April 12–14, 2011
Chicago, Illinois, USA

**Association for
Computing Machinery**

Advancing Computing as a Science & Profession

HSCC'11

Proceedings of the 14th International Conference on
Hybrid Systems: Computation and Control

Sponsored by:
ACM SIGBED

and part of:
CPSWeek

Supported by:
National Science Foundation and NEC

Association for Computing Machinery

Advancing Computing as a Science & Profession

The Association for Computing Machinery
2 Penn Plaza, Suite 701
New York, New York 10121-0701

Notice to Past Authors of ACM-Published Articles
ACM intends to create a complete electronic archive of all articles and/or other material previously published by ACM. If you have written a work that has been previously published by ACM in any journal or conference proceedings prior to 1978, or any SIG Newsletter at any time, and you do NOT want this work to appear in the ACM Digital Library, please inform permissions@acm.org, stating the title of the work, the author(s), and where and when published.

ISBN: 978-1-4503-0629-4

Additional copies may be ordered prepaid from:

ACM Order Department
PO Box 30777
New York, NY 10087-0777, USA

Phone: 1-800-342-6626 (USA and Canada)
+1-212-626-0500 (Global)
Fax: +1-212-944-1318
E-mail: acmhelp@acm.org
Hours of Operation: 8:30 am – 4:30 pm ET

ACM Order Number: 100111

Printed in the USA

Preface

It is our great pleasure to welcome you to the *14th ACM International Conference on Hybrid Systems: Computation and Control (HSCC 2011),* held in Chicago, April 12-14, 2011. This year's conference continues its tradition of being the premier forum for presentation of research results and experience reports on leading edge issues on the theory and practice of embedded reactive systems involving the interplay between discrete and continuous dynamic behaviors.

The mission of the conference is to share information on the latest advancements, both practical and theoretical, in the design, analysis, control, optimization, and implementation of hybrid systems. Previous editions of *HSCC* were held in Berkeley, 1998, Nijmegen, 1999, Pittsburgh, 2000, Rome, 2001, Palo Alto, 2002, Prague, 2003, Philadelphia, 2004, Zurich, 2005, Santa Barbara, 2006, Pisa, 2007, St. Louis, 2008, San Francisco, 2009 and Stockholm, 2010.

The call for papers, which included for the first time a call for tool-presentation papers in addition to regular papers, attracted a large number of very competitive submissions from Asia, Canada, Europe, and the United States. After a very careful reviewing process and in-depth discussions, the program committee accepted 31 regular papers and 4 tool-presentation papers for presentation. All submitted papers were reviewed by at least 3 PC members with support from external reviewers.

HSCC 2011 takes place under the umbrella of the 4th Cyber-Physical Systems Week *(CPSWeek),* which is the collocated cluster of five conferences and workshops: *HSCC*, the International Conference on Cyber-Physical Systems *(ICCPS)*, the International Conference on Information Processing in Sensor Networks *(IPSN)*, the Conference on Languages, Compilers, and Tools for Embedded Systems *(LCTES)*, and the Real-Time and Embedded Technology and Application Symposium *(RTAS)*. Through *CPSWeek*, the five conferences had joint plenary speakers, poster and demo sessions, workshops and tutorials as well as joint social events.

Putting together *HSCC 2011* was a team effort. We thank the authors for providing the content of the conference program. We thank the Program Committee members, the additional reviewers, and the *HSCC* Steering Committee members for their help in composing a strong program. We thank *ACM SIGBED* for being the sponsor of *HSCC*, and the *CPSWeek* for providing an umbrella for *HSCC*. We are grateful for the financial support to *HSCC* from *NEC* and the *NSF*. We thank *EasyChair* for hosting the management service for paper submissions and reviewing, and Ezio Bartocci for also helping to organize the conference web page. We thank Adrienne Griscti from ACM and Sheridan Printing for taking care of processing and printing the papers in a timely manner. Finally, we thank George Pappas for accepting to be a plenary speaker.

We hope you will find this program interesting and thought provoking, and that the conference will provide you with a valuable opportunity to share ideas with other researchers and practitioners from institutions around the world.

Emilio Frazzoli
Program Co-Chair
Massachusetts Institute of Technology
Cambridge, MA, USA

Radu Grosu
Program Co-Chair
State University of New York
Stony Brook, NY, USA

Table of Contents

Session 5: Approximation and Abstraction

Session 6: Applications I

Session 7: Synthesis I

Session 8: Synthesis II

Session 9: Applications II

Session 10: Networked Systems

Session 11: Tool Presentation Papers

Author Index

HSCC 2011 Conference Organization

CPS Week General Chair: Marco Caccamo *(University of Illinois at Urbana-Champaign, USA)*

Program Chairs: Emilio Frazzoli *(Massachusetts Institute of Technology, USA)*
Radu Grosu *(State University of New York at Stony Brook, USA)*

Steering Committee: Rajeev Alur *(University of Pennsylvania, USA)*
Bruce Krogh *(Carnegie Mellon University, USA)*
Oded Maler *(VERIMAG, France)*
Manfred Morari *(ETH Zurich, Switzerland)*
George J. Pappas *(University of Pennsylvania, USA)*
Werner Damm *(Universität Oldenburg, Germany)*

Program Committee: Aaron Ames *(Texas A&M, USA)*
Panos Antsaklis *(University of Notre Dame, USA)*
Eugene Asarin *(University Paris 7, France)*
Ezio Bartocci *(State University of New York at Stony Brook, USA)*
Calin Belta *(Boston University, USA)*
Antonio Bicchi *(Università di Pisa, Italy)*
Francesco Borrelli *(University of California at Berkeley, USA)*
Manfred Broy *(Technical University of Munich, Germany)*
Rance Cleaveland *(University of Maryland, USA)*
Patrick Cousot *(New York University, USA)*
Thao Dang *(French National Center for Scientific Research, France)*
Domitilla Del Vecchio *(Massachusetts Institute of Technology, USA)*
Stefano Di Cairano *(Ford Motor Company, USA)*
Geir Dullerud *(University of Illinois at Urbana-Champaign, USA)*
Magnus Egerstedt *(Georgia Institute of Technology, USA)*
Antoine Girard *(Université Joseph Fourier, France)*
Holger Hermanns *(University of Saarland, Germany)*
Klaus Havelund *(NASA Jet Propulsion Laboratory, USA)*
Franjo Ivančić *(NEC Laboratories, USA)*
T. John Koo *(Chinese Academy of Sciences, China)*
Kim Larsen *(University of Aalborg, Denmark)*
Hai Lin *(National University of Singapore, Singapore)*
Jan Lunze *(Ruhr-University, Bochum, Germany)*
Nancy A. Lynch *(Massachusetts Institute of Technology, USA)*
Rupak Majumdar *(Max Planck Institute for Software Systems, Germany)*
Alexandre Megretski *(Massachusetts Institute of Technology, USA)*
Bud Mishra *(New York University, USA)*
Sayan Mitra *(University of Illinois at Urbana Champaign, USA)*
Carla Piazza *(Università degli Studi di Udine, Italy)*
André Platzer *(Carnegie Mellon University, USA)*
Jean-Francois Raskin *(University Libre de Bruxelles, Belgium)*

Sponsor:

Supporters:

NEC   Empowered by Innovation

x

Wireless Control Networks:
Modeling, Synthesis, Robustness, Security

George J. Pappas
Department of Electrical and Systems Engineering
University of Pennsylvania
Philadelphia, PA
pappasg@ee.upenn.edu

ABSTRACT

Control networks are based on time-triggered wireless substrates for industrial automation control, such as the WirelessHART and Honeywell's OneWireless. Control networks have fundamental di_erences over their sensor network counterparts as they also include actuation and the physical dynamics. A great challenge in such systems is understanding cross-cutting interfaces between computing systems, control systems, sensor networks, and time-triggered communications.

A mathematical framework is _rst proposed for modeling and analyzing multi-hop control networks that use time-triggered communication protocols. We propose formal models for analyzing robustness of multi-hop control networks, where data is exchanged through a multi-hop communication network subject to disruptions. Time-triggered protocols enable our approach to be compositional and hence addresses the problem of designing scalable scheduling and routing policies for multiple control loops closed on the same multi-hop control network. We then present a method to stabilize a plant using just a network of resource constrained wireless nodes. As opposed to traditional networked control schemes where the nodes simply route information to and from a centralized controller, our approach treats the wireless network itself as the controller. The key idea is that each node updates its internal state to be a linear combination of the states of the nodes in its neighborhood. We show that this causes the entire network to behave as a linear dynamical system, with sparsity constraints imposed by the network topology.

We provide a synthesis procedure to program the network controller and present a scheme that can handle node failures while preserving stability. We also consider the design of an intrusion detection system (IDS), which observes the transmissions of certain nodes in the network and uses that information to recover the plant outputs (for diagnostic purposes) and identify malicious behavior by any of the wireless nodes in the network.

Categories and Subject Descriptors

C.3 [Special-Purpose And Application-Based Systems]: Real-time and embedded systems

General Terms

Algorithms, Design, Security, Theory.

BIO

George J. Pappas is the Joseph Moore Professor in the Department of Electrical and Systems Engineering at the University of Pennsylvania. He also holds a secondary appointment in the Departments of Computer and Information Sciences, and Mechanical Engineering and Applied Mechanics. He is member of the GRASP Lab and the PRECISE Center. He currently serves as the Deputy Dean for Research in the School of Engineering and Applied Science. His research focuses on control theory and in particular, hybrid systems, embedded systems, cyber-physical systems, hierarchical and distributed control systems, with applications to unmanned aerial vehicles, distributed robotics, green buildings, and biomolecular networks. He is a Fellow of IEEE, and has received various awards such as the Antonio Ruberti Young Researcher Prize, the George S. Axelby Award, and the National Science Foundation PECASE.

Observability Implies Observer Design for Switched Linear Systems

Aneel Tanwani[*]
Coordinated Science Laboratory
University of Illinois at Urbana-Champaign, USA
tanwani2@illinois.edu

Hyungbo Shim[†]
ASRI, School of Electrical Engineering
Seoul National University, Korea
hshim@snu.ac.kr

Daniel Liberzon[*]
Coordinated Science Laboratory
University of Illinois at Urbana-Champaign, USA
liberzon@illinois.edu

ABSTRACT

This paper presents a characterization of observability and an observer design method for a class of hybrid systems. A necessary and sufficient condition is presented for observability, globally in time, when the system evolves under predetermined mode transitions. A relatively weaker characterization is given for determinability, the property that concerns with recovery of the original state at some time rather than at all times. These conditions are then utilized in the construction of a hybrid observer that is feasible for implementation in practice. The observer, without using the derivatives of the output, generates the state estimate that converges to the actual state under persistent switching.

Categories and Subject Descriptors

J.2 [**Physical Sciences and Engineering**]: Engineering

General Terms

Algorithms, Theory

Keywords

Observability, Observer design, Switched linear systems

1. INTRODUCTION

This paper studies observability conditions and observer construction for a class of hybrid systems where the continuous dynamics are modeled as linear differential equations; the state trajectories exhibit jumps during their evo-

[*]A. Tanwani and D. Liberzon are supported by NSF under the grant ECCS-0821153.

[†]H. Shim is supported by Basic Science Research Program through the National Research Foundation of Korea, funded by Ministry of Education, Science and Technology (grant 2010-0001966).

lution; and discrete dynamics are represented by an exogenous switching signal. Often called *switched systems*, they are described mathematically as:

$$\dot{x}(t) = A_{\sigma(t)}x(t) + B_{\sigma(t)}u(t), \qquad t \neq \{t_q\}, \qquad (1a)$$

$$x(t_q) = E_{\sigma(t_q^-)}x(t_q^-) + F_{\sigma(t_q^-)}v_q, \qquad q \geq 1, \qquad (1b)$$

$$y(t) = C_{\sigma(t)}x(t) + D_{\sigma(t)}u(t), \qquad t \geq t_0, \qquad (1c)$$

where $x(t) \in \mathbb{R}^n$ is the state, $y(t) \in \mathbb{R}^{d_y}$ is the output, $v_q \in \mathbb{R}^{d_v}$ and $u(t) \in \mathbb{R}^{d_u}$ are the inputs, and $u(\cdot)$ is a measurable function. The switching signal $\sigma : \mathbb{R} \mapsto \mathbb{N}$ (set of natural numbers) is a piecewise constant and right-continuous function that changes its value at switching times $\{t_q\}$, $q \in \mathbb{N}$. It is assumed that there are a finite number of switching times in any finite time interval, thus we rule out the Zeno phenomenon in our problem formulation. The switching mode $\sigma(t)$ and the switching times $\{t_q\}$ may be governed by a supervisory logic controller, or determined internally depending on the system state, or considered as an external input. In any case, it is assumed in this paper that the signal $\sigma(\cdot)$ (and thus, the active mode and the switching time $\{t_q\}$ as well) is known over the interval of interest. For estimation of the switching signal $\sigma(t)$, one may be referred to, e.g., [4,7,13,14].

In the past decade, the structural properties of hybrid systems have been investigated by many researchers and observability along with observer construction has been one of them. In hybrid systems, the observability can be studied from various perspectives. If we allow for the use of the differential operator in the observer, then it may be desirable to determine the continuous state of the system instantaneously from the measured output. This in turn requires each subsystem to be observable, however, the problem becomes nontrivial when the switching signal is treated as a discrete state and simultaneous recovery of the discrete and continuous state is required for observability. Some results on this problem are published in [2,6,13].

On the other hand, if the mode transitions are represented by a known switching signal then, even though the individual subsystems are not observable, it is still possible to recover the initial state $x(t_0)$ when the output is observed over an interval $[t_0, T)$ that involves multiple switching instants. This phenomenon is of particular interest for switched systems as the notion of instantaneous observability and observability over an interval[1] coincide for linear time invariant sys-

---

[1]See Definition 1 for precise meaning.

tems. This variant of the observability in switched systems has been studied most notably by [3, 11, 16]. The authors in [8, 9] have studied the observability problem for the systems that allow jumps in the states but they do not consider the change in the dynamics that is introduced by switching to different matrices associated with the active mode. The observer design has also received some attention in the literature [1, 4, 10], where the authors have assumed that each mode in the system is in fact observable admitting a state observer, and have treated the switching as a source of perturbation effect. This approach immediately incurs the need of a common Lyapunov function for the switched error dynamics, or a fixed amount of dwell-time between switching instants, because it is intrinsically a stability problem of the error dynamics.

The approach adopted in this paper is similar to [3, 16] in the sense that we consider observability over an interval. The authors in [3] have presented a coordinate dependent sufficient condition that leads to observer construction; the work of [16] primarily addresses the question whether there exists a switching signal which makes it possible to recover $x(t_0)$ from the knowledge of the output. Whereas, in this paper, similar to our recent work in [12], the switching signal is considered to be known and fixed, so that the trajectory of the system satisfies a time varying linear differential equation. Then for that particular trajectory, we answer the question whether it is possible to recover $x(t_0)$ from the knowledge of the measured output. We present a necessary and sufficient condition for observability over an interval, which is independent of coordinate transformations. Since this condition depends upon the switching times and requires computation of the state transition matrices, we also provide easily verifiable conditions that are either necessary or sufficient for the main condition. Also, with a similar tool set, the notion of determinability, which is more in the spirit of recovering the current state based on the knowledge of inputs and outputs in the past, is developed. Moreover, a hybrid observer for system (1) is designed based on the proposed necessary and sufficient condition which was not the case in [16]. Since the observers are useful for various engineering applications, their utility mainly lies in their online operation method. This thought is essentially rooted in the idea for observer construction adopted in this paper: the idea of combining the partial information available from each mode and collecting them at one instance of time to get the estimate of the state. We show that under mild assumptions, such an estimate converges to the actual state of the plant. We remark that the main contribution of this paper is to present a characterization of observability and an observer design for the systems represented by (1). To the best of authors' knowledge, the considered class of linear systems is the most general for this purpose because it combines both the state jumps and mode switchings.

More emphasis will be given to the case when the individual modes of the system (1) are not observable (in the classical sense of linear time-invariant systems theory) since it is obvious that the system becomes immediately observable when the system is switched to the observable mode. In such cases, switching signal plays a pivotal role as the observability of the switched system not only depends upon the mode sequence but also the switching times. In order to facilitate our understanding of this matter, let us begin with an example.

EXAMPLE 1. Consider a switched system characterized by:

$$A_1 = \begin{bmatrix} 0 & 0 \\ 0 & 0 \end{bmatrix}, \qquad A_2 = \begin{bmatrix} 0 & 1 \\ -1 & 0 \end{bmatrix}$$
$$C_1 = \begin{bmatrix} 1 & 0 \end{bmatrix}, \qquad C_2 = \begin{bmatrix} 0 & 0 \end{bmatrix}$$

with $E_i = I$, $F_i = 0$, $B_i = 0$, and $D_i = 0$ for $i \in \{1, 2\}$. It is noted that neither of the pair (A_1, C_1) or (A_2, C_2) is observable. However, if the switching signal $\sigma(t)$ changes its value in the order of $1 \to 2 \to 1$ at times t_1 and t_2, then we can recover the state. In fact, it turns out that at least two switchings are necessary and the switching sequence should contain the subsequence of modes $\{1, 2, 1\}$. For instance, if the switching happens as $1 \to 2 \to 1$, the output y at time t_1^- (just before the first switching) and t_2 (just after the second switching) are: $y(t_1^-) = C_1 x(t_1^-) = x_1(t_0)$, and $y(t_2) = C_1 e^{A_2 \tau} x(t_0) = \cos \tau \cdot x_1(t_0) + \sin \tau \cdot x_2(t_0)$, where $x(t_0) = [x_1(t_0), x_2(t_0)]^\top$ is the initial condition and $\tau = t_2 - t_1$. Then, it is obvious that $x(t_0)$ can be recovered from two measurements $y(t_1^-)$ and $y(t_2)$ if $\tau \neq k\pi$ with $k \in \mathbb{N}$. On the other hand, any switching signal whose duration for the mode 2 is an integer multiple of π is a 'singular' switching signal (see Remark 1 for the meaning of singular switching signals).

Notation: For a square matrix A and a subspace $\mathcal{V}$, we denote by $\langle A | \mathcal{V} \rangle$ the smallest A-invariant subspace containing $\mathcal{V}$, and by $\langle \mathcal{V} | A \rangle$ the largest A-invariant subspace contained in $\mathcal{V}$. (See Property 7 in the Appendix for their computation.) With a matrix A, $\mathcal{R}(A)$ denotes the column space (range space) of A. For a possibly non-invertible matrix A, the pre-image of a subspace $\mathcal{V}$ through A is given by $A^{-1}\mathcal{V} = \{x : Ax \in \mathcal{V}\}$. Let $\ker A := A^{-1}\{0\}$, then it is seen that $A^{-1} \ker C = \ker(CA)$ for a matrix C. For convenience of notation, let $A^{-\top}\mathcal{V} := (A^\top)^{-1}\mathcal{V}$ where $A^\top$ is the transpose of A, and it is understood that $A_2^{-1} A_1^{-1} \mathcal{V} = A_2^{-1}(A_1^{-1}\mathcal{V})$. Also, we denote the products of matrices A_i as $\prod_{i=j}^{k} A_i := A_j A_{j+1} \cdots A_k$ when $j < k$, and $\prod_{i=j}^{k} A_i := A_j A_{j-1} \cdots A_k$ when $j > k$. The notation $\text{col}(A_1, \ldots, A_k)$ means the vertical stack of matrices $A_1, \cdots, A_k$, that is, $[A_1^\top, \ldots, A_k^\top]^\top$.

2. GEOMETRIC CONDITIONS FOR OBSERVABILITY

To make precise the notions of observability and determinability considered in this paper, let us introduce the formal definitions.

DEFINITION 1. Let $(\sigma^i, u^i, v^i, y^i, x^i)$, for $i = 1, 2$, be the signals that satisfy (1) over an interval[2] $[t_0, T^+)$. We say that the system (1) is $[t_0, T^+)$-*observable* if the equality $(\sigma^1, u^1, v^1, y^1) = (\sigma^2, u^2, v^2, y^2)$ implies that $x^1(t_0) = x^2(t_0)$. Similarly, the system (1) is said to be $[t_0, T^+)$-*determinable*

---

[2] The notation $[t_0, T^+)$ is used to denote the interval $[t_0, T + \varepsilon)$, where $\varepsilon > 0$ is arbitrarily small. In fact, because of the right continuity of the switching signal, the output $y(T)$ belongs to the next mode when T is the switching instant. Then, the point-wise measurement $y(T)$ is insufficient to contain the information for the new mode, and thus, it is imperative to consider the output signal over the interval $[t_0, T + \varepsilon)$ with $\varepsilon > 0$. This definition implicitly implies that the observability property does not change for sufficiently small ε (which is true, and becomes clear shortly).

4

if the equality $(\sigma^1, u^1, v^1, y^1) = (\sigma^2, u^2, v^2, y^2)$ implies that $x^1(T) = x^2(T)$.

Since the initial state $x(t_0)$, the switching signal σ, and the inputs (u, v) uniquely determine $x(t)$ on $[t_0, T^+)$ by (1), observability is achieved if and only if the state trajectory $x(t)$, for each $t \in [t_0, T^+)$, is uniquely determined by the inputs, the output, and the switching signal. Obviously, observability implies determinability by forward integration of (1), but the converse is not true due to the possibility of noninvertible matrices E_σ. In case there is no jump map (1b), or each E_σ is invertible, observability and determinability are equivalent. The notion of determinability has also been called reconstructability in [11].

PROPOSITION 1. For a switching signal σ, the system (1) is $[t_0, T^+)$-observable (or, determinable) if, and only if, zero inputs and zero output on the interval $[t_0, T^+)$ imply that $x(t_0) = 0$ (or, $x(T) = 0$).

PROOF. Since the zero solution with the zero inputs yields the zero output, the necessity follows from the fact that $x(t_0)$ (or, $x(T)$) is uniquely determined from the inputs and the outputs. For the sufficiency, suppose that the system (1) is not $[t_0, T^+)$-observable (or determinable); that is, there exist two different states $x^1(t_0)$ and $x^2(t_0)$ (or, $x^1(T)$ and $x^2(T)$) that yield the same output y under the same inputs (u, v). Let $\tilde{x}(t) := x^1(t) - x^2(t)$, where $x^i(t)$, $i = 1, 2$, is the solution of (1) which takes the value $x^i(t_0)$ at initial time t_0 (or, $x^i(T)$ at terminal time T). Then, by linearity, it follows that $\dot{\tilde{x}} = A_\sigma \tilde{x}$, $\tilde{x}(t_q) = E_\sigma \tilde{x}(t_q^-)$, and $C_\sigma \tilde{x} = C_\sigma x^1 - C_\sigma x^2 = y - y = 0$, but $\tilde{x}(t_0) = x^1(t_0) - x^2(t_0) \neq 0$ (or, $\tilde{x}(T) = x^1(T) - x^2(T) \neq 0$). Hence, zero inputs and zero output do not imply $x(t_0) = 0$ (or, $x(T) = 0$), and the sufficiency holds. $\square$

Because of Proposition 1, we are motivated to introduce the following homogeneous switched system, which has been obtained by setting the inputs (u, v) equal to zero in (1):

$$\dot{x}(t) = A_{\sigma(t)} x(t), \quad y(t) = C_{\sigma(t)} x(t), \quad t \in [t_{q-1}, t_q) \quad \text{(2a)}$$

$$x(t_q) = E_{\sigma(t)} x(t_q^-). \quad \text{(2b)}$$

If this homogeneous system is observable (or, determinable) with a given σ, then $y \equiv 0$ implies that $x(t_0) = 0$ (or, $x(T) = 0$), and in terms of description of system (1), this means that zero inputs and zero output yield $x(t_0) = 0$ (or, $x(T) = 0$); hence, (1) is observable (or, determinable) because of Proposition 1. On the other hand, if the system (1) is observable (or, determinable), then it is still observable (or, determinable) with zero inputs, which is described as system (2). Thus, the observability (or, determinability) of systems (1) and (2) are equivalent.

Before going further, let us rename the switching sequence for convenience. For system (1), when the switching signal $\sigma(t)$ takes the mode sequence $\{q_1, q_2, q_3, \cdots\}$, we rename them as increasing integers $\{1, 2, 3, \cdots\}$, which is ever increasing even though the same mode is revisited; for convenience, this sequence is indexed by q and not by $\sigma(t)$. Moreover, it is often the case that the mode of the system changes without the state jump (1b), or the state jumps without switching to another mode. In the former case, we can simply take $E_q = I$ and $F_q = 0$, and in the latter case, we increase the mode index by one and take $A_q = A_{q+1}$, $B_q = B_{q+1}$ and so on. In this way various situations fit into

the description of (1) with increasing mode sequence. The switching time t_q is the instant when transition from mode q to mode $q + 1$ takes place.

2.1 Necessary and Sufficient Conditions for Observability

In this section, we present a characterization of the unobservable subspace for the system (2) with a given switching signal. Towards this end, let $\mathcal{N}_q^m$ $(m \geq q)$ denote the set of states at $t = t_{q-1}$ for system (2) that generate identically zero output over $[t_{q-1}, t_{m-1}^+)$. Then, it is easily seen that $\mathcal{N}_q^m$ is actually a subspace due to linearity of (2), and we call $\mathcal{N}_q^m$ the *unobservable subspace for* $[t_{q-1}, t_{m-1}^+)$. It can be seen that the system (2) is an LTI system between two consecutive switching times, so that its unobservable subspace on the interval $[t_{q-1}, t_q)$ is simply given by the largest A_q-invariant subspace contained in $\ker C_q$, i.e., $\langle \ker C_q | A_q \rangle = \ker G_q$ where $G_q := \text{col}(C_q, C_q A_q, \cdots, C_q A_q^{n-1})$. So it is clear that $\mathcal{N}_q^q = \ker G_q$. Now, when the measured output is available over the interval $[t_{q-1}, t_{m-1}^+)$ that includes switchings at $t_q, t_{q+1}, \ldots, t_{m-1}$, more information about the state is obtained in general so that $\mathcal{N}_q^m$ gets smaller as the difference $m - q$ gets larger, and we claim that the subspace $\mathcal{N}_q^m$ can be computed recursively as follows:

$$\begin{aligned} \mathcal{N}_m^m &= \ker G_m, \\ \mathcal{N}_q^m &= \ker G_q \cap e^{-A_q \tau_q} E_q^{-1} \mathcal{N}_{q+1}^m, \quad 1 \leq q \leq m-1, \end{aligned} \quad \text{(3)}$$

where $\tau_q = t_q - t_{q-1}$. The following theorem presents a necessary and sufficient condition for observability of the system (1) while proving the claim in the process.

> THEOREM 1. For the system (2) with a switching signal $\sigma_{[t_0, t_{m-1}^+)}$, the unobservable subspace for $[t_0, t_{m-1}^+)$ at t_0 is given by $\mathcal{N}_1^m$ from (3). Therefore, the system (1) is $[t_0, t_{m-1}^+)$-observable if, and only if,
>
> $$\mathcal{N}_1^m = \{0\}. \quad \text{(4)}$$

From (3), it is not difficult to arrive at the following formula for $\mathcal{N}_q^m$:

$$\mathcal{N}_q^m = \ker G_q \cap \left(\bigcap_{i=q+1}^{m} \prod_{j=q}^{i-1} e^{-A_j \tau_j} E_j^{-1} \ker G_i \right) \quad \text{(5a)}$$

$$= \ker G_q \cap \left(\bigcap_{i=q+1}^{m} \ker \left(G_i \prod_{l=i-1}^{q} E_l e^{A_l \tau_l} \right) \right). \quad \text{(5b)}$$

From this expression, it is easily seen that $\mathcal{N}_1^{m_1} \supseteq \mathcal{N}_1^{m_2}$ if $m_1 \leq m_2$. Therefore, in case the interval under consideration is not finite and the switching is persistent, observability of system (1) is determined by whether there exists an $m \in \mathbb{N}$ such that (4) holds.

PROOF OF THEOREM 1. *Sufficiency.* Using the result of Proposition 1, it suffices to show that the identically zero output of (2) implies $x(t_0) = 0$. Assume that $y \equiv 0$ on $[t_0, t_{m-1}^+)$. Then, it is immediate that $x(t_{m-1}) \in \mathcal{N}_m^m = \ker G_m$. We next apply the inductive argument to show that $x(t_{q-1}) \in \mathcal{N}_q^m$ for $1 \leq q \leq m-1$. Suppose that $x(t_q) \in \mathcal{N}_{q+1}^m$, then $x(t_{q-1}) \in e^{-A_q \tau_q} E_q^{-1} \mathcal{N}_{q+1}^m$ since $x(t)$

is the solution of (2). Zero output on the interval $[t_{q-1}, t_q)$ also implies that $x(t_{q-1}) \in \ker G_q$. Therefore,

$$x(t_{q-1}) \in \ker G_q \cap e^{-A_q \tau_q} E_q^{-1} \mathcal{N}_{q+1}^m.$$

From (3), it follows that $x(t_{q-1}) \in \mathcal{N}_q^m$. This induction proves the claim that $\mathcal{N}_q^m$ is given by (3). With $q = 1$, it is seen that $x(t_0) \in \mathcal{N}_1^m = \{0\}$, which proves the sufficiency.

Necessity. Assuming that $\mathcal{N}_1^m \neq \{0\}$, we show that a non-zero initial state $x(t_0) \in \mathcal{N}_1^m$ yields the solution of (2) such that $y \equiv 0$ on $[t_0, t_{m-1}^+)$, which implies unobservability. First, we show the following implication;

$$x(t_{q-1}) \in \mathcal{N}_q^m \quad \Rightarrow \quad x(t_q) \in \mathcal{N}_{q+1}^m, \qquad q < m. \quad (6)$$

Indeed, assuming that $x(t_{q-1}) \in \mathcal{N}_q^m$ with $q < m$, it follows that, $x(t_q) = E_q e^{A_q \tau_q} x(t_{q-1})$, which further gives,

$$\begin{aligned} x(t_q) &\in E_q e^{A_q \tau_q} \mathcal{N}_q^m \\ &= E_q e^{A_q \tau_q} \left(\ker G_q \cap e^{-A_q \tau_q} E_q^{-1} \mathcal{N}_{q+1}^m \right) \\ &\subseteq E_q \ker G_q \cap E_q E_q^{-1} \mathcal{N}_{q+1}^m \\ &= E_q \ker G_q \cap \mathcal{N}_{q+1}^m \cap \mathcal{R}(E_q) \subseteq \mathcal{N}_{q+1}^m \end{aligned}$$

by using (3) and Properties 2, 3, and 11 in the Appendix. Therefore, for $0 \leq q \leq m - 1$, $x(t_q) \in \mathcal{N}_{q+1}^m \subseteq \ker G_{q+1}$, and the solution $x(t) = e^{A_{q+1}(t - t_q)} x(t_q)$ for $t \in [t_q, t_{q+1})$ satisfies that $y(t) = C_{q+1} x(t) = 0$ for $t \in [t_q, t_{q+1})$ due to A_{q+1}-invariance of $\ker G_{q+1}$. $\square$

REMARK 1. The observability condition (4) given in Theorem 1 is dependent upon a particular switching signal under consideration, and it is entirely possible that the system is observable for certain switching signals and unobservable for others (*cf.* Example 1). Note that a switching signal is composed of a mode sequence and switching times. We call a switching signal σ *singular* when the observability condition (4) does not hold with σ, but the condition happens to hold by changing the switching times of σ while preserving the mode sequence.

In order to inspect the observability of the system (2), one can compute $\mathcal{N}_1^m$ by (5) (the formula (5b) may be preferable because the computation of pre-image due to E_j^{-1} is avoided). However, the computation of matrix exponent may be heavy in practice (especially for large dimensional systems) and one may want to resort to the following sufficient, or necessary conditions, which are independent of switching times and only take mode sequence into consideration. Hence, once the sufficient condition in Corollary 1 holds (respectively, the necessary condition in Corollary 2 is violated), then the system is observable (resp. unobservable) for any switching signal that has the same switching mode sequence regardless of the switching times.

COROLLARY 1. Let $\overline{\mathcal{N}}_1^m$ be an over-approximation of $\mathcal{N}_1^m$ that is defined as follows:

$$\begin{aligned} \overline{\mathcal{N}}_m^m &:= \ker G_m, \\ \overline{\mathcal{N}}_q^m &:= \left\langle A_q | \ker G_q \cap E_q^{-1} \overline{\mathcal{N}}_{q+1}^m \right\rangle, \qquad 1 \leq q \leq m - 1. \end{aligned}$$

The system (1) is $[t_0, t_{m-1}^+)$-observable if $\overline{\mathcal{N}}_1^m = \{0\}$.

PROOF. The proof is completed by showing that $\mathcal{N}_q^m \subseteq \overline{\mathcal{N}}_q^m$ for $1 \leq q \leq m$. First, note that $\mathcal{N}_m^m = \overline{\mathcal{N}}_m^m$. Assuming that $\mathcal{N}_{q+1}^m \subseteq \overline{\mathcal{N}}_{q+1}^m$ for $1 \leq q \leq m - 1$, we now claim

that $\mathcal{N}_q^m \subseteq \overline{\mathcal{N}}_q^m$. Indeed, by Properties 3, 9, and 11 in the Appendix, and the recursion equation (3), we obtain

$$\begin{aligned} \mathcal{N}_q^m &= \ker G_q \cap e^{-A_q \tau_q} E_q^{-1} \mathcal{N}_{q+1}^m \\ &= e^{-A_q \tau_q} \left(\ker G_q \cap E_q^{-1} \mathcal{N}_{q+1}^m \right) \\ &\subseteq \left\langle A_q | \ker G_q \cap E_q^{-1} \mathcal{N}_{q+1}^m \right\rangle \\ &\subseteq \left\langle A_q | \ker G_q \cap E_q^{-1} \overline{\mathcal{N}}_{q+1}^m \right\rangle = \overline{\mathcal{N}}_q^m, \quad 1 \leq q \leq m - 1. \end{aligned}$$

Therefore, the condition $\overline{\mathcal{N}}_1^m = \{0\}$ implies (4). $\square$

COROLLARY 2. Let $\underline{\mathcal{N}}_1^m$ be an under-approximation of $\mathcal{N}_1^m$ that is defined as follows:

$$\begin{aligned} \underline{\mathcal{N}}_m^m &:= \ker G_m, \\ \underline{\mathcal{N}}_q^m &:= \left\langle \ker G_q \cap E_q^{-1} \underline{\mathcal{N}}_{q+1}^m | A_q \right\rangle, \qquad 1 \leq q \leq m - 1. \end{aligned}$$

If system (1) is $[t_0, t_{m-1}^+)$-observable, then $\underline{\mathcal{N}}_1^m = \{0\}$.

PROOF. The proof proceeds similar to Corollary 1. With $\mathcal{N}_m^m = \underline{\mathcal{N}}_m^m$, we assume that $\mathcal{N}_{q+1}^m \supseteq \underline{\mathcal{N}}_{q+1}^m$ for $1 \leq q \leq m - 1$, and claim that $\mathcal{N}_q^m \supseteq \underline{\mathcal{N}}_q^m$. Again by Properties 3, 9, and 11 in the Appendix, and employing equation (3), we obtain

$$\begin{aligned} \mathcal{N}_q^m &= e^{-A_q \tau_q} \left(\ker G_q \cap E_q^{-1} \mathcal{N}_{q+1}^m \right) \\ &\supseteq \left\langle \ker G_q \cap E_q^{-1} \mathcal{N}_{q+1}^m | A_q \right\rangle \\ &\supseteq \left\langle \ker G_q \cap E_q^{-1} \underline{\mathcal{N}}_{q+1}^m | A_q \right\rangle = \underline{\mathcal{N}}_q^m, \quad 1 \leq q \leq m - 1. \end{aligned}$$

The condition $\underline{\mathcal{N}}_1^m = \{0\}$ is implied by (4). $\square$

REMARK 2. By taking orthogonal complements of $\mathcal{N}_q^m$, $\overline{\mathcal{N}}_q^m$ and $\underline{\mathcal{N}}_q^m$, respectively, we get dual conditions, using Properties 5, 6, 8, and 10 in the Appendix, as follows. The system (1) is $[t_0, t_{m-1}^+)$-observable if and only if $\mathcal{P}_1^m = \mathbb{R}^n$ where

$$\mathcal{P}_1^m := (\mathcal{N}_1^m)^\perp = \mathcal{R}(G_1^\top) + \sum_{i=2}^{m} \prod_{j=1}^{i-1} e^{A_j^\top \tau_j} E_j^\top \mathcal{R}(G_j^\top).$$

Based on the above definition, one can state Corollary 1 and Corollary 2 in alternate forms. System (1) is $[t_0, t_{m-1}^+)$-observable if $\underline{\mathcal{P}}_1^m = \mathbb{R}^n$, where $\underline{\mathcal{P}}_1^m$ is computed as:

$$\begin{aligned} \underline{\mathcal{P}}_m^m &= (\overline{\mathcal{N}}_m^m)^\perp = \mathcal{R}(G_m^\top) \\ \underline{\mathcal{P}}_q^m &= (\overline{\mathcal{N}}_q^m)^\perp = \left\langle \mathcal{R}(G_q^\top) + E_q^\top \underline{\mathcal{P}}_{q+1}^m | A_q^\top \right\rangle, \quad 1 \leq q \leq m - 1. \end{aligned}$$

Also, if system (1) is $[t_0, t_{m-1}^+)$-observable then $\overline{\mathcal{P}}_1^m = \mathbb{R}^n$, where $\overline{\mathcal{P}}_1^m$ is defined sequentially as:

$$\begin{aligned} \overline{\mathcal{P}}_m^m &= (\underline{\mathcal{N}}_m^m)^\perp = \mathcal{R}(G_m^\top) \\ \overline{\mathcal{P}}_q^m &= (\underline{\mathcal{N}}_q^m)^\perp = \left\langle A_q^\top | \mathcal{R}(G_q^\top) + E_q^\top \overline{\mathcal{P}}_{q+1}^m \right\rangle, \quad 1 \leq q \leq m - 1. \end{aligned}$$

2.2 Necessary and Sufficient Conditions for Determinability

In order to study determinability of the system (1) and arrive at a result parallel to Theorem 1, our first goal is to develop an object similar to $\mathcal{N}_q^m$. So, for system (2) with a given switching signal, let $\mathcal{Q}_q^m$ be the set of states that can be reached at time $t = t_{m-1}$ while producing the zero output on the interval $[t_{q-1}, t_{m-1}^+)$. We call $\mathcal{Q}_q^m$ the *undeterminable subspace for* $[t_{q-1}, t_{m-1}^+)$. Then, it can be shown, similarly to the proof of Theorem 1, that $\mathcal{Q}_q^m$ is computed recursively as follows:

$$\mathcal{Q}_q^q = \ker G_q$$
$$\mathcal{Q}_q^k = \ker G_k \cap E_{k-1} e^{A_{k-1}\tau_{k-1}} \mathcal{Q}_q^{k-1}, \quad q+1 \le k \le m. \tag{7}$$

These sequential definitions lead to following expression for $\mathcal{Q}_q^m$:

$$\mathcal{Q}_q^m = \ker G_m \cap E_{m-1} \ker(G_{m-1}) \cap$$
$$\left(\bigcap_{i=q}^{m-2} \prod_{l=m-1}^{i+1} E_l e^{A_l \tau_l} E_i \ker G_i \right), \tag{8}$$

with $\mathcal{Q}_q^q = \ker G_q$. In the above equation, the subspace $(\Pi_{l=m-1}^{i+1} E_l e^{A_l \tau_l} E_i \ker G_i)$ indicates the set of states at time $t = t_{m-1}$ obtained by propagating the unobservable state of the mode i, that is active during the interval $[t_{i-1}, t_i)$, under the dynamics of system (2). Intersection of these subspaces with $\ker G_m$ shows that $\mathcal{Q}_q^m$ is the set of states that cannot be determined from the zero output at time $t = t_{m-1}$. Then, the determinability can be characterized as in the following theorem (which is given without proof).

THEOREM 2. For the system (2) and a given switching signal $\sigma_{[t_0, t_{m-1}^+)}$, the undeterminable subspace for $[t_0, t_{m-1}^+)$ at t_{m-1} is given by $\mathcal{Q}_1^m$ of (8). Therefore, the system (1) is $[t_0, t_{m-1}^+)$-determinable if and only if

$$\mathcal{Q}_1^m = \{0\}. \tag{9}$$

The condition (9) is equivalent to (4) when all E_q matrices, $q = 1, \ldots, m-1$, are invertible because of the relation

$$\mathcal{Q}_1^m = \prod_{l=m-1}^{1} E_l e^{A_l \tau_l} \mathcal{N}_1^m.$$

On the other hand, if any of the jump maps E_q is a zero matrix, then (9) holds regardless of (4) (which makes sense because we can immediately determine that $x(t_{m-1}) = 0$ in this case).

The recursive expression (7) shows that the sequence $\{\mathcal{Q}_1^1, \mathcal{Q}_1^2, \mathcal{Q}_1^3, \cdots\}$ is moving forward in the sense that $\mathcal{Q}_1^{k+1}$ is computed from $\mathcal{Q}_1^k$ and from the information about the running mode k such as G_{k+1}, E_k, A_k, and τ_k. This fact illustrates that the computation of $\mathcal{Q}_1^m$ is more suitable for online implementation (since m increases as time sets forward), compared to the computation of $\mathcal{N}_1^m$, which requires a backward computation from $\mathcal{N}_m^m$ (see (3)).

COROLLARY 3. The system (1) is $[t_0, t_{m-1}^+)$-determinable if $\overline{\mathcal{Q}}_1^m = \{0\}$, where $\overline{\mathcal{Q}}_1^m$ is computed by

$$\overline{\mathcal{Q}}_1^1 := \ker G_1$$
$$\overline{\mathcal{Q}}_1^q := E_{q-1} \left\langle A_{q-1} | \overline{\mathcal{Q}}_1^{q-1} \right\rangle \cap \ker G_q, \quad 2 \le q \le m.$$

COROLLARY 4. If system (1) is $[t_0, t_{m-1}^+)$-determinable, then $\underline{\mathcal{Q}}_1^m = \{0\}$, where $\underline{\mathcal{Q}}_1^m$ is computed by

$$\underline{\mathcal{Q}}_1^1 := \ker G_1$$
$$\underline{\mathcal{Q}}_1^q := E_{q-1} \left\langle \underline{\mathcal{Q}}_1^{q-1} | A_{q-1} \right\rangle \cap \ker G_q, \quad 2 \le q \le m.$$

The above corollaries are proved by showing that $\underline{\mathcal{Q}}_1^q \subseteq \mathcal{Q}_1^q \subseteq \overline{\mathcal{Q}}_1^q$. It is noted again that the computation of sequential subspaces in Corollary 3 and Corollary 4 proceeds forward in time.

REMARK 3. An alternative dual characterization of determinability is possible by inspecting whether the complete state information is available while going forward in time. This is achieved in terms of the subspace $\mathcal{M}_q^m$, obtained by taking the orthogonal complement of $\mathcal{Q}_q^m$. Using Properties 5, 6, 8, and 10 in the Appendix, the following expression follows from (8):

$$\mathcal{M}_q^m := (\mathcal{Q}_q^m)^\perp = \sum_{i=q}^{m-2} \prod_{l=m-1}^{i+1} E_l^{-\top} e^{-A_l^\top \tau_l} E_i^{-\top} \mathcal{R}(G_i^\top)$$
$$+ E_{m-1}^{-\top} \mathcal{R}(G_{m-1}^\top) + \mathcal{R}(G_m^\top). \tag{10}$$

In other words, $\mathcal{M}_q^m$ is the set of states at time instant $t = t_{m-1}$ that can be identified, modulo the unobservable subspace at t_{m-1}, from the information of y over the interval $[t_{q-1}, t_{m-1}^+)$. Therefore, the dual statement for determinability is that the system (1) is $[t_0, t_{m-1}^+)$-determinable if and only if

$$\mathcal{M}_1^m = \mathbb{R}^n. \tag{11}$$

It is noted that a recursive expression for $\mathcal{M}_1^m$ is given by

$$\mathcal{M}_1^1 = \mathcal{R}(G_1^\top)$$
$$\mathcal{M}_1^q = E_{q-1}^{-\top} e^{-A_{q-1}^\top \tau_{q-1}} \mathcal{M}_1^{q-1} + \mathcal{R}(G_q^\top), \quad 2 \le q \le m,$$

and the dual statements of Corollaries 3 and 4, that are independent of switching times, are given as follows: system (1) is $[t_0, t_{m-1}^+)$-determinable if $\underline{\mathcal{M}}_1^m = \mathbb{R}^n$, where

$$\underline{\mathcal{M}}_1^1 := (\overline{\mathcal{Q}}_1^m)^\perp = \mathcal{R}(G_1^\top),$$
$$\underline{\mathcal{M}}_1^q := (\overline{\mathcal{Q}}_1^q)^\perp = E_{q-1}^{-\top} \left\langle \underline{\mathcal{M}}_1^{q-1} | A_{q-1}^\top \right\rangle + \mathcal{R}(G_q^\top), \ 2 \le q \le m.$$

Similarly, if system (1) is $[t_0, t_{m-1}^+)$-determinable then $\overline{\mathcal{M}}_1^m = \mathbb{R}^n$, where $\overline{\mathcal{M}}_1^m$ is computed as follows:

$$\overline{\mathcal{M}}_1^1 := (\underline{\mathcal{Q}}_1^m)^\perp = \mathcal{R}(G_1^\top),$$
$$\overline{\mathcal{M}}_1^q := (\underline{\mathcal{Q}}_1^q)^\perp = E_{q-1}^{-\top} \left\langle A_{q-1}^\top | \overline{\mathcal{M}}_1^{q-1} \right\rangle + \mathcal{R}(G_q^\top), \ 2 \le q \le m.$$

3. OBSERVER DESIGN

In engineering practice, an observer is designed to provide an estimate of the actual state value at current time. In this regard, determinability (weaker than observability according to Definition 1) is a suitable notion. Based on the conditions obtained for determinability in the previous section, an asymptotic observer is designed for the system (1) in this section. By asymptotic observer, we mean that the estimate $\hat{x}(t)$ converges to the plant state $x(t)$ as $t \to \infty$, and in order to achieve this convergence, we introduce the following assumptions.

ASSUMPTION 1. 1. The switching is persistent in the sense that there exists a $D > 0$ such that a switch occurs at least once in every time interval of length D; that is,

$$t_q - t_{q-1} < D, \qquad \forall q \in \mathbb{N}. \tag{12}$$

2. The system is persistently determinable in the sense that there exists an $N \in \mathbb{N}$ such that

$$\dim \mathcal{M}_{q-N}^q = n, \qquad \forall q \geq N+1. \tag{13}$$

(The integer N is interpreted as the minimal number of switches required to gain determinability.)

3. $\|A_q\|$ is uniformly bounded for all $q \in \mathbb{N}$ (which is always the case when A_q belongs to a finite set).

We disregard the time consumed for computation by assuming that the data processor is fairly fast compared to the plant process. The computation time, however, needs to be considered in real-time application if the plant itself is fast.

The observer we propose is a hybrid dynamical system of the form

$$\dot{\hat{x}}(t) = A_q \hat{x}(t) + B_q u(t), \qquad t \in [t_{q-1}, t_q), \tag{14a}$$

$$\hat{x}(t_q) = E_q(\hat{x}(t_q^-) - \xi_q(t_q^-)) + F_q v_q, \qquad q \geq 1, \tag{14b}$$

$$\xi_q(t_q^-) = \begin{cases} \mathcal{L}_q(y_{[t_{q-N-1}, t_q)}, u_{[t_{q-N-1}, t_q)}, v_{[q-N,q-1]}), & q > N, \\ 0, & 1 \leq q \leq N, \end{cases} \tag{14c}$$

with an arbitrary initial state $\hat{x}(t_0) \in \mathbb{R}^n$, where $v_{[q-N,q-1]}$ denotes the vector $[v_{q-N}, v_{q-N+1}, \ldots, v_{q-1}]^\top$. It is seen that the observer consists of a system copy and an estimate update law by some operator $\mathcal{L}_q$. So the goal is to design the operator $\mathcal{L}_q$ such that $\hat{x}(t) \to x(t)$. It will turn out that the proposed operator $\mathcal{L}_q$ consists of dynamic observers for partial states at each mode, a procedure for accumulating the partial state information, and an inversion formula for recovering full state information. In the sequel, we develop the structure of the operator $\mathcal{L}_q$ and based on that, a procedure for implementation of hybrid observer is outlined in Algorithm 1. It is then shown in Theorem 3 that the state estimates computed according to the parameter bounds in Algorithm 1 indeed converge to the actual state of the system.

With $\tilde{x} := \hat{x} - x$, the error dynamics are described by,

$$\dot{\tilde{x}}(t) = A_q \tilde{x}(t), \qquad t \neq t_q, \tag{15a}$$

$$\tilde{x}(t_q) = E_q(\tilde{x}(t_q^-) - \xi_q(t_q^-)). \tag{15b}$$

The output error can now be defined as $\tilde{y}(t) := C_q \hat{x}(t) + D_q u(t) - y(t) = C_q \tilde{x}(t)$.

Based on the description of error dynamics, we design partial observers for each mode q using the idea similar to Kalman observability decomposition [5]. Choose a matrix Z^q such that its columns are an orthonormal basis of $\mathcal{R}(G_q^\top)$, so that $\mathcal{R}(Z^q) = \mathcal{R}(G_q^\top)$. Further, choose a matrix W^q such that its columns are an orthonormal basis of $\ker G_q$. From the construction, there are matrices $S_q \in \mathbb{R}^{r_q \times r_q}$ and $R_q \in \mathbb{R}^{d_y \times r_q}$, where $r_q = \operatorname{rank} G_q$, such that $Z^{q\top} A_q = S_q Z^{q\top}$ and $C_q = R_q Z^{q\top}$, and that the pair (S_q, R_q) is observable. Let $z^q := Z^{q\top} \tilde{x}$ and $w^q := W^{q\top} \tilde{x}$, so that z^q (resp. w^q) denotes the observable (resp. unobservable) states of mode q. Thus, for the interval $[t_{q-1}, t_q)$, we obtain,

$$\dot{z}^q = Z^{q\top} A_q \tilde{x} = S_q z^q, \quad \tilde{y} = C_q \tilde{x} = R_q z^q, \tag{16a}$$

$$z^q(t_{q-1}) = Z^{q\top} \tilde{x}(t_{q-1}). \tag{16b}$$

Since z^q is observable over the interval $[t_{q-1}, t_q)$, a standard Luenberger observer, whose role is to estimate $z^q(t_q^-)$ at the

end of the interval, is designed as:

$$\dot{\hat{z}}^q = S_q \hat{z}^q + L_q(\tilde{y} - R_q \hat{z}^q), \quad t \in [t_{q-1}, t_q), \tag{17a}$$

$$\hat{z}^q(t_{q-1}) = 0, \tag{17b}$$

where L_q is a matrix such that $(S_q - L_q R_q)$ is Hurwitz. Note that we have fixed the initial condition of the estimator to be zero for each interval.

Let us denote the vector $[\tau_{i+1}, \cdots, \tau_j]$ simply by $\tau_{\{i+1,j\}}$ (where $j > i$), which will be often dropped when used as an argument for succinct presentation. With $j > i$, define the state-transport matrix

$$\Psi_i^j(\tau_{\{i+1,j\}}) := e^{A_j \tau_j} E_{j-1} e^{A_{j-1} \tau_{j-1}} E_{j-2} \cdots e^{A_{i+1} \tau_{i+1}} E_i, \tag{18}$$

and for convenience $\Psi_i^i := I$. We now define a matrix $\Theta_i^q(\tau_{\{i+1,q\}})$ whose columns form a basis of the subspace $\mathcal{R}(\Psi_i^q(\tau_{\{i+1,q\}}) W^i)^\perp$; that is,

$$\mathcal{R}(\Theta_i^q(\tau_{\{i+1,q\}})) = \mathcal{R}(\Psi_i^q(\tau_{\{i+1,q\}}) W^i)^\perp, \quad i = q-N, \cdots, q.$$

By construction, each column of Θ_i^q is orthogonal to the subspace $\ker G_i$ that has been transported from t_i^- to t_q^- along the error dynamics (15). This matrix Θ_i^q will be used for filtering out the unobservable component in the state estimate obtained from the mode i after being transported to the time t_q^-. As a convention, we take Θ_i^q to be a null matrix whenever $\mathcal{R}(\Psi_i^q(\tau_{\{i+1,q\}}) W^i)^\perp = \{0\}$.

Using the determinability of the system (Assumption 1.2), it will be shown later in the proof of Theorem 3 that the matrix

$$\Theta_q := [\Theta_q^q \vdots \cdots \vdots \Theta_{q-N}^q] \tag{19}$$

has rank n. Equivalently, $\Theta_q^\top$ has n independent columns and is left-invertible, so that $(\Theta_q^\top)^\dagger = (\Theta_q \Theta_q^\top)^{-1} \Theta_q$, where $\dagger$ denotes the left-pseudo-inverse. Introduce the notation

$$\begin{aligned} \xi_{\{q-N,q-1\}}^- &:= \operatorname{col}(\xi_{q-N}(t_{q-N}^-), \ldots, \xi_{q-1}(t_{q-1}^-)), \\ \hat{z}_{\{q-N,q\}}^- &:= \operatorname{col}(\hat{z}^{q-N}(t_{q-N}^-), \ldots, \hat{z}^q(t_q^-)), \end{aligned} \tag{20}$$

and define the vector Ξ_q as follows:

$$\Xi_q(\hat{z}_{\{q-N,q\}}^-, \xi_{\{q-N,q-1\}}^-) :=$$
$$\begin{bmatrix} \Theta_q^{q\top} \Psi_q^q Z^q \hat{z}^q(t_q^-) \\ \Theta_{q-1}^{q\top} (\Psi_{q-1}^q Z^{q-1} \hat{z}^{q-1}(t_{q-1}^-) - \Psi_{q-1}^q \xi_{q-1}(t_{q-1}^-)) \\ \vdots \\ \Theta_{q-N}^{q\top} (\Psi_{q-N}^q Z^{q-N} \hat{z}^{q-N}(t_{q-N}^-) - \sum_{l=q-N}^{q-1} \Psi_l^q \xi_l(t_l^-)) \end{bmatrix}.$$

We then compute $\xi_q(t_q^-)$ in (14c) as:

$$\xi_q(t_q^-) = (\Theta_q^\top)^\dagger \Xi_q(\hat{z}_{\{q-N,q\}}^-, \xi_{\{q-N,q-1\}}^-) \tag{21}$$

which corresponds to the operator $\mathcal{L}_q$. Finally, as the last piece of notation, we define the matrices M_j^q, $j = q-N, \cdots, q$, as follows:

$$[M_q^q, M_{q-1}^q, \cdots, M_{q-N}^q] := E_q(\Theta_q^\top)^\dagger \times$$
$$\operatorname{blockdiag}\left(\Theta_q^{q\top} \Psi_q^q, \Theta_{q-1}^{q\top} \Psi_{q-1}^q, \cdots, \Theta_{q-N}^{q\top} \Psi_{q-N}^q\right). \tag{22}$$

Each M_j^q ($j = q-N, \cdots, q$) is an n by n matrix whose argument is $\tau_{\{q-N+1,q\}}$ in general (due to the inversion of $\Theta_q^\top$), while the argument of both Θ_q^q and Ψ_q^q is $\tau_{\{j+1,q\}}$.

With all the quantities defined so far, the proposed observer (14) is implemented using the following algorithm:

Algorithm 1 Implementation of hybrid observer

Input: σ, u, v, y
Initialization: Run (14) for $t \in [t_0, t_{N+1})$ with some $\hat{x}(t_0)$
1: **for all** $q \geq N+1$ **do**
2:    **for** $j = q - N$ **to** q **do**
3:       Compute the injection gain L_j such that

$$\|M_j^q(\tau_{\{q-N+1,q\}})Z^j e^{(S_j - L_j R_j)\tau_j} Z^{j\top}\| \leq c \quad (23)$$

where the constant c is chosen so that

$$0 < c < \frac{1}{N+1}. \quad (24)$$

4:       Obtain $\hat{z}^j(t_j^-)$ by running the individual observer (17) for the j-th mode.
5:    **end for**
6:    Compute $\xi_q(t_q^-)$ from (21), as an implementation of (14c).
7:    Compute $\hat{x}(t_q)$ using (14b) and run (14a) over the interval $[t_q, t_{q+1})$.
8: **end for**

The following theorem shows that the above implementation guarantees the convergence of the estimation error to zero.

THEOREM 3. Under Assumption 1, if the hybrid observer (14) is implemented as in Algorithm 1, with $\mathcal{L}_q$ computed through the observers (17) and the map $(\Theta_q^\top)^\dagger \Xi_q$ in (21), then the estimated state $\hat{x}(t)$ has the property that:

$$\lim_{t \to \infty} |\hat{x}(t) - x(t)| = 0. \quad (25)$$

REMARK 4. Note that the computation of the gains L_j's requires the knowledge of switching times in order to generate converging estimates. Thus, post-processing of the data is required since L_j's cannot be computed a priori. Also, in the operation of the observer, we ignored the time required for computation at time instant t_q. In fact, the outcome $\xi_q(t_q^-)$ becomes available not at t_q but at $t_q + T_{comp}$ where T_{comp} is the time elapsed during computation. It is conjectured that the error caused by this time-delayed update in (14) can be suppressed by taking smaller value of c in (24) while the update is actually performed at $t_q + T_{comp}$ using another state-transport matrix. Detailed analysis on improving the quality of the observer is an ongoing work.

PROOF OF THEOREM 3. Using (15), it follows from Assumptions 1.1 and 1.3 that the estimation error $\tilde{x}(t)$ for the interval $[t_q, t_{q+1})$ is bounded by

$$|\tilde{x}(t)| = |e^{A_{q+1}(t-t_q)}\tilde{x}(t_q)| \leq e^{a(t-t_q)}|\tilde{x}(t_q)|$$

with a constant a such that $\|A_q\| \leq a$ for all $q \in \mathbb{N}$, and thus,

$$|\tilde{x}(t)| \leq e^{aD}|\tilde{x}(t_q)|.$$

Therefore, if $|\tilde{x}(t_q)| \to 0$ as $q \to \infty$, then we achieve that

$$\lim_{t \to \infty} |\tilde{x}(t)| = 0. \quad (26)$$

The remainder of the proof shows that $|\tilde{x}(t_q)| \to 0$ as $q \to \infty$ under the conditions stated in the theorem statement.

Note that, $\tilde{x}(t_q^-)$ can be written as,

$$\tilde{x}(t_q^-) = \begin{bmatrix} Z^{q\top} \\ W^{q\top} \end{bmatrix}^{-1} \begin{bmatrix} z^q(t_q^-) \\ w^q(t_q^-) \end{bmatrix} = Z^q z^q(t_q^-) + W^q w^q(t_q^-). \quad (27)$$

The matrix $\Psi_i^j(\tau_{\{i+1,j\}})$, defined in (18), transports $\tilde{x}(t_i^-)$ to $\tilde{x}(t_j^-)$ along (15) by

$$\tilde{x}(t_j^-) = \Psi_i^j(\tau_{\{i+1,j\}})\tilde{x}(t_i^-) - \sum_{l=i}^{j-1} \Psi_l^j(\tau_{\{l+1,j\}})\xi_l(t_l^-). \quad (28)$$

We now have the following series of equivalent expressions for $\tilde{x}(t_q^-)$:

$$\begin{aligned}
\tilde{x}(t_q^-) &= Z^q z^q(t_q^-) + W^q w^q(t_q^-) \\
&= \Psi_{q-1}^q Z^{q-1} z^{q-1}(t_{q-1}^-) + \Psi_{q-1}^q W^{q-1} w^{q-1}(t_{q-1}^-) \\
&\quad - \Psi_{q-1}^q \xi_{q-1}(t_{q-1}^-) \\
&= \Psi_{q-2}^q Z^{q-2} z^{q-2}(t_{q-2}^-) + \Psi_{q-2}^q W^{q-2} w^{q-2}(t_{q-2}^-) \\
&\quad - \Psi_{q-2}^q \xi_{q-2}(t_{q-2}^-) - \Psi_{q-1}^q \xi_{q-1}(t_{q-1}^-) \\
&\vdots \\
&= \Psi_{q-N}^q Z^{q-N} z^{q-N}(t_{q-N}^-) \\
&\quad + \Psi_{q-N}^q W^{q-N} w^{q-N}(t_{q-N}^-) - \sum_{l=q-N}^{q-1} \Psi_l^q \xi_l(t_l^-).
\end{aligned} \quad (29)$$

To appreciate the implication of this equivalence, we first note that for each $q - N \leq i \leq q$, the term $\Psi_i^q Z^i z^i(t_i^-)$ transports the observable information of the i-th mode from the interval $[t_{i-1}, t_i)$ to the time instant t_q^-. This observable information is corrupted by the unknown term $w^i(t_i^-)$, but since the information is being accumulated at t_q^- from modes $i = q - N, \cdots, q$, the idea is to combine the partial information from each mode to recover $\tilde{x}(t_q^-)$. This is where we use the notion of determinability. By Properties 1, 5, and 6 in the Appendix, and the fact that $\mathcal{R}(W^i)^\perp = (\ker G_i)^\perp = \mathcal{R}(G_i^\top)$ and $e^{-A_q^\top \tau_q}\mathcal{R}(G_q^\top) = \mathcal{R}(G_q^\top)$, it follows under Assumption 1.2 that

$$\begin{aligned}
&\mathcal{R}(W^q)^\perp + \mathcal{R}(\Psi_{q-1}^q W^{q-1})^\perp + \cdots + \mathcal{R}(\Psi_{q-N}^q W^{q-N})^\perp \\
&= e^{-A_q^\top \tau_q}\Big(\mathcal{R}(G_q^\top) + E_{q-1}^{-\top}\mathcal{R}(G_{q-1}^\top) + \\
&\quad \sum_{i=q-N}^{q-2} \Pi_{l=q-1}^{i+1} E_l^{-\top} e^{-A_l^\top \tau_l} E_i^{-\top}\mathcal{R}(G_i^\top) \Big) \\
&= e^{-A_q^\top \tau_q}\mathcal{M}_{q-N}^q = \mathbb{R}^n.
\end{aligned}$$

This equation shows that the matrix Θ_q defined in (19) has rank n, and is left-invertible. Keeping in mind that the range space of each Θ_i^q is orthogonal to $\mathcal{R}(\Psi_i^q W^i)$, each equality in (29) leads to the following relation:

$$\Theta_i^{q\top}\tilde{x}(t_q^-) = \Theta_i^{q\top}\left(\Psi_i^q Z^i z^i(t_i^-) - \sum_{l=i}^{q-1} \Psi_l^q \xi_l(t_l^-) \right), \quad (30)$$

for $i = q - N, \cdots, q$. Stacking (30) from $i = q$ to $i = q - N$, and employing the left-inverse of $\Theta_q^\top$, we obtain that

$$\tilde{x}(t_q^-) = (\Theta_q^\top)^\dagger \Xi_q(z_{\{q-N,q\}}^-, \xi_{\{q-N,q-1\}}^-) \quad (31)$$

9

where $z_{\{q-N,q\}}^-$ is defined similarly as in (20). It is seen from (31) that, if we are able to estimate $z_{\{q-N,q\}}^-$ without error, then the plant state $x(t_q^-)$ is exactly recovered by (31) because $x(t_q^-) = \hat{x}(t_q^-) - \tilde{x}(t_q^-)$ and both entities on the right side of the equation are known. However, since this is not the case, $z_{\{q-N,q\}}^-$ has been replaced with its estimate $\hat{z}_{\{q-N,q\}}^-$ in (21), and $\xi_q(t_q^-)$ is set as an estimate of $\tilde{x}(t_q^-)$ there.

Thanks to the linearity of Ξ_q in its arguments, it is noted that,

$$
\begin{aligned}
\tilde{x}(t_q) &= E_q(\tilde{x}(t_q^-) - \xi_q(t_q^-)) \\
&= E_q(\Theta_q^\top)^\dagger \Big(\Xi_q(z_{\{q-N,q\}}^-, \xi_{\{q-N,q-1\}}^-) \\
&\quad - \Xi_q(\hat{z}_{\{q-N,q\}}^-, \xi_{\{q-N,q-1\}}^-) \Big) \\
&= -E_q(\Theta_q^\top)^\dagger \Xi_q(\tilde{z}_{\{q-N,q\}}^-, 0)
\end{aligned}
\tag{32}
$$

where $\tilde{z}_{\{q-N,q\}}^- := \hat{z}_{\{q-N,q\}}^- - z_{\{q-N,q\}}^- = \mathrm{col}(\tilde{z}^{q-N}(t_{q-N}^-), \dots, \tilde{z}^q(t_q^-))$. It follows from (16) and (17) that

$$
\tilde{z}^i(t_{i-1}) = \hat{z}^i(t_{i-1}) - z^i(t_{i-1}) = 0 - Z^{i\top}\tilde{x}(t_{i-1})
$$

and

$$
\tilde{z}^i(t_i^-) = e^{(S_i - L_i R_i)\tau_i}\tilde{z}^i(t_{i-1}) = -e^{(S_i - L_i R_i)\tau_i}Z^{i\top}\tilde{x}(t_{i-1}).
$$

Plugging this expression in (32), and using the definition of M_j^q ($j = q-N, \dots, q$) from (22), we get

$$
\tilde{x}(t_q) = \sum_{j=q-N}^{q} M_j^q(\tau_{\{q-N+1,q\}})Z^j e^{(S_j - L_j R_j)\tau_j}Z^{j\top}\tilde{x}(t_{j-1}).
$$

Then, from the selection of gains L_j's satisfying (23), we have that

$$
|\tilde{x}(t_q)| \le \sum_{j=q-N}^{q} c|\tilde{x}(t_{j-1})|.
\tag{33}
$$

Finally, the statement of the following lemma, proof of which appears in the Appendix, aids us in the completion of the proof of Theorem 3

LEMMA 1. A sequence $\{a_i\}$ satisfying

$$
|a_i| \le c(|a_{i-1}| + |a_{i-2}| + \cdots + |a_{i-N-1}|), \qquad i > N,
$$

with $0 \le c < 1/(N+1)$ converges to zero: $\lim_{i\to\infty} a_i = 0$.

Applying Lemma 1 to (33), we see that $|\tilde{x}(t_q)| \to 0$ as $q \to \infty$, whence the desired result follows. $\square$

EXAMPLE 2. We demonstrate the operation of the proposed observer for the switched system considered in Example 1. We assume that each mode is activated for τ seconds, so that the persistent switching signal is:

$$
\sigma(t) = \begin{cases} 1 & \text{if } t \in [2k\tau, (2k+1)\tau), \\ 2 & \text{if } t \in [(2k+1)\tau, (2k+2)\tau), \end{cases}
\tag{34}
$$

where $k = 0, 1, 2, \cdots$, and the underlying assumption is that $\tau \neq \kappa\pi$ for any $\kappa \in \mathbb{N}$. As mentioned earlier, the system is observable (and thus, determinable) with this switching signal if the mode sequence $1 \to 2 \to 1$ is contained in a time interval. Hence, we pick $N = 3$ for Assumption 1.2 to hold. For brevity, we call $[2k\tau, (2k+1)\tau)$, the *odd* interval, and

$[(2k+1)\tau, (2k+2)\tau)$, the *even* interval. With an arbitrary initial condition $\hat{x}(0)$, the observer to be implemented is:

$$
\left.\begin{aligned} \dot{\hat{x}}(t) &= A_1\hat{x}(t) \\ \hat{y}(t) &= C_1\hat{x}(t) \end{aligned}\right\}, \quad t \in [2k\tau, (2k+1)\tau),
\tag{35a}
$$

$$
\left.\begin{aligned} \dot{\hat{x}}(t) &= A_2\hat{x}(t) \\ \hat{y}(t) &= C_2\hat{x}(t) \end{aligned}\right\}, \quad t \in [(2k+1)\tau, (2k+2)\tau),
\tag{35b}
$$

$$
\hat{x}(q\tau) = \hat{x}(q\tau^-) - \xi_q(q\tau^-), \qquad q > 3.
\tag{35c}
$$

In order to determine the value of $\xi_q(q\tau^-)$, we start off with the estimators for the observable modes of each subsystem, denoted by z^q in (16). Note that mode 1 has a one-dimensional observable subspace whereas for mode 2, the unobservable subspace is $\mathbb{R}^2$. Since mode 1 is active on every odd interval and mode 2 on every even interval, z^q for every odd q represents the partial information obtained from mode 1, and z^q for every even q is a null vector as no information is extracted from mode 2. So the one-dimensional z-observer in (17) is implemented only for odd intervals. For odd q, we compute

$$
G_q = \begin{bmatrix} 1 & 0 \\ 0 & 0 \end{bmatrix}, \mathcal{R}(G_q^\top) = \mathrm{span}\left\{\begin{bmatrix} 1 \\ 0 \end{bmatrix}\right\}, W^q = \begin{bmatrix} 0 \\ 1 \end{bmatrix}, Z^q = \begin{bmatrix} 1 \\ 0 \end{bmatrix},
$$

so that one may choose $S^q = 0$ and $R^q = 1$, which yields the observer (17) as

$$
\dot{\hat{z}}^q = -l_q\hat{z}^q + l_q\tilde{y}, \qquad t \in [(q-1)\tau, q\tau), \quad q: \text{odd},
$$

with the initial condition $\hat{z}^q((q-1)\tau) = 0$, and $\tilde{y}$ being the difference between the measured output and the estimated output of (35). The gain l_q will be chosen later by (36). For q even, we take $W^q = I_{2\times 2}$, and $G_q = 0_{2\times 2}$, so that Z^q, S^q, R^q are null-matrices.

The next step is to use the value of $\hat{z}^q(q\tau^-)$ to compute $\xi_q(q\tau^-)$, $q > 3$. We use the notation ξ^q to denote $\xi_q(q\tau^-)$, and let ξ_1^q be the first component of the vector ξ^q. Since $N = 3$, it follows from (14c) that $\xi^1 = \xi^2 = \xi^3 = \mathrm{col}(0,0)$. The matrices appearing in the computation of ξ^q are given as follows: for every *odd* $q > 3$:

$$
\Psi_{q-3}^q = \begin{bmatrix} \cos\tau & \sin\tau \\ -\sin\tau & \cos\tau \end{bmatrix} \Rightarrow (\Psi_{q-3}^q I_{2\times 2})^\perp = \{0\},
$$

$$
\Psi_{q-2}^q = \begin{bmatrix} \cos\tau & \sin\tau \\ -\sin\tau & \cos\tau \end{bmatrix} \Rightarrow \left(\Psi_{q-2}^q \begin{bmatrix} 0 \\ 1 \end{bmatrix}\right)^\perp = \left\{\begin{bmatrix} \cos\tau \\ -\sin\tau \end{bmatrix}\right\},
$$

$$
\Psi_{q-1}^q = I_{2\times 2} \Rightarrow (\Psi_{q-1}^q I_{2\times 2})^\perp = \{0\},
$$

$$
\Psi_q^q = I_{2\times 2} \Rightarrow \left(\Psi_q^q \begin{bmatrix} 0 \\ 1 \end{bmatrix}\right)^\perp = \left\{\begin{bmatrix} 1 \\ 0 \end{bmatrix}\right\},
$$

where the braces $\{\cdot\}$ denote the linear combination of the elements it contains. These subspaces directly lead to the expressions for Θ_j^q, $j = q-3, \dots, q$, so that

$$
\Theta_q = \begin{bmatrix} 1 & \cos\tau \\ 0 & -\sin\tau \end{bmatrix}, \quad q = 5, 7, \dots,
$$

where we have used the convention that Θ_i^q is a null matrix whenever $\mathcal{R}(\Psi_i^q(\tau_{\{i+1,q\}})W^i)^\perp = \{0\}$. Hence, the error correction term can be computed recursively for every odd $q > 3$ by the formula:

$$
\xi^q = \Theta_q^{-\top}\left[\begin{matrix} \hat{z}^q(t_q^-) \\ \hat{z}^{q-2}(t_{q-2}^-) - \xi_1^{q-2} - [\cos\tau \ -\sin\tau]\xi^{q-1} \end{matrix}\right].
$$

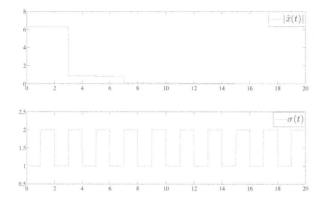
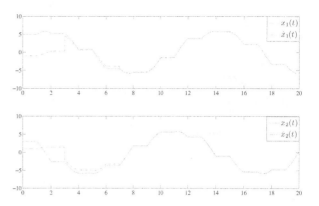

Figure 1: Size of state estimation error and the switching signal

Figure 2: Converging state estimates

Also, for odd q, we obtain that $M_{q-1}^q = 0$, $M_{q-3}^q = 0$,

$$M_q^q = \begin{bmatrix} 1 & 0 \\ \frac{\cos \tau}{\sin \tau} & 0 \end{bmatrix}, \quad \text{and} \quad M_{q-2}^q = \begin{bmatrix} 0 & 0 \\ -\frac{1}{\sin \tau} & 0 \end{bmatrix}.$$

Next, for every *even* $q > 3$, we repeat the same calculations to get:

$$\Psi_{q-3}^q = \begin{bmatrix} \cos 2\tau & \sin 2\tau \\ -\sin 2\tau & \cos 2\tau \end{bmatrix} \Rightarrow \left(\Psi_{q-3}^q \begin{bmatrix} 0 \\ 1 \end{bmatrix} \right)^\perp = \left\{ \begin{bmatrix} \cos 2\tau \\ -\sin 2\tau \end{bmatrix} \right\},$$

$$\Psi_{q-2}^q = \begin{bmatrix} \cos \tau & \sin \tau \\ -\sin \tau & \cos \tau \end{bmatrix} \Rightarrow \left(\Psi_{q-2}^q I_{2\times 2} \right)^\perp = \{0\},$$

$$\Psi_{q-1}^q = \begin{bmatrix} \cos \tau & \sin \tau \\ -\sin \tau & \cos \tau \end{bmatrix} \Rightarrow \left(\Psi_{q-1}^q \begin{bmatrix} 0 \\ 1 \end{bmatrix} \right)^\perp = \left\{ \begin{bmatrix} \cos \tau \\ -\sin \tau \end{bmatrix} \right\},$$

$$\Psi_q^q = I_{2\times 2} \Rightarrow \left(\Psi_q^q I_{2\times 2} \right)^\perp = \{0\}.$$

Once again, using the expressions for Θ_j^q, $j = q-3, \ldots, q$, based on these subspaces, one gets,

$$\Theta_q = \begin{bmatrix} \cos \tau & \cos 2\tau \\ -\sin \tau & -\sin 2\tau \end{bmatrix}, \quad q = 4, 6, 8, \cdots,$$

so that

$$\xi^q = \Theta_q^{-\top} \begin{bmatrix} \hat{z}^{q-1}(t_{q-1}^-) - \xi_1^{q-1} \\ \hat{z}^{q-3}(t_{q-3}^-) - \xi_1^{q-3} - [\cos \tau \ -\sin \tau](\xi^{q-2} + \xi^{q-1}) \end{bmatrix}.$$

Again, we obtain for even q that $M_q^q = M_{q-2}^q = 0$,

$$M_{q-1}^q = \begin{bmatrix} \frac{\sin 2\tau}{\sin \tau} & 0 \\ \frac{\cos 2\tau}{\sin \tau} & 0 \end{bmatrix} \quad \text{and} \quad M_{q-3}^q = \begin{bmatrix} -1 & 0 \\ -\frac{\cos \tau}{\sin \tau} & 0 \end{bmatrix}.$$

By computing the induced 2-norm of a matrix, it is seen that, for any $q > 3$ and $q-3 \leq j \leq q$,

$$\| M_j^q Z^j e^{(S_j - l_j R_j)\tau_j} Z^{j\top} \| = \begin{cases} 0 & \text{if } j \text{ is even} \\ \frac{e^{-l_j \tau}}{|\sin \tau|} & \text{if } j \text{ is odd}. \end{cases}$$

Therefore, for the relation (24) and (23), it is enough to choose l_q (for odd q) such that

$$\frac{1}{|\sin \tau|} e^{-l_q \tau} < \frac{1}{N+1} = \frac{1}{4}$$

or,

$$l_q > \frac{1}{\tau} \ln \frac{4}{|\sin \tau|}. \tag{36}$$

Once again it can be seen that the singularity occurs when τ is an integer multiple of π. Moreover, if τ approaches this singularity, then the gain required for convergence gets arbitrarily large. This also explains why the knowledge of the switching signal is required in general to compute the observer gains.

Some results of simulations with $\tau = 1$ and $l_q = 2$ for odd q, are illustrated in Fig. 1 and Fig. 2. The figures clearly show the hybrid nature of the proposed observer, which is caused by the jump discontinuity in error correction. For this particular example, the error does not grow between the switching times because the subsystem 2 is just a rotating dynamics.

4. CONCLUSION

This paper has presented conditions for observability and determinability of switched linear systems with state jumps. Based on these conditions, an asymptotic observer is constructed that combines the partial information obtained from each mode to get an estimate of the state vector. Under the assumption of persistent switching, the error analysis shows that the estimate converges to the actual state. The proposed method relies on the homogeneity that a linear switched system and linear jump maps provide with. In fact, it is seen in (29) that the transportation of the partially observable state information (represented by z), obtained at each mode, can be computed even with some unobservable information (by w). Since homogeneity guarantees that the observable information is not altered by this transportation process, the unobservable components are simply filtered out after the transportation. We emphasize that this idea may not be transparently applied to nonlinear systems.

Acknowledgement

The authors appreciate the discussions with Stephan Trenn related to observability conditions.

Appendix: Proof of Lemma 1

Let $c = \alpha/(N+1)$ with $0 < \alpha < 1$. Then it is obvious that, for $i > N$,

$$|a_i| \leq \frac{\alpha}{N+1} \sum_{k=i-N-1}^{i-1} |a_k| \leq \alpha \max_{i-N-1 \leq k \leq i-1} |a_k|. \tag{37}$$

Similarly, it follows that

$$|a_{i+1}| \leq \alpha \max_{i-N \leq k \leq i} |a_k|$$

$$\leq \alpha \max \left\{ |a_{i-N-1}|, \max_{i-N \leq k \leq i-1} |a_k|, |a_i| \right\}$$

$$\leq \alpha \max_{i-N-1 \leq k \leq i-1} |a_k|,$$

where the last inequality follows from (37). By induction, this leads to

$$\max_{i \leq k \leq i+N} |a_k| \leq \alpha \max_{i-N-1 \leq k \leq i-1} |a_k|.$$

that is, the maximum value of the sequence $\{a_i\}$ over the length of window $N+1$ is strictly decreasing and converging to zero, which proves the desired result. □

Appendix: Some Useful Facts

Let $\mathcal{V}_1$, $\mathcal{V}_2$, and $\mathcal{V}$ be any linear subspaces, A be a (not necessarily invertible) $n \times n$ matrix, and B, C be matrices of suitable dimension. The following properties can be found in the literature such as [15], or developed with little effort.

1. $A\mathcal{R}(B) = \mathcal{R}(AB)$ and $A^{-1} \ker B = \ker(BA)$.
2. $A^{-1}A\mathcal{V} = \mathcal{V} + \ker A$, and $AA^{-1}\mathcal{V} = \mathcal{V} \cap \mathcal{R}(A)$.
3. $A^{-1}(\mathcal{V}_1 \cap \mathcal{V}_2) = A^{-1}\mathcal{V}_1 \cap A^{-1}\mathcal{V}_2$, and $A(\mathcal{V}_1 \cap \mathcal{V}_2) \subseteq A\mathcal{V}_1 \cap A\mathcal{V}_2$ (with equality if and only if $(\mathcal{V}_1 + \mathcal{V}_2) \cap \ker A = \mathcal{V}_1 \cap \ker A + \mathcal{V}_2 \cap \ker A$, which holds, in particular, for any invertible A).
4. $A\mathcal{V}_1 + A\mathcal{V}_2 = A(\mathcal{V}_1 + \mathcal{V}_2)$, and $A^{-1}\mathcal{V}_1 + A^{-1}\mathcal{V}_2 \subseteq A^{-1}(\mathcal{V}_1 + \mathcal{V}_2)$ (with equality if and only if $(\mathcal{V}_1 + \mathcal{V}_2) \cap \mathcal{R}(A) = \mathcal{V}_1 \cap \mathcal{R}(A) + \mathcal{V}_2 \cap \mathcal{R}(A)$, which holds, in particular, for any invertible A).
5. $(\ker A)^\perp = \mathcal{R}(A^\top)$.
6. $(A^\top \mathcal{V})^\perp = A^{-1}\mathcal{V}^\perp$ and $(A^{-1}\mathcal{V})^\perp = A^\top \mathcal{V}^\perp$.
7. $\langle A|\mathcal{V}\rangle = \mathcal{V} + A\mathcal{V} + A^2\mathcal{V} + \cdots + A^{n-1}\mathcal{V}$ and $\langle \mathcal{V}|A\rangle = \mathcal{V} \cap A^{-1}\mathcal{V} \cap A^{-2}\mathcal{V} \cap \cdots \cap A^{-(n-1)}\mathcal{V}$.
8. $\langle \mathcal{V}_1 \cap \mathcal{V}_2|A\rangle = \langle \mathcal{V}_1|A\rangle \cap \langle \mathcal{V}_2|A\rangle$ and $\langle A|\mathcal{V}_1 \cap \mathcal{V}_2\rangle \subset \langle A|\mathcal{V}_1\rangle \cap \langle A|\mathcal{V}_2\rangle$.
9. $e^{At}\mathcal{V} \subseteq \langle A|\mathcal{V}\rangle$ and $\langle \mathcal{V}|A\rangle \subseteq e^{At}\mathcal{V}$ for any t.
10. $\langle A|\mathcal{V}\rangle^\perp = \langle \mathcal{V}^\perp|A^\top\rangle$.

Now, with $G := \mathrm{col}(C, CA, \ldots, CA^{n-1})$,

11. $e^{At} \ker G = \ker G$ and $e^{A^\top t}\mathcal{R}(G^\top) = \mathcal{R}(G^\top)$ for all t.
12. $\langle \ker G|A\rangle = \ker G$ and $\langle A^\top|\mathcal{R}(G^\top)\rangle = \mathcal{R}(G^\top)$.

5. REFERENCES

[1] A. Alessandri and P. Coletta. Switching observers for continuous-time and discrete-time linear systems. In *American Control Conference*, pages 2516–2521, 2001.

[2] M. Babaali and G. J. Pappas. Observability of switched linear systems in continuous time. In M. Morari and L. Thiele, editors, *Hyb. Sys: Comp. & Control*, volume 3414 of *Lecture Notes in Computer Science*, pages 103–117. Springer Berlin/Heidelberg, 2005.

[3] A. Balluchi, L. Benvenuti, M. D. Di Benedetto, and A. L. Sangiovanni-Vincentelli. Observability for hybrid systems. In *IEEE Conf. on Decision and Control*, pages 1159–1164, 2003.

[4] A. Balluchi, L. Benvenuti, M. D. Di Benedetto, and A. L. Sangiovanni-Vincentelli. Design of observers for hybrid systems. In C. Tomlin and M. Greenstreet,

editors, *Hyb. Sys: Comp. & Control*, volume 2289 of *Lecture Notes in Computer Science*, pages 76–89. Springer Berlin/Heidelberg, 2002.

[5] C. T. Chen. *Linear System Theory and Design*. Oxford University Press, Inc., New York, 3rd edition, 1999.

[6] P. Collins and J. H. van Schuppen. Observability of piecewise-affine hybrid systems. In R. Alur and G. J. Pappas, editors, *Hyb. Sys: Comp. & Control*, volume 2993 of *Lecture Notes in Computer Science*, pages 265–279. Springer Berlin/Heidelberg, 2004.

[7] M. Fliess, C. Join, and W. Perruquetti. Real-time estimation for switched linear systems. In *IEEE Conf. on Decision and Control*, pages 941–946, 2008.

[8] E. A. Medina and D. A. Lawrence. Reachability and observability of linear impulsive systems. *Automatica*, 44(5):1304–1309, 2008.

[9] E. A. Medina and D. A. Lawrence. State estimation for linear impulsive systems. In *American Control Conference*, pages 1183–1188, 2009.

[10] S. Pettersson. Designing switched observers for switched systems using multiple Lyapunov functions and dwell-time switching. In *IFAC Conf. on Analysis and Design of Hybrid Systems*, pages 18–23, 2006.

[11] Z. Sun and S. S. Ge. *Switched Linear Systems: Control and Design*. Springer-Verlag London, 2005.

[12] A. Tanwani and S. Trenn. On observability of switched differential-algebraic equations. In *IEEE Conf. on Decision and Control*, pages 5656–5661, 2010.

[13] R. Vidal, A. Chiuso, S. Soatto, and S. Sastry. Observability of linear hybrid systems. In O. Maler and A. Pnueli, editors, *Hyb. Sys: Comp. & Control*, volume 2623 of *Lecture Notes in Computer Science*, pages 526–539. Berlin, Germany: Springer, 2003.

[14] L. Vu and D. Liberzon. Invertibility of switched linear systems. *Automatica*, 44(4):949–958, 2008.

[15] W. M. Wonham. *Linear Multivariable Control: A Geometric Approach*. Springer-Verlag, New York, 3rd edition, 1985.

[16] G. Xie and L. Wang. Necessary and sufficient conditions for controllability and observability of switched impulsive control systems. *IEEE Trans. on Automatic Control*, 49:960–966, 2004.

Analysis of the Joint Spectral Radius via Lyapunov Functions on Path-Complete Graphs

Amir Ali Ahmadi (a_a_a@mit.edu)
Raphaël M. Jungers (raphael.jungers@uclouvain.be)
Pablo A. Parrilo (parrilo@mit.edu)
Mardavij Roozbehani (mardavij@mit.edu)
Laboratory for Information and Decision Systems
Massachusetts Institute of Technology
Cambridge, MA, USA

ABSTRACT

We study the problem of approximating the joint spectral radius (JSR) of a finite set of matrices. Our approach is based on the analysis of the underlying switched linear system via inequalities imposed between multiple Lyapunov functions associated to a labeled directed graph. Inspired by concepts in automata theory and symbolic dynamics, we define a class of graphs called path-complete graphs, and show that any such graph gives rise to a method for proving stability of the switched system. This enables us to derive several asymptotically tight hierarchies of semidefinite programming relaxations that unify and generalize many existing techniques such as common quadratic, common sum of squares, maximum/minimum-of-quadratics Lyapunov functions. We characterize all path-complete graphs consisting of two nodes on an alphabet of two matrices and compare their performance. For the general case of any set of $n \times n$ matrices we propose semidefinite programs of modest size that approximate the JSR within a multiplicative factor of $1/\sqrt[4]{n}$ of the true value. We establish a notion of duality among path-complete graphs and a constructive converse Lyapunov theorem for maximum/minimum-of-quadratics Lyapunov functions.

Categories and Subject Descriptors

I.1.2 [**Symbolic and algebraic manipulations**]: Algorithms—*Analysis of algorithms*

General Terms

Theory, algorithms

Keywords

Joint spectral radius, Lyapunov methods, finite automata, semidefinite programming, stability of switched systems

1. INTRODUCTION

Given a finite set of matrices $\mathcal{A} := \{A_1, ..., A_m\}$, their *joint spectral radius* $\rho(\mathcal{A})$ is defined as

$$\rho(\mathcal{A}) = \lim_{k \to \infty} \max_{\sigma \in \{1,...,m\}^k} \|A_{\sigma_k} ... A_{\sigma_2} A_{\sigma_1}\|^{1/k}, \qquad (1)$$

where the quantity $\rho(\mathcal{A})$ is independent of the norm used in (1). The joint spectral radius (JSR) is a natural generalization of the spectral radius of a single square matrix and it characterizes the maximal growth rate that can be obtained by taking products, of arbitrary length, of all possible permutations of $A_1, ..., A_m$. This concept was introduced by Rota and Strang [26] in the early 60s and has since been the subject of extensive research within the engineering and the mathematics communities alike. Aside from a wealth of fascinating mathematical questions that arise from the JSR, the notion emerges in many areas of application such as stability of switched linear dynamical systems, computation of the capacity of codes, continuity of wavelet functions, convergence of consensus algorithms, trackability of graphs, and many others. See [16] and references therein for a recent survey of the theory and applications of the JSR.

Motivated by the abundance of applications, there has been much work on efficient computation of the joint spectral radius; see e.g. [1], [5], [4], [21], and references therein. Unfortunately, the negative results in the literature certainly restrict the horizon of possibilities. In [6], Blondel and Tsitsiklis prove that even when the set $\mathcal{A}$ consists of only two matrices, the question of testing whether $\rho(\mathcal{A}) \leq 1$ is undecidable. They also show that unless P=NP, one cannot compute an approximation $\hat{\rho}$ of ρ that satisfies $|\hat{\rho} - \rho| \leq \epsilon\rho$, in a number of steps polynomial in the size of $\mathcal{A}$ and ϵ [27]. It is easy to show that the spectral radius of any finite product of length k raised to the power of $1/k$ gives a lower bound on ρ [16]. Our focus, however, will be on computing good upper bounds for ρ, which requires more elaborate techniques.

There is an attractive connection between the joint spectral radius and the stability properties of an arbitrary switched linear system; i.e., dynamical systems of the form

$$x_{k+1} = A_{\sigma(k)} x_k, \qquad (2)$$

where $\sigma : \mathbb{Z} \to \{1, ..., m\}$ is a map from the set of integers to the set of indices. It is well-known that $\rho < 1$ if and only if system (2) is *absolutely asymptotically stable* (AAS), that is, (globally) asymptotically stable for all switching sequences. Moreover, it is known [18] that absolute asymptotic stability of (2) is equivalent to absolute asymptotic stability of the linear difference inclusion

$$x_{k+1} \in \text{co}\mathcal{A} \; x_k, \qquad (3)$$

where co$\mathcal{A}$ here denotes the convex hull of the set $\mathcal{A}$. Therefore, any method for obtaining upper bounds on the joint spectral radius provides sufficient conditions for stability of systems of type (2) or (3). Conversely, if we can prove absolute asymptotic stability of (2) or (3) for the set $\mathcal{A}_\gamma := \{\gamma A_1, \ldots, \gamma A_m\}$ for some positive scalar γ, then we get an upper bound of $\frac{1}{\gamma}$ on $\rho(\mathcal{A})$. (This follows from the scaling property of the JSR: $\rho(\mathcal{A}_\gamma) = \gamma\rho(\mathcal{A})$.) One advantage of working with the notion of the joint spectral radius is that it gives a way of rigorously quantifying the performance guarantee of different techniques for stability analysis of systems (2) or (3).

Perhaps the most well-established technique for proving stability of switched systems is the use of a *common (or simultaneous) Lyapunov function*. The idea here is that if there is a continuous, positive, and homogeneous (Lyapunov) function $V(x): \mathbb{R}^n \to \mathbb{R}$ that for some $\gamma > 1$ satisfies

$$V(\gamma A_i x) \leq V(x) \quad \forall i = 1, \ldots, m, \quad \forall x \in \mathbb{R}^n, \qquad (4)$$

(i.e., $V(x)$ decreases no matter which matrix is applied), then the system in (2) (or in (3)) is AAS. Conversely, it is known that if the system is AAS, then there exists a *convex* common Lyapunov function (in fact a norm); see e.g. [16, p. 24]. However, this function is not in general finitely constructable. A popular approach has been to try to approximate this function by a class of functions that we can efficiently search for using semidefinite programming. Semidefinite programs (SDPs) can be solved with arbitrary accuracy in polynomial time and lead to efficient computational methods for approximation of the JSR. As an example, if we take the Lyapunov function to be quadratic (i.e., $V(x) = x^T P x$), then the search for such a Lyapunov function can be formulated as the following SDP:

$$\begin{aligned} P &\succ 0 \\ \gamma^2 A_i^T P A_i &\preceq P \quad \forall i = 1, \ldots, m. \end{aligned} \qquad (5)$$

The quality of performance of common quadratic (CQ) Lyapunov functions is a well-studied topic. In particular, it is known [5] that the estimate $\hat{\rho}_{CQ}$ obtained by this method[1] satisfies

$$\frac{1}{\sqrt{n}}\hat{\rho}_{CQ}(\mathcal{A}) \leq \rho(\mathcal{A}) \leq \hat{\rho}_{CQ}(\mathcal{A}), \qquad (6)$$

where n is the dimension of the matrices. This bound is a direct consequence of John's ellipsoid theorem and is known to be tight [3].

In [21], the use of sum of squares (SOS) polynomial Lyapunov functions of degree $2d$ was proposed as a common Lyapunov function for the switched system in (2). The search for such a Lyapunov function can again be formulated as a semidefinite program. This method does considerably better than a common quadratic Lyapunov function in practice and its estimate $\hat{\rho}_{SOS,2d}$ satisfies the bound

$$\frac{1}{\sqrt[2d]{\eta}}\hat{\rho}_{SOS,2d}(\mathcal{A}) \leq \rho(\mathcal{A}) \leq \hat{\rho}_{SOS,2d}(\mathcal{A}), \qquad (7)$$

---

[1]The estimate $\hat{\rho}_{CQ}$ is the reciprocal of the largest γ that satisfies (5) and can be found by bisection.

where $\eta = \min\{m, \binom{n+d-1}{d}\}$. Furthermore, as the degree $2d$ goes to infinity, the estimate $\hat{\rho}_{SOS,2d}$ converges to the true value of ρ [21]. The semidefinite programming based methods for approximation of the JSR have been recently generalized and put in the framework of conic programming [22].

1.1 Contributions and organization

It is natural to ask whether one can develop better approximation schemes for the joint spectral radius by using multiple Lyapunov functions as opposed to requiring simultaneous contractibility of a single Lyapunov function with respect to all the matrices. More concretely, our goal is to understand how we can write inequalities among, say, k different Lyapunov functions $V_1(x), \ldots, V_k(x)$ that imply absolute asymptotic stability of (2) and can be checked via semidefinite programming.

The general idea of using several Lyapunov functions for analysis of switched systems is a very natural one and has already appeared in the literature (although to our knowledge not in the context of the approximation of the JSR); see e.g. [15], [7], [13], [12], [10]. Perhaps one of the earliest references is the work on "piecewise quadratic Lyapunov functions" in [15]. However, this work is in the different framework of constrained switching, where the dynamics switches depending on which region of the space the trajectory is traversing (as opposed to arbitrary switching). In this setting, there is a natural way of using several Lyapunov functions: assign one Lyapunov function per region and "glue them together". Closer to our setting, there is a body of work in the literature that gives sufficient conditions for existence of piecewise Lyapunov functions of the type $\max\{x^T P_1 x, \ldots, x^T P_k x\}$, $\min\{x^T P_1 x, \ldots, x^T P_k x\}$, and conv$\{x^T P_1 x, \ldots, x^T P_k x\}$, i.e, the pointwise maximum, the pointwise minimum, and the convex envelope of a set of quadratic functions [13], [12], [10], [14]. These works are mostly done in continuous time for analysis of linear differential inclusions, but they have obvious discrete time counterparts. The main drawback of these methods is that in their greatest generality, they involve solving bilinear matrix inequalities, which are nonconvex and in general NP-hard. One therefore has to turn to heuristics, which have no performance guarantees and their computation time quickly becomes prohibitive when the dimension of the system increases. Moreover, all of these methods solely provide sufficient conditions for stability with no performance guarantees.

There are several unanswered questions that in our view deserve a more thorough study: (i) With a focus on conditions that are amenable to convex optimization, what are the different ways to write a set of inequalities among k Lyapunov functions that imply absolute asymptotic stability of (2)? Can we give a unifying framework that includes the previously proposed Lyapunov functions and perhaps also introduces new ones? (ii) Among the different sets of inequalities that imply stability, can we identify some that are less conservative than some other? (iii) The available methods on piecewise Lyapunov functions solely provide sufficient conditions for stability with no guarantee on their performance. Can we give converse theorems that guarantee the existence of a feasible solution to our search for a given accuracy?

In this work, we provide the foundation to answer these questions. More concretely, our contributions can be summarized as follows. We propose a unifying framework based on a representation of Lyapunov inequalities with labeled graphs and making some connections with basic concepts in automata theory. This is done in Section 2, where we define the notion of a path-complete graph (Definition 2.2) and prove that any such graph provides an approximation scheme for the JSR (Theorem 2.4). In this section, we also show that many of the previously proposed methods come from particular classes of path-complete graphs (e.g., Corollary 2.5 and Corollary 2.6). In Section 3, we characterize all the path-complete graphs with two nodes for the analysis of the JSR of two matrices. We completely determine how the approximations obtained from these graphs compare. In Section 4, we study in more depth the approximation properties of a particular class of path-complete graphs that seem to perform very well in practice. We prove in Section 4.1 that certain path-complete graphs that are in some sense dual to each other always give the same bound on the JSR (Theorem 4.1). We present a numerical example in Section 4.2. Section 5 includes approximation guarantees for a subclass of our methods, and in particular a converse theorem for the method of max-of-quadratics Lyapunov functions (Theorem 5.1). Finally, our conclusions and some future directions are discussed in Section 6.

2. PATH-COMPLETE GRAPHS AND THE JOINT SPECTRAL RADIUS

In what follows, we will think of the set of matrices $\mathcal{A} := \{A_1, ..., A_m\}$ as a finite alphabet and we will often refer to a finite product of matrices from this set as a *word*. We denote the set of all words $A_{i_t} \ldots A_{i_1}$ of length t by $\mathcal{A}^t$. Contrary to the standard convention in automata theory, our convention is to read a word from right to left. This is in accordance with the order of matrix multiplication. The set of all finite words is denoted by $\mathcal{A}^*$; i.e., $\mathcal{A}^* = \bigcup_{t \in \mathbb{Z}^+} \mathcal{A}^t$.

The basic idea behind our framework is to represent through a graph all the possible occurrences of products that can appear in a run of the dynamical system in (2), and assert via some Lyapunov inequalities that no matter what occurrence appears, the product must remain stable. A convenient way of representing these Lyapunov inequalities is via a directed labeled graph $G(N, E)$. Each node of this graph is assigned to a (continuous, positive definite, and homogeneous) Lyapunov function $V_i(x) : \mathbb{R}^n \to \mathbb{R}$, and each edge is labeled by a finite product of matrices, i.e., by a word from the set $\mathcal{A}^*$. As illustrated in Figure 1, given two nodes with Lyapunov functions $V_i(x)$ and $V_j(x)$ and an arc going from node i to node j labeled with the matrix A_l, we write the Lyapunov inequality:

$$V_j(A_l x) \le V_i(x) \quad \forall x \in \mathbb{R}^n. \tag{8}$$

The problem that we are interested in is to understand which sets of Lyapunov inequalities imply stability of the switched system in (2). We will answer this question based on the corresponding graph.

For reasons that will become clear shortly, we would like to reduce graphs whose arcs have arbitrary labels from the set $\mathcal{A}^*$ to graphs whose arcs have labels from the set $\mathcal{A}$, i.e, labels of length one. This is explained next.

DEFINITION 2.1. *Given a labeled directed graph $G(N, E)$, we define its* expanded graph $G^e(N^e, E^e)$ *as the outcome of the following procedure. For every edge $(i, j) \in E$ with*

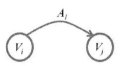

Figure 1: **Graphical representation of Lyapunov inequalities. The graph above corresponds to the Lyapunov inequality** $V_j(A_l x) \le V_i(x)$. **Here,** A_l **can be a single matrix from** $\mathcal{A}$ **or a finite product of matrices from** $\mathcal{A}$.

label $A_{ik} \ldots A_{i1} \in \mathcal{A}^k$, *where* $k > 1$, *we remove the edge* (i, j) *and replace it with k new edges* $(s_q, s_{q+1}) \in E^e \setminus E$: $q \in \{0, \ldots, k-1\}$, *where* $s_0 = i$ *and* $s_k = j$.[2] *(These new edges go from node i through $k-1$ newly added nodes $s_1, \ldots, s_{k-1}$ and then to node j.) We then label the new edges* $(i, s_1), \ldots, (s_q, s_{q+1}), \ldots, (s_{k-1}, j)$ *with* $A_{i1}, \ldots, A_{ik}$ *respectively.*

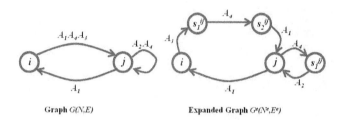

Graph $G(N,E)$ **Expanded Graph** $G^e(N^e, E^e)$

Figure 2: **Graph expansion: edges with labels of length more than one are broken into new edges with labels of length one.**

An example of a graph and its expansion is given in Figure 2. Note that if a graph has only labels of length one, then its expanded graph equals itself. The next definition is central to our development.

DEFINITION 2.2. *Given a directed graph $G(N, E)$ whose arcs are labeled with words from the set $\mathcal{A}^*$, we say that the graph is* path-complete, *if for all finite words* $A_{\sigma_k} \ldots A_{\sigma_1}$ *of any length k (i.e., for all words in $\mathcal{A}^*$), there is a directed path in its expanded graph $G^e(N^e, E^e)$ such that the labels on the edges of this path are the labels A_{σ_1} up to A_{σ_k}.*

In Figure 3, we present eight path-complete graphs on the alphabet $\mathcal{A} = \{A_1, A_2\}$. The fact that these graphs are path-complete is obvious for the graphs in (a), (b), (e), (f), and (h), but perhaps not so obvious for graphs in (c), (d), and (g). One way to check if a graph is path-complete is to think of it as a finite automaton by introducing an auxiliary start node (state) with free transitions to every node and by making all the other nodes be accepting states. Then, there are well-known algorithms (see e.g. [11, Chap. 4]) that check whether the language accepted by an automaton is $\mathcal{A}^*$, which is equivalent to the graph being path-complete. At least for the cases where the automata are deterministic (i.e., when all outgoing arcs from any node have different

---

[2]It is understood that the node index s_q depends on the original nodes i and j. To keep the notation simple we write s_q instead of s_q^{ij}.

labels), these algorithms are very efficient. (They run in $O(|N|^2)$ time.) Similar algorithms exist in the symbolic dynamics literature; see e.g. [19, Chap. 3]. Our interest in path-complete graphs stems from the Theorem 2.4 below that establishes that any such graph gives a method for approximation of the JSR. We introduce one last definition before we state this theorem.

DEFINITION 2.3. *Let $\mathcal{A} = \{A_1, \ldots, A_m\}$ be a set of matrices. Given a path-complete graph $G(N, E)$ and $|N|$ functions $V_i(x)$, we say that $\{V_i(x) \mid i = 1, \ldots, |N|\}$ is a Piecewise Lyapunov Function (PLF) associated with $G(N, E)$ if*

$$V_j(L(e)x) \leq V_i(x) \qquad \forall x \in \mathbb{R}^n, \quad \forall e \in E,$$

where $L(e) \in \mathcal{A}^*$ is the label associated with edge $e \in E$ going from node i to node j.

THEOREM 2.4. *Consider a finite set of matrices $\mathcal{A} = \{A_1, \ldots, A_m\}$. For a scalar $\gamma > 0$, let $\mathcal{A}_\gamma := \{\gamma A_1, \ldots, \gamma A_m\}$. Let $G(N, E)$ be a path-complete graph whose edges are labeled with words from $\mathcal{A}_\gamma^*$. If there exist positive, continuous, and homogeneous[3] Lyapunov functions $V_i(x)$, one per node of the graph, such that $\{V_i(x) \mid i = 1, \ldots, |N|\}$ is a piecewise Lyapunov function associated with $G(N, E)$, then $\rho(\mathcal{A}) \leq \frac{1}{\gamma}$.*

PROOF. We will first prove the claim for the special case where the labels of the arcs of $G(N, E)$ belong to $\mathcal{A}_\gamma$ and therefore $G(N, E) = G^e(N^e, E^e)$. The general case will be reduced to this case afterwards. Let us take the degree of homogeneity of the Lyapunov functions $V_i(x)$ to be d, i.e., $V_i(\lambda x) = \lambda^d V_i(x)$ for all $\lambda \in \mathbb{R}$. (The actual value of d is irrelevant.) By positivity, continuity, and homogeneity of $V_i(x)$, there exist scalars α_i and β_i with $0 < \alpha_i \leq \beta_i$ for $i = 1, \ldots, |N|$, such that

$$\alpha_i \|x\|^d \leq V_i(x) \leq \beta_i \|x\|^d, \tag{9}$$

for all $x \in \mathbb{R}^n$ and for all $i = 1, \ldots, |N|$. Let

$$\xi = \max_{i, j \in \{1, \ldots, |N|\}^2} \frac{\beta_i}{\alpha_j}. \tag{10}$$

Now consider an arbitrary product $A_{\sigma_k} \ldots A_{\sigma_1}$ of length k. Because the graph is path-complete, there will be a directed path corresponding to this product that consists of k arcs, and goes from some node i to some node j. If we write the chain of k Lyapunov inequalities associated with these arcs (cf. Figure 1), then we get

$$V_j(\gamma^k A_{\sigma_k} \ldots A_{\sigma_1} x) \leq V_i(x),$$

which by homogeneity of the Lyapunov functions can be rearranged to

$$\left(\frac{V_j(A_{\sigma_k} \ldots A_{\sigma_1} x)}{V_i(x)} \right)^{\frac{1}{d}} \leq \frac{1}{\gamma^k}. \tag{11}$$

---

[3]The requirement of homogeneity can be replaced by radial unboundedness which is implied by homogeneity and positivity. However, since the dynamical system in (2) is homogeneous, there is no conservatism in asking $V_i(x)$ to be homogeneous [25].

We can now bound the norm of $A_{\sigma_k} \ldots A_{\sigma_1}$ as follows:

$$
\begin{aligned}
\|A_{\sigma_k} \ldots A_{\sigma_1}\| &\leq \max_x \frac{\|A_{\sigma_k} \ldots A_{\sigma_1} x\|}{\|x\|} \\
&\leq \left(\frac{\beta_i}{\alpha_j} \right)^{\frac{1}{d}} \max_x \frac{V_j^{\frac{1}{d}}(A_{\sigma_k} \ldots A_{\sigma_1} x)}{V_i^{\frac{1}{d}}(x)} \\
&\leq \left(\frac{\beta_i}{\alpha_j} \right)^{\frac{1}{d}} \frac{1}{\gamma^k} \\
&\leq \xi^{\frac{1}{d}} \frac{1}{\gamma^k},
\end{aligned}
$$

where the last three inequalities follow from (9), (11), and (10) respectively. From the definition of the JSR in (1), after taking the k-th root and the limit $k \to \infty$, we get that $\rho(\mathcal{A}) \leq \frac{1}{\gamma}$ and the claim is established.

Now consider the case where at least one edge of $G(N, E)$ has a label of length more than one and hence $G^e(N^e, E^e) \neq G(N, E)$. We will start with the Lyapunov functions $V_i(x)$ assigned to the nodes of $G(N, E)$ and from them we will explicitly construct $|N^e|$ Lyapunov functions for the nodes of $G^e(N^e, E^e)$ that satisfy the Lyapunov inequalities associated to the edges in E^e. Once this is done, in view of our preceding argument and the fact that the edges of $G^e(N^e, E^e)$ have labels of length one by definition, the proof will be completed.

For $j \in N^e$, let us denote the new Lyapunov functions by $V_j^e(x)$. It is sufficient to give the construction for the case where $|N^e| = |N| + 1$. The result for the general case with $|N^e| = |N| + l$, $l > 1$, follows by induction. Let $s \in N^e \backslash N$ be the added node in the expanded graph, and $q, r \in N$ be such that $(s, q) \in E^e$ and $(r, s) \in E^e$ with A_{sq} and A_{rs} as the corresponding labels respectively. Define

$$
V_j^e(x) = \begin{cases} V_j(x), & \text{if} \quad j \in N \\ V_q(A_{sq}x), & \text{if} \quad j = s. \end{cases} \tag{12}
$$

By construction, r and q, and subsequently, A_{sq} and A_{rs} are uniquely defined and hence, $\{V_j^e(x) \mid j \in N^e\}$ is well defined. We only need to show that

$$V_q(A_{sq}x) \leq V_s^e(x) \tag{13}$$
$$V_s^e(A_{rs}x) \leq V_r(x). \tag{14}$$

Inequality (13) follows trivially from (12). Furthermore, it follows from (12) that

$$
\begin{aligned}
V_s^e(A_{rs}x) &= V_q(A_{sq}A_{rs}x) \\
&\leq V_r(x),
\end{aligned}
$$

where the inequality follows from the fact that for $i \in N$, the functions $V_i(x)$ satisfy the Lyapunov inequalities of the edges of $G(N, E)$. $\square$

REMARK 2.1. *If the matrix A_{sq} is not invertible, the extended function $V_j^e(x)$ as defined in (12) will only be positive semidefinite. However, since our goal is to approximate the JSR, we will never be concerned with invertibility of the matrices in $\mathcal{A}$. Indeed, since the JSR is continuous in the entries of the matrices [16], we can always perturb the matrices slightly to make them invertible without changing the JSR by much. In particular, for any $\alpha > 0$, there exist $0 < \varepsilon, \delta < \alpha$ such that*

$$\hat{A}_{sq} = \frac{A_{sq} + \delta I}{1 + \varepsilon}$$

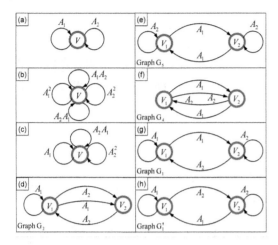

Figure 3: Examples of path-complete graphs for the alphabet $\{A_1, A_2\}$. If Lyapunov functions satisfying the inequalities associated with any of these graphs are found, then we get an upper bound of unity on $\rho(A_1, A_2)$.

is invertible and (12)–(14) are satisfied with $A_{sq} = \hat{A}_{sq}$.

To understand the generality of this framework more clearly, let us revisit the path-complete graphs in Figure 3 for the study of the case where the set $\mathcal{A} = \{A_1, A_2\}$ consists of only two matrices. For all of these graphs if our choice for the Lyapunov functions $V(x)$ or $V_1(x)$ and $V_2(x)$ are quadratic functions or sum of squares polynomial functions, then we can formulate the well-established semidefinite programs that search for these candidate Lyapunov functions.

The graph in (a), which is the simplest one, corresponds to the well-known common Lyapunov function approach. The graph in (b) is a common Lyapunov function applied to all products of length two. This graph also obviously implies stability.[4] But the graph in (c) tells us that if we find a Lyapunov function that decreases whenever A_1, A_2^2, and $A_2 A_1$ are applied (but with no requirement when $A_1 A_2$ is applied), then we still get stability. This is a priori not so obvious and we believe this approach has not appeared in the literature before. We will later prove (Theorem 5.2) a bound for the quality of approximation of path-complete graphs of this type, where a common Lyapunov function is required to decrease with respect to products of different lengths. The graph in (c) is also an example that explains why we needed the expansion process. Note that for the unexpanded graph, there is no path for the word $A_1 A_2$ or any succession of the word $A_1 A_2$, or for any word of the form A_2^{2k-1}, $k \in \mathbb{N}$. However, one can check that in the expanded graph of graph (c), there is a path for every finite word, and this in turn allows us to conclude stability from the Lyapunov inequalities of graph (c).

Let us comment now on the graphs with two nodes and four arcs, which each impose four Lyapunov inequalities. We can show that if $V_1(x)$ and $V_2(x)$ satisfy the inequalities of any of the graphs (d), (e), (f), or (g), then $\max\{V_1(x), V_2(x)\}$ is a common Lyapunov function for the switched system. If

---

[4]By slight abuse of terminology, we say that a graph implies stability meaning (of course) that the associated Lyapunov inequalities imply stability.

$V_1(x)$ and $V_2(x)$ satisfy the inequalities of any of the graphs in (e), (f), and (h), then $\min\{V_1(x), V_2(x)\}$ is a common Lyapunov function. These arguments serve as alternative proofs of stability and in the case where V_1 and V_2 are quadratic functions, they correspond to the works in [13], [12], [10], [14]. The next two corollaries prove these statements in a more general setting.

COROLLARY 2.5. *Consider a set of m matrices and the switched linear system in (2) or (3). If there exist k positive definite matrices P_j such that*

$$\forall \{i, k\} \in \{1, \ldots, m\}^2, \ \exists j \in \{1, \ldots, m\}$$
$$\text{such that} \quad \gamma^2 A_i^T P_j A_i \preceq P_k, \quad (15)$$

for some $\gamma > 1$, then the system is absolutely asymptotically stable. Moreover, the pointwise minimum

$$\min\{x^T P_1 x, \ldots, x^T P_k x\}$$

of the quadratic functions serves as a common Lyapunov function.

PROOF. The inequalities in (15) imply that every node of the associated graph has outgoing edges labeled with all the different m matrices. Therefore, it is obvious that the graph is path-complete. The proof that the pointwise minimum of the quadratics is a common Lyapunov function is easy and left to the reader. □

COROLLARY 2.6. *Consider a set of m matrices and the switched linear system in (2) or (3). If there exist k positive definite matrices P_j such that*

$$\forall \{i, j\} \in \{1, \ldots, m\}^2, \ \exists k \in \{1, \ldots, m\}$$
$$\text{such that} \quad \gamma^2 A_i^T P_j A_i \preceq P_k, \quad (16)$$

for some $\gamma > 1$, then the system is absolutely asymptotically stable. Moreover, the pointwise maximum

$$\max\{x^T P_1 x, \ldots, x^T P_k x\}$$

of the quadratic functions serves as a common Lyapunov function.

PROOF. The inequalities in (16) imply that every node of the associated graph has incoming edges labeled with all the different m matrices. This implies that the associated graph is path-complete. To see this, consider any product $A_{i_k} \ldots A_{i_1}$ and consider a new graph obtained by reversing the directions of all the edges. Since this new graph has now outgoing edges with all different labels for every node, it is clearly path-complete and in particular it has a path for the backwards word $A_{i_1} \ldots A_{i_k}$. If we now trace this path backwards, we get exactly a path in the original graph for the word $A_{i_k} \ldots A_{i_1}$.

The proof that the pointwise maximum of the quadratics is a common Lyapunov function is easy and again left to the reader. □

REMARK 2.2. *The linear matrix inequalities in (15) and (16) are (convex) sufficient conditions for existence of min-of-quadratics or max-of-quadratics Lyapunov functions. The converse is not true. The works in [13], [12], [10], [14] have additional multipliers in (15) and (16) that make the inequalities non-convex but when solved with a heuristic method contain a larger family of min-of-quadratics and max-of-quadratics Lyapunov functions. Even if the non-convex inequalities with multipliers could be solved exactly, except for* special cases where the $\mathcal{S}$-procedure is exact (e.g., the case of two quadratic functions), these methods still do not completely characterize min-of-quadratics and max-of-quadratics functions.

REMARK 2.3. *Two other well-established references in the literature that (when specialized to the analysis of arbitrary switched linear systems) turn out to be particular classes of path-complete graphs are the work in [17] on "path-dependent quadratic Lyapunov functions", and the work in [8] on "parameter dependent Lyapunov functions". In fact, the LMIs suggested in these works are special cases of Corollary 2.5 and 2.6 respectively, hence revealing a connection to the min/max-of-quadratics type Lyapunov functions. We will elaborate further on this connection in an extended version of this work.*

When we have so many different ways of imposing conditions for stability, it is natural to ask which ones are better. This seems to be a hard question in general, but we have studied in detail all the path-complete graphs with two nodes that imply stability for the case of a switched system with two matrices. This is the subject of the next section. The connections between the bounds obtained from these graphs are not always obvious. For example, we will see that the graphs (a), (e), and (f) always give the same bound on the joint spectral radius; i.e, one graph will succeed in proving stability if and only if the other will. So, there is no point in increasing the number of decision variables and the number of constraints and impose (e) or (f) in place of (a). The same is true for the graphs in (c) and (d), which makes graph (c) preferable to graph (d). (See Proposition 3.2.)

3. PATH-COMPLETE GRAPHS WITH TWO NODES

In this section, we characterize and compare all the path-complete graphs consisting of two nodes, an alphabet set $\mathcal{A} = \{A_1, A_2\}$, and arc labels of unit length. We refer the reader to [23], [24] for a more general understanding of how the Lyapunov inequalities associated to certain pairs of graphs relate to each other.

3.1 The set of path-complete graphs

The next lemma establishes that for thorough analysis of the case of two matrices and two nodes, we only need to examine graphs with four or less arcs.

LEMMA 3.1. *Let $G(\{1, 2\}, E)$ be a path-complete graph for $\mathcal{A} = \{A_1, A_2\}$ with labels of length one. Let $\{V_1, V_2\}$ be a piecewise Lyapunov function for G. If $|E| > 4$, then, either*
(i) there exists $\hat{e} \in E$ such that $G(\{1, 2\}, E \backslash \hat{e})$ is a path-complete graph,
(ii) either V_1 or V_2 or both are common Lyapunov functions for $\mathcal{A}$.

PROOF. If node 1 has more than one self-arc, then either these arcs have the same label, in which case one of them can be removed without changing the output set of words, or, they have different labels, in which case V_1 is a Lyapunov function for $\mathcal{A}$. By symmetry, the same argument holds for node 2. It remains to present a proof for the case where no node has more than one self-arc. If $|E| > 4$, then at least one node has three or more outgoing arcs. Without loss of generality let node 1 be as such, $e_1, e_2,$ and e_3 be

the corresponding arcs, and $L(e_1) = L(e_2) = A_1$. Let $\mathcal{D}(e)$ denote the destination node of $e \in E$. If $\mathcal{D}(e_1) = \mathcal{D}(e_2) = 2$, then e_1 (or e_2) can be removed without changing the output set. If $\mathcal{D}(e_1) \neq \mathcal{D}(e_2)$, assume, without loss of generality, that $\mathcal{D}(e_1) = 1$ and $\mathcal{D}(e_2) = 2$. Now, if $L(e_3) = A_1$, then regardless of its destination node, e_3 can be removed. The only remaining possibility is that $L(e_3) = A_2$ and $\mathcal{D}(e_3) = 2$. In this case, it can be verified that e_2 can be removed without affecting the output set of words. $\square$

It can be verified that a path-complete graph with two nodes and less than four arcs must necessarily place two arcs with different labels on one node, which necessitates existence of a single Lyapunov function for the underlying switched system. Since we are interested in exploiting the favorable properties of Piecewise Lyapunov Functions (PLFs) in approximation of the JSR we will focus on graphs with four arcs.

3.2 Comparison of performance

It can be verified that for path-complete graphs with two nodes, four arcs, and two matrices, and without multiple self loops on a single node, there are a total of nine distinct graph topologies to consider (several redundant cases arise which can be shown to be equivalent to one of the nine cases via swapping the nodes). Of the nine graphs, six have the property that every node has two incoming arcs with different labels—we call these *primal graphs*; six have the property that every node has two outgoing arcs with different labels—we call these *dual graphs*; and three are in both the primal and the dual set of graphs—we call these *self-dual graphs*. The self-dual graphs are least interesting to us since, as we will show, they necessitate existence of a single Lyapunov function for $\mathcal{A}$ (cf. Proposition 3.2, equation (19)).

The (strictly) primal graphs are Graph G_1 (Figure 3 (g)), Graph G_2, (Figure 3 (d)), and Graph G_3 which is obtained by swapping the roles of A_1 and A_2 in G_2 (not shown). The self-dual graphs are Graph G_4 (Figure 3 (f)), Graph G_5 (Figure 3 (e)), and Graph G_6 which is obtained by swapping the roles of A_1 and A_2 in G_5 (not shown). The (strictly) dual graphs are obtained by reversing the direction of the arrows in the primals and are denoted by G_1', G_2', G_3' respectively. For instance, G_1' is the graph shown in Figure 3 (h). The rest of the dual graphs are not shown.

Note that all of these graphs perform at least as well as a common Lyapunov function because we can always take $V_1(x) = V_2(x)$. We know from Corollary 2.6 and 2.5 that the primal graphs imply that $\max\{V_1(x), V_2(x)\}$ is a Lyapunov function, whereas, the dual graphs imply that $\min\{V_1(x), V_2(x)\}$ is a Lyapunov function.

Notation: Given a set of matrices $\mathcal{A} = \{A_1, \cdots, A_m\}$, a path-complete graph $G(N, E)$, and a class of functions $\mathcal{V}$, we denote by $\hat{\rho}_{\mathcal{V}, G}(\mathcal{A})$, the upperbound on the JSR of $\mathcal{A}$ that can be obtained by numerical optimization of PLFs $V_i \in \mathcal{V}$, $i \in \{1, \cdots |N|\}$, defined over G. With a slight abuse of notation, we denote by $\hat{\rho}_{\mathcal{V}}(\mathcal{A})$, the upperbound that is obtained by using a common Lyapunov function $V \in \mathcal{V}$.

PROPOSITION 3.2. *Consider $\mathcal{A} = \{A_1, A_2\}$, and let $\mathcal{A}_2 = \{A_1, A_2 A_1, A_2^2\}$, $\mathcal{A}_2' = \{A_1, A_1 A_2, A_2^2\}$, $\mathcal{A}_3 = \{A_2, A_1 A_2, A_1^2\}$, $\mathcal{A}_3' = \{A_2, A_2 A_1, A_1^2\}$. For $\mathcal{S} \in \{\mathcal{A}_i, \mathcal{A}_i' \mid i = 2, 3\}$, let $G_{\mathcal{S}} = G(\{1\}, E_{\mathcal{S}})$ be the graph with one node and three edges such that $\{L(e) \mid e \in E_{\mathcal{S}}\} = \mathcal{S}$.*

Then, we have

$$\hat{\rho}_{\mathcal{V},G_2}(\mathcal{A}) = \hat{\rho}_{\mathcal{V},G_{\mathcal{A}_2}}(\mathcal{A}), \quad \hat{\rho}_{\mathcal{V},G_3}(\mathcal{A}) = \hat{\rho}_{\mathcal{V},G_{\mathcal{A}_3}}(\mathcal{A}), \quad (17)$$

$$\hat{\rho}_{\mathcal{V},G_2'}(\mathcal{A}) = \hat{\rho}_{\mathcal{V},G_{\mathcal{A}_2'}}(\mathcal{A}), \quad \hat{\rho}_{\mathcal{V},G_3'}(\mathcal{A}) = \hat{\rho}_{\mathcal{V},G_{\mathcal{A}_3'}}(\mathcal{A}), \quad (18)$$

$$\hat{\rho}_{\mathcal{V},G_i}(\mathcal{A}) = \hat{\rho}_{\mathcal{V}}(\mathcal{A}), \qquad i = 4,5,6, \quad (19)$$

$$\hat{\rho}_{\mathcal{V},G_1}(\mathcal{A}) = \hat{\rho}_{\mathcal{V},G_1'}(\mathcal{A}). \quad (20)$$

PROOF. We start by proving the left equality in (17). Let $\{V_1, V_2\}$ be a PLF associated with G_2. It can be verified that V_1 is a Lyapunov function associated with $G_{\mathcal{A}_2}$, and therefore, $\hat{\rho}_{\mathcal{V},G_{\mathcal{A}_2}}(\mathcal{A}) \leq \hat{\rho}_{\mathcal{V},G_2}(\mathcal{A})$. Similarly, if $V \in \mathcal{V}$ is a Lyapunov function associated with $G_{\mathcal{A}_2}$, then one can check that $\{V_1, V_2 \mid V_1(x) = V(x), V_2(x) = V(A_2 x)\}$ is a PLF associated with G_2, and hence, $\hat{\rho}_{\mathcal{V},G_{\mathcal{A}_2}}(\mathcal{A}) \geq \hat{\rho}_{\mathcal{V},G_2}(\mathcal{A})$. The proofs for the rest of the equalities in (17) and (18) are analogous. The proof of (19) is as follows. Let $\{V_1, V_2\}$ be a PLF associated with G_i, $i = 4,5,6$. It can be then verified that $V = V_1 + V_2$ is a common Lyapunov function for $\mathcal{A}$, and hence, $\hat{\rho}_{\mathcal{V},G_i}(\mathcal{A}) \geq \hat{\rho}_{\mathcal{V}}(\mathcal{A})$, $i = 4,5,6$. The other direction is trivial: If $V \in \mathcal{V}$ is a common Lyapunov function for $\mathcal{A}$, then $\{V_1, V_2 \mid V_1 = V_2 = V\}$ is a PLF associated with G_i, and hence, $\hat{\rho}_{\mathcal{V},G_i}(\mathcal{A}) \leq \hat{\rho}_{\mathcal{V}}(\mathcal{A})$, $i = 4,5,6$. The proof of (20) is based on similar arguments; the PLFs associated with G_1 and G_1' can be derived from one another via $V_1'(A_1 x) = V_1(x)$, and $V_2'(A_2 x) = V_2(x)$. $\square$

REMARK 3.1. *Proposition 3.2 (20) establishes the equivalence of the bounds obtained from the primal and dual graphs G_1 and G_1' for general class of Lyapunov functions. This, however, is not true for graphs G_2 and G_3 and there exist examples for which*

$$\hat{\rho}_{\mathcal{V},G_2}(\mathcal{A}) \neq \hat{\rho}_{\mathcal{V},G_2'}(\mathcal{A}),$$

$$\hat{\rho}_{\mathcal{V},G_3}(\mathcal{A}) \neq \hat{\rho}_{\mathcal{V},G_3'}(\mathcal{A}).$$

The three primal graphs G_1, G_2, and G_3 can outperform one another depending on the problem data. We ran 100 test cases on random 5×5 matrices with elements uniformly distributed in $[-1, 1]$, and observed that G_1 resulted in the least conservative bound on the JSR in approximately 77% of the test case, and G_2 and G_3 in approximately 53% of the test cases (the overlap is due to ties). Furthermore, $\hat{\rho}_{\mathcal{V},G_1}(\{A_1, A_2\})$ is invariant under (i) relabeling of A_1 and A_2 (obvious), and (ii) transposing of A_1 and A_2 (Corollary 4.2). These are desirable properties which fail to hold for G_2 and G_3 or their duals. Motivated by these observations, we generalize G_1 and its dual G_1' to the case of m matrices and m Lyapunov functions and establish that they have certain appealing properties. We will prove in the next section (cf. Theorem 4.3) that these graphs always perform better than a common Lyapunov function in 2 steps (i.e., for $\mathcal{A}^2 = \{A_1^2, A_1 A_2, A_2 A_1, A_2^2\}$), whereas, this fact is not true for G_2 and G_3 (or their duals).

4. A PARTICULAR FAMILY OF PATH-COMPLETE GRAPHS

The framework of path-complete graphs provides a multitude of semidefinite programming based techniques for the approximation of the JSR whose performance vary with computational cost. For instance, as we increase the number of nodes of the graph, or the degree of the polynomial

Lyapunov functions assigned to the nodes, or the number of arcs of the graph that instead of labels of length one have labels of higher length, we clearly obtain better results but at a higher computational cost. Many of these approximation techniques are asymptotically tight, so in theory they can be used to achieve any desired accuracy of approximation. For example,

$$\hat{\rho}_{\mathcal{V}^{SOS,2d}}(\mathcal{A}) \to \rho(\mathcal{A}) \text{ as } 2d \to \infty,$$

where $\mathcal{V}^{SOS,2d}$ denotes the class of sum of squares homogeneous polynomial Lyapunov functions of degree $2d$. (Recall our notation for bounds from Section 3.2.) It is also true that a common quadratic Lyapunov function for products of higher length achieves the true JSR asymptotically [16]; i.e.[5],

$$\sqrt[t]{\hat{\rho}_{\mathcal{V}^2}(\mathcal{A}^t)} \to \rho(\mathcal{A}) \text{ as } t \to \infty.$$

Nevertheless, it is desirable for practical purposes to identify a class of path-complete graphs that provide a good tradeoff between quality of approximation and computational cost. Towards this objective, we propose the use of m quadratic functions $x^T P_i x$ satisfying the set of linear matrix inequalities (LMIs)

$$\begin{aligned} P_i &\succ 0 \quad \forall i = 1, \ldots, m, \\ \gamma^2 A_i^T P_j A_i &\preceq P_i \quad \forall i,j = \{1, \ldots, m\}^2 \end{aligned} \quad (21)$$

or the set of LMIs

$$\begin{aligned} P_i &\succ 0 \quad \forall i = 1, \ldots, m, \\ \gamma^2 A_i^T P_i A_i &\preceq P_j \quad \forall i,j = \{1, \ldots, m\}^2 \end{aligned} \quad (22)$$

for the approximation of the JSR of a set of m matrices. Observe from Corollary 2.5 and Corollary 2.6 that the first LMIs give rise to max-of-quadratics Lyapunov functions, whereas the second LMIs lead to min-of-quadratics Lyapunov functions. Throughout this section, we denote the path-complete graphs associated with (21) and (22) with G_1 and G_1' respectively. For the case $m = 2$, our notation is consistent with the previous section and these graphs are illustrated in Figure 3 (g) and (h). Note that we can obtain G_1 and G_1' from each other by reversing the direction of the edges. For this reason, we say that these graphs are dual to each other. We will prove later in this section that the approximation bound obtained by these graphs (i.e., the reciprocal of the largest γ for which the LMIs (21) or (22) hold) is always the same and lies within a multiplicative factor of $\frac{1}{\sqrt[4]{n}}$ of the true JSR, where n is the dimension of the matrices.

4.1 Duality and invariance under transposition

In [9], [10], it is shown that absolute asymptotic stability of the linear difference inclusion in (3) defined by the matrices $\mathcal{A} = \{A_1, \ldots, A_m\}$ is equivalent to absolute asymptotic stability of (3) for the transposed matrices $\mathcal{A}^T := \{A_1^T, \ldots, A_m^T\}$. Note that this fact is obvious from the definition of the JSR in (1), since $\rho(\mathcal{A}) = \rho(\mathcal{A}^T)$. It is also well-known that

$$\hat{\rho}_{\mathcal{V}^2}(\mathcal{A}) = \hat{\rho}_{\mathcal{V}^2}(\mathcal{A}^T).$$

---

[5]By $\mathcal{V}^2$ we denote the class of quadratic homogeneous polynomials. We drop the superscript "SOS" because nonnegative quadratic polynomials are always sums of squares.

Indeed, if $x^T P x$ is a common quadratic Lyapunov function for the set $\mathcal{A}$, then $x^T P^{-1} x$ is a common quadratic Lyapunov function for the set $\mathcal{A}^T$. However, this nice property is not true for the bound obtained from some other techniques. For example[6],

$$\hat{\rho}_{VSOS,4}(\mathcal{A}) \neq \hat{\rho}_{VSOS,4}(\mathcal{A}^T). \tag{23}$$

Similarly, the bound obtained by non-convex inequalities proposed in [9] is not invariant under transposing the matrices. For such methods, one would have to run the numerical optimization twice– once for the set $\mathcal{A}$ and once for the set $\mathcal{A}^T$– and then pick the better bound of the two. We will show that by contrast, the bound obtained from the LMIs in (21) and (22) are invariant under transposing the matrices. Before we do that, let us prove a general result, which states that the bounds resulting from a path-complete graph (with quadratic Lyapunov functions as nodes) and its dual are always the same, provided that these bounds are invariant under transposing the matrices.

THEOREM 4.1. *Let $G(N, E)$ be a path-complete graph, and let its dual graph $G'(N, E')$ be the graph obtained by reversing the direction of the edges and the order of the matrices in the labels of each edge. If*

$$\hat{\rho}_{\mathcal{V}^2, G}(\mathcal{A}) = \hat{\rho}_{\mathcal{V}^2, G}(\mathcal{A}^T), \tag{24}$$

then, the two following equations hold:

$$\hat{\rho}_{\mathcal{V}^2, G}(\mathcal{A}) = \hat{\rho}_{\mathcal{V}^2, G'}(\mathcal{A}), \tag{25}$$

$$\hat{\rho}_{\mathcal{V}^2, G'}(\mathcal{A}) = \hat{\rho}_{\mathcal{V}^2, G'}(\mathcal{A}^T). \tag{26}$$

PROOF. For ease of notation, we prove the claim for the case where the labels of the edges of $G(N, E)$ have length one. The proof of the general case is identical.

Pick an arbitrary edge $(i, j) \in E$ going from node i to node j, and let the associated constraint be given by

$$A_l^T P_j A_l \preceq P_i,$$

for some $A_l \in \mathcal{A}$. If this inequality holds for some positive definite matrices P_i and P_j, then because $\hat{\rho}_{\mathcal{V}^2, G}(\mathcal{A}) = \hat{\rho}_{\mathcal{V}^2, G}(\mathcal{A}^T)$, we will have

$$A_l \tilde{P}_j A_l^T \preceq \tilde{P}_i,$$

for some other positive definite matrices $\tilde{P}_i$ and $\tilde{P}_j$. By applying the Schur complement twice, we get that the last inequality implies

$$A_l^T \tilde{P}_i^{-1} A_l \preceq \tilde{P}_j^{-1}.$$

But this inequality shows that $\tilde{P}_i^{-1}$ and $\tilde{P}_j^{-1}$ satisfy the constraint associated with edge $(j, i) \in E'$. Therefore, the claim in (25) is established. The equality in (26) follows directly from (24) and (25). $\square$

COROLLARY 4.2. *For the path-complete graphs G_1 and G_2 associated with the LMIs in (21) and (22), we have*

$$\hat{\rho}_{\mathcal{V}^2, G_1}(\mathcal{A}) = \hat{\rho}_{\mathcal{V}^2, G_1}(\mathcal{A}^T) = \hat{\rho}_{\mathcal{V}^2, G_1'}(\mathcal{A}) = \hat{\rho}_{\mathcal{V}^2, G_1'}(\mathcal{A}^T). \tag{27}$$

---

[6]We have examples that show the statement in (23), which we do not present because of space limitations. See [9] for such an example in the continuous time setting.

PROOF. We prove the leftmost equality. The other two equalities then follow from Theorem 4.1. Let $P_i, i = 1, \ldots, m$ satisfy the LMIs in (21) for the set of matrices $\mathcal{A}$. The reader can check that

$$\tilde{P}_i = A_i P_i^{-1} A_i^T, \quad i = 1, \ldots, m$$

satisfy the LMIs in (21) for the set of matrices $\mathcal{A}^T$. $\square$

We next prove a bound on the quality of approximation of the estimate resulting from the LMIs in (21) and (22).

THEOREM 4.3. *Let $\mathcal{A}$ be a set of m matrices in $\mathbb{R}^{n \times n}$ with JSR $\rho(\mathcal{A})$. Let $\hat{\rho}_{\mathcal{V}^2, G_1}(\mathcal{A})$ and $\hat{\rho}_{\mathcal{V}^2, G_1'}(\mathcal{A})$ be the bounds on the JSR obtained from the LMIs in (21) and (22) respectively. Then,*

$$\frac{1}{\sqrt[4]{n}} \hat{\rho}_{\mathcal{V}^2, G_1}(\mathcal{A}) \leq \rho(\mathcal{A}) \leq \hat{\rho}_{\mathcal{V}^2, G_1}(\mathcal{A}), \tag{28}$$

and

$$\frac{1}{\sqrt[4]{n}} \hat{\rho}_{\mathcal{V}^2, G_1'}(\mathcal{A}) \leq \rho(\mathcal{A}) \leq \hat{\rho}_{\mathcal{V}^2, G_1'}(\mathcal{A}). \tag{29}$$

PROOF. By Corollary 4.2, $\hat{\rho}_{\mathcal{V}^2, G_1}(\mathcal{A}) = \hat{\rho}_{\mathcal{V}^2, G_1'}(\mathcal{A})$ and therefore it is enough to prove (28). The right inequality in (28) is an obvious consequence of G_1 being a path-complete graph (Theorem 2.4). To prove the left inequality, consider the set $\mathcal{A}^2$ consisting of all m^2 products of length two. In view of (6), a common quadratic Lyapunov function for this set satisfies the bound

$$\frac{1}{\sqrt{n}} \hat{\rho}_{\mathcal{V}^2}(\mathcal{A}^2) \leq \rho(\mathcal{A}^2).$$

It is easy to show that

$$\rho(\mathcal{A}^2) = \rho^2(\mathcal{A}).$$

See e.g. [16]. Therefore,

$$\frac{1}{\sqrt[4]{n}} \hat{\rho}_{\mathcal{V}^2}^{\frac{1}{2}}(\mathcal{A}^2) \leq \rho(\mathcal{A}). \tag{30}$$

Now suppose for some $\gamma > 0$, $x^T Q x$ is a common quadratic Lyapunov function for the matrices in $\mathcal{A}_\gamma^2$; i.e., it satisfies

$$Q \succ 0$$
$$\gamma^4 (A_i A_j)^T Q A_i A_j \preceq Q \quad \forall i, j = \{1, \ldots, m\}^2.$$

Then, we leave it to the reader to check that

$$P_i = Q + A_i^T Q A_i, \quad i = 1, \ldots, m$$

satisfy (21). Hence,

$$\hat{\rho}_{\mathcal{V}^2, G_1}(\mathcal{A}) \leq \hat{\rho}_{\mathcal{V}^2}^{\frac{1}{2}}(\mathcal{A}^2),$$

and in view of (30) the claim is established. $\square$

Note that the bounds in (28) and (29) are independent of the number of matrices. Moreover, we remark that these bounds are tighter, in terms of their dependence on n, than the known bounds for $\hat{\rho}_{VSOS,2d}$ for any finite degree $2d$ of the sum of squares polynomials. The reader can check that the bound in (7) goes asymptotically as $\frac{1}{\sqrt{n}}$. Numerical evidence suggests that the performance of both the bound obtained by sum of squares polynomials and the bound obtained by the LMIs in (21) and (22) is much better than the provable bounds in (7) and in Theorem 4.3. The problem of improving these bounds or establishing their tightness is open. It goes without saying that instead of quadratic functions, we can associate sum of squares polynomials to the nodes of G_1 and obtain a more powerful technique for which we can also prove better bounds with the exact same arguments.

4.2 Numerical example

In the proof of Theorem 4.3, we essentially showed that the bound obtained from LMIs in (21) is tighter than the bound obtained from a common quadratic applied to products of length two. The example below shows that the LMIs in (21) can in fact do better than a common quadratic applied to products of *any* finite length.

EXAMPLE 4.1. *Consider the set of matrices* $\mathcal{A} = \{A_1, A_2\}$, *with*

$$A_1 = \begin{bmatrix} 1 & 0 \\ 1 & 0 \end{bmatrix}, \quad A_2 = \begin{bmatrix} 0 & 1 \\ 0 & -1 \end{bmatrix}.$$

This is a benchmark set of matrices that has been studied in [3], [21], [2] because it gives the worst case approximation ratio of a common quadratic Lyapunov function. Indeed, it is easy to show that $\rho(\mathcal{A}) = 1$, *but* $\hat{\rho}_{\mathcal{V}2}(\mathcal{A}) = \sqrt{2}$. *Moreover, the bound obtained by a common quadratic function applied to the set* $\mathcal{A}^t$ *is*

$$\hat{\rho}_{\mathcal{V}2}^{\frac{1}{t}}(\mathcal{A}^t) = 2^{1/2t},$$

which for no finite value of t is exact. On the other hand, we show that the LMIs in (21) give the exact bound; i.e., $\hat{\rho}_{\mathcal{V}2,G_1}(\mathcal{A}) = 1$. *Due to the simple structure of* A_1 *and* A_2, *we can even give an analytical expression for our Lyapunov functions. Given any* $\varepsilon > 0$, *the LMIs in (21) with* $\gamma = 1/(1 + \varepsilon)$ *are feasible with*

$$P_1 = \begin{bmatrix} a & 0 \\ 0 & b \end{bmatrix}, \qquad P_2 = \begin{bmatrix} b & 0 \\ 0 & a \end{bmatrix},$$

for any $b > 0$ *and* $a > b/2\varepsilon$.

5. CONVERSE LYAPUNOV THEOREMS AND MORE APPROXIMATION BOUNDS

It is well-known that existence of a Lyapunov function which is the pointwise maximum of quadratics is not only sufficient but also necessary for absolute asymptotic stability of (2) or (3) [20]. This is a very intuitive fact, if we recall that switched systems of type (2) and (3) always admit a convex Lyapunov function. Indeed, if we take "enough" quadratics, the convex and compact unit sublevel set of a convex Lyapunov function can be approximated arbitrarily well with sublevel sets of max-of-quadratics Lyapunov functions, which are intersections of ellipsoids. An obvious consequence of this fact is that the bound obtained from max-of-quadratics Lyapunov functions is asymptotically tight for the approximation of the JSR. However, this converse Lyapunov theorem does not answer two natural questions of importance in practice: (i) How many quadratic functions do we need to achieve a desired quality of approximation? (ii) Can we search for these quadratic functions via semidefinite programming or do we need to resort to non-convex formulations? Our next theorem provides an answer to these questions. We then prove a similar result for another interesting subclass of our methods. Due to length constraints, we only briefly sketch the common idea behind the two theorems. The interested reader can find the full proofs in the journal version of the present paper.

THEOREM 5.1. *Let* $\mathcal{A}$ *be any set of m matrices in* $\mathbb{R}^{n \times n}$. *Given any positive integer l, there exists an explicit path-complete graph G consisting of* m^{l-1} *nodes assigned to quadratic Lyapunov functions and* m^l *arcs with labels of length one such that the linear matrix inequalities associated with G imply existence of a max-of-quadratics Lyapunov function and the resulting bound obtained from the LMIs satisfies*

$$\frac{1}{\sqrt[2l]{n}} \hat{\rho}_{\mathcal{V}2,G}(\mathcal{A}) \leq \rho(\mathcal{A}) \leq \hat{\rho}_{\mathcal{V}2,G}(\mathcal{A}). \tag{31}$$

THEOREM 5.2. *Let* $\mathcal{A}$ *be a set of matrices in* $\mathbb{R}^{n \times n}$. *Let* $\tilde{G}(\{1\}, E)$ *be a path-complete graph, and l be the length of the shortest word in* $\tilde{\mathcal{A}} = \{L(e) : e \in E\}$. *Then* $\hat{\rho}_{\mathcal{V}2,\tilde{G}}(\mathcal{A})$ *provides an estimate of* $\rho(\mathcal{A})$ *that satisfies*

$$\frac{1}{\sqrt[2l]{n}} \hat{\rho}_{\mathcal{V}2,\tilde{G}}(\mathcal{A}) \leq \rho(\mathcal{A}) \leq \hat{\rho}_{\mathcal{V}2,\tilde{G}}(\mathcal{A}).$$

PROOF. (Sketch of the proof of Theorems 5.1 and 5.2) For the proof of Theorem 5.1, we define the graph G as follows: there is one node v_w for each word $w \in \{1, \ldots, m\}^{l-1}$. For each node v_w and each index $j \in \{1, \ldots, m\}$, there is an edge with the label A_j from v_w to $v_{w'}$ iff $w'j = xw$ for some label $x \in \{1, \ldots, m\}$ (xw' means the concatenation of the label x with the word w').
Now, for both proofs, denoting the corresponding graph by G, we show that if $\mathcal{A}^l$ has a common quadratic Lyapunov function, then

$$\hat{\rho}_{\mathcal{V}2,G} \leq 1,$$

which implies the result. $\square$

REMARK 5.1. *In view of NP-hardness of approximation of the JSR [27], the fact that the number of quadratic functions and the number of LMIs grow exponentially in l is to be expected.*

6. CONCLUSIONS AND FUTURE DIRECTIONS

We studied the use of multiple Lyapunov functions for the formulation of semidefinite programming based approximation algorithms for computing upper bounds on the joint spectral radius of a finite set of matrices (or equivalently establishing absolute asymptotic stability of an arbitrary switched linear system). We introduced the notion of a path-complete graph, which was inspired by well-established concepts in automata theory. We showed that every path-complete graph gives rise to a technique for the approximation of the JSR. This provided a unifying framework that includes many of the previously proposed techniques and also introduces new ones. (In fact, all families of LMIs that we are aware of appear to be particular cases of our method.) We compared the quality of the bound obtained from certain classes of path-complete graphs, including all path-complete graphs with two nodes on an alphabet of two matrices, and also a certain family of dual path-complete graphs. We proposed a specific class of such graphs that appear to work particularly well in practice. We proved that the bound obtained from these graphs is invariant under transposition of the matrices and is always within a multiplicative factor of $1/\sqrt[4]{n}$ from the true JSR. Finally, we presented two converse Lyapunov theorems, one for a new class of methods that propose the use of a common quadratic Lyapunov function for a set of words of possibly different lengths, and the other for the well-known methods of minimum and maximum-of-quadratics Lyapunov functions. These theorems yield explicit and systematic constructions of semidefinite programs that achieve any desired accuracy of approximation.

Some of the interesting questions that can be explored in the future are the following. What is the complexity of recognizing path-complete graphs when the underlying finite automata are non-deterministic? What are some other classes of path-complete graphs that lead to new techniques for proving stability of switched systems? How can we compare the performance of different path-complete graphs in a systematic way? Given a set of matrices, a class of Lyapunov functions, and a fixed size for the graph, can we come up with the least conservative topology of a path-complete graph? Within the framework that we proposed, do all the Lyapunov inequalities that prove stability come from path-complete graphs? What are the analogues of the results of this paper for continuous time switched systems? We hope that this work will stimulate further research in these directions.

7. REFERENCES

[1] A. A. Ahmadi. Non-monotonic Lyapunov functions for stability of nonlinear and switched systems: theory and computation. Master's Thesis, Massachusetts Institute of Technology, June 2008. Available from http://dspace.mit.edu/handle/1721.1/44206.

[2] A. A. Ahmadi and P. A. Parrilo. Non-monotonic Lyapunov functions for stability of discrete time nonlinear and switched systems. In *Proceedings of the 47th IEEE Conference on Decision and Control*, 2008.

[3] T. Ando and M.-H. Shih. Simultaneous contractibility. *SIAM Journal on Matrix Analysis and Applications*, 19:487–498, 1998.

[4] V. D. Blondel and Y. Nesterov. Computationally efficient approximations of the joint spectral radius. *SIAM J. Matrix Anal. Appl.*, 27(1):256–272, 2005.

[5] V. D. Blondel, Y. Nesterov, and J. Theys. On the accuracy of the ellipsoidal norm approximation of the joint spectral radius. *Linear Algebra and its Applications*, 394:91–107, 2005.

[6] V. D. Blondel and J. N. Tsitsiklis. The boundedness of all products of a pair of matrices is undecidable. *Systems and Control Letters*, 41:135–140, 2000.

[7] M. S. Branicky. Multiple Lyapunov functions and other analysis tools for switched and hybrid systems. *IEEE Transactions on Automatic Control*, 43(4):475–482, 1998.

[8] J. Daafouz and J. Bernussou. Parameter dependent Lyapunov functions for discrete time systems with time varying parametric uncertainties. *Systems and Control Letters*, 43(5):355–359, 2001.

[9] R. Goebel, T. Hu, and A. R. Teel. Dual matrix inequalities in stability and performance analysis of linear differential/difference inclusions. In *Current Trends in Nonlinear Systems and Control*, pages 103–122. 2006.

[10] R. Goebel, A. R. Teel, T. Hu, and Z. Lin. Conjugate convex Lyapunov functions for dual linear differential inclusions. *IEEE Transactions on Automatic Control*, 51(4):661–666, 2006.

[11] J. E. Hopcroft, R. Motwani, and J. D. Ullman. *Introduction to automata theory, languages, and computation*. Addison Wesley, 2001.

[12] T. Hu and Z. Lin. Absolute stability analysis of discrete-time systems with composite quadratic Lyapunov functions. *IEEE Transactions on Automatic Control*, 50(6):781–797, 2005.

[13] T. Hu, L. Ma, and Z. Li. On several composite quadratic Lyapunov functions for switched systems. In *Proceedings of the 45th IEEE Conference on Decision and Control*, 2006.

[14] T. Hu, L. Ma, and Z. Lin. Stabilization of switched systems via composite quadratic functions. *IEEE Transactions on Automatic Control*, 53(11):2571 – 2585, 2008.

[15] M. Johansson and A. Rantzer. Computation of piecewise quadratic Lyapunov functions for hybrid systems. *IEEE Transactions on Automatic Control*, 43(4):555–559, 1998.

[16] R. Jungers. *The joint spectral radius: theory and applications*, volume 385 of *Lecture Notes in Control and Information Sciences*. Springer, 2009.

[17] J. W. Lee and G. E. Dullerud. Uniform stabilization of discrete-time switched and Markovian jump linear systems. *Automatica*, 42(2):205–218, 2006.

[18] H. Lin and P. J. Antsaklis. Stability and stabilizability of switched linear systems: a short survey of recent results. In *Proceedings of IEEE International Symposium on Intelligent Control*, 2005.

[19] D. Lind and B. Marcus. *An introduction to symbolic dynamics and coding*. Cambridge University Press, 1995.

[20] A. Molchanov and Y. Pyatnitskiy. Criteria of asymptotic stability of differential and difference inclusions encountered in control theory. *Systems and Control Letters*, 13:59–64, 1989.

[21] P. A. Parrilo and A. Jadbabaie. Approximation of the joint spectral radius using sum of squares. *Linear Algebra and its Applications*, 428(10):2385–2402, 2008.

[22] V. Y. Protasov, R. M. Jungers, and V. D. Blondel. Joint spectral characteristics of matrices: a conic programming approach. *SIAM Journal on Matrix Analysis and Applications*, 31(4):2146–2162, 2010.

[23] M. Roozbehani. *Optimization of Lyapunov invariants in analysis and implementation of safety-critical software systems*. PhD thesis, Massachusetts Institute of Technology.

[24] M. Roozbehani, A. Megretski, E. Frazzoli, and E. Feron. Distributed Lyapunov functions in analysis of graph models of software. *Springer Lecture Notes in Computer Science*, 4981:443–456, 2008.

[25] L. Rosier. Homogeneous Lyapunov function for homogeneous continuous vector fields. *Systems Control Letters*, 19(6):467–473, 1992.

[26] G. C. Rota and W. G. Strang. A note on the joint spectral radius. *Indag. Math.*, 22:379–381, 1960.

[27] J. N. Tsitsiklis and V. Blondel. The Lyapunov exponent and joint spectral radius of pairs of matrices are hard- when not impossible- to compute and to approximate. *Mathematics of Control, Signals, and Systems*, 10:31–40, 1997.

The Earlier the Better: A Theory of Timed Actor Interfaces*

Marc Geilen
Eindhoven University of Technology
m.c.w.geilen@tue.nl

Stavros Tripakis
University of California, Berkeley
stavros@eecs.berkeley.edu

Maarten Wiggers
University of Twente
m.h.wiggers@utwente.nl

ABSTRACT

Programming embedded and cyber-physical systems requires attention not only to functional behavior and correctness, but also to non-functional aspects and specifically timing and performance. A structured, compositional, model-based approach based on stepwise refinement and abstraction techniques can support the development process, increase its quality and reduce development time through automation of synthesis, analysis or verification. Toward this, we introduce a theory of timed actors whose notion of refinement is based on the principle of worst-case design that permeates the world of performance-critical systems. This is in contrast with the classical behavioral and functional refinements based on restricting sets of behaviors. Our refinement allows time-deterministic abstractions to be made of time-non-deterministic systems, improving efficiency and reducing complexity of formal analysis. We show how our theory relates to, and can be used to reconcile existing time and performance models and their established theories.

Categories and Subject Descriptors

D.2.2 [**Software Engineering**]: Design Tools and Techniques—*Modules and Interfaces*; D.2.13 [**Software Engineering**]: Reusable Software

General Terms

Design, Languages, Theory, Verification

Keywords

Actors, Dataflow, Compositionality, Throughput, Latency, Interfaces, Refinement

* This work was supported in part by the Center for Hybrid and Embedded Software Systems (CHESS) at UC Berkeley, which receives support from the National Science Foundation (NSF awards #0720882 (CSR-EHS: PRET) and #0931843 (ActionWebs), the U. S. Army Research Office (ARO #W911NF-07-2-0019), the U. S. Air Force Office of Scientific Research (MURI #FA9550-06-0312), the Air Force Research Lab (AFRL), the Multiscale Systems Center (MuSyC), one of six research centers funded under the Focus Center Research Program, a Semiconductor Research Corporation program, and the following companies: Bosch, National Instruments, Thales, and Toyota.

1. INTRODUCTION

Advances in sensor, actuator and computer hardware currently enable new classes of applications, often described under the terms *embedded* or *cyber-physical systems* (ECPS). Examples of such systems can be found in the domains of robotics, health care, transportation, and energy. ECPS are different from traditional computing systems, because in ECPS the computer is in tight interaction with a physical environment, which it monitors and possibly controls. The requirements of the closed-loop system (computer + environment) are not purely functional. Instead, they often involve timing or performance properties, such as throughput or latency.

Abstraction and compositionality have been two key principles in developing large and complex systems. Although a large number of methods employing these principles exist to deal with functional properties (e.g., see [5, 9, 24, 25]), less attention has been paid to timing and performance. This paper contributes toward filling this gap.

Our approach can be termed *model based*. High-level models that are suitable for analysis are used as specifications or for design-space exploration. Refinement and abstraction steps are used to move between high-level models, lower-level models and implementations. The process guarantees that the results of the analysis (e.g., bounds on throughput or latency) are preserved during refinement. Our paper defines a general model and a suitable notion of abstraction and refinement that support this process. The model is compositional in the sense that refinement between models consisting of many components can be achieved by refining individual components separately.

Our treatment follows *interface theories* [11], which can be seen as type theories focusing on dynamic and concurrent behavior. Our interfaces, called *actor interfaces*, are inspired by *actor-oriented* models of computation such as process networks [20] and data flow [14].

Actors operate by consuming and producing *tokens* on their input and output ports, respectively. Since our primary goal is timing and performance analysis, we completely abstract away from token values, and keep only the times in which these tokens are produced. Actors are then defined as relations between input and output sequences of discrete events occurring in a given time axis. Examples of such event sequences are shown in Figure 2.

The main novelty of our theory lies in its notion of refinement, which is based on the principle *the earlier the better*. In particular, actor A refines actor B if, for the same input, A produces no fewer events and no later, in the worst case, than those produced by B. For example, an actor A that non-deterministically delays its input by some time $t \in [1, 2]$ refines an actor B that deterministically delays its input by a constant time of 3. This is in sharp contrast with most standard notions of refinement which rely on the principle that the implementation should have fewer possible behaviors and thus be "more deterministic" than the specification. With the standard notions, actor A does not refine B, although it would refine an actor B' that non-deterministically delays its input by some time $t \in [0, 3]$.

The earlier-is-better refinement principle is interesting because it allows *deterministic abstractions of non-deterministic systems*. System implementations can often be seen as time-non-deterministic because of high variability in execution and communication delays, dynamic scheduling, and other effects that are expensive or impossible to model precisely. Time-deterministic models, on the other hand, suffer less from state explosion problems, and are also more suitable for deriving analytic bounds.

The main contributions of this work are the following:

- We develop an interface theory of timed actors with a refinement relation that follows the earlier-is-better principle and preserves worst-case bounds on performance metrics (throughput, latency). (Sections 4–7).

- Our framework unifies existing models (SDF and variants, automata, service curves, etc.) by treating actors semantically, as relations on event sequences, rather than syntactically, as defined by specific models such as automata or dataflow. (Section 8).

Omitted proofs can be found in the extended version of this paper [1]. The latter also contains additional material omitted from this version due to space limitations.

2. MOTIVATING EXAMPLE

To illustrate the use of our framework, we present an example of an MP3 play-back application. The application is based on a fragment of the car radio case study presented in [31]. Our goal is to show how such an application can be handled within our framework, using stepwise refinement from specification to implementation, such that performance guarantees are preserved during the process.

The layers of the refinement process are shown in Figure 1. The top layer captures the specification. It consists of a single actor SPEC, with a single output port, and a single event sequence τ at this port, defined by $\tau(n) = 50 + n/44.1$ ms, for $n \in \mathbb{N}$. That is, the n-th event in the sequence occurs at time $\tau(n)$. SPEC specifies the required behavior of an MP3 player where audio samples are produced at a rate of 44.1 kHz, starting with an initial 50 ms delay to allow for processing.

For simplicity, we do not model input tokens, assuming they are always available for consumption. Also note that the system typically includes a component such as a digital-to-analog converter (DAC) which consumes the audio samples produced at port p, buffers them and reproduces them periodically. We omit DAC since it does not take part in the refinement process.

The next layer is an application model consisting of actors DEC (decoder), SRC (sample-rate converter), and actor D1 explained below. DEC and SRC are timed *synchronous data flow* (SDF) [22] actors. SDF actors communicate by conceptually unbounded FIFO queues. They "fire" as soon

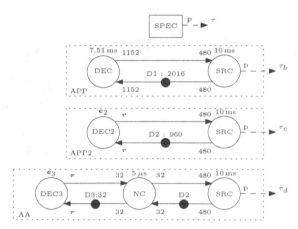

Figure 1: Successive refinements of an MP3 play-back application.

as a fixed number of tokens become available at their input queues and, after a fixed duration, produce a fixed number of tokens at their output queues. For instance, DEC consumes and produces 1152 tokens per firing on the queues from and to the SRC actor. Each firing of DEC takes 7.51 ms. For a formal definition of SDF actors see Section 8.1. D1 is an actor modeling 2016 initial tokens on the queue from SRC to DEC. Formally, it is an instance of parameterized actor I_k defined in Example 8. All dataflow actors in Figure 1 (DEC, SRC, DEC2, DEC3 and NC) implicitly have a self-edge with a single initial token so that firings of the same actor do not overlap (i.e., each firing completes before the next one starts).

The global application model is a single composite actor APP obtained by composing the three actors above, first in series and then in feedback, and then hiding all ports except the output port p of SRC. Section 5 precisely defines the compositions and hiding. Because APP is an SDF model and hence deterministic, APP produces a single event sequence τ_b at p. We have captured APP in the dataflow analysis tool SDF3 (http://www.es.ele.tue.nl/sdf3/) and have used the tool to check that τ_b refines τ, i.e., that each event in τ_b occurs no later than the corresponding event in τ. As a result, APP refines SPEC.

The motivation for the third layer is buffer considerations. In this layer, the SDF actor DEC is replaced by the *cyclo-static data flow* (CSDF) [6] actor DEC2. This substitution results in smaller buffers on the queue from SRC to DEC2 [31]. CSDF actors generalize SDF actors by allowing the token consumption/production rates to vary periodically, as an SDF actor that cycles between a finite set of firing *phases*. In our example, DEC2 has 39 phases, captured by the notation $\boldsymbol{r} = [0, 0, [32]^{18}, 0, [32]^{18}]$. In the first two phases DEC decodes frame headers without consuming nor producing any tokens. The subsequent 18 phases each consume and produce 32 tokens, and are followed by a header decoding phase with no tokens consumed or produced. Finally there are 18 more phases that each consume and produce 32 tokens. This sequence of phases is repeated for each MP3 frame. The durations of these phases are given by $\boldsymbol{e}_2 = [670, 2700, [40]^{18}, 2700, [40]^{18}]\mu s$. That is, phase 1 takes 670 μs, phase 2 takes 2700 μs, and so on.

Using arguments similar to those presented later in Example 3, we can show that DEC2 refines DEC. The composite actor APP2 is produced by first refining DEC to DEC2, and then reducing the number of initial tokens from D1 to D2, while maintaining that APP2 refines APP. The latter is ensured by using SDF3 to compute τ_c for a given D2, and checking that τ_c refines τ_b.

The bottom layer is an *architecture aware* model (AA) that is close to a distributed implementation on a multi-processor architecture with network-on-chip (NoC) communication. In this layer DEC2 is replaced by the composition of DEC3, D3 and NC. DEC3 is identical to DEC2 except for its firing durations which are reduced to $e_3 = [670, 2700, [30]^{18}, 2700, [30]^{18}]\mu s$, because the communication is modeled separately. NC is an SDF actor that models the NoC behavior. It can be shown that the composition of DEC3, NC and D3 refines DEC2. This and our compositionality Propositions 2 and 3 imply that AA refines APP2.

The final implementation (not shown in the figure) can be compositionally shown to refine the AA model. For instance, the NC actor conservatively abstracts the NoC implementation [16]. It is important to mention that although implementations are time-non-deterministic for multiple reasons, e.g., software execution times or run-time scheduling, the models in Figure 1 are time-deterministic.

3. RELATED WORK

Abstraction and compositionality have been extensively studied from an untimed perspective, focusing on functional correctness (e.g., see [5, 9, 24, 25]). Timing has also been considered, implicitly or explicitly, in a number of frameworks. Our treatment has been inspired in particular by interface theories such as interface automata [11] which use game-theoretic interpretations of composition and refinement, that are more appropriate for open systems. Although interface automata have no explicit notion of time, discrete time can be implicitly modeled by adding a special "tick" output. [12] follows [11] but uses timed automata [2] instead of discrete automata. However, a notion of refinement is not defined in [12]. [10] extends [12] with a notion of refinement in the spirit of alternating simulation [3], adapted for timed systems.

The refinement notions used in all works above differ from ours in a fundamental way: in our case, earlier is better, whereas in the above works, if the implementation can produce an output a at some time t, then the (refined) specification must also be able to produce a at the *same* time t. Thus, an implementation that can produce a only at times $t \leq 1$ does not refine a specification that can produce a only at times $t \geq 2$. Another major difference is that performance metrics such as throughput and latency are not considered in any of the above works.

Our work is about non-deterministic models and worst-case performance bounds and as such differs from probabilistic frameworks such as Markov decision processes, or stochastic process algebras or games (e.g., see [26, 19, 18, 13]). Worst-case performance bounds can be derived using techniques from the *network calculus* (NC) [7] or *real-time calculus* (RTC) [27]. Refinement relations have been considered recently in these frameworks [17, 28]. Semantically, these relations correspond to trace containment at the outputs and as such do not follow the earlier-is-better principle. An important feature of NC and RTC is that they can model resources, e.g., available computation power, and therefore be used in applications such as schedulability analysis. We

do not explicitly distinguish resources in our framework. In NC and RTC, behaviors are typically captured by arrival or service curves. Service curves can be seen as a special class of actors [1]. Service curves have limited expressiveness: they cannot generally capture, for instance, languages produced by automata actors. The same can be said of real-time scheduling theory (e.g., see [8]). Automata-based models have been used for scheduling and resource modeling, e.g., as in [30], where tasks are described as ω-regular languages representing sets of admissible schedules. Refinement is not considered in this work, and although it could be defined as language containment, this would not follow the earlier-is-better principle.

$(\max, +)$ algebra and its relatives (e.g., see [4]) are used as an underlying system theory for different discrete event system frameworks, including NC, RTC and SDF. $(\max, +)$ algebra is mostly limited to deterministic, $(\max, +)$-linear systems. Our framework is more general: it can capture non-determinism in time, an essential property in order to be able to relate time-deterministic specification models such as SDF to implementations that have variable timing.

Our work has also been inspired by the work in [32], where task graph implementations are conservatively abstracted to timed dataflow specifications.

4. ACTORS

We consider actor interfaces (in short *actors*) as relations between finite or infinite sequences of input tokens and sequences of output tokens. We abstract from token content, and instead focus on arrival or production times represented as timestamps from some totally ordered, continuous time domain $(\mathcal{T}, \leq)$. $\mathcal{T}$ contains a minimal element denoted 0. We also add to $\mathcal{T}$ a maximal element denoted ∞, so that $t < \infty$ for all $t \in \mathcal{T}$. $\mathcal{T}^\infty$ denotes $\mathcal{T} \cup \{\infty\}$. $\mathbb{N}$ denotes the set of natural numbers and we assume $0 \in \mathbb{N}$. $\mathbb{R}$ denotes the set of real numbers and $\mathbb{R}^{\geq 0}$ the set of non-negative reals.

DEFINITION 1 (EVENT SEQUENCE). *An* event sequence *is a total mapping* $\tau : \mathbb{N} \to \mathcal{T}^\infty$, *such that* τ *is weakly monotone, that is, for every* $k, m \in \mathbb{N}$ *and* $k \leq m$ *we have* $\tau(k) \leq \tau(m)$.

Thus $\tau(n)$ captures the arrival time of the n-th token, with $\tau(n) = \infty$ interpreted as event n being *absent*. Then, all events $n' > n$ must also be absent. Because of this property, an event sequence τ can also be viewed as a finite or infinite sequence of timestamps in $\mathcal{T}$. The *length* of τ, denoted $|\tau|$, is the smallest $n \in \mathbb{N}$ such that $\tau(n) = \infty$, and with somewhat abusive notation $|\tau| = \infty$ if $\tau(n) < \infty$ for all $n \in \mathbb{N}$. We use ϵ to denote the *empty* event sequence, $\epsilon(n) = \infty$ for all n. Given event sequence τ and timestamp $t \in \mathcal{T}$ such that $t \leq \tau(0)$, $t \cdot \tau$ denotes the event sequence consisting of t followed by τ. The set of all event sequences is denoted Tr.

Event sequences are communicated over *ports*. For a set P of ports, $Tr(P)$ denotes $P \to Tr$, the set of total functions that map each port of P to an event sequence. Elements of $Tr(P)$ are called *event traces over P*. We sometimes use the notation $(p, n, t) \in x$ instead of $x(p)(n) = t$.

DEFINITION 2 (EARLIER-THAN AND PREFIX ORDERS). *For* $\tau, \tau' \in Tr$, τ *is said to be* earlier than τ', *denoted* $\tau \leq \tau'$, *iff* $|\tau| = |\tau'|$ *and for all* $n < |\tau|$, $\tau(n) \leq \tau'(n)$. $\leq$ *is called the* earlier-than *relation. In addition we consider the* prefix *relation:* $\tau \preceq \tau'$ *iff* $|\tau| \leq |\tau'|$ *and for every* $n < |\tau|$,

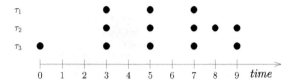

Figure 2: Three event sequences.

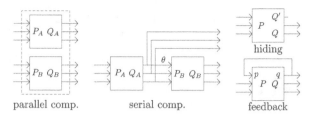

Figure 3: Actor compositions.

$\tau(n) = \tau'(n)$. *We lift $\leq$ and $\preceq$ to event traces $x, x' \in Tr(P)$ in the usual way: $x \leq x'$ iff for all $p \in P$, $x(p) \leq x'(p)$; $x \preceq x'$ iff for all $p \in P$, $x(p) \preceq x'(p)$.*

EXAMPLE 1. *Figure 2 shows three event sequences τ_1, τ_2 and τ_3 visualized as black dots on a horizontal time line: $\tau_1 = 3 \cdot 5 \cdot 7 \cdot \epsilon$, $\tau_2 = 3 \cdot 5 \cdot 7 \cdot 8 \cdot 9 \cdot \epsilon$ and $\tau_3 = 0 \cdot 3 \cdot 5 \cdot 7 \cdot 9 \cdot \epsilon$. τ_1 is a prefix of τ_2: $\tau_1 \preceq \tau_2$; and τ_3 is earlier than τ_2: $\tau_3 \leq \tau_2$. But $\tau_1 \nleq \tau_2$.*

$(Tr, \preceq)$ and $(Tr(P), \preceq)$ are complete partial orders (CPOs). $(Tr, \leq)$ and $(Tr(P), \leq)$ are pre-CPOs (they have no unique minimal element). We use $\bigsqcup_{\preceq} C$ to denote the least upper bound of a chain C in a CPO with partial order $\preceq$.

If x_1 is an event trace over ports P_1 and x_2 is an event trace over ports P_2, and P_1 and P_2 are disjoint, then $x_1 \cup x_2$ denotes the event trace over $P_1 \cup P_2$ such that $(x_1 \cup x_2)(p) = x_1(p)$ if $p \in P_1$ and $(x_1 \cup x_2)(p) = x_2(p)$ if $p \in P_2$. $x \uparrow Q$ is identical to x, but with all ports in Q removed from the domain.

DEFINITION 3 (ACTOR). *An actor is a tuple $A = (P, Q, R_A)$ with a set P of input ports, a set Q of output ports and an event trace relation $R_A \subseteq Tr(P) \times Tr(Q)$. We use xAy to denote $(x, y) \in R_A$ when we leave the three-tuple of A implicit. A is called deterministic if R_A is a partial function. The set of all legal input traces of A is:*
$$\text{in}_A = \{x \in Tr(P) \mid \exists y \in Tr(Q) : xAy\}.$$

Note that an input trace x models the times that input tokens are *produced* by the environment of the actor, and not the times that these tokens are *consumed* by the actor. Token consumption times can be modeled by adding special output ports to the actor as explained in [1].

In order to study composition and refinement later on, we need to investigate actors with respect to different kinds of monotone changes to their input and output. We therefore introduce the following family of definitions.

DEFINITION 4. (INPUT-CLOSURES, MONOTONICITIES AND CONTINUITIES) *Let A be an actor with input ports P and output ports Q. A is called input-complete iff $\text{in}_A = Tr(P)$. Given a partial order $\preceq$ on $Tr(P)$ and $Tr(Q)$, A is called (inverse) $\preceq$-input-closed iff for every $x \in \text{in}_A$ and $x' \in Tr(P)$, $x' \preceq x$ ($x \preceq x'$) implies $x' \in \text{in}_A$. A is called (inverse) $\preceq$-monotone iff for every x, y and x' such that xAy, $x' \in \text{in}_A$ and $x \preceq x'$ ($x' \preceq x$), there exists y' such that $y \preceq y'$ ($y' \preceq y$) and $x'Ay'$. Assuming $\preceq$ yields pre-CPOs, A is called $\preceq$-continuous iff for every pair $\{x_k\}$ and $\{y_k\}$ of chains of event traces w.r.t. $(Tr(P), \preceq)$ and $(Tr(Q), \preceq)$ respectively, if $x_k A y_k$ for all k, then $(\bigsqcup_{\preceq} \{x_k\}) A (\bigsqcup_{\preceq} \{y_k\})$.*

EXAMPLE 2 (DELAY ACTORS). *A variable delay actor $\Delta_{[d_1, d_2]}$ with minimum and maximum delay $d_1, d_2 \in \mathbb{R}^{\geq 0}$,*

where $d_1 \leq d_2$, is an actor with one input port p, one output port q, time domain $\mathcal{T} = \mathbb{R}^{\geq 0}$, and such that
$$x \Delta_{[d_1, d_2]} y \text{ iff } |x(p)| = |y(q)| \wedge \forall n < |x(p)| :$$
$$x(p)(n) + d_1 \leq y(q)(n) \leq x(p)(n) + d_2$$
$$\wedge \ (n > 0 \implies y(q)(n) \geq y(q)(n-1)).$$

$\Delta_{[d_1, d_2]}$ *is input-complete, $\preceq$- and $\leq$-monotone in both directions, and $\preceq$- and $\leq$-continuous, but not deterministic in general. The constant delay actor Δ_d is the deterministic variable delay actor $\Delta_{[d,d]}$.*

5. COMPOSITIONS

Actor interfaces can be composed to yield new actor interfaces. The composition operators defined in this paper are illustrated in Figure 3. Parallel composition composes two interfaces side-by-side without interaction:

DEFINITION 5 (PARALLEL COMPOSITION). *Let A and B be two actors with disjoint input ports P_A and P_B and disjoint output ports Q_A and Q_B respectively. Then the parallel composition $A \| B$ is an actor with input ports $P_A \cup P_B$, output ports $Q_A \cup Q_B$, and relation $A \| B = \{(x_1 \cup x_2, y_1 \cup y_2) \mid x_1 A y_1 \wedge x_2 B y_2\}$.*

Parallel composition is clearly associative and commutative. It is also easy to see that it preserves all monotonicity, continuity and closure properties if both actors have them.

DEFINITION 6 (SERIAL COMPOSITION). *Let A and B be two actors with disjoint input ports P_A and P_B and disjoint output ports Q_A and Q_B respectively. Let θ be a bijective function from Q_A to P_B. Then the serial composition $A\theta B$ is an actor with input ports $P_\theta = P_A$, output ports $Q_\theta = Q_A \cup Q_B$, and whose relation is defined as follows. First, we lift the mapping of ports to event traces: $\theta(y) = \{(\theta(p), n, t) \mid (p, n, t) \in y\}$. The input-output relation of the composite actor $A\theta B$ is then defined as:*
$$A\theta B = \{(x_1, y_1 \cup y_2) \in \text{in}_\theta \times Tr(Q_\theta) \mid x_1 A y_1 \wedge \theta(y_1) B y_2\}$$
where: $\text{in}_\theta = \{x \in \text{in}_A \mid \forall y_1 : xAy_1 \implies \theta(y_1) \in \text{in}_B\}$.

in_θ captures the set of legal inputs of the composite actor $A\theta B$. In the spirit of [5, 11], we adopt a "demonic" interpretation of non-determinism, where an input x is legal in $A\theta B$ only if *any* intermediate output that the first actor A may produce for x is a legal input (after relabeling) of the second actor B. In that case, we say that actor B is *receptive* to actor A w.r.t. θ. An input-complete actor is receptive to any other actor. If B is receptive to A or A is deterministic, then $A\theta B$ reduces to standard composition of relations.

26

Moreover, if both A and B are deterministic (respectively, input-complete) then so is $A\theta B$. Serial composition is associative [1].

The requirement that θ is total and onto is not restrictive. For example, suppose A has two output ports q_1, q_2 and B has two input ports p_1, p_2, but we only want to connect q_1 to p_1. To do this, we can extend A with additional input and output ports p_{p_2} and q_{p_2}, respectively, corresponding to p_2. A acts as the identity function on p_{p_2} and q_{p_2}, that is, for all x, y such that xAy, $y(q_{p_2}) = x(p_{p_2})$. Then we can connect q_{p_2} to p_2. Similarly, we can extend B with additional input and output ports p_{q_2} and q_{q_2}, and connect q_2 to p_{q_2}.

A hiding operator can be used to make internal event sequences unobservable.

DEFINITION 7 (HIDING). *Let $A = (P, Q, R_A)$ be an actor and let $Q' \subseteq Q$. The* hiding *of Q' in A is the actor*

$$A \backslash Q' = (P, Q \backslash Q', \{(x, y \uparrow Q') \mid xAy\}).$$

Note that $\text{in}_{A \backslash Q'} = \text{in}_A$. Hiding preserves all forms of monotonicity and continuity, as well as determinism.

DEFINITION 8 (FEEDBACK). *Let $A(P, Q, R_A)$ be an actor and let $p \in P$ and $q \in Q$. The* feedback composition *of A on (p, q) is the actor*

$$A(p = q) = (P \backslash \{p\}, Q, \{(x \uparrow \{p\}, y) \mid xAy \wedge x(p) = y(q)\}).$$

Feedback is commutative [1].

It is well-known from the study of systems with feedback that the behavior of such a system may not be unique, even if the system is deterministic, or that the behavior may not be constructively computable from the behavior of the actor, depending on the nature of the actor. To effectively apply feedback we will typically require additional constraints on the actor. In the following proposition we describe a case in which a solution can be constructively characterized by a method reminiscent of those used in Kahn Process Networks (KPN) [20]. Our result can also handle non-deterministic actors, however. See also the related Proposition 3.

PROPOSITION 1. *If actor A is input-complete, $\preceq$-monotone and $\preceq$-continuous, then $A(p = q)$ is input-complete, $\preceq$-monotone and $\preceq$-continuous.*

PROOF. Let $A = (P, Q, R_A)$, with $p \in P$ and $q \in Q$. If x is an event trace and p a port of x, then $x[p \to \tau]$ denotes the event trace obtained from x by setting the sequence at p to τ and leaving the sequences at all other ports unchanged. We show here only input completeness because it illustrates the existence of a fixed-point of the feedback. For a detailed proof, see [1]. Let $x \in Tr(P \backslash \{p\})$, then by input-completeness, $x_0 = x[p \to \epsilon] \in \text{in}_A$. Hence there is some y_0 such that $x_0 A y_0$. Now let $x_1 = x[p \to y_0(q)]$. Clearly $x_0 \preceq x_1$. Further by $\preceq$-continuity, there exists y_1 such that $x_1 A y_1$ and $y_0 \preceq y_1$. Repeating the procedure with $x_{k+1} = x[p \to y_k(q)]$ we create two chains $\{x_k\}$ and $\{y_k\}$ in the prefix CPO, such that for all k, $x_k A y_k$. Let $x' = \bigsqcup_{\preceq} \{x_k\}$ and $y' = \bigsqcup_{\preceq} \{y_k\}$, then by construction of the chains and by $\preceq$-continuity, respectively $x'(p) = y'(q)$ and $x' A y'$. Therefore, $x' \uparrow \{p\} = x \in \text{in}_{A(p=q)}$. Thus, $A(p = q)$ is input-complete. $\square$

The assumptions used in the above result may appear strong at first sight. Note, however, that similar assumptions are

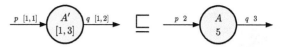

Figure 4: SDF actor A refined by CSDF actor A'.

often used in fixpoint theorems, even for deterministic systems. Although we could have restricted our attention to actors that have such properties by definition, we chose not to do so, since one of our goals is to be as general as possible and to examine the required assumptions on a case-by-case basis. Note that some actor formalisms (e.g., SDF) ensure these properties by definition, however, other formalisms (e.g., automata) don't.

6. REFINEMENT

Refinement is a relation between two actors A and B, allowing one to replace actor A by actor B in a given context and obtain "same or better" results, in the worst case. If τ_A and τ_B are event sequences produced by A and B, respectively, then "τ_B is same or better than τ_A" means the following: τ_B should have at least as many events as τ_A and for every event they have in common, the event should be produced in τ_B no later than in τ_A. We first capture this relation on event sequences and event traces.

DEFINITION 9. *Event sequence τ* refines *event sequence τ', denoted $\tau \sqsubseteq \tau'$, iff for all $n \in \mathbb{N}$, $\tau(n) \leq \tau'(n)$. $\sqsubseteq$ is lifted to event traces $x, x' \in Tr(P)$ in the standard way: $x \sqsubseteq x'$ iff for all $p \in P$, $x(p) \sqsubseteq x'(p)$.*

For example, for the event sequences shown in Figure 2, we have $\tau_3 \sqsubseteq \tau_2$, $\tau_2 \sqsubseteq \tau_1$, but $\tau_1 \not\sqsubseteq \tau_2$. The refinement relations on event sequences and event traces are partial orders, i.e., reflexive, transitive and antisymmetric. Moreover, the set of traces $(Tr(P), \sqsubseteq)$ equipped with the refinement order is a lattice. The supremum and infimum of traces is the point-wise supremum and infimum respectively. The event sequence $\vec{0}$, defined by $\vec{0}(n) = 0$ for all $n \in \mathbb{N}$, is the least element. The empty sequence ϵ, is the greatest element.

Note that for all $x, x' \in Tr(P)$, both $x' \preceq x$ and $x \leq x'$ imply $x \sqsubseteq x'$ and $x \sqsubseteq x'$ iff there exists $x'' \in Tr(P)$ such that $x'' \preceq x$ and $x'' \leq x'$; also, $x \sqsubseteq x'$ implies that for all $p \in P$, $|x'(p)| \leq |x(p)|$ and if both x and x' are infinite, then $x \sqsubseteq x'$ iff $x \leq x'$.

Knowing what refinement of event traces means, we can now define a refinement relation on actors.

DEFINITION 10 (REFINEMENT). *Let $A = (P, Q, R_A)$ and $B = (P, Q, R_B)$ be actors. B* refines *A, denoted $B \sqsubseteq A$, iff*
(1) $\text{in}_A \subseteq \text{in}_B$; and
(2) $\forall x \in \text{in}_A, \forall y : xBy \implies \exists y' : y \sqsubseteq y' \wedge xAy'$.

Condition (1) states that for actor B to refine actor A, B must accept at least all the inputs that actor A accepts. Condition (2) states that any behavior of actor B is no worse than a worst-case behavior of A on the same input. Note that this is where we deviate from the standard notions of refinement that implement the "more output deterministic" principle, which amounts to using the stronger constraint $y = y'$ instead of $y \sqsubseteq y'$ in Condition (2).

The requirement that both A and B have the same sets of input and output ports is not restrictive. Every output port of A (resp. input port of B) must also be an output port of B (resp. input port of A): otherwise replacing A by B in certain contexts may result in open inputs. Any output port of B (resp. input port of A) not originally in A (resp. B) can be added to it as a "dummy" port.

EXAMPLE 3 (CSDF REFINING SDF). *Figure 4 shows an SDF actor A refined by a CSDF actor A'. At each firing, which takes 5 time units to complete, A consumes 2 and produces 3 tokens. A' cycles between two firing phases: in the first, which takes 1 time unit, 1 token is consumed and 1 is produced; in the second, which takes 3 time units, 1 token is consumed and 2 are produced. We observe that for the same number of input tokens, A' produces no fewer (and sometimes strictly more) output tokens than A, because A' can fire on a single input token, whereas A requires two. Moreover, because of the earlier activation, as well as the shorter processing time, A' produces outputs no later than A. Therefore, A' refines A.*

It is worth noting that the refinement in the above example would not hold had we used in Definition 10, $y = y'$ as in traditional refinement relations, or even $y \leq y'$ instead of $y \sqsubseteq y'$. This is because, for the input sequence containing a single token, A' produces strictly more tokens than A. As the example of Section 2 shows, it is important to be able to replace SDF actors by CSDF actors in applications, which partly motivated our novel definition of refinement.

Actor refinement is a pre-order: it is reflexive and transitive [1]. However, it is not antisymmetric. Indeed, for the constant and variable delay actors Δ_d and $\Delta_{[d_1, d_2]}$ (see Example 2), and for $d = d_2$, we have both $\Delta_d \sqsubseteq \Delta_{[d_1, d]}$ and $\Delta_{[d_1, d]} \sqsubseteq \Delta_d$. Yet $\Delta_d \neq \Delta_{[d_1, d]}$ when $d_1 < d$.

It is easy to show that refinement is always preserved by parallel composition and hiding [1]. Refinement is preserved by serial composition under natural conditions, namely, consuming actor B should not refuse better input and should not produce worse output on better input:

PROPOSITION 2. *(1) If $A' \sqsubseteq A$ and B is $\sqsubseteq$-input-closed and $\sqsubseteq$-monotone, then $A'\theta B \sqsubseteq A\theta B$. (2) If $B' \sqsubseteq B$ then $A\theta B' \sqsubseteq A\theta B$.*

Feedback preserves refinement under the following conditions:

PROPOSITION 3. *Let A be an inverse $\sqsubseteq$-input-closed, $\sqsubseteq$-monotone and $\sqsubseteq$-continuous actor, and let A' be an input-complete, $\preceq$-monotone and $\preceq$-continuous actor such that $A' \sqsubseteq A$. Then $A'(p = q) \sqsubseteq A(p = q)$.*

EXAMPLE 4. *Consider actors $A = (\{p\}, \{q\}, R_A)$ and $A' = (\{p\}, \{q\}, R_{A'})$ with input-output relations*

$$R_A = \{(x, y) \mid (\forall n : y(q)(n) = x(p)(n) + 2) \vee$$
$$(y(q)(0) = 0 \wedge \forall n : y(q)(n+1) = x(p)(n))\}, \text{ and}$$
$$R_{A'} = \{(x, y) \mid y(q)(0) = 0 \wedge \forall n : y(q)(n+1) = x(p)(n) + 1\}.$$

Both A and A' are input-complete, $\preceq$-monotone in both directions, $\preceq$-continuous and $\leq$-monotone in both directions. A is non-deterministic but A' is deterministic. A' refines A because the unique output sequence of A' can be matched with the (later) output sequence of A produced by the first disjunct. If we connect A in feedback, we get $A(p = q)$ with a single (output) port q, and producing either the empty sequence $y(q) = \epsilon$ or the zero sequence $y(q)(n) = 0$ for all n.

A' in feedback produces a single sequence $y'(q)(n) = n$. Any sequence refines ϵ, therefore, $A'(p = q) \sqsubseteq A(p = q)$.

7. PERFORMANCE METRICS

We often care about the performance of our systems in terms of specific metrics such as throughput or latency [7, 27, 23, 15]. In this section we show that our notion of refinement is strong enough to provide guarantees on performance under a refinement process. For simplicity we assume in this section, that $\mathcal{T} = \mathbb{R}^{\geq 0}$.

We begin by defining throughput for an infinite event sequence τ. A first attempt is to define throughput as the limit behavior of the average number of tokens appearing in the sequence per unit of time: $T(\tau) = \lim_{n \to \infty} \frac{n}{\tau(n)}$. By the usual definition of the limit, it exists and is equal to T if

$$\forall \varepsilon > 0 : \exists K > 0 : \forall n > K : T - \varepsilon < \frac{n}{\tau(n)} < T + \varepsilon.$$

However, this limit may not always exist for a given τ. Because among all possible behaviors of an actor, there may be some for which it does not exist, we focus instead on throughput *bounds*, which are more robust against this. We consider lower bounds, which are preserved by refinement.

DEFINITION 11. *Given infinite event sequence τ, its lower bound on throughput is*

$$T^{lb}(\tau) = \sup\{T \in \mathbb{R}^{\geq 0} \mid \exists K > 0 : \forall n > K : n > \tau(n) \cdot T\}$$

where by convention we take $\sup \mathbb{R}^{\geq 0} = \infty$.

$T^{lb}(\tau)$ is the greatest lower bound on the asymptotic average throughput of τ (also known as the *limit inferior* of the sequence $n/\tau(n)$). Multiplying both sides of the inequalities by $\tau(n)$ avoids division by zero problems. For a *zeno* sequence τ, where timestamps do not diverge to ∞, i.e., $\exists t \in \mathbb{R}^{\geq 0} : \forall n \in \mathbb{N} : \tau(n) < t$, we have $T^{lb}(\tau) = \sup \mathbb{R}^{\geq 0} = \infty$. This holds in particular for the zero sequence $\vec{0}$ with $\vec{0}(n) = 0$ for all n.

PROPOSITION 4. *For any two infinite event sequences τ_1 and τ_2, if $\tau_1 \sqsubseteq \tau_2$, then $T^{lb}(\tau_1) \geq T^{lb}(\tau_2)$.*

We next define the throughput bound for an actor A. An actor may have multiple output ports with generally different throughputs. For a given port, the throughput at that port may depend on the input trace as well as on non-deterministic choices of the actor. We therefore consider the worst-case scenario.

DEFINITION 12. *Given actor $A = (P, Q, R_A)$, output port $q \in Q$ and input trace $x \in Tr(P)$, the lower bound on throughput of A w.r.t. q, x is:*

$$T^{lb}(A, x, q) = \inf\{T^{lb}(\tau) \mid \exists y : xAy \wedge \tau = y(q)\}.$$

For example, for the actor SPEC of Section 2, which has no inputs and a unique output port, we have $T^{lb}(\text{SPEC}) = T^{lb}(50 + {}^n/_{44.1}) = \sup\{T \in \mathbb{R}^{\geq 0} \mid \exists K > 0 : \forall n > K : n > (50 + {}^n/_{44.1})T\} = \sup\{T \in \mathbb{R}^{\geq 0} \mid T < 44.1\} = 44.1$.

For two actors A and B with the same sets of input and output ports P and Q, respectively, we shall write $T^{lb}(A, x) \leq T^{lb}(B, x)$ to mean $T^{lb}(A, x, q) \leq T^{lb}(B, x, q)$ for all $q \in Q$. This notation is used in Proposition 6 below.

We next turn to latency, another prominent performance metric. We define latency as the smallest upper bound on observed time differences between related input and output events. The pairs of events that we want to relate are explicitly specified as follows:

DEFINITION 13. *An input-output event specification (IOES) for a set P of input ports and a set Q of output ports is a relation $\mathcal{E} \subseteq 2^{P \times \mathbb{N}} \times 2^{Q \times \mathbb{N}}$. $\mathcal{E}$ is called* valid *for $(x,y) \in Tr(P) \times Tr(Q)$ iff for every $(E_P, E_Q) \in \mathcal{E}$, if $x(p)(m) \neq \infty$ for every $(p,m) \in E_P$, then $y(q)(n) \neq \infty$ for every $(q,n) \in E_Q$. $\mathcal{E}$ is called* valid *for an actor $A = (P, Q, R_A)$ iff it is valid for every $(x,y) \in R_A$.*

A pair $(E_P, E_Q) \in \mathcal{E}$ says that we want to measure the maximum latency between an input event in E_P and an output event in E_Q, provided all events in E_P have arrived. See Example 5, given below, for an illustration.

DEFINITION 14. *Let $\mathcal{E}$ be a valid IOES for $(x,y) \in Tr(P) \times Tr(Q)$. The upper bound on latency is defined as:*

$$D^{\mathcal{E}}(x,y) = \sup\{y(q)(n) - x(p)(m) \mid (E_P, E_Q) \in \mathcal{E},$$
$$E_P \subseteq \text{dom}(x), (p,m) \in E_P, (q,n) \in E_Q\}$$

where by convention $\sup \emptyset = 0$ and $\text{dom}(x)$ denotes the set of all pairs (p,n) such that $x(p)(n) \neq \infty$.

$D^{\mathcal{E}}(x,y)$ is the largest among all delays between an input and an output event that occur in x and y and are related by $\mathcal{E}$, provided all other events in the same input group are also in x. Notice that, by the assumption of validity of $\mathcal{E}$ for (x,y), $E_P \subseteq \text{dom}(x)$ implies $E_Q \subseteq \text{dom}(y)$, for every $(E_P, E_Q) \in \mathcal{E}$.

EXAMPLE 5. *Consider a deterministic actor A with two input ports p_1, p_2 and a single output port q. Suppose A consumes one token from each input port, and for every such pair, produces a token at q, after some constant delay, say $d \in \mathbb{R}^{\geq 0}$. Let x_1 and x_2 be two input event traces, with $x_1 = \{(p_1, 2 \cdot \epsilon), (p_2, 4 \cdot \epsilon)\}$ and $x_2 = \{(p_1, 2 \cdot 5 \cdot \epsilon), (p_2, 4 \cdot \epsilon)\}$. For both x_1 and x_2, A produces the same output event trace $y = \{(q, (4+d) \cdot \epsilon)\}$. This is because, in x_2, A waits for a second input to arrive at p_2 before it can produce a second output. To measure the latency of A, we can define $\mathcal{E}$ to be:*

$$\mathcal{E} = \{(\{(p_1, n), (p_2, n)\}, \{(q, n)\}) \mid n \in \mathbb{N}\}.$$

This makes $\mathcal{E}$ a valid IOES for x_1, y, as well as for x_2, y, and gives us $D^{\mathcal{E}}(x_1, y) = D^{\mathcal{E}}(x_2, y) = d$, as is to be expected.

Keeping the reference input trace fixed, refinement of output traces is guaranteed to not worsen latency:

PROPOSITION 5. *Let x, y_1 and y_2 be event traces such that $y_1 \sqsubseteq y_2$. Suppose $\mathcal{E}$ is valid for x and y_2. Then $\mathcal{E}$ is valid for x and y_1 and $D^{\mathcal{E}}(x, y_1) \leq D^{\mathcal{E}}(x, y_2)$.*

DEFINITION 15. *An IOES $\mathcal{E}$ is valid for an actor A iff $\mathcal{E}$ is valid for every (x,y) such that xAy. For a valid $\mathcal{E}$, the worst-case latency of A on input event trace x is*

$$D^{\mathcal{E}}(A, x) = \sup_{y \text{ s.t. } xAy} \{D^{\mathcal{E}}(x, y)\}.$$

EXAMPLE 6. *A suitable IOES for the variable delay actor $\Delta_{[d_1, d_2]}$ from Example 2 is $\mathcal{E} = \{(\{(p, n)\}, \{(q, n)\}) \mid n \in \mathbb{N}\}$. $\mathcal{E}$ is valid for $\Delta_{[d_1, d_2]}$ and $D^{\mathcal{E}}(\Delta_{[d_1, d_2]}, x) = d_2$, for any non-empty input event trace x.*

The following states the main preservation results for throughput and latency performance bounds under refinement:

PROPOSITION 6. *Let $B \sqsubseteq A$ and $\mathcal{E}$ be a valid IOES for A. Then for any $x \in \text{in}_A$, $T^{lb}(B, x) \geq T^{lb}(A, x)$ and $D^{\mathcal{E}}(B, x) \leq D^{\mathcal{E}}(A, x)$.*

Proposition 6 can be used together with Propositions 2 and 3 to guarantee that worst-case performance bounds are preserved during compositional refinement of models, as in the example of Section 2.

8. REPRESENTATIONS & ALGORITHMS

So far, our treatment has been semantical, regarding actors as sets of input-output event traces. In this section, we consider syntactic, finite representations. We show that the semantics commonly associated with these representations can be embedded naturally in our theory. We also provide algorithms to check refinement and compute compositions and performance metrics on such representations. Our intention in this section is not to be complete, but rather to give examples of how our theory can be instantiated and automated.

8.1 Synchronous Data Flow

We have informally used timed SDF actors in previous examples. In this section we formally define them. Typically, in timed SDF models the time domain is the non-negative reals or integers. In the rest of this subsection, we therefore assume that $\mathcal{T} = \mathbb{R}^{\geq 0}$ or $\mathcal{T} = \mathbb{N}$.

DEFINITION 16 (SDF ACTORS). *An actor $A = (P, Q, R_A)$ is a homogeneous SDF actor with firing duration $d \in \mathcal{T}$, iff*

$$R_A = \{(x,y) \mid \forall q \in Q : |y(q)| = \min_{p \in P} |x(p)| \wedge$$
$$\forall n < |y(q)| : y(q)(n) = \max_{p \in P} x(p)(n) + d\}.$$

That is, the n-th firing of A starts as soon as the n-th token has arrived on every input. The firing takes d time units, after which a single output token is produced on each output. A is an SDF actor with token transfer quanta $\boldsymbol{r} : P \cup Q \to \mathbb{N}$ and firing duration $d \in \mathcal{T}$ iff

$$R_A = \{(x,y) \mid \forall q \in Q : |y(q)| = \boldsymbol{r}(q) \cdot \min_{p \in P}\left(|x(p)| \div \boldsymbol{r}(p)\right)$$
$$\wedge \ \forall n < |y(q)| : y(q)(n) = d +$$
$$\max_{p \in P} x(p)\left((n \div \boldsymbol{r}(q) + 1) \cdot \boldsymbol{r}(p) - 1\right)\}.$$

where $\div$ denotes the quotient of the integer division.

A (non-homogeneous) SDF actor can consume respectively produce more than a single token on its inputs and outputs with every firing, using rates according to $\boldsymbol{r}$. SDF actors are deterministic and have constant delays d. In SDF literature they are often implicitly understood to abstract behaviours with varying (non-deterministic) execution times in a conservative way.

EXAMPLE 7. *Consider the SDF actor A shown in Figure 4. A has an input port p (quantum 2) and an output port q (quantum 3). Its firing duration is 5. An example input-output event trace of A is shown below. The firings of A start at times 2, 4 and 5 and overlap in time. Note that the 7-th input token does not lead to any output.*

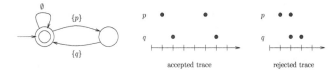

Figure 5: Implicit-tick automaton example.

CSDF actors like the one on the left of Figure 4 can be formalized similarly, taking into account that they periodically cycle through firings with different quanta and firing durations.

An *SDF graph* represents the composition of multiple SDF actors, as in the examples shown in Figure 1. Edges in SDF graphs are often annotated with initial tokens representing the fact that the initial state of some queues is non-empty. To model this, we introduce an explicit actor:

EXAMPLE 8. *The initial token actor with $k \in \mathbb{N}$ tokens is $I_k = (\{p\}, \{q\}, R_{I_k})$, where $(x, y) \in R_{I_k}$ iff for all $n \in \mathbb{N}$:*

$$y(q)(n) = 0 \text{ if } n < k, \text{ and } y(q)(n) = x(p)(n - k) \text{ otherwise.}$$

That is, I_k outputs k initial tokens at time 0, and then behaves as the identity function. I_k satisfies all monotonicity and continuity properties.

An SDF graph cannot always be reduced to an equivalent SDF actor. Indeed, in general, the serial or parallel composition of two SDF actors is not an SDF actor [29] (but of course it is an actor in the sense of this paper).

Let A_1 and A_2 be two SDF actors. We want to check whether $A_1 \sqsubseteq A_2$. Clearly, A_1 and A_2 must have the same sets of input and output ports, say P and Q. Suppose A_1 and A_2 have quanta functions r_1 and r_2, and firing durations d_1 and d_2, respectively.

PROPOSITION 7. $A_1 \sqsubseteq A_2$ iff $d_1 \leq d_2$ and $\forall p \in P, q \in Q, n \leq r_1(p) \cdot r_2(p) : r_1(q) \cdot (n \div r_1(p)) \geq r_2(q) \cdot (n \div r_2(p))$.

The above result is generalized in [1] which discusses how refinement can be checked not only on SDF actors but also on SDF graphs, using (max, +) algebra. [1] also discusses how throughput can be computed on SDF graphs. The proposition that follows summarizes the latter result. For an SDF graph A with external input and output ports P and Q, $r_A : P \cup Q \rightarrow \mathbb{N}$ denotes the *repetition vector* of the graph, which assigns to every port the relative rates at which tokens are consumed and produced [22].

PROPOSITION 8. *Let A be a strongly connected SDF graph with input ports P and output ports Q. Let x be an input trace of A. Then A has a computable internal throughput bound T^A, and*

$$T^{lb}(A, x, q) = r_A(q) \cdot \min(T^A, \min_{p \in P} \frac{T^{lb}(x(p))}{r_A(p)}).$$

Similarly, latency of SDF actors as defined in this work, can be computed using existing analysis techniques from SDF literature [15, 23]. It is natural to specify the related input-output events in patterns which repeat with the periodic behavior of SDF iterations.

8.2 Discrete-Time Automata

An other natural representation of actors is automata. Automata, in contrast with SDF actors, do not have $\sqsubseteq$-monotonicity and input-closure built-in, and such properties have to be explicitly verified when necessary. There are

many automata variants, over finite or infinite words, with various acceptance conditions, finite or infinite-state, and so on. We are not going to propose a single automaton-based model for actors. Instead we will discuss some general ideas as well as some cases for which we have algorithms. We limit our discussion to *discrete-time* automata (DTA) in the sense that time is counted by discrete transitions. DTA generate actors over a discrete time domain, $\mathcal{T} = \mathbb{N}$. The ideas naturally apply also to timed automata [2] and yield actors where $\mathcal{T} = \mathbb{R}^{\geq 0}$.

One possible model is an automaton whose transitions are labeled with subsets of $P \cup Q$, the set of input or output ports. An example is shown in Figure 5. The state drawn with two circles is the *accepting* state. In this *implicit-tick model* each transition corresponds to one time unit. If the transition is labeled by some set of ports $V \subseteq P \cup Q$, then an event at each port in V occurs at the corresponding instant in $\mathbb{N}$. V may be empty, as in the self-loop transition of the automaton shown in the figure. In this case, no events occur at that time instant.

The implicit-tick model cannot capture event traces where more than one event occurs simultaneously at the same port (and therefore an implicit-tick actor cannot be input-complete). An alternative is to dissociate the elapse of time from transitions, by introducing a special label, t, denoting one time unit. Then, we can use automata whose transitions are labeled with single ports or t, that is, whose alphabet is $P \cup Q \cup \{t\}$. We call this the *explicit-tick model*. We do not further discuss this model here: it is explored in [1].

A DTA M defines an actor $A(M) = (P, Q, R_{A(M)})$ as follows. Every (finite or infinite) accepting run of M generates a (finite or infinite) word w in the language of M, denoted $\mathcal{L}(M)$. Every word w can be mapped to a unique event trace pair $Tr(w) \in Tr(P) \times Tr(Q)$, as illustrated in Figure 5. Then, $R_{A(M)}$ is the set of all event trace pairs generated by words in $\mathcal{L}(M)$, i.e., the set $\{ Tr(w) \mid w \in \mathcal{L}(M) \}$, also denoted $Tr(\mathcal{L}(M))$. An *ITA* is an implicit-tick automaton on finite words. An *ITBA* is an implicit-tick Büchi automaton. Finite-word DTA are strictly less expressive than corresponding Büchi versions. Every ITA M can be transformed into an ITBA M' such that $A(M) = A(M')$.

Two distinct words w and w' of an ITA M can result in the same event trace, for instance, if $w' = w \cdot \emptyset^n$, for different $n \geq 1$. To avoid technical complications related to this, we will assume that M is *tick-closed*, that is, for any $w \in \mathcal{L}(M)$, $w \cdot \emptyset^* \subseteq \mathcal{L}(M)$. The ITA of Figure 5 is tick-closed. Any ITA M can be transformed to a tick-closed ITA M' so that $A(M) = A(M')$. In the rest of this section we assume all ITA to be tick-closed. We also assume that automata are finite-state, all their states are reachable from the initial state, and there is no state that cannot reach an accepting state by a non-empty path.

Given actors represented by discrete-time automata, we are interested in answering various questions.

"Given M and M', is $A(M) = A(M')$?" For ITA, there is a bijection between infinite words and event traces, that is, $\forall w, w' \in (2^{P \cup Q})^\omega : w \neq w' \iff Tr(w) \neq Tr(w')$. (Note that $Tr(w)$ may be finite, even though w is infinite, if w ends in $\emptyset^\omega$.) The same bijection does not hold for finite words as explained above. Nevertheless, because ITA are assumed to be tick-closed, we can show:

PROPOSITION 9. *For two ITA (ITBA) M_1 and M_2, we have $A(M_1) = A(M_2)$ iff $\mathcal{L}(M_1) = \mathcal{L}(M_2)$.*

"Given M, is $A(M)$ deterministic?" Determinism of M does not imply determinism of $A(M)$, because of the different role of input and output symbols. $A(M)$ is non-deterministic iff there are two words $w, w' \in \mathcal{L}(M)$, with $(x, y) = Tr(w)$ and $(x', y') = Tr(w')$, such that $x = x'$ but $y \neq y'$.

PROPOSITION 10. *For any ITA or ITBA M it is decidable whether $A(M)$ is deterministic.*

The proof uses a synchronous product of M with itself synchronizing only on input events to check for words with the same input event trace, but a different output event trace [1].

"Given M_1 and M_2, compute M so that $A(M) = A(M_1) \| A(M_2)$." M can be computed as a product of M_1 and M_2, so that $\mathcal{L}(M)$ contains exactly those words w such that $Tr(w) = Tr(w_1) \cup Tr(w_2)$, for $w_i \in \mathcal{L}(M_i)$ and $i = 1, 2$. If M_1 and M_2 belong to the implicit-tick model, M is a synchronous product, so that a pair of transitions $\xrightarrow{P_1 \cup Q_1}$ of M_1 and $\xrightarrow{P_2 \cup Q_2}$ of M_2 yields a transition $\xrightarrow{P_1 \cup P_2 \cup Q_1 \cup Q_2}$ in M.

"Given M_1 and M_2, and a bijection θ from the output ports of M_1 to the input ports of M_2, compute M so that $A(M) = A(M_1) \theta A(M_2)$."

Feeding the output of M_1 (after relabeling) into the input of M_2 can be achieved by a product of both automata synchronizing on the corresponding ports. The main challenge in computing the serial composition is to ensure that the constraint in_θ is satisfied (see Definition 6). We assume M_1 and M_2 are ITA. We construct the composite automaton M as the synchronous product of M_1, M_2 and a third automaton M_{in} capturing the constraint in_θ. We can construct M_{in} as an alternating automaton such that $Tr(\mathcal{L}(M_{in})) = in_\theta$ [1]. M_{in} can then be converted into a non-deterministic automaton using standard techniques.

"Given M, input port p and output port q, compute M' such that $A(M') = A(M)(p = q)$." If M is an ITA or an ITBA, then M' can be easily obtained by removing from M all transitions $\xrightarrow{P' \cup Q'}$ except those that satisfy $p \in P' \iff q \in Q'$.

An important question is to check for actor refinement. "Given M_1 and M_2, is $A(M_1) \sqsubseteq A(M_2)$?" We show that checking actor refinement on ITA can be reduced to checking language containment with respect to an appropriate closure. Given automaton M, we construct an (initially infinite state) automaton $M_{\infty \sqsubseteq}$ that recognizes the refinement closure of M, i.e., it accepts all words of M, but also all words that correspond to traces that refine the traces of M. We define $M_{\infty \sqsubseteq}$ for a single output port q, but the construction can be generalized to multiple output ports. Figure 6 shows an example. We add a counter n to count the surplus of q events. Whenever later in a word M requires a q event, we allow this event to be absent and decrease the

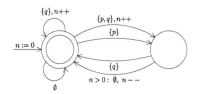

Figure 6: Refinement closure of Figure 5.

counter. The following gives a precise definition of this idea, parameterized with a bound k on the counter.

DEFINITION 17. *Let $M = (S, s_0, E, F)$ be an ITA with a single output port q, states S, initial state $s_0 \in S$, accepting states $F \subseteq S$, and transitions E. For $k \in \mathbb{N}$, the k-bounded refinement closure of M is the automaton $M_{k \sqsubseteq} = (S_{k \sqsubseteq}, s_{k \sqsubseteq, 0}, E_{k \sqsubseteq}, F_{k \sqsubseteq})$ such that $S_{k \sqsubseteq} = \{(s, n) \mid s \in S, 0 \leq n \leq k\}$, $s_{k \sqsubseteq, 0} = (s_0, 0)$, and $F_{k \sqsubseteq} = \{(s, n) \in S_{k \sqsubseteq} \mid s \in F\}$. For every transition $(s_1, \sigma, s_2) \in E$, we have the following transitions in $E_{k \sqsubseteq}$:*

$$
\begin{array}{ll}
((s_1, n), \sigma, (s_2, n)) & \text{if } 0 \leq n \leq k \\
((s_1, n), \sigma \cup \{q\}, (s_2, n+1)) & \text{if } 0 \leq n < k, q \notin \sigma \\
((s_1, n), \sigma \cup \{q\}, (s_2, n)) & \text{if } n = k, q \notin \sigma \\
((s_1, n), \sigma \setminus \{q\}, (s_2, n-1)) & \text{if } 0 < n \leq k, q \in \sigma
\end{array}
$$

The (unbounded) refinement closure of M is the automaton $M_{\infty \sqsubseteq}$ defined in a similar way, but where counter n is unbounded.

LEMMA 1. *Let M_1 and M_2 be ITA with the same input ports P and the same, single output port q. Then $A(M_1) \sqsubseteq A(M_2)$ iff for every $w \in \mathcal{L}(M_1)$ such that $Tr(w) = (x, y)$ and $x \in in_{A(M_2)}$, $w \in \mathcal{L}(M_{2, \infty \sqsubseteq})$.*

Unfortunately, $M_{2, \infty \sqsubseteq}$ is not a finite-state automaton, in fact, $\mathcal{L}(M_{2, \infty \sqsubseteq})$ is not always regular. Let $\mathcal{L}(M) = (\{p, q\} \cdot \emptyset)^*$, where p is the only input port and q the only output port. Then $\mathcal{L}(M_{\infty \sqsubseteq})$ contains all words of the form $(\{p, q\} \cdot \{q\})^n \cdot \{p\}^n$, for any $n \in \mathbb{N}$. Based on this, we can show that $\mathcal{L}(M_{\infty \sqsubseteq})$ is not regular. Despite this difficulty, we can make use of the finite memory property of M_1 and M_2 to find an upper bound on the size of the refinement closure, which proves that checking refinement for ITA is decidable:

PROPOSITION 11. *Let M_1 and M_2 be deterministic ITA with the same input ports P and the same, single output port q. $A(M_1) \sqsubseteq A(M_2)$ iff for every $w \in \mathcal{L}(M_1)$ such that $Tr(w) = (x, y)$ with $x \in in_{A(M_2)}$, $w \in \mathcal{L}(M_{2, N \sqsubseteq})$, where $N = |S_1| \times |S_2|$.*

"Given ITBA M with sets of input and output ports P and Q, respectively, and given an output port $q \in Q$ and an input trace $x \in Tr(P)$, what is $T^{lb}(A(M), x, q)$?" To compute this, we need a finite representation for x. A natural choice is to represent x as a deterministic ITBA M_x that only refers to ports in P. We require that M_x generates a single trace x. These assumptions imply that M_x has the form of a "lasso" (a single path eventually returning to an earlier state).

First, we compute a product of M and M_x such that the two automata synchronize on inputs. We remove from the product all strongly connected components (SCCs) that contain no accepting state of M and denote the result by M'.

We assign a weight to each transition $\xrightarrow{P'/Q'}$ of M': weight 1 if $q \in Q'$ and weight 0 otherwise. With these weights M' can be viewed as a weighted directed graph. We run Karp's algorithm [21] to compute the minimum cycle mean of M', denoted MCM. MCM is the minimum over all simple cycles κ in M'' of the ratio $\frac{w(\kappa)}{|\kappa|}$, where $w(\kappa)$ is the sum of weights of all transitions in κ and $|\kappa|$ is the length of κ (i.e., the number of transitions in κ).

PROPOSITION 12. $T^{lb}(A(M), x, q) = MCM$.

(proof sketch) The tricky part is that Karp's algorithm considers all cycles, including non-accepting (in the Büchi sense) cycles. However, all cycles are guaranteed to belong to an accepting SCC (otherwise the SCC is removed by construction of M'). Then, from any cycle it is possible to reach an accepting state and then return to the cycle. This "detour" may increase the throughput by some amount ε, however, ε can be made arbitrarily small by taking the detour very infrequently (but infinitely often, to be accepting).

9. DISCUSSION AND FUTURE WORK

We have proposed an interface theory for timed actors with a refinement relation based on the earlier-is-better principle, suitable for worst-case performance analysis. Our framework is compositional and unifies existing formalisms, allowing different types of models to be used in the same design process, e.g., automata models could refine SDF models.

The earlier-is-better principle may not seem directly applicable in scenarios where outputs should be produced not too late but not too early either. A possible alternative approach is to combine the earlier-is-better refinement with a corresponding *later-is-better* version, obtained by replacing $y \sqsubseteq y'$ by $y' \sqsubseteq y$ in Condition (2) of Definition 10. Separate specifications could then be used, expressing upper and lower bounds on timing behavior, and refined using the earlier- or later-is-better relation, respectively. Examining in detail this hybrid approach is part of future work.

Other directions for future work include examining the algorithmic complexity of the various computational problems and coming up with practically useful algorithms; implementing the algorithms and performing experiments; integrating new representations under our framework; and having a comparative study of the different representations, for instance, in terms of expressiveness and complexity.

10. REFERENCES

[1] Extended version of this paper available at http://www.eecs.berkeley.edu/Pubs/TechRpts/2010/EECS-2010-130.html.

[2] R. Alur and D. Dill. A theory of timed automata. *Theor. Comput. Sci.*, 126:183–235, 1994.

[3] R. Alur, T. Henzinger, O. Kupferman, and M. Vardi. Alternating refinement relations. In *CONCUR*, 1998.

[4] F. Baccelli, G. Olsder, J.-P. Quadrat, and G. Cohen. *Synchronization and linearity. An algebra for discrete event systems.* Wiley, 1992.

[5] R.-J. Back and J. Wright. *Refinement Calculus.* Springer, 1998.

[6] G. Bilsen, M. Engels, R. Lauwereins, and J. Peperstraete. Cyclo-static dataflow. *IEEE Tran. on Signal Processing*, 44(2), 1996.

[7] J.-Y. L. Boudec and P. Thiran. *Network Calculus: A Theory of Deterministic Queuing Systems for the Internet.* Springer, 2001.

[8] G. Buttazzo. *Hard Real-time computing systems.* Kluwer, 1997.

[9] E. Clarke, O. Grumberg, and D. Peled. *Model Checking.* MIT Press, 2000.

[10] A. David, K. G. Larsen, A. Legay, U. Nyman, and A. Wasowski. Timed I/O automata: a complete specification theory for real-time systems. In *HSCC*, pages 91–100, 2010.

[11] L. de Alfaro and T. Henzinger. Interface automata. In *Foundations of Software Engineering (FSE).* ACM Press, 2001.

[12] L. de Alfaro, T. A. Henzinger, and M. I. A. Stoelinga. Timed interfaces. In *EMSOFT*, pages 108–122, 2002.

[13] L. de Alfaro, R. Majumdar, V. Raman, and M. Stoelinga. Game refinement relations and metrics. *L. Meth. in Comp. Sc.*, 4(3), 2008.

[14] J. B. Dennis. First version of a data flow procedure language. In *Programming Symposium*, pages 362–376. Springer-Verlag, 1974.

[15] A. Ghamarian, S. Stuijk, T. Basten, M. Geilen, and B. Theelen. Latency minimization for synchronous data flow graphs. *Digital Systems Design, Euromicro Symposium on*, pages 189–196, 2007.

[16] A. Hansson, M. Wiggers, A. Moonen, K. Goossens, and M. Bekooij. Enabling application-level performance guarantees in network-based systems on chip by applying dataflow analysis. *IET CDT*, 3(5), 2009.

[17] T. Henzinger and S. Matic. An interface algebra for real-time components. In *RTAS*, pages 253 – 266, 2006.

[18] H. Hermanns, U. Herzog, and J.-P. Katoen. Process algebra for performance evaluation. *Theor. Comput. Sci.*, 274(1-2):43–87, 2002.

[19] B. Jonsson and W. Yi. Testing preorders for probabilistic processes can be characterized by simulations. *Theor. Comput. Sci.*, 282(1):33–51, 2002.

[20] G. Kahn. The semantics of a simple language for parallel programming. In *Inf. Proc. 74*, 1974.

[21] R. Karp. A characterization of the minimum cycle mean in a digraph. *Discr.Math.*, 23(3):309–311, 1978.

[22] E. Lee and D. Messerschmitt. Synchronous data flow. *Proceedings of the IEEE*, 75(9):1235–1245, 1987.

[23] O. M. Moreira and M. J. G. Bekooij. Self-timed scheduling analysis for real-time applications. *EURASIP J. on Adv. in Signal Proc.*, 2007.

[24] F. Nielson, H. Nielson, and C. Hankin. *Principles of Program Analysis.* Springer, 1999.

[25] B. Pierce. *Types and Programming Languages.* MIT Press, 2002.

[26] R. Segala and N. Lynch. Probabilistic simulations for probabilistic processes. *Nordic J. of Computing*, 2(2):250–273, 1995.

[27] L. Thiele, S. Chakraborty, and M. Naedele. Real-time calculus for scheduling hard real-time systems. In *ISCAS*, pages 101–104, 2000.

[28] L. Thiele, E. Wandeler, and N. Stoimenov. Real-time interfaces for composing real-time systems. In *EMSOFT*, pages 34–43, 2006.

[29] S. Tripakis, D. Bui, M. Geilen, B. Rodiers, and E. A. Lee. Compositionality in Synchronous Data Flow: Modular Code Generation from Hierarchical SDF Graphs. ACM TECS (to appear).

[30] G. Weiss and R. Alur. Automata based interfaces for control and scheduling. In *HSCC*, pages 601–613. Springer, 2007.

[31] M. Wiggers, M. Bekooij, P. Jansen, and G. Smit. Efficient computation of buffer capacities for cyclo-static real-time systems with back-pressure. In *RTAS*, pages 281–292, 2007.

[32] M. Wiggers, M. Bekooij, and G. Smit. Monotonicity and run-time scheduling. In *EMSOFT*, pages 177–186, 2009.

Measuring Performance of Continuous-Time Stochastic Processes using Timed Automata

Tomáš Brázdil
brazdil@fi.muni.cz

Jan Krčál
krcal@fi.muni.cz

Jan Křetínský
jan.kretinsky@fi.muni.cz

Antonín Kučera
kucera@fi.muni.cz

Vojtěch Řehák
rehak@fi.muni.cz

Faculty of Informatics, Masaryk University
Botanická 68a, 60200 Brno
Czech Republic

ABSTRACT

We propose deterministic timed automata (DTA) as a model-independent language for specifying performance and dependability measures over continuous-time stochastic processes. Technically, these measures are defined as limit frequencies of locations (control states) of a DTA that observes computations of a given stochastic process. Then, we study the properties of DTA measures over semi-Markov processes in greater detail. We show that DTA measures over semi-Markov processes are well-defined with probability one, and there are only finitely many values that can be assumed by these measures with positive probability. We also give an algorithm which approximates these values and the associated probabilities up to an arbitrarily small given precision. Thus, we obtain a general and effective framework for analysing DTA measures over semi-Markov processes.

Categories and Subject Descriptors

G.3 [**Mathematics of Computing**]: Probability and statistics—*Stochastic processes*

General Terms

Performance, Theory

Keywords

semi-Markov processes, timed automata, performance analysis, general state space Markov chains, stochastic stability

1. INTRODUCTION

Continuous-time stochastic processes, such as continuous-time Markov chains, semi-Markov processes, or generalized semi-Markov processes [21, 6, 19, 17], have been widely used in practice to determine performance and dependability characteristics of real-world systems. The desired behaviour

of such systems is specified by various measures such as mean response time, throughput, expected frequency of errors, etc. These measures are often formulated just semi-formally and chosen specifically for the system under study in a somewhat ad hoc manner. One example of a rigorous and model-independent specification language for performance and dependability properties is Continuous Stochastic Logic (CSL) [3, 5] which allows to specify both steady state and transient measures over the underlying stochastic process. The syntax and semantics of CSL is inspired by the well-known non-probabilistic logic CTL [14]. The syntax of CSL defines *state* and *path* formulae, interpreted over the states and runs of a given stochastic process $\mathcal{M}$. In particular, there are two probabilistic operators, $\mathcal{P}_{\bowtie \varrho}(\cdot)$ and $\mathcal{S}_{\bowtie \varrho}(\cdot)$, which refer to the transient and steady state behaviour of $\mathcal{M}$, respectively. Here $\bowtie$ is a numerical comparison (such as $\leq$) and $\varrho \in [0, 1]$ is a rational constant. If φ is a path formula[1] (which is either valid or invalid for every run of $\mathcal{M}$), then $\mathcal{P}_{\geq 0.7}(\varphi)$ is a state formula which says "the probability of all runs satisfying φ is at least 0.7". If Φ is a state formula, i.e., Φ is either valid or invalid in every state, then $\mathcal{S}_{\geq 0.5}(\Phi)$ is also a state formula which says "the π-weighted sum over all states where Φ holds is at least 0.5". Here π is the steady-state distribution of $\mathcal{M}$. The logic CSL can express quite complicated properties and the corresponding model-checking problem over continuous-time Markov chains is decidable. However, there are also several disadvantages.

(a) The semantics of steady state probabilistic operator $\mathcal{S}_{\bowtie \varrho}(\cdot)$ assumes the existence of invariant distribution which is not guaranteed to exist for all types of stochastic processes with continuous time (the existing works mainly consider CSL as a specification language for ergodic continuous-time Markov chains).

(b) In CSL formulae, all measures are explicitly quantified, and the model-checking algorithm just verifies constraints over these measures. Alternatively, we might wish to *compute* certain measures up to a given precision.

In this paper, we propose *deterministic timed automata (DTA)* [2] as a model-independent specification language

---

[1]In CSL, φ can be of the form $\mathcal{X}_I \Phi$ or $\Phi_1 \mathcal{U}_I \Phi_2$ where Φ, Φ_1, Φ_2 are state formulae, and $\mathcal{X}_I, \mathcal{U}_I$ are the modal connectives of CTL parametrized by an interval I. Boolean connectives can be used to combine just state formulae.

for performance and dependability measures of continuous-time stochastic processes. The "language" of DTA can be interpreted over *arbitrary* stochastic processes that generate timed words, and their expressive power appears sufficiently rich to capture many interesting run-time properties (although we do not relate the expressiveness of CSL and DTA formally, they are surely incomparable because of different "nature" of the two formalisms). Roughly speaking, a DTA $\mathcal{A}$ "observes" runs of a given stochastic process $\mathcal{M}$ and "remembers" certain information in its control states (which are called *locations*). Since $\mathcal{A}$ is deterministic, for every run σ of $\mathcal{M}$ there is a unique computation $\mathcal{A}(\sigma)$ of $\mathcal{A}$, which determines a unique tuple of "frequencies" of visits to the individual locations of $\mathcal{A}$ along σ. These frequencies are the values of "performance measures" defined by $\mathcal{A}$ (in fact, we consider *discrete* and *timed* frequencies which are based on the same concept but defined somewhat differently).

Let us explain the idea in more detail. Consider some stochastic process $\mathcal{M}$ whose computations (or runs) are infinite sequences of the form $\sigma = s_0 t_0 s_1 t_1 \cdots$ where all s_i are "states" and t_i is the time spent by performing the transition from s_i to s_{i+1}. Also assume a suitable probability space defined over the runs of $\mathcal{M}$. Let Σ by a finite alphabet and L a labelling which assigns a unique letter $L(s) \in \Sigma$ to every state s of $\mathcal{M}$. Intuitively, the letters of Σ correspond to collections of predicates that are valid in a given state. Thus, every run $\sigma = s_0 t_0 s_1 t_1 \cdots$ of $\mathcal{M}$ determines a unique *timed word* $w_\sigma = L(s_0) t_0 L(s_1) t_1 \cdots$ over Σ.

A DTA over Σ is a finite-state automaton $\mathcal{A}$ equipped with finitely many internal clocks. Each control state (or location) q of $\mathcal{A}$ has finitely many outgoing edges $q \longrightarrow q'$ labeled by triples (a, g, X), where $a \in \Sigma$, g is a "guard" (a constraint on the current clock values), and X is a subset of clocks that are reset to zero after performing the edge. A configuration of $\mathcal{A}$ is a pair (q, ν), where q and ν are the current location and the current clock valuation, respectively. Every timed word $w = c_0 c_1 c_2 c_3 \cdots$ over Σ (where $c_i \in \Sigma$ iff i is even) then determines a unique run $\mathcal{A}(w) = (q_0, \nu_0)(q_1, \nu_1)(q_2, \nu_2)\cdots$ of $\mathcal{A}$ where q_0 is an *initial* location, ν_0 assigns zero to every clock, and (q_{i+1}, ν_{i+1}) is obtained from (q_i, ν_i) either by performing the only enabled edge $q_i \longrightarrow q_{i+1}$ labeled by (c_i, g, X) if i is even, or by simultaneously increasing all clocks by c_i if i is odd.

As a simple example, consider the following DTA $\hat{\mathcal{A}}$ over the alphabet $\{a\}$ with one clock x and the initial location q_0:

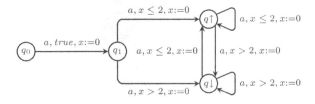

Intuitively, $\hat{\mathcal{A}}$ observes time stamps in a given timed word and enters either $q{\uparrow}$ or $q{\downarrow}$ depending on whether a given stamp is bounded by 2 or not, respectively. For example, a word $w = a\ 0.2\ a\ 2.4\ a\ 2.1 \cdots$ determines the run $\hat{\mathcal{A}}(w) = (q_0, 0)(q_1, 0)(q_1, 0.2)(q{\uparrow}, 0)(q{\uparrow}, 2.4)(q{\downarrow}, 0)(q{\downarrow}, 2.1) \cdots$

Let $w = a_0 t_0 a_1 t_1 \cdots$ be a timed word over Σ and q a location of $\mathcal{A}$. For every $i \in \mathbb{N}_0$, let $T^i(w)$ be the stamp t_i of w, and $Q^i(w)$ the location of $\mathcal{A}$ entered after reading the finite prefix $a_0 t_0 \cdots a_i$ of w. Further, let $1_q^i(w)$ be either 1 or 0 depending on whether $Q^i(w) = q$ or not, respectively. We define the *discrete* and *timed* frequency of visits to q along

$\mathcal{A}(w)$, denoted by $\mathbf{d}_q^{\mathcal{A}}(w)$ and $\mathbf{c}_q^{\mathcal{A}}(w)$, in the following way (the '$\mathcal{A}$' index is omitted when it is clear from the context):

$$\mathbf{d}_q^{\mathcal{A}}(w) = \limsup_{n \to \infty} \frac{\sum_{i=1}^n 1_q^i(w)}{n}$$

$$\mathbf{c}_q^{\mathcal{A}}(w) = \limsup_{n \to \infty} \frac{\sum_{i=1}^n T^i(w) \cdot 1_q^i(w)}{\sum_{i=1}^n T^i(w)}$$

Thus, every timed word w determines the tuple $\mathbf{d}^{\mathcal{A}}(w) = \left(\mathbf{d}_q^{\mathcal{A}}(w)\right)_{q \in Q}$ and the tuple $\mathbf{c}^{\mathcal{A}}(w) = \left(\mathbf{c}_q^{\mathcal{A}}(w)\right)_{q \in Q}$ of *discrete* and *timed* $\mathcal{A}$-*measures*, respectively.

DTA measures can encode various performance and dependability properties of stochastic systems with continuous time. For example, consider again the DTA $\hat{\mathcal{A}}$ above and assume that all states of a given stochastic process $\mathcal{M}$ are labeled with a. Then, the fraction

$$\frac{\mathbf{d}_{q\uparrow}(w_\sigma)}{\mathbf{d}_{q\uparrow}(w_\sigma) + \mathbf{d}_{q\downarrow}(w_\sigma)}$$

corresponds to the percentage of transitions of $\mathcal{M}$ that are performed within 2 seconds along a run σ. If $\mathcal{M}$ is an ergodic continuous-time Markov chain, then the above fraction takes the *same* value for almost all runs σ of $\mathcal{M}$. However, it makes sense to consider this fraction also for non-ergodic processes. For example, we may be interested in the expected value of $\mathbf{d}_{q\uparrow}/(\mathbf{d}_{q\uparrow} + \mathbf{d}_{q\downarrow})$, or in the probability of all runs σ such that the fraction is at least 0.5.

One general trouble with DTA measures is that $\mathbf{d}_q^{\mathcal{A}}(w)$ and $\mathbf{c}_q^{\mathcal{A}}(w)$ faithfully capture the frequency of visits to q along w only if the limits

$$\lim_{n \to \infty} \frac{\sum_{i=1}^n 1_q^i(w)}{n} \quad \text{and} \quad \lim_{n \to \infty} \frac{\sum_{i=1}^n T^i(w) \cdot 1_q^i(w)}{\sum_{i=1}^n T^i(w)}$$

exist, in which case we say that $\mathbf{d}^{\mathcal{A}}$ and $\mathbf{c}^{\mathcal{A}}$ are *well-defined* for w, respectively. So, one general question that should be answered when analyzing the properties of DTA measures over a particular class of stochastic processes is whether $\mathbf{d}^{\mathcal{A}}$ and $\mathbf{c}^{\mathcal{A}}$ are well-defined for almost all runs. If the answer is negative, we might either try to re-design our DTA or accept the fact that the limit frequency of the considered event simply does not exist (and stick to lim sup).

In this paper, we study DTA measures over semi-Markov processes (SMPs). An SMP is essentially a discrete-time Markov chain where each transition is assigned (apart of its discrete probability) a *delay density*, which defines the distribution of time needed to perform the transition. A computation (run) of an SMP $\mathcal{M}$ is initiated in some state s_0, which is also chosen randomly according to a fixed initial distribution over the state space of $\mathcal{M}$. The next transition is selected according to the fixed transition probabilities, and the selected transition takes time chosen randomly according to the density associated to the transition. Hence, each run of $\mathcal{M}$ is an infinite sequence $s_0 t_0 s_1 t_1 \cdots$, where all s_i are states of $\mathcal{M}$ and t_i are time stamps. The probability of (certain) subsets of runs in $\mathcal{M}$ is measured in the standard way (see Section 2).

The main contribution of this paper are general results about DTA measures over semi-Markov processes, which are valid for all SMPs where the employed density functions are bounded from zero on every closed subinterval (see Section 2). Under this assumption, we prove that for every SMP $\mathcal{M}$ and every DTA $\mathcal{A}$ we have the following:

(1) Both discrete and timed $\mathcal{A}$-measures are well defined for almost all runs of $\mathcal{M}$.

(2) Almost all runs of $\mathcal{M}$ can be divided into finitely many pairwise disjoint subsets $\mathcal{R}_1, \ldots, \mathcal{R}_k$ so that $\mathbf{d}^{\mathcal{A}}(w)$ takes the same value for almost all $w \in \mathcal{R}_j$, where $1 \leq j \leq k$. The same result holds also for $\mathbf{c}^{\mathcal{A}}$. (Let us note that k can be larger than 1 even if $\mathcal{M}$ is strongly connected.)

(3) The observations behind the results of (1) and (2) can be used to compute the k and effectively approximate the probability of all $\mathcal{R}_j$ together with the associated values of discrete or timed $\mathcal{A}$-measures up to an arbitrarily small given precision. More precisely, we show that these quantities are expressible using the m-step transition kernel P^m of the product process $\mathcal{M} \times \mathcal{A}$ defined for $\mathcal{M}$ and $\mathcal{A}$ (see Section 3.2), and we give generic bounds on the number of steps m that is sufficient to achieve the required precision. The m-step transition kernel is defined by nested integrals (see Section 3.1) and can be approximated by numerical methods (see, e.g., [16, 9]). This makes the whole framework effective. The design of more efficient algorithms as well as more detailed analysis applicable to concrete subclasses of SMP are left for future work.

To get some intuition about potential applicability of our results (and about the actual power of DTA which is hidden mainly in their ability to accumulate the total time of several transitions in internal clocks), let us start with a simple example. Consider the following itinerary for travelling between Brno and Prague:

	Brno	Kuřim	Tišnov	Čáslav	Prague
arrival		1:15	2:30	3:30	4:50
departure	0:00	1:20	2:40	3:35	

A traveller has to change a train at each of the three intermediate stops, and she needs at least 3 minutes to walk between the platforms. Assume that all trains depart on time, but can be delayed. Further, assume that travelling time between X and Y has density $f_{X\text{-}Y}$. We wonder what is the chance that a traveller reaches Prague from Brno without missing any train and at most 5 minutes after the scheduled arrival. Answering this question "by hand" is not simple (though still possible). However, it is almost trivial to rephrase this question in terms of DTA measures. The itinerary can be modeled by the following semi-Markov process, where the density f is irrelevant and $\Sigma = \{B, K, T, \check{C}, P\}$.

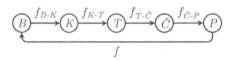

The property of "reaching Prague from Brno without missing any train and at most 5 minutes after the scheduled arrival" is encoded by the DTA $\bar{\mathcal{A}}$ of Figure 1. The automaton uses just one clock x to measure the total elapsed time, and the guards reflect the required timing constraints. Starting in location *init*, the automaton eventually reaches either the location $p{\uparrow}$ or $p{\downarrow}$, which corresponds to satisfaction or violation of the above property, and then it is "restarted". Hence, we are interested in the relative frequency of visits to $p{\uparrow}$ among the visits to $p{\uparrow}$ or $p{\downarrow}$. Using our results, it follows that $\mathbf{d}^{\mathcal{A}}$ is well-defined and takes the same value for almost all runs of $\mathcal{M}$. Hence, the random variable $\mathbf{d}_{p\uparrow}/(\mathbf{d}_{p\uparrow}+\mathbf{d}_{p\downarrow})$ also takes the same value with probability one, and this (unique) value is the quantity of our interest.

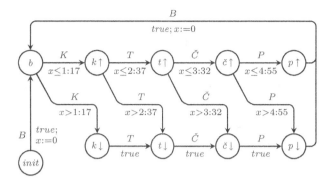

Figure 1: A deterministic timed automaton $\bar{\mathcal{A}}$.

Now imagine we wish to model and analyse the flow of passengers in London metro at rush hours. The SMP states then correspond to stations, transition probabilities encode the percentage of passengers traveling in a given direction, and the densities encode the distribution of travelling time. A DTA can be used to monitor a complex list of timing restrictions such as "there is enough time to change a train", "travelling between important stations does not take more than 30 minutes if one the given routes is used", "trains do not arrive more than 2 minutes later than scheduled", etc. For this we already need several internal clocks. Apart of some auxiliary locations, the constructed DTA would also have special locations used to encode satisfaction/violation of a given restriction (in the DTA $\bar{\mathcal{A}}$ of Figure 1, $(p, \uparrow)$ and $(p, \downarrow)$ are such special locations). Using the results presented in this paper, one may not only study the overall satisfaction of these restrictions, but also estimate the impact of changes in the underlying model (for example, if a given line becomes slower due to some repairs, one may evaluate the decrease in various dependability measures without changing the constructed DTA).

Proof techniques. For a given SMP $\mathcal{M}$ and a given DTA $\mathcal{A}$ we first construct their synchronized product $\mathcal{M} \times \mathcal{A}$, which is another stochastic process. In fact, it turns out that $\mathcal{M} \times \mathcal{A}$ is a discrete-time Markov chain with uncountable state-space. Then, we apply a variant of the standard region construction [2] and thus partition the state-space of $\mathcal{M} \times \mathcal{A}$ into finitely many equivalence classes. At the very core of our paper there are several non-trivial observations about the structure of $\mathcal{M} \times \mathcal{A}$ and its region graph which establish a powerful link to the well-developed ergodic theory of Markov chains with general state-space (see, e.g., [18, 20]). In this way, we obtain the results of items (1) and (2) mentioned above. Some additional work is required to analyze the algorithm presented in Section 4 (whose properties are summarized in item (3) above). Due to space constraints, most of the proofs are omitted and can be found in [10].

Related work. There is a vast literature on continuous-time Markov chains, semi-Markov processes, or even more general stochastic models such as generalized semi-Markov processes (we refer to, e.g., [21, 6, 19, 17]). In the computer science context, most works on continuous-time stochastic models concern model-checking against a given class of temporal properties [3, 5]. The usefulness of CSL model-checking for dependability analysis is advocated in [15]. Timed automata [2] have been originally used as a model of (non-stochastic) real-time systems. Probabilistic semantics of timed automata is proposed in [4, 7]. The idea of using timed automata as a specification language for continuous-time stochastic processes is relatively recent. In [13], the model-checking problem for continuous-time Markov chains and linear-time properties represented by timed automata is considered (the task is to dermine the probability of all timed words that are accepted by a given timed automaton). A more general model of two-player games over generalized semi-Markov processes with qualitative reachability objectives specified by deterministic timed automata is studied in [11].

2. PRELIMINARIES

In this paper, the sets of all positive integers, non-negative integers, real numbers, positive real numbers, and non-negative real numbers are denoted by $\mathbb{N}$, $\mathbb{N}_0$, $\mathbb{R}$, $\mathbb{R}_{>0}$, and $\mathbb{R}_{\geq 0}$, respectively.

Let A be a finite or countably infinite set. A *discrete probability distribution* on A is a function $\alpha : A \to \mathbb{R}_{\geq 0}$ such that $\sum_{a \in A} \alpha(a) = 1$. We say that α is *rational* if $\alpha(a)$ is rational for every $a \in A$. The set of all distributions on A is denoted by $\mathcal{D}(A)$. A *σ-field* over a set Ω is a set $\mathcal{F} \subseteq 2^{\Omega}$ that includes Ω and is closed under complement and countable union. A *measurable space* is a pair $(\Omega, \mathcal{F})$ where Ω is a set called *sample space* and $\mathcal{F}$ is a σ-field over Ω whose elements are called *measurable sets*. A *probability measure* over a measurable space $(\Omega, \mathcal{F})$ is a function $\mathcal{P} : \mathcal{F} \to \mathbb{R}_{\geq 0}$ such that, for each countable collection $\{X_i\}_{i \in I}$ of pairwise disjoint elements of $\mathcal{F}$, $\mathcal{P}(\bigcup_{i \in I} X_i) = \sum_{i \in I} \mathcal{P}(X_i)$, and moreover $\mathcal{P}(\Omega) = 1$. A *probability space* is a triple $(\Omega, \mathcal{F}, \mathcal{P})$, where $(\Omega, \mathcal{F})$ is a measurable space and $\mathcal{P}$ is a probability measure over $(\Omega, \mathcal{F})$. We say that a property $A \subseteq \Omega$ holds *for almost all* elements of a measurable set Y if $\mathcal{P}(Y) > 0$, $A \cap Y \in \mathcal{F}$, and $\mathcal{P}(A \mid Y) = 1$.

All of the integrals used in this paper should be understood as Lebesgue integrals, although we use Riemann-like notation when appropriate.

2.1 Semi-Markov processes

A semi-Markov process (see, e.g., [21]) can be seen as discrete-time Markov chains where each transition is equipped with a density function specifying the distribution of time needed to perform the transition. Formally, let $\mathfrak{D}$ be a set of delay densities, i.e., measurable functions $f : \mathbb{R} \to \mathbb{R}_{\geq 0}$ satisfying $\int_0^\infty f(t)\,dt = 1$ where $f(t) = 0$ for every $t < 0$. Moreover, for technical reasons, we assume that each $f \in \mathfrak{D}$ satisfies the following: There is an interval I either of the form $[\ell, u]$ with $\ell, u \in \mathbb{N}_0, \ell < u$, or $[\ell, \infty)$ with $\ell \in \mathbb{N}_0$, such that

- for all $t \in \mathbb{R} \setminus I$ we have that $f(t) = 0$,

- for all $[c, d] \subseteq I$ there is $b > 0$ such that for all $t \in [c, d]$ we have that $f(t) \geq b$.

The assumption that ℓ, u are natural numbers is adopted only for the sake of simplicity. Our results can easily be generalized to the setting where I is an interval with rational bounds or even a finite union of such intervals.

DEFINITION 2.1. *A* semi-Markov process (SMP) *is a tuple* $\mathcal{M} = (S, \mathbf{P}, \mathbf{D}, \alpha_0)$, *where S is a finite set of states, $\mathbf{P} : S \to \mathcal{D}(S)$ is a transition probability function, $\mathbf{D} : S \times S \to \mathfrak{D}$ is a delay function which to each transition assigns its delay density, and $\alpha_0 \in \mathcal{D}(S)$ is an initial distribution.*

A computation (run) of a SMP $\mathcal{M}$ is initiated in some state s_0, which is chosen randomly according to α_0. In the current state s_i, the next state s_{i+1} is selected randomly according to the distribution $\mathbf{P}(s_i)$, and the selected transition (s_i, s_{i+1}) takes a random time t_i chosen according to the density $\mathbf{D}(s_i, s_{i+1})$. Hence, each run of $\mathcal{M}$ is an infinite timed word $s_0 \, t_0 \, s_1 \, t_1 \cdots$, where $s_i \in S$ and $t_i \in \mathbb{R}_{\geq 0}$ for all $i \in \mathbb{N}_0$. We use $\mathcal{R}_{\mathcal{M}}$ to denote the set of all runs of $\mathcal{M}$.

Now we define a probability space $(\mathcal{R}_{\mathcal{M}}, \mathcal{F}_{\mathcal{M}}, \mathcal{P}_{\mathcal{M}})$ over the runs of $\mathcal{M}$ (we often omit the index $\mathcal{M}$ if it is clear from the context). A *template* is a finite sequence of the form $B = s_0 \, I_0 \, s_1 \, I_1 \cdots s_{n+1}$ such that $n \geq 0$ and I_i is an interval in $\mathbb{R}_{\geq 0}$ for every $0 \leq i \leq n$. Each such B determines the corresponding *cylinder* $\mathcal{R}(B) \subseteq \mathcal{R}$ consisting of all runs of the form $\hat{s}_0 \, t_0 \, \hat{s}_1 \, t_1 \cdots$, where $\hat{s}_i = s_i$ for all $0 \leq i \leq n+1$, and $t_i \in I_i$ for all $0 \leq i \leq n$. The σ-field $\mathcal{F}$ is the Borel σ-field generated by all cylinders. For each template $B = s_0 \, I_0 \, s_1 \, I_1 \cdots s_{n+1}$, let $p_i = \mathbf{P}(s_i)(s_{i+1})$ and $f_i = \mathbf{D}(s_i, s_{i+1})$ for all $0 \leq i \leq n$. The probability $\mathcal{P}(\mathcal{R}(B))$ is defined as follows:

$$\alpha_0(s_0) \cdot \prod_{i=0}^{n} p_i \cdot \int_{t_i \in I_i} f_i(t_i)\,dt_i$$

Then, $\mathcal{P}$ is extended to $\mathcal{F}$ (in the unique way) by applying the extension theorem (see, e.g., [8]).

2.2 Deterministic timed automata

Let $\mathcal{X}$ be a finite set of *clocks*. A *valuation* is a function $\nu : \mathcal{X} \to \mathbb{R}_{\geq 0}$. For every valuation ν and every subset $X \subseteq \mathcal{X}$ of clocks, we use $\nu[X := \mathbf{0}]$ to denote the unique valuation such that $\nu[X := \mathbf{0}](x)$ is equal either to 0 or $\nu(x)$, depending on whether $x \in X$ or not, respectively. Further, for every valuation ν and every $\delta \in \mathbb{R}_{\geq 0}$, the symbol $\nu + \delta$ denotes the unique valuation such that $(\nu + \delta)(x) = \nu(x) + \delta$ for all $x \in \mathcal{X}$. Sometimes we assume an implicit linear ordering on clocks and slightly abuse our notation by identifying a valuation ν with the associated vector of reals.

A *clock constraint* (or *guard*) is a finite conjunction of basic constraints of the form $x \bowtie c$, where $x \in \mathcal{X}$, $\bowtie \in \{<, \leq, >, \geq\}$, and $c \in \mathbb{N}_0$. For every valuation ν and every clock constraint g we have that ν either does or does not satisfy g, written $\nu \models g$ or $\nu \not\models g$, respectively (the satisfaction relation is defined in the expected way). Sometimes we identify a guard g with the set of all valuations that satisfy g and write, e.g., $g \cap g'$. The set of all guards over $\mathcal{X}$ is denoted by $\mathcal{B}(\mathcal{X})$.

DEFINITION 2.2. *A* deterministic timed automaton (DTA) *is a tuple* $\mathcal{A} = (Q, \Sigma, \mathcal{X}, \longrightarrow, q_0)$, *where Q is a nonempty finite set of locations, Σ is a finite alphabet, $\mathcal{X}$ is a finite set of clocks, $q_0 \in Q$ is an initial location, and* $\longrightarrow \subseteq Q \times \Sigma \times \mathcal{B}(\mathcal{X}) \times 2^{\mathcal{X}} \times Q$ *is an edge relation such that for all $q \in Q$ and $a \in \Sigma$ we have the following:*

1. *the guards are deterministic, i.e., for all edges of the form (q, a, g_1, X_1, q_1) and (q, a, g_2, X_2, q_2) such that $g_1 \cap g_2 \neq \emptyset$ we have that $g_1 = g_2$, $X_1 = X_2$, and $q_1 = q_2$;*

2. *the guards are total, i.e., for all $q \in Q$, $a \in \Sigma$, and every valuation ν there is an edge (q, a, g, X, q') such that $\nu \models g$.*

A *configuration* of $\mathcal{A}$ is a pair (q, ν), where $q \in Q$ and ν is a valuation. An *infinite timed word* over Σ is an infinite sequence $w = c_0 \, c_1 \, c_2 \, c_3 \cdots$, where $c_i \in \Sigma$ when i is even, and

$c_i \in \mathbb{R}_{\geq 0}$ when i is odd. The *run* of $\mathcal{A}$ on w is the unique infinite sequence of configurations $\mathcal{A}(w) = (q_0, \nu_0)(q_1, \nu_1) \cdots$ such that q_0 is the initial location of $\mathcal{A}$, $\nu_0(x) = 0$ for all $x \in \mathcal{X}$, and for each $i \in \mathbb{N}_0$ we have that

- if c_i is a time stamp, then $q_{i+1} = q_i$ and $\nu_{i+1} = \nu_i + c_i$;

- if c_i is a letter of Σ, then there is a unique edge (q_i, c_i, g, X, q) such that $\nu_i \models g$, and we require that $q_{i+1} = q$ and $\nu_{i+1} = \nu_i[X := \mathbf{0}]$.

Notice that we do not define any acceptance condition for DTA. Instead, we understand DTA as finite-state *observers* that analyze timed words and report about certain events by entering designated locations. The "frequency" of these events is formally captured by the quantities d_q and c_q defined below.

Let $\mathcal{A} = (Q, \Sigma, \mathcal{X}, \longrightarrow, q_0)$ be a DTA, $q \in Q$ some location, and $w = a_0 t_0 a_1 t_1 \cdots$ a timed word over Σ. For every $i \in \mathbb{N}_0$, let $T^i(w)$ be the stamp t_i of w, and $Q^i(w)$ the unique location of $\mathcal{A}$ entered after reading the finite prefix $a_0 t_0 \cdots a_i$ of w. Further, let $1_q^i(w)$ be either 1 or 0 depending on whether $Q^i(w) = q$ or not, respectively. The *discrete* and *timed* frequency of visits to q along $\mathcal{A}(w)$, denoted by $\mathbf{d}_q^{\mathcal{A}}(w)$ and $\mathbf{c}_q^{\mathcal{A}}(w)$, are defined in the following way (if $\mathcal{A}$ is clear, it is omitted):

$$\mathbf{d}_q^{\mathcal{A}}(w) = \limsup_{n \to \infty} \frac{\sum_{i=1}^n 1_q^i(w)}{n}$$

$$\mathbf{c}_q^{\mathcal{A}}(w) = \limsup_{n \to \infty} \frac{\sum_{i=1}^n T^i(w) \cdot 1_q^i(w)}{\sum_{i=1}^n T^i(w)}$$

Hence, every timed word w determines the tuple $\mathbf{d}^{\mathcal{A}} = \left(\mathbf{d}_q^{\mathcal{A}}(w)\right)_{q \in Q}$ and the tuple $\mathbf{c}^{\mathcal{A}} = \left(\mathbf{c}_q^{\mathcal{A}}(w)\right)_{q \in Q}$ of *discrete* and *timed* $\mathcal{A}$-measures, respectively. The $\mathcal{A}$-measures were defined using $\limsup$, because the corresponding limits may not exist in general. If $\lim_{n \to \infty} \sum_{i=1}^n 1_q^i(w)/n$ exists for all $q \in Q$, we say that $\mathbf{d}^{\mathcal{A}}$ is *well-defined* for w. Similarly, if $\lim_{n \to \infty} (\sum_{i=1}^n T^i(w) \cdot 1_q^i(w))/(\sum_{i=1}^n T^i(w))$ exists for all q, we say that $\mathbf{c}^{\mathcal{A}}$ is *well-defined* for w.

As we already noted in Section 1, a DTA $\mathcal{A}$ can be used to observe runs in a given SMP $\mathcal{M}$ after labeling all states of $\mathcal{M}$ with the letters of Σ by a suitable $L : S \to \Sigma$. Then, every run $\sigma = s_0 t_0 s_1 t_1 \cdots$ of $\mathcal{M}$ determines a unique timed word $w_\sigma = L(s_0) t_0 L(s_1) t_1 \cdots$, and one can easily show that for every timed word w, the set $\{\sigma \in \mathcal{R} \mid w_\sigma = w\}$ is measurable in $(\mathcal{R}, \mathcal{F}, \mathcal{P})$.

3. DTA MEASURES OVER SMPS

Throughout this section we fix an SMP $\mathcal{M} = (S, \mathbf{P}, \mathbf{D}, \alpha_0)$ and a DTA $\mathcal{A} = (Q, \Sigma, \mathcal{X}, \longrightarrow, q_0)$ where $\mathcal{X} = \{x_1, \ldots, x_n\}$. To simplify our notation, we assume that $\Sigma = S$, i.e., every run σ of $\mathcal{M}$ is a timed word over Σ (hence, we do not need to introduce any labeling $L : S \to \Sigma$). This technical assumption does not affect the generality of our results (all of our arguments and proofs work exactly as they are, we only need to rewrite them using less readable notation). Our goal is to prove the following:

THEOREM 3.1.

1. $\mathbf{d}^{\mathcal{A}}$ *is well-defined for almost all runs of $\mathcal{M}$.*

2. *There are pairwise disjoint sets $\mathcal{R}_1, \ldots, \mathcal{R}_k$ of runs in $\mathcal{M}$ such that $\mathcal{P}(\mathcal{R}_1 \cup \cdots \cup \mathcal{R}_k) = 1$, and for every $1 \leq j \leq k$ there is a tuple D_j such that $\mathbf{d}^{\mathcal{A}}(\sigma) =$*

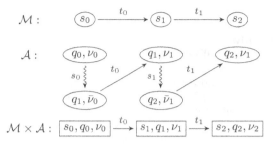

Figure 2: Synchronizing $\mathcal{M}$ and $\mathcal{A}$ in $\mathcal{M} \times \mathcal{A}$. Notice that $\nu_0 = \bar{\nu}_0 = \mathbf{0}$ and $\nu_{i+1} = \bar{\nu}_i + t_i$.

D_j *for almost all $\sigma \in \mathcal{R}_j$ (we use $D_{j,q}$ to denote the q-component of D_j).*

In Section 4, we show how to compute the k and approximate $\mathcal{P}(\mathcal{R}_j)$ and D_j up to an arbitrarily small given precision.

An immediate corollary of Theorem 3.1 is an analogous result for $\mathbf{c}^{\mathcal{A}}$.

COROLLARY 3.2. $\mathbf{c}^{\mathcal{A}}$ *is well-defined for almost all runs of $\mathcal{M}$. Further, there are pairwise disjoint sets $\mathcal{R}_1, \ldots, \mathcal{R}_K$ of runs in $\mathcal{M}$ such that $\mathcal{P}(\mathcal{R}_1 \cup \cdots \cup \mathcal{R}_K) = 1$, and for every $1 \leq j \leq K$ there is a tuple C_j such that $\mathbf{c}^{\mathcal{A}}(\sigma) = C_j$ for almost all $\sigma \in \mathcal{R}_j$.*

Corollary 3.2 follows from Theorem 3.1 simply by considering the discrete $\mathbf{d}^{S \times \mathcal{A}}$ measure, where the DTA $S \times \mathcal{A}$ is obtained from $\mathcal{A}$ in the following way: the set of locations of $S \times \mathcal{A}$ is $\{q_0\} \cup (S \times Q)$, and for every transition (q_0, s, g, X, q') of $\mathcal{A}$ we add a transition $(q_0, s, g, X, (s, q'))$ to $S \times \mathcal{A}$ and for every transition (q, s, g, X, q') and every $s' \in S$ we add a transition $((s', q), s, g, X, (s, q'))$ to $S \times \mathcal{A}$. The initial location of $S \times \mathcal{A}$ is q_0. Intuitively, $S \times \mathcal{A}$ is the same as $\mathcal{A}$ but it explicitly "remembers" the letter which was used to enter the current location. Let k and D_j be the constants of Theorem 3.1 constructed for $\mathcal{M}$ and $S \times \mathcal{A}$. Observe that the expected time of performing a transition from a given $s \in S$, denoted by E_s, is given by $E_s = \sum_{s' \in S} \mathbf{P}(s)(s') \cdot E_{s,s'}$, where $E_{s,s'}$ is the expectation of a random variable with the density $\mathbf{D}(s, s')$. From this we easily obtain that

$$C_{j,q} = \frac{\sum_{s \in S} E_s \cdot D_{j,(s,q)}}{\sum_{p \in Q} \sum_{s \in S} E_s \cdot D_{j,(s,p)}} \qquad (1)$$

for all $q \in Q$ and $1 \leq j \leq k$. The details are given in [10]. Hence, we can also compute the constant K and approximate $\mathcal{P}(\mathcal{R}_j)$ and C_j for every $1 \leq j \leq K$ using Equation (1).

It remains to prove Theorem 3.1. Let us start by sketching the overall structure of our proof. First, we construct a synchronous product $\mathcal{M} \times \mathcal{A}$ of $\mathcal{M}$ and $\mathcal{A}$, which is a Markov chain with an uncountable state space $\Gamma_{\mathcal{M} \times \mathcal{A}} = S \times Q \times (\mathbb{R}_{\geq 0})^n$. Intuitively, $\mathcal{M} \times \mathcal{A}$ behaves in the same way as $\mathcal{M}$ and simulates the computation of $\mathcal{A}$ on-the-fly (see Figure 2). Then, we construct a finite region graph $G_{\mathcal{M} \times \mathcal{A}}$ over the product $\mathcal{M} \times \mathcal{A}$. The nodes of $G_{\mathcal{M} \times \mathcal{A}}$ are the sets of states that, roughly speaking, satisfy the same guards of $\mathcal{A}$. Edges are induced by transitions of the product (note that if two states satisfy the same guards, the sets of enabled outgoing transitions are the same). By relying on arguments presented in [1, 11], we show that almost all runs reach a node of a bottom strongly connected component (BSCC) $\mathcal{C}$ of $G_{\mathcal{M} \times \mathcal{A}}$ (by definition, each run which enters $\mathcal{C}$ remains in $\mathcal{C}$). This gives us the partition of the set of runs of $\mathcal{M}$ into the sets $\mathcal{R}_1, \dots, \mathcal{R}_k$ (each $\mathcal{R}_j$ corresponds to one of the BSCCs of $G_{\mathcal{M} \times \mathcal{A}}$).

Subsequently, we concentrate on a fixed BSCC $\mathcal{C}$, and prove that almost all runs that reach $\mathcal{C}$ have the same frequency of visits to a given $q \in Q$ (this gives us the constant $D_{j,q}$). Here we employ several deep results from the theory of general state space Markov chains (see Theorem 3.6). To apply these results, we prove that assuming aperiodicity of $G_{\mathcal{M} \times \mathcal{A}}$ (see Definition 3.10), the state space of the product $\mathcal{M} \times \mathcal{A}$ is *small* (see Definition 3.5 and Lemma 3.11 below). This is perhaps the most demanding part of our proof. Roughly speaking, we show that there is a distinguished subset of states reachable from each state in a fixed number of steps with probability bounded from 0. By applying Theorem 3.6, we obtain a complete invariant distribution on the product, i.e., in principle, we obtain a constant frequency of any non-trivial subset of states. From this we derive our results in a straightforward way. If $G_{\mathcal{M} \times \mathcal{A}}$ is periodic, we use standard techniques for removing periodicity and then basically follow the same stream of arguments as in the aperiodic case.

3.1 General state space Markov chains

We start by recalling the definition of "ordinary" discrete-time Markov chains with discrete state space (DTMC). A DTMC is given by a finite or countably infinite state space S, an initial probability distribution over S, and a one-step transition matrix P which defines the probability $P(s, s')$ of every transion $(s, s') \in S \times S$ so that $\sum_{s' \in S} P(s, s') = 1$ for every $s \in S$. In the setting of uncountable state spaces, transition probabilities cannot be specified by a transition matrix. Instead, one defines the probabilities of moving from a given state s to a given measurable subset X of states. Hence, the concept of transition matrix is replaced with a more general notion of *transition kernel* defined below.

DEFINITION 3.3. *A* transition kernel *over a measurable space* $(\Gamma, \mathcal{G})$ *is a function* $P : \Gamma \times \mathcal{G} \to [0,1]$ *such that*

1. $P(z, \cdot)$ *is a probability measure over* $(\Gamma, \mathcal{G})$ *for each* $z \in \Gamma$;

2. $P(\cdot, A)$ *is a measurable function for each* $A \in \mathcal{G}$ *(i.e., for every* $c \in \mathbb{R}$, *the set of all* $z \in \Gamma$ *satisfying* $P(z, A) \geq c$ *belongs to* $\mathcal{G}$*).*

A transition kernel is the core of the following definition.

DEFINITION 3.4. *A* general state space Markov chain (GSSMC) *with a state space* $(\Gamma, \mathcal{G})$, *a transition kernel* P *and an initial probability measure* μ *is a stochastic process* $\Phi = \Phi_1, \Phi_2, \dots$ *such that each* Φ_i *is a random variable over a probability space* $(\Omega_\Phi, \mathcal{F}_\Phi, \mathcal{P}_\Phi)$ *where*

- Ω_Φ *is a set of* runs, *i.e., infinite words over* Γ.

- $\mathcal{F}_\Phi$ *is the product* σ-field $\bigotimes_{i=0}^{\infty} \mathcal{G}$.

- $\mathcal{P}_\Phi$ *is the unique probability measure over* $(\Omega_\Phi, \mathcal{F}_\Phi)$ *such that for every finite sequence* $A_0, \cdots, A_n \in \mathcal{F}_\Phi$ *we have that* $\mathcal{P}_\Phi(\Phi_0 \in A_0, \cdots, \Phi_n \in A_n)$ *is equal to*

$$\int_{y_0 \in A_0} \cdots \int_{y_{n-1} \in A_{n-1}} \mu(dy_0) \cdot P(y_0, dy_1) \cdots P(y_{n-1}, A_n).$$
(2)

- *Each* Φ_i *is the projection of elements of* Ω_Φ *onto the* i-th *component.*

A *path* is a finite sequence $z_1 \cdots z_n$ of states from Γ. From Equation (2) we get that Φ also satisfies the following properties which will be used to show several results about the chain Φ by working with the transition kernel only.

1. $\mathcal{P}_\Phi(\Phi_0 \in A_0) = \mu(A_0)$,

2. $\mathcal{P}_\Phi(\Phi_{n+1} \in A \mid \Phi_n, \dots, \Phi_0) = \mathcal{P}_\Phi(\Phi_{n+1} \in A \mid \Phi_n) = P(\Phi_n, A)$ almost surely,

3. $\mathcal{P}_\Phi(\Phi_{n+m} \in A \mid \Phi_n) = P^m(\Phi_n, A)$ almost surely,

where the m-step transition kernel P^m is defined as follows:

$$P^1(z, A) = P(z, A)$$
$$P^{i+1}(z, A) = \int_\Gamma P(z, dy) \cdot P^i(y, A).$$

Notice that the transition kernel and the m-step transition kernel are analogous counterparts to the transition matrix and the k-step transition matrix of a DTMC.

As we mentioned above, our proof of Theorem 3.1 employs several results of GSSMC theory. In particular, we make use of the notion of *smallness* of the state space defined as follows.

DEFINITION 3.5. *Let* $m \in \mathbb{N}$, $\varepsilon > 0$, *and* ν *be a probability measure on* $\mathcal{G}$. *A set* $C \in \mathcal{G}$ *is* (m, ε, ν)-small *if for all* $x \in C$ *and* $B \in \mathcal{G}$ *we have that* $P^m(x, B) \geq \varepsilon \cdot \nu(B)$.

GSSMCs where the whole state space is small have many nice properties, and the relevant ones are summarized in the following theorem.

THEOREM 3.6. *If* Γ *is* (m, ε, ν)-small, then

1. [**Existence of invariant measure**] *There exists a unique probability measure* π *such that for all* $A \in \mathcal{G}$ *we have that*

$$\pi(A) \quad = \quad \int_\Gamma \pi(dx) P(x, A)$$

2. [**Strong law of large numbers**] *If* $h : \Gamma \to \mathbb{R}$ *satisfies* $\int_\Gamma h(x) \pi(dx) < \infty$, *then almost surely*

$$\lim_{n \to \infty} \frac{\sum_{i=1}^n h(\Phi_i)}{n} \quad = \quad \int_\Gamma h(x) \pi(dx)$$

3. [**Uniform ergodicity**] *For all* $x \in \Gamma$, $A \in \mathcal{G}$, *and all* $n \in \mathbb{N}$,

$$\sup_{A \in \mathcal{G}} |P^n(x, A) - \pi(A)| \quad \leq \quad (1 - \varepsilon)^{\lfloor n/m \rfloor}$$

PROOF. The theorem is a consequence of stadard results for GSSMCs. Since Γ is (m, ε, ν)-small, we have

(i) Φ is by definition φ-irreducible for $\varphi = \nu$, and thus also ψ-irreducible by [18, Proposition 4.2.2];

(ii) Γ is by definition also (a, ε, ν)-petite (see [18, Section 5.5.2]), where a is the Dirac distribution on $\mathbb{N}_0$ with $a(m) = 1$, $a(n) = 0$ for $n \neq m$;

(iii) the first return time to Γ is trivially 1.

ad 1. By (iii), Γ is not uniformly transient, hence by (i), (ii) and [18, Theorem 8.0.2], Φ is recurrent. Thus by [18, Theorem 10.0.1], there exists a unique invariant probability measure π.

ad 2. By (i)-(iii) and [18, Theorem 10.4.10 (ii)], Φ is positive Harris. Therefore, we may apply [18, Theorem 17.0.1 (i)] and obtain the desired result.

ad 3. This follows immediately from [20, Theorem 8]. $\square$

3.2 The product process

The *product process* of $\mathcal{M}$ and $\mathcal{A}$, denoted by $\mathcal{M} \times \mathcal{A}$, is a GSSMC with the state space $\Gamma_{\mathcal{M} \times \mathcal{A}} = S \times Q \times (\mathbb{R}_{\geq 0})^n$, where $n = |\mathcal{X}|$ is the number of clocks of $\mathcal{A}$. The σ-field over $\Gamma_{\mathcal{M} \times \mathcal{A}}$ is the product σ-field $\mathcal{G}_{\mathcal{M} \times \mathcal{A}} = 2^S \otimes 2^Q \otimes \mathfrak{B}^n$ where $\mathfrak{B}^n$ is the Borel σ-field over the set $(\mathbb{R}_{\geq 0})^n$. For each $A \in \mathcal{G}_{\mathcal{M} \times \mathcal{A}}$, the initial probability $\mu_{\mathcal{M} \times \mathcal{A}}(A)$ is equal to $\sum_{(s,q_0,\mathbf{0}) \in A} \alpha_0(s)$ (recall that α_0 is the initial distribution of $\mathcal{M}$).

The behavior of $\mathcal{M} \times \mathcal{A}$ is depicted in Figure 2. Each step of the product process corresponds to one step of $\mathcal{M}$ and two steps of $\mathcal{A}$. The step of the product starts by simulating the discrete step of $\mathcal{A}$ that reads the current state of $\mathcal{M}$ and possibly resets some clocks, followed by simulating simultaneously the step of $\mathcal{M}$ that takes time t and the corresponding step of $\mathcal{A}$ which reads the time stamp t.

Now we define the transition kernel $P_{\mathcal{M} \times \mathcal{A}}$ of the product process. Let $z = (s, q, \nu)$ be a state of $\Gamma_{\mathcal{M} \times \mathcal{A}}$, and let $(\bar{q}, \bar{\nu})$ be the configuration of $\mathcal{A}$ entered from the configuration (q, ν) after reading s (note that $\bar{\nu}$ is not necessarily the same as ν because $\mathcal{A}$ may reset some clocks). It suffices to define $P_{\mathcal{M} \times \mathcal{A}}(z, \cdot)$ only for generators of $\mathcal{G}_{\mathcal{M} \times \mathcal{A}}$ and then apply the extension theorem (see, e.g., [8]) to obtain a unique probability measure $P_{\mathcal{M} \times \mathcal{A}}(z, \cdot)$ over $(\Gamma_{\mathcal{M} \times \mathcal{A}}, \mathcal{G}_{\mathcal{M} \times \mathcal{A}})$. Generators of $\mathcal{G}_{\mathcal{M} \times \mathcal{A}}$ are sets of the form $\{s'\} \times \{q'\} \times \mathbf{I}$ where $s' \in S$, $q' \in Q$ and $\mathbf{I}$ is the product $I_1 \times \cdots \times I_n$ of intervals I_i in $\mathbb{R}_{\geq 0}$. If $q' \neq \bar{q}$, then we define $P_{\mathcal{M} \times \mathcal{A}}(z, \{s'\} \times \{q'\} \times \mathbf{I}) = 0$. Otherwise, we define

$$P_{\mathcal{M} \times \mathcal{A}}(z, \{s'\} \times \{q'\} \times \mathbf{I}) = \mathbf{P}(s)(s') \cdot \int_0^\infty f(t) \cdot 1_{\mathbf{I}}(\bar{\nu} + t) dt$$

Here $f = \mathbf{D}(s, s')$ and $1_{\mathbf{I}}$ is the indicator function of the set $\mathbf{I}$.

Since $P_{\mathcal{M} \times \mathcal{A}}(z, \cdot)$ is by definition a probability measure over $(\Gamma_{\mathcal{M} \times \mathcal{A}}, \mathcal{G}_{\mathcal{M} \times \mathcal{A}})$, it remains to check the second condition of Definition 3.3.

LEMMA 3.7. *Let $A \in \mathcal{G}_{\mathcal{M} \times \mathcal{A}}$. Then $P_{\mathcal{M} \times \mathcal{A}}(\cdot, A)$ is a measurable function, i.e., $\mathcal{M} \times \mathcal{A}$ is a GSSMC.*

A proof of this lemma can be found in [10]. Recall that by Definition 3.4, $\mathcal{P}_{\mathcal{M} \times \mathcal{A}}$ is the unique probability measure on the product σ-field $\mathcal{F}_{\mathcal{M} \times \mathcal{A}} = \bigotimes_{i=0}^\infty \mathcal{G}_{\mathcal{M} \times \mathcal{A}}$ induced by $P_{\mathcal{M} \times \mathcal{A}}$ and the initial probability measure $\mu_{\mathcal{M} \times \mathcal{A}}$.

3.2.1 The correspondence between $\mathcal{M} \times \mathcal{A}$ and $\mathcal{M}$

In this subsection we show that $\mathcal{M} \times \mathcal{A}$ correctly reflects the behaviour of $\mathcal{M}$. First, we define the $\mathbf{d}^{\mathcal{A}}$ measure for $\mathcal{M} \times \mathcal{A}$. (As the DTA $\mathcal{A}$ is fixed, we omit them and write $\mathbf{d}$ and $\mathbf{d}_q$ instead of $\mathbf{d}^{\mathcal{A}}$ and $\mathbf{d}_q^{\mathcal{A}}$, respectively.) Let $\sigma = (s_0, q_0, \nu_0)(s_1, q_1, \nu_1) \cdots$ be a run of $\mathcal{M} \times \mathcal{A}$ and $q \in Q$ a location. For every $i \in \mathbb{N}_0$, let $1_q^i(\sigma)$ be either 1 or 0 depending on whether if $q_i = q$ or not, respectively. We put

$$\mathbf{d}_q(\sigma) = \limsup_{n \to \infty} \frac{\sum_{i=1}^n 1_q^i(\sigma)}{n}$$

LEMMA 3.8. *There is a measurable one-to-one mapping ξ from the set of runs of $\mathcal{M}$ to the set of runs of $\mathcal{M} \times \mathcal{A}$ such that*

- *ξ preserves measure, i.e., for every measurable set X of runs of $\mathcal{M}$ we have that $\xi(X)$ is also measurable and $\mathcal{P}_{\mathcal{M}}(X) = \mathcal{P}_{\mathcal{M} \times \mathcal{A}}(\xi(X))$;*

- *ξ preserves $\mathbf{d}$, i.e., for every run σ of $\mathcal{M}$ and every $q \in Q$ we have that $\mathbf{d}_q(\sigma)$ is well-defined iff $\mathbf{d}_q(\xi(\sigma))$ is well-defined, and $\mathbf{d}_q(\sigma) = \mathbf{d}_q(\xi(\sigma))$.*

A formal proof of Lemma 3.8 is given in [10].

3.2.2 The region graph of $\mathcal{M} \times \mathcal{A}$

Although the state-space $\Gamma_{\mathcal{M} \times \mathcal{A}}$ is uncountable, we can define the standard *region relation* $\sim$ [2] over $\Gamma_{\mathcal{M} \times \mathcal{A}}$ with finite index, and then work with finitely many *regions*. For a given $a \in \mathbb{R}$, we use $frac(a)$ to denote the fractional part of a, and $int(a)$ to denote the integral part of a. For $a, b \in \mathbb{R}$, we say that a and b *agree on integral part* if $int(a) = int(b)$ and neither or both a, b are integers.

We denote by $B_{\max}$ the maximal constant that appears in the guards of $\mathcal{A}$ and say that a clock $x \in \mathcal{X}$ is *relevant* for ν if $\nu(x) \leq B_{\max}$. Finally, we put $(s_1, q_1, \nu_1) \sim (s_2, q_2, \nu_2)$ if

- $s_1 = s_2$ and $q_1 = q_2$;

- for all relevant $x \in \mathcal{X}$ we have that $\nu_1(x)$ and $\nu_2(x)$ agree on integral parts;

- for all relevant $x, y \in \mathcal{X}$ we have that $frac(\nu_1(x)) \leq frac(\nu_1(y))$ iff $frac(\nu_2(x)) \leq frac(\nu_2(y))$.

Note that $\sim$ is an equivalence with finite index. The equivalence classes of $\sim$ are called *regions*. Observe that states in the same region have the same behavior with respect to qualitative reachability. This is formalized in the following lemma.

LEMMA 3.9. *Let R and T be regions and $z, z' \in R$. Then $P_{\mathcal{M} \times \mathcal{A}}(z, T) > 0$ iff $P_{\mathcal{M} \times \mathcal{A}}(z', T) > 0$.*

A proof of Lemma 3.9 can be found in [11]. Further, we define a finite *region graph* $G_{\mathcal{M} \times \mathcal{A}} = (V, E)$ where the set of vertices V is the set of regions and for every pair of regions R, R' there is an edge $(R, R') \in E$ iff $P_{\mathcal{M} \times \mathcal{A}}(z, R') > 0$ for some $z \in R$ (due to Lemma 3.9, the concrete choice of z is irrelevant). For technical reasons, we assume that V contains only regions reachable with positive probability in $\mathcal{M} \times \mathcal{A}$.

3.3 Finishing the proof of Theorem 3.1

Our proof is divided into three parts. In the first part we consider a general region graph which is not necessarily strongly connected, and show that we can actually concentrate just on its BSCCs. In the second part we study a given BSCC under the aperiodicity assumption. Finally, in the last part we consider a general BSCC which may be periodic. (The second part is included mainly for the sake of readability.)

Non-strongly connected region graph

Let $\mathcal{C}_1, \ldots, \mathcal{C}_k$ be the BSCCs of the region graph. The set $\mathcal{R}_i$ consists of all runs σ of $\mathcal{M}$ such that $\xi(\omega)$ visits (a configuration in a region of) $\mathcal{C}_i$, where ξ is the mapping of Lemma 3.8. By applying the arguments of [1, 11], it follows that almost runs in $\mathcal{M} \times \mathcal{A}$ visit a configuration of a BSCC. By Lemma 3.8, ξ preserves $\mathbf{d}$ and the probability $\mathcal{P}_{\mathcal{M}}(\mathcal{R}_i)$ is equal to the probability of visiting $\mathcal{C}_i$ in $\mathcal{M} \times \mathcal{A}$. Further, since the value of $\mathbf{d}$ does not depend on a finite prefix of a run, we may safely assume that $\mathcal{M} \times \mathcal{A}$ is initialized in $\mathcal{C}_i$ in such a way that the initial distribution corresponds to the conditional distribution of the first visit to $\mathcal{C}_i$ conditioned on visiting $\mathcal{C}_i$.

In a BSCC $\mathcal{C}_i$, there may be some *growing* clocks that are never reset. Since the values of growing clocks are just constantly increasing, the product process never returns to a state it has visited before. Therefore, there is no invariant distribution. Observe that all runs initiated in C_i eventually reach a configuration where the values of all growing clocks are larger than the maximal constant $B_{\max}$ employed in the guards of $\mathcal{A}$. This means that C_i actually consists *only* of regions where all growing clocks are irrelevant (see Section 3.2.2), because C_i would not be strongly connected otherwise. Hence, we can safely *remove* every growing clock x from C_i, replacing all guards of the form $x > c$ or $x \geq c$ with *true* and all guards of the form $x < c$ or $x \leq c$ with *false*. So, from now on we assume that there are no growing clocks in C_i.

Strongly connected & aperiodic region graph

In this part we consider a given BSCC C_i of the region graph $G_{\mathcal{M} \times \mathcal{A}}$. This is equivalent to assuming that $G_{\mathcal{M} \times \mathcal{A}}$ is strongly connected and $\Gamma_{\mathcal{M} \times \mathcal{A}}$ is equal to the union of all regions of $G_{\mathcal{M} \times \mathcal{A}}$ (recall that $G_{\mathcal{M} \times \mathcal{A}}$ consists just of regions reachable with positive probability in $\mathcal{M} \times \mathcal{A}$). We also assume that there are no growing clocks (see the previous part). Further, in this subsection we assume that $G_{\mathcal{M} \times \mathcal{A}}$ is aperiodic in the following sense.

DEFINITION 3.10. *A period p of the region graph $G_{\mathcal{M} \times \mathcal{A}}$ is the greatest common divisor of lengths of all cycles in $G_{\mathcal{M} \times \mathcal{A}}$. The region graph $G_{\mathcal{M} \times \mathcal{A}}$ is aperiodic if $p = 1$.*

The key to proving Theorem 3.1 in the current restricted setting is to show that the state space of $\mathcal{M} \times \mathcal{A}$ is small (recall Definition 3.5) and then apply Theorem 3.6 (1) and (2) to obtain the required characterization of the long-run behavior of $\mathcal{M} \times \mathcal{A}$.

PROPOSITION 3.11. *Assume that $G_{\mathcal{M} \times \mathcal{A}}$ is strongly connected and aperiodic. Then there exist a region R, a measurable subset $S \subseteq R$, $n \in \mathbb{N}$, $b > 0$, and a probability measure κ such that $\kappa(S) = 1$ and for all measurable $T \subseteq S$ and $z \in \Gamma_{\mathcal{M} \times \mathcal{A}}$ we have that $P^n_{\mathcal{M} \times \mathcal{A}}(z, T) > b \cdot \kappa(T)$. In other words, the set $\Gamma_{\mathcal{M} \times \mathcal{A}}$ of all states of the GSSMC $\mathcal{M} \times \mathcal{A}$ is (n, b, κ)-small.*

PROOF SKETCH. We show that there exist $z^* \in \Gamma_{\mathcal{M} \times \mathcal{A}}$, $n \in \mathbb{N}$, and $\gamma > 0$ such that for an arbitrary starting state $z \in \Gamma_{\mathcal{M} \times \mathcal{A}}$ there is a path from z to z^* of length exactly n that is γ-*wide* in the sense that the waiting time of any transition in the path can be changed by $\pm \gamma$ without ending up in a different region in the end. The target set S then corresponds to a "neighbourhood" of z^* within the region of z^*. Any small enough sub-neighbourhood of z^* is visited by a set of runs that follow the γ-wide path closely enough. The probability of this set of runs then depends linearly on the size of the sub-neighbourhood when measured by κ, where κ is essentially the Lebesgue measure restricted to S.

So, it remains to find suitable z^*, n, and γ. For a given starting state $z \in \Gamma_{\mathcal{M} \times \mathcal{A}}$, we construct a path of fixed length n (independent of z) that always ends in the same state z^*. Further, the path is γ-wide for some $\gamma > 0$ independent of z. Technically, the path is obtained by concatenating five sub-paths each of which has a fixed length independent of z. These sub-paths are described in greater detail below.

In the first sub-path, we move to a δ-*separated* state for some fixed $\delta > 0$ independent of z. A state is δ-separated if the fractional parts of all relevant clocks are approximately equally distributed on the $[0, 1]$ line segment (each two of them have distance at least δ). We can easily build the first sub-path so that it is δ-wide.

For the second sub-path, we first fix some region R_1. Since $G_{\mathcal{M} \times \mathcal{A}}$ is strongly connected and aperiodic, there is a fixed n'

such that R_1 is reachable from an arbitrary state of $\Gamma_{\mathcal{M} \times \mathcal{A}}$ in exactly n' transitions. The second sub-path is chosen as a (δ/n')-wide path of length n' that leads to a (δ/n')-separated state of R_1 (we show that such a sub-path is guaranteed to exist; intuitively, the reason why the separation and wideness may decrease proportionally to n' is that the fractional parts of relevant clock may be forced to move closer and closer to each other by the resets performed along the sub-path).

In the third sub-path, we squeeze the fractional parts of all relevant clocks close to 0. We go through a fixed region path $R_1 \cdots R_k$ (independent of z) so that in each step we shift the time by an integral value minus a small constant c (note that the fractional parts of clocks reset during this path have fixed relative distances). Thus, we reach a state z'_k that is "almost fixed" in the sense that the values of all relevant clocks in z'_k are the same for every starting state z. Note that the third sub-path is c-wide. At this point, we should note that if we defined the product process somewhat differently by identifying all states differing only in the values of irrelevant clocks (which does not lead to any technical complications), we would be done, i.e., we could put $z^* = z'_k$. We have neglected this possibility mainly for presentation reasons. So, we need two more sub-paths to fix the values of irrelevant clocks.

In the fourth sub-path, we act similarly as in the first sub-path and prepare ourselves for the final sub-path. We reach a δ-separated state that is almost equal to a fixed state $z_\ell \in R_\ell$. Again, we do it by a δ-wide path of a fixed length.

In the fifth sub-path, we follow a fixed region path $R_\ell \cdots R_{\ell+m}$ such that each clock not relevant in R_ℓ is reset along this path, and hence we reach a fixed state $z^* \in R_{\ell+m}$. Here we use our assumption that every clock can be reset to zero (i.e., there are no growing clocks). $\square$

Now we may finish the proof of Theorem 3.1. By Theorem 3.6 (1), there is a unique invariant distribution π on $\Gamma_{\mathcal{M} \times \mathcal{A}}$. For every $q \in Q$, we denote by A_q the set of all states of $\mathcal{M} \times \mathcal{A}$ of the form $(s, q, \nu) \in \Gamma_{\mathcal{M} \times \mathcal{A}}$. By Theorem 3.6 (2), for almost all runs σ of $\mathcal{M} \times \mathcal{A}$ we have that $\mathbf{d}(\sigma)$ is well-defined and $\mathbf{d}_q(\sigma) = \sum \pi(A_q)$. By Lemma 3.8, we obtain the same for almost all runs of $\mathcal{M}$.

Strongly connected & periodic region graph

Now we consider a general BSCC C_i of the region graph $G_{\mathcal{M} \times \mathcal{A}}$. Technically, we adopt the same setup as the previous part but remove the aperiodicity condition. That is, we assume that $G_{\mathcal{M} \times \mathcal{A}}$ is strongly connected, $\Gamma_{\mathcal{M} \times \mathcal{A}}$ is equal to the union of all regions of $G_{\mathcal{M} \times \mathcal{A}}$, and there are no growing clocks.

Let p be the period of $G_{\mathcal{M} \times \mathcal{A}}$. In this case, $\mathcal{M} \times \mathcal{A}$ is not necessarily small in the sense of Definition 3.5. By employing standard methods for periodic Markov chains, we decompose $\mathcal{M} \times \mathcal{A}$ into p stochastic processes $\Phi_0, \ldots, \Phi_{p-1}$ where each Φ_k makes steps corresponding to p steps of the original process $\mathcal{M} \times \mathcal{A}$ (except for the first step which corresponds just to k steps of $\mathcal{M} \times \mathcal{A}$). Each Φ_k is aperiodic and hence small (this follows by slightly generalizing the arguments of the previous part; see Proposition 3.13). Thus, we can apply Theorem 3.6 to each Φ_k separately and express the frequency of visits to q in Φ_k in terms of a unique invariant distribution π_k for Φ_k. Finally, we obtain the frequency of visits to q in $\mathcal{M} \times \mathcal{A}$ as an average of the corresponding frequencies in Φ_k.

Let us start by decomposing the set of nodes V of $G_{\mathcal{M} \times \mathcal{A}}$

into p classes that constitute a cyclic structure (see e.g. [12, Theorem 4.1]).

LEMMA 3.12. *There are disjoint sets* $V_0, \ldots, V_{p-1} \subseteq V$ *such that* $V = \bigcup_{k=0}^{p-1} V_k$ *and for all* $u, v \in V$ *we have that* $(u, v) \in E$ *iff there is* $k \in \{0, \ldots, p-1\}$ *satisfying* $u \in V_k$ *and* $v \in V_j$ *where* $j = (k+1) \mod p$.

For each $k \in \{0, \ldots, p-1\}$ we construct a GSSMC Φ_k with state space $\Gamma_{\mathcal{M} \times \mathcal{A}}^k = \bigcup_{R \in V_k} R$, a transition kernel $P^p(\cdot, \cdot)$ restricted to $\Gamma_{\mathcal{M} \times \mathcal{A}}^k$, and an initial probability measure μ_k defined by $\mu_k(A) = \int_{z \in \Gamma_{\mathcal{M} \times \mathcal{A}}} \mu(dz) \cdot P^k(z, A)$. For each k, we define the discrete frequency $\mathbf{d}_q^k$ of visits q in the process Φ_k. Then we show that if $\mathbf{d}^k$ is well-defined in Φ_k, we can express the frequency $\mathbf{d}_q$ in $\mathcal{M} \times \mathcal{A}$.

Note that for every run $z_0 z_1 \cdots$ of $\mathcal{M} \times \mathcal{A}$, the word $z_k z_{p+k} z_{2p+k}$ is a run of Φ_k. For a run $\sigma = (s_0, q_0, \nu_0)(s_1, q_1, \nu_1) \cdots$, $k \in \{0, \ldots, p-1\}$, and a location $q \in Q$, let define $1_q^{i,k}(\sigma)$ to be either 1 or 0 depending on whether $q_{ip+k} = q$ or not, respectively. Further, we put

$$\mathbf{d}_q^k(\sigma) = \limsup_{n \to \infty} \frac{\sum_{i=1}^n 1_q^{i,k}(\sigma)}{n}$$

Assuming that each $\mathbf{d}^k$ is well-defined, for almost all runs σ of $\mathcal{M} \times \mathcal{A}$ we have the following:

$$\mathbf{d}_q(\sigma) = \lim_{n \to \infty} \frac{\sum_{i=1}^n 1_q^i(\sigma)}{n} = \lim_{n \to \infty} \frac{\sum_{i=1}^n \sum_{k=0}^{p-1} 1_q^{i,k}(\sigma)}{np}$$
$$= \frac{1}{p} \sum_{k=0}^{p-1} \lim_{n \to \infty} \frac{\sum_{i=1}^n 1_q^{i,k}(\sigma)}{n} = \frac{1}{p} \sum_{k=0}^{p-1} \mathbf{d}_q^k(\sigma)$$

So, it suffices to concentrate on $\mathbf{d}_q^k$. The following proposition is a generalization of Proposition 3.11 to periodic processes.

PROPOSITION 3.13. *Assume that* $G_{\mathcal{M} \times \mathcal{A}}$ *is strongly connected and has a period* p. *For every* $k \in \{0, \ldots, p-1\}$ *there exist a region* $R_k \in V_k$, *a measurable* $S_k \subset R_k$, $n_k \in \mathbb{N}$, $b_k > 0$, *and a probability measure* κ_k *such that* $\kappa_k(S_k) = 1$ *and for every measurable* $T \subseteq S_k$ *and* $z \in \Gamma_{\mathcal{M} \times \mathcal{A}}^k$ *we have* $P_{\mathcal{M} \times \mathcal{A}}^{n_k \cdot p}(z, T) > b_k \cdot \kappa_k(T)$. *In other words,* Φ_k *is* (n_k, b_k, κ_k)-*small.*

By Theorem 3.6 (1), for every $k \in \{0, \ldots, p-1\}$, there is a unique invariant distribution π_k on $\Gamma_{\mathcal{M} \times \mathcal{A}}$ for the process Φ_k. By Theorem 3.6 (2), each $\mathbf{d}^k$ is well-defined and for almost all runs σ we have that $\mathbf{d}_q^k(\sigma) = \pi_k(A_q)$. Thus, we obtain

$$\mathbf{d}_q(\sigma) = \frac{1}{p} \sum_{k=0}^{p-1} \pi_k(A_q)$$

4. APPROXIMATING DTA MEASURES

In this section we show how to approximate the DTA measures for SMPs using the m-step transition kernel $P_{\mathcal{M} \times \mathcal{A}}^m$ of $\mathcal{M} \times \mathcal{A}$. The procedure for computing $P_{\mathcal{M} \times \mathcal{A}}^m$ up to a sufficient precision is taken as a "black box" part of the algorithm, we concentrate just on developing generic bounds on m that are sufficient to achieve the required precision.

For simplicity, we assume that the initial distribution α_0 of $\mathcal{M}$ assigns 1 to some $s_0 \in S$ (all of the results presented in this section can easily be generalized to an arbitrary initial distribution). The initial state in $\mathcal{M} \times \mathcal{A}$ is $z_0 = (s_0, q_0, \mathbf{0})$.

As we already noted in the previous section, the constant k of Theorem 3.1 is the number of BSCCs of $G_{\mathcal{M} \times \mathcal{A}}$. For the rest of this section, we fix some $1 \leq j \leq k$, and write just $\mathcal{C}$, $\mathcal{R}$ and D instead of $\mathcal{C}_j$, $\mathcal{R}_j$ and D_j, respectively. We slightly abuse our notation by using $\mathcal{C}$ to denote also the set of configurations that belong to some region of $\mathcal{C}$ (particularly in expressions such as $P_{\mathcal{M} \times \mathcal{A}}(z, \mathcal{C})$).

The probability $\mathcal{P}_{\mathcal{M}}(\mathcal{R})$ is equal to the probability of visiting $\mathcal{C}$ in $\mathcal{M} \times \mathcal{A}$. Observe that

$$\mathcal{P}_{\mathcal{M}}(\mathcal{R}) = \lim_{i \to \infty} P_{\mathcal{M} \times \mathcal{A}}^i(z_0, \mathcal{C})$$

Let us analyze the speed of this approximation. First, we need to introduce several parameters. Let $p_{\min}$ be the smallest transition probability in $\mathcal{M}$, and $\mathfrak{D}(\mathcal{M})$ the set of delay densities used in $\mathcal{M}$, i.e., $\mathfrak{D}(\mathcal{M}) = \{\mathbf{D}(s, s') \mid s, s' \in S\}$. Let $|V|$ be the number of vertices (regions) of $G_{\mathcal{M} \times \mathcal{A}}$. Due to our assumptions imposed on delay densities, there is a fixed bound $c_{\mathfrak{D}} > 0$ such that, for all $f \in \mathfrak{D}(\mathcal{M})$ and $x \in [0, B_{\max}]$, either $f(x) > c_{\mathfrak{D}}$ or $f(x) = 0$. Further, $\int_{B_{\max}}^\infty f(x)dx$ is either larger than $c_{\mathfrak{D}}$ or equal to 0.

THEOREM 4.1. *For every* $i \in \mathbb{N}$ *we have that*

$$\mathcal{P}_{\mathcal{M}}(\mathcal{R}) - P_{\mathcal{M} \times \mathcal{A}}^i(z_0, \mathcal{C}) \leq \left(1 - \left(\frac{p_{min} \cdot c_{\mathfrak{D}}}{c}\right)^c\right)^{\lfloor i/c \rfloor}$$

where $c = 4 \cdot |V|$.

PROOF SKETCH. We denote by B the union of all regions that belong to BSCCs of $G_{\mathcal{M} \times \mathcal{A}}$. We show that for $c = 4 \cdot |V|$ there is a lower bound $p_{bound} = (p_{\min} \cdot c_{\mathfrak{D}} \cdot 1/c)^c$ on the probability of reaching B in at most c steps from any state $z \in \Gamma_{\mathcal{M} \times \mathcal{A}}$. Note that then the probability of not hitting B after $i = m \cdot c$ steps is at most $(1 - p_{bound})^m$. However, this means that $P_{\mathcal{M} \times \mathcal{A}}^i(z, \mathcal{C})$ cannot differ from the probability of reaching $\mathcal{C}$ (and thus also from $\mathcal{P}_{\mathcal{M}}(\mathcal{R})$) by more than $(1 - p_{bound})^m$ because $\mathcal{C} \subseteq B$ and the probability of reaching $\mathcal{C}$ from $B \setminus \mathcal{C}$ is 0.

The bound p_{bound} is provided by arguments similar to the proof of Proposition 3.11. From any state z we build a δ-wide path to a state in B that has length bounded by $4 \cdot |V|$ such that $\delta = p_{\min} \cdot c_{\mathfrak{D}} \cdot 1/c$. The paths that follow this δ-wide path closely enough (hence, reach B) have probability p_{bound}. □

Now let us concentrate on approximating the tuple D. This can be done by considering just the BSCC $\mathcal{C}$. Similarly as in Section 3, from now on we assume that $\mathcal{C}$ is the set of nodes of $G_{\mathcal{M} \times \mathcal{A}}$ (i.e., $G_{\mathcal{M} \times \mathcal{A}}$ is strongly-connected) and that $\Gamma_{\mathcal{M} \times \mathcal{A}}$ is equal to the union of all regions of $\mathcal{C}$.

As in Section 3, we start with the aperiodic case. Then, Theorem 3.6 (3.) implies that each D_q can be approximated using $P_{\mathcal{M} \times \mathcal{A}}^i(u, A_q)$ where u is an arbitrary state of $\Gamma_{\mathcal{M} \times \mathcal{A}}$ and A_q is the set of all states of $\mathcal{M} \times \mathcal{A}$ of the form (s, q, ν). More precisely, we obtain the following:

THEOREM 4.2. *Assume that* $G_{\mathcal{M} \times \mathcal{A}}$ *is strongly connected and aperiodic. Then for all* $i \in \mathbb{N}$, $u \in \Gamma_{\mathcal{M} \times \mathcal{A}}$, *and* $q \in Q$

$$\left| D_q - P_{\mathcal{M} \times \mathcal{A}}^i(u, A_q) \right| \leq \left(1 - \left(\frac{p_{\min} \cdot c_{\mathfrak{D}}}{r}\right)^r\right)^{\lfloor i/r \rfloor}$$

where $r = \lfloor |V|^{4 \ln |V|} \rfloor$.

PROOF. From the proof of Proposition 3.13 (for details see [10]), we obtain that $\Gamma_{\mathcal{M} \times \mathcal{A}}$ is (m, ε, κ)-small with $m \leq r$ and $\varepsilon = (\frac{p_{\min} c_{\mathfrak{D}}}{r})^r$, and the result follows from Theorem 3.6 (3.). □

Now let us consider the general (periodic) case. We adopt the same notation as in Section 3, i.e., the period of $G_{\mathcal{M} \times \mathcal{A}}$ is denoted by p, the decomposition of the set V by $V_0, \ldots, V_{p-1}$ (see Lemma 3.12), and $\Gamma^k_{\mathcal{M} \times \mathcal{A}}$ denotes the set $\bigcup_{R \in V_k} R$ for every $k \in \{0, \ldots, p-1\}$.

THEOREM 4.3. *For every $i \in \mathbb{N}$ we have that*

$$\left| D_q - \frac{1}{p} \cdot \sum_{k=0}^{p-1} P^{i \cdot p}_{\mathcal{M} \times \mathcal{A}}(u_k, A_q) \right| \leq \left(1 - \left(\frac{p_{\min} \cdot c_{\mathfrak{D}}}{r} \right)^r \right)^{\lfloor i/r \rfloor}$$

where $u_k \in \Gamma^k_{\mathcal{M} \times \mathcal{A}}$ and $r = \lfloor |V|^{4 \ln |V|} \rfloor$.

PROOF. Due to the results of Section 3 we have that $D_q = \frac{1}{p} \cdot \sum_{k=0}^{p-1} \pi_k(A_q)$, where π_k is the invariant measure for the k-th aperiodic decomposition Φ_k of the product process $\mathcal{M} \times \mathcal{A}$ (i.e. π_k is a measure over $\Gamma^k_{\mathcal{M} \times \mathcal{A}}$). From the proof of Proposition 3.13 (for details see [10]), $\Gamma^k_{\mathcal{M} \times \mathcal{A}}$ is (m, ε, κ)-small with $m \leq r$ and $\varepsilon = (\frac{p_{\min} c_{\mathfrak{D}}}{r})^r$, and the result follows from Theorem 3.6 (3.) applied to each $\Gamma^k_{\mathcal{M} \times \mathcal{A}}$ separately. □

5. CONCLUSIONS

We have shown that DTA measures over semi-Markov processes are well-defined for almost all runs and assume only finitely many values with positive probability. We also indicated how to approximate DTA measures and the associated probabilities up to an arbitrarily small given precision.

Our approximation algorithm is quite naive and there is a lot of space for further improvement. An interesting open question is whether one can design more efficient algorithms with low complexity in the size of SMP (the size of DTA specifications should stay relatively small in most applications, and hence the (inevitable) exponential blowup in the size of DTA is actually not so problematic).

Another interesting question is whether the results presented in this paper can be extended to more general stochastic models such as generalized semi-Markov processes.

Acknowledgement

The authors thank Petr Slovák for his many useful comments. The work has been supported by the Institute for Theoretical Computer Science, project No. 1M0545, the Czech Science Foundation, grant No. P202/10/1469 (T. Brázdil, A. Kučera), No. 201/08/P459 (V. Řehák), and No. 102/09/H042 (J. Krčál), and Brno PhD Talent Financial Aid (J. Křetínský).

6. REFERENCES

[1] R. Alur, C. Courcoubetis, and D.L. Dill. Verifying automata specifications of probabilistic real-time systems. In *Real-Time: Theory in Practice*, volume 600 of *Lecture Notes in Computer Science*, pages 28–44. Springer, 1992.

[2] R. Alur and D. Dill. A theory of timed automata. *Theoretical Computer Science*, 126(2):183–235, 1994. Fundamental Study.

[3] A. Aziz, K. Sanwal, V. Singhal, and R.K. Brayton. Model-checking continuous-time Markov chains. *ACM Transactions on Computational Logic*, 1(1):162–170, 2000.

[4] C. Baier, N. Bertrand, P. Bouyer, T. Brihaye, and M. Größer. Almost-sure model checking of infinite paths in one-clock timed automata. In *Proceedings of LICS 2008*, pages 217–226. IEEE Computer Society Press, 2008.

[5] C. Baier, B.R. Haverkort, H. Hermanns, and J.-P. Katoen. Model-checking algorithms for continuous-time Markov chains. *IEEE Transactions on Software Engineering*, 29(6):524–541, 2003.

[6] V.L. Barbu and N. Limnios. *Semi-Markov Chains and Hidden Semi-Markov Models toward Applications*. Springer, 2008.

[7] N. Bertrand, P. Bouyer, T. Brihaye, and N. Markey. Quantitative model-checking of one-clock timed automata under probabilistic semantics. In *Proceedings of 5th Int. Conf. on Quantitative Evaluation of Systems (QEST'08)*, pages 55–64. IEEE Computer Society Press, 2008.

[8] P. Billingsley. *Probability and Measure*. Wiley, 1995.

[9] D. Bini, G. Latouche, and B. Meini. *Numerical methods for Structured Markov Chains*. Oxford University Press, 2005.

[10] T. Brázdil, J. Krčál, J. Křetínský, A. Kučera, and V. Řehák. Measuring performance of continuous-time stochastic processes using timed automata. *CoRR*, abs/1101.4204, 2011.

[11] T. Brázdil, J. Krčál, J. Křetínský, A. Kučera, and V. Řehák. Stochastic real-time games with qualitative timed automata objectives. In *Proceedings of CONCUR 2010*, volume 6269 of *Lecture Notes in Computer Science*, pages 207–221. Springer, 2010.

[12] P. Brémaud. *Markov chains: Gibbs fields, Monte Carlo simulation, and queues*. Springer, 1998.

[13] T. Chen, T. Han, J.-P. Katoen, and A. Mereacre. Quantitative model checking of continuous-time Markov chains against timed automata specifications. In *Proceedings of LICS 2009*, pages 309–318. IEEE Computer Society Press, 2009.

[14] E.A. Emerson and E.M. Clark. Using branching time temporal logic to synthesize synchronization skeletons. *Science of Computer Programming*, 2:241–266, 1982.

[15] B.R. Haverkort, H. Hermanns, and J.-P. Katoen. On the use of model checking techniques for quantitative dependability evaluation. In *Proceedings of IEEE Symp. Reliable Distributed Systems*, pages 228–238, 2000.

[16] R. Kress. *Numerical Analysis*. Springer, 1998.

[17] K. Matthes. Zur Theorie der Bedienungsprozesse. *Transactions of the Third Prague Conference on Information Theory, Statistical Decision Functions, Random Processes*, pages 513–528, 1962.

[18] S. Meyn and R.L. Tweedie. *Markov Chains and Stochastic Stability*. Cambridge University Press, 2009.

[19] J.R. Norris. *Markov Chains*. Cambridge University Press, 1998.

[20] G.O. Roberts and J.S. Rosenthal. General state space Markov chains and MCMC algorithms. *Probability Surveys*, 1:20–71, 2004.

[21] S.M. Ross. *Stochastic Processes*. Wiley, 1996.

Measurability and Safety Verification for Stochastic Hybrid Systems

Martin Fränzle
Carl von Ossietzky University, Germany

Ernst Moritz Hahn
Saarland University, Germany

Holger Hermanns
Saarland University, Germany

Nicolás Wolovick
National University of Córdoba, Argentina

Lijun Zhang
Technical University of Denmark

ABSTRACT

Dealing with the interplay of randomness and continuous time is important for the formal verification of many real systems. Considering both facets is especially important for wireless sensor networks, distributed control applications, and many other systems of growing importance. An important traditional design and verification goal for such systems is to ensure that unsafe states can never be reached. In the stochastic setting, this translates to the question whether the probability to reach unsafe states remains tolerable. In this paper, we consider stochastic hybrid systems where the continuous-time behaviour is given by differential equations, as for usual hybrid systems, but the targets of discrete jumps are chosen by probability distributions. These distributions may be general measures on state sets. Also non-determinism is supported, and the latter is exploited in an abstraction and evaluation method that establishes safe upper bounds on reachability probabilities. To arrive there requires us to solve semantic intricacies as well as practical problems. In particular, we show that measurability of a complete system follows from the measurability of its constituent parts. On the practical side, we enhance tool support to work effectively on such general models. Experimental evidence is provided demonstrating the applicability of our approach on three case studies, tackled using a prototypical implementation.

Categories and Subject Descriptors: I.6.4 [Computing Methodologies]: Simulation and Modelling - Model Validation and Analysis; C.1.m [Computer Systems Organization]: Processor Architectures - Hybrid Systems; G.3 [Mathematics of Computing]: Probability and Statistics

General Terms: Reliability, Verification.

1. INTRODUCTION

In many modern application areas of hybrid systems, *random phenomena* occur. This is especially true for wireless sensing and control applications, where message loss probabilities and other random effects (node placement, node failure, battery drain, measurement imprecision) turn the overall control problem into a problem that can only be managed with a certain, hopefully sufficiently large, probability. The need to integrate probabilities into hybrid systems formalisms has led to several different notions of *stochastic hybrid systems*, each from a distinct perspective [2, 3, 9, 22]. They differ in the point of attack where to introduce randomness. One option is to replace deterministic jumps by probability distributions over deterministic jumps. Another option is to generalise the differential equation components inside a mode by a stochastic differential equation component. More general models can be obtained by blending the above two choices, and by combining them with memoryless timed probabilistic jumps [8], and with non-determinism. Piecewise-deterministic Markov processes [12] are a prominent example, constituting deterministic hybrid system models augmented with memoryless timed probabilistic jumps.

An important problem in hybrid systems theory is that of reachability analysis. In general terms, a reachability analysis problem consists in evaluating whether a given system will reach certain unsafe states, starting from certain initial states. This problem is associated with the safety verification problem: if the system cannot reach any unsafe state, then the system is declared to be safe. In the probabilistic setting, the safety verification problem can be formulated as that of checking whether the probability that the system trajectories reach an unsafe state from its initial states can be bounded by some given probability threshold.

Recently [23], we have introduced a technique that piggybacks a quantitative probabilistic reachability analysis for *probabilistic* hybrid automata on a qualitative reachability checker for *non-probabilistic* hybrid automata. It does so to compute upper bounds on maximal reachability probabilities, but is restricted to probability distributions with finite support. In this paper, we extend the approach to stochastic hybrid automata which feature *continuous* measures over states, for instance given by a density function, as well as non-deterministic behaviour. We show that the well-definedness of the individual automata parts leads to the well-definedness of the model semantics, such that we can define meaningful probabilities when resolving the non-determinism. For this we harvest recent results on non-deterministic labelled Markov processes [11] to overcome intricate measurability issues. To handle reachability problems for this model class, we over-approximate the stochastic hybrid automaton by a probabilistic hybrid automaton, in which the probability to reach unsafe states can not be lower than in the original model. Because this transformation is done on the high-level description of the automata, we can combine it with our previous bounding technique and algorithmic implementation. We are thus able to tackle a broad class of stochastic hybrid systems. Due to the presence of continuous non-determinism, but absence of diffusion in the differential equation components, the model is distinguished from more classical stochastic hybrid systems representations for which reachability computations have been tackled in different ways [1, 2, 9, 19, 20]. Among them, grid-based methods [1] are a promising approach with some similarities to our work. In contrast to those, we do not compute a fixed state-space abstraction prior to the actual analysis. Instead, an abstraction is computed by a solver for reachability in a non-probabilistic version of the hybrid systems under consideration. From the abstraction obtained this way, we compute a probabilistic model which faithfully over-approximates the actual maximal reachability probability.

The paper is organised as follows: Section 2 describes the model which will later on form the semantics of stochastic hybrid automata. Then, in Section 3, we give the high-level specification model of stochastic hybrid automata. Afterwards, in Section 4 we describe how to compute over-approximating reachability probabilities in an abstracted model of the automata. The practical applicability of the method is demonstrated in Section 5. Finally, we conclude the paper in Section 6.

2. SEMANTIC MODELS

We define non-deterministic Markov processes, which shall later appear as the semantics of stochastic hybrid automata. Subsequently, we show how to abstract these models, which potentially feature continuous measures, to probabilistic automata with measures of only finite support. For the specification of these models, we need some preliminary definitions of measure theoretic concepts.

Measure Theory Background.

A family Σ of subsets of the set S is a σ-algebra provided it is closed under complement and σ-union (denumerable union), a set $A \in \Sigma$ is then called measurable. We denote by $\sigma(\mathcal{A})$ the smallest σ-algebra containing the sets of $\mathcal{A}$, and it is said to be generated by this set. The Borel σ-algebra over $\mathbb{R}$ is generated by intervals of rational endpoints, and it is denoted $\mathcal{B}(\mathbb{R}) := \sigma(\{[p, q] \mid p, q \in \mathbb{Q}\})$. For dimensions greater than one, it is generated by rectangles with rational endpoints and denoted by $\mathcal{B}(\mathbb{R}^n)$. The pair (S, Σ) is called a measurable space, and where convenient we will use $(\mathbb{R}^n, \mathcal{B}(\mathbb{R}^n))$ to denote the particular Borel case.

Given two measurable spaces (S_0, Σ_0), (S_1, Σ_1), the product space is given by $(S_0 \times S_1, \Sigma_0 \otimes \Sigma_1)$, where the product σ-algebra is generated by rectangles $A_0 \times A_1$, with $A_i \in \Sigma_i$. Given $M \in \Sigma_0 \otimes \Sigma_1$, the section at s_0 is the measurable set $M_{|s_0} := \{s_1 \mid (s_0, s_1) \in M\}$.

A function $\mu : \Sigma \to [0, 1]$ is called σ-additive if $\mu(\biguplus_{i \in I} A_i) = \sum_{i \in I} \mu(A_i)$ for countable index sets I. We speak of a probability measure if $\mu(S) = 1$. The support of a measure $Supp(\mu)$ is a measurable set A such that $\mu(S \setminus A) = 0$, and for the Borel σ-algebra there is a (unique) smallest closed set C^0 with this property. The Dirac probability measure δ_s is 1 only in $\{s\}$. A function is measurable, denoted $f : (\mathbb{R}, \mathcal{B}(\mathbb{R})) \to (\mathbb{R}, \mathcal{B}(\mathbb{R}))$ in the real-valued case, if every backwards image of a generator is measurable,

$f^{-1}([p, q)) \in \mathcal{B}(\mathbb{R})$ for all rational p and q. The indicator function $\chi_A : S \to \{0, 1\}$ is $\chi_A(x) := 1$ iff $x \in A$, and it is measurable if $A \in \Sigma$. A function f is simple if it is of the form $f(x) = \sum_{i=1}^{n} c_i \chi_{A_i}(x)$, and if $c_i \in \mathbb{R}$ and $A_i \in \Sigma$ then f is also measurable. Without loss of generality, we can assume pairwise disjoint A_i. The set of probability measures $\Delta(S)$ on (S, Σ) can be endowed with σ-algebra $\Delta(\Sigma)$ [14] generated by the set $\Delta^{>q}(A) := \{\mu \mid \mu(A) > q\}$, i.e. the measures such that when applied to $A \in \Sigma$ give a value greater than $q \in \mathbb{Q} \cap [0, 1]$. In order to define measurable functions $f : S \to \Delta(\Sigma)$, we need to define the σ-algebra on $\Delta(\Sigma)$. We use [11] $H(\Delta(\Sigma)) := \sigma(\{H_\Phi \mid \Phi \in \Delta(\Sigma)\})$, where the hit set is $H_\Phi := \{\xi \in \Delta(\Sigma) \mid \xi \cap \Phi \neq \emptyset\}$, such that f is measurable if $f^{-1}(H_\Phi) \in \Sigma$, and we write $f : (S, \Sigma) \to (\Delta(\Sigma), H(\Delta(\Sigma)))$.

We let $\Delta_f(S)$ denote the set of finite measures, which contains all μ such that $|Supp(\mu)| < \infty$. For a singleton set $\{a\} \in Supp(\mu)$ we write $a \in Supp(\mu)$, $\mu(a) = \mu(\{a\})$, and so on.

We can now define our stochastic models.

DEFINITION 1. A non-deterministic Markov process (NMP) $\mathcal{M}$ is a tuple $(S, \Sigma, Init, Steps, UnSafe)$ where

- S denotes the (possibly uncountable) set of states,

- Σ is a σ-algebra on S. We use the subset $Init \in \Sigma$ to specify the set of initial states and $UnSafe \in \Sigma$ for the set of unsafe states.

- $Steps : (S, \Sigma) \to (\Delta(\Sigma), H(\Delta(\Sigma)))$ is a measurable transition function. We require that $Steps(s) \neq \emptyset$ for all $s \in S$.

A probabilistic automaton (PA) is an NMP where transition functions are restricted to finite support measures, $Steps : S \to 2^{\Delta_f(S)}$. For PA, measurability considerations are not needed to specify meaningful probabilities.

This definition is essentially the same as the one of D'Argenio et al. [11], with the difference that we do not use action labels, as they are not needed in our setting. If $\mu \in Steps(s)$, we call μ a successor probability measure of s. Measurability considerations are indeed necessary for general NMP, as we demonstrate in the following example.

EXAMPLE 1. Let $(S, \mathcal{B}(S), \{s_0\}, Steps, \{s_b\})$ be an NMP with $S := \{s_0\} \uplus [0, 1] \uplus \{s_b\}$. We define $Steps(s_0) := \{\mu\}$, where μ is the uniform distribution on $[0, 1]$, and let $Steps(s_b) := \{\delta_{s_b}\}$. Further, for $s \in \mathcal{V}$ we let $Steps(s) := \{\delta_{s_b}\}$ and for $s \notin \mathcal{V}$ we define $Steps(s) := \{\delta_s\}$, where $\mathcal{V} \subset [0, 1]$ is a Vitali set, which is known not to be Borel-measurable. All $Steps(\cdot)$ are singleton sets, so there is no non-determinism in the example. Thus, the probability to reach s_b from s_0 should be uniquely defined. However, this probability does not exist, because it depends on the measure of the non-measurable Vitali set. Discrete measures (which are the only ones allowed in PA) eliminate the Vitali set problem, since they discard all but a finite set of points out of it. For instance, if instead of using the uniform distribution we set $\mu(s_1) = \mu(s_2) = \mu(s_3) = \frac{1}{3}$ for $s_1, s_2 \in \mathcal{V}$ and $s_3 \notin \mathcal{V}$ and $\mu(\cdot) = 0$ else, we can specify the reachability probability as $\frac{2}{3}$, although formally the model still does not fulfil the measurability requirements.

PAs have been introduced by Segala and Lynch [21]. For PAs, we can always assign a probability larger than zero to each individual state which is possibly chosen as successor. This is not the case for general NMPs: all individual probabilities of moving to a successor state may be zero.

For an NMP $\mathcal{M} = (S, \Sigma, Init, Steps, UnSafe)$, we specify the maximal n-step probability to reach the unsafe states, starting in a state $s \in S$, as $Reach_{\leq 0}^{\mathcal{M}}(s) = 1$ if $s \in UnSafe$ and 0 else,

$$Reach_{\leq n+1}^{\mathcal{M}}(s) := \begin{cases} 1 \text{ if } s \in UnSafe, \\ \sup_{\mu \in Steps(s)} \int Reach_{\leq n}^{\mathcal{M}}(s') \, d\mu(s') \text{ else,} \end{cases}$$

where $\int$ denotes Lebesgue integration. For a PA $\mathcal{M} = (S, \Sigma, Init, Steps, UnSafe)$, we can simplify the latter formula to

$$Reach_{\leq n+1}^{\mathcal{M}}(s) = \begin{cases} 1 \text{ if } s \in UnSafe, \\ \sup_{\mu \in Steps(s)} \sum_{s' \in Supp(\mu)} Reach_{\leq n}^{\mathcal{M}}(s') \mu(s') \text{ else.} \end{cases}$$

The maximal unbounded reachability probability is

$$Reach^{\mathcal{M}}(s) := \lim_{n \to \infty} Reach_{\leq n}^{\mathcal{M}}(s).$$

These definitions indeed define functions in n and s:

LEMMA 1. *Let $\mathcal{M}$ be an NMP. Then $Reach_{\leq n}^{\mathcal{M}}(s)$ is well-defined for all non-negative n, as is $Reach^{\mathcal{M}}(s)$.*

We now specify approximations between general NMPs and PAs operating on the same state set.

DEFINITION 2. *Let $\mathcal{M}_c = (S, \Sigma, Init_c, Steps_c, UnSafe_c)$ be an NMP, and $\mathcal{M}_f = (S, \Sigma, Init_f, Steps_f, UnSafe_f)$ be a PA. We say $\mathcal{M}_f$ is an abstraction of $\mathcal{M}_c$, if $Init_c \subseteq Init_f$, $UnSafe_c \subseteq UnSafe_f$, and moreover, for each $s \in S$ and $\mu_c \in Steps_c(s)$,*

- *there exist pairwise disjoint $A_1, \ldots, A_n \in \Sigma$ such that $\mu_c(\bigcup_{i=1}^{n} A_i) = 1$, and*

- *in the PA, for each $(s_1, \ldots, s_n) \in A_1 \times \cdots \times A_n$ there exists $\mu_f \in Steps_f(s)$ such that $\mu_f(s_i) = \mu_c(A_i)$ for all $1 \leq i \leq n$.*

We write $\mathcal{M}_c \preceq_{cf} \mathcal{M}_f$ if $\mathcal{M}_f$ is an abstraction of $\mathcal{M}_c$.

Notably, and in contrast to related abstraction methods [1], we do not fix a representative of the probability measure, but instead introduce uncountable non-determinism over the possible successors. While this may seem unfamiliar and impractical, indeed it is not. The models under consideration may anyway have an uncountably large state-space, such that a direct analysis is impossible. Later on, semantics of stochastic hybrid automata will be given as uncountably large NMPs. However, we will show how to compute abstractions of the semantics directly from the high-level description of the model. Thereby, we will construct a finitely large state-space, and represent the uncountable non-determinism by a finite number of transitions.

For PAs, simulation preorders have been introduced [21], and subsequently exploited [23], to analyse safety properties. When restricted to this subclass, our notion of abstraction establishes a special case of such simulation preorders.

Applying the abstraction from Definition 2 does not decrease the maximal reachability probability.

LEMMA 2. *Let $\mathcal{M}_c = (S, \Sigma, Init_c, Steps_c, UnSafe_c)$ be an NMP and $\mathcal{M}_f = (S, Init_f, Steps_f, UnSafe_f)$ be a PA such*

that $\mathcal{M}_c \preceq_{cf} \mathcal{M}_f$. *For all $s \in S$ and all non-negative n, we have*

$$Reach_{\leq n}^{\mathcal{M}_c}(s) \leq Reach_{\leq n}^{\mathcal{M}_f}(s)$$

and

$$Reach^{\mathcal{M}_c}(s) \leq Reach^{\mathcal{M}_f}(s).$$

Thus, if we can show that reachability probabilities in $\mathcal{M}_f$ are below a certain threshold, this is also the case in $\mathcal{M}_c$.

3. STOCHASTIC HYBRID AUTOMATA

In this section, we provide definitions for the fragment of stochastic hybrid automata addressed in this paper. We define the underlying semantics of these high-level models in terms of the models of Section 2 and show that the measurability of the semantics of a complete automaton follows from the measurability of its constituent parts.

3.1 Model Description

Probabilistic hybrid automata as considered in our previous work [23] require probability measures to have finite support and also require that only a finite number of non-deterministic choices occurs in each state. In the following, we describe an extension to *stochastic* hybrid automata, in which we allow continuous probability distributions and uncountable non-determinism in discrete assignments, yet not over continuous distributions.

Let m denote a variable ranging over a finite set of modes $M := \{m_1, \ldots, m_n\}$, and let $\boldsymbol{x} := (x_1, \ldots, x_k)$ be a vector of variables ranging over real numbers $\mathbb{R}$. For denoting the derivatives of $\boldsymbol{x}$ we use $\dot{\boldsymbol{x}} := (\dot{x}_1, \ldots, \dot{x}_k)$, ranging over $\mathbb{R}$ correspondingly. With m' and $\boldsymbol{x}' := (x'_1, \ldots, x'_k)$ we denote primed versions of m and $\boldsymbol{x}$ respectively, as subsequently used to specify values resulting from discrete jumps of a hybrid automaton.

Later on, $S := M \times \mathbb{R}^k$ will denote the state-space of the semantics of the hybrid automaton. We let $\Sigma := \mathcal{B}(S)$ denote the Borel σ-algebra on the state-space. Further, let $H(\Sigma)$ denote the σ-algebra generated by all $H_A := \{B \in \Sigma \mid A \cap B \neq \emptyset\}$ where $A \in \Sigma$. (This construction is similar to the ones used in non-probabilistic NLMP [10].) A *state-space constraint* is a constraint $\mathbf{s} \subseteq M \times \mathbb{R}^k$ over modes and variables. A *flow constraint* is a constraint $\mathbf{f} \subseteq M \times \mathbb{R}^k \times \mathbb{R}^k$ over the variables m, $\boldsymbol{x}$, $\dot{\boldsymbol{x}}$.

A *probabilistic guarded command* $\mathbf{c}$ shall be defined as

$$condition \to p_1 : update_1 + \ldots + p_n : update_n$$

where $n \geq 1$ denotes the cardinality of the probabilistic branching of $\mathbf{c}$ with $p_i > 0$ for $i = 1, \ldots, n$ and $\sum_{i=1}^{n} p_i = 1$. We demand that $condition \in \Sigma$ is a measurable constraint over $(m, \boldsymbol{x})$, and that $update_i : (S, \Sigma) \to (\Sigma, H(\Sigma))$ is a measurable function denoting a reset mapping for m and $\boldsymbol{x}$ for all $i = 1, \ldots, n$. Observe that for different $i \neq j$, it could be the case that $update_i(m, \boldsymbol{x}) \cap update_j(m, \boldsymbol{x}) \neq \emptyset$. In our notation, if we do not mention a variable in a guarded command, it remains unchanged.

EXAMPLE 2.

$$\begin{aligned} m = m_1 \quad \to \quad & 0.2 : m' = m_2 \land x'_1 \leq x_2 - 0.84 \\ + \quad & 0.2 : m' = m_2 \land x_2 - 0.85 \leq x'_1 \leq x_2 - 0.25 \\ + \quad & 0.2 : m' = m_2 \land x_2 - 0.26 \leq x'_1 \leq x_2 + 0.26 \\ + \quad & 0.2 : m' = m_2 \land x_2 + 0.25 \leq x'_1 \leq x_2 + 0.85 \\ + \quad & 0.2 : m' = m_2 \land x'_1 \geq x_2 + 0.84 \end{aligned}$$

is a probabilistic guarded command. It can be executed when being in mode m_1. With probability 1, we move to mode m_2. With a probability of 0.2 each, a certain interval is chosen, and the variable x_1 is non-deterministically set to an arbitrary value within this interval. The endpoints of the intervals depend on the value of x_2. Other variables remain unchanged.

While previously [23] we restricted to commands where the updates $update_i$ map a state to a unique successor, we here allow the updates to be predicates over successor states. This leads to a possibly uncountable non-determinism, as in Example 2.

To model continuous measures, we introduce an additional form of guarded commands. Let $M : (S, \Sigma) \to (\Delta(S), \Delta(\Sigma))$ be a measurable function mapping states to probability measures. A *stochastic guarded command* is of the form

$$condition \to M.$$

EXAMPLE 3. *We specify $M(m_1, x_1, x_2, \ldots, x_n)$ as*

$$M(m_1, x_1, x_2, \ldots, x_n) \left(\{m_2\} \times [a, b] \times \bigtimes_{i=2}^{n} \{x_i\} \right)$$

$$:= \frac{1}{\sqrt{2\pi}} \int_a^b \exp\left(-\frac{1}{2}(x - x_2)^2 \right) \mathrm{d}x,$$

with the unique extension of this measure to other Borel sets. Then $m = m_1 \to M$ is a stochastic guarded command, which we denote by c. It can execute under the same conditions as the probabilistic guarded command from Example 2, and has the same target mode. However, x_1 is set according to the normal distribution $\mathcal{N}(x_2, 1)$ with expected value x_2 and standard deviation 1. Due to the properties of the normal distribution, such a command is suited to set a variable according to a probability distribution which is centred around an ideal value from which the variable may deviate into both directions. In practice, such perturbations arise from inexact measurements, deviations of production parameters in a production line, etc.

With these preparations, we can define stochastic hybrid automata.

DEFINITION 3. *A stochastic hybrid automaton is a tuple $\mathcal{H} = (\mathrm{M}, \boldsymbol{x}, Init, Flow, \mathrm{C}, UnSafe)$ where*

- M *is a finite set of modes and $\boldsymbol{x}$ is a set of k variables,*

- $Init \subseteq \mathrm{M} \times \mathbb{R}^k$ *is a constraint on the initial states,*

- $UnSafe \subseteq \mathrm{M} \times \mathbb{R}^k$ *describes the unsafe states,*

- $Flow \subseteq \mathrm{M} \times \mathbb{R}^k \times \mathbb{R}^k$ *is a flow constraint and*

- C *denotes a finite set of guarded commands. We denote the subset of probabilistic guarded commands as C_f and the subset of stochastic guarded commands as C_c.*

We require $Flow$ to be measurable in the following sense: for each $m \in \mathrm{M}$, the pre-post-relation $T :=$

$$\left\{ (\boldsymbol{x}, \boldsymbol{y}) \in \mathbb{R}^k \times \mathbb{R}^k \left| \begin{array}{c} \exists e \geq 0, f : [0, e] \to \mathbb{R}^k \ differentiable : \\ \left(\begin{array}{c} f(0) = \boldsymbol{x} \\ \wedge f(e) = \boldsymbol{y} \\ \wedge \forall t \in [0, e] : \\ (m, f(t), \dot{f}(t)) \in Flow \end{array} \right) \end{array} \right. \right\}$$

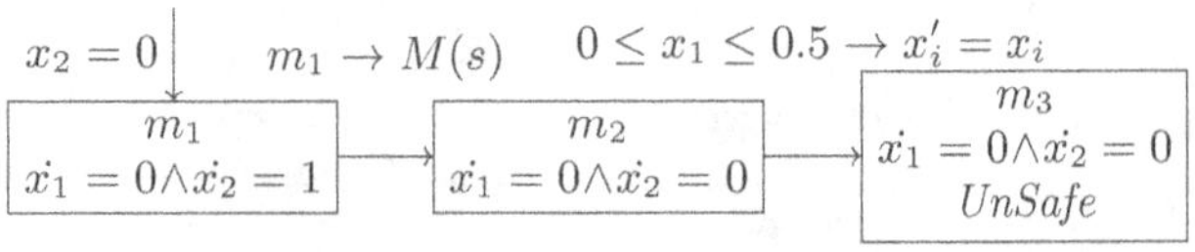

Figure 1: Example stochastic hybrid automaton. Here, $M(s)$ with $s = (m, x_1, x_2)$ is a Dirac distribution with respect to x_2 and a normal distribution over x_1, combined such that x_2 keeps its value while x_1 has expected value x_2 with standard deviation 1.

mediated by the continuous flow is a measurable set in $\mathcal{B}(\mathbb{R}^k \times \mathbb{R}^k)$, and we require $post^m(x) := T_{|\boldsymbol{x}}$ to be measurable, i.e. $post^m : (\mathbb{R}^k, \mathcal{B}(\mathbb{R}^k)) \to (\mathcal{B}(\mathbb{R}^k), H(\mathcal{B}(\mathbb{R}^k)))$. Furthermore, we require $Init$ and $UnSafe$ to be measurable sets.

Measurability of most of the above model constituents can be guaranteed by considering o-minimal definable sets. General results connecting o-minimal definability with measurability [6, 7] show that a sufficient criterion for the above pre-post relation T being Borel-measurable is that it is definable in some o-minimal theory over the reals. In practice, this holds for the pre-post relations manipulated by hybrid model checkers, as all current hybrid model checkers tackle differential equations by providing descriptions or approximations of their reach set and thus of the above pre-post relation via sets definable in o-minimal theories over the reals, such as finite unions of rectangular boxes, zonotopes, polyhedra, ellipsoids, or by differential invariants. In a nutshell, the general results connecting o-minimality with measurability considers the standard parts [6] of o-minimal theories and shows them to be Borel measurable. This, together with the fact that the standard part $\mathrm{st}(A)$ satisfies $\mathrm{st}(A) = A$ provided $A \subseteq \mathbb{R}^k$ [7] implies that relations definable by o-minimal theories over the reals are Borel measurable [6]. Hence, T is Borel measurable if described in some o-minimal theory. The function $post^m$ [16] is defined as section of relation T. Although this definition ensures that $post^m(x)$ gives measurable sets for every x, it does not imply measurability of $post^m$ itself, which is why we need to require it.

We talk of a *probabilistic* hybrid automaton if $\mathrm{C}_c = \emptyset$. We still allow uncountable non-determinism for this model class. In this subclass, measurability restrictions are not needed to guarantee that probability measures are well-defined in the low-level semantical model.

EXAMPLE 4. *Consider the model in Figure 1. We assume that μ is the normal distribution of the command c from Example 3. We are interested in the maximal probability to finally reach the target mode m_3. To obtain the maximal probability, we wait in m_1 until $x_2 = 0.25$. Then, we jump to m_2. Let $f(x)$ be the density of $\mathcal{N}(0.25, 1)$. With probability $\int_0^{0.5} f(x) \, \mathrm{d}x \approx 0.197$ we can finally jump from m_2 to m_3.*

3.2 Semantics of Stochastic Hybrid Automata

The semantics of a stochastic hybrid automaton $\mathcal{H} = (\mathrm{M}, \boldsymbol{x}, Init, Flow, \mathrm{C}, UnSafe)$ is the tuple $Sem(\mathcal{H}) := (S, \Sigma, Init, Steps, UnSafe)$ where $S := \mathrm{M} \times \mathbb{R}^k$, $\Sigma := \mathcal{B}(S)$ and we define $Steps$ as the union of two transition relations $Steps_{\mathcal{T}}, Steps_{\mathcal{J}} : S \to \Delta(\Sigma)$. The semantics of timed steps is defined as

$$Steps_{\mathcal{T}}((m, \boldsymbol{x})) := \{\delta_{(m, \boldsymbol{x}')} \mid \boldsymbol{x}' \in post^m(\boldsymbol{x})\}.$$

Now we define the semantics of guarded commands. To start with, we define the semantics of a probabilistic guarded command $\mathsf{c} = con \rightarrow p_1 : u_1 + \ldots + p_n : u_n$ by: $Steps_{\mathsf{c}}(s) := \emptyset$ if $s \notin con$, and otherwise

$$Steps_{\mathsf{c}}(s) := \left\{ \sum_{i=1}^{n} p_i \delta_{s_i} \mid (s_1, \ldots, s_n) \in u_1(s) \times \cdots \times u_n(s) \right\}.$$

Inside the above formula, we have weighted sums of Dirac probability measures. A step induced by a probabilistic guarded command has as successors all measures, such that with probability p_i we choose a state of the ith update. For measures in which we have a state which is the successor of different updates, the probabilities of these updates are added up.

Next, for a stochastic guarded command $\mathsf{c} = con \rightarrow M$ we define

$$Steps_{\mathsf{c}}(s) := \left\{ \begin{array}{ll} \{M(s)\} & \text{if } s \in con, \\ \emptyset & \text{else.} \end{array} \right.$$

Then for $s \in S$, we let

$$Steps_{\mathcal{J}}(s) := \bigcup_{\mathsf{c} \in \mathbb{C}} Steps_{\mathsf{c}}(s)$$

and

$$Steps(s) := \left\{ \begin{array}{l} Steps_{\mathcal{T}}(s) \cup Steps_{\mathcal{J}}(s), \\ \qquad \text{if } Steps_{\mathcal{T}}(s) \cup Steps_{\mathcal{J}}(s) \neq \emptyset, \\ \{\delta_s\} \text{ else.} \end{array} \right.$$

The possible steps in the semantics are thus all possible transitions induced by jumps or timed transitions. The self-loops introduced using Dirac distributions are necessary to guarantee that each state has at least one successor measure.

It remains to show that the semantics is well-defined. Because of this, for $s \in S$ it must be the case that $Steps(s)$ is an element of $\Delta(\Sigma)$ and that $Steps$ is a measurable function. If this holds, then $Sem(\mathcal{H})$ is indeed an NMP.

LEMMA 3. *Let $Sem(\mathcal{H}) = (S, \Sigma, Init, Steps, UnSafe)$ be a tuple that is the semantics of a stochastic hybrid automaton $\mathcal{H}$. Then, $Steps : (S, \Sigma) \rightarrow (\Delta(\Sigma), H(\Delta(\Sigma)))$ is a measurable function mapping states to elements of $\Delta(\Sigma)$. In turn, $Sem(\mathcal{H})$ is an NMP. In case $\mathcal{H}$ is purely probabilistic, $Sem(\mathcal{H})$ is a PA.*

4. OVER-APPROXIMATING STOCHASTIC HYBRID AUTOMATA

Over-approximation of the semantics of a *probabilistic* hybrid automaton $\mathcal{H}$ by a finite PA $\mathcal{M}$ with an abstract state-space, written $\mathcal{M} \in Abs_f(\mathcal{H})$, implies that the reachability probability of unsafe states in $\mathcal{M}$ is no lower than in $\mathcal{H}$ [22, 23].

Now we describe how we can over-approximate a *stochastic* hybrid automaton by a *probabilistic* hybrid automaton. At first, we describe how to abstract a single stochastic guarded command into a probabilistic command.

DEFINITION 4. *Consider a stochastic guarded command c defined by condition $\rightarrow M$. Fix $p_i \in [0, 1]$ such that $\sum_{i=1}^{n} p_i = 1$. Let $\hat{g}_1, \ldots, \hat{g}_n : S \rightarrow \Sigma$ be functions such that $M(s)(\hat{g}_i(s)) = p_i$, and $M(s)\left(\bigcup_{i=1}^{n} \hat{g}_i(s)\right) = 1$, for all $s \in S$. Further, we require the sets $\hat{g}_1(s), \ldots, \hat{g}_n(s)$ to be*

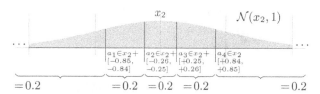

pairwise disjoint. Let $g_1, \ldots, g_n : S \rightarrow 2^S$ be functions satisfying $\hat{g}_i(s) \subseteq g_i(s)$, for all $s \in S$. We define the probabilistic guarded command $Abs_{\mathsf{c}}(\mathsf{c}, g_1, \ldots, g_n, p_1, \ldots, p_n)$ as

$$condition \rightarrow p_1 : g_1 + \ldots + p_n : g_n,$$

and call g_i abstraction functions.

By abstracting a command this way, we may introduce uncountable additional non-determinism. Overlapping sets $g_i(s), g_j(s)$, $i \neq j$ are allowed. This feature can be used for instance, if the exact $\hat{g}_i(s), \hat{g}_j(s)$ corresponding to probabilities p_i, p_j cannot be computed. As already stated in Section 2, this is no drawback of our method. In the final abstraction that we compute, the non-determinism will be over-approximated by a finite number of transitions. Notice that the abstraction of a single command is done symbolically in the high-level description of the probabilistic hybrid automaton instead of the low-level model, as in grid-based methods.

EXAMPLE 5. *Consider the stochastic guarded command from Example 3. Let $p_1 := \ldots := p_5 := 0.2$ and consider $a_1 \in x_2 + [-0.85, -0.84]$, $a_2 \in x_2 + [-0.26, -0.25]$, $a_3 \in x_2 + [0.25, 0.26]$, $a_4 \in x_2 + [0.84, 0.85]$. We define*

$$I_1 := \{m_2\} \times (-\infty, a_1] \times \bigtimes_{i=2}^{n} \{x_i\},$$

$$I_2 := \{m_2\} \times (a_1, a_2] \times \bigtimes_{i=2}^{n} \{x_i\},$$

$$\ldots$$

Assume that by a precomputation (which will later on be described in Subsection 4.1) we known that for states $s = (m_1, x_1, \ldots, x_n)$ it is

$$M(s)(I_1) = M(s)(I_2) = M(s)(I_3) = \ldots = p_i = 0.2,$$

as illustrated in Figure 2. For each point s' of the support of $M(s)$, we have at least one I_i which contains s'. We also know that $M(s)(\{m_2\} \times (-\infty, \infty) \times \bigtimes_{i=2}^{n} \{x_i\}) = 1$. Thus, we can define

$$\hat{g}_1(s) := \{m_2\} \times (-\infty, a_1] \times \bigtimes_{i=2}^{n} \{x_i\}$$

$$\subseteq g_1(s) := \{m_2\} \times (-\infty, x_2 - 0.84] \times \bigtimes_{i=2}^{n} \{x_i\},$$

$$\hat{g}_2(s) := \{m_2\} \times (a_1, a_2] \times \bigtimes_{i=2}^{n} \{x_i\}$$

$$\subseteq g_2(s) := \{m_2\} \times [x_2 - 0.85, x_2 - 0.25] \times \bigtimes_{i=2}^{n} \{x_i\},$$

$$\ldots$$

Because of this, the probabilistic guarded command of Example 2 is an abstraction of the stochastic guarded command of Example 3.

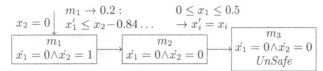

Figure 3: Probabilistic hybrid automaton abstraction of the stochastic hybrid automaton of Figure 1.

In the abstraction of a complete stochastic hybrid automaton, all stochastic guarded commands are abstracted by probabilistic guarded commands.

DEFINITION 5. *Let $\mathcal{H} = (\mathtt{M}, \boldsymbol{x}, Init, Flow, \mathcal{C}, UnSafe)$ be a stochastic hybrid automaton, and consider a family of abstraction functions $F = ((g_{c,1}, \ldots, g_{c,n}), (p_{c,1}, \ldots, p_{c,n}))_{c \in \mathcal{C}_c}$ with corresponding probabilities. Then we define the probabilistic hybrid automaton abstraction of $\mathcal{H}$ as*

$$Abs_c(\mathcal{H}, F) := (\mathtt{M}, \boldsymbol{x}, Init, Flow, Abs_c(\mathcal{C}), UnSafe)$$

where $Abs_c(\mathcal{C}) := \mathcal{C}_f \cup \{Abs_c(\mathtt{c}, g_{c,1}, \ldots, g_{c,n}, p_{c,1}, \ldots, p_{c,n}) \mid \mathtt{c} \in \mathcal{C}_c\}$.

We state the correctness of the over-approximation.

LEMMA 4. *Consider a stochastic hybrid automaton $\mathcal{H}$ and a family F of abstraction functions with corresponding probabilities. Then $Sem(\mathcal{H}) \preceq_{cf} Sem(Abs_c(\mathcal{H}, F))$.*

EXAMPLE 6. *In Figure 3 we give a possible over-approximation of the automaton of Figure 1. We use the abstraction from Example 2 (denoted $\mathtt{c}'$) of the stochastic guarded command of Example 3 (denoted by $\mathtt{c}$) (see Example 5) for the only stochastic guarded command in this model. In the abstraction, the reachability probability is higher than it was originally. If in m_1 we wait until $x_2 = 0.2$, for two branches of $\mathtt{c}'$ we may enter m_3, thus the reachability probability is $2 \cdot 0.2 = 0.4$. By splitting M into more equidistant parts, we could decrease this probability.*

4.1 Obtaining Abstraction Functions

For the abstraction to be applicable in practice, it is crucial to compute the family of abstraction functions. As there exist quite diverse forms of random variables, we cannot give an algorithm to handle all cases. Instead, we sketch how to obtain over-approximation functions for certain classes of random variables.

At first, consider a probability measure $\mu : \mathcal{B}(\mathbb{R}) \to [0, 1]$ given by a density function $f(x)$, for instance the normal distribution. Using numerical methods, we can then compute bounds for a_i such that $\mu((-\infty, a_1]) = p_1, \mu((a_1, a_2]) = p_2, \ldots, \mu((a_{n-1}, \infty)) = p_n$, for some fixed $p_1, \ldots, p_n$. Following Example 5, for $\mathcal{N}(0, 1)$ and $n = 5, p_i = 0.2$, we could obtain $a_1 \in [-0.85, -0.84], a_2 \in [-0.26, -0.25], \ldots$. We transform the random variable to get state-dependent intervals. For the example, we use that $\mathcal{N}(x, y) = \frac{\mathcal{N}(0,1) - x}{y}$. Thus, we can transform corresponding interval endpoints b_i to $b_i(x, y) = b_i(0, 1) \cdot y + x$. When setting $x = x_2, y = 1$, we obtain the same intervals as given in Example 5.

If the cumulative distribution function $F(x)$ of a random variable is known and we can compute a closed-form of F^{-1}, we can use a method similar to the inverse transform method. Consider the exponential distribution with

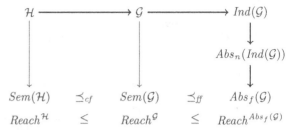

Figure 4: Scheme of abstraction. $\mathcal{G}$ denotes the abstraction $Abs_c(\mathcal{H}, F)$ and $Abs_n(.)$ denotes abstraction of a non-probabilistic hybrid automata, and $\preceq_{ff}$ denotes the simulation relation on PAs [21].

state-dependent λ. We have that if $F_\lambda(a_i) = p_i$ then $a_i = -\ln(1 - p_i)\frac{1}{\lambda}$. We can then obtain adjoint intervals which have a certain probability by precomputing $[b_i, b_i'] \ni -\ln(1 - p_i)$ and thus we specify command branches $p_i : \frac{b_{i-1}}{\lambda} \le x \le \frac{b_i'}{\lambda}$.

For probability measures in two variables, we consider $f(\cdot, (-\infty, \infty))$ first, and then split each $f([a_i, a_{i+1}], \cdot)$ again. This technique extends to any finite number k of variables. If we split each of them into a number of n parts, the support of the abstracting distribution has a size of n^k. Thus, the worst-case complexity of this method is rather bad. However, the case that only one or few variables change appears to be the practically relevant case for us. It occurs in settings where the environment can be observed only with limited accuracy, as the ones discussed in Section 5.

4.2 Finite Abstractions

In the previous sections, we have seen that a stochastic hybrid automaton can be abstracted by a probabilistic hybrid automaton with uncountably many states and transitions. To effectively obtain probability bounds, we abstract this automaton by a finite state probabilistic automaton, which can then be submitted to a probabilistic model checker for further analysis. We can do so by harvesting previous work [23], which proceeds via a non-probabilistic version $Ind(\mathcal{G})$ of the original probabilistic hybrid automaton $\mathcal{G}$ to arrive at a finite abstraction $Abs_f(\mathcal{G})$. A slight adaptation is needed since we thus far did not consider non-determinism within one command, which we now require in order to over-approximate stochastic guarded commands. Nevertheless, the correctness proofs stay unchanged when allowing it. Due to space limitations, we cannot give a complete description of the previous work here.

An overview of the entire approach is depicted in Figure 4. The computation of maximal reachability probabilities in the resulting finite-state PA is done via well-established numerical recipes, and is the capstone ingredient in this effective computation of safe upper bounds for reachability properties for this general class of stochastic hybrid automata.

5. EXPERIMENTS

We experiment with the approach outlined thus far using the tool PROHVER (probabilistic hybrid automata verifier) [23] on a selection of stochastic hybrid automata case studies. In each case, we first abstract stochastic guarded commands (so far manually) to probabilistic guarded commands, which PROHVER can handle. Thus, we obtain a

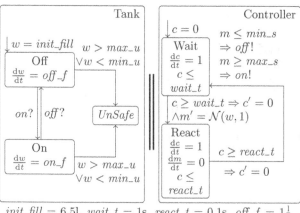

$init\_fill = 6.5l$, $wait\_t = 1s$, $react\_t = 0.1s$, $off\_f = 1\frac{1}{s}$, $on\_f = -2\frac{1}{s}$, $max\_s = 8l$, $min\_s = 5l$, $max\_u = 12l$, $min\_u = 1l$

Figure 5: Water level control with perturbed measurements, modelling measurement deviation by normal distribution $\mathcal{N}(w, 1)$.

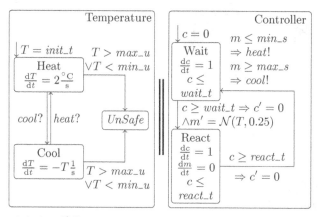

$init\_t = 8°C$, $wait\_t = 1s$, $react\_t = 0.1s$, $max\_s = 9°C$, $min\_s = 6°C$, $max\_u = 12°C$, $min\_u = 3°C$

Figure 6: Temperature control with perturbed measurements, a variant of the model from Figure 5 exhibiting more complex continuous dynamics.

probabilistic hybrid automaton which over-approximates the original stochastic hybrid automaton. Then, our tool uses a modified version of PHAVER [13] to obtain the transition relation of a finite-state abstraction of a non-probabilistic projection of the hybrid automaton. PROHVER then reintroduces the probabilities to this abstraction and constructs a corresponding finite-state Markov decision process. The maximal reachability probabilities we can obtain herein over-approximate the ones which can be obtained in the semantics of the original stochastic hybrid automaton.

To show the applicability of our approach, we applied PROHVER on three case studies, which are small but diverse in the nature of their behaviour. In the examples considered, we focus on reachability probabilities with upper time bounds (obtained by using an additional clock to measure time), because these correspond to very natural verification problems for the settings considered. Notably, our method is not restricted to time-bounded reachability. Actually, time-unbounded problems are simpler (because no additional clock is needed). In the examples considered, time-unbounded reachability probabilities would always be 1. Experiments were run on a Pentium 4 with 2.67 GHz and 4 GB RAM. Models and tools can be found on http://depend.cs.uni-saarland.de/tools/prohver/.

5.1 Water Level Control

We consider a model of a water level control system (extended from the one of Alur et al. [4] and our previous paper [23]). In particular, we use this case study to demonstrate the influence which different abstractions of the same continuous stochastic command have. A water tank is filled by a constant stream of water, and is connected to a pump which is used to avoid overflow of the tank. A control system operates the pump in order to keep the water level within predefined bounds. The controller is connected to a sensor measuring the level of water in the tank. A sketch of the model is given in Figure 5. The state "Tank" models the tank and the pump, and w is the water level. Initially, the tank contains a given amount of water. Whenever the pump is turned off in state "Off", the tank fills with a constant rate

due to the inflow. Conversely, more water is pumped out than flows in when the pump is on.

The controller is modelled by automaton "Controller". In state "Wait", the controller waits for a certain amount of time. Upon the transition to "React", the controller measures the water level. To model the uncertainties in measurement, we set the variable m to a normal distribution with expected value w (the actual water level) and standard deviation 1. According to the measurement obtained, the controller switches the pump off or on.

We are interested in the probability that within a given time bound, the water level leaves the legal interval. In Table 1, we give upper bounds for this probability for different time bounds as well as the number of states in the abstraction computed by PHAVER and the time needed for the analysis. For the stochastic guarded command simulating the measurement, we consider different abstractions by probabilistic guarded commands of different precision, for which we give the abstraction functions in the table caption. When we refine the abstraction A to a more precise B, the probability bound decreases. If we introduce additional non-determinism as in abstraction C, probabilities increase again. If we refine B again into D, we obtain even lower probability bounds. The price to be paid for increasing precision, however, is in the number of abstract states computed by PHAVER as well as a corresponding increase in the time needed to compute the abstraction.

Manual analysis shows that in this case study, the over-approximation of the probabilities only results from the abstraction of the stochastic guarded command into a probabilistic guarded command and is not increased further by the state-space abstraction.

5.2 Temperature Control

We consider a temperature control system extended from a previous case study [23], originally studied by Alur et al. [5]. In Figure 6 we depict the system structure. We ask whether using an air conditioning control system we are able to keep the temperature of a room within a certain range. In contrast to the water level case, the model features dynamics governed by linear rather than piecewise constant ODE, and

time bound	Abstraction A			Abstraction B			Abstraction C			Abstraction D		
	prob.	build (s)	states	prob.	build (s)	states	prob.	build (s)	states	prob.	build (s)	states
20s	0.1987	3	999	0.0982	3	1306	0.1359	3	1306	0.0465	5	1920
30s	0.2830	6	2232	0.1433	8	2935	0.1870	8	2935	0.0693	15	4341
40s	0.3580	16	3951	0.1860	18	5212	0.2547	18	5212	0.0916	47	7734
50s	0.4250	34	6156	0.2264	42	8137	0.3024	43	8137	0.1134	108	12099
60s	0.4848	67	8847	0.2647	86	11710	0.3577	85	11710	0.1347	219	17436

Table 1: Water level control results. We round probabilities to four decimal places. Abstractions used are $A = w + \{[-2, 2], (-\infty, 1.9] \cup [1.9, \infty)\}$, $B = w + \{[-2, 2], (-\infty, 1.9], [1.9, \infty)\}$, $C = w + \{[-2.7, 2.7], (-\infty, 1.2), [1.2, \infty)\}$, $D = w + \{[-1.5, 1.5], [-1.5, -2], [1.5, 2], (-\infty, 1.9), [1.9, \infty)\}$.

instead of varying the splitting of the normal distribution, we vary the refine interval used by PHAVer to analyse such systems. Smaller intervals lead to more precise abstractions.

In Table 2, we give probability bounds and performance statistics. We used a refine interval on the variable T which models the temperature. The interval lengths are given in the table. For all instances there is an interval length small enough to obtain a probability bound that is the best possible using the given abstraction of the normal distribution. Smaller intervals were of no use in this case. The drastic discontinuities in probability bounds obtained are a consequence abstraction by PHAVer.

5.3 Moving-block Train Control

As a more complex example of a hybrid system implementing a safety-critical control policy, we present a model of headway control in the railway domain (Figure 7). A more extensive description of the setting plus a closely related case study containing a sampling-related bug not present in the current model appeared in a different publication [17]. In contrast to fully automated transport, which is in general simpler to analyse (as the system is completely under control of the embedded systems) our sample system implements safe-guarding technology that leaves trains under full human control provided safety is not at risk. It is thus an open system, giving rise to the aforementioned analysis problems.

Our model implements safe interlocking of railway track segments by means of a "moving block" principle of operation. While conventional interlocking schemes in the railway domain lock a number of static track segments in full, the moving block principle enhances traffic density by reserving a "moving block" ahead of the train which moves smoothly with the train. This block is large enough to guarantee safety even in cases requiring emergency stops, i.e. has a dynamically changing block-length depending on current speed and braking capabilities. There are two variants of this principle, namely train separation in relative braking distance, where the spacing of two successive trains depends on the current speeds of both trains, and train separation in absolute braking distance, where the distance between two trains equals the braking distance of the second train plus an additional safety distance (here given as $sd = 400$m). We use the second variant, as also employed in the European Train Control System (ETCS) Level 3.

Our simplified model consists of a leader train, a follower train, and a moving-block control regularly measuring the leader train position and communicating a related *movement authority* to the follower. The leader train is freely controlled by its operator within the physical limits of the

train, while the follower train may be forced to controlled braking if coming close to the leader. The control principle is as follows:

1. 8 seconds after communicating the last movement authority, the moving-block control takes a fresh measurement m of the leader train position s_l. This measurement may be noisy.

2. Afterwards, a fresh movement authority derived from this measurement is sent to the follower. The movement authority is the measured position m minus the length l of the leader train and further reduced by the safety distance sd. Due to an unreliable communication medium, this value may reach the follower (in which case its movement authority $auth$ is updated to $m - l - sd$) or not. In the latter case, which occurs with probability 0.1, the follower's movement authority stays as is.

3. Based on the movement authority, the follower continuously checks the deceleration required to stop exactly at the movement authority. Due to PHAVer being confined to linear arithmetic, this deceleration is conservatively approximated as $a_{req} = \frac{v \cdot v_{max}}{2(s - auth)}$, where v is the actual speed, v_{max} the (constant) top speed, and s the current position of the follower train, rather than the physically more adequate, yet nonlinear, $a_{req} = \frac{v^2}{2(s - auth)}$ of the original model [17].

4. The follower applies automatic braking whenever the value of a_{req} falls below a certain threshold b_{on}. In this case, the follower's brake controller applies maximum deceleration a_{min}, leading to a stop before the movement authority as $a_{min} < b_{on}$. Automatic braking ends as soon as the necessary deceleration a_{req} exceeds a switch-off threshold $b_{off} > b_{on}$. The thresholds b_{on} and b_{off} are separate to prevent the automatic braking system from repeatedly engaging and disengaging in intervals of approximately 8 seconds when the leading train is moving.

We consider the probability to reach the state "Crash" in which the follower train has collided with the leader train. In Table 3, we give probability bounds and performance results. We modelled the measurement error using a normal distribution with expected value s_l, i.e. the current position of the leader train. In the table, we considered different standard deviations of the measurement. The abstraction used for each of them can be obtained using structurally equal Markov decision processes, only with different probabilities. Thus, we only needed to compute the abstraction

time bound	interval length ∞			interval length 2			interval length 1			interval length 0.5		
	prob.	build (s)	states	prob.	build (s)	states	prob.	build (s)	states	prob.	build (s)	states
2s	1	0.03	7	0	0.17	16	0	0.21	21	0	0.30	31
4s	1	0.05	23	1	1.26	269	0.284643	1.61	316	0.284643	2.97	546
6s	1	0.07	39	1	5.79	1518	0.360221	8.66	2233	0.360221	17.39	3797
8s	1	0.10	55	1	19.27	4655	1	35.62	8261	0.488265	81.39	16051
10s	1	0.12	71	1	53.25	10442	1	119.33	20578	0.590683	507.12	44233

Table 2: Temperature control results. To abstract $\mathcal{N}(T, 0.25)$, we used $T + \{[-0.25, 0.25], (-\infty, -0.25], [0.25, \infty)\}$.

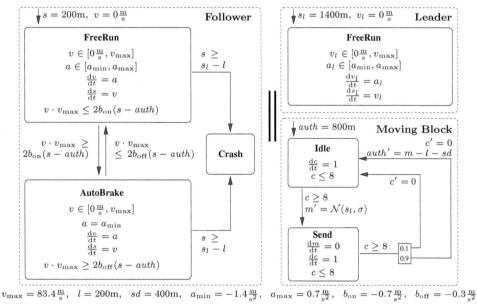

$v_{\max} = 83.4 \frac{m}{s}, \quad l = 200\text{m}, \quad sd = 400\text{m}, \quad a_{\min} = -1.4 \frac{m}{s^2}, \quad a_{\max} = 0.7 \frac{m}{s^2}, \quad b_{\text{on}} = -0.7 \frac{m}{s^2}, \quad b_{\text{off}} = -0.3 \frac{m}{s^2}$

Figure 7: Moving-block train distance control with perturbed measurement of leader train position (using normal distribution $\mathcal{N}(s_l, \sigma)$ centred around actual value, with standard deviation σ) and unreliable communication of resultant movement authorities (failure probability 0.1). "Crash" represents collision of trains.

once for all deviations, and just had to change the transition probabilities before obtaining probability bounds from the abstraction. It was sufficient to split the normal distribution into two parts. Depending on where we set the split-point, we obtained probability bounds of different quality. Although this hybrid automaton is not piecewise constant, such that PHAVER needs to over-approximate the set of reachable states, we are still able to obtain useful probability bounds when using an adequate abstraction, without refine intervals.

The graph in Figure 8 reveals the expected positive correlation between measurement error and risk, but also the effectiveness of the fault-tolerance mechanism handling communication loss. We see that crashes due to communication losses are effectively avoided, rooted in the principle of maintaining the last received movement authority whenever no fresh authority is at hand. In fact, risk correlates negatively with the likelihood of communication loss. The function correlating risk to measurement error and probability of communication loss has been computed by the tool PARAM [15].

6. CONCLUSION

In this paper, we have defined a notion of stochastic hybrid systems which supports both non-determinism and continuous probability distributions in discrete jumps. We have discussed well-definedness of the semantics and have developed

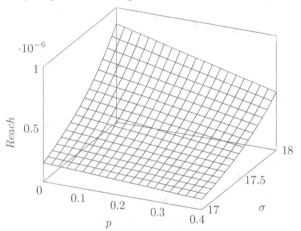

Figure 8: Bounds for probability of crash as a function of probability of movement authority loss p and standard deviation σ of distance measurement. A time bound of 100s and Abstraction A was used.

means to safely over-approximating reachability probabilities for such systems. As the underlying state-space abstraction which we exploit for computing probabilities is the one computed with the help of model checkers for non-stochastic hybrid systems, improvements in efficiency of such tools directly carry over to the technique we describe. The applica-

time bound	Abstraction A				Abstraction B			
	probability ($\sigma = 10, 15, 20$)			build (s)	states	probability ($\sigma = 10, 15, 20$)			build (s)	states
60s	7.110E-19	6.215E-09	2.141E-05	65	571	1.806E-06	2.700E-03	3.847E-02	62	571
80s	1.016E-18	8.879E-09	3.058E-05	201	1440	2.580E-06	3.855E-03	5.450E-02	183	1440
100s	1.219E-18	1.066E-08	3.669E-05	470	2398	3.096E-06	4.624E-03	6.504E-02	472	2392
120s	1.524E-18	1.332E-08	4.587E-05	1260	4536	3.870E-06	5.777E-03	8.063E-02	1210	4524
140s	1.727E-18	1.509E-08	5.198E-05	2541	6568	4.386E-06	6.544E-03	9.088E-02	2524	6550
160s	2.031E-18	1.776E-08	6.116E-05	5764	10701	5.160E-06	7.695E-03	1.060E-01	5700	10665

Table 3: **Train control results. For abstraction A we use a division of the normal distribution into $s_l + \{(-\infty, 91], [89, \infty)\}$. For B, we split the distribution into $s_l + \{(-\infty, 51], [49, \infty)\}$. We give probabilities for different values σ of the standard deviation of the measurement.**

bility of our approach has been demonstrated on three case studies, tackled using a prototypical implementation.

As future work, we want to extend our techniques to reason about the loss of precision introduced by the abstraction and consider the question how to split continuous distributions in an optimal way. We have assumed a finite number of modes and commands. Using additional abstractions [18] for the discrete part, our technique can be extended to models which have a very large or infinite number of modes.

Acknowledgements. This work was supported by the SFB/TR 14 AVACS, by ANPCyT PICT 02272, by SeCyT-UNC 2010-2011, by the VKR Centre of Excellence project MT-LAB, the DAAD-MinCyT project QTDDS, FP7-ICT MoVeS and FP7-ICT Quasimodo.

We would like to thank Pedro Sánchez Terraf from the National University of Córdoba and Stefan Ratschan from the Academy of Sciences of the Czech Republic for fruitful discussions on the measurability problems.

7. REFERENCES

[1] A. Abate, J. Katoen, J. Lygeros, and M. Prandini. Approximate model checking of stochastic hybrid systems. *European Journal of Control*, 2010.

[2] A. Abate, M. Prandini, J. Lygeros, and S. Sastry. Probabilistic reachability and safety for controlled discrete time stochastic hybrid systems. *Automatica*, 44(11):2724–2734, 2008.

[3] E. Altman and V. Gaitsgory. Asymptotic optimization of a nonlinear hybrid system governed by a Markov decision process. *SIAM Journal of Control and Optimization*, 35(6):2070–2085, 1997.

[4] R. Alur, C. Courcoubetis, N. Halbwachs, T. A. Henzinger, P.-H. Ho, X. Nicollin, A. Olivero, J. Sifakis, and S. Yovine. The algorithmic analysis of hybrid systems. *TCS*, 138:3–34, 1995.

[5] R. Alur, T. Dang, and F. Ivancic. Predicate abstraction for reachability analysis of hybrid systems. *ACM Transactions on Embedded Computing Systems*, 5(1):152–199, 2006.

[6] Y. Baisalov and B. Poizat. Paires de structures o-minimales. *J. Symb. Log.*, 63(2):570–578, 1998.

[7] A. Berarducci and M. Otero. An additive measure in o-minimal expansions of fields. *The Quarterly Journal of Mathematics*, 55(4):411–419, 2004.

[8] H. Blom and J. Lygeros. *Stochastic Hybrid Systems: Theory and Safety Critical Applications*, volume 337 of *LNCIS*. Springer, 2006.

[9] M. L. Bujorianu. Extended stochastic hybrid systems and their reachability problem. In *HSCC*, pages 234–249, 2004.

[10] P. R. D'Argenio, P. S. Terraf, and N. Wolovick. Bisimulations for nondeterministic labeled Markov processes. *Math. Struct. in Comp. Science*, 2010. Under consideration for publication.

[11] P. R. D'Argenio, N. Wolovick, P. S. Terraf, and P. Celayes. Nondeterministic labeled Markov processes: Bisimulations and logical characterization. In *QEST*, pages 11–20. IEEE Computer Society, 2009.

[12] M. H. A. Davis. Piecewise-deterministic Markov processes: A general class of non-diffusion stochastic models. *Journal of the Royal Statistical Society*, 46(3):353–388, 1984.

[13] G. Frehse. PHAVer: Algorithmic verification of hybrid systems past HyTech. In *HSCC*, pages 258–273, 2005.

[14] M. Giry. A categorical approach to probability theory. In *Categorical Aspects of Topology and Analysis*, pages 68–85. Springer, 1982.

[15] E. M. Hahn, H. Hermanns, B. Wachter, and L. Zhang. PARAM: A model checker for parametric Markov models. In *CAV*, pages 660–664, 2010.

[16] T. A. Henzinger. The theory of hybrid automata. In *LICS*, pages 278–292, 1996.

[17] C. Herde, A. Eggers, M. Fränzle, and T. Teige. Analysis of hybrid systems using HySAT. In *ICONS*, pages 196–201. IEEE Computer Society, 2008.

[18] M. Kwiatkowska, G. Norman, and D. Parker. A framework for verification of software with time and probabilities. In *FORMATS*, volume 6246 of *LNCS*, pages 25–45. Springer, 2010.

[19] S. Prajna, A. Jadbabaie, and G. J. Pappas. A framework for worst-case and stochastic safety verification using barrier certificates. *IEEE TAC*, 52(8):1415–1429, 2007.

[20] M. Prandini and J. Hu. A stochastic approximation method for reachability computations. In Blom and Lygeros [8], pages 107–139.

[21] R. Segala and N. Lynch. Probabilistic simulations for probabilistic processes. *NJC*, 2(2):250–273, 1995.

[22] J. Sproston. Decidable model checking of probabilistic hybrid automata. In *FTRTFT*, pages 31–45, 2000.

[23] L. Zhang, Z. She, S. Ratschan, H. Hermanns, and E. M. Hahn. Safety verification for probabilistic hybrid systems. In *CAV*, pages 196–211, 2010.

Stochastic Non Sequitur Behavior Analysis of Fault Tolerant Hybrid Systems

Manuela L. Bujorianu and Marius C. Bujorianu
Centre for Interdisciplinary Computational and Dynamical Analysis, University of Manchester, Oxford Street,
Manchester, UK
(Manuela, Marius).Bujorianu@manchester.ac.uk

ABSTRACT

In this paper, we introduce a new stochastic analysis method for the uncontrollable behaviour of hybrid systems triggered by faults. Models like manoeuvre automata can be used for controllable transitions between stable modes. However, the normal behaviour can become easily unpredictable when a fault occurs, and a human or an automated supervisor needs to take very quickly the right actions to drive the system back into a controllable state. The system behaviour while transiting two controllable states is called non sequitur. Using stochastic analysis we investigate how to extract information about non sequitur that can be used in stochastic control.

Categories and Subject Descriptors

B.3.4 [**Reliability, Testing and Fault-Tolerance**]: Diagnostics

General Terms

Reliability

Keywords

non sequitur behavior, manoeuvre automata, Markov models, stochastic hybrid systems, fault tolerant control.

1. INTRODUCTION

Operating in physical environments that are often very hazardous, an open hybrid system can face situations of malfunctioning components. If, in the past, the control of malfunctioning was done by a human operator, now there is a strong emphasis on automatic control for fault tolerant systems (*fault tolerant control* [15], [13], [35] [30]). Hybrid systems have proved themselves to be a useful conceptual vehicle for fault tolerant control.

According to some authors, until now, the fault tolerant control for hybrid systems has not yet been intensively studied [36]. Such studies deserve sustain investigations for both

academic and practical merits. Among these, the safety is the driving motivation of this research.

The stochastic hybrid systems offer more primitives for fault tolerant control because of the versatile randomization techniques. The general model of stochastic hybrid systems developed in [2] offers an intricate concept for fault tolerant control in the form of so-called spontaneous transitions. These transitions are not guarded and they are triggered according to a Poisson like distribution. This statistical assumption is motivated by the interpretation that these transitions are actually triggered by the apparition of faults. Many system properties have been developed supposing the existence of such spontaneous transitions. Therefore these transitions can be used for problems related to fault tolerant control.

Many authors distinguish between continuous and discrete faults. This separation is useful especially for the design of reconfigurable control. However, the focus, in this paper, is on the analysis of system behaviour at the occurrence of faults. The results of such analysis can then be used for improving the control design process. Simulations based on Markov Chain Monte Carlo and other sampling techniques [3] can offer a vital information on how the random trajectories generated by faults do cluster or how they behave on the field surfaces of equiprobability to return in a controllable mode or not.

The principle, we adopt in this work, is that analytical models and facts should be investigated in the first place. The outcome of these investigations should be then applied to simulations and to the derivation of practical control algorithms.

In this paper, we focus on the analytical side of the stochastic analysis of hybrid fault tolerant systems. We apply standard concepts of mathematical physics, like probability density functions and probability current, to the random trajectories between controllable working regimes. We develop an analytic framework and investigate its probabilitic properties.

2. PROBLEM FORMULATION

The departing point in our study is based on the interpretation of hybrid systems as manoeuvre automata [17]. These models have been fruitfully applied to aerospace systems like aircraft and helicopters. In this interpretation, a discrete transition is actually an abstraction of certain activities like decision making processes, or changes of operating regime. For a manoeuvre automaton, it is supposed that each discrete transition has associated a continuous be-

HSCC'11, April 12–14, 2011, Chicago, Illinois, USA.
Copyright 2011 ACM 978-1-4503-0629-4/11/04 ...$10.00.

haviour that realizes a smooth change between two continuous modes. Applying this interpretation to the general model of stochastic hybrid systems, there is an issue about what sort of continuous behaviour corresponds to a manoeuvre. The major difference is that a manoeuvre is a continuous behaviour corresponding to a controllable and smooth motion. In contrast, the continuous behaviour triggered by the apparition of a fault is rather uncontrollable and random. Another major difference is that the system evolves not into a single mode but into the modes of a new hybrid system. For example, let consider an aircraft with three engines, but it is usually designed to operate also with two engines. Consider the scenario when a fault shuts an engine down and possibly makes some structural damage to the aircraft. In most cases, the aircraft will behave randomly until the pilot will stabilize the system that will evolve according to the hybrid system model corresponding to the new dynamics. This sort of behaviour corresponds to a discrete spontaneous transition in the general model of stochastic hybrid systems.

In order to study the random behaviour provoked by faults, we need a new mathematical model for defining and investigating specific problems. The spontaneous discrete transitions can be defined as mode changes, but in a continuous interpretation one needs to define randomized continuous transitions between hybrid systems. This level of detail would be indeed suitable for studying problems of stability, controllability and design. For analysis, we need to abstract away some aspects of the hybrid behaviours. We take inspiration from an approach from physics called metastability. We will abstract the dynamics of a hybrid system by a so-called metastable set of states. In physics, a metastable state set is a macroscopic view of a dynamical system, which at microscopic level consists of switching between different continuous dynamics. Applying this interpretation to our problem, we define a configuration space. In this configuration space, the trajectories of a given hybrid system are modelled by state set called metastable in order to distinguish it from other sets. A point from the boundary of a metastable set could have one of the following interpretations: a starting state of a hybrid system, or a state of a hybrid system where a fault occurs. A dynamics in the configuration space consists of trajectories that can enter and exit a metastable set only through points of its boundary. The dynamics inside and outside of the metastable sets have the same nature, i.e. the global process is a switching diffusion process [18]. Switching diffusion processes represent a particular class of stochastic hybrid systems where the discrete transitions are governed by a Markov chain and the continuous evolutions by a set of stochastic differential equations..

The problem of interest is then the stochastic analysis of system behaviour between two such metastable state sets. The results of this analysis will provide information on how the systems tend to behave when certain types of faults occur and certain types of partial control are applied. Of course, the mathematical models that consider explicitly faults and control will be very complex, and their stochastic analysis would be extraordinarily difficult to set up and interpret. Consequently, some simplifications will be initially considered. The major assumption is that the effects of faults and the partial control are incorporated in the mathematical model of the dynamics. Again, for the sake of simplicity, we will consider models described by diffusion processes which have been widely used in science and engineering.

Formally, we consider a stochastic hybrid system modelled by a switching diffusion process (x_t) with the state space $\mathbf{X}$ (subspace of an appropriate Euclidean space). This stochastic hybrid system has in modes the continuous evolution described by some diffusion processes and the "discrete transitions" are governed by some rates without changing the continuous state. This stochastic hybrid realization can be thought of as a stochastic process whose trajectories are obtained by the concatenation of the component diffusions in a continuous way.

The two metastable sets will be denoted by $\mathbf{A}$ and $\mathbf{B}$. Some measurability assumptions will be also necessary. The problem proposed is to study the trajectories that leave the boundary of $\mathbf{A}$ and go to $\mathbf{B}$. These trajectories might have some undesired properties, therefore the study of their statistical/probabilistic parameters will be of great interest. First problem we need to solve is to identify the mathematical objects that illustrate remarkable properties of such trajectories. First idea is to define their probability densities and their time evolution. A very useful concept will be the so-called state constrained reachability probabilities developed in [7]. Then, we would like to understand how these trajectories are "spred" between $\mathbf{A}$ and $\mathbf{B}$. This can be achieved by using some physical probabilistic concepts related to the Fokker Planck Kolmogorov equations. Such problems will be developed in the following sections.

We call non-sequitur outline the mathematical model that describes the fault tolerant hybrid system abstracted away using the principles described before. In Latin, Non-sequitur is a slogan meaning either "it does not follow" or an abrupt non-smooth change. In logic, the phrase is used to denoted a fault reasoning. Such non-sequitur outline is depicted in Fig. 2. The trajectory in the configuration space will be described at macroscopic level as the trajectory of a stochastic process that visits infinitely often the two metastable configurations. However, *non-sequitur trajectories* represent the truncations of the global trajectory between the two metastable configurations. The non-sequitur trajectories are drawn in bold style.

Revisiting the aircraft example, the long trajectory represents the all aircraft flights over a long period of time (possibly the entire life time). The transition between the metastable configurations could be in large number (possible infinite), but they still represent rare events. This statistical hypotheses will be formalized in the mathematical model.

In the next section, we define the underlying model of stochastic hybrid systems used in this paper, namely switching diffusion processes. In Section 4, we briefly present the state constrained reachability problem and its main properties. Then, in Section 5, we define formally *non-sequitur trajectories*. Section 6 will provide the main results of this paper that represent characterizations of the non-sequitur trajectories. The paper ends with the final remarks.

3. SWITCHING DIFFUSION PROCESSES

Switching Diffusion Processes (SDP) were introduced by Ghosh et. al. in 1991. They are a class of non-linear, continuous-time, stochastic hybrid processes that have been used to model a large number of applications such as fault tolerant control systems, multiple target tracking, flexible manufacturing systems, etc. SDP involve a hybrid state space, with both continuous and discrete states. The continuous state evolves according to a stochastic differential equation (SDE), while the discrete state is a controlled Markov chain. Both the dynamics of the SDE and the transition matrix of the Markov chain depend on the hybrid state. The continuous hybrid state evolves without jumps, i.e. the evolution of the continuous state can be assumed to be a continuous function of time.

In the following, we introduce formally SDP following [18, 19, 20].

A *switching diffusion process* is a collection

$$H = (Q, X, u, Init, \sigma, \lambda_{ij})$$

where:

- $Q = \{1, 2, ..., N\}$ is a finite set of discrete variables, $N \in \mathbb{N}$;

- $X = \mathbb{R}^n$ is the continuous state space;

- $u : Q \times X \to \mathbb{R}^n$ is a vector field;

- $Init : \mathcal{B}(Q \times X) \to [0, 1]$ is an initial probability measure on $(Q \times X, \mathcal{B}(Q \times X))$;

- $\sigma : Q \times X \to \mathbb{R}^{n \times n}$ is a state dependent real-valued matrix;

- $\lambda_{ij} : X \to \mathbb{R}$, $i, j \in Q$ are a set of z-dependent transition rates, with $\lambda_{ij}(.) \geq 0$ if $i \neq j$ and

$$\sum_{j \in Q} \lambda_{ij}(z) = 0$$

for all $i \in Q$, $z \in X$.

We denote with $x = (q, z)$ the hybrid state of an SDP. To ensure the SDP model is well-defined [18, 19, 20] introduce the following assumption.

ASSUMPTION 1. *The functions $u(i, z)$, $\sigma_{kj}(i, z)$ and $\lambda_{kj}(z)$ are bounded and Lipschitz continuous in z.*

Assumption 1 ensures that for any $i \in Q$, the solution to the SDE where W_t is an n-dimensional standard Wiener process, exists and is unique.

For $i, j \in Q$ and $z \in \mathbb{R}^n$ let $\Delta(i, j, z)$ be consecutive, with respect to the lexicographic ordering on $Q \times Q$, left closed, right open intervals of the real line, each having length $\lambda_{ij}(z)$ (for details see [18, 19, 20]). Now define a function $h : \mathbb{R}^n \times Q \times \mathbb{R} \to \mathbb{R}$ by setting

$$h(z, i, y) = \begin{cases} j - i & \text{if } y \in \Delta(i, j, z) \\ 0 & \text{otherwise.} \end{cases}$$

The SDP executions can be defined using the function h.

A stochastic process $x_t = (q(t), z(t))$ is called an SDP execution if it is the solution of the following SDE and stochastic integral:

$$dz(t) = u(q(t), z(t))dt + \sigma(q(t), z(t))dW_t,$$
$$dq(t) = \int_{\mathbb{R}} h(z(t), q(t^-), z)\varphi(dt, dz)$$

for $t \geq 0$ with

$$z(0) = z_0, \quad q(0) = q_0,$$

where

$x_0 = (q(0), z(0))$ is a random variable extracted according to the probability measure $Init$;

- W_t is a n-dimensional standard Wiener process;

- $\varphi(dt, dz)$ is an $\mathcal{M}(\mathbb{R}^+ \times \mathbb{R})$-valued Poisson random measure with intensity $dt \times m(dz)$, where m is the Lebesgue measure on $\mathbb{R}$ (see [22]);

- $\varphi(., .)$, W_t, and $(q(0), z(0))$ are independent.

It is known that an SDP $M = (x_t)$ is a strong Markov process, and its trajectories are continuous on $[0, \infty)$. The hybrid state space is

$$\mathbf{X} = Q \times X.$$

The underlying probability space is denoted by $(\Omega, \mathcal{F})$. $\{\mathcal{F}_t\}$ is the natural filtration of the process (the 'history' of the process). The *(Wiener) probability*

$$\mathbf{P}_x : (\Omega, \mathcal{F}) \to [0, 1]$$

is a probability measure such that $\mathbf{P}_x(x_t \in \mathbf{A})$ is $\mathcal{B}$-measurable in $x \in \mathbf{X}$, for each $t \in [0, \infty)$ and $\mathbf{A} \in \mathcal{B}$, and

$$\mathbf{P}_x(x_0 = x) = 1.$$

The meaning of these elements associated to (x_t) are standard for continuous-parameter Markov processes [10].

Let $\mathbf{B}(\mathbf{X})$ be the set of bounded measurable real valued function defined on the state space $\mathbf{X}$. The measurability is considered w.r.t. the Borel σ-algebra of $\mathbf{X}$. The Markov property of the process M (in particular, the Chapman-Kolmogorov equations) allows us to introduce the *semigroup of operators* associated to M, denoted by

$$\mathcal{P} := (P_t)_{t > 0}$$

Each P_t $(t > 0)$ maps $\mathbf{B}(\mathbf{X})$ into itself, and it is given as

$$P_t f(x) := \mathbf{E}_x f(x_t), \forall x \in \mathbf{X}, \quad (1)$$

where $\mathbf{E}_x$ is the expectation with respect to $\mathbf{P}_x$.

The infinitesimal generator $\mathcal{L}$ is the derivative of P_t at $t = 0$. Let $D(\mathcal{L}) \subset \mathbf{B}(\mathbf{X})$ be the set of functions f for which the following limit exists

$$\lim_{t \searrow 0} \frac{1}{t}(P_t f - f). \quad (2)$$

In case of existence, this limit is denoted by $\mathcal{L}f$.

For an SDP, the infinitesimal generator is a generalization of a diffusion generator and can be expressed as follows (see, for example [29]):

$$\begin{aligned} \mathcal{L}f(q, z) &= \sum_{i=1}^{n} u_i(q, z)\frac{\partial f(q, z)}{\partial z_i} + \quad (3) \\ &+ \frac{1}{2}\sum_{i,j=1}^{n} [\sigma(q, z)\sigma(q, z)^\mathsf{T}]_{ij}\frac{\partial^2 f(q, z)}{\partial z_i \partial z_j} + \\ &+ \sum_{k=1}^{N} \lambda_{qk}(z)f(k, z). \end{aligned}$$

This generator maps functions twice differentiable on $\mathbf{X}$ with compact support to continuous functions with compact support on the same state space.

4. STATE CONSTRAINED REACHABILITY

State-constrained reachability analysis denotes a reachability problem with additional conditions (constraints) on the system trajectories. Let us consider $\mathbf{A}, \mathbf{B}$ two Borel measurable sets of the state space X with disjoint closures, i.e.

$$\mathbf{A}, \mathbf{B} \in \mathcal{B}(X) \text{ and } \overline{\mathbf{A}} \cap \overline{\mathbf{B}} = \emptyset.$$

We consider two fundamental situations. Suppose that the system paths start from a given initial state x and we are interested in a target state, let say $\mathbf{B}$. These trajectories can hit the state set $\mathbf{A}$ or not.

The state constrained reachability (waypoint reachability) problem has been introduced in [7]. This problem aims to compute the probability, denoted by $p_{\mathbf{A} \to \mathbf{B}}(x)$, of all trajectories that hit $\mathbf{B}$ only after hitting $\mathbf{A}$ (as illustrated in Fig.1).

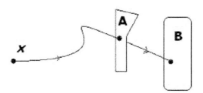

Figure 1: Waypoint reachability

Now we consider the executions (paths) of the stochastic hybrid process that start in $x = (q, z) \in \mathbf{X}$. When we investigate the state-constrained reachability, we ask the probability that these trajectories visit $\mathbf{A}$ before visiting eventually $\mathbf{B}$. Mathematically, this is the probability of

$$\{\omega | x_t(\omega) \notin \mathbf{B}, \ t \leq T_{\mathbf{A}}\}.$$

where $T_{\mathbf{A}}$ is the first hitting time of A

$$T_{\mathbf{A}} = \inf\{t > 0 | x_t \in \mathbf{A}\} \quad (4)$$

Moreover, using the first hitting time $T_{\mathbf{B}}$ of $\mathbf{B}$, we are interested to compute

$$p_{\mathbf{A} \to \mathbf{B}}(x) = \mathbf{P}_x[T_{\mathbf{A}} < T_{\mathbf{B}}]. \quad (5)$$

where $\mathbf{P}_x$ consider only the trajectories that start in $x = (q, z) \in \mathbf{X}$.

THEOREM 1. *[7] State-constrained reachability probability $p_{\mathbf{A} \to \mathbf{B}}$ solves the following Dirichlet boundary value problem:*

$$\begin{cases} \mathcal{L}p(x) = 0 & x \in \mathbf{X} \backslash (\mathbf{A} \cup \mathbf{B}) \\ p(x) = 1 & x \in \mathbf{A} \\ p(x) = 0 & x \in \mathbf{B}. \end{cases} \quad (6)$$

where $\mathcal{L}$ is the infinitesimal generator of the underlying stochastic hybrid process.

In particular, we can apply this result to SDP with the generator given by (3). For SDP, numerical methods for the estimation of the solutions of such problem can be easily obtained extending the methods available for diffusion processes [14]. Moreover, for SDP, some good properties of the component diffusions (like path continuity, Feller property) are preserved by the switching mechanism. Because of this, we expect that we may be able to calculate classical solutions for (6) if some differentiability properties are imposed w.r.t. transition rates (λ_{ij}).

5. NON-SEQUITUR TRAJECTORIES

In this section we start with an SDP (x_t) that is supposed to be *ergodic*, i.e. its operator semigroup satisfies the following condition

$$P_t g = g \text{ iff } g = const.$$

This means that its generator $\mathcal{L}$ has only constants in its null space.

Moreover, we suppose that the process has an invariant probability measure μ, i.e.

$$\mu(\mathbf{A}) = \int_{\mathbf{X}} p_t(x, \mathbf{A}) \mu(dx)$$

for all $t \geq 0$ and $\mathbf{A} \in \mathcal{B}(\mathbf{X})$. This is equivalent with the fact that

$$\lim_{t \to \infty} \frac{1}{t} \int_0^t g(x_s) ds = \mathbf{E}_\mu g \quad (7)$$

with respect to any measurable bounded function $g : \mathbf{X} \to \mathbb{R}$. Here, $\mathbf{E}_\mu$ is the expectation with respect to μ. In this case, we say that the process is ergodic with respect to the invariant measure μ. This ergodicity means that the time averages equal space averages.

On the invariant measure μ we impose that is absolutely continuous with respect to the Lebesgue measure, i.e.

$$d\mu(x) = \rho(x) dx.$$

Another useful assumption about the underlying process is that its joint probabilities (of the discrete q_t and continuous part z_t) have a density function ρ w.r.t. the Lebesgue measure, i.e.

$$\mathbf{P}(q_t = q, \ z_t \in A) = \int_A \rho_t(q, z) dz$$

for all $t \geq 0$ and $A \in \mathcal{B}(X)$.

The ergodicity assumption implies that

$$\rho_t(x) \to \rho(x)$$

for all $x = (q, z) \in \mathbf{X}$.

Conditions for the existence of an invariant measure and problems related to ergodic for SDP have been studied in [20]. This is why, we consider that SDP represent a modelling framework for stochastic hybrid systems for which the problems proposed in this paper make sense. Since, we have imposed the existence of an invariant measure, our SDP should be recurrent (i.e. a state set can be visited infinitely many times by the process) [20].

We are interested not in the statistical properties of the long run trajectories, but rather in the properties that characterize these trajectories when they connect two metastable configurations. Let us formalize the problem. Suppose that $\mathbf{A}, \mathbf{B} \in \mathcal{B}(\mathbf{X})$ are two disjoint metastable measurable sets in the SDP hybrid state space (eventually two operational modes). Let consider an arbitrary trajectory (x_t) that has nonempty intersection with both of them. Since the process is ergodic, this path makes transitions between $\mathbf{A}$ and $\mathbf{B}$ infinitely often. Then we consider only the "pieces" of this trajectory that connect the boundary of $\mathbf{A}$ (denoted by $\partial \mathbf{A}$) with the boundary of $\mathbf{B}$ (denoted by $\partial \mathbf{B}$). Such pieces of trajectory, we will call *non-sequitur trajectories*. Then we are interested to characterize probabilistically the entire set of such trajectories. The ergodicity of the process ensures that these characterizations are independent of the particular trajectory used to generate the non-sequitur trajectories.

In defining such trajectories, we have been inspired by transition path simulation and rare events problems [11, 12]

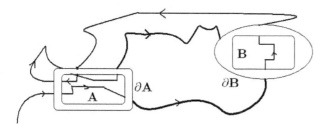

Figure 2: Non-sequitur trajectories

Let us denote by $\mathcal{T}_{\mathbf{AB}}^c$ the time interval when the trajectory x_t evolves outside of $\mathbf{A}$ and $\mathbf{B}$, i.e.

$t \in \mathcal{T}_{\mathbf{AB}}^c$ iff $x_t \notin \mathbf{A} \cup \mathbf{B}$, and $x_{T_{\mathbf{AB}}^{\rightarrow}(t)} \in \mathbf{B}$ and $x_{T_{\mathbf{AB}}^{\leftarrow}(t)} \in \mathbf{A}$,

where

$$
\begin{aligned}
T_{\mathbf{AB}}^{\rightarrow}(t) &= \inf\{s \geq t | x_s \in \mathbf{A} \cup \mathbf{B}\} \\
T_{\mathbf{AB}}^{\leftarrow}(t) &= \sup\{s \leq t | x_s \in \mathbf{A} \cup \mathbf{B}\}.
\end{aligned}
$$

The the collection of non-sequitur trajectories is

$$\Omega_{\mathbf{A}\rightarrow\mathbf{B}} := \{x_t | t \in \mathcal{T}_{\mathbf{AB}}^c\}.$$

In the remaining of the paper, we aim to characterize the set $\Omega_{\mathbf{A}\rightarrow\mathbf{B}}$ in terms of process parameters and of state constrained reachability probabilities.

6. NON-SEQUITUR PROBABILITIES

The aim of this section is to characterize the densities of the non-sequitur trajectories using the following state constrained reach probabilities:

- $p_{\mathbf{B}\rightarrow\mathbf{A}}(x)$ - the probability that the trajectory starting from a state x that is situated outside of $\mathbf{A} \cup \mathbf{B}$ reaches $\mathbf{B}$ before $\mathbf{A}$;

- $\overleftarrow{p}_{\mathbf{B}\rightarrow\mathbf{A}}(x)$ - the probability that the trajectory arriving in state x outside of $\mathbf{A} \cup \mathbf{B}$ has visited $\mathbf{A}$ only after visiting $\mathbf{B}$.

Remark that the probability $p_{\mathbf{B}\rightarrow\mathbf{A}}(x)$ can be expressed as a conditional expectation, i.e.

$$p_{\mathbf{B}\rightarrow\mathbf{A}}(x) = \mathbf{E}_x[1_{\mathbf{B}}(x_{T_{\mathbf{A}\cup\mathbf{B}}})]$$

where $T_{\mathbf{A}\cup\mathbf{B}}$ is the first hitting of the process with respect to $\mathbf{A} \cup \mathbf{B}$. Note, since we work with an SDP, this will be the classical solution for the Dirichlet boundary value problem (6).

It can be observed that $\overleftarrow{p}_{\mathbf{B}\rightarrow\mathbf{A}}(x)$ is the dual of $p_{\mathbf{B}\rightarrow\mathbf{A}}(x)$. This means that if we consider the time reverse process (x_t^*) (the time on the trajectories goes backward), we have

$$\overleftarrow{p}_{\mathbf{B}\rightarrow\mathbf{A}}(x) = p_{\mathbf{B}\rightarrow\mathbf{A}}^*(x)$$

where $p_{\mathbf{B}\rightarrow\mathbf{A}}^*$ is defined for (x_t^*) in the same way like $p_{\mathbf{B}\rightarrow\mathbf{A}}$ is defined for (x_t).

For the characterization of the probabilities $\overleftarrow{p}_{\mathbf{B}\rightarrow\mathbf{A}}(x)$, we need to use the reverse time, and this will be solution for the following Dirichlet boundary value problem:

$$
\begin{cases}
\mathcal{L}^* \overleftarrow{p}_{\mathbf{B}\rightarrow\mathbf{A}} = 0 & \text{in } \mathbf{X}\backslash(\mathbf{A} \cup \mathbf{B}), \\
\overleftarrow{p}_{\mathbf{B}\rightarrow\mathbf{A}} = 1 & \text{on } \partial\mathbf{A}, \\
\overleftarrow{p}_{\mathbf{B}\rightarrow\mathbf{A}} = 0 & \text{on } \partial\mathbf{B},
\end{cases}
$$

where $\mathcal{L}^*$ is the adjoint operator of $\mathcal{L}$. It is interesting to note that if the process is reversible in time, the infinitesimal generator is Hermitian $\mathcal{L} = \mathcal{L}^*$ and

$$\overleftarrow{p}_{\mathbf{B}\rightarrow\mathbf{A}} = 1 - p_{\mathbf{B}\rightarrow\mathbf{A}}.$$

For studying non-sequitur trajectories with respect to $\mathbf{A}$ and $\mathbf{B}$, we are interested to compute their measure, which is a probability denoted by $\mu_{\mathbf{AB}}$, with the density function denoted by $\phi_{\mathbf{AB}}$. This can be thought as the conditional probability density to observe an arbitrary trajectory at the state $x \notin \mathbf{A} \cup \mathbf{B}$ at an instant time t, provided that it is non-sequitur at t. Mathematically, this is equal with the probability density to observe an arbitrary trajectory (that is $\rho(x)$) multiplied with the product of the state constrained reach probabilities

$$p_{\mathbf{B}\rightarrow\mathbf{A}}(x) \cdot \overleftarrow{p}_{\mathbf{B}\rightarrow\mathbf{A}}(x)$$

corresponding to $\mathbf{A}$ and $\mathbf{B}$.

The probability $\mu_{\mathbf{AB}}$ of the non-sequitur trajectories (w.r.t. the metastable sets $\mathbf{A}$ and $\mathbf{B}$) is defined such that it is supported in

$$\mathbf{X}_{\mathbf{AB}} = \mathbf{X}\backslash(\mathbf{A} \cup \mathbf{B})$$

and these trajectories have to be ergodic with respect to $\mu_{\mathbf{AB}}$. Then $\mu_{\mathbf{AB}}$ must satisfy the following condition:

$$
\begin{aligned}
\mu_{\mathbf{AB}}(E) \quad : \quad &= \lim_{T \to \infty} \frac{\int_0^T 1_E(x_t) \cdot 1_{\mathcal{T}_{\mathbf{AB}}^c}(t) dt}{\alpha(\mathcal{T}_{\mathbf{AB}}^c \cap [0, T])} \quad (8) \\
&= \int_E \phi_{\mathbf{AB}}(x) dx,
\end{aligned}
$$

for all measurable set $E \subset \mathbf{X}_{\mathbf{AB}}$, where α is the Lebesgue measure on $\mathbb{R}$ and $1_E(x_t)$ is the indicator function of $\{t | x_t \in E\}$.

THEOREM 2. *The probability density function $\phi_{\mathbf{AB}}$ of the non-sequitur trajectories can be express as follows*

$$\phi_{\mathbf{AB}}(x) = \frac{\eta_{\mathbf{AB}}(x)\rho(x).}{\int_{\mathbf{X}_{\mathbf{AB}}} \eta_{\mathbf{AB}}(x)\rho(x) dx}.$$

where

$$\eta_{\mathbf{AB}}(x) := p_{\mathbf{B}\rightarrow\mathbf{A}}(x) \cdot \overleftarrow{p}_{\mathbf{B}\rightarrow\mathbf{A}}(x). \quad (9)$$

PROOF. We start with the definition (8), and transform the right hand side using a suitable partition of the time interval $\mathcal{T}_{\mathbf{AB}}^c$. The methodology is based on the Prop. 3 from [7]. Suppose that the following notations are in forced:

$$
\begin{aligned}
T_{\mathbf{A}}^{\rightarrow}(t) &= \inf\{s \geq t | x_s \in \mathbf{A}\} \\
T_{\mathbf{B}}^{\rightarrow}(t) &= \inf\{s \geq t | x_s \in \mathbf{B}\}
\end{aligned}
$$

are the first hitting times of $\mathbf{A}$ (resp. $\mathbf{B}$) after time t, and

$$
\begin{aligned}
T_{\mathbf{A}}^{\leftarrow}(t) &= \sup\{s \leq t | x_s \in \mathbf{A}\} \\
T_{\mathbf{B}}^{\leftarrow}(t) &= \sup\{s \leq t | x_s \in \mathbf{B}\}
\end{aligned}
$$

are the last exit times from $\mathbf{A}$ (resp. $\mathbf{B}$) before time t.

$T_{\mathbf{A}}^{\rightarrow}(t)$ describes the times when the process visits $\mathbf{A}$ forward in time (after time t), and $T_{\mathbf{A}}^{\leftarrow}(t)$ describes the process visits in $\mathbf{A}$ "backward" in time (before time t). The first one characterizes the direct process (x_t) evolution, and the latter one describes the dual process (x_t^*).

Then

$$
\begin{aligned}
\mu_{\mathbf{AB}}(E) &= \lim_{T \to \infty} \frac{\int_0^T 1_E(x_t) \cdot 1_{\mathcal{T}_{\mathbf{AB}}^c}(t)dt}{\alpha(\mathcal{T}_{\mathbf{AB}}^c \cap [0,T])} = \\
&= \lim_{T \to \infty} \frac{\int_0^T 1_E(x_t) \cdot 1_{[T_{\mathbf{A}}^{\rightarrow} < T_{\mathbf{B}}^{\rightarrow}]} 1_{[T_{\mathbf{B}}^{\leftarrow} < T_{\mathbf{A}}^{\leftarrow}]} dt}{\int_0^T 1_{\mathbf{X_{AB}}}(x_t) \cdot 1_{[T_{\mathbf{A}}^{\rightarrow} < T_{\mathbf{B}}^{\rightarrow}]} 1_{[T_{\mathbf{B}}^{\leftarrow} < T_{\mathbf{A}}^{\leftarrow}]} dt}
\end{aligned}
$$

and using the ergodicity condition (7), we have

$$
\mu_{\mathbf{AB}}(E) = \frac{\int_E \mathbf{P}_x[T_{\mathbf{A}}^{\rightarrow} < T_{\mathbf{B}}^{\rightarrow} \text{ and } T_{\mathbf{B}}^{\leftarrow} < T_{\mathbf{A}}^{\leftarrow}] \cdot \rho(x)dx}{\int_{\mathbf{X_{AB}}} \mathbf{P}_x[T_{\mathbf{A}}^{\rightarrow} < T_{\mathbf{B}}^{\rightarrow} \text{ and } T_{\mathbf{B}}^{\leftarrow} < T_{\mathbf{A}}^{\leftarrow}] \cdot \rho(x)dx}
$$

Since every SDP is a strong Markov process, the Markov property holds with respect to the stopping times, and we can write

$$
\mathbf{P}_x[T_{\mathbf{A}}^{\rightarrow} < T_{\mathbf{B}}^{\rightarrow} \text{ and } T_{\mathbf{B}}^{\leftarrow} < T_{\mathbf{A}}^{\leftarrow}] = p_{\mathbf{B} \to \mathbf{A}}(x) \cdot \overleftarrow{p}_{\mathbf{B} \to \mathbf{A}}(x).
$$

This conducts to the expression of the density $\phi_{\mathbf{AB}}$ and ends the proof. ∎

Clearly, the computation of the probability density $\phi_{\mathbf{AB}}$ will be based, in the end, on the infinitesimal generator of the underlying stochastic hybrid process. In large dimensional systems, the main problem is how to solve the boundary value problem (6) that gives the state constrained reachability probabilities. This is not a trivial problem and there exists a variety of computational methods (see [7] and the references therein).

EXAMPLE 1. *Let $\{X_t\}_{t \in \mathbb{R}}$ be the Markov diffusion process on $\mathbb{R}^n$ defined by the stochastic differential equation*

$$
dX_t = \nabla Y(X_t)dt + \sqrt{2}dW_t
$$

where dW_t is the Wiener process, and Y is a twice differentiable function such that

$$
V = \int_{\mathbb{R}^n} e^{-Y(x)}dx
$$

is finite.
The diffusion infinitesimal generator

$$
\mathcal{L} = -\nabla Y \cdot \nabla + \triangle
$$

is ergodic with respect to distribution

$$
d\mu(x) = V^{-1}e^{-Y(x)}dx
$$

We have that

$$
p_{\mathbf{B} \to \mathbf{A}}(x) = 1 - \overleftarrow{p}_{\mathbf{B} \to \mathbf{A}}(x)
$$

because the processes $\{X_t\}_{t \in \mathbb{R}}$ and $\{X_{-t}\}_{t \in \mathbb{R}}$ are statistically (which is because $\{X_t\}_{t \in \mathbb{R}}$ is time reversible). $p_{\mathbf{B} \to \mathbf{A}}(x)$ can be calculated as the solution of the backward Kolmogorov equation

$$
\begin{aligned}
\nabla Y \cdot \nabla p &= \triangle p \text{ on } \mathbb{R}^n \backslash (\mathbf{A} \cup \mathbf{B}) \\
p &= 0 \text{ on } \mathbf{A} \\
p &= 1 \text{ on } \mathbf{B}
\end{aligned}
$$

Suppose that Y is ogiven by

$$
Y(x,y) = \frac{5}{2}(1 - x^2)^2 + 5y^2
$$

and

$$
\mathbf{A} = \{x < -0.8\} \text{ and } \mathbf{B} = \{x > 0.8\}
$$

Then

$$
p_{\mathbf{B} \to \mathbf{A}}(x) = \frac{\int_{-0.8}^x e^{\frac{5}{2}(1-z^2)^2}dz}{\int_{-0.8}^{0.8} e^{\frac{5}{2}(1-z^2)^2}dz}
$$

The probability density function associated with $\mu_{\mathbf{AB}}$ is to be compared with

$$
\int_{-0.8}^{0.8} e^{\frac{5}{2}(1-z^2)^2}dz \cdot \int_{\mathbb{R}} e^{-5y^2}dy
$$

The level sets of the density coincide with the level sets of $Y(x,y)$. The statistical analysis indicates that when a trajectory leaves the sets $\mathbf{A}$ and $\mathbf{B}$ it preferably visits the regions near the two minima of $Y(x,y)$ located at $(x,y) = (\pm 1, 0)$. These trajectories are very unlikely to produce non-sequitur trajectories in the future.

Since the sets $\mathbf{A}$ and $\mathbf{B}$ are supposed metastable, the system will have a higher concentration of the non-sequitur trajectories in some specific zones in the space $\mathbf{X_{AB}}$. These zones will be of interest for our study. For this study, we may define a new stochastic reachability problem for non-sequitur trajectories. We may also investigate the properties of some occupation time distributions for the non-sequitur trajectories.

The probability density $\phi_{\mathbf{AB}}$ is indeed a useful concept for studying the non-sequitur trajectories, but it may not give enough information about the displacement of these trajectories. Therefore, to capture such characteristics, other useful "physical" concepts should be deploit.

In the following, we are going to introduce some probabilistic objects that can offer more insights in the study of non-sequitur trajectories.

First recall that the reachability function for a stochastic hybrid process (x_t) corresponding to a target set $E \in \mathcal{B}(\mathbf{X})$ can be expressed as follows:

$$
\varphi^{\to E}(x) = \mathbf{E}_x(\max_{t \geq 0} 1_E(x_t)), \forall x \in \mathbf{X}.
$$

where $\mathbf{E}_x$ is the expectation w.r.t. $\mathbf{P}_x$.

Clearly, the reachability function concept can be extended to non-sequitur trajectories, for a target set $E \in \mathcal{B}(\mathbf{X_{AB}})$ we have to consider only the times instances from $\mathcal{T}_{\mathbf{AB}}^c$, i.e.

$$
\varphi_{\mathbf{AB}}^{\to E}(x) = \mathbf{E}_x(\max_{t \in \mathcal{T}_{\mathbf{AB}}^c} 1_E(x_t)), \forall x \in \mathbf{X_{AB}}.
$$

This definition is expressive, but the characterization of such function might be difficult to be expressed analytically and challenging. The equality

$$
\mathbf{E}_x(\max_{t \in \mathcal{T}_{\mathbf{AB}}^c} 1_E(x_t)) = \sup_{t \in \mathcal{T}_{\mathbf{AB}}^c} p_t(x, E)
$$

holds only in some special cases (e.g. the family $\{1_E(x_t)\}_{t \in \mathcal{T}_{\mathbf{AB}}^c}$ has the lattice property). Here, $p_t(x, E)$ is the transition probability of (x_t) to "transit" from x to E.

Therefore, some other options have to be considered. We may choose another measure for the reachability like

$$
\nu_{\mathbf{AB}}^{\to E} := \int_0^\infty \mathbf{E}_x[1_E(x_t)] \cdot 1_{\mathcal{T}_{\mathbf{AB}}^c}(t)dt
$$

that we expect to be computable via dynamic programming methods.

Going further, let us recall that for a sample path ω, the *occupation time distribution* up to the time $t > 0$ for the process (x_t) can be defined as follows:

$$L_t(\omega, G) = \frac{1}{t} \int_0^t 1_{\mathbf{B}}(x_s(\omega)) ds, \; G \in \mathcal{B}(X). \qquad (10)$$

If the process is conservative, then $L_t(\omega, \cdot)$ is a probability measure on $(\mathbf{X}, \mathcal{B})$. This measure is intensively studied in the context of large deviation theory [34].

For a given measurable set $G \subset \mathbf{X}_{\mathbf{AB}}$, when we extend this concept to the case of non-sequitur trajectories between $\mathbf{A}$ and $\mathbf{B}$ that enter G, we get

$$L_{\mathbf{AB}}(G) = \lim_{s \searrow 0} \frac{1}{s} \lim_{T \to \infty} \frac{1}{T} \int_0^T [1_G(x_t) \cdot 1_{G^c}(x_{t+s}) 1_{\mathcal{T}^c_{\mathbf{AB}}}(t) dt \qquad (11)$$

where $G^c = \mathbf{X} \backslash G$.

Moreover, we may define the traffic corresponding to G considering only the trajectories (i.e. measure of those trajectories that enter G coming from $\mathbf{A}$ and going to $\mathbf{B}$ minus the measure of those trajectories that enter G from the opposite direction). Therefore, we can define

$$\kappa_{\mathbf{AB}}(G) = L_{\mathbf{AB}}(G) - L_{\mathbf{AB}}(G^c). \qquad (12)$$

For studying the "concentration" of the non-sequitur trajectories we can use the so-called probability current or flux. To deal with such a flux, we can use the Fokker-Planck-Kolmogorov equation associated to the local part (diffusion part) of the SDP. Then Fokker-Planck equation is

$$\frac{\partial \rho}{\partial t} + div(\mathbf{j}_t) = 0 \qquad (13)$$

where div denotes the divergence operator w.r.t. the continuous variables of the state space. $\mathbf{j}$ is the *probability current (or flux)*, that is defined as the following vector field:

$$\mathbf{j}_t = \rho_t u - \frac{1}{2} \sum_{r=1}^n div(\rho_t \sigma_r) \sigma_r$$

where $(\sigma_r)_{r=1,..,n}$ are the rows of σ.

Equation (13) is a local conservation equation, which accounts for the fact that, between its switchings, the process evolves continuously according to the SDE (3). Note that this single equation actually hides a system of PDEs - one for each component diffusion. Due to the ergodicity assumption, we have

$$\mathbf{j}_t(q, z) \to \mathbf{j}(q, z) \text{ as } t \to \infty,$$

where $\mathbf{j}$ is the equilibrium probability current (steady-state current) defined by

$$\mathbf{j} = \rho u - \frac{1}{2} \sum_{r=1}^n div(\rho \sigma_r) \sigma_r.$$

Then we can restrict this probability current to the non-sequitur trajectories (w.r.t. $\mathbf{A}$ and $\mathbf{B}$) only, and to obtain a new object denoted by $\mathbf{j}_{\mathbf{AB}}$. Note that the probability current for a general stochastic hybrid system (when there exist also guarded discrete transitions) has been discussed in [1]. But, in our case, since we have supposed that the underlying process is only an SDP, we do not need the expression of $\mathbf{j}$ correlated with the jumps. Moreover, we can

suppose that the non-sequitur trajectories obey only to one stochastic differential equation.

When the process (x_t) has only a diffusion component (degenerates into a diffusion), the following result can be proved.

THEOREM 3. *The probability current* $\mathbf{j}_{\mathbf{AB}}$ *of the non-sequitur trajectories (w.r.t.* $\mathbf{A}$ *and* $\mathbf{B}$*) for a diffusion process* (z_t) *can be expressed as follows:*

$$\mathbf{j}_{\mathbf{AB}} : \mathbf{X}_{\mathbf{AB}} \to \mathbb{R}^n$$
$$\mathbf{j}_{\mathbf{AB}} = \eta_{\mathbf{AB}} \cdot \mathbf{j} + \mathbf{j}_{\mathbf{A} \to \mathbf{B}} - \mathbf{j}_{\mathbf{B} \to \mathbf{A}} \qquad (14)$$

where $\eta_{\mathbf{AB}}$ *is given by (9), and*

$$\mathbf{j}_{\mathbf{A} \to \mathbf{B}} := \rho \cdot \overleftarrow{p}_{\mathbf{B} \to \mathbf{A}} \cdot \sigma \cdot div[p_{\mathbf{B} \to \mathbf{A}}]$$
$$\mathbf{j}_{\mathbf{B} \to \mathbf{A}} := \rho \cdot p_{\mathbf{B} \to \mathbf{A}} \cdot \sigma \cdot div[\overleftarrow{p}_{\mathbf{B} \to \mathbf{A}}].$$

This result can be obtained by straightforward calculations for the probability flux between $\mathbf{A}$ and $\mathbf{B}$, using the ergodicity of the underlying process and the fact that $\mathbf{j}_{\mathbf{AB}}$ can be obtained as the restriction of total probability flux $\mathbf{j}$ to $\mathcal{T}^c_{\mathbf{AB}}$.

For the understanding of these concepts, we may think at the case of a simple continuous time Markov chain whose transition probability distributions are governed by the master equation. The probability $P_t(x)$, to find the system in the state x at time t, is governed by the master equation

$$\partial_t P_t(x) = \sum_{y \neq x} \mathbf{j}_x^y(t) \qquad (15)$$

with the net probability current defined by

$$\mathbf{j}_x^y = \lambda_x^y P_t(y) - \lambda_y^x P_t(x)$$

where λ_x^y and λ_y^x are the rates of the transition from x to y (resp. y to x).

This equation illustrates how the probability flux goes between two state x and y. Since, the process is supposed to be ergodic, any equilibrated trajectory $\{x_t\}$ induces an equilibrium probability current $\mathbf{j}_{xy}$ between any pair (x, y) of states. In other words, $\mathbf{j}_{xy}$ is the average number of jumps from x to y per time unit observed in an infinitesimal time interval dt. The master equation (15) expresses the time derivative of the probability as the balance of the currents flowing in and out the state x. Defining

$$\mathbf{j}_x^+ := \sum_{y \neq x} \lambda_x^y P_t(y) \text{ and } \mathbf{j}_x^- := \sum_{y \neq x} \lambda_y^x P_t(x)$$

the equation (15) simply states the conservation of probability through the expression

$$\partial_t P_t(x) = \mathbf{j}_x^+ - \mathbf{j}_x^-.$$

In the steady state, the current appears to be globally balanced on each state so that

$$\mathbf{j}_x^+ = \mathbf{j}_x^-.$$

It should be clear that probability current induced by the ensemble of non-sequitur trajectories corresponding to the sets $\mathbf{A}$ and $\mathbf{B}$ differs from equilibrium probability current. Intuitively, that current is given by the equilibrium probability current

$$\mathbf{j}_{xy} = \lambda_x^y \rho_y$$

weighted with the probability that the process visits first **A** and then **B** before jumping from x to y and weighted by the probability that the process will continue to **B** not to **A** after jumping from x to y.

The above discussions show and the formula (14) that the study of such probability flux is closely related with the underlying process (x_t), but also with the reversed time (dual) process (x_t^*). Their operator semigroups (P_t) and (P_t^*) are related by

$$\int P_t(f) \cdot g d\mu = \int f \cdot P_t^*(g) d\mu$$

where μ is the invariant measure.

If the process is symmetric in time, then the two semigroups coincide and the things are easier to deal with. For the general case, in order to obtain an elegant formula for the probability flux of the non-sequitur trajectories (like in the case of Markov chain), one needs to handle the Fokker Planck Kolmogorov equations of the processes (x_t) and (x_t^*). Practically, we need to consider the equilibrium probability currents for the two processes.. This analysis is not straightforward, and might be necessary to study in the first instance the connections between the infinitesimal generator and the equilibrium probability flux. This will constitute the subject of an upcoming paper.

The probability flux $\mathbf{j_{AB}}$ can be thought of as a candidate map (via a suitably chose rate for the transitions) that describes the clustering of the non-sequitur trajectories. This map might give an appropriate partition of state space between **A** and **B**, that can be further used to define a discretization of space $\mathbf{X_{AB}}$. Moreover, an interesting problem would be to consider obstacles in this space and to look for estimations of the set of non-sequitur trajectories that avoid these obstacles. However, the probability flux and the occupation time measure (11) are closely related.

7. ROADMAP TO APPLICATIONS

A promising application area for the analysis techniques presented in this paper is the *prevention of dangerous falling* for vulnerable people such the elderly, or people with physical handicap. The falling behaviour can be characterized as a set of random trajectories that depart from a metastable state (like standing, walking, or running) and reach a forbidden/unwanted metastable state (like hitting a hard surface). Obviously, some trajectories could reach a metastable state corresponding to human balance when human control is applied. Other trajectories correspond to fallings that can not be prevented anymore by humans. In this case, an external hybrid system consisting of discrete (intelligent) controller and a physical effector like an exoskeleton could intervene and stop the person from falling. The intelligent controller should be constantly monitoring the trajectory of a part of the human body and anticipate its membership to certain trajectory probabilistic clusters, corresponding to uncontrollable falling. Then an electronic device can very quickly activate an exoskeleton, or another suitable physical facility. This topic has a significant medical, economic and societal impact, since hundred of thousands of big injuries worldwide are due to preventable fallings.

Important applications for the probabilistic analysis of the non-sequitur behaviour are the *fault detection* and the improvement of *training* programmes for pilots [33]. When the non-sequitur behaviour is caused by a fault, the trajec-

tory probabilistic densities could provide a characterization of that fault. An example of such densities characterized by the trajectory shape is described for a toy example in [23]. An area where such analysis would be very useful is the aircraft flight control [4]. The term aircraft upset is used to describe a dangerous condition, which may result in the loss of control, or the total loss of the aircraft. The loss of control may be due to system failures, but also to adverse weather or pilot disorientation. Obviously, these are examples of non-sequitur behaviour. The US NASA aviation safety programme has introduced (in 2000) the concepts of upset prevention and upset recovery to prevent the air crashes provoked by aircraft upset. These operations could benefit from the probabilistic analysis of the non-sequitur behaviour. In 2009, the Royal Aeronautical Society has formed a working group for researching simulations of aircraft upset conditions. An NTSB statistics for the period 1994-2003 indicates that aircraft upset (i.e. non-sequitur behaviour that reached a forbidden region of the state space) caused the loss of 2100 lives in 32 accidents. A more precise statistics provided by Boeing indicates the loss of 2051 lives in 22 accidents within the period 1998-2007. A list of examples of non-sequitur behaviour during flight comprises the following accidents:

- Colgan Air Flight 3407 [21]. The plane stalled and crashed 5.0 nmi short of the runway threshold. The cause was pilot error in following stall recovery procedures.

- China Airlines Flight 006 [9]. The plane rolled to the right and entered a high dive attitude. Fortunately, a recovery from the dive has been produced just below 11,000 ft.

- United Flight B-720 [32] doved until recovery at 14,000 ft. The flight stalled when encountering severe turbulence, downdrafts and updrafts.

- Pan Am B-707 [27] went into a high dive when cruising over the Atlantic. Luckily, the control was recovered just over 6,000 ft. The cause was an accidental disconnection of the autopilot. Under the pressure of significant G-forces, the aircraft suffered extensive structural damages.

- TWA Flight 841 did high dive from 39,000 ft to 5,000 ft in 63 seconds. The cause was improper manipulation of flaps/slats by pilots.

- The A330 [5] test flight in 1994: During a test flight, the control was lost while flying to close to the ground.

An illustrative example for the use of the analysis of the non-sequitur behaviour of flights for detecting mechanical faults is the NTSB investigation of the air crashes of USAir Flight 427 [26], the United Airlines Flight 585 [24], Copa Airlines 737, and the upset of Eastwind Airlines Flight 517, MetroJet Flight 2710 [25], and SilkAir Flight 185 [28]. The non-sequitur behaviours of these flights are parts of the same probabilistic trajectory cluster, and this fact suggests a common cause in a part failure. The investigators discovered that the rudders were doing a "hardover" and jamming due to thermal shock. On November, 2002, the FAA ordered Boeing to upgrade the rudder control system on the 737 type aircrafts.

Faults are difficult to predict with accuracy and in time. Even more, faults are very difficult, if not impossible to be prevented. The analysis of faults and of the non-sequitur behaviour that are provoked shows the consequences of the faults could be avoided. At least, it could be minimized the severity of these consequences in terms of casualties and economic losses. When faults are detected and diagnosed quickly enough, it is possible to subsequently reconfigure the control such that the system will continue its operation under safety until it will be switched off. In the case of safety critical systems, the apparition of faults that are not detected and acted upon can have catastrophic consequences. Here, there are two example for the magnitude of disasters faults can produce:

- The explosion from 4th June 1996 of the Ariane 5 space shuttle, causing lives tall and the loss of a 7 billion USD worth project: The fault appeared in the internal reference unit, and major conceptual problems have risen like the interaction between software and its physical shell. This prompted out the birth of a new modelling paradigm called cyber physical systems.

- The explosion from 26th April 1986 at the nuclear power plant at Chernobyl in Ukraine: The human victims counted 15000 lives immediately after the accident and 50000 handicapped persons. More than 5 million people were exposed to radiation and some of them died because of the induced diseases such as cancer and leukemia.

The state constrained reach probabilities can be used also measures for *system reliability*. The current measures [31] are far much simpler, in the form of first passage time probabilities. The reliability is measured by evaluating the probability that the system will not reach a fatally dangerous state within a finite horizon time. These methods for reliability assessment do not consider the system capability to achieve a certain objective, while the measures we consider account this capability via target state sets.

8. CONCLUSIONS

In this paper, we set up an analytic framework for studying random behaviours generated by the apparitions of faults. We have focused on the study of those system trajectories that depart from a controllable functioning mode and end up in a new controllable mode. We have shown that such trajectories tend to form clusters and we modelled probabilistically the density of such clusters. For these, we have used a new concept of stochastic reachability analysis called state constrained reachability, which was introduced in a recent paper. We have highlighted the role of equiprobability surfaces where the chances to getting into a controllable or uncontrollable mode are equal.

The analytical techniques that we have developed could be fruitfully applied for flight control and guidance systems. For an aircraft, the short time random behaviour, which is the result of a fault, could mean reaching a fatal unstable behaviour or saving the aircraft. The development of autopilots that could use the statistical properties of the analytical studies and simulations can improve dramatically the flight dependability and safety. The following developments of this paper will focus on a case study from aerospace.

The approach we propose can be thought of as a dualization of the manoeuvre automata philosophy. The manoeuvres are controllable continuous trajectories that realize smooth mode transitions. In contrast, we study the continuous trajectories that are only partially controllable and that eventually end into a controllable mode. Equally, we may have introduced the uncontrollable trajectories that depart from a controllable mode and end into the cemetery point. Obviously, the study of one set of trajectories reduces to the study of the other. We prefer a problem formulation, which is more optimistic.

One major advantage of using stochastic analysis consists of the possibility to abstract away hybrid trajectories into diffusion paths. The technique we used is inspired by the multi-scale modelling in mathematical physics. In our interpretation, the hybrid behaviour corresponds to the microscopic scale in physics. The diffusion based abstraction corresponds to the macro scale. The lifting from micro to macro can be very elegantly done in the framework of partial differential equations and diffusion like stochastic processes. These theories have been axiomatically presented in a uniform manner within a theory called Hilbertean formal methods [8].

The major departure compared with the existing approaches on stochastic analysis of fault tolerant systems (see, for example, [37], [35] and the references therein) is that the existing research focuses on the stochastic stability, while we investigate the unstable behaviours generated by faults.

Simulations will be carried out to illustrate the trajectory clustering phenomena. Also, there are many analytic properties that need to be investigated like the last exit time from a surface of equal probability, and the net average flux of the non-sequitur trajectories that cross such surface. Another statistical parameter that should be considered might be the occupation time measure for of non-sequitur trajectories that is fundamental in the large deviation problems (used for studying of rare events probabilities). Obviously, the most related work is the stochastic simulation of rare events in free flight control [3]. In that approach, a set of events of very small probability is considered for simulation in order to produce a very high confidence system. Our approach shares the rare events modelling ideas, but with some differences. For a long run, the probabilities might not be always small. However, considering the entire system life time, these events (faults) can be considered rare.

Acknowledgments
This work was funded by the EPSRC project CICADA, ref. EP/E050441/1.

9. REFERENCES

[1] Bect, J.: *A Unifying Formulation of the Fokker-Planck-Kolmogorov Equation for General Stochastic Hybrid Systems.* Proc. of IFAC World Congress (2008).

[2] Blom, H.A.P., Lygeros, J. (Eds.): "*Stochastic Hybrid Systems: Theory and Safety Critical Applications*". Springer Verlag LNCIS series **337** (2006).

[3] Blom, H.A.P. and Bakker, G.J. and Krystul, J. *Rare Event Estimation for a Large-Scale Stochastic Hybrid System With Air Traffic Application.* In: Rare Event Simulation using Monte Carlo Methods. John Wiley & Sons, (2009): 193-214.

[4] Boeing Company *Aerodynamic Principles of Large-Airplane Upsets* http://www.boeing.com/commercial/aeromagazine/ aero_03/textonly/fo01txt.html

[5] A330 Test Flight (1994) http://aviation-safety.net/database/record.php?id=19940630-0&lang=fr

[6] Bujorianu, M.L., Lygeros, J.: *Towards Modelling of General Stochastic Hybrid Systems.* In [2]: 3-30.

[7] Bujorianu, M.L., Bujorianu, M.C.: *State-Constrained Reachability for Stochastic Hybrid Systems.* IFAC Conference on Analysis and Design of Hybrid Systems (2009).

[8] Bujorianu M.C., Bujorianu M.L. *Towards Hilbertian Formal Methods* in Proc. of ACSD, IEEE Computer Society Press (2007).

[9] China Airlines Flight 006 http://www.rvs.uni-bielefeld.de/publications/Incidents/DOCS/ComAndRep/ ChinaAir/AAR8603.html

[10] Davis, M.H.A.: *"Markov Models and Optimization"* Chapman & Hall, London, (1993).

[11] Dellago, C., Bolhuis, P. G., Geissler, P. L. *Transition Path Sampling Methods.* In Computer Simulations in Condensed Matter: from Materials to Chemical Biology, Springer Verlag (2006).

[12] Dellago, C.; Bolhuis, P. G.: *Transition Path Sampling and Other Advanced Simulation Techniques for Rare Events.* Springer Verlag (2010).

[13] Ducard, G.J.J. *"Fault Tolerant Flight Control Systems. Practical Methods for Small Unmanned Aerial Vehicles Challenge"* Springer Verlag, Advances in Industrial Control Series (2009).

[14] Douglas, C.C., Haase, G., Langer, U.: *A Tutorial on Elliptic PDE Solvers and their Parallelization.* SIAM (2003).

[15] Edwards, C., Lombaerts, T., Smaili, H. *"Fault Tolerant Flight Control: A Benchmark Challenge"* Springer Verlag, Lectures Notes in Control and Information Sciences 399 (2010).

[16] Frazzoli, E.: *Robust Hybrid Control for Autonomous Vehicle Motion Planning.* PhD thesis, MIT. Cambridge, MA (2001).

[17] Frazzoli, E., Dahleh, M.A., Feron, E.: *A Maneuver-Based Hybrid Control Architecture for Autonomous Vehicle Motion Planning.* In T. Samad, G. Balas, eds., Software Enabled Control: Information Technology for Dynamical Systems. Wiley-IEEE Press (2003).

[18] Ghosh, M.K., Arapostathis, A., Marcus, S.I.: *An Optimal Control Problem Arising in Flexible Manufacturing Systems.* in Proc. 30th IEEE Conf. on Decision and Control, (1991): 1884-1849.

[19] Ghosh, M.K., Arapostathis, A., Marcus, S.I.: *Optimal Control of Switching Diffusions with Application to Flexible Manifacturing Systems.* SIAM J. Control Optim. **31**(5), (1993): 1183-1204.

[20] Ghosh, M.K., Arapostathis, A., Marcus, S.I.: *Ergodic Control of Switching Diffusions.* SIAM J. Control Optim. **35**(6) (1997): 1952-1988.

[21] Hradecky, S.: *Crash: Colgan DH8D at Buffalo on Feb 12th 2009, impacted home while on approach.* Aviation Herald, February (2010).

[22] Jacod, J., Shiryayev, A.N.: *"Limit Theorems for Stochastic Processes."* Springer Verlag (1987).

[23] Lunze J.: *Fault Diagnosis of Discretely Controlled Continuous Systems by means of Discrete Event Models* Discrete Event Dyn. Syst. 18 (2008): 181-2010.

[24] NTSB *Aircraft Accident Report 01/01: United Airlines Flight 585* http://www.ntsb.gov/Publictn/2001/AAR0101.pdf

[25] NTSB *MetroJet B-737 In-Flight Event.* February 24, (1999) http://www.ntsb.gov/Pressrel/1999/990223.htm

[26] NTSB *Aircraft Accident Report Uncontrolled Descent and Collision with Terrain USAir Flight 427* http://www.ntsb.gov/publictn/1999/AAR9901.pdf

[27] NTSB *Aircraft Accident Report Pan Am B707, Over The Atlantic, between London and Gander,* February 3, (1959).

[28] National Transportation Safety Committee *Aircraft Accident Report* Jakarta (2000) http://www.dephub.go.id/knkt/ntsc_aviation/Revised-MI185%20Final%20Report%20(2001)%20.pdf

[29] Raffard, R.L., Hu, J., Tomlin, C.J.: *Adjoint-based Optimal Control of the Expected Exit Time for Stochastic Hybrid Systems.* in Proc. of HSCC: Hybrid Systems: Computation and Control, Springer Verlag (2005).

[30] Riganelli, O., Grosu, R., Ramakrishnan, C.R., Smolka, S.A.: *Power Optimization in Fault-Tolerant MANETs.* In Proc. of Hase'08, the 11th IEEE High Assurance Systems Engineering Symposium, IEEE Computer Society, (2008): 362-370.

[31] Shaked M., Shanthikumar J.G.: *Reliability and Maintainability* in "Handbook of Operations Research and Management Science" vol. 2, North Holland, (1990)

[32] Flight United B-720 *Turbulence Penetration Study,* UAL B720, Upset near O'Neal Nebraska, July 12, (1963):18-19.

[33] U.S. FAA *Airplane Upset Recovery Training Aid,* Revision 2, 443 pages, http://www.faa.gov/other_visit/aviation_industry/ airline _opera-tors/training/media/AP_UpsetRecovery_Book.pdf

[34] Varadhan, S.R.S.: *"Large Deviations and Applications".* SIAM (1984).

[35] Mahmoud, M.M., Jiang, J., Zhang, Y. *"Active Fault Tolerant Control Systems. Stochastic Analysis and Synthesis."* Springer Verlag, Lectures Notes in Control and Information Sciences 287, (2003).

[36] Yang, H., Jiang, B., Cocquempot, V.: *"Fault Tolerant Control Design for Hybrid Systems."* Springer Verlag, Lectures Notes in Control and Information Sciences 397 (2010).

[37] Yang, H., Jiang, B., Cocquempot, V.: *Fault Tolerance Analysis for Stochastic Systems Using Switching Diffusion Processes* International Journal of Control 82(8), (2009): 1516-1525.

Quantified Differential Invariants*

André Platzer
Carnegie Mellon University
Computer Science Department
Pittsburgh, PA, USA
aplatzer@cs.cmu.edu

ABSTRACT

We address the verification problem for distributed hybrid systems with nontrivial dynamics. Consider air traffic collision avoidance maneuvers, for example. Verifying dynamic appearance of aircraft during an ongoing collision avoidance maneuver is a longstanding and essentially unsolved problem. The resulting systems are not hybrid systems and their state space is not of the form $\mathbb{R}^n$. They are distributed hybrid systems with nontrivial continuous and discrete dynamics in distributed state spaces whose dimension and topology changes dynamically over time. We present the first formal verification technique that can handle the complicated nonlinear dynamics of these systems. We introduce quantified differential invariants, which are properties that can be checked for invariance along the dynamics of the distributed hybrid system based on differentiation, quantified substitution, and quantifier elimination in real-closed fields. This gives a computationally attractive technique, because it works without having to solve the infinite-dimensional differential equation systems underlying distributed hybrid systems. We formally verify a roundabout maneuver in which aircraft can appear dynamically.

Categories and Subject Descriptors

F.3.1 [**Logics and Meanings of Programs**]: Specifying and Verifying and Reasoning about Programs; D.2.4 [**Software Engineering**]: Software/Program Verification

General Terms

Verification, Theory, Algorithms

Keywords

Distributed hybrid systems, verification logic, quantified differential equations, quantified differential invariants

*This material is based upon work supported by the National Science Foundation under Grant Nos. CNS-0926181, CNS-0931985, and CNS-1035800.

1. INTRODUCTION

Hybrid systems are everywhere [1, 9]. But several hybrid systems applications are really more than just hybrid systems. They involve multiple participants that communicate with each other to achieve their respective control objectives. The set of participants may even change over time whenever participants join or leave the networked control region. The resulting system dynamics has several aspects of hybrid systems, including their joint discrete and continuous dynamics. But that's not all there is to it.

When multiple participants are involved, the system has a high-dimensional hybrid dynamics. Then, the overall system behavior depends on the effects of how the system participants communicate and on structural properties like the communication topology, which may change over time. The most tricky part is when participants can join or leave the system at runtime, leading to reconfigurable system behavior and dynamics with varying dimensionality (appearance and disappearance add or remove state variables during a system evolution). The continuous system dynamics can no longer be described by ordinary differential equations. It already requires arbitrary-dimensional quantified differential equations [17]. Even without appearance and disappearance, flat representations of the system as a fixed set of all participants in a gigantic hybrid automaton becomes computationally infeasible for moderately-sized systems already.

The class of systems that share the above aspects is called *distributed hybrid systems* [6, 17], based on models for reconfigurable hybrid systems [5, 12]. Distributed hybrid systems evolve according to a joint discrete, continuous, structural, and dimensionality-changing dynamics. Hybrid systems [20, 21, 24, 18, 16] cannot represent the distributed aspects nor those of dynamic appearance and disappearance of participants. They would need infinitely many variables and differential equations to describe the system. Distributed systems [2] cannot represent the continuous system dynamics. We need to join both worlds to faithfully represent the system. But we also need verification techniques to handle the complicated resulting dynamics.

Prominent examples of distributed hybrid systems arise, e.g., in air traffic control, where multiple aircraft are flying to their respective goals. They use collision avoidance maneuvers as needed to prevent collisions resulting from traffic conflicts in imperfect flight trajectory planning, which have gone unnoticed by pilots and air traffic controllers.

As an example, consider the roundabout collision avoidance maneuver [19, 26] that aircraft x and y are following in Fig. 1. Here, the hybrid system dynamics comes from the continuous movement of the aircraft and the discrete flight decisions by pilots and autopilots. Even though this has usually been ignored in formal analysis, air traffic control systems are really distributed hybrid systems, because the set of aircraft participating in a coordinated flight maneuver may be large and may change over time

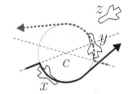

Figure 1: New appearance during collision avoidance

when more aircraft come near a traffic conflict or safely leave the dense area. Aircraft z may suddenly come too close to the collision avoidance maneuver that x and y are still performing in Fig. 1. Then z appears new into the horizon of relevance for x and y. This can happen when z changes its flight direction unaware of the avoidance intentions of x and y, when the collision avoidance of x and y takes longer than expected, or when z was forced to do collision avoidance with another aircraft a and now approaches the x, y roundabout. This is especially common in crowded airspace where global planning of all trajectories is computationally infeasible. Verifying the dynamic appearance of aircraft in the horizon of relevance during an air traffic control maneuver is an essentially unsolved problem. One major reason why distribution effects have been ignored in analysis so far is that formal verification techniques did not previously support distributed hybrid systems, especially not with such complicated dynamics as required in air traffic control. Previous techniques needed ad-hoc tricks to discuss the problem away but could not verify it.

This paper is set out to correct this analytic deficiency by developing verification techniques for distributed hybrid systems with complicated nonlinear dynamics. For this purpose, we present a generalization of differential invariants [15, 18, 16]. These symbolic safety certificates, which we call *quantified differential invariants* can be checked for invariance along the dynamics of the distributed hybrid system based on differentiation, quantified substitution, and quantifier elimination in real-closed fields. This gives a computationally attractive technique, especially because it works without having to solve the infinite- or arbitrary-dimensional quantified differential equation systems underlying distributed hybrid systems. Our approach bears some resemblance to Lyapunov functions. But it works for safety instead of stability, it supports arbitrary logical formulas instead of just a single function, and we have extended it appropriately to cover distributed hybrid systems.

Differential invariants [15, 18, 16] are formulas F that do not change their truth-value along the dynamics of a differential equation describing a continuous system transition in a hybrid system; see Fig. 2. To prove that F is a differential invariant, it is sufficient to check a condition on the directional derivatives of all terms of the formula [18]. Like almost all other verification techniques, differential invariants only work for hybrid systems, not for distributed hybrid systems. Here we present an extension to distributed hybrid systems. In a sense made precise by our complete axiomatization relative to quantified

Figure 2: Differential invariant F

differential equations [17], we can focus mostly on verification techniques for quantified differential equations, here, because our previous proof calculus lifts any such verification technique completely to distributed hybrid systems.

Our main contributions are as follows. We introduce a generalization of differential invariants (called *quantified differential invariants*), which can verify quantified differential equations. In combination with our previous work [17], this is the first formal verification technique for distributed hybrid systems with nontrivial continuous dynamics. We prove that our verification technique is formally sound. As an application scenario, we develop and verify a collision avoidance maneuver in air traffic control, where dynamic appearance of aircraft has been a major unsolved problem before. We do not claim to solve all issues in how to fly the best maneuver. Yet, we present the first verification technique with which these maneuvers can be verified. We believe this to be an important step in the development of formally assured air traffic control maneuvers and other distributed designs of cyber-physical systems.

2. RELATED WORK

The importance of understanding dynamic / reconfigurable distributed hybrid systems was recognized in modeling languages SHIFT [5] and R-Charon [12]. They focused on simulation and compilation [5] or the development of a semantics [12], so that no verification is possible yet. Stochastic simulation has been proposed [13], but soundness has not been proven, because ensuring coverage is difficult by simulation.

For distributed hybrid systems, even giving a formal semantics is very challenging [3, 22, 12, 27]! A formal semantics has been defined for a hybrid version of CSP [3] and a hybrid version of the π-calculus [22]. Rounds [22] also presented a semantics for a spatial logic for these processes. But from the semantics alone, no verification is possible in these approaches, except perhaps by manual semantic reasoning.

Other process-algebraic approaches, like χ [27], have been developed for modeling and simulation. Verification is still limited to small fragments that can be translated directly to other verification tools for (non-distributed) hybrid systems.

Our focus, instead, is on sound formal verification, not on simulation. And it is on distributed hybrid systems.

Approaches for distributed systems [2] do not cover hybrid systems, because the addition of differential equations to distributed systems is even more challenging than the addition of differential equations to discrete dynamics.

Multi-party distributed control has been suggested for air traffic control [26, 25, 14, 10]. Due to limits in previous verification technology, no full formal verification of the distributed hybrid dynamics has been possible for these systems yet. Random simulation has been proposed [14], but that does not guarantee that the maneuver always works safely.

Ad-hoc informal arguments like "the distributed effects can be ignored when we hope that at most 5 aircraft are close" are neither general nor formal verification. They also do not show whether the system stays safe if, nevertheless, a sixth aircraft a approaches. If each 5-tuple of aircraft around the sixth aircraft u assumes that u cannot appear and has to go some place else, then the system will still be unsafe, because u would have to stop moving (and drop out of the sky) to comply with all those assumptions.

Lyapunov-style verification has been used successfully for hybrid systems, including barrier certificates [20, 21], template equations [24, 23], differential invariants [15, 18, 16] and a constraint-based template approach [7]. Here, we present an extension for distributed hybrid systems.

3. DISTRIBUTED HYBRID SYSTEMS

We review and extend a system model that we have recently introduced for modeling distributed hybrid systems [17]. The basic idea is to parameterize hybrid system models by agents and provide a means of quantifying over all affected agents (of a certain type, e.g., A for aircraft). Hence, instead of a primitive state variable like position $x : \mathbb{R}$, distributed hybrid systems have a state function $x : A \to \mathbb{R}$ so that $x(i)$ will be the position of aircraft i and $x(j)$ denotes the position of aircraft j. In order to express that all aircraft i evolve simultaneously, we will use quantified differential equations like $\forall i \, x(i)' = \theta$ with a universal quantifier $\forall i$ (i.e., for all aircraft i) and some term θ for the flight equation for each i. For the purpose of simplifying the presentation, we ignore typing information (like A for aircraft, C for cars, and $\mathbb{R}$ for reals), because it will be clear from the context. Typing is needed for modelling multi-agent hybrid systems with agents of different kinds (a mixed systems with both aircraft and cars, not just all aircraft). We refer to [17] for details on typing.

Quantified Hybrid Programs.

As a system model for distributed hybrid systems, we use *quantified hybrid programs* (QHP) [17]. QHPs are a Kleene algebra with tests [11] based on quantified assignments and quantified differential equation systems for describing distributed hybrid dynamics. QHPs are defined by the following grammar (α, β are QHPs, θ a term, i a variable, f is a function symbol, and H is a formula of first-order logic):

$$\alpha, \beta \ ::= \ \forall i \, f(i) := \theta \mid \forall i \, f(i)' = \theta \, \& \, H \mid i := \mathsf{new}$$
$$\mid \ ?H \mid \alpha \cup \beta \mid \alpha; \beta \mid \alpha^*$$

The effect of *quantified assignment* $\forall i \, f(i) := \theta$ is an instantaneous discrete jump assigning θ to $f(i)$ simultaneously for all objects i. Usually variable i occurs in the term θ. The effect of *quantified differential equation* $\forall i \, f(i)' = \theta \, \& \, H$ is a continuous evolution where, for all objects i, all differential equations $f(i)' = \theta$ hold *and* (written &) formula H holds throughout the evolution (the state remains in the region described by H). Again, variable i usually occurs in the term θ. The dynamics of QHPs changes the interpretation of terms over time: $f(i)'$ is intended to denote the derivative of the interpretation of the term $f(i)$ over time during continuous evolution, not the derivative of $f(i)$ by its argument i. For $f(i)'$ to be defined, we assume f is an $\mathbb{R}$-valued function symbol. Time itself is implicit and can be axiomatized by a differential equation when needed. The effect of $i := \mathsf{new}$ is to add a new object i (say a new aircraft of type A) to the system. Then $x(i)$ could denote its position, $d(i)$ its direction, and so on.

The effect of *test* $?H$ is a *skip* (i.e., no change) if formula H is true in the current state and *abort* (blocking the system run by a failed assertion), otherwise. Tests can be used to define if-then-else. *Nondeterministic choice* $\alpha \cup \beta$ is for alternatives in the behavior of the distributed hybrid system. In the *sequential composition* $\alpha; \beta$, QHP β starts after α finishes (β never starts if α continues indefinitely). *Nondeterministic repetition* α^* repeats α an arbitrary number of times, possibly zero times.

QHPs can be extended [17] to systems of quantified differential equations, simultaneous assignments to multiple functions f, g, or statements with multiple quantifiers ($\forall i \, \forall j \, \dots$).

We use logical formulas with operators $\wedge$ (and), $\vee$ (or), $\to$ (implies), $\forall$ (forall) and $\exists$ (exists) to express properties of QHPs. In addition to all formulas of first-order real arithmetic, we allow formulas of the form $[\alpha]\phi$ with a QHP α and formula ϕ. Formula $[\alpha]\phi$ is true in a state ν iff ϕ is true in all states that are reachable from ν by following the transitions of α. We only need to define the reachability relation of QHPs in order to define a semantics for these formulas [17].

Case Study: Distributed Roundabout Maneuvers.

Before we discuss the semantics in detail, we first give an intuitive example that we will also use as a case study for distributed hybrid systems verification in Section 7. We consider the roundabout collision avoidance maneuver for air traffic control [26, 19, 16]. In the literature, formal verification of the hybrid dynamics of air traffic control focused on a fixed number of aircraft, usually two. In reality, many more aircraft are in the same flight corridor, even if not all of them participate in the same maneuver. They may be involved in multiple distributed maneuvers at the same time, however. Perfect global trajectory planning quickly becomes infeasible then. The verification itself also becomes much more complicated for three aircraft already. Explicit replication of the system dynamics n times is computationally infeasible for larger n. Yet, collision avoidance maneuvers need to work for an (essentially) unbounded number of aircraft. Because global trajectory planning is infeasible, the appearance of other aircraft into a local collision avoidance maneuver always has to be expected and managed safely. See Fig. 1 for a general illustration of roundabout-style collision avoidance maneuvers and the phenomenon of dynamic appearance of some new aircraft z into the horizon of relevance.

The resulting flight control system has several characteristics of hybrid dynamics. But it is not a hybrid system and does not even have a fixed finite number of variables in a fixed finite-dimensional state space. The system forms a distributed hybrid system, in which all aircraft fly at the same time and new aircraft may appear from remote areas into the local flight scenario. Each aircraft i has a position $x(i) = (x_1(i), x_2(i))$ and a velocity vector $d(i) = (d_1(i), d_2(i))$. We model the continuous dynamics of an aircraft i that follows a flight curve with an angular velocity $\omega(i)$ by the (function) differential equation:

$$x_1(i)' = d_1(i), x_2(i)' = d_2(i),$$
$$d_1(i)' = -\omega(i)d_2(i), d_2(i)' = \omega(i)d_1(i) \quad (\mathcal{F}_{\omega(i)}(i))$$

This differential equation, which we denote by $\mathcal{F}_{\omega(i)}(i)$, is the standard equation for curved flight from the literature [26, 19, 16], but lifted to function symbols that are parameterized by aircraft i. Now, the quantified differential equation $\forall i \, \mathcal{F}_{\omega(i)}(i)$ characterizes that *all* aircraft i fly along their respective (function) differential equation $\mathcal{F}_{\omega(i)}(i)$ according to their respective angular velocities $\omega(i)$ at the same time. This quantified differential equation captures what no finite-dimensional differential equation system could ever do. It characterizes the simultaneous movement of an unbounded, arbitrary, and even growing or shrinking set of aircraft.

$$\psi \equiv \forall i, j \, \mathcal{P}(i,j) \rightarrow [drm] \forall i, j \, \mathcal{P}(i,j)$$

$$drm \equiv (\text{free; } \quad newac; \quad entry; \quad \forall i \, \mathcal{F}_\omega(i))^*$$

$$free \equiv \forall i \, \mathcal{F}_{\omega(i)}(i) \, \& \, \forall i, j \, \mathcal{P}(i,j)$$

$$newac \equiv (n := \text{new}; \; ? \forall i \, \mathcal{P}(i,n)) \cup (? true)$$

$$entry \equiv c_1 := *; c_2 := *; \omega := *;$$
$$\forall i \, d_1(i) := -\omega(x_2(i) - c_2); \; \forall i \, d_2(i) := \omega(x_1(i) - c_1)$$

Figure 3: Distributed roundabout collision avoidance maneuver for air traffic control

Two aircraft i and j have violated the safe separation property if they falsify the following formula

$$\mathcal{P}(i,j) \equiv i = j \vee (x_1(i) - x_1(j))^2 + (x_2(i) - x_2(j))^2 \geq p^2$$

which says that aircraft i and j are either identical or separated by at least the protected zone p (usually 5mi). For the aircraft control system to be safe, all aircraft have to be safely separated, i.e., need to satisfy $\forall i, j \, \mathcal{P}(i,j)$. Yet, it is not enough that this formula holds in the beginning of the system evolution. It has to hold always. We express this safety property for collision avoidance by the logical formula ψ shown in Fig. 3. It uses the formula $[drm] \forall i, j \, \mathcal{P}(i,j)$ to say that safe separation $\mathcal{P}(i,j)$ holds for all aircraft i, j *always* after following the dynamics of the distributed hybrid system drm arbitrarily long, if separation holds in the beginning of the evolution (the assumption before $\rightarrow$).

We present a QHP for the distributed roundabout maneuver drm in Fig. 3. First, all aircraft perform free flight (*free*). They fly arbitrarily with their respective angular velocity $\omega(i)$, but can only do so as long as the system stays safe in the evolution domain $\forall i, j \, \mathcal{P}(i,j)$. That is, free flight is only permitted when the aircraft are still far enough apart. It is easy to generalize the maneuver by allowing the aircraft to repeatedly change their angular velocity $\omega(i)$ in *free*. For space reasons, we do not pursue this here. Secondly, a new aircraft may appear by the operation $n := \text{new}$ in *newac*. Yet, it may only appear at positions satisfying the subsequent test $? \forall i \, \mathcal{P}(i,n)$ of n having sufficient distance to all other aircraft i. The system cannot possibly be safe if the sensors and communication is so bad that an aircraft could suddenly appear 5 feet before a collision. If no aircraft appears (the choice after operator $\cup$ in *newac*), there is no further test (trivial test $? true$). Thirdly, the aircraft (including the new one, if created during *newac*) coordinate their flight paths locally and initiate collision avoidance as necessary (*entry*). For *entry*, we use a simplified procedure based on what has been used in classical roundabouts [26, 15]. Operation *entry* chooses a center $c = (c_1, c_2)$ like in Fig. 1 and a common angular velocity ω for the roundabout. It then directs all aircraft ($\forall i$) into the appropriate directions $d(i)$. Such random assignments $c_1 := *$ that choose an arbitrary value for c_1 are definable [17]. In the last step of drm, all aircraft follow coordinated roundabout circles by evolving along $\forall i \, \mathcal{F}_\omega(i)$ with the common ω. Overall, drm repeats collision avoidance maneuvers when necessary, as indicated by the repetition $*$.

It might look as if new aircraft were only allowed to appear immediately before a collision avoidance maneuver happens, not in free flight. That is not the case, though, because it is perfectly okay to follow the subsequent collision avoidance circle $\forall i \, \mathcal{F}_\omega(i)$ for zero time units if the system is still safe, because it can then re-enter *free* immediately since it still satisfies its evolution domain restriction $\forall i, j \, \mathcal{P}(i,j)$.

Semantics and Differential State Flows.

What is the behavior of a QHP? A system *state* ν associates a function $\nu(x)$ of appropriate type with each function symbol x, which associates the x-value $\nu(x)(o)$ to each object o. The set of all states is denoted by $\mathcal{S}$. For representing appearance and disappearance of objects in our semantic model, we use an existence function $\mathsf{E}()$ that has value $\mathsf{E}(o) = 1$ if object o exists. For an elaboration of how we represent (dis)appearance and the new operator, we refer to previous work [17].

For a formula ϕ of first-order logic, we write $\nu \models \phi$ iff ϕ is true at state ν. We write $\nu \models [\alpha] \phi$ iff $\tau \models \phi$ for all states τ reachable from ν by a transition of QHP α. For this, we subsequently define the reachability relation $(\nu, \tau) \in \rho(\alpha)$.

We write $\nu \llbracket \theta \rrbracket$ to denote the value of term θ in state ν. Further, $\nu_i^e \llbracket \theta \rrbracket$ denotes the value of term θ in state ν_i^e, i.e., in a state like ν, but with the interpretation of variable i changed to e. Discrete assignments can be given a semantics based on the effect they have on the state, i.e., how they transform the current state ν to the next state τ. We define the behavioral semantics of quantified discrete assignments as taking an effect on all objects (say aircraft) at once. For differential equations, however, this is more difficult, because a single state is not sufficient for giving a semantics to differential equations, let alone quantified differential equations.

For a semantics for quantified differential equations, we first have to give a meaning to differential function symbols like $x(i)'$. At isolated states, differential symbols do not have a well-defined semantics, because derivatives are not defined. Thus, we consider a flow φ of the system and define a semantics for $x(i)'$ at every state along this flow.

Definition 1. (DIFFERENTIAL STATE FLOW). A function $\varphi : [0, r] \rightarrow \mathcal{S}$ is called *(differential) state flow* of duration $r \geq 0$ if φ is componentwise continuous on $[0, r]$, i.e., for all $x \in \Sigma$ and all u, $\varphi(\zeta)(x)(u)$ is continuous in ζ. Then, the *differentially augmented state* $\bar{\varphi}(\zeta)$ of φ at $\zeta \in [0, r]$ agrees with $\varphi(\zeta)$ except that it assigns values to some of the extra differential function symbols $x^{(1)}$: If $\varphi(t)(x)(u)$ is continuously differentiable in t at ζ for all u, then $\bar{\varphi}(\zeta)(x^{(1)})$ is defined as the function $u \mapsto \frac{d\varphi(t)(x)(u)}{dt}(\zeta)$ that maps u to the time derivative $\frac{d\varphi(t)(x)(u)}{dt}(\zeta)$ at ζ; otherwise the value of $x^{(1)}$ is not defined for $\bar{\varphi}(\zeta)$. The value of $x(i)'$ at $\bar{\varphi}(\zeta)$ is defined to be the same as the value of $x^{(1)}(i)$ at $\bar{\varphi}(\zeta)$.

A state flow φ of duration r is called *state flow of the order of quantified differntial equation* $\forall i \, f(i)' = \theta \, \& \, H$ iff the value of each differential symbol occurring in it is defined on $[0, r]$. For a formula F with differential function symbols, and a state flow φ of the order of F (all differential symbols are defined), we write $\varphi \models F$ iff for all $\zeta \in [0, r]$, $\bar{\varphi}(\zeta) \models_\mathbb{R} F$ using the standard semantics $\models_\mathbb{R}$ of first-order real arithmetic. In particular, $\varphi \models \forall i \, f(i)' = \theta \, \& \, H$ iff, at each time $\zeta \in [0, r]$ and for each interpretation e of i:

- All differential equations hold and derivatives exist (trivial for $r = 0$):

$$\frac{d\left(\varphi(t)_i^e \llbracket f(i) \rrbracket\right)}{dt}(\zeta) = (\varphi(\zeta)_i^e \llbracket \theta \rrbracket)$$

- And the evolution domain is respected: $\varphi(\zeta)_i^e \models H$.

The *transition relation*, $\rho(\alpha) \subseteq \mathcal{S} \times \mathcal{S}$, of QHP α specifies which state $\tau \in \mathcal{S}$ is reachable from $\nu \in \mathcal{S}$ by running QHP α. It is defined inductively:

1. $(\nu, \tau) \in \rho(\forall i\, f(i) := \theta)$ iff state τ is identical to ν except that for each interpretation e of i, the value of f changes as follows: $\tau(f)(e) = \nu_i^e[\![\theta]\!]$.

2. $(\nu, \tau) \in \rho(\forall i\, f(i)' = \theta \,\&\, H)$ iff there is a differential state flow $\varphi{:}[0, r] \to \mathcal{S}$ for some $r \geq 0$ with $\varphi(0) = \nu$ and $\varphi(r) = \tau$ such that $\varphi \models \forall i\, f(i)' = \theta \,\&\, H$.

3. $\rho(?H) = \{(\nu, \nu) \; : \; \nu \models H\}$

4. $\rho(\alpha \cup \beta) = \rho(\alpha) \cup \rho(\beta)$

5. $\rho(\alpha; \beta) = \{(\nu, \tau) \; : \; (\nu, z) \in \rho(\alpha) \text{ and } (z, \tau) \in \rho(\beta)$ for a state $z\}$

6. $(\nu, \tau) \in \rho(\alpha^*)$ iff there is an $n \in \mathbb{N}$ with $n \geq 0$ and states $\nu = \sigma_0, \ldots, \sigma_n = \tau$ such that $(\sigma_i, \sigma_{i+1}) \in \rho(\alpha)$ for all $0 \leq i < n$.

4. DERIVATIONS & DIFFERENTIATION

Now that we have defined a semantics, we know what the behavior of the distributed hybrid system models of QHPs is. The next important question is how we can prove properties of this behavior, e.g., that a QHP never leaves a safe region. The trouble is that we cannot rely on having good solutions of the quantified differential equations that distributed hybrid systems follow. First of all, quantified differential equations do not have a fixed dimension and are thus more complicated than ordinary differential equations. Secondly, "most" differential equations (quantified or not) either have solutions that are outside decidable classes of arithmetic or have no closed-form solutions at all. The flight equation $\mathcal{F}_{\omega(i)}(i)$, for instance, has trigonometric solutions with first-order functions, which is not even semi-decidable. In either case, we cannot rely on using the solutions for verification purposes. Instead, we are looking for a different way of proving properties about the behavior of quantified differential equations by working with the quantified differential equations, not their solutions.

In preparation for that, we define the notions that we will use as crucial reasoning primitives in Section 5. We will define quantified differential invariants of formulas using syntactic total derivations and their relation to analytic differentiation. We first introduce these notions and generalize them to quantified differential equations of distributed hybrid systems. We generally assume all formulas to be given in *prefix disjunctive normal form*, that is they are of the form $Q \bigvee_i \bigwedge_j F_{i,j}$ with a prefix Q of quantifiers and atomic formulas $F_{i,j}$ that have no quantifiers or logical operators.

Derivations.

Hybrid systems evolve along trajectories that are defined in terms of the effects of discrete transitions and in terms of solutions of their differential equations. A (vectorial) function f is a solution of a differential equation system if the function f satisfies the differential equation, in which we replace f' by its analytic differentiation. This is a good mathematical definition, but not computational enough for

verification purposes, because analytic differentiation is defined as a limit process at infinitely many points in time. This becomes computationally even worse for quantified differential equations, because those amount to an essentially infinite-dimensional differential equation system, which we cannot easily solve simultaneously for all positions at once.

In order to turn this mathematically precise definition into a computationally tractable algorithm, we, instead, define an entirely syntactic and algebraic total derivation and then prove that its valuation along differential flows coincides with analytic differentiation. We can easily compute the syntactic total derivation algebraically. And then we show that its value coincides with the result of analytic differentiation. In addition, we generalize total derivations to quantified formulas and constraints with first-order function symbols, as necessary for quantified differential equations.

Definition 2. (DERIVATION). The operator D that is defined as follows on terms is called *syntactic (total) derivation*:

$$D(r) = 0 \qquad \text{for } r \in \mathbb{Q} \tag{1a}$$

$$D(x(s)) = x(s)' \quad \text{for function symbol } x : C \to \mathbb{R}$$
$$\text{with } C \neq \mathbb{R} \text{ discrete} \tag{1b}$$

$$D(a + b) = D(a) + D(b) \tag{1c}$$

$$D(a - b) = D(a) - D(b) \tag{1d}$$

$$D(a \cdot b) = D(a) \cdot b + a \cdot D(b) \tag{1e}$$

$$D(a/b) = (D(a) \cdot b - a \cdot D(b))/b^2 \tag{1f}$$

We extend D to first-order formulas F in prefix disjunctive normal form as follows:

$$D(\forall i\, F) \equiv \forall i\, D(F)$$
$$D(\exists i\, F) \equiv \forall i\, D(F)$$
$$D(F \wedge G) \equiv D(F) \wedge D(G)$$
$$D(F \vee G) \equiv D(F) \wedge D(G)$$
$$D(a \geq b) \equiv D(a) \geq D(b) \quad \text{accordingly for } <, >, \leq, =.$$

In the aircraft example, consider the safe separation formula $\forall i, j\, \mathcal{P}(i, j)$. We compute its syntactic total derivation $D(\forall i, j\, \mathcal{P}(i, j))$ to be the following (differential) expression:

$$\forall i, j\, \big(i' = j' \wedge$$
$$2(x_1(i) - x_1(j))(x_1(i)' - x_1(j)')$$
$$+ 2(x_2(i) - x_2(j))(x_2(i)' - x_2(j)') \geq 0\big)$$

For making use of this syntactic total derivation for verification purposes, we need to define how we can understand its differential function symbols $x_1(i)', x_2(j)', i'$ and so on. These differential function symbols do not even have a well-defined semantics if we try to evaluate or prove the above expression at an isolated state. But the first thing we need to do is to understand what the above expression could mean at all. Def. 2 is entirely syntactical (the $x_1(i)'$ are just symbols), which is good, because then we can compute it algebraically during verification. But what is its meaning? What is its relationship to the results of real analytic differentiation $\frac{d}{dt}$ that define the behavioral system semantics?

Differentiation.

In the following key lemma, we show that the syntactic derivation D directly coincides with analytic differentiation

$\frac{d}{dt}$, even for terms with differential function symbols, as occurring in quantified differential equations.

LEMMA 1 (DERIVATION LEMMA). *The valuation of terms is a differential homomorphism: Let θ be a term and let φ : $[0, r] \to \mathcal{S}$ be any state flow of the order of $D(\theta)$ and of duration $r > 0$ along which the value of θ is defined (as no divisions by zero occur). Then we have for all $\zeta \in [0, r]$ that*

$$\frac{\mathsf{d}\,\varphi(t)[\![\theta]\!]}{\mathsf{d}t}(\zeta) = \bar\varphi(\zeta)[\![D(\theta)]\!].$$

In particular, $\varphi(t)[\![\theta]\!]$ is continuously differentiable (where θ is defined) and its derivative exists on $[0, r]$.

PROOF. The proof is an inductive consequence of the correspondence of the semantics of differential symbols and analytic derivatives in state flows (Def. 1). It uses the assumption that the flow φ remains within the domain of definition of θ and is continuously differentiable in all variables of θ. In particular, all denominators are nonzero during φ.

- If θ is a function term $x(s)$, the lemma holds by Def. 1:

$$\frac{\mathsf{d}\,\varphi(t)[\![x(s)]\!]}{\mathsf{d}t}(\zeta) = \frac{\mathsf{d}\,\varphi(t)(x)(\varphi(t)[\![s]\!])}{\mathsf{d}t}(\zeta)$$
$$\overset{(!)}{=} \frac{\mathsf{d}\,\varphi(t)(x)(\varphi(\zeta)[\![s]\!])}{\mathsf{d}t}(\zeta) = \bar\varphi(\zeta)(x')(\varphi(\zeta)[\![s]\!])$$
$$= \bar\varphi(\zeta)[\![x'(s)]\!] = \bar\varphi(\zeta)[\![D(x(s))]\!].$$

  The equation marked (!) holds, because the argument s of x has type $C \neq \mathbb{R}$, which is equipped with the discrete topology. Consequently, only the neighborhood $\{\zeta\}$ is relevant in the limit process of the derivative, because s only has constant local dynamics. The derivative exists because the state flow is of order 1 in x and, thus, (continuously) differentiable for x.

- If θ is of the form $a + b$, the desired result can be obtained by using the properties of derivatives, derivations (Def. 2), and evaluation $\nu[\![\cdot]\!]$ of terms:

$$\frac{\mathsf{d}}{\mathsf{d}t}(\varphi(t)[\![a + b]\!])(\zeta)$$
$$= \frac{\mathsf{d}}{\mathsf{d}t}(\varphi(t)[\![a]\!] + \varphi(t)[\![b]\!])(\zeta) \qquad \nu[\![\cdot]\!] \text{ homomorphic}$$
$$= \frac{\mathsf{d}}{\mathsf{d}t}(\varphi(t)[\![a]\!])(\zeta) + \frac{\mathsf{d}}{\mathsf{d}t}(\varphi(t)[\![b]\!])(\zeta) \quad \frac{\mathsf{d}}{\mathsf{d}t} \text{ is linear}$$
$$= \bar\varphi(\zeta)[\![D(a)]\!] + \bar\varphi(\zeta)[\![D(b)]\!] \qquad \text{induction hyp.}$$
$$= \bar\varphi(\zeta)[\![D(a) + D(b)]\!] \qquad \nu[\![\cdot]\!] \text{ homomorphic}$$
$$= \bar\varphi(\zeta)[\![D(a + b)]\!] \qquad D(\cdot) \text{ derivation}$$

- The case where θ is of the form $a \cdot b$ or $a - b$ is similar, using Leibniz product rule (1e) or subtractivity (1d) of Def. 2, respectively.

- The case where θ is of the form a/b uses (1f) of Def. 2 and depends on the assumption that $b \neq 0$ along φ. This assumption holds as the value of θ is assumed to be defined all along state flow φ.

- The values of numbers $r \in \mathbb{Q}$ do not change during a state flow (in fact, they are not affected by the state at all); hence their derivative is $D(r) = 0$. $\square$

With Lemma 1, syntactic total derivations are directly related to the behavior during continuous flows of the system, which is good. But how can we use them in a proof? Expressions like $D(\forall i, j\, \mathcal{P}(i, j))$ are related to expressions with a meaning along a continuous flow. But we do not want to reason explicitly about what happens at a point in time during a continuous flow, otherwise we would need to know their solutions for verification and be back at square one. What prevents us from making sense of an expression like $D(\forall i, j\, \mathcal{P}(i, j))$ is the occurrence of differential function symbols like $x_2(i)'$ in it, which are only well-defined along a flow. So instead, we find a way to get rid of the differential function symbols without changing the meaning, i.e., the link to the behavioral semantics of the distributed hybrid system.

Consider some quantified differential equation $\forall i\, f(i)' = \theta$. What is the relationship between this quantified differential equation and a quantified assignment $\forall i\, f(i) := \theta$ to the function symbol $f(i)$? Obviously that the quantified differential equation takes effect continuously where term θ describes the rate of change of $f(i)$ along a flow φ, yet the quantified assignment just has an instant effect at a single state ν of changing $f(i)$ to the new value θ once and then leaving $f(i)$ alone. This is quite a fundamental difference.

But now, what is the relationship between the quantified differential equation $\forall i\, f(i)' = \theta$ and a quantified assignment $\forall i\, f(i)' := \theta$ to the *differential function symbol $f(i)'$*? This question is more tricky. We cannot really understand the latter quantified assignment at a single state ν. First of all, the semantics of differential function symbols is only well-defined along a flow φ, not at an isolated state ν. The semantics is further defined locally per (differentially augmented) state $\bar\varphi(\zeta)$ for each time ζ. So, instead, we consider a flow φ and one of its local differential states $\bar\varphi(\zeta)$. At this local differential state, we perform the quantified assignment $\forall i\, f(i)' := \theta$ and consider its effect on a differential term v. That is we consider $[\forall i\, f(i)' := \theta]v$ at $\bar\varphi(\zeta)$. Now the interesting point is that the effect of this operation corresponds directly to the local effect of the quantified differential equation. That is, if the flow φ respects the quantified differential equation $\forall i\, f(i)' = \theta$, then the quantified assignment $\forall i\, f(i)' := \theta$ does not alter the value of any differential terms.

That is, we show that a quantified assignment of the right-hand side θ of the differential equation to the differential term $f(i)'$ on its left-hand side does not change the value of arbitrary differential terms along a flow φ that already respects this quantified differential equation. This is an interesting generalized differential substitution property. It shows that quantified differential equations have consequences that correspond to substitutions by quantified assignments along their flows. Hence, locally, there is a way of understanding the effects of a quantified differential equations expressed as quantified assignments. We make this formally precise in the following lemma.

LEMMA 2 (DIFFERENTIAL SUBSTITUTION PROPERTY). *If φ is a state flow satisfying $\varphi \models \forall i\, f(i)' = \theta\, \&\, H$, then the property $\varphi \models v = [\forall i\, f(i)' := \theta]v$ holds for all (differential) terms v that include only differential symbols of the form $f(i)'$ for some variable i.*

PROOF. The proof is by induction on the structure of v. Consider a point in time ζ during the flow φ.

1. If v is a differential symbol, then, by assumption, it is

$$(DI) \quad \frac{H \to [\forall i\, f(i)' := \theta] D(F)}{F \to [\forall i\, f(i)' = \theta \,\&\, H]F}$$

$$(DC) \quad \frac{F \to [\forall i\, f(i)' = \theta \,\&\, H]G \qquad F \to [\forall i\, f(i)' = \theta \,\&\, H \wedge G]F}{F \to [\forall i\, f(i)' = \theta \,\&\, H]F}$$

$$([:=]) \quad \frac{\text{if } \exists i\, i = [\mathcal{A}]u \text{ then } \forall i\, (i = [\mathcal{A}]u \to \phi(\theta)) \text{ else } \phi(f([\mathcal{A}]u)) \text{ fi}}{\phi([\forall i\, f(i) := \theta]f(u))} \quad 1$$

$$([\cup]) \quad \frac{[\alpha]\phi \wedge [\beta]\phi}{[\alpha \cup \beta]\phi}$$

---

[1] Occurrence $f(u)$ in $\phi(f(u))$ is not in scope of a modality and we abbreviate quantified assignment $\forall i\, f(i) := \theta$ by $\mathcal{A}$.

Figure 4: Proof rules using quantified differential invariants for distributed hybrid systems

of the form $f(i)'$. Then

$$\bar\varphi(\zeta)[\![f(i)']\!] = \bar\varphi(\zeta)[\![\theta]\!] = \bar\varphi(\zeta)[\![[\forall i\, f(i)' := \theta]f(i)']\!]$$

because $f(i)'$ and θ have the same value along any φ satisfying the assumption $\varphi \models \forall i\, f(i)' = \theta \,\&\, H$.

2. If v is a non-differential function symbol $g(i)$ with a variable i, then it is not affected by assigning to differential symbols. Thus, $\varphi \models g(i) = [\forall i\, f(i)' := \theta]g(i)$.

3. If v is a function term of the form $f(s)$ for a function symbol f and (possibly vectorial) term s, then f itself is not affected by the assignment, but s might be. Then

$$\bar\varphi(\zeta)[\![[\forall i\, f(i)' := \theta]f(s)]\!] = \bar\varphi(\zeta)[\![f([\forall i\, f(i)' := \theta]s)]\!]$$
$$= \bar\varphi(\zeta)[\![f(s)]\!]$$

The last equation holds, because, by induction hypothesis, $\varphi \models s = [\forall i\, f(i)' := \theta]s$, which directly implies that $\bar\varphi(\zeta) \models s = [\forall i\, f(i)' := \theta]s$. $\qquad\square$

5. QUANTIFIED DIFFERENTIAL INVARIANTS

Based on the notions introduced in the last sections, we can now describe our new verification approach for distributed hybrid systems with nontrivial continuous dynamics. Our verification approach is based on logic and automated theorem proving techniques. For each kind of operator that can occur in a QHP describing a distributed hybrid system, we need to give a proof rule that takes care of it. That is, for the operators ; , $\cup$, *, ? and, most importantly, for quantified assignments and quantified differential equations. We list our proof rules for verifying distributed hybrid systems in Fig. 4 and explain them subsequently. For the operators ; , $\cup$, *, ? of Kleene algebras with tests [11], there are classical proof rules, which we have previously shown to apply to hybrid systems [15]. It can be shown easily [17] that we can still use the rules for Kleene algebras in distributed hybrid systems. For instance, the proof rule [$\cup$] axiomatizes non-deterministic choice $\cup$. This rule expresses that, when we want to prove the formula $[\alpha \cup \beta]\phi$ below the inference bar (*conclusion*), it is sufficient to prove the formula $[\alpha]\phi \wedge [\beta]\phi$ above the inference bar (*premiss*). And this makes sense, because if the premiss holds, that is, if all behavior of QHP α safely stays in the region described by formula ϕ (i.e., $[\alpha]\phi$ holds) and, independently, all behavior of β stays in ϕ

(i.e., $[\beta]\phi$ holds), then all behavior of the compound system $\alpha \cup \beta$, which can choose between following any behavior of α and any behavior of β, stays safely in ϕ (i.e., the conclusion holds). This proof rule, like all of our other proof rules, decomposes a property of a compound system $\alpha \cup \beta$ into properties of simpler subsystems. This compositional verification principle is beneficial for scalability purposes, because it helps taming the system complexity by recursively reducing the system to its parts during verification.

Most importantly, we need proof rules for quantified differential equations in order to be able to prove properties of the form $[\forall i\, f(i)' = \theta \,\&\, H]F$. One option is to assume that we know an explicit closed-form solution of the quantified differential equation and use that solution as a quantified assignment $\forall i\, f(i) := \theta$ to verify that property F holds at all times when following the solution, while staying in the evolution domain H. This is the option we have pursued in previous work [17]. It is correct, but the problem with that approach is that it only works if we can find a simple closed-form solution of the quantified differential equation. It does not work, however, if the quantified differential equation has no closed-form solution that we can write down, not if it has one but we cannot compute it, and not if it only has solutions that fall into undecidable classes of arithmetic. Flight equations for curved flight, for instance, have solutions with undecidable arithmetic [16].

Here, we thus follow an entirely different approach that is not limited to working with closed-form solutions of differential equations. We present an approach that has some resemblance to Lyapunov functions. But it works for safety instead of stability, and it supports arbitrary formulas instead of just a single function. Most importantly, we have extended the approach appropriately to cover distributed hybrid systems and their arbitrary-dimensional dynamics, including appearance and disappearance of participants.

The primary insight is that, when we want to verify a property of a differential equation, we do not have to know the solution for proving a property about it. If we want to know whether formula $[\forall i\, f(i)' = \theta \,\&\, H]F$ holds, i.e., whether we always safely stay in the region described by formula F when following that continuous dynamics, then we do not need to know global solutions of where exactly each point of the state space will evolve to when following the dynamics. All we need to know is whether we can possibly ever go from somewhere safe to somewhere unsafe. This is the intuition illustrated in Fig. 2. We check if the local continuous dynamics always pushes the system state in a direction where F is becoming "more" true, not in a direction where it could become false. Then, if the system also starts safe (in F), it will always stay safe no matter where we go.

For quantified differential equations, one of the extra challenges is that the system does not have a fixed finite dimension but can be arbitrary-dimensional. Consequently, there is not even a finite vector space in which the local directions of the vector field of the differential equations can be described and checked. Instead we need a criterion that captures our verification approach based on implicit properties of the local dynamics at uncountably infinitely many points in an essentially infinite-dimensional vector field. The cardinality of this set is at least that of $\mathbb{R}^{\mathbb{N}}$ (vaguely: "∞^∞").

In Section 4, we have introduced entirely symbolic notions of syntactic total derivations and the differential substitution property, which we use to turn the above intuitions into the formally precise and rigorous proof rule DI. For a formula F if we can prove the premiss of rule DI, i.e., that, after a differential substitution $[\forall i\, f(i)' := \theta]$, its total derivative $D(F)$ is valid in the evolution domain region H, then the conclusion of DI is valid, i.e., that the system stays in region F when it starts in F (left assumption in conclusion). It is important that we add the quantified assignment $\forall i\, f(i)' := \theta$ in the premiss, because, otherwise, the premiss of DI is not even a logical formula that would have a well-defined semantics when evaluated in a state. Unlike F, the total derivative $D(F)$ will contain differential function symbols like $f(i)'$, which do not have a semantics in isolated states but only along a flow. The quantified assignment, however, defines a value for those differential function symbols, which has a well-defined correspondence to the local dynamics of quantified differential equations by way of Lemma 2. The quantified assignment resulting in the premiss of rule DI can be handled subsequently by rule $[:=]$.

We call formula F in rule DI a *quantified differential invariant* for quantified differential equation $\forall i\, f(i)' = \theta \,\&\, H$. Note that stronger assumptions than H are generally unsound for the premiss of DI; see previous work for details [15, 16]. Even though stronger assumptions than H have been proposed for hybrid systems [20, 7], they are generally unsound even there. That is the reason why we take extra care in this paper to make sure our proof system is actually sound (cannot prove invalid formulas). Instead, we use a proof rule called differential cut (DC) that can be used to accumulate more knowledge and extra assumptions about the dynamics successively in a sound way. The right premiss proves that the property holds when assuming G as an additional restriction on the evolution domain region, and the left premiss proves that G is actually an invariant so that restricting the dynamics to G on the right branch is just a pseudo-restriction.

Rule $[:=]$ handles quantified assignments [17]. Their effect depends on whether $\forall i\, f(i) := \theta$ *matches* $f(u)$, i.e., there is a choice for i such that $f(u)$ is affected by the assignment, because u is of the form i for some i. If it matches, the premiss uses the term θ assigned to $f(i)$ instead of $f(u)$. Otherwise, the occurrence of f in $\phi(f(u))$ will be left unchanged. Rule $[:=]$ makes a case distinction on matching by if-then-else. In either case, the original quantified assignment $\forall i\, f(i) := \theta$, which we abbreviate by $\mathcal{A}$, will be applied to u in the premiss, because the value of argument u may also be affected by $\mathcal{A}$, recursively. The side condition on rule $[:=]$ makes sure that we use rule $[:=]$ in the appropriate order.

We use classical proof rules for the operators of the Kleene algebra with tests, and refer to the literature for details [11, 17]. We also use a proof rule (written ℝ) for real arithmetic based on quantifier elimination in real-closed fields [4]. Because we do not need the details of real arithmetic for the purpose of this paper, we consider it as a black box and refer to previous work for an elaboration of real arithmetic [17].

As a simple example, consider the following proof:

$$
\dfrac{\dfrac{\dfrac{true}{\forall i\, 3(x(i)^2 + x(i)^4 + 2) \geq 0}}{[\forall i\, x(i)' := x(i)^2 + x(i)^4 + 2]\forall i\, 3x(i)' \geq 0}}{\forall i\, 3x(i) \geq 1 \rightarrow [\forall i\, x(i)' = x(i)^2 + x(i)^4 + 2]\forall i\, 3x(i) \geq 1}
\begin{array}{l}\text{ℝ}\\[18pt]\text{[:=]}\\[14pt]\text{}^{DI}\end{array}
$$

This simple proof shows that $\forall i\, 3x(i) \geq 1$ is a (quantified differential) invariant of the quantified differential equation

$\forall i\, x(i)' = x(i)^2 + x(i)^4 + 2$. Note that this differential equation is difficult to solve and the solution falls into undecidable classes of arithmetic. At the bottom the proof starts with rule DI that reduces verification to a check on the total differential of the formula after an assignment of the differential function symbol to the right hand side of the quantified differential equation. The proof then uses rule $[:=]$ to handle the quantified assignment by substitution and finally can be proven by quantifier elimination for real arithmetic (marked by ℝ). In this simple example, the quantifier for i can be handled in a very simple modular way. We consider a more complicated example in our case study in Section 7.

6. SOUNDNESS

The proof rules in Fig. 4 would be entirely useless if they were unsound, because they could then be used to claim counterfactual properties as "proven" that do not hold in reality. In order to show the appropriateness of the proof rules, we, thus, prove that all provable properties are actually true.

THEOREM 1 (SOUNDNESS). *The quantified differential invariant proof rules in Fig. 4 are sound, i.e., every formula they prove is a valid property (of the distributed hybrid system that the formula refers to), i.e., it is true in all states.*

PROOF. We prove soundness of each proof rule.

DC Rule DC can be proven sound using the fact that the left premise implies that every flow φ that satisfies the quantified differential equation and evolution domain restriction H also satisfies G *all along* the flow. Thus, $\varphi \models \forall i\, f(i)' = \theta \,\&\, H$ implies $\varphi \models \forall i\, f(i)' = \theta \,\&\, H \wedge G$ so that the right premise entails the conclusion.

DI Assume that the premise is valid, i.e., true in all states. We have to show that the conclusion is valid too. Let ν be a state that satisfies the assumption F of the conclusion as, otherwise, there is nothing to show. We prove soundness by induction on the structure of F. First, we assume F to be quantifier-free in disjunctive normal form and consider any disjunct G of F that is true at ν. In order to show that F is invariant during the continuous evolution, it is sufficient to show that each conjunct of G is. We can assume these conjuncts to be of the form $c \geq 0$ (or $c > 0$ where the proof is similar). Now let $\varphi : [0, r] \to \mathcal{S}$ be any state flow with $\varphi \models \forall i\, x(i)' = \theta \,\&\, H$ beginning in $\varphi(0) = \nu$. By antecedent, $\nu \models F$. We assume duration $r > 0$, because the other case is immediate ($\nu \models F$ already holds). We show that F holds all along the flow φ, i.e., $\varphi \models F$.

Consider the case where F is of the form $c \geq 0$ (or $c > 0$ where the proof is similar). Suppose there was a $\zeta \in [0, r]$ where $\varphi(\zeta) \models c < 0$; this will lead to a contradiction. Then the function $h : [0, r] \to \mathbb{R}$ defined as $h(t) = \varphi(t)[\![c]\!]$ satisfies $h(0) \geq 0 > h(\zeta)$, because the antecedent shows $\nu \models c \geq 0$. Now, φ is of the order of $D(c)$, because: φ is of order 1 for all symbols $x(i)$, and trivially of order ∞ for variables that do not change during the differential equation. The value of c is defined all along φ, because we have assumed H to guard against zeros of denominators. Thus, by Lemma 1, h is continuous on $[0, r]$ and differentiable at every $\xi \in (0, r)$. By mean value theorem there is a $\xi \in (0, \zeta)$ such that $\frac{dh(t)}{dt}(\xi) \cdot (\zeta - 0) = h(\zeta) - h(0) < 0$.

In particular, since $\zeta \geq 0$, we can conclude $\frac{dh(t)}{dt}(\xi) < 0$. Lemma 1 implies that $\frac{dh(t)}{dt}(\xi) = \bar{\varphi}(\xi)[\![D(c)]\!] < 0$. And $\bar{\varphi}(\xi)[\![D(c)]\!] = \bar{\varphi}(\xi)[\![[\forall i\, x(i)' := \theta]D(c)]\!]$ by Lemma 2, as $\varphi \models \forall i\, x(i)' = \theta \,\&\, H$. This, however, is now a contradiction, because the premise actually implies that $\varphi \models H \to [\forall i\, x(i)' := \theta]D(c) \geq 0$. In particular, since $\bar{\varphi}(\xi) \models H$ holds by definition of the semantics, we have $\bar{\varphi}(\xi) \models [\forall i\, x(i)' := \theta]D(c) \geq 0$.

Second, consider the case where F is of the form $\forall j\, G$ for a fresh variable j that we can assume to occur only in G by renaming. Again let $\varphi : [0, r] \to \mathcal{S}$ be any state flow with $\varphi \models \forall i\, x(i)' = \theta \,\&\, H$ beginning in $\varphi(0) = \nu$ with $\nu \models F$. Consider *any* value e for the quantified variable j, then $\nu_j^e \models G$, i.e., $\nu \models G_j^e$. Premiss $H \to [\forall i\, x(i)' := \theta]D(F)$ is provable and, thus, by induction hypothesis valid. Then $H \to [\forall i\, x(i)' := \theta]D(G_j^e)$ is valid too, because $F \equiv \forall j\, G$ implies $D(F) \equiv \forall j\, D(G)$ and $[\forall i\, x(i)' := \theta]\forall j\, D(G)$ entails $[\forall i\, x(i)' := \theta]D(G_j^e)$ by the definition of the semantics. Consequently, G_j^e satisfies all assumptions of rule DI and the induction hypothesis implies that $G_j^e \to [\forall i\, x(i)' = \theta \,\&\, H]G_j^e$ is valid. Since assumption $\nu \models G_j^e$ holds, we know that G_j^e holds at all times when following $\forall i\, x(i)' = \theta \,\&\, H$. Now e was arbitrary, thus $\forall j\, G \to [\forall i\, x(i)' = \theta \,\&\, H]\forall j\, G$ is valid. If F is of the form $\exists j\, G$, the proof is similar, except for the last step.

[:=] For a proof of the soundness of rule [:=] and [∪], we refer to earlier work [17]. □

7. DISTRIBUTED ROUNDABOUT FLIGHT VERIFICATION

As an example of a distributed hybrid system, we have verified collision freedom in a roundabout flight collision avoidance maneuver; see Fig. 3. Unlike classical versions considered in the literature, we verify the roundabout maneuver for arbitrarily many aircraft and for an unbounded number of new aircraft that may appear into the horizon of relevance during the collision avoidance maneuver; see Fig. 1. In Fig. 5, we show the most important part of the proof for the collision-freedom property defined in Fig. 3. This part of the proof shows that the circle phase of the roundabout maneuver stays collision-free indefinitely for an arbitrary number of aircraft. That is the most crucial part, because we have to know that the aircraft remain safe during the actual roundabout collision avoidance circle. In other flight modes (e.g., *free*), the aircraft are safe by construction, because the evolution domain $\forall i, j\, \mathcal{P}(i, j)$ forces them to switch to a roundabout collision avoidance circle when the aircraft come too close. The condition $\forall i, j\, \mathcal{T}(i, j)$ characterizes compatible tangential maneuvering choices and can be proven to hold after *entry*. Without a condition like $\mathcal{T}(i, j)$, roundabouts can be unsafe [15, 16] so it is crucial that *entry* establishes it. For a systematic derivation of how to construct $\mathcal{T}(i, j)$, we refer to previous work [15, 16].

Note that the maneuver cannot be proven using any hybrid systems verification technique, because the dimension is parametric and unbounded and may even change dynamically during the remainder of the maneuver. The single proof in Fig. 5 corresponds to infinitely many proofs for systems with n aircraft for all n (plus unbounded dynamic changes of n in the proof for flight phase *newac*).

Our proof shows that the distributed roundabout maneuver safely avoids collisions for arbitrarily many aircraft (even with dynamic appearance of new aircraft). The above maneuver still requires all aircraft in the horizon of relevance to participate in the collision avoidance maneuver. In fact, we can show that this is unnecessary for aircraft that are far enough away and that may be engaged in other roundabouts. For space reasons, a discussion of these phenomena is beyond the scope of this paper, however.

8. CONCLUSIONS

Many cyber-physical systems are really distributed hybrid systems, with joint discrete, continuous, structural, and dimensional dynamics. This makes them challenging for formal verification. With hybrid systems verification, we cannot understand the distributed aspects of these systems nor aspects of dynamic appearance and disappearance of participants. With distributed systems verification, we cannot understand the continuous system dynamics. We present the first verification technique for distributed hybrid systems with nontrivial dynamics, which captures all these kinds of dynamics at once. We introduce quantified differential invariants for verifying properties of quantified differential equations. These quantified differential invariants are computationally attractive, because they can be used to verify distributed hybrid systems without having to solve their quantified differential equation systems. In particular, quantified differential invariants can be used even if the solutions cannot be computed, fall into undecidable classes of arithmetic, or do not even exist in closed form. We prove soundness of our verification approach and formally verify collision-freedom in a distributed roundabout maneuver in which new aircraft can appear dynamically at runtime.

Future work includes a more detailed study of the distributed roundabout maneuver and improving automation. In particular a number of assumptions in our distributed roundabout maneuver would be interesting to relax in future work (e.g., overly simplistic entry procedure, perfect communication, and synchronicity). Our verification approach does not depend on these assumptions, but the example does.

9. ACKNOWLEDGMENTS

I would like to thank the anonymous referees for their helpful comments.

10. REFERENCES

[1] R. Alur and G. J. Pappas, editors. *Hybrid Systems: Computation and Control*, volume 2993 of *LNCS*. Springer, 2004.

[2] P. C. Attie and N. A. Lynch. Dynamic input/output automata: A formal model for dynamic systems. In K. G. Larsen and M. Nielsen, editors, *CONCUR*, volume 2154 of *LNCS*, pages 137–151. Springer, 2001.

[3] Z. Chaochen, W. Ji, and A. P. Ravn. A formal description of hybrid systems. In R. Alur, T. A. Henzinger, and E. D. Sontag, editors, *Hybrid Systems*, volume 1066 of *LNCS*, pages 511–530. Springer, 1995.

[4] G. E. Collins and H. Hong. Partial cylindrical algebraic decomposition for quantifier elimination. *J. Symb. Comput.*, 12(3):299–328, 1991.

[5] A. Deshpande, A. Göllü, and P. Varaiya. SHIFT: A formalism and a programming language for dynamic

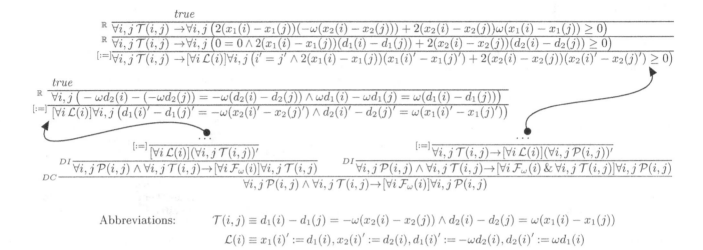

Abbreviations: $\mathcal{T}(i,j) \equiv d_1(i) - d_1(j) = -\omega(x_2(i) - x_2(j)) \wedge d_2(i) - d_2(j) = \omega(x_1(i) - x_1(j))$

$\mathcal{L}(i) \equiv x_1(i)' := d_1(i), x_2(i)' := d_2(i), d_1(i)' := -\omega d_2(i), d_2(i)' := \omega d_1(i)$

Figure 5: Proof for collision freedom of roundabout collision avoidance maneuver circle

networks of hybrid automata. In P. J. Antsaklis, W. Kohn, A. Nerode, and S. Sastry, editors, *Hybrid Systems*, volume 1273 of *LNCS*, pages 113–133. Springer, 1996.

[6] S. Gilbert, N. Lynch, S. Mitra, and T. Nolte. Self-stabilizing robot formations over unreliable networks. *ACM Trans. Auton. Adapt. Syst.*, 4(3):1–29, 2009.

[7] S. Gulwani and A. Tiwari. Constraint-based approach for analysis of hybrid systems. In Gupta and Malik [8], pages 190–203.

[8] A. Gupta and S. Malik, editors. *Computer Aided Verification*, volume 5123 of *LNCS*. Springer, 2008.

[9] J. P. Hespanha and A. Tiwari, editors. *Hybrid Systems: Computation and Control*, volume 3927 of *LNCS*. Springer, 2006.

[10] I. Hwang, J. Kim, and C. Tomlin. Protocol-based conflict resolution for air traffic control. *Air Traffic Control Quarterly*, 15(1):1–34, 2007.

[11] D. Kozen. Kleene algebra with tests. *ACM Trans. Program. Lang. Syst.*, 19(3):427–443, 1997.

[12] F. Kratz, O. Sokolsky, G. J. Pappas, and I. Lee. R-Charon, a modeling language for reconfigurable hybrid systems. In Hespanha and Tiwari [9], pages 392–406.

[13] J. Meseguer and R. Sharykin. Specification and analysis of distributed object-based stochastic hybrid systems. In Hespanha and Tiwari [9], pages 460–475.

[14] L. Pallottino, V. G. Scordio, E. Frazzoli, and A. Bicchi. Decentralized cooperative policy for conflict resolution in multi-vehicle systems. *IEEE Trans. on Robotics*, 23(6):1170–1183, 2007.

[15] A. Platzer. Differential-algebraic dynamic logic for differential-algebraic programs. *J. Log. Comput.*, 20(1):309–352, 2010.

[16] A. Platzer. *Logical Analysis of Hybrid Systems: Proving Theorems for Complex Dynamics*. Springer, Heidelberg, 2010.

[17] A. Platzer. Quantified differential dynamic logic for distributed hybrid systems. In A. Dawar and H. Veith,

editors, *CSL*, volume 6247 of *LNCS*, pages 469–483. Springer, 2010.

[18] A. Platzer and E. M. Clarke. Computing differential invariants of hybrid systems as fixedpoints. In Gupta and Malik [8], pages 176–189.

[19] A. Platzer and E. M. Clarke. Formal verification of curved flight collision avoidance maneuvers: A case study. In A. Cavalcanti and D. Dams, editors, *FM*, volume 5850 of *LNCS*, pages 547–562. Springer, 2009.

[20] S. Prajna and A. Jadbabaie. Safety verification of hybrid systems using barrier certificates. In Alur and Pappas [1], pages 477–492.

[21] S. Prajna, A. Jadbabaie, and G. J. Pappas. A framework for worst-case and stochastic safety verification using barrier certificates. *IEEE T. Automat. Contr.*, 52(8):1415–1429, 2007.

[22] W. C. Rounds. A spatial logic for the hybrid π-calculus. In Alur and Pappas [1], pages 508–522.

[23] S. Sankaranarayanan. Automatic invariant generation for hybrid systems using ideal fixed points. In K. H. Johansson and W. Yi, editors, *HSCC*, pages 221–230. ACM, 2010.

[24] S. Sankaranarayanan, H. Sipma, and Z. Manna. Constructing invariants for hybrid systems. In Alur and Pappas [1], pages 539–554.

[25] C. Tomlin, G. Pappas, J. Košecká, J. Lygeros, and S. Sastry. Advanced air traffic automation: A case study in distributed decentralized control. In B. Siciliano and K. Valavanis, editors, *Control Problems in Robotics and Automation*, volume 230 of *Lecture Notes in Control and Information Sciences*, pages 261–295. Springer, 1998.

[26] C. Tomlin, G. J. Pappas, and S. Sastry. Conflict resolution for air traffic management: a study in multi-agent hybrid systems. *IEEE T. Automat. Contr.*, 43(4):509–521, 1998.

[27] D. A. van Beek, K. L. Man, M. A. Reniers, J. E. Rooda, and R. R. H. Schiffelers. Syntax and consistent equation semantics of hybrid Chi. *J. Log. Algebr. Program.*, 68(1-2):129–210, 2006.

Decidability and Complexity for the Verification of Safety Properties of Reasonable Linear Hybrid Automata

Werner Damm
Carl von Ossietzky
University Oldenburg,
Oldenburg, Germany
damm@offis.de

Carsten Ihlemann
Max-Planck-Institut für
Informatik,
Saarbrücken, Germany
ihlemann@mpi-
inf.mpg.de

Viorica
Sofronie-Stokkermans
Max-Planck-Institut für
Informatik,
Saarbrücken, Germany
sofronie@mpi-inf.mpg.de

ABSTRACT

This paper identifies an industrially relevant class of linear hybrid automata (LHA) called *reasonable LHA* for which parametric verification of safety properties with exhaustive entry conditions can be done in polynomial time and time-bounded reachability with exhaustive entry conditions can be decided in nondeterministic polynomial time for non-parametric verification and in exponential time for parametric verification. Deciding whether an LHA is reasonable is shown to be decidable in polynomial time.

Categories and Subject Descriptors

D.2.4 [**Software/Program verification**]; J.7 [**Computers in other systems**]

General Terms

Theory, Verification

1. INTRODUCTION

Parametric verification asks for the derivation of a constraint on parameters of a hybrid system specification such that a given set of requirements is satisfied. From an industrial perspective, it is a key instrument supporting the concept phase in the design of new control applications, such as for the automotive or industrial domain, in order to determine key system requirements such as required reaction times, required rate of changes of system variables, required bounds on disturbances, etc. In typical industrial design flows, Matlab-Simulink/Stateflow models are used for the specification of the underlying controllers, and safety and stability are verified informally, using simulation and testing. Often, the implementation of such controllers is based on code generated automatically from such specification models, employing tools such as the Embedded Encoder or Target Link. For use in safety critical applications, modeling guidelines are imposed to ease testability and verification.

Giving such specification models a formal semantics in terms of hybrid automata makes them amenable to formal verification techniques. The challenge in making these applicable in practice is in identifying industrially relevant classes of hybrid automata, for which verification of safety properties is truly feasible, that is, were safety properties can be decided in polynomial time. This paper proposes *reasonable linear hybrid automata* as such a model: we show that for (non-parametric) reasonable linear hybrid automata:

- safety properties can be decided in polynomial time

- time-bounded reachability properties can be checked in nondeterministic polynomial time

provided that the entry conditions cover the inner envelopes of the modes. For the parametric verification of such properties we identify conditions when the complexity is in PTIME, NP, or EXPTIME. Finally, we show that the property of *being reasonable* can be decided in polynomial time.

Linear hybrid automata (LHA) [2] allow for linear constraints on the values of the continuous variables and on their rate of growth of the form $\sum_{i=1}^{n} a_i x_i \leq a$ (resp. $\sum_{i=1}^{n} c_i \dot{x}_i \leq c$). As in [2], we assume that guards and invariants are given as conjunctions of linear constraints, and that updates are expressed in linear arithmetic. This class of hybrid automata has been shown to be expressive enough to give sufficiently concise conservative abstractions for non-linear hybrid systems (see e.g. [12]), and comes with symbolic semi-decision procedures, such as originally proposed in [11] and subsequently often refined, such as in [8, 14, 9, 5]. We propose a novel subclass of LHA for which a particular class of safety properties can be decided in polynomial time. First, we require each mode additionally to be equipped with what is often called an inner envelope of the mode invariant. Intuitively, these serve to assure chatter-freedom, that is to assure each state is visted with a minimal dwelling time. We then restrict ourselves to what we call safety properties with exhaustive entry conditions of the form:

$$\Box(\phi_{\mathsf{ExhEntry}} \rightarrow \Box\phi_{\mathsf{safe}})$$

and timed-reachability properties with exhaustive entry conditions:

$$\Box(\phi_{\mathsf{ExhEntry}} \rightarrow \Diamond_{\leq t}\phi_{\mathsf{safe}})$$

where ϕ_{ExhEntry} is the disjunction of all inner envelopes of the modes, and both ϕ_{safe} and the inner envelopes are given by convex linear constraints. Our reasonable LHA exploit modeling guidelines used in safety critical system designs, enforcing the following restrictions on LHA:

Input determinism: Guards of transitions originating from one and the same state are mutually exclusive, and initial states of modes are disjoint.

Invariant compatability: Mode invariants are safe (i.e. are all subsets of ϕ_{safe}). Moreover, whenever the invariants becomes false, then there is at least one transition leaving the mode whose guard is enabled.

Chatter-freedom: All transitions enter a mode within its inner envelope. Moreover, there is a minimal dwelling time s.t. when a mode is entered in its inner envelope no guard of a transition leaving the mode will be enabled before the minimal dwelling time is expired.

Our approach is based on a reduction of such verification problems to checking satisfiability of formulae in fragments of linear arithmetic, which can be decided in PTIME for verification of safety properties, and in NP for time-bounded reachability properties. We use as a running example a simple chemical process control, where substances to be mixed must maintain constant ratios within margins. Here ϕ_{safe} encodes that indeed the ratio of concentrations is maintained, and that the tank is never overflowing. As an example of a time bound reachability property, we are interested in proving that the tank is drained within 200 ms if the critical ratio of concentrations is exceeded. We have used the SMT solver Z3 to verify the above safety and time-bounded reachability properties within fragments of a second.

Related work. A considerable amount of work was dedicated to identifying classes of hybrid automata for which checking safety is decidable. One of the first decidability results is for initialized rectangular hybrid automata [11], a restricted class of rectangular hybrid automata (i.e. hybrid automata for which at each control location the flow is described by differential inclusions of the form $\dot{x} \in [a,b]$) which require resets for continous variables upon mode transitions, unless the newly entered and left mode share the same dynamics. These results have been extended in [20] which identifies classes of LHA for which reachability is decidable and in [19] which uses translations of verification problems into satisfiability problems for theories of real numbers (possibly with exponentiation). Decidability of *finite-precision* hybrid automata was studied in [1]. In [18, 3] *o-minimal hybrid systems* are studied; it is shown that for such hybrid systems reachability can be reduced to bounded reachability (or, in other words that such systems always allow finite bisimulations) – this is used to give decidability results. The restrictions imposed when defining o-minimal hybrid systems – in particular the requirement that only constant resets are allowed – are quite severe. In contrast, the restrictions we impose (input determinism, invariant compatibility, chatter freedom) are much more likely to be observed in real applications. Parametric verification of linear hybrid automata was first studied in [2]. In [23, 24] a dynamic hybrid logic is developed allowing non-linear differential equations within modalities, and possibilities of deriving constraints on parameters guaranteeing satisfaction of system requirements are discussed. While [23, 24] support significantly richer classes of dynamics, the underlying logic is highly undecidable, and there is no investigation into decidable subclasses of the verification problem. We improve on previous work by Wang [31], in that our approach gives an effective procedure, while his back-reachability based symbolic execution approach is not guaranteed to terminate. [8] incrementally learns constraints on good parameters from counterexamples; this approach is only terminating if there are only finitely many counterexamples. In [8] parameters are not allowed as bounds for derivatives, and system variables and parameters cannot be combined multiplicatively. In our work we could in principle allow this, at the price of passing from linear constraints to non-linear constraints. First steps towards the verification of parametric systems, including inferring constraints on the parameters which ensure safety, were taken in [27]. Here we extend these results.

To the best of our knowledge no systematic study exists in which classes of properties which can be checked in PTIME are identified, and conditions on LHA which allow for efficient checking of safety properties and/or time-bounded reachability are given. This is the aim of our paper.

The paper is structured as follows. Sect. 2 introduces the main concepts on hybrid automata used in the paper, our running example, and a summary of complexity results for fragments of linear real arithmetic. In Sect. 3 we show that (for convex formulae) invariant checking is decidable in PTIME, and bounded model checking in NP. In Sect. 4 reasonable LHA are defined and the complexity of checking this property is studied; we then show that for reasonable LHA checking safety properties with exhaustive entry conditions can be reduced to invariant checking (and hence can be checked in PTIME) and checking time-bounded reachability properties can be reduced to bounded model checking (hence is in NP). The complexity results are summarized in Sect. 5. Sect. 6 contains some experimental results.

2. PRELIMINARIES

We introduce the classes of hybrid automata studied in this paper, as well as the verification tasks we consider, then give a summary of complexity results for linear arithmetic.

2.1 Hybrid Automata

DEFINITION 1. *A hybrid automaton (HA) is a tuple*

$$S = (X, Q, \mathsf{flow}, \mathsf{Inv}, \mathsf{Init}, E, \mathsf{jump})$$

consisting of:

(1) *A finite set $X = \{x_1, \ldots, x_n\}$ of real valued variables and a finite set Q of control modes, that together define the state space of the system.*

(2) *A family $\{\mathsf{flow}_q \mid q \in Q\}$ specifying the continuous dynamics in each control mode, where flow_q is a predicate over the variables in $X \cup \dot{X}$, where $\dot{X} = \{\dot{x}_1, \ldots, \dot{x}_n\}$, where $\dot{x}_i$ is the derivative of x_i.*

(3) *A family $\{\mathsf{Inv}_q \mid q \in Q\}$ defining the invariant conditions for each control mode, where for every $q \in Q$, Inv_q is a predicate over the variables in X.*

(4) *A family $\{\mathsf{Init}_q \mid q \in Q\}$ defining the initial states for each control mode, where for every $q \in Q$, Init_q is a predicate over the variables in X.*

(5) *The control switches, modeled by a finite multiset E with elements in $Q \times Q$. Every $(q, q') \in E$ is a directed edge between q (source mode) and q' (target mode).*

(6) A family of guards for every control switch $\{\text{guard}_e \mid e \in E\}$, where guard_e is a predicate over X.

(7) A family of jump conditions $\{\text{jump}_e \mid e \in E\}$, where jump_e is a predicate over $X \cup X'$, where $X' = \{x'_1, \ldots, x'_n\}$ is a copy of X consisting of "primed" variables.

DEFINITION 2. *An HA is* input-deterministic *if:*

(i) $\text{Init}_{q_1} \wedge \text{Init}_{q_2} \models \perp$ *for all* $q_1 \neq q_2, q_1, q_2 \in Q$

(ii) $\text{guard}_{(q,q_1)} \wedge \text{guard}_{(q,q_2)} \models \perp$ *for all* $(q, q_1), (q, q_2) \in E, q_1 \neq q_2$.

At any time instant, the state of a hybrid automaton specifies a control location and values for all data variables.

DEFINITION 3. *A* state *of the hybrid automaton S is a pair (q, a) consisting of a control mode $q \in Q$ and a vector $a = (a_1, \ldots, a_n)$ that represents a value $a_i \in \mathbb{R}$ for each variable $x_i \in X$. A state (q, a) is* admissible *if Inv_q is true when each x_i is replaced by a_i. A state (q, a) is* initial *if Init_q is true when each x_i is replaced by a_i.*

There are the following types of state change:

(Jump) The state can change by an instantaneous transition that changes the control location and the values of data variables according to the jump conditions, or

(Flow) The state can change due to the evolution in a given control mode over an interval of time: the values of the data variables change in a continuous manner according to the flow rules of the current control location.

DEFINITION 4. *A* run *of the hybrid automaton S is a finite sequence $s_0 s_1 \ldots s_k$ of admissible states s_j such that*

- *the first state s_0 is an initial state of S,*

- *each pair (s_j, s_{j+1}) of consecutive states in the sequence is either a jump of S or a flow of S.*

2.1.1 Linear Hybrid Automata

An atomic linear predicate is a linear inequality (strict or non-strict inequality between a rational constant and a linear combination of variables with rational coefficients, such as $3x_1 - x_2 + 7x_5 \leq 4$). A convex linear predicate is a finite conjunction of linear inequalities. A state assertion s for S is a family $\{s(q) \mid q \in Q\}$, where $s(q)$ is a predicate over X (expressing constraints which hold in state s for mode q).

DEFINITION 5. *[2] A* linear hybrid automaton (LHA) *is a hybrid automaton which satisfies the following requirements:*

1. **Linearity:** *For every control mode $q \in Q$, the flow condition flow_q, the invariant condition Inv_q, and the initial condition Init_q are convex linear predicates. For every control switch $e = (q, q') \in E$, the jump condition jump_e and the guard guard_e are convex linear predicates. In addition, we assume that the flow conditions flow_q are conjunctions of non-strict inequalities.*

2. **Flow independence:** *For every control mode $q \in Q$, the flow condition flow_q is a predicate over the variables in $\dot{X}$ only (and does not contain any variables from X). This requirement ensures that the possible flows are independent from the values of the variables, and depend only on the control mode.*

EXAMPLE 1. We consider the following (very simplified) chemical plant example, modeling the situation in which we control the reaction of two substances, and the separation of the substance produced by the reaction.

Let x_1, x_2 and x_3 be variables which describe the evolution of the volume of substances 1 and 2, and the substance 3 generated from their reaction, respectively. Assume that $\epsilon_a > 0, \delta_a > 0, L_f > 0, \min > 0, \text{dmin} > 0$. The plant is described by a hybrid automaton with four modes:

Mode 1: Fill In this mode the temperature is low, and hence the substances 1 and 2 do not react. In this mode the substances 1 and 2 (possibly mixed with a very small quantity of substance 3) are filled in the tank in equal quantities up to a certain margin of error. This is described by the following invariants and flow conditions:

$\text{Inv}_1 \quad x_1 + x_2 + x_3 \leq L_f \ \wedge \ \bigwedge_{i=1}^{3} x_i \geq 0 \ \wedge$
$\quad\quad -\epsilon_a \leq x_1 - x_2 \leq \epsilon_a \ \wedge \ 0 \leq x_3 \leq \min$
$\text{flow}_1 \quad \dot{x}_1 \geq \text{dmin} \wedge \dot{x}_2 \geq \text{dmin} \wedge \dot{x}_3 = 0 \wedge -\delta_a \leq \dot{x}_1 - \dot{x}_2 \leq \delta_a$

If the proportion is not kept the system jumps into mode 4 (**Dump**); if the total quantity of substances exceeds level L_f (tank filled) the system jumps into mode 2 (**React**).

Mode 2: React In this mode the temperature is high, and the substances 1 and 2 react. The reaction consumes equal quantities of substances 1 and 2 and produces substance 3.

$\text{Inv}_2 \quad L_f \leq x_1 + x_2 + x_3 \leq L_{\text{overflow}} \ \wedge \ \bigwedge_{i=1}^{3} x_i \geq 0 \ \wedge$
$\quad\quad -\epsilon_a \leq x_1 - x_2 \leq \epsilon_a \ \wedge \ 0 \leq x_3 \leq \max$
$\text{flow}_2 \quad \dot{x}_1 \leq -\text{dmin} \wedge \dot{x}_2 \leq -\text{dmin} \wedge \dot{x}_3 > \text{dmin}$
$\quad\quad \wedge \dot{x}_1 = \dot{x}_2 \wedge \dot{x}_3 + \dot{x}_1 + \dot{x}_2 = 0$

If the proportion between substances 1 and 2 is not kept the system jumps into mode 4 (**Dump**); if the total quantity of substances 1 and 2 is below some minimal level min the system jumps into mode 3 (**Filter**).

Mode 3: Filter In this mode the temperature is low again and the substance 3 is filtered out.

$\text{Inv}_3 \quad x_1 + x_2 + x_3 \leq L_{\text{overflow}} \ \wedge \ \bigwedge_{i=1}^{3} x_i \geq 0 \wedge$
$\quad\quad -\epsilon_a \leq x_1 - x_2 \leq \epsilon_a \ \wedge \ x_3 \geq \min$
$\text{flow}_3 \quad \dot{x}_1 = 0 \wedge \dot{x}_2 = 0 \wedge \dot{x}_3 \leq -\text{dmin}$

If the proportion between substances 1 and 2 is not kept the system jumps into mode 4 (**Dump**). Otherwise, if the concentration of substance 3 is below some minimal level min the system jumps into mode 1 (**Fill**).

Mode 4: Dump In this mode the content of the tank is emptied. For simplicity we assume that this happens instantaneously, i.e. $\text{Inv}_4 : \bigwedge_{i=1}^{3} x_i = 0$ and $\text{flow}_4 : \bigwedge_{i=1}^{3} \dot{x}_i = 0$.

Jumps. The automaton has the following jumps:

- $e = (1, 2)$ with $\text{guard}_e(x_1, x_2, x_3) = x_1 + x_2 + x_3 \geq L_f$ which leaves the variables unchanged.

- $e = (2, 3)$ with $\text{guard}_e(x_1, x_2, x_3) = x_1 + x_2 \leq \min$ which leaves the variables unchanged.

- $e = (3, 1)$ with $\text{guard}_e(x_1, x_2, x_3) = -\epsilon_a \leq x_1 - x_2 \leq \epsilon_a \wedge 0 \leq x_3 \leq \min$ which leaves the variables unchanged.

- Two edges e_1^1, e_2^1 from 1 to 4, and two edges e_1^2, e_2^2 from 2 to 4, with $\text{guard}_{e_1^j}(x_1, x_2, x_3) = x_1 - x_2 \geq \epsilon_a$, $\text{guard}_{e_2^j}(x_1, x_2, x_3) = x_1 - x_2 \leq -\epsilon_a$; and $\text{jump}_{e_i^j}(x_1, x_2, x_3, x'_1, x'_2, x'_3) = \bigwedge_{i=1}^{3} x'_i = 0$, $(j, i = 1, 2)$;

- Two edges e_1^3, e_2^3 from 3 to 4, with $\text{guard}_{e_1^3}(x_1, x_2, x_3) = x_3 \leq \min \wedge x_1 - x_2 \geq \epsilon_a$; $\text{guard}_{e_2^3}(x_1, x_2, x_3) = x_3 \leq$

$\min \wedge x_1 - x_2 \leq -\epsilon_a$, and $\mathsf{jump}_{e_i^3}(x_1, x_2, x_3, x_1', x_2', x_3') = \bigwedge_{i=1}^{3} x_i' = 0$ for $i = 1, 2$.

2.1.2 Extended Hybrid Automata

We also consider extended hybrid automata (EHA), in which a set of "inner envelopes" of modes are specified.

DEFINITION 6. *An extended hybrid automaton is a tuple:*

$$S = (X, Q, \mathsf{flow}, \mathsf{Inv}, \mathsf{InEnv}, \mathsf{Init}, E, \mathsf{jump}),$$

with the property that $(X, Q, \mathsf{flow}, \mathsf{Inv}, \mathsf{Init}, E, \mathsf{jump})$ *is a hybrid automaton and in which for every mode* $q \in Q$ *an inner envelope* InEnv_q *is specified, which is "contained" in the mode invariant, i.e. it satisfies the following condition:*

$$\models \forall x_1, \ldots, x_n (\mathsf{InEnv}_q(x_1, \ldots, x_n) \rightarrow \mathsf{Inv}_q(x_1, \ldots, x_n)).$$

An extended linear hybrid automaton (ELHA) is a LHA in which each InEnv_q *can be described as a conjunction of linear inequalities (as a convex predicate).*

EXAMPLE 2. *An inner envelope for mode 1 in Ex. 1 could be:* $\mathsf{InEnv}_1(x_1, x_2, x_3) = 0 \leq x_1 + x_2 + x_3 \leq L_{\mathsf{safe}} \wedge \bigwedge_{i=1}^{3} x_i \geq 0$
$\wedge -\epsilon_a \leq x_1 - x_2 \leq \epsilon_a \wedge 0 \leq x_3 \leq \min$
where $L_{\mathsf{safe}} < L_f$.

The following property will play an important rôle:

DEFINITION 7. *Let* (s_1, s_2) *be a jump in* S *under mode switch* (q, q'), *where* $s_1 = (q, a_1, \ldots, a_n), s_2 = (q', a_1', \ldots, a_n')$. *We say that the jump* (s_1, s_2) *is* mode reachable *(w.r.t. q) if there exists a state* $s_0 = (q, a_1^0, \ldots, a_n^0)$ *with* $\mathsf{InEnv}_q(a_1^0, \ldots, a_n^0)$ *and there exists a flow in mode* q *from* s_0 *to* s_1.

2.1.3 Verification Problems

We consider the following verification problems:

Invariant checking, i.e. the problem of checking whether a formula Ψ is invariant in a hybrid automaton S, i.e.:

(1) $\mathsf{Init}_q \models \Psi$ for all $q \in Q$;

(2) Ψ is invariant under jumps and flows:

(Flow) For every flow in a mode q, the continuous variables satisfy Ψ both during and at the end of the flow.

(Jump) For every jump according to a control switch e, if the values of the continuous variables satisfy Ψ before the jump, they satisfy Ψ after the jump.

Bounded model checking, i.e. the problem of checking whether a formula Safe is preserved under runs of length bounded by k, i.e.:

(1) $\mathsf{Init}_q \models \mathsf{Safe}$ for every $q \in Q$;

(2) Safe is preserved under runs of length j for all $1 \leq j \leq k$.

Safety properties. We consider properties of the form:

$$\phi = \Box(\phi_{\mathsf{entry}} \rightarrow \Box \phi_{\mathsf{safe}})$$

stating that for every run, if the predicate ϕ_{entry} holds at the beginning of the run, ϕ_{safe} is always true during the run.

Time-bounded reachability problems, i.e. checking properties stating that for every run σ in the automaton S, if ϕ_{entry} holds at the beginning of the run, then there exists a run σ of total duration bounded by t such that ϕ_{safe} is true at the end of the run, i.e. of the form:

$$\phi = \Box(\phi_{\mathsf{entry}} \rightarrow \Diamond_{\leq t} \phi_{\mathsf{safe}}).$$

ASSUMPTION. The safety and time-bounded reachability conditions we analyze in the paper are assumed to have *exhaustive entry conditions* (i.e. $\phi_{\mathsf{entry}} = \phi_{\mathsf{ExhEntry}} := \bigvee_{q \in Q} \mathsf{InEnv}_q$) and ϕ_{safe} is a convex linear predicate over X.

2.2 Complexity for Linear Arithmetic

We summarize the complexity results for satisfiability checking in fragments of linear arithmetic used in this paper:

THEOREM 8 (LINEAR ARITHMETIC). *The following hold:*

(1) *Satisfiability of any conjunction of linear inequalities can be checked in polynomial time [15, 30].*

(2) *The complexity of checking the satisfiability of sets of clauses in linear arithmetic is in NP [28].*

We present some classes of sets of clauses in linear arithmetic for which satisfiability can be checked in PTIME.

Horn disjunctive linear constraints. A Horn-disjunctive linear constraint (or HDL for short) is a disjunction $d_1 \vee \cdots \vee d_n$ where each $d_i, i = 1, \ldots, n$ is a linear inequality or a linear disequation, and the number of inequalities among $d_1, \ldots, d_n$ does not exceed one.

THEOREM 9 ([16]). *The satisfiability of any conjunction* C *of Horn disjunctive linear constraints (over* $\mathbb{R}$ *or* $\mathbb{Q}$*) can be decided in PTIME. Moreover, we can eliminate* n *variables from a set* C *of Horn disjunctive linear constraints in time* $O(|C|^n)$. [1]

$UTVPI^{\neq}$ **constraints**[2] are conjunctions of constraints of the form $ax + by \leq c$ or $ax + by \neq c$, with $a, b \in \{-1, 0, 1\}$).

THEOREM 10 ([17]). *The satisfiability of any set* C *of* $UTVPI^{\neq}$ *constraints (over* $\mathbb{R}$ *or* $\mathbb{Q}$*) can be decided in time* $O(n^3 + d)$, *where* d *is the number of disequations and* n *is the number of variables in* C. *Any number of variables can be eliminated in time* $O(dn^4)$, *where* n *and* d *are as before.*

Ord-Horn constraints. An Ord-Horn constraint is a constraint of the form $\bigwedge_{i=1}^{n} x_i \leq y_i \rightarrow x \leq y$, where x_i, y_i, x, y are variables.

THEOREM 11 ([22]). *The satisfiability of any conjunction of Ord-Horn constraints (over* $\mathbb{R}$ *or* $\mathbb{Q}$*) can be decided in PTIME.*

2.3 Notation

We use the following notation. If $x_1, \ldots, x_n$ are continuous variables (varying over time) we will denote the sequence $x_1, \ldots, x_n$ with $\overline{x}$ and the sequence of values $x_1(t), \ldots, x_n(t)$ of these variables at a time t with $\overline{x}(t)$. We denote the sequence $\dot{x}_1, \ldots, \dot{x}_n$ with $\dot{\overline{x}}$. A similar notation is used also for variables with superscripts: If $x_1^k, \ldots, x_n^k$ are continuous variables we denote the sequence $x_1^k, \ldots, x_n^k$ with $\overline{x}^k$ and the sequence $\dot{x}_1^k, \ldots, \dot{x}_n^k$ with $\dot{\overline{x}}^k$.

---

[1] As mentioned in [16], this is a PTIME result under the assumption that the number of variables n which are eliminated is fixed (then the degree of the polynomial is directly proportional to the number of variables to be eliminated).
[2] $UTVPI^{\neq}$ stands for "Unit two variables per inequality with disequalities".

3. SIMPLE VERIFICATION PROBLEMS

The verification problems we consider in this section are invariant checking and bounded model checking for linear hybrid automata. We show that invariant checking is decidable in polynomial time, and bounded model checking is in general decidable in nondeterministically polynomial time, also for parametric systems if the constraints on parameters are expressed by linear inequalities, and parameters occur only as bounds of linear constraints over X. We also analyze the complexity of generating constraints on parameters.

In Section 4.2 we consider more general safety properties and bounded time reachability and show that for certain *reasonable* (extended) linear hybrid automata checking safety properties can be reduced to invariant checking and checking time-bounded reachability properties can be reduced to problems very similar to bounded model checking.

3.1 Complexity of Invariant Checking

Let S be an LHA with continuous variables $x_1, \ldots, x_n$. We first consider the problem of checking whether a formula $\Psi(x_1, \ldots, x_n)$ is invariant (under jumps and flows) in S. To efficiently solve this problem we use a reduction to the problem of checking the satisfiability of systems of linear inequalities. By definition, the mode invariants, initial states and guards of mode switches are described as conjunctions of strict and non-strict linear inequalities. These constraints can also be expressed referring to the time moment t. The fact that $\mathsf{Inv}_q = \bigwedge_{j=1}^{m_q}(\sum_{i=1}^{n} a_{ij}^q x_i \leq a_j^q)$ can be expressed by:

$$\mathsf{Inv}_q(x_1(t), \ldots, x_n(t)) = \bigwedge_{j=1}^{m_q}(\sum_{i=1}^{n} a_{ij}^q x_i(t) \leq a_j^q)$$

and similarly for guard_e and Init_q. The flow conditions are expressed by non-strict linear inequalities of the form

$\mathsf{flow}_q = \bigwedge_{j=1}^{n_q}(\sum_{i=1}^{n} c_{ij}^q \dot{x}_i \leq c_j^q)$, i.e.

$\mathsf{flow}_q(t) = \bigwedge_{j=1}^{n_q}(\sum_{i=1}^{n} c_{ij}^q \dot{x}_i(t) \leq c_j^q)$.

We also can express the flow conditions $\mathsf{flow}_q(t)$ without referring to the values of the derivative, using the following formulae (where $0 \leq t \leq t'$)[3]:

$$\underline{\mathsf{flow}}_q(t, t') = \bigwedge_{j=1}^{n_q}(\sum_{i=1}^{n} c_{ij}^q (x_i(t') - x_i(t)) \leq c_j^q(t' - t)).$$

We axiomatize a flow in control mode q in the time interval $[t_0, t_1]$ (where $0 \leq t_0 \leq t_1$) as follows:

$$\mathsf{Flow}_q(t_0, t_1) = \begin{aligned} &\forall t(t_0 \leq t \leq t_1 \rightarrow \mathsf{Inv}_q(\overline{x}(t))) \wedge \\ &\forall t, t'(t_0 \leq t \leq t' \leq t_1 \rightarrow \underline{\mathsf{flow}}_q(t, t')). \end{aligned}$$

Assume that the jump update jump_e and guard guard_e at moment t for the control switch $e = (q, q') \in E$ are expressed by the following convex linear predicate:

$$\begin{aligned} \mathsf{guard}_e(\overline{x}(t)) &= \bigwedge_{j=1}^{m_e}(\sum_{i=1}^{n} g_{ij}^e x_i(t) \leq g_j^e) \\ \mathsf{jump}_e(\overline{x}(t), \overline{x}'(0)) &= \bigwedge_{j=1}^{h_e}(\sum_{i=1}^{n} b_{ij}^e x_i(t) + c_{ij}^e x_i'(0) \leq d_j^e). \end{aligned}$$

For $e = (q, q') \in E$, we axiomatize the jump condition by

$$\mathsf{Jump}_e(\overline{x}, \overline{x}') := \mathsf{guard}_e(\overline{x}) \wedge \mathsf{jump}_e(\overline{x}, \overline{x}').$$

We want to analyze whether certain properties are preserved during jumps and flows. Let Ψ be a state assertion. Ψ is preserved under a flow from time t_0 to t (where $0 \leq t_0 < t$) in state q iff the following formula is unsatisfiable:

$$\mathsf{Inv}_q(\overline{x}(t_0)) \wedge \Psi(\overline{x}(t_0)) \wedge \mathsf{Flow}_q(t_0, t) \wedge \neg\Psi(\overline{x}(t)).$$

---
[3]By Thm. 12, no information is lost with this encoding.

Note that this is a satisfiability problem for a formula with free variables t_0 and t (implicitly existentially quantified) with the subformula $\mathsf{Flow}_q(t_0, t)$ containing a universal quantifier, i.e. for a formula containing an alternation of quantifiers of the form $\exists\forall$. In what follows we will show that in certain cases a simpler encoding is possible.

An optimized translation. We present a considerably simpler encoding of the runs in linear hybrid automata, in which for $t_0 \leq t_1$, $\mathsf{Flow}_q(t_0, t_1)$ is replaced with:

$$\mathsf{Inv}_q(x(t_0)) \wedge \mathsf{Inv}_q(x(t_1)) \wedge \underline{\mathsf{flow}}_q(t_0, t_1).$$

THEOREM 12. *The following are equivalent for any LHA:*

(1) Ψ *is an invariant of the automaton.*

(2) *For every* $q \in Q$ *and* $e = (q, q') \in E$, *the formulae* $F_{\mathsf{Flow}}(q)$ *and* $F_{\mathsf{jump}}(e)$ *are unsatisfiable, where:*

$$F_{\mathsf{Flow}}(q) = \Psi(\overline{x}(t_0)) \wedge \mathsf{Flow}_q(t_0, t) \wedge \neg\Psi(\overline{x}(t)) \wedge t \geq t_0$$

$$\begin{aligned} F_{\mathsf{jump}}(e) = &\Psi(\overline{x}(t)) \wedge \mathsf{Jump}_e(\overline{x}(t), \overline{x}'(0)) \wedge \\ &\mathsf{Inv}_{q'}(\overline{x}'(0)) \wedge \neg\Psi(\overline{x}'(0)). \end{aligned}$$

(3) *For every* $q \in Q$ *and* $(q, q') \in E$, *the formulae* $F_{\mathsf{flow}}(q)$ *and* $F_{\mathsf{jump}}(e)$ *are unsatisfiable, where:*

$$\begin{aligned} F_{\mathsf{flow}}(q) = \ &\Psi(\overline{x}(t_0)) \wedge \mathsf{Inv}_q(\overline{x}(t_0)) \wedge \mathsf{Inv}_q(\overline{x}(t)) \wedge \\ &\wedge \underline{\mathsf{flow}}_q(t_0, t) \wedge \neg\Psi(\overline{x}(t)) \wedge t \geq t_0. \end{aligned}$$

Proof: The equivalence is a consequence of the properties of continuous and differentiable functions and of the convexity of the mode invariants. $\square$

An important consequence of Thm. 12 is the fact that for LHA checking invariance of safety properties expressible as convex linear predicates (i.e. conjunctions of strict and non-strict inequalities) over X can be done in polynomial time.

COROLLARY 13. *Let S be a LHA and Ψ be a convex linear predicate over X. The satisfiability of any of the formulae $F_{\mathsf{flow}}(q)$ and $F_{\mathsf{jump}}(e)$ in Thm. 12 can be checked in PTIME.*

Proof: All formulae of type $\mathsf{Inv}_q(\overline{x}(t))$ are conjunctions of linear inequalities in $x_1(t), \ldots, x_n(t)$. All formulae of the form $\mathsf{Jump}_e(\overline{x}(t), \overline{x}'(0))$ are conjunctions of linear inequalities in $x_1(t), \ldots, x_n(t), x_1'(0), \ldots, x_n'(0)$. The formula $\underline{\mathsf{flow}}_q(t_0, t_1)$ is a conjunction of non-strict linear inequalities in $x_1(t_0)$, $x_1(t_1), \ldots, x_n(t_0), x_n(t_1), t_0, t_1$; $\Psi(\overline{x}(t))$ is a conjunction of linear inequalities in $x_1(t), \ldots, x_n(t)$, and $\neg\Psi(\overline{x}, t)$ is a disjunction of linear inequalities in $x_1(t), \ldots, x_n(t)$. The PTIME complexity is based on the fact that any conjunction of linear inequalities and strict inequalities can be checked in polynomial time [15] (cf. Thm. 8). $\square$

Parametric linear hybrid automata. We consider two problems: (1) parametric verification (assuming that constraints on the parameters are given) and (2) synthesis of constraints on parameters. The following results are consequences of Theorems 12, 8 and 9:

THEOREM 14. *Consider a weak notion of parametricity, in which we allow the bounds in the linear inequalities over the values of the continuous variables to be parameters. The following hold:*

(1) Parametric verification. *Assume that relationships between the parameters – expressed as a conjunction Γ of linear inequalities – are given. Then checking whether a property Ψ expressed as convex linear predicate over X is an invariant is decidable in PTIME.*

(2) Constraint synthesis. *The problem of deriving constraints on parameters which guarantee that a convex linear predicate over X, Ψ, is an invariant has polynomial complexity, with the degree of the polynomial equal with the number of continuous variables of the system.*

Remark 1. If all constraints are in $UTVPI^{\neq}$ then by Theorem 10 quantifier elimination can be done in time $O(dn^4)$, where d is the number of disequations and n is the number of variables in the set of constraints in Theorem 12.

Remark 2. If the description of the automaton also contains parameters as coefficients and/or bounds for the linear inequalities describing the flows then Theorem 12 provides a reduction to checking the satisfiability of a family of non-linear constraints and parametric verification and the complexity is in general exponential.

EXAMPLE 3. *We illustrate the ideas on Example 1. The invariance property we study is:*

$$\Psi(x_1, x_2, x_3) = x_1 + x_2 + x_3 \leq L_{\text{overflow}} \wedge -\epsilon \leq x_1 - x_2 \leq \epsilon.$$

We assume that $L_f < L_{\text{overflow}}$ and $\epsilon_a < \epsilon$. By Thm. 12, Ψ is an invariant iff for every mode $q \in \{1, 2, 3, 4\}$ the following formula $F_{\text{flow}}(q)$ is unsatisfiable:

$$\Psi(\overline{x}(0)) \wedge \neg\Psi(\overline{x}(t)) \wedge \text{Inv}_q(\overline{x}(0)) \wedge \text{Inv}_q(\overline{x}(t)) \wedge \underline{\text{flow}}_q(\overline{x}, t)$$

and $F_{\text{Jump}}(e)$ is unsatisfiable for all $e \in E$. We ignore the redundant formulae corresponding to jumps which do not change the values of the variables and only analyze the jumps with reset, namely $e_{i_j} = (i_j, 4), i = 1, 2, 3, j = 1, 2.$[4]

As an illustration we present the formula $F_{\text{flow}}(q)$ for $q = 2$ (invariance under the flow in reaction mode):

$$(x_1(0)+x_2(0)+x_3(0) \leq L_{\text{overflow}} \wedge -\epsilon \leq x_1(0)-x_2(0) \leq \epsilon) \wedge$$
$$\neg(x_1(t)+x_2(t)+x_3(t) \leq L_{\text{overflow}} \wedge -\epsilon \leq x_1(t)-x_2(t) \leq \epsilon) \wedge$$

$$(L_f \leq x_1(0) + x_2(0) + x_3(0) \leq L_{\text{overflow}} \wedge x_3(0) \leq \text{max} \wedge$$
$$L_f \leq x_1(t) + x_2(t) + x_3(t) \leq L_{\text{overflow}} \wedge x_3(t) \leq \text{max} \wedge$$
$$x_1(t) - x_1(0) \leq -\text{dmin} \wedge x_2(t) - x_2(0) \leq -\text{dmin} \wedge$$
$$x_3(t) - x_3(0) \geq \text{dmin} \wedge x_1(t) - x_1(0) - (x_2(t) - x_2(0)) = 0 \wedge$$
$$x_1(t) - x_1(0) + x_2(t) - x_2(0) + x_3(t) - x_3(0) = 0)$$

This is the disjunction of two conjunctions of strict and non-strict linear inequalities, hence its satisfiability can be checked in PTIME. We used H-PILoT [13] for checking the satisfiability of $F_{\text{flow}}(q)$ (assuming $L_f < L_{\text{overflow}} \wedge \epsilon_a < \epsilon$) or to generate constraints on the parameters which guarantee (un)satisfiability of $F_{\text{flow}}(q)$. We also used H-PILoT directly on $F_{\text{Flow}}(q)$ (the complete instantiation facilities of H-PILoT allowed to construct $F_{\text{flow}}(q)$ from $F_{\text{Flow}}(q)$ automatically).

3.2 Bounded Model Checking

We now consider the problem of checking whether a formula Safe is preserved under runs of length bounded by k.

THEOREM 15. *Assume that Safe is a state assertion expressed by a convex linear predicate and that the LHA S is input-deterministic. The following are equivalent:*

(1) There exists no run σ of S with length at most k starting with a state satisfying Init such that condition Safe does not hold at the end of σ.

---

[4]For these, the verification tasks are trivial.

(2) The formulae F_i are unsatisfiable for all $1 \leq i \leq k$, where F_i states that there exists a run with end point satisfying $\neg$Safe passing through exactly i states, i.e. is the following formula (where InState is a predicate which indicates the state the system is in):

$$\bigvee_{q \in Q} [\text{Init}_q(\overline{x}^1(0))$$
$$\wedge \text{InState}^1 = q \wedge \text{Inv}_q(\overline{x}^1(0)) \wedge \underline{\text{flow}}_q(\overline{x}^1, t_1) \wedge \text{Inv}_q(\overline{x}^1(t_1)) \wedge t_1 \geq 0$$
$$\wedge \bigvee_{(q,q') \in E} \text{guard}_{(q,q')}(\overline{x}^1(t_1))] \wedge$$
$$\bigwedge_{\substack{e_1 = \\ (q,q') \in E}} [\text{InState}^1 = q \wedge \text{guard}_{e_1}(\overline{x}^1(t_1)) \rightarrow (\text{jump}_{e_1}(\overline{x}^1(t_1), \overline{x}^2(0))$$
$$\wedge \text{InState}^2 = q' \wedge \text{Inv}_{q'}(\overline{x}^2(0)) \wedge \underline{\text{flow}}_{q'}(\overline{x}^2, t_2) \wedge \text{Inv}_{q'}(\overline{x}^2(t_2)) \wedge t_2 \geq 0$$
$$\wedge \bigvee_{(q',q'') \in E} \text{guard}_{(q',q'')}(\overline{x}^2(t_2)))] \wedge$$
$$\cdots$$
$$\bigwedge_{\substack{e_{i-1} = \\ (q,q') \in E}} [\text{InState}^{i-1} = q \wedge \text{guard}_{e_{i-1}}(\overline{x}^{i-1}(t_{i-1})) \rightarrow \text{jump}_{e_{i-1}}(\overline{x}^{i-1}(t_{i-1}), \overline{x}^i(0))$$
$$\wedge \text{InState}^i = q' \wedge \text{Inv}_{q'}(\overline{x}^i(0)) \wedge \underline{\text{flow}}_{q'}(\overline{x}^i, t_i) \wedge \text{Inv}_{q'}(\overline{x}^i(t_i)) \wedge t_i \geq 0] \wedge$$
$$\neg\text{Safe}(\overline{x}^i(t_i))$$

Hence, the bounded model checking problem is in NP.

Proof: The equivalence of (1) and (2) is a consequence of the convexity assumptions in the definition of a LHA. The NP complexity of the bounded model problem is a consequence of the results in [28] (cf. Thm. 8). □

We identify situations in which checking unsatisfiability of the formulae in (2) above can be performed in PTIME.

COROLLARY 16. *Assume that for every $e \in E$, guard_e contains only equalities, and from each mode there is at most one switch to another mode. Then checking the satisfiability of all formulae F_i can be done in PTIME.*

Proof: The premise ensures that each F_i is equivalent to a disjunction of $|Q| * \text{length}(\text{Safe})$ conjunctions of Horn disjunctive constraints. The result follows from Thm. 9. □

We study some alternative situations in which PTIME satisfiability can be guaranteed.

THEOREM 17. *Assume that Init_q, Inv_q, Safe, Jump, guard_e can be expressed as conjunctions of inequalities between variables, $\underline{\text{flow}}$ can be expressed as a conjunction of non-strict inequalities between variables, and that from each mode there is at most one switch to another mode. We can adapt the encoding of the formulae F_i by accumulating the overall time (such that the last constraint is $t_i \leq t$). Hence, the satisfiability of all formulae F_i can be checked in PTIME.*

Proof: The premise ensures that F_i is a conjunction of Ord-Horn constraints for all i, so we can use Thm. 11. □

Remark: The restriction that the constraint in $\underline{\text{flow}}$ can be expressed as a non-strict inequality between variables holds iff the only constraints allowed in $\underline{\text{flow}}$ are monotonicity or antitonicity conditions of the form:

$$\text{if } t_1 \leq t_2 \text{ then } x(t_1) \leq x(t_2) \qquad (\text{or } x(t_1) \geq x(t_2)).$$

Parametric hybrid automata. The results in Theorem 14 lift in a natural way to yield complexity results for bounded model checking of parametric hybrid automata (in this case EXPTIME).[5]

---

[5]The remark following Theorem 14 will not hold in this case because the complexity results for quantifier elimination for $UTVPI^{\neq}$ refer only to unit clauses.

EXAMPLE 4. *Consider the LHA in Example 1, and the safety property in Example 3. We want to check whether an unsafe state can be reached from an initial state in at most two steps (where $\mathsf{Init}_q = \mathsf{Inv}_q$ for every $q \in Q$). This can be reduced to checking whether the following formula is unsatisfiable (where the disjunction in the first part of the formula is taken over all modes, and the conjunction in the middle part of the formula is taken over all mode switches):*

$[(\mathsf{InState}^1{=}1 \wedge \mathsf{Inv}_1(x_1^1(0), x_2^1(0), x_3^1(0)) \wedge$
$\quad \underline{\mathsf{flow}_1}(x_1^1, x_2^1, x_3^1, t_1) \wedge \mathsf{Inv}_1(x_1^1(t_1), x_2^1(t_1), x_3^1(t_1)) \wedge t_1 {\geq} 0 \wedge$
$\quad (\mathsf{guard}_{(1,2)}(x_1^1(t_1), x_2^1(t_1), x_3^1(t_1)) \vee \mathsf{guard}_{(1,4)}(x_1^1(t_1), x_2^1(t_1), x_3^1(t_1))$
$\quad \vee \mathsf{guard}_{(1,4)'}(x_1^1(t_1), x_2^1(t_1), x_3^1(t_1)))$
$\vee \ldots$
$(\mathsf{InState}^1{=}4 \wedge \mathsf{Inv}_4(x_1^1(0), x_2^1(0), x_3^1(0)) \wedge$
$\quad \underline{\mathsf{flow}_4}(x_1^1, x_2^1, x_3^1, t_1) \wedge \mathsf{Inv}_4(x_1^1(t_1), x_2^1(t_1), x_3^1(t_1)) \wedge t_1 {\geq} 0)]$
$\wedge$
$[(\mathsf{InState}^1{=}1 \wedge \sum_{i=1}^3 x_i^1(t_1) \geq L_f \rightarrow \bigwedge_{i=1}^3 x_i^2(0) = x_i^1(t_1) \wedge$
$\quad \mathsf{InState}^2{=}2 \wedge \mathsf{Inv}_2(\overline{x}^2(0)) \wedge \underline{\mathsf{flow}_2}(\overline{x}^2, t_2) \wedge \mathsf{Inv}_2(\overline{x}^2(t_2)) \wedge t_2 {\geq} 0)$
$\wedge (\mathsf{InState}^1{=}1 \wedge x_1^1(t_1) - x_2^1(t_1) \geq \epsilon_a \rightarrow \bigwedge_{i=1}^3 x_i^2(0) = 0 \wedge$
$\quad \mathsf{InState}^2{=}4 \wedge \bigwedge_{i=1}^3 x_i^2(0) = 0 \wedge \bigwedge_{i=1}^3 x_i^2(t_2) = 0 \wedge t_2 {\geq} 0)$
$\wedge (\mathsf{InState}^1{=}1 \wedge x_1^1(t_1) - x_2^1(t_1) \leq -\epsilon_a \rightarrow \bigwedge_{i=1}^3 x_i^2(0) = 0 \wedge$
$\quad \mathsf{InState}^2{=}4 \wedge \bigwedge_{i=1}^3 x_i^2(0) = 0 \wedge \bigwedge_{i=1}^3 x_i^2(t_2) = 0 \wedge t_2 {\geq} 0)]$
$\wedge \ldots$
$[(\mathsf{InState}^1{=}3 \wedge -\epsilon_a \leq x_1^1(t_1) - x_2^1(t_1) \leq \epsilon_a \wedge 0 \leq x_3^1(t_1) \leq \min \rightarrow$
$\quad \bigwedge_{i=1}^3 x_i^2(0) = x_i^1(t_1) \wedge \mathsf{InState}^2{=}1 \wedge \mathsf{Inv}_1(x_1^2(0), x_2^2(0), x_3^2(0)) \wedge$
$\quad \underline{\mathsf{flow}_1}(\overline{x}^2, t_2) \wedge \mathsf{Inv}_1(x_1^2(t_2), x_2^2(t_2), x_3^2(t_2)) \wedge t_2 {\geq} 0)$
$\wedge (\mathsf{InState}^1{=}3 \wedge x_1^1(t_1) - x_2^1(t_1) \geq \epsilon_a \rightarrow \bigwedge_{i=1}^3 x_i^2(0) = 0 \wedge$
$\quad \mathsf{InState}^2{=}4 \wedge \bigwedge_{i=1}^3 x_i^2(0) = 0 \wedge \bigwedge_{i=1}^3 x_i^2(t_2) = 0 \wedge t_2 {\geq} 0)$
$\wedge (\mathsf{InState}^1{=}3 \wedge x_1^1(t_1) - x_2^1(t_1) \leq -\epsilon_a \rightarrow \bigwedge_{i=1}^3 x_i^2(0) = 0 \wedge$
$\quad \mathsf{InState}^2{=}4 \wedge \bigwedge_{i=1}^3 x_i^2(0) = 0 \wedge \bigwedge_{i=1}^3 x_i^2(t_2) = 0 \wedge t_2 {\geq} 0)] \wedge$
$(x_1^2(t_2) + x_2^2(t_2) + x_3^2(t_2) > L_{\mathsf{overflow}} \vee x_1^2(t_2) - x_2^2(t_2) > \epsilon \vee x_1^2(t_2) - x_2^2(t_2) < -\epsilon)$.

It can be seen that every component in the disjunction in the first part of the formula is a conjunction of linear inequalities. The disjunction in the last part of the formula is a disjunction of strict linear inequalities. Writing $a < b$ as $a \leq b \wedge a \neq b$ it is easy to see that this last formula can be seen as a disjunction of Horn disjunctive linear constraints.

The second part of the formula can be written as a conjunction of implications of the form $\bigwedge_{i=1}^n p_i \rightarrow p$, where $p_1, \ldots, p_n$ and p are linear inequalities. In conclusion, using distributivity, the formula above can be written as the disjunction of $4 * 3$ conjunctions of implications of the form $\bigwedge_{i=1}^n p_i \rightarrow p$, where $p_1, \ldots, p_n$ and p are linear inequalities.

If all the guards in the automaton are expressed using equalities (for our example, only using the limit conditions instead of the full inequalities, e.g. $x_1 + x_2 + x_3 = L_f$ instead of $x_1 + x_2 + x_3 \geq L_f$ for the transition from mode 1 to mode 2) then the formula is the disjunction of 12 conjunctions of Horn disjunctive linear constraints, hence its satisfiability is decidable in PTIME.

4. REASONABLE HYBRID AUTOMATA

We are interested in a class of controllers for which mode entry conditions are chosen with sufficient safety margin such that it is guaranteed that the system remains in a node at least a given time, and for which mode invariants are compatible with the safety condition(s).

Invariant compatibility. We consider hybrid automata in which the mode invariants are compatible with the safety condition Safe and with the guards of the transitions.

DEFINITION 18. *The mode invariants of a hybrid automaton S are compatible with the safety condition Safe and with the guards of the transitions if:*

(1a) All mode invariants are safe: For every mode $q \in Q$,
$\models \forall \overline{x}(\mathsf{Inv}_q(\overline{x}) \rightarrow \mathsf{Safe}(\overline{x}))$;

(1b) There exists $\delta > 0$ such that if a mode invariant is true at time t and becomes false at time $t + \delta$, then one of the guards of a jump from that mode becomes true at a time moment t' with $t \leq t' \leq t + \delta$:

$\forall t[\mathsf{Inv}_q(\overline{x}(t)) \wedge \mathsf{flow}_q(t, t + \delta) \wedge \neg \mathsf{Inv}_q(\overline{x}(t+\delta)) \rightarrow$
$\exists t'(t \leq t' \leq t + \delta \wedge \mathsf{Inv}_q(\overline{x}(t)) \wedge \bigvee_{(q,q') \in E} \mathsf{guard}_{(q,q')}(\overline{x}(t')))]$.

Chatter-freedom. A hybrid automaton S is *chatter-free* if mode entry conditions are chosen with sufficient safety margin such that it is guaranteed that the system remains in a node at least a given time ϵ_t. Formally:

DEFINITION 19. *A hybrid automaton S is* chatter-free *if:*

(2a) all transitions lead to an inner envelope, i.e. for all $q \in Q$, $(q, q') \in E$ the following formula is valid:
$\forall \overline{x}(\mathsf{Inv}_q(\overline{x}) \wedge \mathsf{guard}_{(q,q')}(\overline{x}) \wedge \mathsf{jump}_{(q,q')}(\overline{x}, \overline{x}') \rightarrow \mathsf{InEnv}_{q'}(\overline{x}'))$;

(2b) for any flow starting in the inner envelope of a mode q, no guard of a mode switch (q, q') will become true in a time interval smaller than ϵ_t.

Reasonable hybrid automata. We say that an extended hybrid automaton S is *reasonable* w.r.t. a safety property Safe, if (1) S is input deterministic; (2) the mode invariants are compatible with Safe and with the guards of the transitions; (3) S is chatter-free.

4.1 Checking the property of being reasonable

We show that checking the property of an extended LHA of being reasonable can be done in PTIME.

Checking input determinism reduces to checking satisfiability of linear constraints, so can be done in PTIME.

Compatibility of the mode invariants with safety conditions expressed as convex linear predicates can be checked in PTIME:

LEMMA 20. *Let S be an LHA and let Safe be a convex linear predicate. The problem of checking whether for all $q \in Q$, $\models \forall \overline{x}(\mathsf{Inv}_q(\overline{x}) \rightarrow \mathsf{Safe}(\overline{x}))$ is decidable in PTIME.*

Proof: For each $q \in Q$, $\models \forall x(\mathsf{Inv}_q(x) \rightarrow \mathsf{Safe}(x))$ iff $\mathsf{Inv}_q(x) \wedge \neg \mathsf{Safe}(x)$ is unsatisfiable. As $\mathsf{Inv}_q(x)$ and $\mathsf{Safe}(x)$ are sets of linear constraints this problem is decidable in PTIME. □

Compatibility with the guards. We present a condition which implies compatibility of invariants with the guards of the transitions. We use the following notation: If C is a conjunction of linear inequalities let $C^b = \{\sum_{j=1}^n a_{ij} x_j = a_i \mid \sum_{j=1}^{n_i} a_{ij} x_j \leq a_i \in C\}$. We say that a set of values $(v_1, \ldots, v_n)$ satisfies C^b if it satisfies at least one equality in C^b.

THEOREM 21. *Let S be an LHA. Assume that there exists $\delta > 0$ s.t. for every $q \in Q$ the following holds:*
$\forall \overline{z}[(\exists t'(0 < t' < \delta \wedge \underline{\mathsf{flow}}(\overline{x}, 0, t') \wedge \mathsf{Inv}_q(\overline{x}(0)) \wedge \bigwedge_{i=1}^n x_i(t') = z_i))$
$\wedge \mathsf{Inv}_q^b(\overline{z}) \rightarrow \bigvee_{(q,q') \in E} \mathsf{guard}_{(q,q')}(z_1, \ldots, z_n)]$.

Then condition (1b) in Def. 18 holds.

Proof: Assume that there exists an evolution according to the flow in mode q s.t. $\mathsf{Inv}_q(\overline{x}(t))$ holds and $\mathsf{Inv}_q(\overline{x}(t + \delta))$ does not hold. Since $x_1, \ldots, x_n$ are all continuous, any linear combination has the intermediate value property, so (since Inv_q is convex) there exists t' with $t \leq t' < t + \delta$ s.t. $\overline{x}(t')$ satisfies at least one of the equalities in Inv_q^b. By the assumption it then follows that $\mathsf{guard}_{(q,q')}(\overline{x}(t'))$ for some $(q, q') \in E$. □

LEMMA 22. *Checking the premise of Thm. 21 can be reduced to satisfiability checking of $|Q|\prod_{e\in E}\text{length}(\text{guard}_e)$ linear constraints (each decidable in PTIME).*

EXAMPLE 5. *Consider a refinement of our example in which e.g. the flow in mode 1 is described by $\dot{x}_1 = \dot{x}_2 = k > 0$, and $\dot{x}_3 = 0$. We check whether the assumption in Thm. 21 holds as follows:*

$$\text{Inv}_1^b = \{x_1 = 0, x_2 = 0, x_3 = 0, x_1 + x_2 + x_3 = 0,$$
$$x_1 + x_2 + x_3 = L_f, x_1 - x_2 = \epsilon_a, x_1 - x_2 = -\epsilon_a\}$$

Let (x_1, x_2, x_3) be a state satisfying some equality in Inv_1^b and such that there exists a flow (of positive length) leading from a state (x_1', x_2', x_3') satisfying Inv_1 to (x_1, x_2, x_3). Because of the flow rules in mode 1, it can be proved that this state cannot satisfy any of $\{x_1 = 0, x_2 = 0, x_1 + x_2 + x_3 = 0\}$.

If $x_1 + x_2 + x_3 = L_f$ then the guard of mode switch (1,2) is true. If $x_1 - x_2 = \epsilon_a$ or $x_1 - x_2 = -\epsilon_a$ then the guard of one of the mode switches from 1 to 4 is true.

Chatter-freedom can also be checked in PTIME.

THEOREM 23. *Let S be an LHA. Checking whether S is chatter-free can be done in PTIME.*

Proof: (2a) reduces to a satisfiability check for linear constraints. (2b) holds iff for every $(q, q') \in E$, $\text{Inv}_q(\overline{x}(0)) \wedge \underline{\text{flow}}_q(0, t) \wedge \text{guard}_{(q,q')}(\overline{x}(t)) \wedge t \leq \epsilon_t$ is unsatisfiable. $\square$

EXAMPLE 6. *Consider a further refinement of Ex. 5 in which e.g. the inner envelope of mode 1 is:*

$$\text{InEnv}_1(x_1, x_2, x_3) = 0 \leq x_1 + x_2 + x_3 \leq L_{\text{safe}} \wedge \bigwedge_{i=1}^{3} x_i \geq 0$$
$$\wedge -\epsilon_a \leq x_1 - x_2 \leq \epsilon_a \wedge 0 \leq x_3 \leq \min$$

The minimal dwelling time in mode 1 (provided the mode is entered in the inner envelope) is ϵ_t iff the following formulae are unsatisfiable:

$$\text{InEnv}_1(x_1(0), x_2(0), x_2(0)) \wedge \underline{\text{flow}}_1(0, t) \wedge$$
$$\text{guard}_e(x_1(t), x_2(t), x_2(t)) \wedge t \leq \epsilon_t$$

where e is one of the transitions from mode 1 into mode 2 or 4. Analyze for instance the switch change (1,2). The corresponding formula is F_e:

$$0 \leq x_1 + x_2 + x_3 \leq L_{\text{safe}} \wedge \text{Inv}_1 \wedge x_3(t) = x_3(0) \wedge$$
$$x_1(t) - x_1(0) \geq k * (t - 0) \wedge x_2(t) - x_2(0) \geq k * (t - 0) \wedge$$
$$L_f \leq x_1(t) + x_2(t) + x_2(t) \wedge t > \epsilon_t.$$

It can be seen (using QE) that F_e is unsatisfiable iff $L_f > L_{\text{safe}} + 2k\epsilon_t$. In our tests we considered: (i) a non-parametric version of the running example in which $L_f, L_{\text{safe}}, k, \epsilon_t, \ldots$ are instantiated such that the constraint above (and similar constraints) are satisfied; (ii) a parametric version of the example in which such constraints are additionally specified; (iii) a parametric version of the example in which we use the equivalence above (and analogous for other modes/switches) to derive constraints on the parameters.

4.2 Verification of Reasonable LHA

We show that for reasonable LHA checking convex safety properties with exhaustive entry conditions is decidable in PTIME and checking time-bounded reachability properties with exhaustive entry conditions is in NP.

4.2.1 Safety Properties

THEOREM 24. *Let $\phi = \Box(\phi_{\text{ExhEntry}} \rightarrow \Box\phi_{\text{safe}})$, and let S be a LHA satisfying the invariant compatibility condition w.r.t. ϕ_{safe}. The following are equivalent:*

(i) $S \models \Box(\phi_{\text{ExhEntry}} \rightarrow \Box\phi_{\text{safe}})$.

(ii) The following invariance conditions hold:

(a) $\phi_{\text{ExhEntry}} \models \phi_{\text{safe}}$.
(b) ϕ_{safe} *is invariant under all flows and all mode-reachable jumps of the hybrid automaton.*

LEMMA 25. *A jump (s_1, s_2) where s_1, s_2 are states with $s_1 = (q, x_1, \ldots, x_n)$ and $s_2 = (q', x_1', \ldots, x_n')$ is mode reachable iff the following formula holds:*

$$\text{Inv}_q(\overline{x}) \wedge \text{guard}_e(\overline{x}) \wedge \text{jump}_e(\overline{x}, \overline{x}') \wedge$$
$$\exists x_1^0, \ldots, x_n^0, t(\text{InEnv}_q(\overline{x}^0) \wedge \bigwedge_{j=1}^{n_q} \sum_{i=1}^{n} c_{ij}^q(x_i^1 - x_i^0) \leq c_j^q * t).$$

where $\underline{\text{flow}}_q = \bigwedge_{j=1}^{n_q} \sum_{i=1}^{n} c_{ij}^q(x_i(t) - x_i(0)) \leq c_j^q * t.$

THEOREM 26. *Let $\phi = \Box(\phi_{\text{ExhEntry}} \rightarrow \Box\phi_{\text{safe}})$ where ϕ_{safe} is a convex linear predicate and S be an LHA satisfying the invariant compatibility condition w.r.t. ϕ_{safe}. Then (i) and (ii) in Thm. 24 are equivalent to:*

(iii) The following formulae are unsatisfiable:

(a) $\phi_{\text{ExhEntry}}(\overline{x}) \wedge \neg\phi_{\text{safe}}(\overline{x})$;

(b) The formulae F_e for every $e = (q, q') \in E$, where:

$$F_e = \phi_{\text{safe}}(\overline{x}) \wedge (\text{Inv}_q(\overline{x}) \wedge \text{guard}_e(\overline{x}) \wedge \text{jump}(\overline{x}, \overline{x}') \wedge$$
$$\exists \overline{x}^0(\text{InEnv}_q(\overline{x}^0) \wedge \bigwedge_{j=1}^{n_q} \sum_{i=1}^{n} c_{ij}^q(x_i^1 - x_i^0) \leq c_j^q * t)) \wedge$$
$$\wedge \neg\phi_{\text{safe}}(\overline{x}')$$

where $\underline{\text{flow}}_q = \bigwedge_{j=1}^{n_q} \sum_{i=1}^{n} c_{ij}^q(x_i(t) - x_i(0)) \leq c_j^q * t.$

Proof: The equivalence of (ii) and (iii) follows as in the proof of Thm. 12, using Lemma 25 and the fact that, by condition (1a) in Def. 18, each formula $F_{\text{flow}}(q)$ in Thm. 12 is unsatisfiable. $\square$

COROLLARY 27. *Let $\phi = \Box(\phi_{\text{ExhEntry}} \rightarrow \Box\phi_{\text{safe}})$ where ϕ_{safe} is a convex linear predicate and S be a reasonable ELHA w.r.t. ϕ_{safe}. The problem of checking whether $S \models \phi$ is decidable in PTIME.*

Note. In Corollary 27 only the invariant compatibility condition in the definition of reasonable ELHA is needed.

4.2.2 Time-Bounded Reachability Properties

We consider time-bounded reachability properties of the form:

$$\phi = \Box(\phi_{\text{ExhEntry}} \rightarrow \Diamond_{\leq t}\phi_{\text{safe}}).$$

THEOREM 28. *Let S be an ELHA satisfying the condition that there is a lower bound ϵ_t for the minimal dwell time for each mode. The following are equivalent:*

(1) $S \models \Box(\phi_{\text{ExhEntry}} \rightarrow \Diamond_{\leq t}\phi_{\text{safe}})$.

(2) There exists a run σ of S with total time length at most t starting with a state satisfying ϕ_{ExhEntry}, s.t. the safety condition ϕ_{safe} holds at the end of σ.

(3) There exists a constant $k(t, \epsilon_t)$ and a run σ of S with at most $k(t, \epsilon_t)$ states (and total time length at most t), starting from a state satisfying ϕ_{ExhEntry} such that ϕ_{safe} holds at the end of σ.

For all input-deterministic ELHA, if ϕ_{safe} is a convex linear predicate then (3) is equivalent to (4):

(4) There exists a constant $k(t, \epsilon_t)$ such that at least one of the formulae $F'_i \wedge \sum_{j=1}^i t_i \leq t$, $1 \leq i \leq k(t, \epsilon_t)$ is satisfiable (where the F'_i are like the formulae F_i in item (2) of Thm. 15, except for the fact that Init is replaced with ϕ_{ExhEntry} and $\neg\mathsf{Safe}$ with ϕ_{safe}).

Proof: (1) and (2) are equivalent by definition. (3) obviously implies (2). Assume that (2) holds. Let $k_1(t, \epsilon_t) = \frac{t}{\epsilon_t}$. Since S must remain at least time ϵ_t in every continuous mode, any run of cumulated time $\leq t$ can have at most $k_1(t, \epsilon_t)$ flows, hence it can have length at most $k(t, \epsilon_t) = 2 * k_1(t, \epsilon_t)$. Hence (3) holds. Under the assumptions in Thm. 15, the equivalence of (3) and (4) can be proved as Thm. 15. $\square$

COROLLARY 29. *Let $\square(\phi_{\mathsf{ExhEntry}} \rightarrow \lozenge_{\leq t}\phi_{\mathsf{safe}})$ be a bounded-time reachability condition with exhaustive entry conditions, where ϕ_{safe} is a convex linear predicate. Let S be a reasonable ELHA, w.r.t. ϕ_{safe}. Then the problem of checking whether $S \models \square(\phi_{\mathsf{entry}} \rightarrow \lozenge_{\leq t}\phi_{\mathsf{safe}})$ is in NP.*

The problem is decidable in PTIME if (i) all guards of mode switches can be expressed as equalities and from each mode there is at most one control switch to another mode, or (ii) all constraints in F'_i are in the OrdHorn class.

Note. In Corollary 29 only the input determinism and chatter-freedom conditions of S are needed. The results also hold for time-bounded reachability properties for which $\phi_{\mathsf{entry}} \models \phi_{\mathsf{ExhEntry}}$.

EXAMPLE 7. *Consider the variant of our running example[6] in Example 3, in which bounds on the rate of change of the substances x_1, x_2, x_3 in modes 2 (React) and 4 (Dump) are additionally specified. Consider the following properties:*

(1) $\phi_1 = \square(\phi_{\mathsf{entry}} \rightarrow \lozenge_{\leq t} x_1 = x_2 = x_3 = 0)$, where $\phi_{\mathsf{entry}} = 0 \leq x_1 + x_2 + x_3 \leq L_{\mathsf{overflow}} \wedge \bigwedge x_i \geq 0 \wedge \neg(-\epsilon_a \leq x_1 - x_2 \leq \epsilon_a)$.

(2) $\phi_2 = \square(\mathsf{InEnv}_1 \rightarrow \lozenge_{\leq t}(x_3 \geq \mathsf{max} \vee x_1 = x_2 = x_3 = 0))$

(1) ϕ_1 states that if x_1, x_2 are mixed in the wrong proportion then the tank is emptied in at most t time units. Since there is no mode switch from state 4 into another state, we only need to check runs of length at most 2, i.e. $\sigma_1 = 14$, $\sigma_2 = 24$ and $\sigma_3 = 34$ or prefixes thereof.

(2) ϕ_2 states that if we start in an entry state in the inner envelope of mode 1 (Fill), then there exists a run of the system such that, in at most t units of time, either a sufficient amount of reaction product is produced or the substances are dumped. Assume that $t = 200s$, and the minimal dwelling time in each mode is $100s$. Then we only need to check runs with two states and one jump, namely $\sigma_1 = 14$ and $\sigma_2 = 12$.

5. SUMMARY OF RESULTS

We can summarize our results as follows. Assume that S is an LHA and the safe states are conjunctions of linear inequalities. The following complexity results hold for non-parametric verification, and also for parametric verification

---

− if the constraints on the parameters are linear inequalities (invariant checking) resp. sets of clauses in linear arithmetic in the classes mentioned in the header (BMC), and parameters are allowed only as bounds for linear inequalities over X (for non-linear constraints, parametric coefficients or parametric bounds in flows the complexity is exponential) [7]:

	Inv. checking	BMC		
		general	eq.guards + det.	ord Horn + det.
LHA	PTIME	NP	PTIME	PTIME

The complexity of the problem of generating constraints on parameters (Inv(1): only bounds on linear combinations of variables are parametric; Inv(2): also coefficients and/or bounds on rates of growth are allowed to be parametric):

	Inv(1)	Inv(2)	BMC
LHA	$O(|C|^n)$	EXPTIME	EXPTIME
LHA ($UTVPI^{\neq}$)	PTIME	EXPTIME	EXPTIME

Complexity of the verification of reasonability conditions:

Inv. comp. (1a)	PTIME
Test implying (1b)	$\prod_{(q,q') \in E}$ length(guard$_{(q,q')}$)
(mode q)	PTIME problems
Chatter-free (2a), (2b)	PTIME

Complexity of verification problems for reasonable ELHA:

$\square(\phi_{\mathsf{ExhEntry}} \rightarrow \square\phi_{\mathsf{safe}})$	$\square(\phi_{\mathsf{ExhEntry}} \rightarrow \lozenge_{\leq t} \phi_{\mathsf{safe}})$	
Invariant Compatibility	Input determinism + Chatter-freeness	+ (equality guards or OrdHorn) + det.
PTIME	NP	PTIME

6. EXPERIMENTAL RESULTS

We checked Example 1 with H-PILoT [13] and the SMT-solver Z3 [21]; for generating constraints between parameters we used H-PILoT together with Redlog [6]. For the parametric examples presented in this paper, some ability to handle non-linear arithmetical constraints was essential; Z3 is one of the few SMT checkers which supports such constraints. In our tests, Z3 proved to be very efficient: For bounded model checking it verified a 100 steps run in slightly more than 1 second. The results are presented in the table below (n is the number of steps, the first column (LHA) refers to BMC for LHA where the time t spent inside of a mode satisfies condition $t \geq 0$; the second column (CF) refers to BMC for chatter-free LHA, so the time t spent inside of a mode satisfies the condition $t \geq \delta$, where δ is a parameter for which we only specified that $\delta > 0$). The time difference is explained by the fact that for LHA a jump may take place after a zero time delay in a mode q, so we need to consider the congruence axioms for the functions x_i, whereas for chatter-free systems this situation cannot occur, so congruence axioms are not needed.

n	LHA	CF	n	LHA	CF	n	LHA	CF
2	0.02s	0.01s	5	0.04s	0.03s	20	0.15s	0.10s
3	0.03s	0.02s	6	0.05s	0.04s	50	0.43s	0.40s
4	0.03s	0.02s	7	0.06s	0.04s	100	1.07s	1.00s

The running times are given in User + sys times (in s). The experiments were carried out on an Intel Xeon 3 GHz, 512 kB cache. Z3 version 2.13 was used.

---

[6] We also considered a refinement of the example in Sect. 1 in which mode 4 (**Dump**) is replaced with three modes: **Add** x_1 (in which the inflow of substance x_2 is stopped and only x_1 is added), **Add** x_2 (in which the inflow of x_1 is stopped and only x_2 is added) and **Dump'** described as Dump. Due to space limitations we do not present this refinement here.

[7]"det." is condition: "$|\{(q, q')|(q, q') \in E\}| \leq 1$ for all $q \in Q$".

7. CONCLUSION

In this paper we proved that verification of safety properties for reasonable ELHA can be reduced to invariant checking resp. to problems very similar to bounded model checking and, ultimately, to checking the validity of certain formulae (obtained using a polynomial reduction). We showed that the problem of checking the validity of such formulae is typically in NP, and identified verification tasks which can be performed in PTIME. For the parametric verification of such properties we identified conditions when the complexity is in PTIME, NP, or EXPTIME. We also analyzed the complexity of checking the property of being reasonable for ELHA.

Acknowledgments. This work was partly supported by the German Research Council (DFG) as part of the Transregional Collaborative Research Center "Automatic Verification and Analysis of Complex Systems" (SFB/TR 14 AVACS). See www.avacs.org for more information.

8. REFERENCES

[1] M. Agrawal, P. S. Thiagarajan. The Discrete Time Behavior of Lazy Linear Hybrid Automata. *Proc. HSCC 2005, LNCS 3414*, 55-69, Springer 2005.

[2] R. Alur, T.A. Henzinger, P.H. Ho. Automatic symbolic verification of embedded systems. *IEEE Trans. Software Eng.* 22(3): 181-201, 1996.

[3] T. Brihaye, Ch. Michaux. On the expressiveness and decidability of o-minimal hybrid systems. *Journal of Complexity* 21(4): 447-478, 2005.

[4] W. Damm, G. Pinto, S. Ratschan. Guaranteed termination in the verification of LTL properties of non-linear robust discrete time hybrid systems. *Int. J. Found. Comput. Sci.* 18(1): 63-86, 2007.

[5] W. Damm, H. Dierks, S. Disch, W. Hagemann, F. Pigorsch, C. Scholl, U. Waldmann, B. Wirtz. Exact and Fully Symbolic Verification of Linear Hybrid Automata with Large Discrete State Spaces. *Science of Computer Programming. Special Issue on Automated Verification of Critical Systems*, Editor M. Roggenbach, Accepted for publication, 2011.

[6] A. Dolzmann and T. Sturm. Redlog: Computer algebra meets computer logic. *ACM SIGSAM Bulletin* 31(2):2-9, 1997.

[7] G.E. Fainekos, G.J. Pappas. Robustness of temporal logic specifications. *Proc. FATES/RV 2006, LNCS 4262*, pp. 178-192, Springer, 2006.

[8] G. Frehse, S.K. Jha, B.H. Krogh. A counterexample guided approach to parameter synthesis for linear hybrid automata. *Proc. HSCC 2008, LNCS 4981*, pp. 187-200, Springer, 2008.

[9] G. Frehse. Tools for the verification of linear hybrid automata models, *Handbook of Hybrid Systems Control, Theory – Tools – Applications*. Cambridge University Press, Cambridge, 2009.

[10] S. Gulwani and A. Tiwari. Constraint-based approach for analysis of hybrid systems. In *Proc. CAV 2008*, LNCS 5123, pp. 190-203, Springer, 2008.

[11] T.A. Henzinger, P.W. Kopke, A. Puri, P. Varaiya. What's decidable about hybrid automata? *Journal of Computer and System Sciences* 57(1): 94-124, 1998.

[12] T.A. Henzinger, P.-H. Ho, and H. Wong-Toi Algorithmic Analysis of Nonlinear Hybrid Systems *IEEE Trans. on Automatic Control* 43:540-554, 1998.

[13] C. Ihlemann and V. Sofronie-Stokkermans. System description: H-PILoT. In *Proc. CADE 2009*, LNAI 5663, pp. 131-139, Springer, 2009.

[14] S. Jha, B.A. Brady, and S.A. Seshia Symbolic Reachability Analysis of Lazy Linear Hybrid Automata *Proceedings of FORMATS 2007*, 2007.

[15] L. Khachian. A polynomial time algorithm for linear programming. *Soviet Math. Dokl.* 20:191-194, 1979.

[16] M. Koubarakis. Tractable disjunctions of linear constraints: basic results and applications to temporal reasoning. *Theor. Comput. Sci.* 266: 311-339, 2001.

[17] M. Koubarakis and S. Skiadopoulos. Querying temporal and spatial constraint networks in PTIME. *Artificial Intelligence* 123: 223-263, 2000.

[18] G. Lafferriere, G.J. Pappas, S. Sastry. O-Minimal hybrid systems. Mathematics of Control, Signals, and Systems, 13(1):1-21, 2000.

[19] G. Lafferriere, G.J. Pappas, S. Yovine. A new class of decidable hybrid systems. *Proc. HSCC 1999*, LNCS 1569, pp.137-151, Springer, 1999.

[20] J.S. Miller. Decidability and complexity results for timed automata and semi-linear hybrid automata. *Proc. HSCC 2000*, LNCS 1790, pp. 296-309, 2000.

[21] L.M. de Moura and N. Bjørner. Z3: An Efficient SMT Solver. *Proc. TACAS 2008*, LNCS 4963, pp. 337-340, 2008.

[22] B. Nebel and H.-J. Bürckert. Reasoning about temporal relations: A maximal tractable subclass of Allen's interval algebra. *Journal of the ACM* 42(1): 43-66, 1995.

[23] A. Platzer and J.-D. Quesel. Logical verification and systematic parametric analysis in train control. *Proc. HSCC 2008*, LNCS 4981, pp. 646-649, Springer, 2008.

[24] A. Platzer and J.-D. Quesel. European train control system: A case study in formal verification. *Proc. ICFEM 2009*, LNCS 5885, pp. 246-265, Springer, 2009.

[25] V. Sofronie-Stokkermans. Hierarchic reasoning in local theory extensions. *Proc. CADE-20*, LNAI 3632, pp. 219-234, Springer, 2005.

[26] V. Sofronie-Stokkermans. Efficient hierarchical reasoning about functions over numerical domains. In *Proc. KI 2008*, LNAI 5243, pp.135-143, Springer, 2008.

[27] V. Sofronie-Stokkermans. Hierarchical reasoning for the verification of parametric systems. *Proc. IJCAR 2010*, LNAI 6173, pp. 171-187, Springer, 2010.

[28] E.D. Sontag. Real addition and the polynomial hierarchy. *Inf. Proc. Letters* 20(3):115-120, 1985.

[29] M. Swaminathan, M. Fränzle. A symbolic decision procedure for robust safety of timed systems. *Proc. TIME 2007*, p. 192, IEEE Computer Society, 2007.

[30] G. J. Tee. Khachian's efficient algorithm for linear inequalities and linear programming. *ACM SIGNUM Newsletter Archive* 15(1):13-15, 1980.

[31] F. Wang. Symbolic Parametric Safety Analysis of Linear Hybrid Systems with BDD-Like Data-Structures. *IEEE Trans. Software Eng.* 31(1): 38-51, 2005.

Quantitative Automata Model Checking of Autonomous Stochastic Hybrid Systems*

Alessandro Abate
Delft Center for Systems & Control
Mekelweg 2, 2628 CD Delft
The Netherlands
a.abate@tudelft.nl

Joost-Pieter Katoen
Software Modeling and Verification Group
Ahornstraße 55
D-52056 Aachen, Germany
katoen@cs.rwth-aachen.de

Alexandru Mereacre
Software Modeling and Verification Group
Ahornstraße 55
D-52056 Aachen, Germany
mereacre@cs.rwth-aachen.de

ABSTRACT

This paper considers the quantitative verification of discrete-time stochastic hybrid systems (DTSHS) against linear time objectives. The central question is to determine the likelihood of all the trajectories in a DTSHS that are accepted by an automaton on finite or infinite words. This verification covers regular and ω-regular properties, and thus comprises the linear temporal logic LTL. This work shows that these quantitative verification problems can be reduced to computing reachability probabilities over the product of an automaton and the DTSHS under study. The computation of reachability probabilities can be performed in a backward-recursive manner, and quantitatively approximated by procedures over discrete-time Markov chains. A case study shows the feasibility of the approach.

Categories and Subject Descriptors

H.4 [**Information Systems Applications**]: Miscellaneous

General Terms

Theory

Keywords

Stochastic hybrid systems, finite state automata, LTL

1. INTRODUCTION

Stochastic hybrid systems (SHS, for short) are pivotal in application areas such as systems biology, air traffic control,

*This research is funded by the DFG research training group 1295 AlgoSyn, by the European Commission under the MoVeS project, FP7-ICT-2009-257005, by the European Commission under Marie Curie grant MANTRAS 249295, and by NWO under VENI grant 016.103.020.

finance, and telecommunication systems [6]. Their main analytical challenge is to treat the intricate intertwining of discrete dynamics, continuous phenomena, and randomness. An important class of properties for such systems are probabilistic invariance (i.e., what is the likelihood to stay in a set of "safe" states for some period of time?) and reachability (i.e., what is the likelihood to reach a certain set of "goal" states within a number of steps?). Variants thereof, such as reach-avoid objectives, have also been considered [18]. A broad palette of techniques has been developed to compute these quantities, e.g., measure-theoretical approaches [5], Monte Carlo simulations [14], coupled Hamilton Jacobi Bellman equations [13], approximation methods that turn an infinite-state problem into a finite-state one [15], and dynamic programming [3]. These approaches exist for controllable as well as autonomous models (i.e. "deterministic" SHS), where in the former case maximal – or dually minimal – probabilities are determined [3].

On the other hand, probabilistic reachability, invariance, and reach-avoid measures have been studied extensively in the field of probabilistic model checking. In fact, these properties can all naturally be expressed in PCTL [11], the probabilistic variant of the branching-time temporal logic CTL. Thanks to the effective support of powerful software tools such as PRISM [1] and MRMC [12], the verification of this class of properties has been successfully applied to numerous models for systems biology, security protocols, hardware circuits and reliability analysis. This insight has recently culminated into the generalization of CTL model checking to continuous state space [7], the application of PCTL model checking to discrete-time SHS (DTSHS) [2], and a study on the relationship between PCTL and dynamic programming for SHS models [17].

The use of PCTL allows for the expression of a significant class of properties that can be analyzed over SHS. This paper takes a somewhat orthogonal direction by considering probabilistic *linear-time* objectives. An example of such an objective is a more advanced reach-avoid property, such as the following: determine the likelihood, starting from some initial point, to reach a set of "goal" states, while avoiding "bad" states, but conditional on visiting some "trigger" states prior to reaching the goal states. Other examples include repeated reachability objectives (certain goal states must be visited repeatedly), conjuncted with a persistence property – from some point on, the system should only obey goal states. Such objectives can naturally be provided as finite-state automata, either on finite or on infinite words. This includes regular and ω-regular properties, and thus covers the linear temporal logic LTL and several properties that cannot be expressed in PCTL (like the ones above).

The central question that we address in this paper is how to determine the probability that a given discrete-time SHS satisfies an automaton. Stated differently, we investigate what is the likelihood of the trajectories in the DTSHS that are accepted by the automaton. In order to cope with the quantitative verification of DTSHS against automata objectives, we resort to a well-known technique from model checking: product construction. We consider the synchronous product of a DTSHS $\mathcal{H}$ and a (Büchi) automaton $\mathcal{A}$. Technically this is achieved by adopting deterministic finite-state automata (for finite words) and separated generalized Büchi automata [8] (for infinite words). The key result is that the probability of all paths in $\mathcal{H}$ that are accepted by $\mathcal{A}$ can be reduced to computing reachability probabilities in the product $\mathcal{H} \otimes \mathcal{A}$. These reachability probabilities can in principle be determined by any available technique for DTSHS. We then consider a finite abstraction of the DTSHS as a discrete-time Markov chain (DTMC), and provide an approximation (with explicit errors) of the time-bounded or -unbounded reachability objectives. Aa a numerical case study, a two-room heating benchmark [2] is used to illustrate the application and feasibility of our approach.

Automated verification of stochastic hybrid systems is *en vogue*, and several quite recent works have appeared focusing on safety properties. Zhang *et al.* [20] propose a technique for verifying probabilistic safety problems by adopting abstraction techniques from the verification of hybrid systems. Fränzle *et al.* apply stochastic satisfiability modulo theory (SSMT) to the symbolic analysis of probabilistic bounded reachability problems of probabilistic hybrid automata. This SSMT-approach has recently been extended to computing expected values of probabilistic hybrid systems, for instance mean-times to failure [10]. This paper complements the verification techniques using PCTL [2, 17], and includes safety as well as liveness properties.

2. PRELIMINARIES

We consider the model of discrete time stochastic hybrid systems (DTSHS) from [2], which is an autonomous (uncontrolled) version of that in [3]. A DTSHS is a stochastic model with a hybrid state space $\mathbb{S} = \cup_{\ell \in Loc}\{\ell\} \times S_\ell, S_\ell \subseteq \mathbb{R}^{d(\ell)}$, given by the disjoint union of continuous domains S_ℓ (each of which with its own dimension, specified by $d : Loc \to \mathbb{N}$) associated to discrete locations Loc, also referred to as the "modes". A point in the hybrid state space $s = (\ell, x)$ is thus made up of two components: a discrete one $\ell \in Loc$ and a continuous one $x \in \mathbb{R}^{d(\ell)}$. Unlike [2, 3], here discrete labels are associated with locations via a function L. Let $\mathcal{B}(\mathbb{S})$ denote the σ-algebra generated by the subsets A of $\mathbb{S}$ of the form $A = \cup_{\ell \in Loc}\{\ell\} \times A_\ell$, where $A_\ell \in \mathcal{B}(\mathbb{R}^{d(\ell)})$ is a Borel set in $\mathbb{R}^{d(\ell)}$. We use the notation $\mathbf{dom}(\ell)$ to denote the domain associated to location ℓ.

DEFINITION 1. *[DTSHS] A DTSHS is a structure* $\mathcal{H} = (Loc, \mathrm{AP}, L, d, \alpha, T_x, T_\ell, T_r)$, *where:*

- *Loc - is a finite set of locations;*

- *AP - is a finite set of atomic propositions;*

- $L : Loc \to 2^{\mathrm{AP}}$ - *is the labeling function, which acts on the discrete locations;*

- $d : Loc \to \mathbb{N}$ - *is the dimension assigned to the continuous domain $\mathbb{R}^{d(\ell)}$ of each location $\ell \in Loc$;*

- $\alpha : \mathcal{B}(\mathbb{S}) \to [0, 1]$ - *is the initial probability distribution;*

- $T_\ell : Loc \times \mathbb{S} \to [0, 1]$ - *is a conditional discrete stochastic kernel, which assigns to each $s \in \mathbb{S}$ a probability distribution, $T_\ell(\cdot|s)$, over Loc;*

- $T_x : \mathcal{B}(\mathbb{R}^{d(\cdot)}) \times \mathbb{S} \to [0, 1]$ - *is a continuous stochastic kernel on $\mathbb{R}^{d(\cdot)}$, conditional on $\mathbb{S}$. It assigns to each $s = (\ell, x) \in \mathbb{S}$ a probability measure, $T_x(\cdot|s)$, on the Borel space $(\mathbb{R}^{d(\ell)}, \mathcal{B}(\mathbb{R}^{d(\ell)}))$. The function $T_x(A_\ell|(\ell, \cdot))$ is assumed to be Borel measurable, for all $\ell \in Loc$ and all $A_\ell \in \mathcal{B}(\mathbb{R}^{d(\ell)})$;*

- $T_r : \mathcal{B}(\mathbb{R}^{d(\cdot)}) \times \mathbb{S} \times Loc \to [0, 1]$ - *is a stochastic kernel on $\mathbb{R}^{d(\cdot)}$, conditional on $\mathbb{S} \times Loc$. It assigns to each $s \in \mathbb{S}$ and $\ell' \in Loc$, a probability measure, $T_r(\cdot|s, \ell')$, on the Borel space $(\mathbb{R}^{d(\ell')}, \mathcal{B}(\mathbb{R}^{d(\ell')}))$. The function $T_r(A_{\ell'}|(\ell, \cdot), \ell')$ is assumed to be Borel measurable for all $\ell, \ell' \in Loc, \ell \neq \ell'$, and all $A_{\ell'} \in \mathcal{B}(\mathbb{R}^{d(\ell')})$.*

EXAMPLE 1. *Fig. 1 depicts the DTSHS $\mathcal{H}_1$ with the set of locations $Loc = \{\ell_0, \ell_1, \ell_2, \ell_3\}$ and the set of atomic propositions $\mathrm{AP} = \{\mathrm{ON}_1, \mathrm{ON}_2, \mathrm{OFF}_1, \mathrm{OFF}_2\}$. Each location ℓ_i is associated with a continuous two-dimensional bounded rectangular domain $S_{\ell_i} = [0, x'_1] \times [0, x'_2] \subset \mathbb{R}^2$ (thus $d(l_i) = 2$), and is labeled with an element from 2^{AP}, for instance $L(\ell_0) = \{\mathrm{ON}_1, \mathrm{ON}_2\}$. The initial distribution is $\alpha(\cdot) = \delta_{(\ell_0, 0)}(\cdot)$. Here $\delta_{(\ell_0, 0)}(\cdot)$ is the Dirac delta function. Each continuous domain S_{ℓ_i} is partitioned into (the same) four non-overlapping sub-regions G_0, G_1, G_2 and G_3 (see Fig. 2). The conditional discrete stochastic kernel T_ℓ is given by $T_\ell(\ell_i|(\ell, x)) = \frac{Leb(G_i)}{Leb(S_\ell)}$, where Leb is the Lebesgue measure. The discrete graphical structure of $\mathcal{H}_1$ is represented in Fig. 1. An edge represents a positive transition probability between pairs of modes. In particular, each self-loop denotes the likelihood of dwelling within G_i, for any location ℓ_i. The conditional stochastic kernel T_x corresponds to a Gaussian distribution $T_x(\cdot|(\ell, x)) = \mathcal{N}(\cdot; \mu(\ell, x), \Sigma(\ell, x))$, where $\mu(\ell, x)$ and $\Sigma(\ell, x)$ are the mean and the covariance, respectively, and are functions of the hybrid state (ℓ, x). For $\ell \neq \ell'$ the stochastic kernel T_r, conditional on a point (ℓ, x), is given by $T_r(\cdot|(\ell, x), \ell') = \delta_{(\ell, x)}(\cdot)$, which denotes a (deterministic) identity map.* $\square$

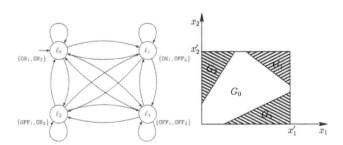

Figure 1: The discrete structure of $\mathcal{H}_1$.

Figure 2: The (partitioned) continuous domains S_{ℓ_i} of $\mathcal{H}_1$.

Semantics.

To simplify the notation, let us introduce a conditional stochastic kernel $T : \mathcal{B}(\mathbb{S}) \times \mathbb{S} \to [0,1]$ defined by

$$T(\{\ell'\} \times A_{\ell'} \,|\, (\ell, x)) = \begin{cases} T_x(A_{\ell'} | (\ell, x)) T_\ell(\ell' | (\ell, x)), & \text{if } \ell' = \ell \\ T_r(A_{\ell'} | (\ell, x), \ell') T_\ell(\ell' | (\ell, x)), & \text{if } \ell' \neq \ell, \end{cases} \quad (1)$$

for all sets $A_{\ell'} \in \mathcal{B}(\mathbb{R}^{d(\ell')})$, $\ell' \in Loc$, and $(\ell, x) \in \mathbb{S}$. We consider the evolution of the DTSHS either over a finite time horizon $\mathbb{T} \subset \mathbb{N}$, or over an infinite one $\mathbb{T} = \mathbb{N}$. The underlying stochastic process of a DTSHS is $\{\mathbf{s}(k), k \in \mathbb{T}\}$, where $\mathbf{s}(k) = (\mathbf{l}(k), \mathbf{x}(k))$ represents the process at step $k \in \mathbb{T}$ (we denote processes with bold font, in order to emphasize the difference from sample points over the state space). The executions of $\{\mathbf{s}(k), k \in \mathbb{T}\}$ are obtained according to the following procedure [3, Definition 3]: the conditional discrete stochastic kernel T_ℓ gives the probability to any location ℓ', given the current state (ℓ, x). If T_ℓ samples location $\ell' = \ell$, the conditional stochastic kernel T_x characterizes the probability for the next point inside the continuous domain S_ℓ of ℓ. If instead T_ℓ samples location $\ell' \neq \ell$, the conditional stochastic kernel T_r induces a probability distribution for the process over domain $S_{\ell'}$ for location ℓ'.

DEFINITION 2. *[Paths] Let $\mathcal{H}$ be a DTSHS. An infinite path starting at state (ℓ_0, x_0) is a sequence $\rho = (\ell_0, x_0) \to (\ell_1, x_1) \to (\ell_2, x_2) \cdots$, such that for every $k \in \mathbb{N}$, $s_k = (\ell_k, x_k) \in \mathbb{S}$. A finite path is a prefix (ending in a state) of an infinite path.*

We define $Paths_*^{\mathcal{H}}$ and $Paths_\omega^{\mathcal{H}}$ as the set of all finite paths and infinite paths in $\mathcal{H}$, respectively. Let also the sets $Paths^{\mathcal{H}}$ and $Paths^{\mathcal{H}}(\ell, x)$ denote all finite and infinite paths in $\mathcal{H}$ and those starting from state (ℓ, x), respectively. For any k less than the length of path ρ, let $\rho[k] := (\ell_k, x_k)$ be the k-th state of ρ. Given a path ρ, the function $\mathbf{lab}(\rho)$ returns the word $w = a_1 a_2 a_3 \ldots$ (sequence of state labels corresponding to path ρ) such that $a_k = L(\rho[k](1))$.

A DTSHS $\mathcal{H}$ with initial probability distribution α is associated to a probability measure $\Pr_\alpha^{\mathcal{H}}$ on paths over a time horizon $[0, k], k \in \mathbb{N}$, as follows. Consider the canonical sample space $\Omega = \mathbb{S}^{k+1}$, endowed with its product topology. Let $C(G(0), G(1), \ldots, G(k))$ denote the cylinder set consisting of all paths $\rho \in Paths^{\mathcal{H}}$ such that $\rho[i] = s_i$, where $G(i) \in \mathcal{B}(\mathbb{S})$, $s_i \in G(i)$ for any $i \leq k$. The probability measure $\Pr_\alpha^{\mathcal{H}}$ on $\mathcal{B}(Paths^{\mathcal{H}})$ is the unique measure defined as: $\Pr_\alpha^{\mathcal{H}}(C(G(0), G(1), \ldots, G(k))) =$

$$\int_{G(0)} \int_{G(1)} \cdots \int_{G(k)} T(da_k | a_{k-1}) \cdots T(da_1 | a_0) \alpha(da_0).$$

Notice that $\Pr_\alpha^{\mathcal{H}}(C(G(0))) = \int_{G(0)} \alpha(da_0)$. Further details on the topological and semantical properties of the DTSHS model can be found in [3].

Automata and LTL specifications.

Here we will distinguish two types of specifications: finite state automata and linear temporal logic (LTL) specifications.

DEFINITION 3. *[DFA] A deterministic finite state automaton (DFA) is a structure $\mathcal{A} = (Q, q_0, \Sigma, F, \Delta)$, where: Q - is a finite set of locations; $q_0 \in Q$ - is the initial location; Σ*

- is a finite alphabet; $F \subseteq Q$ - is a set of accept locations; $\Delta : Q \times \Sigma \to Q$ - is a transition function.

From here on we assume that $\Sigma = 2^{AP}$, and let Σ^* and Σ^ω denote the set of all finite and infinite words over Σ, respectively. A finite word $w \in \Sigma^*$ is accepted by a DFA $\mathcal{A}$, if there exists a finite *run* (or a path) $\theta \in Q^*$ such that $\theta[0] = q_0$, $\Delta(\theta[i], w[i]) = \theta[i+1]$ for $i \geq 0$ and there exists a $j \in \mathbb{N}, j < \infty$, such that $\theta[j] \in F$. Note that $w[i]$ (resp. $\theta[i]$) denotes the i-th letter (resp. state) on w (resp. θ). The accepted language of $\mathcal{A}$, denoted $\mathcal{L}_*(\mathcal{A})$, is the set of all words accepted by $\mathcal{A}$. Notice that one could define Δ as a transition relation (as opposed to a function), which results in a nondeterministic finite state automaton (NFA). It is well known that DFAs are equally expressive as NFA and that for any NFA a canonical minimal DFA exists [16].

DEFINITION 4. *[GBA] A generalized Büchi automaton (GBA) is a structure $\mathcal{A} = (Q, Q_0, \Sigma, \mathcal{F}, \Delta)$, where Q - is a finite set of locations; $Q_0 \subseteq Q$ - is a set of initial locations; Σ - is a finite alphabet; $\mathcal{F} \subseteq 2^Q$ - is a set of acceptance sets; $\Delta \subseteq Q \times \Sigma \times Q$ - is a transition relation.*

We sometimes write $q \xrightarrow{\sigma} q'$ if $(q, \sigma, q') \in \Delta$ for simplicity. An infinite *word* $w \in \Sigma^\omega$ is accepted by $\mathcal{A}$, if there exists an infinite *run* $\theta \in Q^\omega$ such that $\theta[0] \in Q_0$, $(\theta[i], w[i], \theta[i+1]) \in \Delta$ for all $i \geq 0$ and for each $F \in \mathcal{F}$, there exist infinitely many indices $j \in \mathbb{N}$ such that $\theta[j] \in F$. The accepted language of $\mathcal{A}$, denoted $\mathcal{L}_\omega(\mathcal{A})$, is the set of all infinite words accepted by $\mathcal{A}$. Given a GBA $\mathcal{A}$ and location q, we denote by $\mathcal{A}[q]$ the GBA $\mathcal{A}$ with q as the unique initial location. Note that $\mathcal{L}_\omega(\mathcal{A}) = \bigcup_{q \in Q_0} \mathcal{L}_\omega(\mathcal{A}[q])$.

DEFINITION 5. *[Separated GBA] A GBA $\mathcal{A}$ is separated if, for any locations $q', q'' \in Q$, $\mathcal{L}_\omega(\mathcal{A}[q']) \cap \mathcal{L}_\omega(\mathcal{A}[q'']) = \emptyset$.*

The set of LTL formulae over the set AP of atomic propositions is defined as follows:

$$\varphi ::= a \mid \varphi \wedge \varphi \mid \neg \varphi \mid \mathsf{X}\,\varphi \mid \varphi \,\mathsf{U}\, \varphi,$$

where $a \in AP$. We interpret LTL formulae over DTSHS $\mathcal{H}$.

DEFINITION 6. *[LTL semantics] For an LTL formula φ, a path $\rho \in Paths^{\mathcal{H}}$, and a step $i \in \mathbb{N}$, the satisfaction relation $\models$ is defined by:*

$$
\begin{array}{lll}
(\rho, i) \models a & \Longleftrightarrow & a \in L(\rho[i](1)) \\
(\rho, i) \models \varphi_1 \wedge \varphi_2 & \Longleftrightarrow & (\rho, i) \models \varphi_1 \text{ and } (\rho, i) \models \varphi_2 \\
(\rho, i) \models \neg \varphi & \Longleftrightarrow & \text{not } (\rho, i) \models \varphi \\
(\rho, i) \models \mathsf{X}\,\varphi & \Longleftrightarrow & (\rho, i+1) \models \varphi \\
(\rho, i) \models \varphi_1 \,\mathsf{U}\, \varphi_2 & \Longleftrightarrow & \exists j \in \mathbb{N}. \infty > j \geq i, (\rho, j) \models \varphi_2 \text{ and} \\
& & \forall k \in \mathbb{N}. i \leq k < j, (\rho, k) \models \varphi_1
\end{array}
$$

An example LTL formula is $a \,\mathsf{U}\, (\neg b \wedge (c \,\mathsf{U}\, d))$. Using the until operator we can define the temporal modalities $\Diamond$ and $\Box$ as $\Diamond\varphi := true \,\mathsf{U}\, \varphi$ and $\Box\varphi := \neg \Diamond \neg\varphi$. The operator $\Diamond\varphi$ is satisfied on all paths where eventually in the future φ holds. The operator $\Box\varphi$ characterizes all the paths that only contain states satisfying φ. The formula $\Box \Diamond \varphi$ means that φ holds infinitely often, whereas $\Diamond \Box \varphi$ means that from some moment on the formula φ will always hold.

Given a DTSHS $\mathcal{H}$, let $\mathcal{L}_\omega(\varphi) = \{\rho \in Paths^{\mathcal{H}} \mid (\rho, 0) \models \varphi\}$ be the *language* of φ in $\mathcal{H}$. The measurability of $\mathcal{L}_\omega(\varphi)$ can be proven in a similar way as in [9].

THEOREM 1. [[8]] *For any LTL formula φ over* AP*, there exists a separated GBA $\mathcal{A}_\varphi = (Q, Q_0, \Sigma, \mathcal{F}, \Delta)$, where $\Sigma = 2^{AP}$ and $|Q| \leqslant 2^{\mathcal{O}(|\varphi|)}$, such that $\mathcal{L}_\omega(\mathcal{A}_\varphi)$ is the set of computations satisfying the formula φ.*

Here $|\varphi|$ denotes the length of the LTL formula φ in terms of the number of operators in φ.

3. REACHABILITY ANALYSIS

This section formally introduces the following problem over a DTSHS $\mathcal{H}$: determine the probability of reaching a certain "goal" or "target" set within a given time horizon, starting from any state in $\mathbb{S}$. More precisely, select any compact Borel set $G \in \mathcal{B}(\mathbb{S})$, representing the goal set. We are interested in determining the probability that the execution associated with the initial condition $s_0 \in \mathbb{S}$ will intersect G within the time horizon $\mathbb{T}$:

$$\mathrm{p}_{s_0}(\Diamond G) := P_{s_0}\{\mathbf{s}(k) \in G \text{ for some } k \in \mathbb{T}\}, \qquad (2)$$

where P_{s_0} denotes the probability measure for an event over the solution of $\mathcal{H}$, conditional on $\mathbf{s}(0) = s_0$: the value of $\mathrm{p}_{s_0}(\Diamond G)$ depends on the initial state s_0. (Notice that the measure P_{s_0} can also be expressed by $\mathrm{Pr}^{\mathcal{H}}$, where $G(0) = s_0$.) If $\mathrm{p}_{s_0}(\Diamond G) \geq \epsilon$, $\epsilon \in (0, 1]$, we say that the system initialized at s_0 reaches G with an ϵ probabilistic guarantee (the case $\epsilon = 0$ is trivially satisfied by all states in $\mathbb{S}$). For a given $\epsilon \in (0, 1]$, we define the *ϵ-probabilistic reachability set* by

$$R(\epsilon, G) = \{s_0 \in \mathbb{S} : \mathrm{p}_{s_0}(\Diamond G) \geq \epsilon\} \qquad (3)$$

of those initial states s_0 that are associated with a process that reaches set G with an ϵ probabilistic guarantee. We show that the problem of computing $\mathrm{p}_{s_0}(\Diamond G)$ can be solved through a backward iterative procedure by representing $\mathrm{p}_{s_0}(\Diamond G)$ as a max function.

3.1 Characterizing Probabilistic Reachability

Step-bounded reachability probability.

Let us consider $\mathbb{T} = [0, N] \subset \mathbb{N}$. Let $1_C : \mathbb{S} \to \{0, 1\}$ denote the indicator function of the set $C \subseteq \mathbb{S}$: $1_C(s) = 1$ if and only if $s \in C$. Observe that $\max_{k \in [0,N]} 1_G(s_k) = 1$, if $\exists k \in [0, N] : s_k \in G$, and 0 otherwise, where $s_k \in \mathbb{S}$. Then, the quantity $\mathrm{p}_{s_0}(\Diamond G)$ in (2) can be expressed as the expectation with respect to the probability measure P_{s_0} of the Bernoulli random variable $\max_{k \in \mathbb{T}} 1_G(\mathbf{s}(k))$, conditional on $\mathbf{s}(0) = s_0$ [3]:

$$\mathrm{p}_{s_0}(\Diamond G) = E_{s_0}\left[\max_{k \in [0,N]} 1_G(\mathbf{s}(k))\right]. \qquad (4)$$

Denote with $\overline{G} = \mathbb{S} \setminus G$, the complement of G over $\mathbb{S}$. Consider the sequence of functions $W_k : \mathbb{S} \times \mathcal{B}(\mathbb{S}) \to [0, 1]$, $k \in [0, N]$, defined for $s \in \mathbb{S}$ and $G \in \mathcal{B}(\mathbb{S})$ by $W_N(s, G) = 1_G(s)$, and for $k < N$:

$$W_k(s, G) = 1_G(s) + 1_{\overline{G}}(s) \int_{\mathbb{S}^{N-k}} \max_{n=k+1,\ldots,N} 1_G(s_n) \cdot$$
$$\prod_{m=k+1}^{N-1} T(ds_{m+1}|s_m)T(ds_{k+1}|s). \qquad (5)$$

It is easily seen that for any $k \in [0, N]$, $W_k(s, G)$ represents the probability that an execution of the DTSHS enters the target set G over the residual time horizon $[k, N]$,

starting from s at time instant k [3]: we name $W_k(s, G)$ the value function at time k. In particular, $W_0(s, G) = E_s[\max_{k \in [0,N]} 1_G(\mathbf{s}(k))]$, evaluated at $s = s_0 \in \mathbb{S}$ returns the quantity of interest $\mathrm{p}_{s_0}(\Diamond G)$, and the ϵ-probabilistic reachability set defined in (3): $R(\epsilon, G) = \{s_0 \in \mathbb{S} : W_0(s_0, G) \geq \epsilon\}$.

The following result states that the value functions can be determined through a backward-recursive procedure.

THEOREM 2. [[3], Lemma 2] *The value functions $W_k : \mathbb{S} \times \mathcal{B}(\mathbb{S}) \to [0, 1]$, defined in (5) can be computed for $s \in \mathbb{S}$ through the following backward recursion for $k < N$:*

$$W_k(s, G) = 1_G(s) + 1_{\overline{G}}(s) \int_{\mathbb{S}} W_{k+1}(s_{k+1}, G)T(ds_{k+1}|s) \tag{6}$$

initialized with $W_N(s, G) = 1_G(s)$.

In conclusion, given an initial distribution α, the related step-bounded reachability probability is simply $\int_{\mathbb{S}} \alpha(ds)p_s(\Diamond G)$.

Step-unbounded reachability probability.

Let us consider now the case $\mathbb{T} = \mathbb{N}$. We denote by

$$\mathrm{p}_{s_0}^\infty(\Diamond G) := P_{s_0}\{\mathbf{s}(k) \in G \text{ for some } k \geq 0\} \qquad (7)$$

the step-unbounded reachability probability. It can be computed as the fixpoint $W(s, G)$ of the following system of integral equations:

$$W(s, G) = 1_G(s) + 1_{\overline{G}}(s) \int_{\mathbb{S}} W(s', G)T(ds'|s). \qquad (8)$$

In this case $\mathrm{p}_{s_0}^\infty(\Diamond G) = W(s_0, G)$. Notice that the map $W : \mathbb{S} \times \mathcal{B}(\mathbb{S}) \to [0, 1]$ is related to W_k as follows $W(s, G) = \lim_{k \to \infty} W_k(s, G)$, for any $s \in \mathbb{S}$.

3.2 Discretization

In most cases, the solution of Equations (6) or (8) is not analytic. In this paper we will use discretization techniques in order to approximate the solution for the time-bounded and time-unbounded reachability probability.

Consider $\mathbb{S} = \bigcup_{\ell \in Loc}\{\ell\} \times S_\ell$ and assume each S_ℓ is compact. We introduce a finite partition for each domain $S_\ell \subset \mathbb{R}^{d(\ell)}$, $\ell \in Loc$, by taking $S_\ell = \bigcup_{i=1}^{m_\ell} S_{\ell,i}$, where $S_{\ell,i} \in \mathcal{B}(\mathbb{R}^{d(\ell)})$ with $S_{\ell,i} \cap S_{\ell,j} = \emptyset$, for all $i \neq j$. Here m_ℓ represents the finite number of partitions for the domain in location ℓ. Denote by $h_{\ell,i}$ the diameter of the set $S_{\ell,i}$ as $h_{\ell,i} = \sup\{\|x - x'\| : x, x' \in S_{\ell,i}\}$ (here we are using the Euclidean norm), and define the grid size parameter by $h := \max_{i=1,\ldots,m_\ell; \ell \in Loc} h_{\ell,i}$. Let us additionally introduce a function $r_\ell : \mathcal{B}(\mathbb{R}^{d(\ell)}) \to \mathbb{R}^{d(\ell)}$ which, given a partition set $S_{\ell,i} \in \mathcal{B}(\mathbb{R}^{d(\ell)})$ and location $\ell \in Loc$, returns a randomly chosen point in $S_{\ell,i}$, denoted with $r_\ell(S_{\ell,i})$, which is also named the "representative point" of the partition set $S_{\ell,i}$. Notice that the discretization can in general be tailored to the target set G, so that $G = \bigcup_{i=1}^{m_\ell^g} S_{\ell,i}^g$, where $\forall \ell \in Loc, m_\ell^g \leq m_\ell$. Using the grid size parameter h we can define the discretized DTMC of a DTSHS as follows:

DEFINITION 7. *[DTMC approximation of DTSHS] For the DTSHS $\mathcal{H} = (Loc, AP, L, d, \alpha, T_x, T_\ell, T_r)$, the DTMC $\mathcal{D}_h = (\mathbb{S}_h, AP, L_h, \alpha_h, P_h)$ is defined as follows: $\mathbb{S}_h = \{(\ell, i)|\ell \in Loc, i \in \{1, \ldots, m_\ell\}\}$ - is the state space; $L_h(\ell, i) = L(\ell)$ - is the labeling function; $\alpha_h(\ell, i) = \int_{S_{\ell,i}} \alpha(\ell, x)dx$ - is the* initial probability distribution; $P_h((\ell, i), (\ell', i')) = T(\ell' \times S_{\ell',i'}|(\ell, r_\ell(S_{\ell,i})))$ - is the transition probability matrix.

Notice that the state space $\mathbb{S}_h$ of the discretized DTMC $\mathcal{D}_h$ is given by pairs (location, partition index). The probability $P_h((\ell, i), (\ell', i'))$ to jump from state (ℓ, i) to state (ℓ', i') is the probability to jump from $r_\ell(S_{\ell,i})$, the representative point of partition $S_{\ell,i}$, to the partition set $S_{\ell',i'}$ in location ℓ'.

Let us denote with $G_h \in \mathbb{S}_h$ the set of states of $\mathcal{D}_h$ corresponding to partitions of $\mathbb{S}$ overlapping with the original target set G of $\mathcal{H}$. Similarly, $\overline{G_h} = \mathbb{S}_h \setminus G_h$. We define the step-bounded and step-unbounded reachability probabilities by introducing functions W_k^h and W^h respectively, both of which are defined on $\mathbb{S}_h \times \mathcal{B}(\mathbb{S}_h)$ and take value in $[0, 1]$:

$$W_k^h(v, G_h) = 1_{G_h}(v) + 1_{\overline{G_h}}(v) \sum_{v_{k+1} \in \mathbb{S}_h} W_{k+1}^h(v_{k+1}, G_h) P_h(v, v_{k+1}), \tag{9}$$

for $k < N$, initialized with $W_N^h(v, G_h) = 1_{G_h}(v)$ and

$$W^h(v, G_h) = 1_{G_h}(v) + 1_{\overline{G_h}}(v) \sum_{v' \in \mathbb{S}_h} W^h(v', G_h) P_h(v, v'). \tag{10}$$

W_k^h and W^h approximate the original functions W_k and W – in the next Section we derive explicit approximation bounds.

Notice that step-bounded and step-unbounded reachability probability is given by a system of linear equations for which solutions can be computed efficiently. If $\alpha_h(v_0) = 1$, the solutions to Eq. (9) and (10) will be denoted as $\widehat{p}_{v_0}(\Diamond G_h)$ and $\widehat{p}_{v_0}^\infty(\Diamond G_h)$, respectively, whereas for an arbitrary initial distribution α_h we get $\sum_{v \in \mathbb{S}_h} \alpha_h(v)\widehat{p}_v(\Diamond G_h)$ and $\sum_{v \in \mathbb{S}_h} \alpha_h(v)\widehat{p}_v^\infty(\Diamond G_h)$, respectively.

3.3 Error Bounds

The quantities $W_k^h(v, G_h)$, $W^h(v, G_h)$, $\widehat{p}_v(\Diamond G_h)$, and $\widehat{p}_v^\infty(\Diamond G_h)$ are all defined on $\mathbb{S}_h$. We can extend them over $\mathbb{S}$ by piecewise constant interpolation – for instance, $W_k^h(s, G_h) = W_k^h(r_\ell(S_{\ell,i}), G_h), \forall s \in S_{\ell,i}, i = 1, \ldots, m_\ell, \ell \in Loc$. In the remaining of this section, we shall refer to the quantities extended over $\mathbb{S}$. We now derive explicit error bounds between the quantities in Equations (6)-(8) and the corresponding quantities in Equations (9)-(10) (again, extended over $\mathbb{S}$).

We assume that the kernels T_ℓ, as well as the densities t_x and t_r of T_x and T_r, satisfy the following Lipschitz continuity assumptions:

ASSUMPTION 1. *For any* $(\ell, x), (\ell, x'), (\ell, x''), (\ell', x'') \in \mathbb{S}$:

$$|T_\ell(\ell'|(\ell, x)) - T_\ell(\ell'|(\ell, x'))| \leq h_1||x - x'||,$$

$$|t_x(x''|(\ell, x)) - t_x(x''|(\ell, x'))| \leq h_2||x - x'||,$$

$$|t_r(x''|(\ell, x), \ell') - t_r(x''|(\ell, x'), \ell')| \leq h_3||x - x'||, \ \ell \neq \ell',$$

where h_1, h_2 *and* h_3 *are finite Lipschitz constants.*

We define $\mathcal{K} \doteq |Loc|h_1 + \lambda(\overline{G}) \cdot (h_2 + (|Loc| - 1)h_3)$, where $\lambda(\overline{G})$ is the Lebesgue measure of the set $\overline{G}$ and $|Loc|$ is the number of discrete locations.

THEOREM 3. *Given a DTSHS* $\mathcal{H}$, *the DTMC* $\mathcal{D}_h$ *obtained with discretization step* h, *a finite time horizon* N *and a target set* $G \in \mathcal{B}(\mathbb{S})$, *the following holds:*

1) $|p_{s_0}(\Diamond G) - \widehat{p}_{v_0}(\Diamond G_h)| \leq N\mathcal{K}h,$

2) $|p_{s_0}^\infty(\Diamond G) - \widehat{p}_{v_0}^\infty(\Diamond G_h)| \leq 2\mathcal{K}h,$

where $v_0 = (\ell, i)$ *and* $s_0 \in G_{\ell,i}$.

4. AUTOMATA MODEL CHECKING

In this section we study the problem of model checking a property specified as a DFA or as a separated GBA against a DTSHS. Recall that the main difference between a DFA-property and a separated GBA-property is that the former reasons over the finite paths whereas the latter reasons over infinite paths. Since every LTL-formula φ can be expressed as a separated GBA (see Section 2), LTL model checking boils down to automata model checking. Let the quantity $\Pr^\mathcal{H}(\mathcal{L}(\mathcal{A})) := \Pr^\mathcal{H}(\rho \in Paths_f^\mathcal{H}|\mathbf{lab}(\rho) \in \mathcal{L}(\mathcal{A}))$ (and $\Pr^\mathcal{H}(\mathcal{L}_\omega(\mathcal{B})) := \Pr^\mathcal{H}(\rho \in Paths_\omega^\mathcal{H}|\mathbf{lab}(\rho) \in \mathcal{L}_\omega(\mathcal{B}))$) denote the probability that the DTSHS $\mathcal{H}$ satisfies the DFA $\mathcal{A}$ (and GBA $\mathcal{B}$, respectively). The measurability of the sets $\{\rho \in Paths_f^\mathcal{H}|\mathbf{lab}(\rho) \in \mathcal{L}(\mathcal{A})\}$ and $\{\rho \in Paths_\omega^\mathcal{H}|\mathbf{lab}(\rho) \in \mathcal{L}_\omega(\mathcal{B})\}$ can be shown as in [19]. We will show that the probability of these events can be computed over the product between $\mathcal{H}$ and $\mathcal{A}$ (and $\mathcal{B}$).

DFA Specifications.

We start considering properties expressed as DFA.

DEFINITION 8. *[Product between DTSHS and DFA] Consider a DTSHS* $\mathcal{H} = (Loc, \mathrm{AP}, L, d, \alpha, T_x, T_\ell, T_r)$ *and a DFA* $\mathcal{A} = (Q, q_0, \Sigma, F, \Delta)$. *Let* $\mathcal{H} \otimes \mathcal{A} = (V, \mathrm{AP}, \widehat{L}, \widehat{\alpha}, \widehat{d}, \widehat{T}_x, \widehat{T}_\ell, \widehat{T}_r)$ *be the product DTSHS, where* $V := Loc \times Q$, $\widehat{L}(\langle \ell, q_0 \rangle) = L(\ell)$, $\widehat{\alpha}(\langle \ell, q_0 \rangle, x) := \alpha(\ell, x)$, $\widehat{d}(\langle \ell, q \rangle) := d(\ell)$ *and the kernels are defined by:*

$$\frac{T_\ell(\ell'|(\ell, x)) = p \ \wedge \ \Delta(q, L(\ell)) = q'}{\widehat{T}_\ell(\langle \ell', q' \rangle|(\langle \ell, q \rangle, x)) = p},$$

$$\frac{T_x(A_\ell|(\ell, x)) = p \ \wedge \ \Delta(q, L(\ell)) = q'}{\widehat{T}_x(A_{\langle \ell, q' \rangle}|(\langle \ell, q \rangle, x)) = p},$$

$$\frac{T_r(A_{\ell'}|(\ell, x), \ell') = p \ \wedge \ \Delta(q, L(\ell)) = q'}{\widehat{T}_r(A_{\langle \ell', q' \rangle}|(\langle \ell, q \rangle, x), \langle \ell', q' \rangle) = p}(\ell \neq \ell').$$

Here $A_{\langle \ell, q' \rangle}$ and $A_{\langle \ell', q' \rangle}$ denote the Borel-measurable sets in $S_{\langle \ell, q' \rangle}$ and $S_{\langle \ell', q' \rangle}$, respectively. The definition of the conditional stochastic kernel $\widehat{T}$ for the product $\mathcal{H} \otimes \mathcal{A}$ is the same as in Eq. (1). We define the set of final locations of $\mathcal{H} \otimes \mathcal{A}$ as $V_F := Loc \times F$.

THEOREM 4. *For any DTSHS* $\mathcal{H}$ *and DFA* $\mathcal{A}$,

$$\Pr^\mathcal{H}(\mathcal{L}(\mathcal{A})) = \Pr^{\mathcal{H} \otimes \mathcal{A}}(\Diamond V_F),$$

where $\Pr^{\mathcal{H} \otimes \mathcal{A}}(\Diamond V_F)$ *is the probability to reach the set of final locations* V_F *in the product* $\mathcal{H} \otimes \mathcal{A}$.

Given the fact that $\mathcal{H} \otimes \mathcal{A}$ is a DTSHS, the probability $\Pr^{\mathcal{H} \otimes \mathcal{A}}(\Diamond V_F)$ can be computed via Eq. (6). In this case, $\Pr^{\mathcal{H} \otimes \mathcal{A}}(\Diamond V_F)$ is the fixpoint of Eq. (8). Notice that in order to make the computation of $\Pr^{\mathcal{H} \otimes \mathcal{A}}(\Diamond V_F)$ more efficient we can make all the states corresponding to the set of locations V_F absorbing.

GBA Specifications.

In order to compute the probability that a DTSHS $\mathcal{H}$ satisfies a separated GBA-property $\mathcal{A}$ we consider two steps. First, we construct the product between $\mathcal{H}$ and the separated GBA $\mathcal{A}$. Second, in the product $\mathcal{H} \otimes \mathcal{A}$ we compute the probability $\Pr^\mathcal{H}(\mathcal{L}_\omega(\mathcal{A}))$.

DEFINITION 9. *[Product between* DTSHS *and* GBA*]* Consider a DTSHS $\mathcal{H} = (Loc, \mathrm{AP}, L, d, \alpha, T_x, T_\ell, T_r)$ and a GBA $\mathcal{A} = (Q, Q_0, \Sigma, \mathcal{F}, \Delta)$. *Their product is defined as* $\mathcal{H} \otimes \mathcal{A} = (V, \mathrm{AP}, \widehat{L}, \widehat{\alpha}, \widehat{d}, \widehat{T}_x, \widehat{T}_\ell, \widehat{T}_r)$ *and is constructed as in Definition 8, except that* $\widehat{\alpha}(\langle \ell, q_0 \rangle, x) := \alpha(\ell, x)$ *for all* $q_0 \in Q_0$.

In general when one takes the product between a generalized *deterministic* Büchi automaton (GDBA) and a DTSHS, the resulting product is a DTSHS. The product between a generalized (nondeterministic) Büchi automaton (GBA) and a DTSHS is not always a DTSHS. This can be seen from the fact that for a location q in $\mathcal{A}$ and a symbol $\sigma \in \Sigma$, $\{(q, \sigma, q'), (q, \sigma, q'')\} \subseteq \Delta$ for $q' \neq q''$, it follows that for a transition $\ell \to \ell'$ in $\mathcal{H}$ with $T_\ell(\ell'|(\ell, x)) = 1$ the product will contain two transitions: $\langle \ell, q \rangle \to \langle \ell', q' \rangle$ and $\langle \ell, q \rangle \to \langle \ell', q'' \rangle$. In this case we get that $\widehat{T}_\ell(\langle \ell', q' \rangle|(\langle \ell, q \rangle, x)) + \widehat{T}_\ell(\langle \ell', q'' \rangle|(\langle \ell, q \rangle, x)) = 2$. In this paper we consider GBA, which are strictly more expressive than GDBA [4]. The separability property will give us the possibility to transform the product into a DTSHS.

EXAMPLE 2. *Fig. 5 shows the product* $\mathcal{H} \otimes \mathcal{A}$ *between the DTSHS* $\mathcal{H}$ *of Fig. 3 and the separated GBA* $\mathcal{A}$ *of Fig. 4. (For each location of the DTSHS* $\mathcal{H}$ *we pick a continuous kernel* T_x *and kernel* T_r *– they can for instance be a conditional exponential or Gaussian distribution.) Notice that the product in its original form does not define a DTSHS as the automaton* $\mathcal{A}$ *is nondeterministic (all dashed transitions in Fig. 5 are nondeterministic). For instance in the product location* v_0 *there are two transitions to the dashed product locations* v_1 *and* v_2. *To each product locations* v_1 *and* v_2 *corresponds the location* ℓ_0 *from the DTSHS* $\mathcal{H}$.

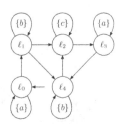

Figure 3: DTSHS $\mathcal{H}$ for Example 2.

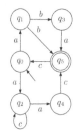

Figure 4: Separated GBA $\mathcal{A}$ for Ex. 2.

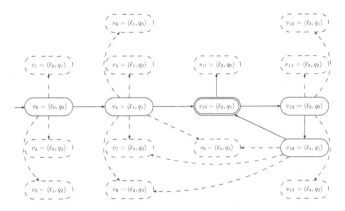

Figure 5: Product $\mathcal{H} \otimes \mathcal{A}$ for Example 2.

In order to compute the probability $\mathrm{Pr}^{\mathcal{H}}(\mathcal{L}_\omega(\mathcal{A}))$, we consider the probability to reach an *accepting bottom strongly connected component* (aBSCC) in the product $\mathcal{H} \otimes \mathcal{A}$. A *strongly connected component* (SCC) is a strongly connected set of discrete locations such that no proper superset is strongly connected.(The notion of SCC is related to that of irreducible class for classical Markov chains.) A *bottom strongly connected component* (BSCC) is an SCC from which no location outside the SCC is reachable.

DEFINITION 10. *[aBSCC] Given the product* $\mathcal{H} \otimes \mathcal{A}$, *a* BSCC $B \subseteq V = Loc \times Q$ *is accepting if for all* $F \in \mathcal{F}$ *of* $\mathcal{A}$, *there exists some* $\langle \ell, q \rangle \in B$ *such that* $q \in F$.

Let the set of final locations be $V_F^\omega = \{v \in B \mid B$ in the set of all *aBSCC* in $\mathcal{H} \otimes \mathcal{A}\}$.

Now the task is to compute $\mathrm{Pr}^{\mathcal{H}}(\mathcal{L}_\omega(\mathcal{A}))$. Using the separability property of GBA $\mathcal{A}$ we obtain a DTSHS out of $\mathcal{H} \otimes \mathcal{A}$. The following lemma asserts that for each accepted word of the separated GBA $\mathcal{A}$ there exists a single accepting path in $\mathcal{H} \otimes \mathcal{A}$.

We say that from location $\langle \ell, q \rangle$ there is a path leading to a *aBSCC* B, if there is a sequence $\langle \ell_0, q_0 \rangle, \langle \ell_1, q_1 \rangle, \ldots, \langle \ell_n, q_n \rangle$ such that $\langle \ell, q \rangle = \langle \ell_0, q_0 \rangle$, $\langle \ell_i, q_i \rangle$ and $\langle \ell_{i+1}, q_{i+1} \rangle$ are connected (if there exists $G' \subseteq \mathbf{dom}(\langle \ell_i, q_i \rangle) \backslash \{\emptyset\}$ such that for all $x \in G'$, $\widehat{T}((\langle \ell_{i+1}, q_{i+1} \rangle, \cdot)|(\langle \ell_i, q_i \rangle, x)) > 0)$ for $0 \leqslant i < n$ and $\langle \ell_n, q_n \rangle \in B$.

LEMMA 1. *Consider the product* $\mathcal{H} \otimes \mathcal{A}$, *where* $\mathcal{A}$ *is a separated GBA. For any aBSCC* B *of the product* $\mathcal{H} \otimes \mathcal{A}$, *it holds that*

1. $\langle \ell, q \rangle \to \langle \ell', q' \rangle$ *and* $\langle \ell, q \rangle \to \langle \ell', q'' \rangle$ *implies* $q' = q''$, *for any* $\langle \ell, q \rangle, \langle \ell', q' \rangle, \langle \ell', q'' \rangle$ *in* B;

2. *if* $\langle \ell, q \rangle$ *and* $\langle \ell, q' \rangle$ *with* $q \neq q'$ *have a path leading to* B *then* $q = q'$.

Using the above lemmas we can conclude that each location of $\mathcal{H} \otimes \mathcal{A}$ that does not lead to an aBSCC can be safely removed. The resulting product is denoted $\mathcal{H}\underline{\otimes}\mathcal{A}$. With reference to Example 2, by removing all dashed transitions and dashed locations in Fig. 5 we obtain the DTSHS $\mathcal{H}\underline{\otimes}\mathcal{A}$.

In general, when searching for an *aBSCC* one relies on the topological discrete structure (which hinges on the conditional discrete stochastic kernels) of $\mathcal{H}\underline{\otimes}\mathcal{A}$. Still, an *aBSCC* in $\mathcal{H}\underline{\otimes}\mathcal{A}$ might not be accepting. To illustrate this fact, consider the product $\mathcal{H}\underline{\otimes}\mathcal{A}$ from Fig. 6 and the set of accepting conditions $\mathcal{F} = \{\{q_0\}, \{q_1\}\}$. It is easy to see that when the conditions $\widehat{T}((v_1, \cdot)|(v_0, x)) > 0$ and $\widehat{T}((v_0, \cdot)|(v_1, y)) > 0$ are satisfied for every $x \in \mathbb{R}^{d(v_0)}$ and $y \in \mathbb{R}^{d(v_1)}$ then the set $B = \{v_0, v_1\}$ is an *aBSCC*. However, let us now assume that the domain S_0 from Fig. 7 is associated to location v_0. The domain S_0 contains two subdomains G_1 and G_2, which are such that $\widehat{T}(v_0 \times S|(v_0, x)) = 0$ and $\widehat{T}((v_1, \cdot)|(v_0, x)) = 0$, for all $x \in G_1 \cup G_2$ and $S \subseteq S_0 \backslash (G_1 \cup G_2)$. This means that when we are in the subdomain G_1 or G_2 there is no way to jump back to $S_0 \backslash (G_1 \cup G_2)$, nor to jump to location v_1. As a result we get that the aBSCC B is not accepting. This example suggests that when searching for accepting BSCCs it is not enough to look at the conditional discrete stochastic kernels — one has to consider also the continuous stochastic kernels.

Given an aBSCC B in $\mathcal{H}\underline{\otimes}\mathcal{A}$ and a set of accepting conditions $\mathcal{F}$ we introduce $acc(B) = \{\langle \ell, q \rangle \in B| q \in F \in \mathcal{F}\}$,

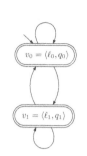

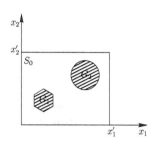

Figure 6: Product $\mathcal{H} \underline{\otimes} \mathcal{A}$ with aBSCC.

Figure 7: Domain with two absorbing subregions: the aBSCC may not be accepting.

the function returning the set of accepting locations in the aBSCC B. For a given set of states $G \subseteq \mathbb{S}$ we define the random variable $\eta(G) = \sum_{k=0}^{\infty} 1_G(\mathbf{s}(k))$, where $\mathbf{s}(k)$ is the stochastic process associated with $\mathcal{H} \underline{\otimes} \mathcal{A}$, and $\mathbb{E}_s(\eta(G))$ is the expected value of $\eta(G)$ over all executions of $\mathbf{s}(k)$ starting from $\mathbf{s}(0) = s$.

DEFINITION 11. *[Recurrent aBSCC] An aBSCC B is recurrent if for the set $G = \{v \times G' | v \in acc(B), G' \subseteq \mathbf{dom}(v)\}$, $\mathbb{E}_s(\eta(G)) = \infty$, for all $s \in G$.*

Here recall that $\mathbf{dom}(v)$ denotes the domain associated to location v. The above definition says that every state from G can reach all other states in G infinitely often.

In order to check whether an aBSCC is recurrent one has to look at the conditional stochastic kernel $\widehat{T}$. For instance, in Fig. 7 one has to find all possible subsets G_i, $i > 0$, of the domain S_0, such that $\widehat{T}(v_0 \times S_0 \backslash (\cup_{i>0} G_i) | (v_0, x)) = 0$, $x \in G_i$. This is equivalent to searching for all absorbing regions S' of the domain S_0. In case one such region S' does exist, one can assign a new location v' to the absorbing region S' and a transition $v \to v'$, such that v' has the domain S' and v has the new absorbing domain $S_0 \backslash S'$. In general searching for absorbing regions is hard as one has to analyse the kernel $\widehat{T}$ for every x (uncountable many) in a continuous domain S. We propose to solve the problem of absorbing regions by discretizing the aBSCC into a DTMC and then searching for absorbing states. Notice that the discretization approach will not guarantee the absence of absorbing states as it relies on the size of the discretization step.

THEOREM 5. *For any separated GBA $\mathcal{A}$ and DTSHS $\mathcal{H}$:*

$$\mathrm{Pr}^{\mathcal{H}}(\mathcal{L}_\omega(\mathcal{A})) = \mathrm{Pr}^{\mathcal{H} \underline{\otimes} \mathcal{A}}(\lozenge \, V_F^\omega).$$

Notice that for the above theorem we only need to compute the probability to reach a set of absorbing final locations V_F^ω. This is enough due to the fact that as long as we are in an aBSCC the DTSHS $\mathcal{H}$ satisfies the GBA-property $\mathcal{A}$ with probability one.

5. CASE STUDY

In this section, we will show the applicability of our theoretical results to a case study.

5.1 Model Description

The following computational study [2] considers a model for the temperature evolution in a building with two rooms. Both rooms are equipped with a heater and each heater switches between the ON and OFF conditions depending on the temperature in the corresponding room. The state of the system is hybrid, with the discrete state component representing the status of the two heaters and the continuous state component representing the temperature in each of the two rooms. The discrete state space is given by $Loc = \{\mathtt{ON}, \mathtt{OFF}\}^2$. The allowed transitions between the locations are depicted in Fig. 1. The continuous state space is $\mathbb{R}^2$, irrespectively of the discrete state value (that is, $d(\ell) = 2, \forall \ell \in Loc$).

We suppose that the temperature of each room, say room i, evolves according to the following stochastic difference equation (SDE):

$$\mathbf{x}_i(k+1) = \mathbf{x}_i(k) + b_i(x_a - \mathbf{x}_i(k)) + a_{ij, j \neq i}(\mathbf{x}_j(k) - \mathbf{x}_i(k)) + c_i 1_{Loc_i}(\ell(k)) + \mathbf{w}_i(k),$$

where x_a represents the ambient temperature (assumed to be constant and equal for both rooms) and $1_{Loc_i}(\cdot)$ is the indicator function of set $Loc_i = \{(\ell_1, \ell_2) \in Loc : \ell_i = \mathtt{ON}\}$. The quantities b_i, a_{ij}, and c_i are non-negative real constants representing the heat transfer rate from room i to the ambient (b_i) and to room $j \neq i$ (a_{ij}), and the heat rate supplied to room i by the heater in room i (c_i). The disturbance $\{\mathbf{w}_i(k), k = 0, \dots, N\}$ affecting the temperature evolution in room i is assumed to be a sequence of independent identically distributed Gaussian random variables with zero mean and variance ν^2. Furthermore, with no loss of generality we suppose that the disturbances $\mathbf{w}_i$ and $\mathbf{w}_j$ affecting the temperature of different rooms ($i \neq j$) are independent.

The continuous transition kernel T_x describing the evolution of the continuous state $x = (x_1; x_2)$ can be easily derived from the SDE above. $T_x : \mathcal{B}(\mathbb{R}^2) \times \mathcal{S} \to [0, 1]$ can be expressed as

$$T_x(\cdot | (\ell, x)) = \mathcal{N}(\cdot; x + Zx + \Gamma(\ell), \nu^2 I), \qquad (11)$$

where $Z \in \mathbb{R}^{2 \times 2}$, $\Gamma(\ell) \in \mathbb{R}^2$, and $I \in \mathbb{R}^{2 \times 2}$ is the identity matrix. For $i = 1, 2$, the element in row i and column j of matrix Z is given by $[Z]_{ij} = a_{ij}$, if $j \neq i$, and $[Z]_{ij} = -(b_i + \sum_{k \neq i, k \in Loc} a_{ik})$, if $j = i$. For $i = 1, 2$, the ith element of vector $\Gamma(\ell)$, $\ell = (\ell_1, \ell_2) \in Loc$, is given by $[\Gamma(\ell)]_i = b_i x_a + c_i$, if $\ell_i = \mathtt{ON}$, and $[\Gamma(\ell)]_i = b_i x_a$, if $\ell_i = \mathtt{OFF}$. The reset kernel is set to coincide with the transition kernel in the current mode, irrespectively of the status to which the heaters possibly switch: $T_r(\cdot | (\ell, x), \ell') = T_x(\cdot | (\ell, x))$, for any $\ell, \ell' \in Loc$, and any $x \in \mathbb{R}^2$.

As for the discrete state evolution, we suppose that each heater switches status based on the temperature of the corresponding room, and independently of the other heater. This is modeled taking the discrete transition kernel $T_\ell : Loc \times \mathcal{S} \to [0, 1]$ as the product of two conditional stochastic kernels $T_{\ell, i} : \{\mathtt{ON}, \mathtt{OFF}\} \times (\{\mathtt{ON}, \mathtt{OFF}\} \times \mathbb{R}) \to [0, 1]$ governing the switching of each heater i. More precisely, we set

$$T_\ell(\ell' | (\ell, x)) = \prod_{i=1}^{2} T_{\ell, i}(\ell'_i | (\ell_i, x_i)), \qquad (12)$$

$\ell = (\ell_1, \ell_2), \ell' = (\ell'_1, \ell'_2) \in Loc, x = (x_1, x_2) \in \mathbb{R}^2$, where

$$T_{\ell, i}(\ell'_i | (\ell_i, x_i)) = \begin{cases} \sigma_i(x_i), & \ell'_i = \mathtt{OFF}, \\ 1 - \sigma_i(x_i), & \ell'_i = \mathtt{ON} \end{cases} \qquad (13)$$

with $\sigma_i : \mathbb{R} \to [0,1]$ a sigmoidal function given by

$$\sigma_i(y) = \frac{y^{d_i}}{\alpha_i^{d_i} + y^{d_i}}, \ y \in \mathbb{R}. \tag{14}$$

Function $\sigma_i(y)$, $y \in \mathbb{R}$, is parameterized by a "threshold" parameter α_i and a "steepness" parameter $d_i > 0$. α_i is the value of y at which the probability of the heater changing status becomes equal to 0.5, whereas d_i is related to the slope of the sigmoidal function at $y = \alpha_i$ (which amounts to $d_i/(4\alpha_i)$). We shall refer to the three possible values for the steepness parameter d_i respectively as $d_i = 1$ (*flat*), $d_i = 10$ (*gradual*), and $d_i = 100$ (*steep*), in increasing order. The values for the threshold α_i are determined as a convex combination of the temperatures x_i^l and x_i^u, $x_i^l < x_i^u$, defining the desired temperature range $[x_i^l, x_i^u]$ in room i.

5.2 Property Specification

We will consider two properties. The first one is a DFA and the second one is an LTL-formula expressed as a GBA. Recall that the difference between a DFA property and an LTL-formula is that the former reasons over the finite paths whereas the latter reasons over the infinite paths.

DFA property.

The property specified as a DFA $\mathcal{A}$ is depicted in Fig. 8.

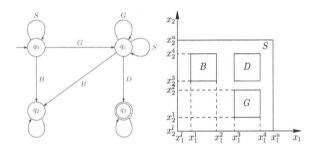

Figure 8: DFA $\mathcal{A}$.

Figure 9: Domains for DFA $\mathcal{A}$ of Fig. 8.

Intuitively, $\mathcal{A}$ describes all the paths, the continuous part of which can reach the region labeled with D (see Fig. 9) by first visiting the region labeled with G while avoiding the regions labeled with B. Region S is given by $([x_1^l, x_1^u] \times [x_2^l, x_2^u]) \backslash (G \cup B \cup D)$. Notice that no equivalent CTL formula can be formulated for property $\mathcal{A}$.

We specify the heating system as a DTSHS $\mathcal{H}$ with 16 locations: to every subset S, G, D and B of each continuous domain we assign a location, each of which has the conditional discrete stochastic kernel T_ℓ specified as in Fig. 1 and Eq. (13). The parameter d_i is taken to be equal to 10 (gradual) and the parameter α_i is equal to $\frac{1}{4}x_i^l + \frac{3}{4}x_i^u$ for $i \in \{1,2\}$. The regions within the continuous domains are specified by the parameters from Table 1. The set of atomic propositions is $\mathrm{AP} = \{S, G, D, B\}$. Every location is labeled with a single element from the set AP. The continuous transition kernels T_x and R are given by Eq. (11), and depend on the parameters $a_{12} = a_{21} = 0.25$, $b_1 = b_2 = 0.1$, $c_1 = 2.6$, $c_2 = 2.4$, $x_a = 6$ and $\nu = 0.5$. We partition the continuous domains $[x_1^l, x_1^u] \times [x_2^l, x_2^u]$ into square regions, uniformly dividing each interval $[x_1^l, x_1^u]$ into l slots. We leverage the discretization technique from Section 3.2 in order to obtain the discretized DTMC from the product $\mathcal{H} \otimes \mathcal{A}$. The DTMC

$x_1^l \backslash x_2^l$	$x_1^1 \backslash x_2^1$	$x_1^2 \backslash x_2^2$	$x_1^3 \backslash x_2^3$	$x_1^4 \backslash x_2^4$	$x_1^u \backslash x_2^u$
$10 \backslash 10$	$15 \backslash 15$	$20 \backslash 20$	$25 \backslash 25$	$30 \backslash 30$	$35 \backslash 35$

Table 1: Parameters characterizing continuous domains.

is highly connected, namely most of the transition probabilites are non zero. The results reported in this section refer to computations performed on a AMD Athlon 64 Dual Core Processor with 2GB RAM. The product construction and the discretization algorithm were implemented in MAT-LAB. Table 2 shows the verification time and the DTMC

Slots l	5	10	20
DTMC states	400	1600	6400
Time (sec)	29.5	466.7	5694.6

Table 2: Verification time for the DFA $\mathcal{A}$ in (Fig. 8) over the DTMC obtained from the DTSHS $\mathcal{H}$.

size for different number of slots. The obtained verification times critically depend on the discretization procedure, rather than the model checking algorithms: the time spent on the product construction and solving the system of linear equations is much smaller compared to the time spent for the generation of the DTMC. Fig. 10 displays the probabil-

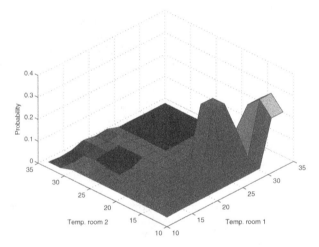

Figure 10: Satisfiability probability for the DFA $\mathcal{A}$ over the DTSHS $\mathcal{H}$ (through its DTMC discretization), with the first set of parameters.

ity that the two-room DTSHS satisfies the DFA property $\mathcal{A}$ given that the initial location is (OFF, OFF) and the continuous state is chosen in any of the 4 domains S, G, B and D. (The surface is obtained at the representative points.) The number of discretization slots l is 10. A similar plot is reported on Fig. 11 in 2D for a parameter choice of d_i of 100 (steep) and of α_i of $\frac{1}{2}x_i^l + \frac{1}{2}x_i^u$, respectively — all other parameters are as before. Here warmer colors denote higher probabilities. In both the described instances, the probability is higher for all the states starting from the domain G or nearby. This is due to the fact that the property $\mathcal{A}$ is satisfied only for the paths of DTSHS that reach D by starting anywhere in G or S and having crossed G.

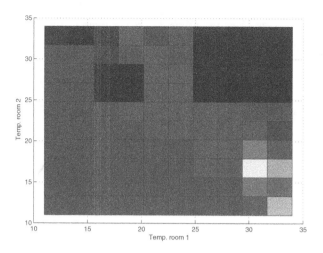

Figure 11: 2D plot for the satisfiability probability for the DFA $\mathcal{A}$ over the DTSHS $\mathcal{H}$ (through its DTMC discretization), with the second set of parameters.

LTL - formula.

We consider the formula $\varphi = \Diamond \Box (D \wedge \neg S \wedge \neg F)$ on the set of atomic propositons $\text{AP} = \{S, D, F\}$ and sets $S = [x_1^l, x_1^u] \times [x_2^l, x_2^u]$ and $D = [x_1^l, x_1^m] \times [x_2^l, x_2^m]$, where $x_i^m = \frac{x_i^u + x_i^l}{2}$, $i \in \{1, 2\}$. The formula signifies that all paths should eventually reach domain D and then stay there forever.

We compute the satisfiability probability of the formula φ on the DTSHS $\mathcal{H}_2$ modeling the two-room heating benchmark, where we consider a slightly different discrete structure, as specified in Fig. 12. For all locations Loc of the

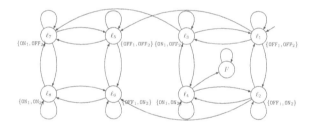

Figure 12: Discrete structure of the DTSHS $\mathcal{H}_2$.

DTSHS $\mathcal{H}_2$, the behavior of the discrete stochastic kernel T_ℓ is defined in Eq. (12). The kernels T_ℓ, T_x and T_r for locations $Loc \backslash \{\ell_4\}$ are normalized, whereas in ℓ_4 we introduce a new location F: this location models a failure mode, which is possibly attained when both heaters in the two rooms are switched on. F is an abstract locations containing four sublocations F_1, F_2, F_3 and F_4 denoting the following hybrid set of states $\{\text{ON}_1, \text{ON}_2\} \times D$, $\{\text{ON}_1, \text{OFF}_2\} \times D$, $\{\text{OFF}_1, \text{ON}_2\} \times D$ and $\{\text{OFF}_1, \text{OFF}_2\} \times (S \cup D)$. The transitions kernels T_ℓ and T_x to the four sublocations are defined accordingly to Eq. (11) and (12), as in $T_\ell(F_i | (\ell_4, x))$ for $F_i \in \{\{\text{ON}_1, \text{ON}_2\}, \{\text{ON}_1, \text{OFF}_2\}, \{\text{OFF}_1, \text{ON}_2\}, \{\text{OFF}_1, \text{OFF}_2\}\}$. The reset transition kernel is defined as $T_r(x' | (\ell_4, x), F_i) = T_x(x' | (\ell_4, x))$ for two cases $F_i \in \{F_1, F_2, F_3\}$ and $x' \in D$, or $F_i = F_4$ and $x' \in S \cup D$. All locations ℓ_1, ℓ_2, ℓ_3 and ℓ_4 are labeled with S (domain S), locations ℓ_5, ℓ_6, ℓ_7 and ℓ_8 are labeled with D (domain D) and locations F_i are labeled

with F. We select the boundary for the continuous domains as $x_1^l = x_2^l = 5$ and $x_1^u = x_2^u = 45$. Table 3 displays the

Slots l	4	8	16
DTMC states	49	193	769
Time (sec)	66.4	142.4	1723.8

Table 3: Verification time for the LTL-formula φ over the DTMC obtained from the DTSHS $\mathcal{H}_2$.

verification time and the DTMC size for different choices of partitioning slots l. Fig. 13 depicts the probability that the

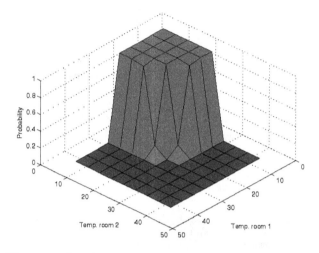

Figure 13: Satisfiability probability for the LTL-formula φ over the DTSHS $\mathcal{H}_2$ (through its DTMC discretization).

DTSHS $\mathcal{H}_2$ satisfies the LTL-formula φ given that the initial location is $\{\text{OFF}, \text{OFF}\}$ and the continuous state is chosen anywhere within the sets S, D of the continuous domains. Notice that the probability is higher for continuous states that are closer to the domain D. All continuous states in domain D satisfy the formula φ with probability one.

6. CONCLUSIONS

In this paper, we have considered the quantitative verification of DTSHS against linear time objectives, specified either as a DFA or as an LTL-formula (Büchi automaton). We have shown that the probability that a DTSHS satisfies a linear time property can be reduced to computing reachability probabilities in the product of the DFA (or the Büchi automaton) and the DTSHS. Future work will include verification of nonautonomous DTSHS and the development of more efficient techniques for the general verification of DTSHS.

7. REFERENCES

[1] PRISM website. http://www.prismmodelchecker.org.
[2] A. Abate, J.-P. Katoen, J. Lygeros, and M. Prandini. Approximate model checking of stochastic hybrid systems. *European Journal of Control*, 16(6):1–18, 2010.
[3] A. Abate, M. Prandini, J. Lygeros, and S. Sastry. Probabilistic reachability and safety for controlled

discrete time stochastic hybrid systems. *Automatica*, 44(11):2724–2734, 2008.

[4] C. Baier and J.-P. Katoen. *Principles of Model Checking*. MIT Press, 2008.

[5] M. Bujorianu and J. Lygeros. Reachability questions in piecewise deterministic Markov processes. In O. Maler and A. Pnueli, editors, *HSCC*, volume 2623 of *LNCS*, pages 126–140. Springer-Verlag, 2003.

[6] C. Cassandras and J. Lygeros, editors. *Stochastic Hybrid Systems*. CRC Press, 2007.

[7] P. Collins and I. S. Zapreev. Computable CTL* for discrete-time and continuous-space dynamic systems. In *RP '09: Proceedings of the 3rd International Workshop on Reachability Problems*, pages 107–119, Berlin, Heidelberg, 2009. Springer-Verlag.

[8] J.-M. Couvreur, N. Saheb, and G. Sutre. An optimal automata approach to LTL model checking of probabilistic systems. In *LPAR*, volume 2850 of *LNCS*, pages 361–375, 2003.

[9] J. Desharnais and P. Panangaden. Continuous stochastic logic characterizes bisimulation of continuous-time Markov processes. *J. Log. Algebr. Program.*, 56(1-2):99–115, 2003.

[10] M. Fränzle, T. Teige, and A. Eggers. Satisfaction meets expectations - computing expected values of probabilistic hybrid systems with SMT. In *IFM*, pages 168–182, 2010.

[11] H. Hansson and B. Jonsson. A logic for reasoning about time and reliability. *Formal Aspects of Computing*, 6:512–535, 1994. 10.1007/BF01211866.

[12] J.-P. Katoen, E. M. Hahn, H. Hermanns, D. N. Jansen, and I. Zapreev. The ins and outs of the probabilistic model checker MRMC. In *QEST*, pages 167–176, 2009.

[13] X. D. Koutsoukos and D. Riley. Computational methods for verification of stochastic hybrid systems. *IEEE Trans. on Systems, Man, and Cybernetics, Part A*, 38(2):385–396, 2008.

[14] A. Lecchini, W. Glover, J. Lygeros, and J. Maciejowski. Monte Carlo optimization for conict resolution in air traffic control. *IEEE Transactions on Intelligent Transportation Systems*, 7:470–482, 2006.

[15] M. Prandini and J. Hu. Stochastic reachability: Theory and numerical approximation. In C. Cassandras and J. Lygeros, editors, *Stochastic hybrid systems*, Automation and Control Engineering Series, pages 107–138. Taylor & Francis Group/CRC Press, 2006.

[16] M. O. Rabin and D. Scott. Finite automata and their decision problems. *IBM Journal of Research and Development*, 3(2):114 –125, 1959.

[17] F. Ramponi, D. Chatterjee, S. Summers, and J. Lygeros. On the connections between PCTL and dynamic programming. In *HSCC*, pages 253–262. ACM, 2010.

[18] S. Summers and J. Lygeros. Verification of discrete time stochastic hybrid systems: A stochastic reach-avoid decision problem. *Automatica*, 46(12):1951–1961, 2010.

[19] M. Y. Vardi and P. Wolper. An automata-theoretic approach to automatic program verification. In *LICS*, pages 332–344, 1986.

[20] L. Zhang, Z. She, S. Ratschan, H. Hermanns, and E. M. Hahn. Safety verification for probabilistic hybrid systems. In *CAV*, volume 6174 of *LNCS*, pages 196–211. Springer Verlag, 2010.

Reachable Set Computation for Uncertain Time-Varying Linear Systems

Matthias Althoff
Carnegie Mellon University
5000 Forbes Ave
Pittsburgh, PA 15213
malthoff@ece.cmu.edu

Colas Le Guernic
New York University
251 Mercer Street
New York, NY 10012
colas@cs.nyu.edu

Bruce H. Krogh
Carnegie Mellon University
5000 Forbes Ave
Pittsburgh, PA 15213
krogh@ece.cmu.edu

ABSTRACT

This paper presents a method for using set-based approximations to the Peano-Baker series to compute overapproximations of reachable sets for linear systems with uncertain, time-varying parameters and inputs. Alternative representations for sets of uncertain system matrices are considered, including matrix polytopes, matrix zonotopes, and interval matrices. For each representation, the computational efficiency and resulting approximation error for reachable set computations are evaluated analytically and empirically. As an application, reachable sets are computed for a truck with hybrid dynamics due to a gain-scheduled yaw controller. As an alternative to computing reachable sets for the hybrid model, for which switching introduces an additional overapproximation error, the gain-scheduled controller is approximated with uncertain time-varying parameters, which leads to more efficient and more accurate reachable set computations.

Categories and Subject Descriptors

G.1.0 [**Numerical Analysis**]: General; I.6.4 [**Simulation and Modeling**]: Model Validation and Analysis

General Terms

Algorithms, Theory, Verification

Keywords

Reachability Analysis, Linear Systems, Uncertain Parameters, Peano-Baker Series, Safety

1. INTRODUCTION

Reachable set computation offers an alternative to numerical simulation for evaluating the correctness of system models with respect to safety specifications, such as limits on system state variables. For hybrid dynamic systems, the principal difficulty is computing the reachable sets for the continuous dynamics. Research in this area is extensive, and, as summarized in the following brief literature review, a number of approaches have been developed for computing and estimating reachable states for various classes of continuous and hybrid dynamic systems. This paper presents a new method for computing overapproximations of reachable sets for linear systems with uncertain, time-varying parameters and bounded inputs using set-based approximations to the Peano-Baker series. Various methods for representing uncertain matrices can be used, allowing for a trade-off between approximation accuracy and computation time.

The method developed in this paper represents reachable sets using zonotopes, a class of polyhedra, so we focus most of our literature review on polyhedra-based methods. For so-called linear hybrid automata in which the continuous dynamics are given as constant convex polyhedral bounds on the derivative of the state vector, the reachable set can be computed directly using operations on polyhedra [13]. Reachable sets of such systems can be used as a basis for the reachability analysis of linear or even more complex systems, such as nonlinear and hybrid systems [7, 14]. The approximation of more complex continuous dynamics, and even linear dynamics, can lead to a computationally prohibitive explosion of discrete states, however. Consequently, methods that deal more directly with the continuous dynamics are often more efficient when the systems of interest cannot be modeled immediately using linear hybrid automata.

Since the set of states that can be reached over a time interval are in general nonconvex for linear dynamic systems, effective approximations of the reachable sets are constructed as unions of convex polyhedra. A number of methods and tools have been developed to approximate reachable sets for continuous dynamic systems [4, 5]. In all of these methods, a sequence of approximations to the reachable sets at discrete points in time (which *are* convex if the set of initial states and bounds on the inputs are convex) are computed and then consecutive sets in the sequence are wrapped with an enclosing convex polyhedron that accounts for the state trajectories between the discrete points in time. When the systems have bounded input signals, the approximation error can grow significantly over time due to the compounded accumulation of overapproximation added at each step to account for the effect of the inputs. In [12] it has been shown that the reachable set of linear time invariant (LTI) systems can be computed without this wrapping effect. This wrapping-free method can be implemented using different set representations, such as ellipsoids [18], polytopes [5], oriented rectangular hulls [26], zonotopes [9], or support functions [10], and is generally superior to methods based on approximating the system dynamics with derivative bounds. Level-set methods [27] are often more accurate than the method developed in [12], but level set methods do not scale to systems with more than a few continuous state variables (typically about four).

If uncertain parameters are considered, most algorithms work with interval methods and multidimensional intervals (axis-oriented boxes) to represent reachable sets [15, 23, 24]. Interval methods are also applied in [21] for the rigorous numerical solution of initial value problems for ordinary differential equations. The algorithms in [24] take advantage of special properties of the system dynamics, such as monotonicity. Interval methods are generally more conservative than other set representations, however. In [1], reachable sets for linear systems with uncertain time-invariant parameters are computed using zonotopes, which can be much more accurate than multidimensional intervals. The procedure presented in this paper to compute reachable sets of linear systems with time-varying parameters is a non-trivial extension of [1], since the state-transition matrix is no longer the matrix exponential for the time-varying case.

Reachability computations for linear systems with uncertain parameters can be applied to analyze nonlinear systems [2] and hybrid systems [3, 11]. Thus, the reachability analysis of linear systems can be seen as a basic module for the reachability analysis of more complicated system classes. In this paper, we illustrate how linear systems with time-varying parameters can be used to compute reachable sets for nonlinear systems. In addition, we show how time-varying parameters can be used to eliminate discrete transitions for some classes of hybrid systems, a procedure we call *continuization*, which makes it possible to compute reachable sets without the errors introduced by set intersections required to deal with transition guards in hybrid system models.

The following section reviews the overall approach to computing reachable sets for continuous dynamic systems and formulates the specific problem addressed. Section 3 develops the overapproximation of the state transition matrix based on overapproximations of matrix operations. This result is used to compute the reachable sets for systems without inputs and with bounded inputs in Sec. 4. Then we show how to perform computations with sets of matrices, using different representations. Section 5 describes matrix-matrix operations and Sec. 6 focuses on matrix-vector operations. The usefulness, efficiency, and accuracy of the proposed overapproximations are illustrated with numerical examples in Sec. 7.

2. PROBLEM FORMULATION

The basic principle of many reachability algorithms, including the approach in this paper, is to compute the reachable set for consecutive time intervals $R([t_{k-1}, t_k])$, where $t_k = k \cdot r$, $r \in \mathbb{R}^+$ is the time increment, and $k \in \mathbb{N}$ is the time step; see [5,6,9,26]. The final reachable set is then given by $R([0, t_f]) = \bigcup_{k=1}^{t_f/r} R([t_{k-1}, t_k])$, where t_f is a multiple of r. Since this finite union can be represented as an enumeration, this paper focuses on the computation of the reachable set for a single time interval $[0, r]$. The basic steps for the computation of $R([0, r])$, shown in Fig. 1, are summarized as follows.

1. Compute the reachable set at $t = r$, neglecting the input (the homogeneous solution, $R^h(r)$);

2. Generate the convex hull of the solution at $t = r$ and the initial set; and

3. Enlarge the convex hull to ensure enclosure of all trajectories for the time interval $t \in [0, r]$, including the effects of inputs.

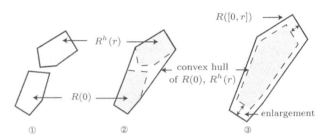

Figure 1: Steps in the computation of an overapproximation of the reachable set for a given time interval.

In this paper we consider time-varying linear systems of the form

$$\dot{x}(t) = A(t)x(t) + u(t), \quad x(0) \in \mathbb{R}^n, \ t \in [0, r], \quad (1)$$

where for given sets $\mathcal{A} \subset \mathbb{R}^{n \times n}$ and $\mathcal{U} \subset \mathbb{R}^n$, $A : \mathbb{R}^+ \to \mathcal{A}$ and $u : \mathbb{R}^+ \to \mathcal{U}$ are piecewise continuous.

Note that the commonly used input formulation $B(t)\tilde{u}(t)$ with $\tilde{u}(t) \in \tilde{\mathcal{U}}$ and $B(t) \in \mathcal{B}$ is accommodated by defining $\mathcal{U} = \{B\tilde{u} | \tilde{u} \in \tilde{\mathcal{U}}, B \in \mathcal{B}\}$.

Let $\chi(t; x_0, A(\cdot), u(\cdot))$, $t \in [0, r]$, denote the solution to (1) for given $x(0) = x_0$, $A(\cdot)$, and $u(\cdot)$. Given a set of initial states, $R(0) \subset \mathbb{R}^n$, our objective is to compute the set of reachable states

$$R^e([0, r]) = \Big\{ \chi(t; x_0, A(\cdot), u(\cdot)) \, \Big| \, x_0 \in R(0), t \in [0, r],$$
$$\forall \tau \in [0, t] A(\tau) \in \mathcal{A}, u(\tau) \in \mathcal{U} \Big\}.$$

The superscript e on $R^e([0, r])$ denotes the exact reachable set. The exact reachable set for time-varying linear systems cannot be computed exactly [19]. Therefore, we aim to compute overapproximations $R([0, r]) \supseteq R^e([0, r])$ that are as accurate as possible, while at the same time ensuring that the computations are efficient and scale well with the system dimension n.

3. OVERAPPROXIMATING THE STATE TRANSITION MATRIX

We first consider the case when there is no input signal in (1). In this case, when $A(\cdot)$ is known, the solution is given by

$$x(t) = \Phi(t, 0)x_0,$$

where the *state transition matrix* $\Phi(t, 0)$ can be computed by the Peano-Baker series

$$\Phi(t, 0) = I + \int_0^t A(\sigma_1)d\sigma_1 + \int_0^t A(\sigma_1) \int_0^{\sigma_1} A(\sigma_2) \, d\sigma_2 \, d\sigma_1$$
$$+ \int_0^t A(\sigma_1) \int_0^{\sigma_1} A(\sigma_2) \int_0^{\sigma_2} A(\sigma_3) \, d\sigma_3 \, d\sigma_2 \, d\sigma_1 + \dots$$
$$(2)$$

This series converges uniformly and absolutely [25, Chap. 3]. When the system matrix is time invariant $A(\cdot) = A$, the solution of $\Phi(t,0)$ is the well-known matrix exponential e^{At}. Since we are only interested in solutions starting at $t = 0$, we write $\Phi(t)$ as a short form of $\Phi(t,0)$ from now on.

To obtain a numerical solution to the Peano-Baker series, the integrals in (2) can be approximated by Riemann sums of the form $\int_0^t A(\sigma_i)d\sigma_i \approx \sum_{l_i=1}^k A(l_i\Delta)\Delta$, where $\Delta \in \mathbb{R}^+$ is a fixed step size. For conciseness throughout the remainder of the paper, $A(m)$ denotes $A(m\Delta)$ for any integer m. Applying the Riemann sum approximation to (2) yields

$$\tilde{\Phi}(t,\Delta) := I + \sum_{l_1=1}^k A(l_1)\Delta + \sum_{l_2=1}^k \sum_{l_1=1}^{l_2} A(l_2)A(l_1)\Delta\,\Delta$$
$$+ \sum_{l_3=1}^k \sum_{l_2=1}^{l_3} \sum_{l_1=1}^{l_2} A(l_3)A(l_2)A(l_1)\Delta\Delta\Delta + \dots$$

Therefore, the matrix $\Phi(t)$ can be approximated iteratively by $\tilde{\Phi}_i(t,\Delta)$, where

$$\tilde{\Phi}_1(t,\Delta) = I + \sum_{l_1=1}^k A(l_1)\Delta,$$

$$\tilde{\Phi}_i(t,\Delta) = \tilde{\Phi}_{i-1}(t,\Delta) + \underbrace{\sum_{l_i=1}^k \cdots \sum_{l_1=1}^{l_2} \left(\prod_{q=1}^i A(l_q)\right)\Delta^i}_{\triangleq \tilde{\Phi}_i^\delta(t,\Delta)}. \quad (3)$$

We now consider the computation of *sets* that overapproximate the range of state transition matrices that result when $A(t) \in \mathcal{A}$. Towards this end, we introduce the following notation and matrix operators. For a set of matrices $\mathcal{A}$, $\mathrm{CH}(\mathcal{A})$ denotes the closed convex hull of $\mathcal{A}$. Given two sets of matrices, $\mathcal{A}$ and $\mathcal{B}$, we denote by $\mathcal{A} \oplus \mathcal{B}$ and $\mathcal{A} \otimes \mathcal{B}$ the sets resulting from the point-wise sums and products of elements of $\mathcal{A}$ and $\mathcal{B}$, respectively; that is,

$$\mathcal{A} \oplus \mathcal{B} = \{A + B \mid A \in \mathcal{A}, B \in \mathcal{B}\}, \text{ and}$$
$$\mathcal{A} \otimes \mathcal{B} = \{A \times B \mid A \in \mathcal{A}, B \in \mathcal{B}\}$$

The point-wise product also holds for the special case when A or B is a scalar. By an abuse of notation, $\otimes$ is sometimes omitted. Considering equation (3), we only know that every $A(l_q)$ belongs to some set $\mathcal{A}$. Thus, $\prod_{q=1}^m A(l_q)$ belongs to $\bigotimes_{q=1}^m \mathcal{A}$, which we will denote by $\mathcal{A}^m$ in the following,[1] and

$$\tilde{\Phi}_i^\delta(t,\Delta) \in \bigoplus_{l_i=1}^k \cdots \bigoplus_{l_1=1}^{l_2} \mathcal{A}^i \Delta^i. \quad (4)$$

The following proposition on distributivity of multiplication by positive scalars over addition for convex matrix sets is useful to simplify the above expression,

PROPOSITION 1. (Distributivity of Matrix Sets) *If $\mathcal{A}$ is convex and $a,b \in \mathbb{R}^+$, then*

$$a\mathcal{A} \oplus b\mathcal{A} = (a+b)\mathcal{A}.$$

PROOF. It is always true that $(a+b)\mathcal{A} \subseteq a\mathcal{A} \oplus b\mathcal{A}$, even if $\mathcal{A}$ is not convex. Further, due to the convexity it follows

---

[1] Note that $\mathcal{A}^m$ is $\{\prod_{q=1}^m A_i \mid A_i \in \mathcal{A}\}$, which is different from and larger than $\{A^m \mid A \in \mathcal{A}\}$.

for $X_1, X_2 \in \mathcal{A}$ and $\alpha \in [0,1]$ that $\alpha X_1 + (1-\alpha)X_2 \in \mathcal{A}$. Making use of $a,b \geq 0$, let $\alpha = \frac{a}{a+b}$, which gives

$$\frac{a}{a+b}X_1 + \frac{b}{a+b}X_2 \in \mathcal{A}.$$

Thus, $aX_1 + bX_2 \in (a+b)\mathcal{A}$ and consequently $a\mathcal{A} \oplus b\mathcal{A} \subseteq (a+b)\mathcal{A}$. $\square$

Using this proposition, the following theorem provides an expression for an overapproximation for the set of possible state transition matrices $\Phi(t)$.

THEOREM 1. (Set of State Transition Matrices) *Let $\mathcal{M}(t)$ denote the set of state transition matrices $\Phi(t)$ when $A(\tau) \in \mathcal{A}$ for $\tau \in [0,t]$. Then $\mathcal{M}(t) \subseteq \overline{\mathcal{M}}(t)$, where*

$$\overline{\mathcal{M}}(t) = \bigoplus_{i=0}^\infty \overline{\mathcal{M}}_i(t), \qquad \overline{\mathcal{M}}_i(t) = \frac{t^i}{i!}\mathrm{CH}(\mathcal{A}^i).$$

PROOF. Proposition 1 implies the set in equation (4) is contained in $\Delta^i \left(\sum_{l_i=1}^k \cdots \sum_{l_1=1}^{l_2} 1\right)\mathrm{CH}(\mathcal{A}^i)$. The summation of ones is computed using the formula

$$\sum_{l=1}^k l^m = \frac{k^{m+1}}{m+1} + \mathcal{O}(k^m)$$

from which the auxiliary results ξ_m are obtained:

$$\xi_1 = \sum_{l_1=1}^{l_2} 1 = l_2 + \mathcal{O}(l_2^0)$$

$$\xi_2 = \sum_{l_2=1}^{l_3} \xi_1 = \sum_{l_2=1}^{l_3} l_2 + \mathcal{O}(l_2^0) = \frac{l_3^2}{2} + \mathcal{O}(l_3^1)$$

$$\cdots$$

$$\xi_i = \sum_{l_i=1}^k \cdots \sum_{l_1=1}^{l_2} 1 = \frac{k^i}{i!} + \mathcal{O}(k^{i-1}).$$

From the relation $k\Delta = t$ it follows that $\tilde{\Phi}_i^\delta(t,\Delta)$ is in $\left(\frac{t^i}{i!} + \Delta\mathcal{O}(t^{i-1})\right)\mathrm{CH}(\mathcal{A}^i)$. Taking the limit as $\Delta \to 0$, the Riemann sums used to approximate the integrals in (2) converge and therefore

$$\Phi_i^\delta(t) \subseteq \frac{t^i}{i!}\mathrm{CH}(\mathcal{A}^i). \quad \square$$

Note that the computation of $\overline{\mathcal{M}}(t)$ resembles the computation of the set of Taylor expansions of e^{At}, namely, $\{\sum_{i=0}^\infty A^i t^i/i! \mid A \in \mathcal{A}\}$, except the relationships between the different occurrences of A are forgotten in Theorem 1 and the convex hull of $\mathcal{A}^i$ has to be computed. We will see in Sec. 5 that this computation requires no additional work in our multiplication procedure since we only represent convex sets.[2]

Defining the norm of a set of matrices $\mathcal{M}$ as

$$\|\mathcal{M}\| = \sup\{\|M\| \mid M \in \mathcal{M}\},$$

---

[2] Even if $\mathcal{A}$ is convex, $\mathcal{A}^2$ might not be, but our approximate multiplication gives a convex overapproximation of $\mathcal{A}^2$, and thus of $\mathrm{CH}(\mathcal{A}^2)$.

where $\|M\|$ denotes a particular matrix norm, the norm of $\overline{\mathcal{M}}(t)$ is bounded as

$$\|\overline{\mathcal{M}}(t)\| \leq 1 + \|\mathcal{A}\|t + \frac{1}{2!}\|\mathcal{A}\|^2 t^2 + \frac{1}{3!}\|\mathcal{A}\|^3 t^3 + \ldots = e^{\|\mathcal{A}\|t}$$

which implies that $\|\overline{\mathcal{M}}(t)\| < \infty$ for $\|\mathcal{A}\| < \infty$ and $t < \infty$.

To approximate $\overline{\mathcal{M}}(t)$, the following proposition shows that we can replace the infinite sum $\overline{\mathcal{M}}(t) = \sum_{i=0}^{\infty} \overline{\mathcal{M}}_i(t)$ with a finite sum of η terms, $\sum_{i=0}^{\eta} \overline{\mathcal{M}}_i(t)$, plus a set that bounds the remaining terms in the infinite sum. For a set of matrices $\mathcal{M}$, the notation $|\mathcal{M}|$ denotes the matrix in which each element is equal to the supremum of the absolute value of the corresponding element in each matrix in $\mathcal{M}$. That is, $|\mathcal{M}|(i,j) = \sup\{|M(i,j)| \mid M \in \mathcal{M}\}$.

PROPOSITION 2. (State Transition Remainder) *The set of remainder matrices $\mathcal{E}(t)$ is an overapproximation of $\bigoplus_{i=\eta+1}^{\infty} \overline{\mathcal{M}}_i(t)$ computed as*

$$\mathcal{E}(t) = [-W(t), W(t)], \quad W(t) = e^{|\mathcal{A}|t} - \sum_{i=0}^{\eta} \frac{t^i}{i!}|\mathcal{A}|^i.$$

PROOF. By induction it follows that $|\mathcal{A}^n| \leq |\mathcal{A}|^n$ element wise. Thus,

$$\left| \bigoplus_{i=\eta+1}^{\infty} \frac{t^i}{i!} \mathrm{CH}(\mathcal{A}^i) \right| \leq \sum_{i=\eta+1}^{\infty} \frac{t^i}{i!}|\mathcal{A}|^i = e^{t|\mathcal{A}|} - \sum_{i=0}^{\eta} \frac{t^i}{i!}|\mathcal{A}|^i. \quad \square$$

4. OVERAPPROXIMATING THE REACHABLE SET

For autonomous uncertain time-varying systems, the set of state transition matrices makes it possible to bound the state of an autonomous system by $x(r) \in \overline{\mathcal{M}}(r)x(0)$ so that the reachable set is obtained by $R(r) = \overline{\mathcal{M}}(r)R(0)$. In this section, we first derive an overapproximation for the reachable set for a time interval, $R([0, r])$, for autonomous systems and then show how to incorporate the effects of uncertain inputs.

The set $R^e([0, r])$ can be approximated by the convex hull of $R(0)$ and $R(r)$. To ensure this approximation is an overapproximation we add an error term $\mathcal{F}(r) \otimes R(0)$ to this convex hull.

For a given trajectory starting from x_0, we know that for any t in $[0, r]$, $x(t)$ is in $\overline{\mathcal{M}}(t)x_0$. Therefore, Theorem 1 implies there exists a sequence of matrices $A_{t,i} \in \mathrm{CH}(A^i)$ such that

$$x(t) = \sum_{i=0}^{\infty} \frac{t^i}{i!} A_{t,i} x_0$$

We approximate $x(t)$ by a point $\hat{x}(t)$ in the convex hull of x_0 and $\overline{\mathcal{M}}(r)x_0$ defined as

$$\hat{x}(t) = (1 - \frac{t}{r})x_0 + \frac{t}{r}\left(\sum_{i=0}^{\infty} \frac{r^i}{i!} A_{t,i}\right) x_0.$$

Because of its dependence on t, $\hat{x}(t)$ may not describe a straight line from x_0 and $x(r)$ as t varies from 0 to r, but it will always stay in the convex hull of x_0 and $\overline{\mathcal{M}}(r)x_0$.

We now evaluate the error made when applying this approximation.

$$x(t) - \hat{x}(t) = \sum_{i=0}^{\infty} \frac{t^i}{i!} A_{t,i} x_0 - (1 - \frac{t}{r})x_0 - \frac{t}{r}\left(\sum_{i=0}^{\infty} \frac{r^i}{i!} A_{t,i}\right) x_0$$

$$= \left(\sum_{i=2}^{\infty} \frac{(t^i - tr^{i-1})}{i!} A_{t,i}\right) x_0$$

For $i > 1$, one can show [1] that

$$\left\{ t^i - tr^{i-1} \mid t \in [0, r] \right\} = \left[\left(i^{\frac{-i}{i-1}} - i^{\frac{-1}{i-1}}\right) r^i, 0 \right].$$

Thus, we can define $\mathcal{F}(r)$ as

$$\mathcal{F}(r) = \bigoplus_{i=2}^{\infty} \frac{r^i}{i!} \mathrm{CH}\left(\{0\} \cup \left(i^{\frac{i}{1-i}} - i^{\frac{1}{1-i}}\right) \mathcal{A}^i\right).$$

And we have $R([0, r]) \subseteq \mathrm{CH}\left(R(0) \cup \overline{\mathcal{M}}(r)R(0)\right) \oplus \mathcal{F}(r)R(0)$.

Similarly to $\overline{\mathcal{M}}(r)$, $\mathcal{F}(r)$ can be overapproximated by considering the sum up to η and overapproximating the remaining terms by $\mathcal{E}(r)$. For an efficient evaluation, $\mathcal{A}$ is overapproximated by an interval matrix (specified later) and interval arithmetic is applied to obtain an overapproximation of $\mathcal{F}(r)$.

We now consider the additional reachable set due to uncertain inputs. Since the superposition principle for linear systems can be applied, the reachable set due to the input can be computed independently of the reachable set for the autonomous system.

THEOREM 2 (INPUT SOLUTION). *The set of reachable states due to the uncertain input $u(t) \in \mathcal{U}$ can be overapproximated by*

$$\mathcal{P}(t) = \bigoplus_{i=0}^{\eta} \left(\frac{t^{i+1}}{(i+1)!} \mathrm{CH}(\mathcal{A}^i \mathcal{U}) \right) \oplus \frac{t}{\eta + 2} \mathcal{E}(t)\{|\mathcal{U}|\}.$$

PROOF. The differential equation $\dot{x}(t) = A(t)x(t) + u(t)$ can be rewritten as

$$\frac{d}{dt}\begin{pmatrix} x(t) \\ 1 \end{pmatrix} = \underbrace{\begin{pmatrix} A(t) & u(t) \\ 0 & 0 \end{pmatrix}}_{A_u(t)} \begin{pmatrix} x(t) \\ 1 \end{pmatrix}$$

Based on Theorem 1, the set of points reachable at time t from $(x_0, 1)^\top$ is included in $\bigoplus_{i=0}^{\infty} \frac{t^i}{i!} \mathrm{CH}(\mathcal{A}_u^i)\{(x_0, 1)^\top\}$ where

$$\mathcal{A}_u = \left\{ \begin{pmatrix} A & u \\ 0 & 0 \end{pmatrix} \middle| A \in \mathcal{A}, u \in \mathcal{U} \right\}.$$

One can show by induction that, for $i > 0$,

$$\mathcal{A}_u^i = \left\{ \begin{pmatrix} A_{i-1}A & A_{i-1}u \\ 0 & 0 \end{pmatrix} \middle| A \in \mathcal{A}, A_{i-1} \in \mathcal{A}^{i-1}, u \in \mathcal{U} \right\}.$$

Thus, taking $x_0 = (0, \ldots, 0)$, we have

$$\mathcal{P}(t) = \bigoplus_{i=0}^{\infty} \frac{t^{i+1}}{(i+1)!} \mathrm{CH}(\mathcal{A}^i \otimes \mathcal{U}).$$

Similarly to Proposition 2, we can compute this infinite sum up to η and bound the remainder by

$$\left| \bigoplus_{i=\eta+1}^{\infty} \frac{t^{i+1}}{(i+1)!} \mathrm{CH}(\mathcal{A}^i \otimes \mathcal{U}) \right| \leq \frac{t}{\eta + 2} \sum_{i=\eta+1}^{\infty} \left(\frac{\eta + 2}{i + 1}\right) \frac{t^i}{i!}|\mathcal{A}|^i|\mathcal{U}|$$

$$\leq \frac{t}{\eta + 2} W(t)|\mathcal{U}| \quad \square$$

If the origin is contained in the set of possible inputs ($0 \in \mathcal{U}$), it holds that $\mathcal{P}([0, r]) = \mathcal{P}(r)$; see [1]. If this is not the case, some correction measures have to be applied [1]. Algorithm 1 summarizes the steps for computing $R([0, t_f])$ under the assumption $0 \in \mathcal{U}$. Note that the error of the computations in Algorithm 1 can be made arbitrarily small when $r \to 0$ while the computational effort grows. When enforcing an upper bound on the number of parameters describing the reachable set, the error cannot be made arbitrarily small since overapproximative order reduction techniques have to be applied to bound the growing number of set parameters.

Algorithm 1 Compute $R([0, t_f])$

Require: Initial set $R(0)$, set of state transition matrices $\overline{\mathcal{M}}(r)$, input set $\mathcal{U}$, set of correction matrices $\mathcal{F}$, time horizon t_f, time step r.
Ensure: $R([0, t_f])$

$\mathcal{H}_0 = \mathrm{CH}(R(0) \cup \overline{\mathcal{M}}(r)R(0)) \oplus \mathcal{F}(r)R(0)$
$\mathcal{P}_0 = \mathcal{P}(r)$
$R_0 = \mathcal{H}_0 \oplus \mathcal{P}_0$
for $k = 1 \ldots t_f/r - 1$ **do**
$\quad R_k = \overline{\mathcal{M}}(r)R_{k-1} \oplus \mathcal{P}_0$
end for
$R([0, t_f]) = \bigcup_{k=1}^{t_f/r} R_{k-1}$

The proof of Theorem 2 uses the fact that we can express a d-dimensional system with inputs as a $(d+1)$-dimensional autonomous system. By not using this transformation, we can use different representations and algorithms for the set of matrices $\mathcal{A}$ and the set of inputs $\mathcal{U}$. On a side note, one can use a similar transformation to ensure that $0 \in \mathcal{U}$.

5. COMPUTING WITH SETS OF MATRICES

The computation of $\overline{\mathcal{M}}(t)$ requires representations of sets of matrices and methods for computing sums and products of sets of matrices using these representations. We note that the space of matrices is itself a vector space, where the inner product of two matrices A and B in $\mathbb{R}^{m \times n}$ can be defined as

$$\langle A \mid B \rangle = \sum_{i=1}^{m} \sum_{j=1}^{n} a_{ij}b_{ij} = trace(AB^{\top}),$$

A number of representations can be used to characterize sets of matrices. Here we consider polytopes in vertex representation, zonotopes, and interval products. Algorithms are known for computing the Minkowski sum for each of these representations; see, e.g., [9]. We describe here how to over-approximate the product of two sets using each representation.

5.1 Matrix Polytopes

We define a matrix polytope, designated by the superscript $[p]$, as the convex hull of a set of matrix vertices $V^{(i)}$; that is,

$$\mathcal{A}^{[p]} = \Big\{ \sum_{i=1}^{r_A} \alpha_i V^{(i)} \Big| V^{(i)} \in \mathbb{R}^{n \times n}, \alpha_i \geq 0, \sum_i \alpha_i = 1 \Big\}.$$

The multiplication of two matrix polytopes $\mathcal{A}^{[p]}\mathcal{B}^{[p]}$, where

the matrix vertices of $\mathcal{B}^{[p]}$ are denoted by $W^{(i)}$, can be over-approximated by another matrix polytope $\mathcal{C}^{[p]}$ given by

$$\mathcal{A}^{[p]} \otimes \mathcal{B}^{[p]}$$
$$= \Big\{ \big(\sum_{i=1}^{r_A} \alpha_i V^{(i)}\big)\big(\sum_{j=1}^{r_B} \beta_j W^{(j)}\big) \Big| \alpha_i, \beta_j \geq 0, \sum_{i=1}^{r_A} \alpha_i = 1, \sum_{j=1}^{r_B} \beta_j = 1 \Big\}$$
$$= \Big\{ \sum_{i=1}^{r_A} \sum_{j=1}^{r_B} \alpha_i \beta_j V^{(i)} W^{(j)} \Big| \alpha_i, \beta_j \geq 0, \sum_{i=1}^{r_A} \alpha_i = 1, \sum_{j=1}^{r_B} \beta_j = 1 \Big\}$$
$$\subseteq \Big\{ \sum_{i=1}^{r_C} \gamma_i X^{(i)} \Big| \gamma_i \geq 0, \sum_{i=1}^{r_C} \gamma_i = 1 \Big\} = \mathcal{C}^{[p]}, \tag{5}$$

where

$$\begin{array}{ll} \gamma_1 = \alpha_1 \beta_1, & \gamma_2 = \alpha_1 \beta_2, \quad \ldots \\ X^{(1)} = V^{(1)}W^{(1)}, & X^{(2)} = V^{(1)}W^{(2)}, \quad \ldots \end{array} \tag{6}$$

In order to show that $\mathcal{C}^{[p]}$ is an overapproximation, as indicated in (5), it has to be shown that for each α_i and β_j value it follows that $\gamma_i \geq 0$ and that $\sum_{i=1}^{r_C} \gamma_i = 1$, i.e. for each matrix $A \in \mathcal{A}^{[p]}$, $B \in \mathcal{B}^{[p]}$, it is true that $AB \in \mathcal{C}^{[p]}$. In (6) it can be immediately seen that the first property $\gamma_i \geq 0$ is always fulfilled. The second property is always true since

$$\sum_{i=1}^{r_C} \gamma_i = \sum_{i=1}^{r_A} \sum_{j=1}^{r_B} \alpha_i \beta_j = \underbrace{\sum_{i=1}^{r_A} \alpha_i}_{=1} \underbrace{\sum_{j=1}^{r_B} \beta_j}_{=1} = 1.$$

The reciprocal property that for each $C \in \mathcal{C}^{[p]}$, there exists matrices $A \in \mathcal{A}^{[p]}$, $B \in \mathcal{B}^{[p]}$ such that $AB = C$, is not always true, meaning the $\mathcal{C}^{[p]}$ is an overapproximation.

The disadvantage of matrix polytopes is their combinatorial complexity. For m vertices, the number of vertices of the lth power is m^l. The Minkowski addition of polytopes up to order l can be done by adding each vertex of the solution up to the $(l-1)^{th}$ order with each vertex of the lth power resulting in $\prod_{i=1}^{l} m^i = m^{\sum_{i=1}^{l} i} = m^{0.5(l(l+1))} = \mathcal{O}(m^{l^2})$ matrices.

5.2 Matrix Zonotopes

Zonotopes, which are polytopes defined as a sum of segments, are useful for computing reachable sets for high-dimensional systems; see e.g. [9]. In a vector space $\mathbb{V}$, zonotopes are specified by a center $c \in \mathbb{V}$ and generators $g^{(i)} \in \mathbb{V}$ as

$$Z = \Big\{ x = c + \sum_{i=1}^{e} \beta_i g^{(i)} \Big| \beta_i \in [-1, 1] \Big\}. \tag{7}$$

Zonotopes are always centrally symmetric to the center c and their order is defined by $\rho = \frac{e}{n}$ (e: number of generators, n: dimension). Zonotopes are denoted in a short form as $Z = (c, g^{(1)}, \ldots, g^{(e)})$. One can interpret the above definition as the Minkowski addition of line segments $l_i = \beta_i g^{(i)}$, $\beta_i \in [-1, 1]$ as illustrated step-by-step in Fig. 2, with $\mathbb{V} = \mathbb{R}^2$.

The use of zonotopes, usually in $\mathbb{R}^n$, is justified by the fact that linear transformation and Minkowski addition, which are both of great importance for reachability analysis, can be computed efficiently; see [9, 12]. Given two zonotopes $Z_1 = (c_1, g^{(1)}, \ldots, g^{(e)})$ and $Z_2 = (c_2, h^{(1)}, \ldots, h^{(u)})$,

$$L Z_1 = (Lc_1, Lg^{(1)}, \ldots, Lg^{(e)}), \quad L \in \mathbb{R}^{n \times n}$$
$$Z_1 \oplus Z_2 = (c_1 + c_2, g^{(1)}, \ldots, g^{(e)}, h^{(1)}, \ldots, h^{(u)}). \tag{8}$$

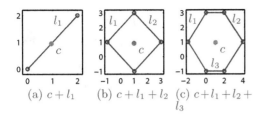

(a) $c + l_1$ (b) $c + l_1 + l_2$ (c) $c + l_1 + l_2 + l_3$

Figure 2: Step-by-step construction of a zonotope from left to right via Minkowski addition of line segments.

The Minkowski sum of two zonotopes is simply an addition of their centers and a concatenation of their generators.

Here we are interested in zonotopes in the space of matrices. We will call them matrix zonotopes and denote them with a superscript $[z]$. A matrix zonotope is defined as

$$\mathcal{A}^{[z]} = \left\{ G^{(0)} + \sum_{i=1}^{\kappa_A} p_i G^{(i)} \,\middle|\, p_i \in [-1,1], G^{(i)} \in \mathbb{R}^{n \times n} \right\},$$

and written in short form as $(G^{(0)}, G^{(1)}, \ldots, G^{(\kappa_A)})$, where the first matrix is referred to as the *matrix center* and the other matrices as *matrix generators*.

The multiplication of two matrix zonotopes $\mathcal{A}^{[z]}\mathcal{B}^{[z]}$, where the matrix generators of $\mathcal{B}^{[z]}$ are denoted by $H^{(i)}$, can be overapproximated by another matrix zonotope $\mathcal{C}^{[z]}$ given by

$$\mathcal{A}^{[z]}\mathcal{B}^{[z]} = \left(G^{(0)} \oplus \bigoplus_{i=1}^{\kappa_A} [-1,1] G^{(i)} \right) \left(H^{(0)} \oplus \bigoplus_{j=1}^{\kappa_B} [-1,1] H^{(j)} \right)$$

$$\subseteq G^{(0)} H^{(0)} \oplus \bigoplus_{\substack{i=0 \\ (i,j) \neq (0,0)}}^{\kappa_A} \bigoplus_{j=0}^{\kappa_B} [-1,1][-1,1] G^{(i)} H^{(j)}$$

$$= I^{(0)} \oplus \bigoplus_{i=1}^{\kappa_C} [-1,1] I^{(i)} = \mathcal{C}^{[z]},$$

(9)

where $I^{(1)} = G^{(0)} H^{(1)}, I^{(2)} = G^{(0)} H^{(2)}, \ldots$

It can be directly seen that $\mathcal{C}^{[z]} \supseteq \mathcal{A}^{[z]}\mathcal{B}^{[z]}$. However, the resulting set of matrices is not exact.

An advantage of matrix zonotopes over matrix polytopes is that they are much more compact in their representation, a property inherited from zonotopes; see [3]. Another advantage is that the Minkowski addition of the two matrix zonotopes $\mathcal{A}^{[z]}$ and $\mathcal{B}^{[z]}$ is computationally cheap since one only has to add their matrix centers and concatenate their matrix generators. That is,

$$\mathcal{A}^{[z]} \oplus \mathcal{B}^{[z]} = (G^{(0)} + H^{(0)}, G^{(1)}, \ldots, G^{(\kappa_A)}, H^{(1)}, \ldots, H^{(\kappa_B)}).$$

The number of matrix generators for the lth power is $(\kappa + 1)^l - 1$. Thus, the number of matrix generators of $\overline{\mathcal{M}}(t)$ up to order l is $\sum_{k=0}^{l} (\kappa+1)^k - 1 = (1 - (\kappa+1)^{l+1})/(-\kappa) - (l+1) = \mathcal{O}(\kappa^l)$. The computational complexity can be drastically reduced by applying order reduction techniques developed for zonotopes; see [9,17]. Equally powerful reduction techniques for polytopes are not known to the authors.

5.3 Interval Matrices

An interval matrix is a special case of a matrix zonotope specified by intervals for each element; that is

$$\mathcal{A}^{[i]} = [\underline{A}, \overline{A}], \quad \forall i,j : \underline{A}_{ij} \leq \overline{A}_{ij}, \quad \underline{A}, \overline{A} \in \mathbb{R}^{n \times n}.$$

The matrix $\underline{A}$ is referred to as the *lower bound* and $\overline{A}$ as the *upper bound* of $\mathcal{A}$.

Interval matrix multiplications are performed using interval arithmetic [16]. The addition and multiplication rules for two real-valued intervals $a^I = [\underline{a}, \overline{a}]$ and $b^I = [\underline{b}, \overline{b}]$ are given by

$$a^I + b^I = [\underline{a} + \underline{b}, \overline{a} + \overline{b}],$$
$$a^I \cdot b^I = [\min(\underline{a}\,\underline{b}, \underline{a}\,\overline{b}, \overline{a}\,\underline{b}, \overline{a}\,\overline{b}), \max(\underline{a}\,\underline{b}, \underline{a}\,\overline{b}, \overline{a}\,\underline{b}, \overline{a}\,\overline{b})].$$

(10)

Using the rules in (10), the multiplication of an interval matrix $\mathcal{A}^{[i]}$ with another interval matrix $\mathcal{B}^{[i]}$ is computed elementwise: $\mathcal{C}^{[i]}_{ij} = \sum_{k=1}^{n} \mathcal{A}^{[i]}_{ik} \mathcal{B}^{[i]}_{kj}$. Clearly, the result is an overapproximation. In contrast to matrix polytopes and matrix zonotopes, where the number of vertices and generators grow after addition and multiplication, the representation does not grow for interval matrices.

6. COMPUTATION OF THE REACHABLE SET

This section shows how the representations defined in the previous section can be use to compute reachable set approximations. In order to compute $\overline{\mathcal{M}}(r) R_k$ in Algorithm 1, the multiplication of a matrix zonotope/polytope or an interval matrix with a zonotope has to be computed. For a small enough time t (as it is typically the case for reachability analysis), the terms $\overline{\mathcal{M}}_i(t)$ for large i values, which are referred to as higher order terms, contribute less to the computation of $\overline{\mathcal{M}}(t)$. Thus, one should use sophisticated computations for the first terms and switch to coarser and more efficient computations for higher order terms. For instance, the first two terms of a matrix zonotope could be computed using matrix zonotope computations, while the other terms are computed via interval matrix overapproximations. Denoting the set of transition matrices using matrix zonotopes by $\overline{\mathcal{M}}^{[z]}(t)$ and the set using interval matrices by $\overline{\mathcal{M}}^{[i]}(t)$, the reachable set is computed as $R_k = \overline{\mathcal{M}}^{[z]}(r) R_{k-1} \oplus \overline{\mathcal{M}}^{[i]}(r) R_{k-1} \oplus \mathcal{P}_0$. This technique is preferred over the transformation of the interval matrix to a matrix zonotope, making it possible to obtain $\overline{\mathcal{M}}(t)$, since the transformation results in too many generators, especially in high dimensional spaces.

In order to tightly overapproximate an interval matrix multiplication, the interval matrix $\mathcal{A}^{[i]}$ is split into a real valued part $A^{[n]} \in \mathbb{R}^{n \times n}$ and a symmetric interval matrix $\mathcal{S} = [-S, S]$: $\mathcal{A}^{[i]} Z \subseteq A^{[n]} Z \oplus \mathcal{S} Z$. The following proposition shows how to compute the symmetric interval matrix part in the zonotope multiplication (8).

PROPOSITION 3 (INTERVAL MATRIX MULTIPLICATION). *The multiplication of a symmetric interval matrix $\mathcal{S} = [-S, S]$ with a zonotope $Z = (c, g^{(1)}, \ldots, g^{(e)})$ can be overapproximated by a hyperrectangle (in zonotope notation) with center* 0: $\mathcal{S} Z = (0, v^{(1)}, \ldots, v^{(n)})$, *where*

$$v_j^{(i)} = \begin{cases} 0, & i \neq j \\ \overline{S}_j (|c| + \sum_{k=1}^{e} |g^{(k)}|), & i = j, \end{cases}$$

and the subscript j of $v_j^{(i)}$ denotes the jth element of $v^{(i)}$ and S_j denotes the jth row of S.

PROOF. The multiplication of a symmetric interval matrix $\mathcal{S}$ with a zonotope Z is overapproximated by $\mathcal{S} Z \subseteq \mathcal{S}\,\texttt{box}(Z)$ and $\texttt{box}$ returns an enclosing axis-aligned box which is computed as proposed in [9] as

$$\texttt{box}(Z) = [c - \Delta g, c + \Delta g], \quad \Delta g = \sum_{i=1}^{e} |g^{(i)}|.$$

It remains to compute $\mathcal{S}[c - \Delta g, c + \Delta g]$, which returns a symmetric interval vector due to the symmetry of $\mathcal{S}$. The upper bound is obtained by $\mathcal{S}\max(|c - \Delta g|, |c + \Delta g|)$. Since $\max(|c - \Delta g|, |c + \Delta g|) = |c| + |\Delta g|$, it follows that

$$\mathcal{S}\,\texttt{box}(Z) = [-S(|c| + |\Delta g|), S(|c| + |\Delta g|)].$$

Rewriting this result in zonotope notation with generators $v^{(i)}$ yields the result of the proposition. $\square$

We now consider the application of a set of linear transformations, represented as a matrix zonotope, to a set of vectors represented as a zonotope.

PROPOSITION 4 (MATRIX ZONOTOPE MULTIPLICATION). *The product of a matrix zonotope $\mathcal{L}^{[z]} = \{L^{(0)} + \sum_{i=1}^{\kappa} p_i L^{(i)} \mid p_i \in [-1, 1]\}$ and a zonotope $Z = (c, g^{(1)}, \ldots, g^{(e)})$ is overapproximated by*

$$
\begin{aligned}
\mathcal{L}^{[z]} Z &= \bigcup_{p_i \in [-1,1]} \left(L^{(0)} Z \oplus \bigoplus_{i=1}^{\kappa} p_i L^{(i)} Z \right) \\
&\subseteq (L^{(0)} c, L^{(0)} g^{(1)}, \ldots, L^{(0)} g^{(e)}, \\
&\quad L^{(1)} c, L^{(1)} g^{(1)}, \ldots, L^{(1)} g^{(e)}, \ldots, \\
&\quad L^{(\kappa)} c, L^{(\kappa)} g^{(1)}, \ldots, L^{(\kappa)} g^{(e)}).
\end{aligned}
$$

PROOF. The result follows directly from the addition and multiplication rule of zonotopes; see (8). $\square$

The multiplication of a matrix polytope with matrix vertices $V^{(1)}, \ldots, V^{(r)}$ and a zonotopes can be performed as $\mathcal{L}^{[p]} Z = \texttt{CH}(V^{(1)} Z, V^{(2)} Z, \ldots, V^{(r)} Z)$. The result is no longer a zonotope in general, so that it has to be overapproximated by a zonotope. The overapproximation of polytopes by zonotopes is computationally expensive (see, e.g., [3]), so that matrix polytopes should be overapproximated by matrix zonotopes beforehand, making it possible to apply Prop. 4.

In order to quickly estimate the size of the error in the state transition matrix overapproximation, it is often helpful to compute with norms instead of applying the previously introduced computational techniques. In order to obtain a tight norm bound, the matrix set $\mathcal{A}$ is overapproximated by an interval matrix $\mathcal{A}^{[i]}$ which is split into a nominal and a symmetric[3] part: $\mathcal{A}^{[i]} = A^{[n]} + [-S, S]$. The norm of the distance of the set of state transition matrices to the exponential matrix of the nominal matrix is computed for

---

[3] Symmetric refers here to the set and not to the matrices it contains.

$\||A^{[n]}| + S\| < \frac{2}{t}$ as

$$\|\overline{\mathcal{M}}(t) - e^{A^{[n]}t}\|$$

$$\leq \frac{\|A^{[n]}\|\,\|S\|\frac{t^2}{2}}{\|A^{[n]}\| \cdot \||A^{[n]}| + S\|\frac{t^2}{4} - (\|A^{[n]}\| + \||A^{[n]}| + S\|)\frac{t}{2} + 1}$$

$$+ \frac{\|S\| t}{1 - \||A^{[n]}| + S\|\frac{t}{2}}.$$

The proof is neglected due to space limitations.

7. NUMERICAL EXAMPLES

In this section we illustrate the proposed methods for computing reachable sets for uncertain time-varying linear systems. The first example demonstrates the difference in accuracy when computing with matrix zonotopes or interval matrices. The scalability of the approach is also demonstrated. The second example shows the usefulness of the approach for computing reachable sets for a hybrid system with nonlinear continuous dynamics.

7.1 Five Dimensional Example

We consider a standard example from the literature [9, 22]. In [9], there are no uncertain system matrices, in [22] the system matrices are bounded by an interval matrix. Here, the system matrices are bounded by a matrix zonotope, whose enclosing interval matrix is exactly the one in [22]: $\dot{x} = A(t)x + u(t)$, $A(t) \in \mathcal{A}^{[z]} = (G^{(0)}, G^{(1)})$, $u(t) \in \mathcal{U} = [-\overline{u}, \overline{u}]$, where $\overline{u} = \begin{bmatrix} 0.1 & 0.1 & 0.1 & 0.1 & 0.1 \end{bmatrix}^T$ and

$$G^{(0)} = \begin{bmatrix} -1 & -4 & 0 & 0 & 0 \\ 4 & -1 & 0 & 0 & 0 \\ 0 & 0 & -3 & 1 & 0 \\ 0 & 0 & -1 & -3 & 0 \\ 0 & 0 & 0 & 0 & -2 \end{bmatrix}, \; G^{(1)} = \begin{bmatrix} 0.1 & 0.1 & 0 & 0 & 0 \\ 0.1 & 0.1 & 0 & 0 & 0 \\ 0 & 0 & 0.1 & 0.1 & 0 \\ 0 & 0 & 0.1 & 0.1 & 0 \\ 0 & 0 & 0 & 0 & 0.1 \end{bmatrix}.$$

The reachable set is computed for a step size of $r = 0.05$ and a time horizon of $t_f = 5$. The number of transition matrix terms is chosen as $\eta = 4$ (the first two using matrix zonotopes, the other 2 using interval matrices), and the order of the zonotopes is limited to $\rho = 20$ using the reduction technique in [9]. The computation time is 0.16 s in MATLAB on an i7 Processor and 6GB memory. Plots of selected projections are shown in Fig. 3. It can be seen that the computation is tighter when using matrix zonotopes instead of a tightly enclosing interval matrix. It has not been considered to separate the uncoupled states of this specific example due to the efficiency of the algorithm.

The scalability of the algorithm is shown by computing reachable sets for several linear systems generated with random parameters. Computation times for system matrices bounded by interval matrices and matrix zonotopes are shown in Table 1. Even if the examples contain uncoupled subsystems, the system is computed as if all states are coupled.

7.2 Rollover Verification of a Truck

We consider the problem of determining if a truck will roll over during a set of maneuvers, where the truck has a yaw controller to improve cornering performance. The truck dynamics is velocity dependent, so the yaw controller is a gain scheduling controller that switches among several controllers depending on the velocity. The switching between different controllers is instantaneous rather than cross-fading [20]. Thus, the overall system becomes a hybrid system which is modeled as a hybrid automaton in [3].

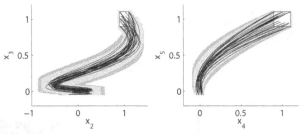

(a) Projection onto x_1, x_2.   (b) Projection onto x_3, x_4.

Figure 3: Reachable set of the five-dimensional example. The dark gray region shows the reachable set when computing with the interval matrix $\mathcal{A}^{[i]}$, while the light gray region shows the result when computing with the original matrix zonotope $\mathcal{A}^{[z]}$. Black lines show exemplary trajectories and the white region is the initial set.

Table 1: Computation times.

Dimension n	5	10	20	50	100
Interval matrix					
CPU-time [s]	0.10	0.12	0.33	0.82	3.64
Matrix zonotope: Nr of generator matrices $\kappa = 1$					
CPU-time [s]	0.13	0.17	0.60	2.65	8.72
Matrix zonotope: Nr of generator matrices $\kappa = 2$					
CPU-time [s]	0.18	0.30	1.13	4.73	18.77
Matrix zonotope: Nr of generator matrices $\kappa = 4$					
CPU-time [s]	0.34	0.68	2.60	18.07	98.70

We consider two approaches to compute the reachable set. First, the standard approach to hybrid system reachability is applied, where the reachable set computation is continued across discrete transitions using intersections with guard sets. These intersections can introduce significant overapproximation errors; see [3, 11]. Second, as an alternative to the standard approach, the reachable set is computed under a larger set of parameter uncertainties when intersecting several invariant sets. The enlarged set of parameter uncertainties is the union of uncertainties within the invariants the reachable set is intersecting. This makes it possible to compute the reachable set without any intersection operation. This approach is referred to as *continuization* and is beneficial when the intersection operation is dominant in the enlargement of the reachable set, while the effect on computing with a larger set of parameter uncertainties is small as in gain scheduling. This approach is applicable only if the hybrid automaton has no jumps.

The truck dynamics is described by the following continuous state variables (see Fig. 4): the side-slip angle at center of mass β, yaw rate $\dot{\Psi}$, sprung mass roll angle Φ, sprung mass roll angle rate $\dot{\Phi}$, unsprung mass roll angle of the front axle $\Phi_{t,f}$ and the rear axle $\Phi_{t,r}$, and velocity v. The input to the system is the steering angle δ and the longitudinal

acceleration a_x. The dynamic equations from [8] are

$$mv(\dot{\beta} + \dot{\Psi}) - m_S h \ddot{\Phi} = Y_\beta \beta + Y_{\dot{\Psi}}(v) \dot{\Psi} + Y_\delta \delta$$

$$-I_{xz} \ddot{\Phi} + I_{zz} \ddot{\Psi} = N_\beta \beta + N_{\dot{\Psi}}(v) \dot{\Psi} + N_\delta \delta$$

$$(I_{xx} + m_S h^2) \ddot{\Phi} - I_{xz} \ddot{\Psi} = m_S g h \Phi + m_S v h (\dot{\beta} + \dot{\Psi}) - k_f(\Phi - \Phi_{t,f})$$

$$- b_f(\dot{\Phi} - \dot{\Phi}_{t,f}) - k_r(\Phi - \Phi_{t,r}) - b_r(\dot{\Phi} - \dot{\Phi}_{t,r})$$

$$-r(Y_{\beta,f}\beta + Y_{\dot{\Psi},f}\dot{\Psi} + Y_\delta \delta) = m_{u,f}v(r - h_{u,f})(\dot{\beta} + \dot{\Psi}) + m_{u,f}gh_{u,f}\Phi_{t,f}$$

$$- k_{t,f}\Phi_{t,f} + k_f(\Phi - \Phi_{t,f}) + b_f(\dot{\Phi} - \dot{\Phi}_{t,f})$$

$$-r(Y_{\beta,r}\beta + Y_{\dot{\Psi},r}\dot{\Psi}) = m_{u,r}v(r - h_{u,r})(\dot{\beta} + \dot{\Psi}) - m_{u,r}gh_{u,r}\Phi_{t,r}$$

$$- k_{t,r}\Phi_{t,r} + k_r(\Phi - \Phi_{t,r}) + b_r(\dot{\Phi} - \dot{\Phi}_{t,r})$$

$$\dot{v} = a_x.$$

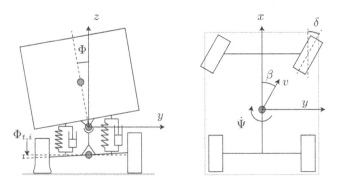

Figure 4: Truck model.

The parameters are chosen as in [8]. In order to obtain a controlled system without steady state error, a PI controller is designed to control the yaw rate $\dot{\Psi}$. Different controllers are active in the intervals $[10, 20 + \Delta v]$, $[20, 30 + \Delta v]$, and $[30, \infty[$ m/s, where $\Delta v > 0$ models the velocity measurement uncertainty. Below 10 m/s, no controller is active. Note that due to the velocity measurement uncertainty, the dynamics is switched in the intervals $[10, 10 + \Delta v]$, $[20, 20 + \Delta v]$, $[30, 30 + \Delta v]$.

The control error is denoted by $e = \dot{\Psi}_d - \dot{\Psi}$, where $\dot{\Psi}_d$ is the desired yaw rate. The desired yaw rate is computed by the steady state solution of $\dot{\Psi}$ when a desired steering angle δ_d or desired lateral acceleration $a_{y,d}$ is used, where $\dot{\Psi}_d = a_{y,d}/v$ in the latter case. The PI controller is written as $\delta = k_1 e + k_2 \int e(t)\,dt$, where k_1 and k_2 are the gains for the proportional and integral part, respectively, which are listed in Table 2.

Table 2: Yaw controller gains.

$v \in$	$[10, 20]$ m/s	$[20, 30]$ m/s	$[30, \infty[$ m/s
controller	$k_1 = 0.4$	$k_1 = 0.5$	$k_1 = 0.6$
gains	$k_2 = 1.5$	$k_2 = 2$	$k_2 = 2.5$

The set of possible system matrices is modeled as a matrix zonotope $\mathcal{A}^{[z]}$ and the set of inputs $\mathcal{U}$ as a zonotope. In order to obtain a tight overapproximation of the reachable set, the sets of possible matrices $\mathcal{A}^{[z]}([t_k, t_{k+1}])$ and inputs $\mathcal{U}([t_k, t_{k+1}])$ are updated for each time interval $[t_k, t_{k+1}]$. After introducing the state vector $x = [\beta, \dot{\Psi}, \Phi, \dot{\Phi}, \Phi_{t,f}, \Phi_{t,r}, v,$ $\int e(t)\,dt]^T$ and grouping the terms of the controlled truck dynamics, one can write them in the form

$$\dot{x} = (p_1 Q^{(1)} + p_2 Q^{(2)} + p_3 Q^{(3)} + p_4 Q^{(4)})x \\ + (p_1 R^{(1)} + p_2 R^{(2)})a_{y,d}, \quad (11)$$

where for $t \in [t_k, t_{k+1}]$

$$p_1 \in \left[\frac{1}{(\overline{v})^2}, \frac{1}{(\underline{v})^2}\right], \quad p_2 \in \left[\frac{1}{\overline{v}}, \frac{1}{\underline{v}}\right], \quad p_3 = 1 \quad p_4 \in [\underline{v}, \overline{v}],$$

and $\underline{v}, \overline{v}$ are the lower and upper bound of the velocity for $t \in [t_k, t_{k+1}]$. The formulation in (11) gets rid of the nonlinearities and makes it possible to obtain the generators $G^{(i)}$ of the matrix zonotope $\mathcal{A}^{[z]}$ as

$$G^{(0)} = \sum_{i=1}^{4} \texttt{center}(p_i)Q^{(i)}, \text{ for } i = 1..4 : G^{(i)} = \texttt{rad}(p_i)Q^{(i)}$$

and analogously for $\mathcal{B}^{[z]}$, where the operators $\texttt{center}()$ and $\texttt{rad}()$ return the center and radius of an interval. The set of inputs is obtained as

$$\mathcal{U}([t_k, t_{k+1}]) = \mathcal{B}^{[z]}([t_k, t_{k+1})) \begin{bmatrix} [\underline{a}_{y,d}, \overline{a}_{y,d}] & [\underline{a}_x, \overline{a}_x] \end{bmatrix}^T.$$

In order to compute the reachable set under the changing parameter intervals, the computation for $\mathcal{H}_0, \mathcal{P}_0$ in Alg. 1 have to be repeated for each time interval instead of only once as presented in Alg. 1.

The reachable set is computed for a deceleration maneuver with $a_x = 0.7g$, where g is the gravity constant. Due to limited tire friction, the truck may still perform steering maneuvers that are uncertain within the corresponding set of lateral accelerations $[-a_{y,d}, a_{y,d}]$ and $a_{y,d} = 0.4g$. The reachable set is computed until it has left the half-space of velocities above 10 m/s. Below this velocity, no controller is active anymore. Parameters of the reachable set computation are specified in Table 3. The set of initial states is $x(0) \in [0, 0.04] \times [0, 0.2] \times [-0.1, 0.1] \times [-0.1, 0.1] \times [-0.01, 0.01] \times [-0.01, 0.01] \times [32.75, 33.25] \times [-0.1, 0.1]$.

Table 3: Parameters of the reachable set computation of the truck.

time step size	$r = 0.01$
maximum zonotope order	$\rho = 60$
Taylor series order	$\eta = 4$
velocity measurement uncertainty	$\Delta v = [0, 0.5]$

The reachable set approximations for the deceleration maneuver using the continuization and the hybrid approach are shown in Fig. 5. It can be observed that an enlargement takes place for x_8 after passing the 30, 20, and 10 m/s borders when using the hybrid approach. The reachable set of x_1 and x_2 is almost the same, while the one projected onto $x_3 - x_6$ is much tighter for the continuization approach. The computational time for the hybrid approach is 85 s and 38 s for the continuization approach. The computations have been performed on an Intel i7 Processor with 6GB memory in MATLAB. The truck starts to rollover when the dynamic forces on the rear inner wheel (which is the critical wheel) overcompensate the force due to gravity, which is the case for $x_6 > 0.55$. Thus, the continuization approach verifies the safety of the maneuvers, while the classical approach fails.

The intersection with guard sets has been computed as presented in [3], where zonotopes are transformed to a half-space representation in order to perform the intersections which are later enclosed by a zonotope to continue the computation with zonotopes. A good compromise between accuracy and efficiency that works surprisingly well for the truck example is to overapproximate the zonotopes by boxes to accelerate the halfspace conversion. The conversion back to zonotopes is done by enclosing the intersected set with two zonotopes in order to increase the accuracy (see [3]). One zonotope is obtained by a principal component technique presented in [26] and the other one by an axis-aligned box.

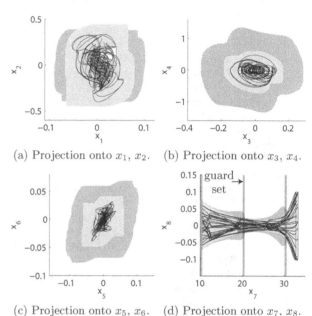

(a) Projection onto x_1, x_2. (b) Projection onto x_3, x_4.

(c) Projection onto x_5, x_6. (d) Projection onto x_7, x_8.

Figure 5: Reachable set of the yaw controlled truck. The light gray region shows the reachable set for the continuization approach. The dark gray region shows the reachable set for the classical hybrid approach. Black lines show exemplary trajectories of the system.

8. CONCLUSIONS

Previous methods for the reachability analysis of uncertain linear time-invariant systems have been extended to uncertain linear time-varying systems. The presented approach can cope with uncertain system matrices, as well as with arbitrary input trajectories whose values are bounded. In addition, the proposed algorithms scale well with the number of continuous state variables. It has also been demonstrated that the approach can be used for reachability analysis of nonlinear and hybrid systems. The continuization approach is promising for hybrid systems with similar continuous dynamics in adjacent locations. Extensions of the approach to better handle nonlinear dynamics are currently being investigated.

Acknowledgments

This research was supported in part by U.S. National Science Foundation grant number CCF-0926181 and the U.S. Air

Force Office of Scientific Research grant number FA9550-06-1-0312.

9. REFERENCES

[1] M. Althoff. *Reachability Analysis and its Application to the Safety Assessment of Autonomous Cars*. PhD thesis, TU München, 2010.

[2] M. Althoff, O. Stursberg, and M. Buss. Reachability analysis of nonlinear systems with uncertain parameters using conservative linearization. In *Proc. of the 47th IEEE Conference on Decision and Control*, pages 4042–4048, 2008.

[3] M. Althoff, O. Stursberg, and M. Buss. Computing reachable sets of hybrid systems using a combination of zonotopes and polytopes. *Nonlinear Analysis: Hybrid Systems*, 4:233–249, 2010.

[4] E. Asarin, T. Dang, and O. Maler. d/dt: A verification tool for hybrid systems. In *Proc. of the Conference on Decision and Control*, pages 2893–2898, 2001.

[5] A. Chutinan and B. H. Krogh. Computational techniques for hybrid system verification. In *IEEE Transactions on Automatic Control*, volume 48, pages 64–75, 2003.

[6] T. Dang. *Vérification et synthèse des systèmes hybrides*. PhD thesis, Institut National Polytechnique de Grenoble, 2000.

[7] Goran Frehse. PHAVer: Algorithmic verification of hybrid systems past HyTech. In *Hybrid Systems: Computation and Control*, LNCS 3413, pages 258–273. Springer, 2005.

[8] P. Gaspar, I. Szaszi, and J. Bokor. The design of a combined control structure to prevent the rollover of heavy vehicles. *European Journal of Control*, 10:1–15, 2004.

[9] A. Girard. Reachability of uncertain linear systems using zonotopes. In *Hybrid Systems: Computation and Control*, LNCS 3414, pages 291–305. Springer, 2005.

[10] A. Girard and C. Le Guernic. Efficient reachability analysis for linear systems using support functions. In *Proc. of the 17th IFAC World Congress*, pages 8966–8971, 2008.

[11] A. Girard and C. Le Guernic. Zonotope/hyperplane intersection for hybrid systems reachability analysis. In *Proc. of Hybrid Systems: Computation and Control*, LNCS 4981, pages 215–228. Springer, 2008.

[12] A. Girard, C. Le Guernic, and O. Maler. Efficient computation of reachable sets of linear time-invariant systems with inputs. In *Hybrid Systems: Computation and Control*, LNCS 3927, pages 257–271. Springer, 2006.

[13] T. Henzinger. *The theory of hybrid automata*, volume 170 of *NATO ASI Series F: Computer and Systems Sciences*, pages 265–292. Springer, 2000.

[14] T. A. Henzinger, P.-H. Ho, and H. Wong-Toi. Algorithmic analysis of nonlinear hybrid systems. *IEEE Transactions on Automatic Control*, 43:540–554, 1998.

[15] T. A. Henzinger, B. Horowitz, R. Majumdar, and H. Wong-Toi. Beyond HyTech: Hybrid systems analysis using interval numerical methods. In *Hybrid Systems: Computation and Control*, LNCS 1790, pages 130–144. Springer, 2000.

[16] L. Jaulin, M. Kieffer, and O. Didrit. *Applied Interval Analysis*. Springer, 2006.

[17] W. Kühn. Rigorously computed orbits of dynamical systems without the wrapping effect. *Computing*, 61:47–67, 1998.

[18] A. B. Kurzhanskiy and P. Varaiya. Ellipsoidal techniques for reachability analysis of discrete-time linear systems. *IEEE Transactions on Automatic Control*, 52:26–38, 2007.

[19] G. Lafferriere, G. J. Pappas, and S. Yovine. Symbolic reachability computation for families of linear vector fields. *Symbolic Computation*, 32:231–253, 2001.

[20] D. J. Leith and W. E. Leithead. Survey of gain-scheduling analysis and design. *International Journal of Control*, 73:1001–1025, 2000.

[21] N. S. Nedialkov and K. R. Jackson. *Perspectives on Enclosure Methods*, chapter A New Perspective on the Wrapping Effect in Interval Methods for Initial Value Problems for Ordinary Differential Equations, pages 219–264. Springer-Verlag, 2001.

[22] N. Ramdani, N. Meslem, and Y. Candau. Reachability analysis of uncertain nonlinear systems using guaranteed set integration. In *Proc. of the 17th IFAC World Congress*, pages 8972–8977, 2008.

[23] N. Ramdani, N. Meslem, and Y. Candau. Reachability of uncertain nonlinear systems using a nonlinear hybridization. In *Hybrid Systems: Computation and Control*, LNCS 4981, pages 415–428. Springer, 2008.

[24] N. Ramdani, N. Meslem, and Y. Candau. Computing reachable sets for uncertain nonlinear monotone systems. *Nonlinear Analysis : Hybrid Systems*, 4:263–278, 2010.

[25] W. J. Rugh. *Linear System Theory*. Prentice Hall, 1996.

[26] O. Stursberg and B. H. Krogh. Efficient representation and computation of reachable sets for hybrid systems. In *Hybrid Systems: Computation and Control*, LNCS 2623, pages 482–497. Springer, 2003.

[27] C. Tomlin, I. Mitchell, A. Bayen, and M. Oishi. Computational techniques for the verification and control of hybrid systems. In *Proceedings of the IEEE*, volume 91, pages 986–1001, 2003.

Scalable Calculation of Reach Sets and Tubes for Nonlinear Systems with Terminal Integrators[*]

A Mixed Implicit Explicit Formulation

Ian M. Mitchell
Department of Computer Science
University of British Columbia
2366 Main Mall, Vancouver, BC, Canada
mitchell@cs.ubc.ca

ABSTRACT

The solution of a particular Hamilton-Jacobi (HJ) partial differential equation (PDE) provides an implicit representation of reach sets and tubes for continuous systems with nonlinear dynamics and can treat inputs in either worst-case or best-case fashion; however, it can rarely be determined analytically and its numerical approximation typically requires computational resources that grow exponentially with the state space dimension. In this paper we describe a new formulation—also based on HJ PDEs—for reach sets and tubes of systems where some states are *terminal integrators*: states whose evolution can be written as an integration over time of the other states. The key contribution of this new mixed implicit explicit (MIE) scheme is that its computational cost is linear in the number of terminal integrators, although still exponential in the dimension of the rest of the state space. Application of the new scheme to four examples of varying dimension provides empirical evidence of its considerable improvement in computational speed.

Categories and Subject Descriptors

J.2 [**Physical Sciences & Engineering**]: Engineering; I.6.4 [**Simulation & Modeling**]: Model Validation & Analysis

General Terms

Verification, Algorithms

Keywords

nonlinear systems, continuous reachability, Hamilton-Jacobi equations, optimal control

1. INTRODUCTION

Reachability has proved a powerful tool in the verification of discrete systems, but it is still impractical for most hybrid and continu-

---

[*]Research supported by a Discovery Grant from the National Science and Engineering Research Council of Canada

ous systems because there is as yet no scalable method of computing reachability for nonlinear continuous systems in high dimensions. It was shown in previous work [8, 10] that an implicit representation of the backwards reach tube of nonlinear systems with adversarial inputs is the viscosity solution of a particular Hamilton-Jacobi (HJ) partial differential equation (PDE). While this formulation permits the representation of nonconvex sets and can treat inputs in either a best-case (for control) or worst-case (for robustness) manner, its practical implementation so far requires computational resources exponential in the dimension of the state space.

The contribution of this paper is a new formulation based on HJ PDEs that leads to a significant reduction in computational complexity for systems with dynamics of a certain form. We divide the state space into two sets of states:

- *Coupled* states, which have the same weak restrictions on their dynamics and are treated in the same manner as in the traditional implicit formulation, but consequently have computational cost that scales exponentially with dimension.
- *Terminal integrator* states, whose dynamics can only depend on the coupled states but which can be treated in a more efficient manner that requires computational resources linear in dimension. The key restriction on terminal integrator states is that no other states may depend on their value(s).

Terminal integrator states often appear in dynamic models of mechanical systems, where position states are simply integrators for velocity states. We call the new scheme a *mixed implicit explicit* (MIE) formulation because the boundary of the reach set or tube is represented implicitly in the coupled dimensions but explicitly (in the form of intervals) in the terminal integrator dimensions.

The remainder of the paper is organized as follows. Section 2 discusses the continuous reachability problem that we seek to solve, the traditional HJ PDE based implicit formulation upon which we build, and related work. Section 3 provides the new formulation for the case of a single terminal integrator state, proves that it can be used to determine the backwards maximal reach set, and outlines how it can be modified to determine other reach sets and tubes. Section 4 demonstrates the application of the new scheme on two examples. Section 5 discusses a generalization of the terminal integrator's dynamics which has less theoretical support, but which is demonstrated on a third example. Section 6 extends the simpler dynamics to the vector case, and demonstrates it on a fourth example. Finally, section 7 compares the implicit and MIE formulations.

2. CONTINUOUS REACHABILITY

This paper concerns itself with the calculation of backwards reach sets and tubes for deterministic nonlinear continuous systems with dynamics given by the ordinary differential equation (ODE)

$$\dot{z} = f(z, u) \qquad (1)$$

for state variables $z \in \mathbb{R}^{d_z}$, input $u \in \mathcal{U}$ where $\mathcal{U}$ is compact and convex, and nonlinear dynamics $f : \mathbb{R}^{d_z} \times \mathcal{U} \to \mathbb{R}^{d_z}$ which are bounded and Lipschitz continuous in z for fixed u. Backwards reachability seeks to determine the set of states which are initial conditions to trajectories of (1) that intersect a specified target set $\mathcal{R}_0$. The reach set is the set of initial states which give rise to trajectories whose endpoints at a specified time lie within the target set, while the reach tube is the set of initial states which give rise to trajectories which arrive at or pass through the target set during a specified time interval; typically the interval is from time zero to a specified time.

Two methods of treating inputs to the system dynamics are outlined in [9]: Maximal reachability uses the inputs to make the reach set or tube as large as possible, while minimal reachability uses the inputs to make the reach set or tube as small as possible. For example, we can formalize the backward maximal reach set as

$$\mathcal{R}(\mathcal{R}_0, t) = \{z_s \mid \exists u(\cdot), \exists z_f \in \mathcal{R}_0, z(t) = z_f\}, \qquad (2)$$

and the backward minimal reach tube as

$$\mathcal{R}(\mathcal{R}_0, t) = \{z_s \mid \forall u(\cdot), \exists z_f \in \mathcal{R}_0, \exists \sigma \in [t, 0], z(\sigma) = z_f\}, \qquad (3)$$

where $t < 0$ and trajectory $z(\cdot)$ solves (1) with initial conditions $z(t) = z_s$ and input signal $u(\cdot)$. It is also possible to formulate versions with adversarial inputs, in which some inputs seek to maximize the size of the reach set or tube, while others seek to minimize it. Formalization of these versions is complicated by the need to consider the knowledge available to the competing players when choosing their inputs; consequently, we do not further pursue the adversarial case here.

2.1 Related Work

Computation of reachable sets for deterministic nonlinear continuous systems remains a challenge despite several decades of research. The methods discussed in this paper are based on the Hamilton-Jacobi(-Bellman)(-Isaacs) equation [8, 7, 10], as explained in the next section. These schemes, as well as the very closely related schemes arising from viability theory, such as [3, 13], share several important characteristics. On the positive side, they are designed to deal directly with nonlinear dynamics and are able to automatically determine optimal inputs in both the best-case and the worst-case senses. On the negative side, their implementation typically requires computational resources exponential in the number of state space dimensions.

There have been a number of attempts to reduce this computational complexity. For systems in which the dynamics decouple, the high dimensional PDE can be broken naturally into multiple, lower-dimensional PDEs. Complete decoupling of the dynamics occurs rarely, so a projection-based scheme whereby coupling terms could be treated as disturbances was proposed in [12] (see section 6 for a discussion of how that approach can be applied to the class of dynamics studied in this paper), and a time-based decoupling where the system dynamics could be separated into fast and slow components was proposed in [6]. The exponential cost of HJ PDE based schemes arises from the requirement in traditional PDE solvers to grid the state space, so [4] proposes to reduce that cost by using a neural network representation of the solution instead.

There is also a vast body of research on other schemes for approximating reachability and/or solving related verification problems in continuous and hybrid systems; so vast that we cannot survey it here. Previous proceedings of the *Hybrid Systems: Computation and Control* workshop provide many suitable entries into the literature for the interested reader.

2.2 The Traditional Implicit Formulation

A method for determining an implicit representation of the backward reach tube of nonlinear dynamic systems with adversarial inputs was proposed in [8]. An implicit representation of the reach set or tube takes the form of a function $\psi : \mathbb{R} \times \mathbb{R}^{d_z} \to \mathbb{R}$ whose zero sublevel set is the reach set or tube:

$$\mathcal{R}(\mathcal{R}_0, t) = \{z \mid \psi(t, z) \leq 0\}. \qquad (4)$$

In [7] it was shown for a minimal reach tube (3) that ψ is the solution of the terminal value HJ PDE

$$D_t \psi(t, z) + H(t, z, D_z \psi(t, z)) = 0 \qquad (5)$$

with Hamiltonian

$$H_{\text{tube}}(t, z, p) = \min\left[0, \max_{u \in \mathcal{U}}(p \cdot f(z, u))\right], \qquad (6)$$

and terminal conditions

$$\psi(0, z) = \psi_0(z) \qquad (7)$$

such that $\mathcal{R}_0 = \{z \mid \psi_0(z) \leq 0\}$.

This result was extended to reach tubes for systems with two adversarial inputs (and therefore maximal reach tubes as well) in [10]. It is straightforward to adapt these PDEs to compute reach sets with various treatments of the inputs. For example, an implicit representation of the maximal reach set (2) is given by the solution of the HJ PDE (5) with Hamiltonian

$$H_{\text{set}}(t, z, p) = \min_{u \in \mathcal{U}}(p \cdot f(z, u)).$$

and terminal conditions (7).

For most systems of interest, (5) cannot be solved analytically. Typical approximation schemes involve creating a grid over the state space $\mathbb{R}^{d_z}$, and hence are only practical for $d_z \leq 4$.

3. THE MIXED IMPLICIT EXPLICIT FORMULATION

In this section we outline a new formulation for computing reachability that involves solving PDEs of dimension lower than (5). We derive the results for the backward maximal reach set (2), although the modifications to compute the other sets and tubes are straightforward.

Consider a system where the state can be decomposed as $z = (y, x)$ and the dynamics (1) as

$$\dot{y} = f(y, u) \qquad (8)$$
$$\dot{x} = b(y) \qquad (9)$$

where $y \in \mathbb{R}^{d_{cp}}$ are the *coupled* state variables, $x \in \mathbb{R}$ is the scalar *terminal integrator* state variable, and $u \in \mathcal{U}$ is the input as above. We adopt this particular choice of variables because the terminal integrator state(s) often represent positions and are hence typically plotted horizontally. The functions f and b are assumed to be Lipschitz continuous and bounded.

Instead of the fully implicit representation of the reach set (4), we will adopt the MIE representation

$$\mathcal{R}(\mathcal{R}_0, t) = \{(y, x) \mid \underline{\phi}(t, y) \leq x \leq \bar{\phi}(t, y)\} \qquad (10)$$

with target set

$$\mathcal{R}_0 = \{(y, x) \mid \underline{\phi}_0(y) \leq x \leq \bar{\phi}_0(y)\}. \qquad (11)$$

Consider first the upper boundary of the reach set. Let $\overline{x}(t, y)$ be the upper boundary of $\mathcal{R}(\mathcal{R}_0, t)$ at time t and coupled state y; in other words $\overline{x}(t, y) = \overline{\phi}(t, y)$. Formally by (9) and the chain rule

$$b(y) = \frac{d}{dt}\overline{x}(t, y) = \frac{d}{dt}\overline{\phi}(t, y) = D_t\overline{\phi}(t, y) + D_y\overline{\phi}(t, y) \cdot f(y, u). \quad (12)$$

Rearranging, we arrive at a Hamilton-Jacobi equation

$$D_t\overline{\phi}(t, y) + D_y\overline{\phi}(t, y) \cdot f(y, u) - b(y) = 0,$$

whose form is identical to those that arise in finite horizon optimal control problems.

The derivation above is entirely formal, but the finite horizon optimal control interpretation provides rigourous support for the resulting formulation. The terminal integrator's dynamics in (9) can be rewritten for $t \leq 0$ as

$$x(0, y(0)) = x(t, y(t)) + \int_t^0 b(y(s))ds \quad (13)$$

which is simply rearranged into the form of a finite horizon cost function

$$x(t, y(t)) = \int_t^0 -b(y(s))ds + x(0, y(0)) \quad (14)$$

with running cost $-b(y)$ and terminal cost $x(0, y)$ for trajectories $y(\cdot)$ generated by (8). Setting the terminal cost to be $x(0, y) = \overline{\phi}_0(y)$, the value function representing the maximum cost to go from each coupled state y at time $t \leq 0$ is the viscosity solution to the terminal value Hamilton-Jacobi-Bellman equation

$$D_t\overline{\phi}(t, y) + H\left(t, y, D_y\overline{\phi}(t, y)\right) = 0 \quad (15)$$

with Hamiltonian

$$\overline{H}_{\text{set}}(t, y, p) = \max_{u \in \mathcal{U}} \left(p \cdot f(y, u) - b(y)\right), \quad (16)$$

and terminal condition

$$\overline{\phi}(0, y) = \overline{\phi}_0(y). \quad (17)$$

PROPOSITION 1. *For a coupled state $\hat{y}$ in the reach set, let $\overline{x}(t, \hat{y})$ be the upper boundary of the interval which is the reach set's projection at time $t < 0$ onto the terminal integrator dimension at that $\hat{y}$*

$$\overline{x}(t, \hat{y}) = \max\{x \mid (\hat{y}, x) \in \mathcal{R}(\mathcal{R}_0, t)\}$$

The solution $\overline{\phi}(t, y)$ of (15)–(17) provides $\overline{x}$

$$\overline{x}(t, \hat{y}) = \overline{\phi}(t, \hat{y}). \quad (18)$$

Note that $\overline{x}(\cdot, \cdot)$ is not assumed to be a trajectory of the system.

PROOF. For a system with dynamics (8), initial condition $y(\tau_0) = \hat{y}$ and cost function

$$\gamma(\tau_0, \hat{y}, u(\cdot)) = \int_{\tau_0}^{\tau_f} \beta(y(\sigma), u(\sigma))d\sigma + \rho(y(\tau_f))$$

we can define the value function

$$\nu(\tau_0, \hat{y}) = \sup_{u(\cdot)} \gamma(\tau_0, \hat{y}, u(\cdot)).$$

The value function is the viscosity solution of the terminal value Hamilton-Jacobi-Bellman PDE (see, for example, [5, 2])

$$D_\tau\nu(\tau, y) + H(\tau, y, \nu(\tau, y)) = 0 \quad (19)$$

with Hamiltonian

$$H(t, y, p) = \max_{u \in \mathcal{U}} \left(p \cdot f(y, u) + \beta(y, u)\right) \quad (20)$$

and terminal condition

$$\nu(\tau_f, y) = \rho(y). \quad (21)$$

Set

$$\begin{aligned} \beta(y, u) &= -b(y), & \rho(y) &= \overline{\phi}_0(y), \\ \tau_0 = t &< 0, & \tau_f &= 0, \end{aligned} \quad (22)$$

so that (19)–(21) are the same as (15)–(17) and $\nu(t, y) = \overline{\phi}(t, y)$.

For any $\epsilon > 0$, we can pick a $\hat{u}(\cdot)$ such that

$$\gamma(t, \hat{y}, \hat{u}(\cdot)) \leq \nu(t, \hat{y}) < \gamma(t, \hat{y}, \hat{u}(\cdot)) + \epsilon \quad (23)$$

and let $y(\cdot)$ be the solution of (8) arising from $\hat{u}(\cdot)$ with initial condition $y(t) = \hat{y}$. Consider a trajectory $x^{(1)}(\cdot)$ solving (9) along $y(\cdot)$ with initial conditions $x^{(1)}(t)$ (we drop the dependence of the terminal integrator's trajectory $x(\cdot)$ on the coupled states' trajectory $y(\cdot)$ because only one coupled state trajectory is considered in the remainder of the proof). Let $x^{(1)}(0) = \overline{x}(0, y(0)) = \overline{\phi}_0(y(0))$. By (14) and (22),

$$\begin{aligned} x^{(1)}(t) &= \int_t^0 -b(y(\sigma))d\sigma + x^{(1)}(0), \\ &= \int_{\tau_0}^{\tau_f} \beta(y(\sigma), u(\sigma))d\sigma + \rho(y(\tau_f)), \\ &= \gamma(t, \hat{y}, u(\cdot)), \end{aligned}$$

and consequently by (23)

$$x^{(1)}(t) \leq \nu(t, \hat{y}) < x^{(1)}(t) + \epsilon.$$

Letting $\epsilon \to 0$, we find $x^{(1)}(t) = \nu(t, \hat{y}) = \overline{\phi}(t, \hat{y})$.

Now consider a second trajectory $x^{(2)}(\cdot)$ solving (9) along $y(\cdot)$ with initial conditions $x^{(2)}(t) > x^{(1)}(t) = \overline{\phi}(t, \hat{y})$. By (13)

$$\begin{aligned} x^{(2)}(0) &= x^{(2)}(t) + \int_t^0 b(y(s))ds \\ &> x^{(1)}(t) + \int_t^0 b(y(s))ds = x^{(1)}(0) = \overline{\phi}(0, y(0)), \end{aligned} \quad (24)$$

which implies $(x^{(2)}(0), y(0)) \notin \mathcal{R}_0$, which implies that $(x^{(2)}(t), \hat{y}) \notin \mathcal{R}(\mathcal{R}_0, t)$.

Conversely, consider that trajectory $x^{(2)}(\cdot)$ starts from $x^{(2)}(t) < x^{(1)}(t)$. Reversing the inequality in (24), it can be shown that $(x^{(2)}(t), \hat{y}) \in \mathcal{R}(\mathcal{R}_0, t)$.

Since states $(x^{(2)}(t), \hat{y})$ are in the reach set if and only if $x^{(2)}(t) \leq x^{(1)}(t)$, we conclude that $\overline{x}(t, \hat{y}) = x^{(1)}(t) = \overline{\phi}(t, \hat{y})$. $\square$

Following the derivation in [10], the reach tube can be computed by adjusting the Hamiltonian so that the computed set only grows; in this case solve (15) with Hamiltonian

$$\overline{H}_{\text{tube}}(t, y, p) = \max\left[0, \max_{u \in \mathcal{U}} (p \cdot f(y, u) - b(y))\right].$$

The derivation for the lower boundary function $\underline{\phi}(t, y)$ of the reach set is similar except that using the input to maximize the size of the reach set requires using the input to minimize the height of the lower boundary, and in the case of a reach tube the lower boundary should only decrease as time goes backwards. Therefore, we need to use Hamiltonians

$$\underline{H}_{\text{set}}(t, y, p) = \min_{u \in \mathcal{U}} (p \cdot f(y, u) - b(y))$$

$$\underline{H}_{\text{tube}}(t, y, p) = \min\left[0, \min_{u \in \mathcal{U}} (p \cdot f(y, u) - b(y))\right]$$

in (15) for the reach set and tube respectively, and terminal condition $\underline{\phi}(0, y) = \underline{\phi}_0(y)$ instead of (17).

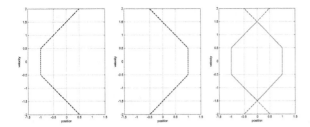

Figure 1: Terminal conditions for the double integrator. The axes are position x and velocity y. Left: $\underline{\phi}_0(y)$. Middle: $\overline{\phi}_0(y)$. Right: $\mathcal{R}_0$ is the region *outside* the solid curve.

Following [10], we can similarly compute minimal backward reachability or even backward reachability under adversarial inputs by appropriate adjustment of the sense of the optimization over inputs in the Hamiltonian.

4. SCALAR EXAMPLES

We demonstrate the MIE scheme on some examples in which the terminal integrator state is a scalar.

4.1 Computational Setting

In the examples that follow, the solution of the HJ PDEs are approximated in MATLAB with the Toolbox of Level Set Methods [11] (ToolboxLS). The numerical schemes used by ToolboxLS require that solutions remain continuous (although not necessarily differentiable). Under the assumptions placed on the dynamics, viscosity solution theory ensures that the solutions of (5) and (15) will remain continuous provided that their terminal condition functions are continuous; consequently, we restrict the target sets in the examples to ensure that $\psi_0(y, x)$, $\overline{\phi}_0(y)$ and $\underline{\phi}_0(y)$ are continuous.

While the software requires continuity, viscosity solution theory allows for more general semicontinuous terminal conditions for finite horizon optimal control problems [2, section V.5.2]. We plan to adapt numerical schemes from conservation laws to permit approximation of solutions with these more general terminal conditions.

All computations were done with ToolboxLS version 1.1 in MATLAB version 7.11 under 64-bit Windows 7 on a Lenovo x200 tablet with 4GB RAM and an Intel Core2 Duo L9400 CPU at 1.86 GHz. MATLAB code for all examples is available at the author's web site.

4.2 The Double Integrator

To demonstrate the technique, consider the simplest possible dynamics with input and a terminal integrator: the traditional double integrator

$$\dot{y} = f(y, u) = u, \qquad \dot{x} = b(y) = y \qquad (25)$$

with $y \in \mathbb{R}$ and $u \in \mathcal{U} = [-u_{\max}, +u_{\max}]$ for some constant $u_{\max} \geq 0$.

The usual double integrator target set is the complement of a rectangle in position x cross velocity y space, which can be translated as finding the set of states such that the system will not exceed specified upper and lower bounds on position and velocity, given upper and lower bounds on acceleration u. To represent the complement, we adjust the interpretation of $\underline{\phi}$ and $\overline{\phi}$ to

$$\mathcal{R}(\mathcal{R}_0, t) = \{(y, x) \mid x \leq \underline{\phi}(t, y) \text{ or } x \geq \overline{\phi}(t, y)\},$$

and similarly for $\mathcal{R}_0$. However, using the complement of a rectangle as $\mathcal{R}_0$ requires piecewise continuous $\overline{\phi}_0(y)$ and $\underline{\phi}_0(y)$, since the maximum and minimum position are discontinuous functions of

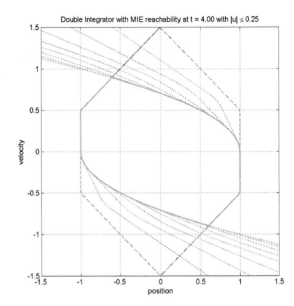

Figure 2: The reach tube for the double integrator computed with the MIE formulation. The reach tube is everything outside the solid curve. The dashed curve shows the target set, while the dotted curves show the evolution of $\overline{\phi}$ and $\underline{\phi}$.

velocity at the upper and lower bounds on velocity. Because ToolboxLS requires continuous functions, we instead use the hexagonal shape shown in figure 1 as the target set. The initial functions in figure 1 are given by

$$\overline{\phi}_0(y) = \min\left[+1, y + \frac{3}{2}, -y + \frac{3}{2}\right],$$

$$\underline{\phi}_0(y) = \max\left[-1, y - \frac{3}{2}, -y - \frac{3}{2}\right]. \qquad (26)$$

Since these functions are the minimum or maximum of three linear functions, they are continuous.

For this reach tube, we choose Hamiltonians to minimize the lower boundary $\underline{\phi}$ subject to the restriction that it can only increase, and maximize the upper boundary $\overline{\phi}$ subject to the restriction that it can only decrease. Plugging (25) into (15) with appropriate Hamiltonians yields the PDEs

$$0 = D_t\overline{\phi}(t, y) + \min\left[0, \max_{u \in \mathcal{U}}\left(D_y\overline{\phi}(t, y) \cdot u - y\right)\right]$$
$$= D_t\overline{\phi}(t, y) + \min\left[0, \left(+u_{\max}|D_y\overline{\phi}(t, y)| - y\right)\right]$$

and

$$0 = D_t\underline{\phi}(t, y) + \max\left[0, \min_{u \in \mathcal{U}}\left(D_y\underline{\phi}(t, y) \cdot u - y\right)\right]$$
$$= D_t\underline{\phi}(t, y) + \max\left[0, \left(-u_{\max}|D_y\underline{\phi}(t, y)| - y\right)\right].$$

Using ToolboxLS we approximate the solutions $\overline{\phi}$ and $\underline{\phi}$ of these PDEs. The results are shown in figure 2, where the final reach tube is everything outside the solid lines.

For comparison purposes, the same double integrator problem can be solved using the traditional implicit formulation from section 2.2. The single HJ PDE in this case is

$$0 = D_t\psi(t, y, x) + \min\left[0, \min_{u \in \mathcal{U}}\left(D_y\psi(t, y, x) \cdot u + D_x\psi(t, y, x) \cdot y\right)\right]$$
$$= D_t\psi(t, y, x) + \min\left[0, \left(-u_{\max}|D_y\psi(t, y, x)| + D_x\psi(t, y, x) \cdot y\right)\right].$$

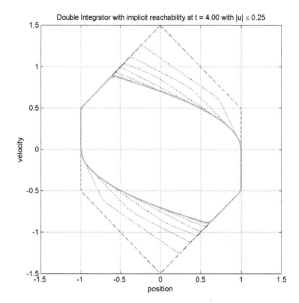

Figure 3: The reach tube for the double integrator computed with the implicit formulation. The reach tube is everything outside the solid curve. The dashed curve shows the target set, while the dotted curves show the evolution of the zero level set of ψ.

The target set is still the complement of a hexagon, so the terminal conditions $\psi_0(y, x)$ are constructed by taking the complement of the intersection of six half-spaces (with implicit surface functions, these operations can be accomplished through some linear algebra and maximum operations). The results are shown in figure 3.

The MIE formulation requires solving two PDEs over a one dimensional grid, while the implicit formulation requires solving a single PDE over a two dimensional grid. Consequently the latter requires at least an order of magnitude more computational effort. In this case the MIE formulation took less than 0.4 seconds on a grid of size 101 (although most of that time is probably consumed by generating the plots), while the implicit formulation took less than 4 seconds on a grid of size 101^2.

4.3 The Rotating Double Integrator

We modify the double integrator so that $y \in \mathbb{R}^2$ and the optimal input is no longer constant along trajectories. The dynamics are

$$\begin{bmatrix} \dot{y}_1 \\ \dot{y}_2 \end{bmatrix} = \dot{y} = f(y, u) = \begin{bmatrix} -y_2 \\ +y_1 \end{bmatrix} + \mu(\|y\|_2) \begin{bmatrix} u_1 \\ u_2 \end{bmatrix}, \qquad (27)$$
$$\dot{x} = b(y) = \|y\|_2$$

where $u \in \mathcal{U} = \{u \in \mathbb{R}^2 \mid \|u\|_2 \leq u_{\max}\}$ for some constant $u_{\max} \geq 0$ and $\mu : \mathbb{R} \to \mathbb{R}$.

If $\mu(\alpha) \equiv 1$, the system behaves radially like the first quadrant of a traditional double integrator. For this example, we choose

$$\mu(\alpha) = 2\sin(4\pi\alpha)$$

so that the optimal vector input not only changes direction and sign, but has variable effect depending on the current state.

For this example we compute only the lower boundary of a backwards minimal reach tube

$$\mathcal{R}(\mathcal{R}_0, t) = \{(y, x) \mid x \geq \underline{\phi}(t, y)\},$$

where $\underline{\phi}_0$, the lower boundary of $\mathcal{R}_0$, is given by

$$\underline{\phi}_0(y) = \min\left[1, 5(1.2 - y_1), 5(1.2 + y_1), 5(1.2 - y_2), 5(1.2 + y_2)\right];$$

and is shown on the left of figure 4. Note that the "position" variable x is the vertical axis (unlike the previous example, where it was shown on the horizontal axis). We use this truncated pyramid shape for the target set boundary to ensure that $\underline{\phi}_0$ is continuous; however, note that the slopes of the sides are much steeper than those in figure 1.

Plugging in the dynamics (27), the Hamiltonian for (15) is

$$H(t, y, p) = \min\left[0, \max_{u \in \mathcal{U}}(p \cdot f(y, u) - b(y))\right]$$
$$= \min\left[0, (-y_2 p_1 + y_1 p_2 + u_{\max}|\mu(\|y\|_2)|\|p\|_2 - \|y\|_2)\right]$$

The results are shown in the middle of figure 4, where the reach tube is everything above the solid surface.

For comparison, the same reach tube computed with the standard implicit scheme uses PDE (5) with Hamiltonian

$$H(t, z, p) = \min\left[0, \left(-y_2 p_{y_1} + y_1 p_{y_2} + u_{\max}|\mu(\|y\|_2)|\|p\|_2 + \|y\|_2 p_x\right)\right]$$

and target set constructed by taking the complement of the intersection of six half-spaces. Results are shown on the right of figure 4.

The MIE formulation requires a single PDE on a two dimensional grid and took 18 seconds on a grid of size 121^2. The implicit formulation requires a single PDE on a three dimensional grid and took around 3000 seconds on a grid of size $121^2 \times 61$.

5. GENERALIZING THE TERMINAL INTEGRATOR

Consider a generalization of the terminal integrator's dynamics

$$\dot{x} = a(y)x + b(y, v) \qquad (28)$$

where $v \in \mathcal{V}$ is an input with $\mathcal{V}$ compact convex. Following the same formal derivation as in (12) for $\overline{\phi}$ for a maximal reach set, for example, we arrive at a terminal value PDE of the form

$$D_t\overline{\phi}(t, y) + H\left(t, y, \overline{\phi}(t, y), D_y\overline{\phi}(t, y)\right) = 0 \qquad (29)$$

with Hamiltonian

$$\overline{H}_{\text{set}}(t, y, q, p) = \max_{u \in \mathcal{U}} \max_{v \in \mathcal{V}}(p \cdot f(y, u) - a(y)q - b(y, v)), \qquad (30)$$

and terminal conditions (17). The major difference in these equations compared to (15) and (16) is that the Hamiltonian now depends on the function value $\overline{\phi}$.

Hamilton-Jacobi PDEs of the form (29) and (30) arise in optimal control theory for discounted finite horizon problems [2, section III.3]. If the linear term is a positive constant $a(y) \equiv a > 0$, then a unique viscosity solution to (29) and (30) exists. Unfortunately, the cost function corresponding to (14) is in this case

$$\int_t^0 -b(y(s), v)e^{a(s-t)}ds + e^{-at}x(0, y(0)) \neq x(t, y(t)),$$

so there is no simple extension of Proposition 1 to make a rigorous argument. Furthermore, the case of $a < 0$ is more likely to occur in practice, since $a > 0$ gives rise to unstable growth in state variable x; however, $a < 0$ in (30) breaks a key monotonicity property assumed of Hamiltonians in viscosity solution theory. The theoretical basis for the MIE formulation with terminal integrator dynamics of the form (28) therefore remains an area of future research.

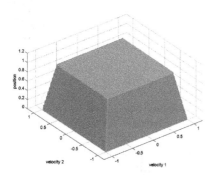 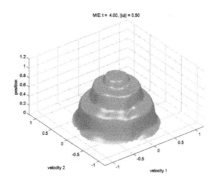 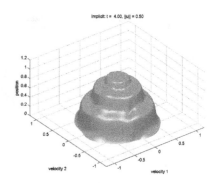

Figure 4: The rotating double integrator example. Note that the "position" variable x is the vertical axis, and in every case the relevant set is *above* the surface shown. Left: The lower boundary $\underline{\phi}_0$ of the target set. Middle: The lower boundary $\underline{\phi}$ of the minimal backwards reach tube computed with the MIE formulation. The ragged bottom edge is a visualization artifact arising from the truncation of the surface for $x < 0$. Right: The lower boundary of the minimal backwards reach tube computed with the implicit formulation.

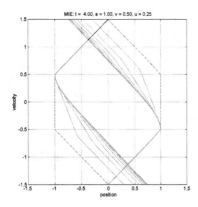

Figure 5: The reach tube for the modified-beyond-recognition double integrator (a double integrator with terminal integrator dynamics of the form (28)) computed with the MIE formulation. The reach tube is everything outside the solid curve. The dashed curve shows the target set, while the dotted curves show the evolution of $\overline{\phi}$ and $\underline{\phi}$.

5.1 The Modified-Beyond-Recognition Double Integrator

Despite the theoretical issues outlined above, it is trivial to adapt the numerical schemes in TOOLBOXLS to handle (29) and (30). To demonstrate, we again modify the double integrator example, this time by adding a linear term and an input to the position variable's dynamics

$$\dot{y} = f(y, u) = u, \qquad \dot{x} = ax + b(y, v) = ax + y + v \qquad (31)$$

with inputs $u \in \mathcal{U} = [-u_{\max}, +u_{\max}]$ and $v \in \mathcal{V} = [-v_{\max}, +v_{\max}]$ for some constants $u_{\max} \geq 0$ and $v_{\max} \geq 0$. We seek to approximate the minimal reach tube for the same target set as in section 4.2, so the Hamiltonian for $\overline{\phi}$ will be

$$H(t, y, q, p) = \min \left[0, \max_{u \in \mathcal{U}} \max_{v \in \mathcal{V}} (p \cdot f(y, u) - aq - b(y, v)) \right]$$
$$= \min \left[0, (+u_{\max}|p| - aq - y + v_{\max}) \right]$$

with PDE (29) and terminal conditions (26). This terminal value HJ PDE is known to have a viscosity solution for $a > 0$. The results computed for $a = +1$, $u_{\max} = 0.25$ and $v_{\max} = 0.5$ are shown

in figure 5, and took less than one second to compute on a grid of size 151. The results using a fully implicit scheme are similar, and so are not shown, although they take about twenty seconds to complete on a grid of size 151^2. While the implicit formulation is much more computationally intensive, it is fully supported by viscosity solution theory for any bounded and Lipschitz continuous $a(y)$ on any bounded domain of x.

6. MULTIPLE TERMINAL INTEGRATORS

We extend to systems with independent terminal integrator states $x \in \mathbb{R}^{d_{\text{ti}}}$ for $d_{\text{ti}} > 1$ whose dynamics are

$$\dot{x}_i = b_i(y) \qquad \text{for } i = 1, 2, \ldots, d_{\text{ti}}.$$

There are three straightforward adaptations of the procedures discussed above to systems of this form. First, one can solve an HJ PDE of the form (5) (with appropriate Hamiltonians and terminal conditions) in the full state space $z \in \mathbb{R}^{d_z}$ to get an implicit representation of the full dimensional reach set or tube. Second, because the terminal integrator variables are fully decoupled from one another, one can use the projection ideas from [12] (without any need to resort to artificial disturbance inputs) and solve d_{ti} separate HJ PDEs of the form (5) in the state spaces $(y, x_i) \in \mathbb{R}^{d_{\text{cp}}+1}$ for $i = 1, 2, \ldots, d_{\text{ti}}$ to get d_{ti} separate implicit representations of the projections of the reach set or tube. Finally, one can solve $2d_{\text{ti}}$ separate HJ PDEs of the form (15) in the state space $y \in \mathbb{R}^{d_{\text{cp}}}$ to get d_{ti} separate MIE representations of the projections of the reach set or tube.

While the decoupled implicit and MIE schemes are obviously appealing because of the considerably lower dimension in which the PDEs need to be solved, there is a challenge: the Hamiltonians (such as (16)) of the resulting vector HJ PDE are coupled through their choice of input u. For maximal reach sets and tubes choosing the input independently for each component of the solution is sound if potentially pessimistic—it may be that to achieve the maximal reach set or tube for x_i one must choose a $\hat{u}_i$ which produces a reach set or tube for x_j that is much smaller than the maximal $\hat{u}_j$. However, this independent choice is unlikely to be too pessimistic, because in general there will be states $(\hat{x}_i, \hat{x}_j)$ such that $\hat{x}_i$ is in the interior of its reachable interval but $\hat{x}_j$ is at one of the boundaries of its reachable interval, and thus $\hat{u}_j$ is the appropriate maximal choice of input.

The situation for minimal reach sets and tubes is more complicated. Simply choosing the inputs independently risks introducing leaky corners into the reach set or tube: There may be states $(\hat{x}_i, \hat{x}_j)$

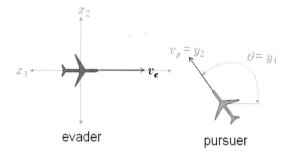

x_2

x_1 — v_e

$v_p = y_2$

$\theta = y_1$

evader   pursuer

Figure 6: The relative coordinate system for the pursuit of an oblivious vehicle problem. The (oblivious) evader is fixed at the origin facing right and cannot modify its speed or heading. The pursuer can modify its speed and heading.

such that $\hat{u}_i$ must be chosen to avoid the target set in dimension i, $\hat{u}_j$ must be chosen to avoid the target set in dimension j, and $\hat{u}_i \neq \hat{u}_j$. In the fully implicit formulation with a single Hamiltonian such as (6), $D_z \psi(t,z)$ (the costate of the corresponding optimal control problem) provides a definitive choice of u. In the MIE formulation, only $D_y \bar{\phi}_i(t,y)$ and $D_y \underline{\phi}_i(t,y)$ are available for $i = 1, 2, \ldots, d_y$. A similar issue exists for the decoupled implicit formulation. Fortunately, there is again a sound if potentially pessimistic solution: Use any single $u \in \mathcal{U}$; for example,

$$\arg\min_{u \in \mathcal{U}} \left(\sum_{i=1}^{d_x} D_y \bar{\phi}_i(t,y) - \sum_{i=1}^{d_x} D_y \underline{\phi}_i(t,y) \right) \cdot f(y,u). \quad (32)$$

Because the choice may be suboptimal in some or all dimensions, the resulting set may not be the true minimal reach set or tube; however, it will not have leaky corners.

6.1 Pursuit of an Oblivious Vehicle

To demonstrate these approaches for a system with multiple terminal integrators, consider a pursuit problem played in a planar workspace. A pursuer vehicle wishes to collide with an evader vehicle. The evader is oblivious to the pursuer in the sense that it travels at constant linear speed v_e and at constant heading, which we choose without loss of generality to be zero. The pursuer travels at speed v_p and heading θ, and may modify its speed through linear acceleration input $a_p \in A_p$ and its heading through angular velocity input $\omega_p \in \Omega_p$. Because collision only depends on relative spatial position, we fix the evader at the origin and use the spatial variables x_1 and x_2 to represent the relative position of the pursuer. See figure 6 for a diagram of the problem. The dynamics can be written as

$$\frac{d}{dt} \begin{bmatrix} \theta \\ v_p \\ x_1 \\ x_2 \end{bmatrix} = \begin{bmatrix} \omega_p \\ a_p \\ -v_e + v_p \cos\theta \\ v_p \sin\theta \end{bmatrix}$$

which we can decompose into coupled dynamics

$$\dot{y} = \frac{d}{dt} \begin{bmatrix} y_1 \\ y_2 \end{bmatrix} = \begin{bmatrix} \theta \\ v_p \end{bmatrix} = \begin{bmatrix} \omega_p \\ a_p \end{bmatrix} = f(y,u)$$

and terminal integrators

$$\dot{x} = \frac{d}{dt} \begin{bmatrix} x_1 \\ x_2 \end{bmatrix} = \begin{bmatrix} -v_e + y_2 \cos y_1 \\ y_2 \sin y_1 \end{bmatrix} = b(y).$$

Note that the coupled dynamics do not depend on the coupled variables and are linear in the inputs, while the terminal integrators are

nonlinear in y_1.

Traditionally the target set for collision problems would be circular in the position variables and independent of other variables; however, in order to treat the position variables as terminal integrators we must decouple their target set components. As a consequence, we use an interval as the target set for each position variable. We also constrain the pursuer's speed such that $0 < v_p^{\min} \leq v_p \leq v_p^{\max}$. Speed constraints of this sort would be appropriate, for example, if the vehicles are fixed-wing aircraft. The resulting target set is a square in $x_1 \times x_2$ space for all headings $\theta = y_1$ and all valid speeds $v_p = y_2$.

While the constraint on v_p certainly affects the target set, it goes further than that. For the backward reach tube in this problem we are only interested in states which can reach the target set within the specified time without violating the constraints on v_p. The construction of the target set ensures that the state within the target set which the trajectory reaches satisfies the constraints, but we need additional effort to ensure that all other states along the trajectory also satisfy the constraints. In the implicit formulation state constraints are enforced by applying a constraint on the solution of the HJ PDE $\psi(t,z) \geq \zeta_{\text{implicit}}(z)$ for all t, where ζ_{implicit} is an implicit surface function for the state constraints; equivalently, $\psi(t,z)$ becomes the solution of a variational inequality. For state constraints on the coupled variables (such as the constraint on $v_p = y_2$ in this problem), the MIE formulation can also apply a constraint of the form $\bar{\phi}(t,y) \leq \zeta_{\text{indicator}}^{\text{upper}}(y)$ (and a corresponding lower bound on $\underline{\phi}$); however, $\zeta_{\text{indicator}}^{\text{upper}}$ is now a discontinuous function with value $+\infty$ for states which satisfy the constraint and value 0 for those which do not.

As mentioned earlier, our current implementation of MIE requires that $\bar{\phi}_i$ and $\underline{\phi}_i$ are continuous with respect to y. That is trivially true with respect to $y_1 = \theta$ (the target set is constant with respect to this variable), but not true with respect to $y_2 = v_p$ because of the constraints on v_p. In order to maintain continuity, the size of the target square is shrunk gradually to zero as $y_2 = v_p$ approaches $v_p^{\min}$ or $v_p^{\max}$. In order to maintain consistency across all formulations, this same smoothed target set is used in all cases. With regard to the constraint functions, no smoothing is necessary for ζ_{implicit} in the implicit formulations because it is chosen as a signed distance function and is hence continuous already, and no smoothing is applied to $\zeta_{\text{indicator}}$ in the MIE formulation beyond the effect of grid discretization.

We approximate the reach set for parameters $v_e = 1$, $A_p = [-0.2, +0.2]$, $\Omega_p = [-0.2, +0.2]$, $v_p^{\min} = 1.0$, $v_p^{\max} = 3.0$ and target set $x \in [-1, +1]^2$ for valid v_p. We compute the backward reach tube out to $t = 2.0$ in each of the three formulations: full dimensional implicit (one PDE in four dimensions), decoupled implicit (two PDEs in three dimensions) and decoupled MIE (four PDEs in two dimensions). For the decoupled calculations, inputs are chosen independently in each PDE, and thus the computed tube is conservative. The results are somewhat difficult to visualize because the reach tube lies in a four dimensional space.

Figure 7 shows the projections of the reach tube into the (x_1, v_p, θ) and (x_2, v_p, θ) subspaces. These projections are computed directly for the two decoupled formulations, and for the full dimensional implicit case we simply take a minimum over the missing dimension of the implicit surface function ψ.

Of course, projections almost always lose some information. Figure 8 shows a sample slice of the reach tube in the (x_1, x_2, θ) subspace for $v_p = 2.0$. This slice is a subset of the data computed by the full dimensional implicit formulation. For the decoupled formulations, this slice can be recreated by backprojecting the lower dimensional reach tubes into full dimensional prisms and then in-

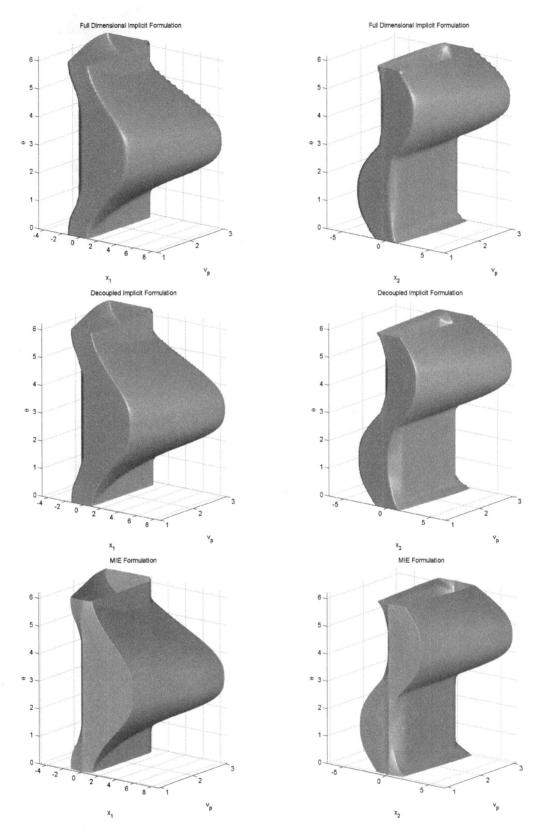

Figure 7: **Projections of the reach tube for the pursuit of an oblivious vehicle problem, computed with the three formulations. Left column:** (x_1, v_p, θ) **projection. Right column:** (x_2, v_p, θ) **projection. Top row: full dimensional implicit formulation. Middle row: decoupled implicit formulation. Bottom row: decoupled MIE formulation.**

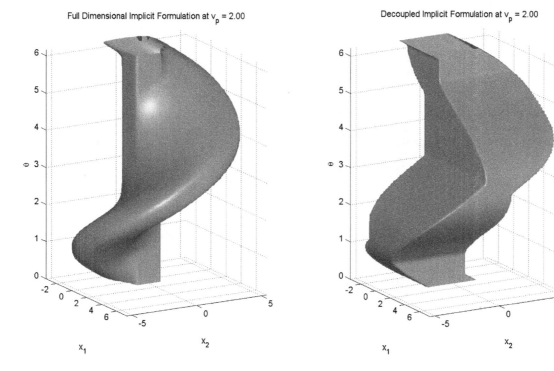

Full Dimensional Implicit Formulation at $v_p = 2.00$

Decoupled Implicit Formulation at $v_p = 2.00$

Figure 8: A slice of the reach tube in (x_1, x_2, θ) space for $v_p = 2.0$. Left: The reach tube as computed by the full dimensional implicit formulation. Right: The reach tube as computed by backprojection of the decoupled implicit formulation (the MIE result would be the same). Because the decoupled formulations (implicit or MIE) work in subsets of the state space, there is inevitably some overapproximation of the reach tube.

tersecting the prisms; the result from the decoupled implicit calculation was easier to backproject and is shown, although the result from the MIE calculation would be virtually identical. Note how the reach tube computed by projections is an overapproximation of the reach tube computed in the full dimensional state space.

While the decoupled formulations effectively result in an overapproximation of the reach tube, the trade-off is their speed. On a grid of size 65×100 the MIE formulation required 3.1 seconds (a nontrivial fraction of which is I/O) to solve four PDEs. On a grid of size $151 \times 65 \times 100$ the decoupled implicit formulation required 541 seconds to solve two PDEs. On a grid of size $151^2 \times 65 \times 100$ the full dimensional implicit formulation required too much memory, but extrapolation from a grid of half that size in each dimension provides a rough estimate of 30 *hours* to finish.

There are two obvious extensions of this problem: permit the evader to modify its speed and/or heading. By including v_e as a (coupled) state variable, it is straightforward to augment this model to permit the evader to modify its speed. The additional state dimension would make the full dimensional implicit formulation five dimensional and hence impractical to compute, but the MIE formulation would be only three dimensional and still quite feasible. It is possible to permit the evader to modify its heading without augmenting the state space—the pursuer heading state θ is replaced with a relative heading state, and the result is a version of the game of two identical vehicles (see, for example, [10] or [12]). Unfortunately, in this model the relative position variables are coupled and hence cannot be treated with the MIE formulation. The alternative is to augment the state space with the evader's heading, which has the same dimensional costs as adding v_e to the state space.

7. COMPARING THE IMPLICIT AND MIE FORMULATIONS

For a system with $y \in \mathbb{R}^{d_{cp}}$ and $x \in \mathbb{R}^{d_{ti}}$, the traditional full dimensional implicit formulation described in section 2.2 requires solving a single PDE (5) over the entire state space. Approximating the solution of such a PDE is typically accomplished by gridding the entire state space and hence requires $O(n^{(d_{cp}+d_{ti})})$ memory for a grid of size n in each dimension, and computational time $O(n^{(d_{cp}+d_{ti})})$ or greater. As described in section 6, because the dynamics of the terminal integrator states are decoupled, the decoupled implicit formulation requires solving d_{ti} separate PDEs in $\mathbb{R}^{d_{cp}+1}$, so the cost would be $O(d_{ti}n^{d_{cp}+1})$. Finally, the MIE formulation for the same system requires solving two PDEs of the form (15) for each terminal integrator state, but these PDEs are solved over only the coupled state space $\mathbb{R}^{d_{cp}}$; thus, the cost is $O(2d_{ti}n^{d_{cp}})$.

Despite their higher computational cost, the implicit formulations (both full and decoupled) do have a few features which the MIE formulation lacks. First, in the implicit formulations the reach set or tube is defined by the zero level set of the solution. The implicit formulation can therefore represent sets with sharp features and some types of discontinuities while maintaining a continuous solution, which makes both theory and numerics easier. Such continuity benefits also accrue when applying constraints to the reach set or tube (such as the constraints on v_p in section 6.1), since those constraints can be represented by continuous implicit functions for the implicit formulations but may require discontinuous functions for the MIE formulation. Furthermore, since only the zero level set is of interest, narrowband or local level set schemes can be used to slightly reduce the memory and cost (although TOOLBOXLS does not implement such schemes). Finally, artificial boundary conditions for the edges of the computational domain are easier to construct when only the zero level set is of interest, because errors caused by such boundary conditions are easily kept away from the zero level set through reinitialization or similar procedures. In the MIE formulation, the entire solution of the PDE is relevant, so none of these benefits apply.

Second, in the implicit formulations the gradient of the solution is the costate of the corresponding finite horizon optimal control problem (see [10] for details). Consequently, this costate can be used to choose optimal inputs along the boundary of the reach set or tube (or throughout the state space under appropriate circumstances). In the MIE formulation, only $D_y\bar{\phi}_i$ and $D_y\underline{\phi}_i$ are available; we are presently investigating how they might be used to deduce optimal inputs.

Given these benefits of implicit schemes, the decoupled implicit formulation might appear to be a "best of both worlds," since it requires only one more dimension than MIE. One dimension, however, is often quite significant computationally; for example, consider the reach tube in section 6.1 which took more than 150 times as long to approximate with the three dimensional decoupled implicit formulation as it did with the two dimensional MIE formulation. Furthermore, the decoupling of the terminal integrator states (whether in an implicit or MIE formulation) introduces two additional weaknesses when compared with the full dimensional implicit formulation. The first is the unresolved coupling through the inputs discussed in section 6. The second is the likely overapproximation of the reach set or tube caused by working in subspaces of the state space, as demonstrated in figure 8. Only the full dimensional implicit formulation can avoid these issues, and its dimension is often impractically large.

8. CONCLUSIONS AND FUTURE WORK

We have demonstrated the new mixed implicit explicit (MIE) formulation for the computation of reach sets and tubes using Hamilton-Jacobi type PDEs for a class of systems with terminal integrators. ODE models of mechanical systems often take this form, since position states are simple integrations of velocity states. While the traditional full dimensional implicit formulation requires computational resources exponential in the number of state space dimensions, the MIE formulation requires resources linear in the number of terminal integrators and exponential in the number of remaining state space dimensions. We have also discussed an intermediate, decoupled implicit formulation for these systems which almost matches the MIE formulation's asymptotic complexity, but which also shares some of its weaknesses.

In the examples, the solutions of the resulting PDEs were approximated using ToolboxLS, but other numerical methods could certainly be applied. In particular, ToolboxLS provides no guarantee on the sign of the error in the approximation. For rigorous computation of reach sets and tubes, schemes providing such guarantees would be desirable; for example, the numerical algorithms for approximating HJ PDEs developed in the viability community, such as those described in [1, 3].

We are currently investigating several aspects of the new formulation. Section 5 raised some theoretical questions regarding an extension of the terminal integrator's dynamics. As discussed in section 6, there are questions regarding the treatment of inputs to the coupled states in the case of vector terminal integrators. Section 7 raised a related issue regarding the inputs: The relationship between the gradients of the MIE solution and the costate of the implicit formulation, since the latter can be used to construct optimal feedback policies. As mentioned in section 4.1, we are planning to implement new schemes in ToolboxLS which will permit compu-

tation with piecewise continuous functions. Finally, the examples used in this paper were all toys; we are working on more realistic problems in five and higher dimensions, and seeking extensions of the MIE formulation which will reduce the computational costs even further.

9. REFERENCES

[1] J.-P. Aubin, A. Bayen, and P. Saint-Pierre. Dirichlet problems for some Hamilton-Jacobi equations with inequality constraints. *SIAM Journal of Control and Optimization*, 47(5):2348–2380, 2008.

[2] M. Bardi and I. Capuzzo-Dolcetta. *Optimal Control and Viscosity Solutions of Hamilton-Jacobi-Bellman equations*. Birkhäuser, Boston, 1997.

[3] P. Cardaliaguet, M. Quincampoix, and P. Saint-Pierre. Set-valued numerical analysis for optimal control and differential games. In M. Bardi, T. E. S. Raghavan, and T. Parthasarathy, editors, *Stochastic and Differential Games: Theory and Numerical Methods*, volume 4 of *Annals of International Society of Dynamic Games*, pages 177–247. Birkhäuser, 1999.

[4] B. Djeridane and J. Lygeros. Neural approximation of PDE solutions: An application to reachability computations. In *Proceedings of the IEEE Conference on Decision and Control*, pages 3034–3039, San Diego, CA, 2006.

[5] L. C. Evans. *Partial Differential Equations*. American Mathematical Society, Providence, Rhode Island, 1998.

[6] I. Kitsios and J. Lygeros. Final glide-back envelope computation for reusable launch vehicle using reachability. In *Proceedings of the IEEE Conference on Decision and Control*, pages 4059–4064, Seville, Spain, 2005.

[7] J. Lygeros. On reachability and minimum cost optimal control. *Automatica*, 40(6):917–927, 2004.

[8] J. Lygeros, C. Tomlin, and S. Sastry. Controllers for reachability specifications for hybrid systems. *Automatica*, 35(3):349–370, 1999.

[9] I. M. Mitchell. Comparing forward and backward reachability as tools for safety analysis. In A. Bemporad, A. Bicchi, and G. Buttazzo, editors, *Hybrid Systems: Computation and Control*, number 4416 in Lecture Notes in Computer Science, pages 428–443. Springer Verlag, 2007.

[10] I. M. Mitchell, A. M. Bayen, and C. J. Tomlin. A time-dependent Hamilton-Jacobi formulation of reachable sets for continuous dynamic games. *IEEE Transactions on Automatic Control*, 50(7):947–957, 2005.

[11] I. M. Mitchell and J. A. Templeton. A toolbox of Hamilton-Jacobi solvers for analysis of nondeterministic continuous and hybrid systems. In M. Morari and L. Thiele, editors, *Hybrid Systems: Computation and Control*, number 3414 in Lecture Notes in Computer Science, pages 480–494. Springer Verlag, 2005.

[12] I. M. Mitchell and C. J. Tomlin. Overapproximating reachable sets by Hamilton-Jacobi projections. *Journal of Scientific Computing*, 19(1–3):323–346, 2003.

[13] P. Saint-Pierre. Hybrid kernels and capture basins for impulse constrained systems. In C. J. Tomlin and M. R. Greenstreet, editors, *Hybrid Systems: Computation and Control*, number 2289 in Lecture Notes in Computer Science, pages 378–392. Springer Verlag, 2002.

Computing Bounded ε-Reach Set with Finite Precision Computations for a Class of Linear Hybrid Automata*

Kyoung-Dae Kim   Sayan Mitra   P. R. Kumar

Department of Electrical and Computer Engineering and Coordinated Science Laboratory
University of Illinois at Urbana-Champaign
Urbana, IL 61801 USA
{kkim50, mitras, prkumar}@illinois.edu

ABSTRACT

In a previous paper [7] we have identified a special class of linear hybrid automata, called *Deterministic Transversal Linear Hybrid Automata*, and shown that an ε-reach set up to a finite time, called a *bounded ε-reach set*, can be computed using infinite precision calculations. However, given the linearity of the system and the consequent presence of matrix exponentials, numerical errors are inevitable in this computation. In this paper we address the problem of determining a bounded ε-reach set using variable finite precision numerical approximations. We present an algorithm for computing it that uses only such numerical approximations. We further develop an architecture for such bounded ε-reach set computation which decouples the basic algorithm for an ε-reach set with given parameter values from the choice of several runtime adaptation needed by several parameters in the variable precision approximations.

Categories and Subject Descriptors

G.M [**Mathematics of Computing**]: Miscellaneous

General Terms

Theory, Algorithm, Verification

Keywords

Linear hybrid automata, reachability, transversal discrete transition, deterministic discrete transition

1. INTRODUCTION

*This material is based upon work partially supported by NSF under Contract Nos. CNS-1035378, CNS-1035340, CNS-1016791, and CCF-0939370, AFOSR under Contract FA9550-09-0121, USARO under Contract Nos. W911NF-08-1-0238 and W-911-NF-0710287.

It is well known that computing the exact reach set of general Hybrid Automaton (HA) is undecidable. In [6, 11] several class of decidable HA with restricted expressive power have been identified. In recent years, research in hybrid system verification has focused on algorithms computing over-approximations of the reachable states of various classes of HAs [2, 10]. In [5], for examples, two techniques, called *clock translation* and *linear phase-portrait approximation*, are proposed to compute an over-approximation of the reach set when the continuous dynamics of a HA is more general than a rectangular HA. In [4], a conservative over-approximation of a reach set of a HA is computed through on-the-fly over-approximation of the phase portrait, which is a variation of the approximation in [5]. In [3], to solve a verification problem of a class of HA, called a polyhedral-invariant HA (PIHA), a finite state transition system, which is a conservative approximation of the original HA, is constructed through a polyhedral approximation of each sampled segment of the continuous state evolution between switching planes.

In [7], we have identified a class of hybrid automata, called *Deterministic Transversal Linear Hybrid Automata (DTLHA)*[1]. We also have shown a new approach to compute an over-approximation of the reach set, with arbitrarily small approximation error ε, up to a finite time, from an initial state. We refer to such a set as a *bounded ε-reach set*. The class of DTLHA consists of linear systems with constant inputs (i.e., where the right hand sides of the differential equations consist of the superposition of a term that is linear in the state and a constant input), for which the linear dynamics as well as the constant input switch along the boundaries of polyhedra, and for which the discrete transitions involved are *deterministic* and *transversal* at each discrete transition time. Since the solutions of linear systems involve matrix exponentials, one however needs to carefully take into account the issue of numerical approximations. In this paper we address the problem of computing with variable finite precision numerical schemes and show that one

[1]Abbreviated simply as DLHA in [7]. In the hybrid system literature [1, 5] the word "linear automaton" has been used to denote a system where the differential equations and inequalities involved have constant right hand sides. However, this does not conform to the standard notion of linearity where the right hand side is allowed to be a function of state. We use the term "linear" in this latter more mathematically standard way that therefore encompasses a larger class of systems.

can still compute a bounded ϵ-reach set. We also present an algorithm that decouples the basic scheme for bounded ϵ-reach set from the numerical approximation issues. This algorithm is additionally more flexible in comparison to the algorithm presented in [7] in terms of allowing more efficient computational strategies.

2. PRELIMINARIES

We consider the problem of the computation of an approximate reach set of a special class of Hybrid Automaton (HA) under some assumptions on the discrete transitions. More precisely, given an initial state x_0, an approximation parameter ϵ, a time bound T, and a jump bound N, we would like to compute a set S such that it contains the actual set of states that are reachable from x_0 in T time or N jumps (whichever happens earlier), and does not contain any states which are more than ϵ away from the actual reachable states. We now describe the class of automata in greater detail.

We assume that the continuous state space $\mathcal{X} \subset \mathbb{R}^n$ is closed and bounded, and is partitioned into a collection of polyhedral regions $\mathcal{C} := \{\mathcal{C}_1, \cdots, \mathcal{C}_m\}$, that is

$$\bigcup_{i=1}^{m} \mathcal{C}_i = \mathcal{X}, \quad \text{s.t.} \quad \mathcal{C}_i^\circ \cap \mathcal{C}_j^\circ = \emptyset \quad \text{for } i \neq j, \quad (1)$$

where m is the size of the partition, each $\mathcal{C}_i \in \mathcal{C}$ is a polyhedron, called *cell*, such that $\mathcal{C}_i^\circ \neq \emptyset$, where $\mathcal{C}_i^\circ$ is the interior of $\mathcal{C}_i$. Two cells $\mathcal{C}_i$ and $\mathcal{C}_j$ are said to be *adjacent* if the affine dimension of $\partial\mathcal{C}_i \cap \partial\mathcal{C}_j$ is $(n-1)$, or, equivalently, cells $\mathcal{C}_i$ and $\mathcal{C}_j$ intersect in an $(n-1)$-dimensional facet. Here $\partial\mathcal{C}_i$ denotes the boundary of $\mathcal{C}_i$. Two cells $\mathcal{C}_i$ and $\mathcal{C}_j$ are said to be *connected* if there exists a sequence of adjacent cells between $\mathcal{C}_i$ and $\mathcal{C}_j$.

DEFINITION 1. *An n-dimensional* Linear Hybrid Automaton (LHA) *is a tuple* $(\mathbb{L}, Inv, A, u)$ *satisfying the following properties. (a) $\mathbb{L}$ is a finite set of locations or discrete states; The state space is $\mathbb{L} \times \mathbb{R}^n$, and an element $(l, x) \in \mathbb{L} \times \mathbb{R}^n$ is called a state. (b) $Inv : \mathbb{L} \to 2^{\mathcal{C}}$ is a function that maps each location to a set of cells [2], called an invariant set of a location, such that (i) for each $l \in \mathbb{L}$, all the cells in $Inv(l)$ are connected, (ii) for any two locations $l, l' \in \mathbb{L}$, $Inv(l)^\circ \cap Inv(l')^\circ = \emptyset$, and (iii) $\cup_{l \in \mathbb{L}} Inv(l) = \mathcal{X}$. (c) $A : \mathbb{L} \to \mathbb{R}^{n \times n}$ is a function that maps each location to an $n \times n$ matrix, and (d) $u : \mathbb{L} \to \mathbb{R}^n$ is a function that maps each location to an n-dimensional vector.*

In the sequel, for each $l_i \in \mathbb{L}$, we use A_i, u_i, Inv_i to denote $A(l_i)$, $u(l_i)$, and $Inv(l_i)$, respectively.

DEFINITION 2. *For a location $l_i \in \mathbb{L}$, a trajectory of duration $t \in \mathbb{R}_{\geq 0}$ for an LHA $\mathcal{A}$ with n continuous dimensions (or variables) is a continuous map η from $[0, t]$ to $\mathbb{R}^n$, such that (a) $\eta(\tau)$ satisfies the differential equation*

$$\dot{\eta}(\tau) = A_i \eta(\tau) + u_i, \quad (2)$$

(b) $\eta(\tau) \in Inv_i$ for every $\tau \in [0, t]$.

For such a trajectory η, its *duration* is t, and it is denoted by $\eta.dur$.

---

[2]Actually, to be precise, the invariant of a location is the union of such cells; however, we abuse the terminology slightly for ease of reading.

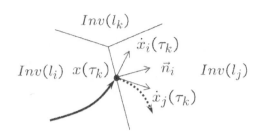

Figure 1: A deterministic and transversal discrete transition from a location l_i to a location l_j occurring at $x(\tau_k) \in \partial Inv(l_i) \cap \partial Inv(l_j)$.

DEFINITION 3. *An execution x of an LHA $\mathcal{A}$ from a starting state $(l_0, x_0) \in \mathbb{L} \times \mathbb{R}^n$ is defined as a continuous map $x : [0, t] \to \mathbb{R}^n$ which is the concatenation of a finite or infinite sequence of trajectories $x = \eta_0 \eta_1 \eta_2 \ldots$ such that (a) $t = \sum_k \eta_k.dur$, (b) $x(0) = \eta_0(0) = x_0 \in Inv_0$, (c) $x(\tau_k) = \eta_k(0) = \eta_{k-1}(\eta_{k-1}.dur)$ for $k \geq 1$, (d) $x(\tau) = \eta_{k-1}(\tau - \tau_{k-1})$ for $\tau \in [\tau_{k-1}, \tau_k)$, where $\tau_0 = 0$, and $\tau_k = \sum_{i=0}^{k-1} \eta_i.dur$ for $k \geq 1$. Note that τ_k for $k \geq 1$ represents the time at the k-th discrete transition between locations and the continuous state is not reset during discrete transitions.*

DEFINITION 4. *For $l_i, l_j \in \mathbb{L}$, a discrete transition from l_i to l_j occurs at a continuous state $x(\tau')$ at time τ', whenever $x(\tau') \in Inv_i \cap Inv_j$ and $x(\tau') = \lim_{\tau \nearrow \tau'} x(\tau)$ where $x(\tau) \in (Inv_i)^\circ$ for $\tau \in (\tau' - \delta, \tau')$ for some $\delta > 0$.*

DEFINITION 5. *A discrete transition is called a* deterministic *discrete transition if there is only one location $l_j \in \mathbb{L}$ to which a discrete transition state $x(\tau_k)$ can make a discrete transition from l_i. Furthermore, for $\epsilon > 0$, we call a discrete transition a* transversal *discrete transition if the following condition is satisfied at $x(\tau_k)$:*

$$\langle \dot{x}_i(\tau_k), \vec{n}_i \rangle \geq \epsilon \quad \wedge \quad \langle \dot{x}_j(\tau_k), \vec{n}_i \rangle \geq \epsilon, \quad (3)$$

where $\vec{n}_i$ is an outward normal vector of ∂Inv_i at $x(\tau_k)$, and $\dot{x}_i(\tau_k) = A_i x(\tau_k) + u_i$, and $\dot{x}_j(\tau_k) = A_j x(\tau_k) + u_j$ are the vector fields at $x(\tau_k)$ evaluated with respect to the continuous dynamics of location l_i and l_j, respectively.

Fig. 1 illustrates a case when $x(\tau_k)$ satisfies such a deterministic and transversal discrete transition conditions. Note that if $x(\tau_k)$ satisfies a deterministic and transversal discrete transition condition, then $x(\tau_k)$ must make a discrete transition from a location l_i to the other unique location l_j. Furthermore, the *Zeno behavior* does not occur if a discrete transition is transversal discrete transition.

DEFINITION 6. *Given an LHA $\mathcal{A}$, a starting state $(l_0, x_0) \in \mathbb{L} \times \mathbb{R}^n$, a time bound T, and a jump bound N, we call an LHA $\mathcal{A}$ as a* Deterministic and Transversal Linear Hybrid Automaton (DTLHA) *if all discrete transitions in the execution starting from x_0 up to time T or up to N transitions (whichever is earlier) are deterministic and transversal.*

DEFINITION 7. *A continuous state in $\mathcal{X}$ is* reachable *if there exists some time t at which it is reached by some execution x.*

DEFINITION 8. *Given a time t, the* bounded reach set *up to time t, denoted as* $\mathcal{R}_t(x_0)$, *of a DTLHA $\mathcal{A}$ is defined to be the set of continuous states that are reachable for some time $\tau \in [0, t]$ by some execution x starting from $x_0 \in Inv_0$.*

DEFINITION 9. *Given $\epsilon > 0$, a set of continuous states S is called a* bounded ϵ-reach set *of a DTLHA $\mathcal{A}$ over a time interval $[0, t]$ from an initial state x_0 if $\mathcal{R}_t(x_0) \subseteq S$ and*

$$d_H(\mathcal{R}_t(x_0), S) \le \epsilon. \qquad (4)$$

where $d_H(\mathcal{P}, \mathcal{Q})$ denotes the Hausdorff distance between two sets $\mathcal{P}$ and $\mathcal{Q}$.

The specific norm that we use in (4) as well as the sequel is the ℓ_∞-norm. Its advantage is that the neighborhoods it induces are polyhedra, in fact hypercubes.

Our results and algorithm also address the following *Safety Problem*[3]: Does the state enter the "unsafe" set of cells corresponding to a specified location within a specified finite time T? Our results show that except for the degenerate case where the state hits the boundary of the unsafe set for the first time at exactly T, the problem is decidable, and that our algorithm resolves this question.

Throughout this paper, we use $\mathcal{D}_t(\mathcal{P})$ to denote the set of states reached at time t from a set $\mathcal{P}$ at time 0. We also use $\mathcal{D}_t(\mathcal{P}, \gamma)$ to denote an over-approximation of $\mathcal{D}_t(\mathcal{P})$ with an approximation parameter $\gamma > 0$, and calling it a γ-approximation of $\mathcal{D}_t(\mathcal{P})$ if it satisfies (i) $\mathcal{D}_t(\mathcal{P}) \subset \mathcal{D}_t(\mathcal{P}, \gamma)$ and (ii) $d_H(\mathcal{D}_t(\mathcal{P}), \mathcal{D}_t(\mathcal{P}, \gamma)) \le \gamma$. Note that $\mathcal{D}_0(\mathcal{P}, \gamma)$ is simply a γ-approximation of the set $\mathcal{P}$.

3. THEORY

In this section, we first present the theoretical results for bounded ϵ-reachability of a DTLHA under the assumption of infinite precision calculation made in [7]. We then derive a set of conditions that can be used to determine the event of a deterministic and transversal discrete transition in computing a bounded ϵ-reach set of a DTLHA which is discussed in more detail in Section 5. In the last part of this section, these results are extended to show that a bounded ϵ-reach set of a DTLHA can be computed without infinite precision calculation capability.

3.1 Bounded ϵ-Reachability of a DTLHA

The approach to compute a bounded ϵ-reach set of a DTLHA from an initial state x_0 in [7] is to over-approximate the bounded reach set through sampling and polyhedral over-approximation. More precisely, for given parameters δ and γ, and a sampling period h, the bounded reach set of a DTLHA from x_0 up to time t_f is over-approximated by

$$\bigcup_{m=0}^{M} \mathcal{D}_{mh}(\mathcal{B}_\delta(x_0), \gamma) \qquad (5)$$

where $\mathcal{B}_\delta(x_0)$ is a polyhedral δ-neighborhood of x_0, γ is a parameter which defines the size of over-approximation of $\mathcal{D}_\tau(\mathcal{B}_\delta(x_0))$ for $\tau \in [0, t_f]$, and $M := \lceil t_f/h \rceil$.

In this approach, the existence of appropriate values for parameters δ, γ, and h is in fact critical in computing a bounded ϵ-reach set of a DTLHA from x_0. In [7], we showed

that for any given $\epsilon > 0$, there exist values for these parameters such that the set in (5) is indeed a bounded ϵ-reach set of a DTLHA from x_0 if every discrete transition is deterministic and transversal.

We present the main results from [7] as follows:

LEMMA 1. *Given $\gamma > 0$, if a sampling period h satisfies the following inequality in (6), then $\mathcal{D}_\tau(\mathcal{B}_\delta(x_0)) \subset \mathcal{D}_t(\mathcal{B}_\delta(x_0), \gamma)$ for $\tau \in [t, t+h]$ for each sample time t:*

$$h < \frac{\gamma}{\bar{v}} \qquad (6)$$

where $\bar{v} := \max_{l_i \in \mathbb{L}}\{\|A_i\|\bar{x} + \|u_i\|\}$ and $\bar{x} := \max_{x \in \mathcal{X}}\|x\|$

LEMMA 2. *Given $\epsilon > 0$, a DTLHA $\mathcal{A}$, an initial state $(l_0, x_0) \in \mathbb{L} \times \mathbb{R}^n$, and a time bound t_f, there exist $\delta > 0$, $\gamma > 0$, and $h > 0$ such that the following hold:*

(i) $\mathcal{R}_{t_f}(x_0) \subset \bigcup_{m=0}^{M} \mathcal{D}_{mh}(\mathcal{B}_\delta(x_0), \gamma)$,

(ii) $dia(\mathcal{D}_{mh}(\mathcal{B}_\delta(x_0), \gamma)) < \epsilon \quad \forall m \in \{0, 1, \cdots, M\}$,

(iii) Suppose $x(\tau_k) \in \partial Inv_i$, $x(\tau_k) = \lim_{\tau \to \tau_k} x(\tau)$, and $\tau_k < t_f$ where $x(\tau) \in (Inv_i)^\circ \ \forall \tau \in (\tau_k - \eta, \tau_k)$ for some $l_i \in \mathbb{L}$ and $\eta > 0$. Then $\mathcal{D}_{t-h}(\mathcal{B}_\delta(x_0)) \subset (Inv_i)^\circ$, $\mathcal{D}_t(\mathcal{B}_\delta(x_0)) \subset (Inv_i)^C$, $\tau_k \in (t-h, t)$, and $t < t_f$ where $\mathcal{D}_\tau(\mathcal{B}_\delta(x_0))$ is computed under the LTI dynamics of $l_i \in \mathbb{L} \ \forall \tau \in [t-h, t]$, and

(iv) Suppose (iii) holds and $x(\tau_k)$ makes a discrete transition from a location l_i to some other location $l_j \in \mathbb{L}$. Then $(\mathcal{D}_{\tau_k}(\mathcal{B}_\delta(x_0), \gamma) \cap Inv_i \cap Inv_j) \subset \mathcal{J}_{i,j}$ and $h < \Delta$ for some appropriate $\delta' > 0$ and $\Delta > 0$ such that $\mathcal{B}_{2\delta'}(x(\tau_k)) \subset (Inv_i \cup Inv_j)$ and

$$\bigcup_{y \in \mathcal{J}_{i,j}} \mathcal{D}_\tau(y) \subset (Inv_j)^\circ \quad \forall \tau \in (\tau_k, \tau_k + \Delta), \quad (7)$$

where $\mathcal{J}_{i,j} := \mathcal{B}_{\delta'}(x(\tau_k)) \cap Inv_i \cap Inv_j$.

where $\mathcal{R}_{t_f}(x_0)$ is the bounded reach set of $\mathcal{A}$, h is determined by (6), $dia(\mathcal{P})$ denotes the diameter of a polyhedron $\mathcal{P}$, and $M := \lceil t_f/h \rceil$.

In summary, for a given bounded reach set $\mathcal{R}_{t_f}(x_0)$ of a DTLHA $\mathcal{A}$ from x_0, the above results state the following: (1) A sampling period $h > 0$ can be determined for any given $\gamma > 0$ so that the bounded reach set can be over-approximated. (2) If there is a discrete transition, then this event can be determined through the over-approximation of sampled states with an appropriate values of δ and h. (3) If a discrete transition is deterministic and transversal, then an over-approximation of the discrete transition state can computed with an appropriate values of δ, γ, and h. (4) If every discrete transition state $x(\tau_k)$ is deterministic and transversal, then a bounded ϵ-reach set of $\mathcal{A}$ can be computed by an appropriate values of δ, γ, and h.

3.2 Conditions for Determination of Deterministic and Transversal Discrete Transition

Now, we elaborate in more detail on the result given in Lemma 2, especially on (iii) and (iv), to develop some conditions which are used in Section 5.

---

[3]We can prove safety as long as there exists a minimal separation between the unsafe set and the actual bounded reach set.

LEMMA 3. *Given a location l_c, if $\mathcal{D}_{t-h}(\mathcal{B}_\delta(x_0)) \subset (Inv_c)^\circ$ and $\mathcal{D}_t(\mathcal{B}_\delta(x_0)) \subset Inv_c^C$ for some $\delta > 0$ and $h > 0$ where $\mathcal{B}_\delta(x_0)$ is a δ-neighborhood of the initial state x_0, then there is a discrete transition from the location l_c within time $(t - h, t)$.*

PROOF. Note $\mathcal{D}_t(x_0) \in \mathcal{D}_t(\mathcal{B}_\delta(x_0))$, where $\mathcal{D}_t(x_0)$ is the reached state at time t from x_0. Similarly, $\mathcal{D}_{t-h}(x_0) \in \mathcal{D}_{t-h}(\mathcal{B}_\delta(x_0))$. From the hypothesis, $\mathcal{D}_t(x_0) \in Inv_c^C$ and $\mathcal{D}_{t-h}(x_0) \in (Inv_c)^\circ$. This implies that there exists $\tau \in (t - h, t)$ such that $\mathcal{D}_s(x_0) \in Inv_c^\circ$ for $s \in [t - h, \tau)$ and $\mathcal{D}_s(x_0) \in Inv_c^C$ for $s \in (\tau, t]$. Hence there is a discrete transition at some time $\tau \in (t - h, t)$. $\square$

LEMMA 4. *Given a polyhedron $\mathcal{P}_t$ at time t, suppose that there is a discrete transition from a location l_c to some other locations, i.e., $\mathcal{P}_{t-h} \subset (Inv_c)^\circ$ and $\mathcal{P}_t \subset Inv_c^C$ for some $h > 0$. Then the discrete transition is deterministic if there exists a location l_n such that $l_n \neq l_c$ and $\mathcal{P}_t \subset (Inv_n)^\circ$.*

PROOF. By Definition 5, the result is trivially true. $\square$

LEMMA 5. *Given polyhedron $\mathcal{P}_t$ at time t, $\gamma > 0$, and $h > 0$ satisfying (6), suppose that there is a deterministic discrete transition from a location l_c to a location l_n, i.e., $\mathcal{P}_{t-h} \subset (Inv_c)^\circ$ and $\mathcal{P}_t \subset (Inv_n)^\circ$ for some $h > 0$. Then for any $\epsilon > 0$, the discrete transition is transversal if the following conditions hold.*

(i) $h < (dia(\mathcal{J}_{c,n})/2)/(2\bar{v})$,

(ii) $\mathcal{D}_0(\mathcal{J}_{c,n}, dia(\mathcal{J}_{c,n})/2) \subset (Inv_c \cup Inv_n)$,

(iii) $\langle \dot{x}_c, \vec{n} \rangle \geq \epsilon \wedge \langle \dot{x}_n, \vec{n} \rangle \geq \epsilon, \quad \forall x \in \mathcal{V}(\mathcal{J}'_{c,n})$,

where $\mathcal{J}_{c,n} := \mathcal{D}_0(\mathcal{P}_t, \gamma) \cap Inv_c \cap Inv_n$, $\mathcal{J}'_{c,n} := \mathcal{D}_0(\mathcal{J}_{c,n}, dia(\mathcal{J}_{c,n})/2) \cap Inv_c \cap Inv_n$, $\bar{v}$ is as defined in (6), $\mathcal{V}(\mathcal{P})$ is a set of vertices of a polyhedron $\mathcal{P}$, $\vec{n}$ is an outward normal vector of ∂Inv_c, and $\dot{x}_i$ is the vector flow evaluated with respect to the LTI dynamics of location $l_i \in \mathbb{L}$.

PROOF. First note that $\mathcal{P}_{t-h} \subset (Inv_c)^\circ$ and $\mathcal{P}_t \subset (Inv_n)^\circ$, since there is a deterministic discrete transition from l_c to l_n. Since γ and h satisfy (6), $\mathcal{P}_{t-h} \subset \mathcal{D}_0(\mathcal{P}_t, \gamma)$. In fact, $\cup_{z \in \mathcal{P}_{t-h}} x(\tau; z) \subset \mathcal{D}_0(\mathcal{P}_t, \gamma)$ for $\tau \in [0, h]$ where $x(\tau; z) := e^{A_c \tau} z + \int_0^\tau e^{A_c s} u_c ds$. Since $\mathcal{D}_{t-h}(x_0) \in \mathcal{P}_{t-h}$ and $\mathcal{D}_t(x_0) \in \mathcal{P}_t$, $\mathcal{D}_{\tau'}(x_0) \in \mathcal{J}_{c,n} := \mathcal{D}_0(\mathcal{P}_t, \gamma) \cap Inv_c \cap Inv_n$ for some $\tau' \in (t - h, t)$ where $\mathcal{D}_{\tau'}(x_0)$ is a discrete transition state from l_c to l_n at time τ'. Thus $\mathcal{J}_{c,n} \neq \emptyset$ (more precisely, $\mathcal{J}_{c,n}^\circ \neq \emptyset$) and it is in fact an over-approximation of the deterministic discrete transition state $x_{\tau'} \in Inv_c \cap Inv_n$.

Notice that if (i) holds, then $\|x(h; z) - z\| < dia(\mathcal{J}_{c,n})/4 < dia(\mathcal{J}_{c,n})/2$ for any $z \in \mathcal{J}_{c,n}$ since $\|x(h; z) - z\| \leq \bar{v}h$ where $x(h; z)$ is the state reached from z at time h under the LTI dynamics of the location l_n and $\bar{v}$ is as defined in (6). Also notice that if (ii) and (iii) hold, then for any $z' \in \mathcal{J}'_{c,n}$, z' satisfies the deterministic and transversal discrete transition condition in Definition 5. If we now consider the fact that $dia(\mathcal{J}'_{c,n}) \geq 2 \cdot dia(\mathcal{J}_{c,n})$, then $x(\tau; z) \in Inv_n^\circ$ for $\tau \in (0, h)$. Since $z \in \mathcal{J}_{c,n}$ is arbitrary, it is easy to see that $\mathcal{D}_\tau(\mathcal{J}_{c,n}) \in Inv_n^\circ$ for $\tau \in (0, h)$. $\square$

3.3 Bounded ϵ-Reachability of a DTLHA with Finite Precision Calculations

The results in Section 3.1 and 3.2 rely on the assumption that the following quantities can be computed exactly:

- $x(t; x_0) = e^{At}x_0 + \int_0^t e^{As} u ds$.

- $\mathcal{H} \cap \mathcal{P}$ where $\mathcal{H}$ is a hyperplane and $\mathcal{P}$ is a polyhedron.

- $hull(\mathcal{V})$ where $hull(\mathcal{V})$ is the convex hull of $\mathcal{V}$ which is a finite set of points in $\mathbb{R}^n$.

However, these exact computation assumptions cannot be satisfied in practice and we can only compute each of these with possibly arbitrarily small computation error. In this section, we extend the theory to incorporate the numerical computation errors.

3.3.1 Approximate Numerical Computations

In the sequel, we use $a(x, y)$ to denote an approximate computation of x with $y \in \mathbb{R}^+$ as an upper bound on the approximation error. The precise definition depends on the types of x:

- If x is a vector or a matrix, then $\|x - a(x, y)\| \leq y$.

- If x is a set, then $d_H(x, a(x, y)) \leq y$ where $d_H(x, z)$ is the Hausdorff distance.

We assume that a set of subroutines or functions are available for approximately computing these quantities, which we use to compute a bounded ϵ-reach set. More precisely, for given μ_c and μ_h, $a(\mathcal{H} \cap \mathcal{P}, \mu_c)$ and $a(hull(\mathcal{V}), \mu_h)$ are available. Moreover, we also assume that a set of approximate computations, specifically, of $a(e^{At}, \sigma_e)$, $a(\int_0^t e^{A\tau} d\tau, \sigma_i)$, $a(A \cdot b, \sigma_p)$, and $a(u + v, \sigma_a)$, are available in computing $x(t; x_0)$ for given approximation errors $\sigma_e, \sigma_i, \sigma_p$, and σ_a. From these approximate computational capabilities, we can derive an upper bound on the approximation error, denoted as μ_x, for $x(t; x_0)$. We first note that, for all approximate computations $a(x, y)$ that are used for computing $x(t; x_0)$, we have $(x - y \cdot \mathbf{1}_{n \times m}) \leq a(x, y) \leq (x + y \cdot \mathbf{1}_{n \times m})$ where $x \in \mathbb{R}^{n \times m}$ and $\mathbf{1}_{n \times m}$ is an n by m matrix whose every element is 1. With this, we derive μ_x as follows.

$$e^{At} - \sigma_e \cdot \mathbf{1}_{n \times n} \leq a(e^{At}, \sigma_e) \leq e^{At} + \sigma_e \cdot \mathbf{1}_{n \times n},$$

$$e^{At}x_0 - (\sigma_e|x_0| + \sigma_p) \cdot \mathbf{1}_{n \times 1} \leq a(e^{At}x_0, \sigma_p)$$
$$\leq e^{At}x_0 + (\sigma_e|x_0| + \sigma_p) \cdot \mathbf{1}_{n \times 1}.$$

Similarly,

$$\int_0^t e^{As} ds \cdot u - (\sigma_i|u| + \sigma_p) \cdot \mathbf{1}_{n \times 1} \leq a(\int_0^t e^{As} ds \cdot u, \sigma_p)$$
$$\leq \int_0^t e^{As} ds \cdot u + (\sigma_i|u| + \sigma_p) \cdot \mathbf{1}_{n \times 1}.$$

Hence, we have

$$x(t; x_0) - \delta_x \leq a(x(t; x_0), \delta_x) \leq x(t; x_0) + \delta_x,$$

where $\delta_x := (2\sigma_p + \sigma_a + \sigma_e|x_0| + \sigma_i|u|) \cdot \mathbf{1}_{n \times 1}$.

Now, we define μ_x as the maximum of $|\delta_x|$ over the continuous state space $\mathcal{X}$ and the control input domain $\mathcal{U}$,

$$\mu_x := \max_{x \in \mathcal{X}, u \in \mathcal{U}} |\delta_x|. \tag{8}$$

3.3.2 Incorporation of Finite Precision Calculations in Bounded ϵ-Reachability of a DTLHA

In this section, we extend the result given in Sections 3.1 and 3.2 to relax the infinite precision computation assumption. Especially, we extend the results in Lemmas 1, 3, 4, and 5. We first discuss how the relation between h and γ in Lemma 1 can be changed under finite precision computation.

LEMMA 6. *Let $\rho > 0$ be an upper bound on the approximation errors of $a(x(t), \rho)$ for some $x(t) \in \mathbb{R}^n$ and some time $t > 0$. Then for a given LTI system $\dot{x} = Ax + u$, if h satisfies $h < (\gamma - \rho)/(\|A\|\bar{x} + \|u\|)$ for any given $\gamma > \rho$, where $\bar{x}$ is as defined in (6), then the following property holds:*

$$\bigcup_{z \in \mathcal{B}_\rho(x(t))} x(\tau; z) \subset \mathcal{B}_\gamma(x(t)), \quad \forall \tau \in [0, h], \qquad (9)$$

where $x(\tau; z) = e^{A\tau}z + \int_0^\tau e^{As} u \, ds$ and $\mathcal{B}_y(x)$ is a y-neighborhood around x.

PROOF. Notice that $a(x(t), \rho) \in \mathcal{B}_\rho(x(t))$ and for any $x(t) \in \mathcal{X}$,

$$\begin{aligned}
\|x(t+h) - x(t)\| &\leq \int_t^{t+h} \|\dot{x}(s)\| ds \\
&\leq (\|A\|\bar{x} + \|u\|)h.
\end{aligned}$$

Since $h < (\gamma - \rho)/(\|A\|\bar{x} + \|u\|)$, $\|x(t+h) - x(t)\| < \gamma - \rho$ for any $x(t) \in \mathcal{X}$. In fact, $\|x(t+s) - x(t)\| < \gamma - \rho$ for all $s \in [0, h]$. Hence for any $z \in \mathcal{B}_\rho(x(t))$, $x(s; z) \in \mathcal{B}_{\gamma-\rho}(z)$ for $s \in [0, h]$. This implies that for any $z \in \mathcal{B}_\rho(x(t))$, $\|x(t) - x(s; z)\| \leq \|x(t) - z\| + \|z - x(s; z)\| \leq \gamma$. $\square$

LEMMA 7. *Given $\rho > 0$ for $a(\mathcal{D}_t(\mathcal{B}_\delta(x_0)), \rho)$, let $\mathcal{P}_t := \mathcal{D}_t(\mathcal{B}_\delta(x_0)), \rho)$. Then if h satisfies the inequality in (10) for a given $\gamma > \rho$, then $\mathcal{D}_\tau(\mathcal{P}_t) \subset \mathcal{D}_t(\mathcal{B}_\delta(x_0), \gamma)$ for all $\tau \in [0, h]$:*

$$h < \frac{\gamma - \rho}{\bar{v}} \qquad (10)$$

where $\bar{v}$ is as defined in (6).

PROOF. Let $\mathcal{V}$ and $\mathcal{V}'$ be the set of extreme points of $\mathcal{D}_t(\mathcal{B}_\delta(x_0))$ and $\mathcal{P}_t$, respectively. Since (10) hold, we know from Lemma 6 that for each $x(t) \in \mathcal{V}$, $\mathcal{D}_\tau(\mathcal{B}_\rho(x(t))) \subset \mathcal{B}_\gamma(x(t))$ for all $\tau \in [0, h]$. Since for each $z \in \mathcal{V}'$, there exists $x(t) \in \mathcal{V}$ such that $\|x(t) - z\| \leq \rho$, Notice that for each $z \in \mathcal{V}'$, $z \in \mathcal{B}_\rho(x(t))$ for some $x(t) \in \mathcal{V}$. Therefore, $\mathcal{D}_\tau(\mathcal{P}_t) \subset \mathcal{D}_t(\mathcal{B}_\delta(x_0), \gamma)$ for all $\tau \in [0, h]$. $\square$

In the sequel, we use $\hat{x}$ to denote $a(x, \rho)$ for a given approximation error bound $\rho > 0$ for simplicity of notation.

The condition (ii) in Lemma 2 enforces the size of an over-approximation of each sampled state along $\mathcal{R}_{t_f}(x_0)$ to be less than the given $\epsilon > 0$. Under the finite precision calculations, it is straightforward to extend the result of (ii) in Lemma 2 as shown in the following Lemma.

LEMMA 8. *Given $\epsilon > 0$, $\rho > 0$, a polyhedron $\mathcal{P}$, and $\hat{\mathcal{P}}$, if $dia(\hat{\mathcal{P}}) < \epsilon - \rho$, then $dia(\mathcal{P}) < \epsilon$.*

PROOF. Recall that $\hat{\mathcal{P}} := a(\mathcal{P}, \rho)$. This implies $d_H(\mathcal{P}, \hat{\mathcal{P}}) < \rho$. Hence $\mathcal{P} \subset \mathcal{D}_0(\hat{\mathcal{P}}, \rho)$. Notice that $dia(\mathcal{D}_0(\hat{\mathcal{P}}, \rho)) \leq dia(\hat{\mathcal{P}}) + \rho$. Hence $dia(\hat{\mathcal{P}}) < \epsilon - \rho$ implies $dia(\mathcal{P}) < \epsilon$. $\square$

Now we address the issue of numerical computation error in determining a deterministic and transversal discrete transition. The conditions developed in the following lemmas are sufficient in that if they are satisfied by a given polyhedron $\hat{\mathcal{P}}$ with a given approximation error ρ at time t, then there is a deterministic and transversal discrete transition at some time $\tau \in [t - h, t]$.

LEMMA 9. *Given $\rho > 0$, a location l_c, and $\hat{\mathcal{D}}_t(\mathcal{B}_\delta(x_0))$ at time t, if $\hat{\mathcal{D}}_{t-h}(\mathcal{B}_\delta(x_0), \rho) \subset (Inv_c)^\circ$ and $\hat{\mathcal{D}}_t(\mathcal{B}_\delta(x_0), \rho) \subset Inv_c^C$ for some $\delta > 0$ and $h > 0$, then there is a discrete transition from the location l_c.*

PROOF. Since $d_H(\mathcal{D}_t(\mathcal{B}_\delta(x_0)), \hat{\mathcal{D}}_t(\mathcal{B}_\delta(x_0))) \leq \rho$, $\mathcal{D}_t(\mathcal{B}_\delta(x_0)) \subset \hat{\mathcal{D}}_t(\mathcal{B}_\delta(x_0), \rho)$. Similarly, $\mathcal{D}_{t-h}(\mathcal{B}_\delta(x_0)) \subset \hat{\mathcal{D}}_{t-h}(\mathcal{B}_\delta(x_0), \rho)$. Hence $\mathcal{D}_t(\mathcal{B}_\delta(x_0)) \subset Inv_c^C$ and $\mathcal{D}_{t-h}(\mathcal{B}_\delta(x_0)) \subset (Inv_c)^\circ$. Then the result follows immediately from Lemma 3. $\square$

LEMMA 10. *Given $\rho > 0$, a location l_c, and a polyhedron $\mathcal{P}_t$ at time t, suppose that there is a discrete transition from a location l_c to some other locations, i.e., $\mathcal{D}_0(\hat{\mathcal{P}}_{t-h}, \rho) \subset (Inv_c)^\circ$ and $\mathcal{D}_0(\hat{\mathcal{P}}_t, \rho) \subset Inv_c^C$ for some $h > 0$. Then there is a deterministic discrete transition from l_c to l_n if there exists a location l_n such that $l_n \neq l_c$ and $\mathcal{D}_0(\hat{\mathcal{P}}_t, \rho) \subset (Inv_n)^\circ$.*

PROOF. Since $\mathcal{P}_{t-h} \subset \mathcal{D}_0(\hat{\mathcal{P}}_{t-h}, \rho)$, $\mathcal{P}_{t-h} \subset (Inv_c)^\circ$. Similarly, $\mathcal{P}_t \subset (Inv_n)^\circ$ since $\mathcal{P}_t \subset \mathcal{D}_0(\hat{\mathcal{P}}_t, \rho)$. Then by Lemma 4, the conclusion holds. $\square$

LEMMA 11. *Given $\rho > 0$, $\gamma > 0$ and $h > 0$ satisfying (10), and a polyhedron $\mathcal{P}_t$ at time t, suppose that there is a deterministic discrete transition from a location l_c to a location l_n, i.e., $\mathcal{D}_0(\hat{\mathcal{P}}_{t-h}, \rho) \subset (Inv_c)^\circ$ and $\mathcal{D}_0(\hat{\mathcal{P}}_t, \rho) \subset (Inv_n)^\circ$ for some $h > 0$. Then for any $\epsilon > 0$, the discrete transition is transversal if the following conditions hold:*

(i) $h < (dia(\hat{\mathcal{J}}_{c,n})/2)/(2\bar{v})$,

(ii) $\mathcal{D}_0(\hat{\mathcal{J}}_{c,n}, dia(\hat{\mathcal{J}}_{c,n})/2) \subset (Inv_c \cup Inv_n)$,

(iii) $\langle \dot{x}_c, \vec{n} \rangle \geq \epsilon \wedge \langle \dot{x}_n, \vec{n} \rangle \geq \epsilon, \quad \forall x \in \mathcal{V}(\hat{\mathcal{J}}'_{c,n})$,

where $\hat{\mathcal{J}}_{c,n} := \mathcal{D}_0(\hat{\mathcal{P}}_t, \gamma + \rho) \cap Inv_c \cap Inv_n$, $\hat{\mathcal{J}}'_{c,n} := \mathcal{D}_0(\hat{\mathcal{J}}_{c,n}, dia(\hat{\mathcal{J}}_{c,n})/2) \cap Inv_c \cap Inv_n$, and $\dot{x}_i$ and $\vec{n}$ are as defined in Lemma 5.

PROOF. Notice that $\mathcal{D}_0(\mathcal{P}_t, \gamma) \subset \mathcal{D}_0(\hat{\mathcal{P}}_t, \gamma + \rho)$ since $d_H(\mathcal{P}_t, \hat{\mathcal{P}}_t) \leq \rho$. Then, by the definition of $\mathcal{J}_{c,n}$ given in Lemma 5 and $\hat{\mathcal{J}}_{c,n}$, we know $\mathcal{J}_{c,n} \subset \hat{\mathcal{J}}_{c,n}$. Hence, $\hat{\mathcal{J}}_{c,n} \neq \emptyset$ and in fact it is an over-approximation of the deterministic discrete transition state as is $\mathcal{J}_{c,n}$ in Lemma 5.

By the same argument used in the proof of Lemma 5, if (i) holds, then $\mathcal{D}_\tau(\hat{\mathcal{J}}_{c,n}) \subset \mathcal{D}_0(\hat{\mathcal{J}}_{c,n}, dia(\hat{\mathcal{J}}_{c,n})/2)$ for $\tau \in (0, h)$. Hence, (ii) and (iii) imply that $\mathcal{D}_\tau(\hat{\mathcal{J}}_{c,n}) \subset Inv_n^\circ$ for $\tau \in [0, h]$. Therefore, the conclusion holds since $\mathcal{J}_{c,n} \subset \hat{\mathcal{J}}_{c,n}$. $\square$

4. ARCHITECTURE

In [7], we have shown that an approximate bounded reach set of a DTLHA from an initial continuous state x_0 in an initial location l_0 can be computed with arbitrarily small approximation error ϵ under the assumption that every discrete transition is deterministic and transversal. We also have proposed an algorithm for such bounded ϵ-reach set computation. Algorithm 1 shows the main computation steps of the proposed algorithm where $\mathcal{B}_\delta(x_0)$ is an δ-neighborhood of a given initial state x_0, $\mathcal{D}_t(\mathcal{B}_\delta(x_0), \gamma)$ is a γ-approximation of $\mathcal{D}_t(\mathcal{B}_\delta(x_0))$, and h is a sampling period corresponding to the value of γ satisfying an over-approximation condition in (6). Note that $\mathcal{R}$ returned from the algorithm is in fact a bounded ϵ-reach set as stated in the following theorem.

THEOREM 1. *Given input $(\mathcal{A}, N, T, l_0, x_0, \epsilon)$, Algorithm 1 terminates in a finite number of iterations and returns $\mathcal{R}$, a* bounded ϵ-reach set of $\mathcal{A}$ from $x_0 \in Inv_0$ up to time $t_f :=$ $\min\{\tau_N, T\}$, if $\mathcal{A}$ is a DTLHA up to t_f:

$$\mathcal{R} := \bigcup_{m=0}^{M} \mathcal{D}_{mh}(\mathcal{B}_\delta(x_0), \gamma) \qquad (11)$$

where δ, γ, h are the values when Algorithm 1 returns and M is a value such that $Mh \in [t_f, t_f + h]$.

Algorithm 1: An algorithm proposed in [7] for a bounded ϵ-reach set of a DTLHA $\mathcal{A}$ from an initial state $x_0 \in Inv_0$.

Input: $\mathcal{A}, N, T, l_0, x_0, \epsilon$

Initialize δ and γ with arbitrary positive real values.
• Initialize $t = 0$, $jump = 0$, and $\mathcal{R} = \emptyset$.
while true **do**
    Compute $\mathcal{B}_\delta(x_0)$ and h from γ.
    Compute $\mathcal{D}_t(\mathcal{B}_\delta(x_0))$ and $\mathcal{D}_t(\mathcal{B}_\delta(x_0), \gamma)$.
    if $dia(\mathcal{D}_t(\mathcal{B}_\delta(x_0), \gamma)) > \epsilon$ **then**
      | Reduce δ, γ and goto •
    if discrete transition **then**
      if deterministic $\wedge$ transversal **then**
        | $t \leftarrow t + h$
        | $jump \leftarrow jump + 1$
        | $\mathcal{R} \leftarrow \mathcal{R} \cup \mathcal{D}_t(\mathcal{B}_\delta(x_0), \gamma)$
      else Reduce δ, γ and goto •
    else $t \leftarrow t + h$ and $\mathcal{R} \leftarrow \mathcal{R} \cup \mathcal{D}_t(\mathcal{B}_\delta(x_0), \gamma)$.
    if $(t \geq T) \vee (jump \geq N)$ **then return** $\mathcal{R}$
end

In the above algorithm and accompanying theoretical results for a bounded ϵ-reach set computation in [7], we have not addressed the issue of computation with finite precision. Moreover, even though the proposed algorithm can compute a bounded ϵ-reach set correctly, it is far from being computationally efficient since the algorithm restarts its ϵ-reach set computation from an initial state x_0 at time $t = 0$ whenever the values of δ and γ are changed. Moreover, the algorithm does not provide any flexibility in choosing the values for δ and γ whenever the algorithm needs to be continued with different δ and γ values, since one specific decision rule resulting in δ and γ which are monotonically decreasing is tightly embedded within the algorithm. We address all these issues in the remainder of this section.

It is helpful to take a top-down approach since we want to address *modularity*, *flexibility*, and *architecture*. Hence we begin by presenting an architecture for a bounded ϵ-reach set computation followed by a new bounded ϵ-reach set algorithm which is based on the theoretical results developed in Section 3 so that the overall computation process can be better optimized in terms of computational efficiency and flexibility.

One of the main objectives of the architecture design is to provide flexibility. We argue that this can be achieved by decoupling the part where decisions are made, called *Policy*, and the part where some specific steps of computation are performed, which is called *Algorithm* in our context (but called *Mechanism* in some other contexts). Fig. 2 shows a proposed architecture based on this design principle. As mentioned above, the proposed architecture consists of roughly four different parts which are Policy (*Policy* module), Algorithm (*Main Algorithm* and *Condition Checking*

modules), Data (*System Description* and *Data* modules), and *Numerical Calculation* module. A more detail explanation of each of these modules is given in below.

Policy.

This module contains a user-defined rule to choose appropriate values of the parameters, especially δ and γ as shown in Algorithm 1, that are needed to continue to compute a bounded ϵ-reach set of a DTLHA when an ambiguous situation is encountered in the Main Algorithm module. Furthermore, this module can make a decision about the choice of numerical calculation algorithms which affect to the computational accuracy for each approximate numerical function defined in Section 3.3.1.

System Description.

The System Description module contains information about the system, described by a modeling language; it consists of $\mathcal{X}$ the domain of continuous state space, a DTLHA $\mathcal{A}$, and an initial continuous state x_0, an initial location l_0 (i.e., discrete state) where x_0 is contained. Also, to specify the required computation, an upper bound T on terminal time, an upper bound N on the total number of discrete transitions, and an approximation parameter ϵ, are described in the System Description module. In short, all information required to describe a problem of a bounded ϵ-reach set computation of a DTLHA is contained in the System Description module.

Data.

The data generated by the System Description module, called `SystemData`, is stored in the Data module which can then be used by the rest of the modules in the architecture. Furthermore, the data which are generated on-the-fly in a bounded ϵ-reach set computation by the Main Algorithm module, called `ReachSetHistory` and `TransitionHistory`, are also stored in this module.

Condition Checking.

To ensure a correct bounded ϵ-reach set computation, a bounded ϵ-reach set algorithm needs to correctly (i) detect a deterministic and transversal discrete transition if there is one, (ii) determine whether the size of the set computed as an over-approximation of the reach set between samples is smaller than the specified parameter ϵ, and (iii) check whether a sampling period h and an over-approximation parameter γ satisfy the relation for over-approximation guarantee. All functions which implement these condition checkings are contained in this module. More detail on this module is given in Section 5.

Main Algorithm.

With the inputs from the Policy module, the Main Algorithm computes a bounded ϵ-reach set utilizing Sub-functions and functions from the Condition Checking module until it either successfully finishes its computation, or cannot make further progress which happens when some required conditions are not met. If the algorithm encounters the latter situation, then it returns to the Policy module indicating the problems that the Policy module has to resolve so as to continue the computation. During a bounded ϵ-reach set computation, the Main Algorithm stores its computational state in two data structures, called `ReachSetHistory`

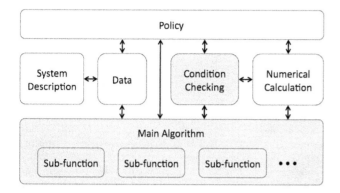

Figure 2: A architecture for bounded ϵ-reach set computation

and `TransitionHistory`, which are in turn stored in Data module. `ReachSetHistory` contains the computation results and the information used to produce the results at each step of computation, described in more detail in Section 5. One of the most important benefits of maintaining this information is that the computational efficiency can be improved significantly since the computation does not need to be restarted from the initial time and state whenever new parameter values, such as a smaller δ and γ, have to be used to continue the computation. Under this architecture, the Policy module can go back to any past computational step and make the Main Algorithm restart the computation from that point. In `TransitionHistory`, the information about the discrete transition is stored. Maintaining this information in `TransitionHistory`, also contributes to improving efficiency of the overall bounded ϵ-reach set computation process. More detail on this module is given in Section 5.

Numerical Calculation.

This module contains a collection of numerical functions for computing a matrix exponential, an integral of a matrix exponential, the intersection between polyhedra, a convex hull of a finite set of points, and so on. Each of these functions is in fact an implementation of some computational algorithms. As an example, $a(e^{At}, \sigma_e)$ can be computed in many different ways as shown in [9]. Each of the different algorithms can computes e^{At} with different accuracy. Hence, the computational accuracy of a bounded ϵ-reach set computation inevitably depends on the choice of the algorithms for computation of each of the $a(x, y)$'s assumed in above. We decouple such issues arising in the low level numerical calculations from our proposed bounded ϵ-reach set algorithm, which is the reason for the separate module for numerical calculation in our architecture.

5. ALGORITHM

The proposed algorithm for a bounded ϵ-reach set computation is decomposed into roughly two parts, the Main Algorithm module and the Condition Checking module. In this section, we discuss these modules in more detail. Recall that we use $\hat{x}$ to denote $a(x, \rho)$ for some given approximation error bound $\rho \in \mathbb{R}^+$. In particular, a polyhedron $\hat{P}$ in the sequel should be understood as an approximation of a polyhedron P, i.e., $\hat{P} := a(P, \rho)$.

5.1 Condition Checking Module

In computing a bounded ϵ-reach set, the following set of questions needs to be answered at each step of computation in the Main Algorithm to produce a correct result:

1. Given δ and γ, is the diameter of $\mathcal{D}_t(\mathcal{B}_\delta(x_0), \gamma)$ at current sample time t less than the given ϵ?

2. Given δ, γ, and h, is $\mathcal{D}_\tau(\mathcal{B}_\delta(x_0)) \subset \mathcal{D}_t(\mathcal{B}_\delta(x_0), \gamma)$ for all $\tau \in [t, t+h]$ at current sample time t?

3. Given h, δ, and $\mathcal{D}_t(\mathcal{B}_\delta(x_0))$, can we conclude that a discrete transition has occurred between $t - h$ and t?

4. If there is a discrete transition as above, is it a deterministic discrete transition?

5. If there is a deterministic discrete transition above, is it a transversal discrete transition?

Corresponding to these questions, the Condition Checking module consists of the following set of functions which are based on the results in Section 3.3.2.

IsEpsilonSmall($\cdot$).

Given $\epsilon > 0$ and a polyhedron $\hat{P}$, this function determines whether $dia(P) < \epsilon$ or not, where $dia(P)$ denotes the diameter of a polyhedron P. As shown in Lemma 8, $dia(P) < \epsilon$ if $dia(\hat{P}) < \epsilon - \rho$. Hence, this function returns `true` if $dia(\hat{P}) < \epsilon - \rho$.

IsOverApproximate($\cdot$).

Given γ and ρ, this function determines whether a sampling period h and γ satisfy the condition (10). Hence, if $h < (\gamma - \rho)/\bar{v}$, this function returns `true` where $\bar{v}$ is as defined (6).

IsTransition($\cdot$).

Given a sampling period h, a location l_c, a polyhedron $\hat{P}_t$ at time t, this function checks if there is a discrete transition from a location l_c at some time in between $t - h$ and t. In Lemma 9, it is shown that if $\mathcal{D}_0(\hat{P}_{t-h}, \rho) \subset (Inv_c)^\circ$ and $\mathcal{D}_0(\hat{P}_t, \rho) \subset Inv_c^C$, then there is indeed a discrete transition at some time in $(t - h, t)$. Assuming that $\mathcal{D}_0(\hat{P}_{t-h}, \rho) \subset (Inv_c)^\circ$ is satisfied at time $t - h$, this function returns `true` if $\mathcal{D}_0(\hat{P}_t, \rho) \subset Inv_c^C$. If $\mathcal{D}_0(\hat{P}_t, \rho) \subset (Inv_c)^\circ$, then this function returns `false`. In the other cases that $(\mathcal{D}_0(\hat{P}_t, \rho) \cap Inv_c \neq \emptyset) \wedge (\mathcal{D}_0(\hat{P}_t, \rho) \cap Inv_c^C \neq \emptyset)$, this function returns `error` to inform that other values of δ or h need to be used to resolve the ambiguity.

IsDeterministic($\cdot$).

Given a location l_c and a polyhedron $\hat{P}$, this function checks if a discrete transition from l_c is a deterministic transition to some other location l_n. Based on the result in Lemma 10, this function returns the location l_n if there is a location $l_n \in \mathbb{L}$ such that $l_n \neq l_c$ and $\mathcal{D}_0(\hat{P}, \rho) \subset (Inv_n)^\circ$. Otherwise it returns `error`.

IsTransversal($\cdot$).

Given h, γ, and a polyhedron $\hat{P}$, this function checks if a discrete transition from a location l_c to other location l_n is a transversal discrete transition or not, using the conditions (i), (ii), and (iii) in Lemma 11. If it is a transversal discrete transition, this function returns $\hat{\mathcal{D}}_h(\hat{\mathcal{J}}_{c,n}, \rho')$, where $\hat{\mathcal{D}}_h(\hat{\mathcal{J}}_{c,n})$ is an approximation of $\mathcal{D}_h(\hat{\mathcal{J}}_{c,n})$ which is the image of $\hat{\mathcal{J}}_{c,n}$ at time h under the linear dynamics of a location l_n. Otherwise, it returns error. Note that ρ' is a numerical calculation error which is introduced during the computation of $\mathcal{D}_h(\hat{\mathcal{J}}_{c,n})$ from $\hat{\mathcal{J}}_{c,n}$.

5.2 Main Algorithm Module

Roughly, the Main Algorithm module consists of two parts. The first part is a function called ReachSet($\cdot$) which is the main function to compute a bounded ϵ-reach set, and the second part is a set of functions called *Sub-functions* which are called by ReachSet($\cdot$) during its computation. We first describe the functions defined as Sub-functions.

ReachNext($\cdot$).

Given h, γ, and a polyhedron $\hat{\mathcal{P}}$, this function returns $\hat{\mathcal{D}}_h(\hat{\mathcal{P}})$, an approximation of the linear image of a polyhedron $\hat{\mathcal{P}}$ at time h under a linear dynamics, and $\hat{\mathcal{D}}_h(\hat{\mathcal{P}}, \gamma)$, an over-approximation of $\hat{\mathcal{D}}_h(\hat{\mathcal{P}})$ for a given over-approximation parameter γ. This function also returns estimates of the upper bound of computation errors ρ' and ρ'' along with $\hat{\mathcal{D}}_h(\hat{\mathcal{P}})$ and $\hat{\mathcal{D}}_h(\hat{\mathcal{P}}, \gamma)$, so that ReachSet($\cdot$) function can keep track of the numerical errors accumulated from the initial time up to the current time t. Notice that ρ' and ρ'' are defined via $d_H(\mathcal{D}_h(\mathcal{P}), \hat{\mathcal{D}}_h(\hat{\mathcal{P}})) \leq \rho'$ and $d_H(\mathcal{D}_h(\mathcal{P}, \gamma), \hat{\mathcal{D}}_h(\hat{\mathcal{P}}, \gamma)) \leq \rho''$, respectively.

To compute $\hat{\mathcal{D}}_h(\hat{\mathcal{P}})$ and $\hat{\mathcal{D}}_h(\hat{\mathcal{P}}, \gamma)$, this function exploits the fact that the polyhedral structure is preserved under a linear dynamics in the following way. Given polyhedron $\hat{\mathcal{P}}$, this function first computes $\mathcal{V}(\hat{\mathcal{P}})$ which is a set that contains the vertices of $\hat{\mathcal{P}}$, and possibly some other points in $\hat{\mathcal{P}}$. (The reason for allowing some other points that are possibly not vertices is because $\hat{\mathcal{P}}$ is itself computed as the linear image of a finite number of points, and we would like to avoid the need to computationally determine precisely which remain extreme points under the linear map). Then for each $v_i \in \mathcal{V}(\hat{\mathcal{P}})$, it computes $v_i(h) := e^{Ah}v_i + \int_0^h e^{As}uds$ where A and u are given by the linear dynamics of a location on which the linear image of $\hat{\mathcal{P}}$ is computed. If we let $\mathcal{V}_h(\hat{\mathcal{P}}) := \{v_i(h) : v_i \in \mathcal{V}(\hat{\mathcal{P}})\}$, then $\mathcal{D}_h(\hat{\mathcal{P}}) := hull(\mathcal{V}_h(\hat{\mathcal{P}}))$ where $hull(\mathcal{V}_h(\hat{\mathcal{P}}))$ is the convex hull of $\mathcal{V}_h(\hat{\mathcal{P}})$. Notice that what we really have here is $\hat{\mathcal{D}}_h(\hat{\mathcal{P}})$, and not $\mathcal{D}_h(\hat{\mathcal{P}})$, since there is a numerical calculation error in the $hull(\mathcal{V}_h(\hat{\mathcal{P}}))$ computation. From $\mathcal{V}_h(\hat{\mathcal{P}})$, this function can also compute $\hat{\mathcal{D}}_h(\hat{\mathcal{P}}, \gamma)$ easily. For each $v_i(h) \in \mathcal{V}_h(\hat{\mathcal{P}})$, it first constructs a hypercubic γ-neighborhood of $v_i(h)$. If we denote such a neighborhood by $\mathcal{B}_\gamma(v_i(h))$, then the convex hull of the set of vertices of $\mathcal{B}_\gamma(v_i(h))$ for all $v_i(h) \in \mathcal{V}_h(\hat{\mathcal{P}})$ defines a $\hat{\mathcal{D}}_h(\hat{\mathcal{P}}, \gamma)$.

AtTransition($\cdot$).

This function is called by ReachSet($\cdot$) when a discrete transition from a given location l_c is detected by IsTransition ($\cdot$). Then this function internally calls IsDeterministic ($\cdot$) and IsTransversal($\cdot$) functions to check if this discrete transition is deterministic and transversal. If it is, then this function returns a location l_n which is returned by IsDeterministic($\cdot$) and $\hat{\mathcal{D}}_h(\hat{\mathcal{J}}_{c,n}, \rho')$ which is returned by IsTransversal($\cdot$). However, if any of these functions returns error, this function returns the same error to in-

dicate the necessity of a decision in the Policy module to resolve the erroneous situation.

ImageAt($\cdot$).

Even though the overall computational efficiency of a bounded ϵ-reach set computation can be improved by the proposed architecture, it is unavoidable to restart the computation from an initial state when the value of parameter δ which defines an initial neighborhood around an initial state is changed. If the algorithm encounters such a situation, ImageAt($\cdot$) can be used to reduce the number of computational steps. Given t and δ, the goal of this function is to compute $\mathcal{D}_t(\mathcal{B}_\delta(x_0))$. More precisely, this function computes $\hat{\mathcal{D}}_t(\mathcal{B}_\delta(x_0))$ and a corresponding numerical calculation error ρ such that $d_H(\mathcal{D}_t(\mathcal{B}_\delta(x_0)), \hat{\mathcal{D}}_t(\mathcal{B}_\delta(x_0))) \leq \rho$. To compute $\hat{\mathcal{D}}_t(\mathcal{B}_\delta(x_0))$ from $\mathcal{B}_\delta(x_0)$, what this function needs to know is the computational history of ReachSet($\cdot$) containing the time τ_k when a discrete transition is detected and the values of the parameters h, γ that were used at the time τ_k. Note that all of these informations are stored in TransitionHistory by ReachSet($\cdot$).

Now, we describe the main function, called ReachSet($\cdot$), in the Main Algorithm.

ReachSet($\cdot$).

Given an input (k, δ, γ, h) from the Policy module, where k indicates one of the past computation steps of ReachSet($\cdot$) from which this function starts its computation, this function computes a bounded ϵ-reach set by utilizing all other functions in the Main Algorithm and the Condition Checking modules. This function first retrieves the computation data at the $(k-1)$-th computation step from the ReachSetHistory and starts its k-th computation step using this data. As shown in Algorithm 2, it continues its computation until it either successfully computes a bounded ϵ-reach set or encounters some error. If there is an error from any of the functions that are called, then this function returns the same error to the Policy module to indicate the cause of the error. For each types of error, ReachSet($\cdot$) expects to have a new input from the Policy module to continue its computation. Besides the input (k, δ, γ, h), an additional set of inputs $(\sigma_e, \sigma_i, \sigma_p, \sigma_a, \mu_c, \mu_h)$ can also be provided by the Policy module when there are numerical computational algorithms with better computational accuracies in the Numerical Calculation modules to resolve an erroneous situation occurring in ReachSet($\cdot$).

As mentioned in Section 4, ReachSet($\cdot$) stores its computation results (or states) in ReachSetHistory data structure at every step of its computation. The information stored in ReachSetHistory includes k which is the step of its computation, and the time t_k at the k-th computation step, $(\delta_k, \gamma_k, h_k)$ that are used in the k-th computation step without causing any error, and $\hat{\mathcal{D}}_{t_k}(\mathcal{B}_{\delta_k}(x_0))$ and $\hat{\mathcal{D}}_{t_k}(\mathcal{B}_{\delta_k}(x_0), \gamma_k)$ along with their corresponding numerical computation errors, ρ_k' and ρ_k''. In addition to ReachSetHistory, ReachSet($\cdot$) maintains another data structure, called TransitionHistory which contains computation information of ReachSet($\cdot$) only at the time of discrete transition between locations, intended to be used in ImageAt($\cdot$).

We now have the following overall main result:

THEOREM 2. *For a given* SystemData $:= (\mathcal{X}, \mathcal{A}, l_0, x_0, T, N, \epsilon)$, *if* ReachSet($\cdot$) *in Algorithm 2 returns* done, *then a*

Algorithm 2: Algorithm of ReachSet(·).

Input: $k, \delta_k, \gamma_k, h_k, \sigma_e, \sigma_i, \sigma_p, \sigma_a, \mu_c, \mu_h$
Result: ReachSetHistory, TransitionHistory

compute μ_x from $(\sigma_e, \sigma_i, \sigma_p, \sigma_a)$
while true do
  Get $(k-1)$-th computation data from
  ReachSetHistory
  if $\delta_k \neq \delta_{k-1}$ then
    call ImageAt() $\rightarrow \hat{\mathcal{D}}_{t_{k-1}}(\mathcal{B}_{\delta_k}(x_0))$
    if error then return error
  end
  if IsOverApproximate() = false then return error
  $t_k = t_{k-1} + h_k$
  call
  ReachNext() $\rightarrow \{\hat{\mathcal{D}}_{t_k}(\mathcal{B}_{\delta_k}(x_0)), \hat{\mathcal{D}}_{t_k}(\mathcal{B}_{\delta_k}(x_0), \gamma_k)\}$
  compute ρ_k s.t.
  $d_H(\mathcal{D}_{t_k}(\mathcal{B}_{\delta_k}(x_0), \gamma_k), \hat{\mathcal{D}}_{t_k}(\mathcal{B}_{\delta_k}(x_0), \gamma_k)) \leq \rho_k$
  if IsEpsilonSmall() = false then return error
  call IsTransition() $\rightarrow$ out
  if out = error then return error
  else if out = false then $l_k \leftarrow l_{k-1}$
  else if out = true then
    call AtTransition() $\rightarrow \{l_k, \hat{\mathcal{P}}\}$
    if error then return error
    $\hat{\mathcal{D}}_{t_k}(\mathcal{B}_{\delta_k}(x_0)) \leftarrow \hat{\mathcal{P}}$
    $\hat{\mathcal{D}}_{t_k}(\mathcal{B}_{\delta_k}(x_0), \gamma_k) \leftarrow \hat{\mathcal{D}}_0(\hat{\mathcal{P}}, \gamma_k)$
    update ρ_k
    jump $\leftarrow$ jump + 1
    store $\{t_k, l_k, h_k\}$ to TransitionHistory
  end
  store to ReachSetHistory the data of
  $\{k, t_k, l_k, \delta_k, \gamma_k, h_k, \rho_k, \hat{\mathcal{D}}_{t_k}(\mathcal{B}_{\delta_k}(x_0)), \hat{\mathcal{D}}_{t_k}(\mathcal{B}_{\delta_k}(x_0), \gamma_k)\}$
  $k \leftarrow k + 1$
  if $(t_k \geq T) \vee (\text{jump} \geq N)$ then return done
end

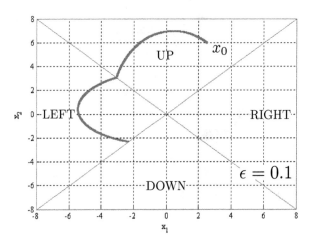

Figure 3: A bounded ϵ-reach set of $\mathcal{A}$ with $\epsilon = 0.1$.

bounded ϵ-reach set of a DTLHA $\mathcal{A}$ over the continuous domain $\mathcal{X}$ from an initial state $x_0 \in Inv_0$, denoted as $\mathcal{R}_{t_f}(x_0, \epsilon)$, is the following:

$$\mathcal{R}_{t_f}(x_0, \epsilon) := \bigcup_{k=1}^{K} \hat{\mathcal{D}}_{t_k}(\mathcal{B}_{\delta_k}(x_0), \gamma_k), \quad (12)$$

where K is the number of data elements in ReachSetHistory, $t_f := \min\{T, \tau_N\}$, and τ_N is the N-th discrete transition.

PROOF. For each $k \leq K$, (i) (γ_k, h_k) satisfies Lemma 7, and (ii) $\hat{\mathcal{D}}_{t_k}(\mathcal{B}_{\delta_k}(x_0), \gamma_k)$ satisfies Lemma 8. These imply that $\hat{\mathcal{D}}_{t_k}(\mathcal{B}_{\delta_k}(x_0), \gamma_k)$ is guaranteed to be a correct γ_k-approximation of $\mathcal{D}_{t_k}(\mathcal{B}_{\delta_k}(x_0))$ by γ_k and h_k, i.e.,

$$\bigcup_{\tau \in [0, h_k]} \mathcal{D}_{t_k + \tau}(\mathcal{B}_{\delta_k}(x_0)) \subset \hat{\mathcal{D}}_{t_k}(\mathcal{B}_{\delta_k}(x_0), \gamma_k). \quad (13)$$

Moreover $dia(\hat{\mathcal{D}}_{t_k}(\mathcal{B}_{\delta_k}(x_0), \gamma_k)) < \epsilon - \rho_k$. If a deterministic and transversal discrete transition is detected at the k-th step by $\hat{\mathcal{D}}_{t_k}(\mathcal{B}_{\delta_k}(x_0))$, then (iii) by Lemmas 9, 10, and 11, there is in fact a deterministic and transversal discrete transition in (t_{k-1}, t_k). This implies that a deterministic and transversal discrete transition event is correctly determined

by ReachSet(·). Finally, the fact that done is returned by ReachSet(·) implies that either $t_k > T$ or jump $> N$. Hence, t_f is $\min\{T, \tau_N\}$. Therefore, we conclude that $\mathcal{R}_{t_f}$ is a bounded ϵ-reach set of $\mathcal{A}$ from x_0. $\square$

6. IMPLEMENTATION

A prototype implementation of the proposed algorithm for a bounded ϵ-reach set of a DTLHA has been developed on Matlab. We use the Multi-Parametric Toolbox [8] for polyhedral operations.

We now illustrate the results of the implementation on an example. The SystemData $:= (\mathcal{X}, \mathcal{A}, l_0, x_0, T, N, \epsilon)$ of this example is the following: (i) $\mathcal{X} := [-8, 8] \times [-8, 8] \subset \mathbb{R}^2$, (ii) $\epsilon = 0.1$ and 0.5, (iii) $T = 10$ sec., (iv) $N = 5$, (v) $\mathcal{A} := (\mathbb{L}, Inv, A, u)$ where $\mathbb{L} = \{UP, DOWN, LEFT, RIGHT\}$, and for each $l \in \mathbb{L}$, $A(l)$ and $u(l)$ are defined as shown in Table 1 and $Inv(l)$ is defined as shown in Fig. 3 and Fig. 4. As an example, the invariant set for the location UP, $Inv(UP)$, is defined as $\mathcal{X} \cap (x_1 - x_2 \leq 0) \cap (x_1 + x_2 \geq 0)$, (vi) $x_0 = (2.5, 6)^T$, and (vii) $l_0 = UP$.

Table 1: $A(l)$ and $u(l)$ for each $l \in \mathbb{L}$ of $\mathcal{A}$

l	$A(l)$	$u(l)$
UP	$\begin{pmatrix} -0.2 & -1 \\ 3 & -0.2 \end{pmatrix}$	$\begin{pmatrix} 0.1 \\ 0.1 \end{pmatrix}$
$DOWN$	$\begin{pmatrix} -0.2 & -1 \\ 3 & -0.2 \end{pmatrix}$	$\begin{pmatrix} -0.2 \\ -0.2 \end{pmatrix}$
$LEFT$	$\begin{pmatrix} -0.2 & -3 \\ 1 & -0.2 \end{pmatrix}$	$\begin{pmatrix} 0.15 \\ 0.15 \end{pmatrix}$
$RIGHT$	$\begin{pmatrix} -0.2 & -3 \\ 1 & -0.2 \end{pmatrix}$	$\begin{pmatrix} 0.3 \\ 0.3 \end{pmatrix}$

To compute a bounded ϵ-reach set of $\mathcal{A}$, we use a policy that (i) uses a fixed value of $\delta = 10^{-5}$ which defines a sufficiently small $\mathcal{B}_\delta(x_0)$, (ii) chooses the value of h and γ on-the-fly to resolve erroneous situations, and (iii) chooses k in nondecreasing manner, and (iv) sets a fixed value of 10^{-7} for $\sigma_e, \sigma_i, \sigma_p, \sigma_a, \mu_c$, and μ_h. We also set 10^{-7} as the minimum value for h and γ.

The bounded ϵ-reach set for two different values of ϵ com-

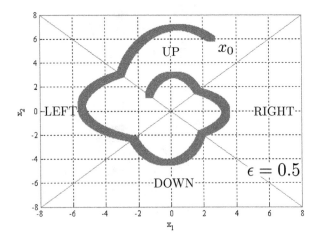

Figure 4: A bounded ϵ-reach set of $\mathcal{A}$ with $\epsilon = 0.5$.

puted by the implementation are shown in Fig. 3 and Fig. 4. For the case of $\epsilon = 0.1$, the algorithm terminates at the computational step $k = 2613$ at which the time $t = 2.2153$ sec. and jump $= 1$ at the location *LEFT*. The reason for this early termination is that the maximum sampling period h, which is determined by the value of ϵ and the accumulated numerical calculation errors ρ, is not large enough to separate $\mathcal{D}_t(\mathcal{B}_\delta(x_0))$ and $\mathcal{D}_{t+h}(\mathcal{B}_\delta(x_0))$ at the time of discrete transition. Hence the algorithm fails to determine the discrete transition from the location *LEFT* to the location *DOWN*. On the contrary, the algorithm successfully returns a bounded ϵ-reach set of $\mathcal{A}$ for the case with $\epsilon = 0.5$ as shown in Fig. 4. In this case, the algorithm terminates at the computational step $k = 1364$ at which the time $t = 5.8496$ sec. and jump $= 5$ at the location *LEFT*.

7. CONCLUSIONS

In this paper, we have extended the theoretical results presented in [7] to compute bounded ϵ-reach sets of Deterministic and Transversal Linear Hybrid Automata with subroutines that only provide finite precision elementary computations. We have also proposed an architecture which separates the elementary subroutines from the policies that adapt the various parameters. This makes the overall algorithm flexible and amenable to different optimizations. An example of a bounded ϵ-reach set computation using a prototype implementation of the proposed algorithm has also been shown in the last section.

8. REFERENCES

[1] R. Alur, C. Courcoubetis, T. A. Henzinger, and P.-H. Ho. Hybrid automata: An algorithmic approach to the specification and verification of hybrid systems. In *HSCC*, pages 250–271, 1993.

[2] R. Alur, T. Dang, and F. Ivančić. Counterexample-guided predicate abstraction of hybrid systems. *Theoretical Computer Science*, 354(2):250–271, 2006.

[3] A. Chutinan and B. H. Krogh. Computational techniques for hybrid system verification. *ITAC*, 48(1):64–75, 2003.

[4] G. Frehse. Phaver: Algorithmic verification of hybrid systems past hytech. *Int. Journal on Software Tools for Technology Transfer*, 10(3):263–279, 2008.

[5] T. A. Henzinger, P.-H. Ho, and H. Wong-Toi. HYTECH: A model checker for hybrid systems. *Int. Journal on Software Tools for Technology Transfer*, 1(1–2):110–122, 1997.

[6] T. A. Henzinger, P. W. Kopke, A. Puri, and P. Varaiya. What's decidable about hybrid automata? In *ACM Symposium on Theory of Computing*, pages 373–382, 1995.

[7] K.-D. Kim, S. Mitra, and P. R. Kumar. Bounded ϵ-reachability of linear hybrid automata with a deterministic and transversal discrete transition condition. In *IEEE CDC*, pages 6177–6182, 2010.

[8] M. Kvasnica, P. Grieder, and M. Baotić. Multi-Parametric Toolbox (MPT), 2004.

[9] C. Moler and C. V. Loan. Nineteen dubious ways to compute the exponential of a matrix, twenty-five yeasr later. *SIAM Review*, 20(4):801–836, 1978.

[10] P. Tabuada, G. J. Pappas, and P. U. Lima. Composing abstractions of hybrid systems. In *HSCC*, pages 136–147, 2002.

[11] V. Vladimerou, P. Prabhakar, M. Viswanathan, and G. E. Dullerud. Stormed hybrid systems. In *ICALP (2)*, LNCS, pages 136–147. Springer, 2008.

Hybridization Domain Construction using Curvature Estimation[*]

Thao Dang
CNRS/VERIMAG
Centre Equation, 2 av de Vignate
38610 Gières, France
Thao.Dang@imag.fr

Romain Testylier
VERIMAG
Centre Equation, 2 av de Vignate
38610 Gières, France
Romain.Testylier@imag.fr

ABSTRACT

This paper is concerned with the reachability computation for non-linear systems using hybridization. The main idea of hybridization is to approximate a non-linear vector field by a piecewise-affine one. The piecewise-affine vector field is defined by building around the set of current states of the system a simplicial domain and using linear interpolation over its vertices. To achieve a good time-efficiency and accuracy of the reachability computation on the approximate system, it is important to find a simplicial domain which, on one hand, is as large as possible and, on the other hand, guarantees a small interpolation error. In our previous work [8], we proposed a method for constructing hybridization domains based on the curvature of the dynamics and showed how the method can be applied to quadratic systems. In this paper we pursue this work further and present two main results. First, we prove an optimality property of the domain construction method for a class of quadratic systems. Second, we propose an algorithm of curvature estimation for more general non-linear systems with non-constant Hessian matrices. This estimation can then be used to determine efficient hybridization domains. We also describe some experimental results to illustrate the main ideas of the algorithm as well as its performance.

Categories and Subject Descriptors

G.1 [**Numerical Analysis**]: Initial value problems

General Terms

Algorithms, Reliability, Verification

Keywords

Hybrid systems, formal verification, reachability analysis

---

[*]Research supported by ANR project VEDECY.

1. INTRODUCTION

Hybrid systems are systems that combine continuous and discrete dynamics and can be used as a mathematical model as well as a computation methodology for many important application domains, such as embedded and cyber-physical systems. Reachability analysis is a central problem in verification and synthesis. For these reasons, numerous reachability computation techniques for hybrid systems have been developed (for example, [12, 2, 17, 5, 20, 16, 7, 15, 14, 11]). In particular, the recently-developed techniques specialized for linear dynamical systems [10, 15, 1, 11] can handle high dimensional systems. Nevertheless, reachability computation for non-linear systems is still a challenging problem. *Hybridization* is an approach to analyze non-linear systems by approximating it by a system with simpler dynamics, such as piecewise-constant or piecewise-affine, which can be handled more efficiently. This idea was first proposed in [9] for numerical simulation of large non-linear systems. It was later adapted to reachability computation in [3, 4, 6]. The principle of hybridization is rather simple: in order to analyze the behavior of a system from some states, one constructs a domain, which we call "hybridization domain", around these states, and in this domain the original non-linear derivative function f is approximated by a simpler, such as affine, function l and an error set U which accounts for the approximation error, that is

$$\forall x \in \Delta \; \exists u \in U \; \text{s.t.} \; f(x) = l(x) + u. \tag{1}$$

Then, the system

$$\dot{x} = Ax + u$$

where u is the uncertain input taking values in U can be used as a conservative approximation of the original non-linear system inside Δ. By "conservative approximation", we mean that any trajectory of the original system is also a trajectory of the approximate system. Other properties of this approximation, such as convergence and qualitative behavior preservation, are described in [4]. Various implementations of hybridization have been developed and used for interesting examples, such as biological systems [6, 8]. While affine dynamics can be handled by relatively efficient methods, the application of the hybridization approach to non-linear systems is still limited by the difficulty in determining good hybridization domains so that the error in the approximate dynamics is not too large. Indeed, large errors in the dynamics approximation might reduce significantly the computation speed since this requires small time steps or a large number of small hybridization domains. In our previous work [8] we proposed a method for constructing efficient hybridization domains using the curvature of the dynamics and showed how the method can be applied to quadratic systems. In this paper we pursue this work further and present two main results. First, we prove an optimality property of the domain construction method for a class of quadratic systems. Second, we propose an algorithm of curvature estimation for more general non-linear systems with non-constant Hessian matrices. This estimation can then be used to determine efficient hybridization domains.

Before presenting our results, we summarize the basic principles of hybridization and its use in the computation of reachable sets of non-linear systems. We also emphasize that although this paper addresses only continuous systems, its extension to hybrid systems is direct. Indeed, the underlying techniques we use to handle approximate piecewise-affine systems generated by hybridization are based on the techniques for hybrid systems [2].

2. HYBRIDIZATION: BASIC IDEAS

In this section, we describe affine hybridization where the approximate vector field in each domain is affine. The use of piecewise-affine systems is motivated by a large choice of available methods for their verification (see for instance [2, 5, 16, 15, 11]). However, other classes of functions can be used, such as constant or multi-affine.

2.1 Principle

We consider a non-linear system:

$$\dot{x}(t) = f(x(t)), \; x \in \mathcal{X} \subseteq \mathbb{R}^n. \tag{2}$$

where the function f is Lipschitz over the state space $\mathcal{X}$.

Given a set of all current states of the system, the trajectories from which we want to explore, we construct a domain which is a neighborhood of this set and then assign an affine vector field to that domain. When the trajectories of the system leave the domain, a new domain is created. This way the non-linear system is approximated by a piecewise-affine system.

In this work, to define approximate affine vector fields, we use linear interpolation over simplicial domains. An important advantage of this approximation method is that using the vertices of a simplex, the affine interpolant is uniquely determined, since each simplex has exactly $(n + 1)$ vertices.

We denote by $\Delta \subseteq \mathbb{R}^n$ a simplex[1] which is used as a hybridization domain. The approximate affine system associated with Δ is defined as follows:

$$\dot{x}(t) = s(x(t), u(t)) = l(x(t)) + u(t) \tag{3}$$

where $x \in \Delta$, $u(\cdot) \in \mathcal{U}_\mu$; l is an affine map of the form:

$$l(x) = Ax + b$$

where A is a matrix of size $n \times n$ and $b \in \mathbb{R}^n$ such that l interpolates the function f at the vertices of Δ, that is for each vertex v of Δ we have $l(v) = f(v)$. The input u is introduced in order to account for the approximation error. More concretely, let μ be the bound of $\|f - l\|$, i.e. for all $x \in \Delta$

$$\|f(x) - l(x)\| \leq \mu$$

---

[1]We recall that a simplex in $\mathbb{R}^n$ is the convex hull of $(n + 1)$ affinely independent points in $\mathbb{R}^n$.

where $\|\cdot\|$ is some norm on $\mathbb{R}^n$. In this work we will consider the norm $\|\cdot\|_\infty$ which is defined as

$$\|x\|_\infty = \max(|x_1|, \ldots, |x_n|).$$

Thus, the error set U contains all the points $u \in \mathbb{R}^n$ such that $\|u\| \leq \mu$. The set $\mathcal{U}_\mu$ is the set of piecewise-continuous input functions of the form $u : \mathbb{R}^+ \to U$.

To determine the error set U, one needs to estimate the error bound μ which depends on the domain geometry and size. It is however important to note that determining the exact maximal error μ is difficult. Suppose that we can find an upper bound of μ, denoted by $\overline{\mu}$. Then, we can choose the error set U to be the ball (thats is a hypercube for the infinity norm) centered at the origin and has radius $\overline{\mu}$. This will be discussed in detail later.

To define the reachable set of the system (3), let $\Phi_s(t, x, u(\cdot))$ be the trajectory starting from a state x under input $u(\cdot) \in \mathcal{U}$. The reachable sets of the system (3) from a set of initial states $X_0 \subseteq \mathcal{X}$ during the interval $[0, t]$ is defined as:

$$Reach_s(t, X_0) =$$
$$\{ y = \Phi_s(\tau, x, u(\cdot)) \mid \tau \in [0, t], x \in X_0, u(\cdot) \in U_\mu \}.$$

The reachable set of the original system can be defined similarly.

It is easy to see that the size of the error set directly impacts the distance between the trajectories of the original and the approximate systems. It is therefore of great interest to construct hybridization domains with a small error bound. An intuitive observation is that to one can achieve a better interpolation accuracy by using smaller hybridization domains. This makes the staying time within this domain shorter, which results in a higher frequency of domain construction and more computation effort. However, when computing the reachable set of a hybrid system, accuracy becomes important since the error accumulation can be drastically aggravated by discrete dynamics in such systems, which induces spurious trajectories the exploration of which might be expensive and lead to erroneous results. In other words, for highly "sensitive" systems, requiring high accuracy can indeed be a way of reducing computation time in the long run. It is also crucial to exploit the structure of the dynamics in order to improve accuracy without causing too much additional computation time. In our previous work [8] we showed that this can be done by considering the relation between the interpolation error and the curvature of the dynamics. This enabled us to construct larger hybridization domains while maintaining the same accuracy. This result was applied to the class of quadratic functions. In this paper we continue this work by extending its application to more general non-linear systems. In addition, the error bound we used in [8] is valid for rather general systems, for more specific systems optimal bounds and domain construction can be considered. This is another problem we address in this paper. Before describing our new contributions, we first recap the result presented in the previous paper [8].

2.2 Hybridization error bound

It is possible to show an error bound that depends on the radius of the smallest containment ball of the simplex. The smallest containment ball of a simplex is the smallest ball that contains the simplex (which should not be confused with its circumcirle). In addition, using the curvature of the functions f one can map the original space (where the functions f are defined) to an "isotropic" space by shrinking the domain along the direction with small curvature. As a result, in the isotropic space, the new domain becomes more "regular" and smaller. It is then possible to derive a tighter error bound depending on the radius of the smallest containment ball of the new domain. In the opposite direction, using the inverse of this transformation, we can extend a domain along the direction with small curvature while preserving the same error bound.

In order to explain this more formally, we first introduce some notation. From now on we write f^i with $i \in \{1, 2, \ldots, n\}$ to indicate the ith component of f.

For a vector $d \in \mathbb{R}^n$, the first-order directional derivative of f^i with $i \in \{1, \ldots, n\}$ along the vector d is

$$\partial f^i(x, d) = \sum_{j=1}^{n} \frac{\partial f^i}{\partial x_j}(x) d_j$$

The Hessian matrix associated with each function f^i for $i \in \{1, \ldots, n\}$ is:

$$H^i(x) = \begin{pmatrix} \frac{\partial^2 f^i}{\partial x_1^2} & \cdots & \frac{\partial^2 f^i}{\partial x_1 x_n} \\ & \cdots & \\ \frac{\partial^2 f^i}{\partial x_1 x_n} & \cdots & \frac{\partial^2 f^i}{\partial x_n^2} \end{pmatrix}. \quad (4)$$

Then, the second-order directional derivative of f^i along d is defined as:

$$\partial^2 f^i(x, d) = d^T H^i(x) d.$$

Given a set $X \subseteq \mathcal{X}$, let $\gamma_X \in \mathbb{R}$ be the smallest real number such that f satisfies the following condition for all unit vector $d \in \mathbb{R}^n$ and for all $x \in X$:

$$\forall i \in \{1, \ldots, n\} : \max_{x \in X, ||d||=1} |\partial^2 f^i(x, d)| \leq \gamma_X. \quad (5)$$

The value γ_X is called the *maximal curvature* of f in X.

Intuitively, we are interested in finding an "isotropic" coordinate transformation so that when mapping a domain back to the original space, we obtain a larger domain while preserving the same interpolation error bound. To this end, we assume the boundedness of the directional curvature of f as follows. Given a set $X \subseteq \mathcal{X}$, we assume that there exists a positive-definite matrix C such that

$$\forall i \in \{1, \ldots, n\} \; \forall x \in X \; \forall d \in \mathbb{R}^n :$$
$$||d|| = 1 \; \wedge \; |\partial^2 f^i(x, d)| \leq d^T C d. \quad (6)$$

The matrix C corresponds to a bound on the directional curvature of f in the set X, and we call C a *curvature tensor matrix* of f in X. Geometrically speaking, for each $x \in X$ the ellipsoid defined by $d^T C d = 1$ is included in the level set defined by $|\partial^2 f^i(x, d)| = 1$ for all $i \in \{1, \ldots, n\}$.

We now derive from C a isotropic transformation as follows. Since C is a symmetric matrix with real entries, we can write its eigen-decomposition:

$$C = S \Xi S^T = S \begin{pmatrix} \xi_1 & 0 & \ldots & 0 \\ 0 & \xi_2 & \ldots & 0 \\ & & \ldots & \\ 0 & 0 & \ldots & \xi_n \end{pmatrix} S^T$$

where ξ_j with $j \in \{1, \ldots, n\}$ are the eigenvalues of C, and S is an orthonormal square matrix containing the eigenvectors of C, that is

$$S = [v_1 v_2 \ldots v_n].$$

Let $\xi_{max}(C)$ and $\xi_{min}(C)$ be the largest and smallest eigenvalues of the matrix C. Then, we consider the linear transformation defined by the matrix

$$T = S W_s S^T = S \begin{pmatrix} \sqrt{\frac{\xi_1}{\xi_{max}(C)}} & \cdots & 0 \\ & \cdots & \\ 0 & \cdots & \sqrt{\frac{\xi_n}{\xi_{max}(C)}} \end{pmatrix} S^T \quad (7)$$

and denote $\hat{x} = Tx$. We call this an "isotropic" transformation. Indeed, the linear transfomation associated with the matrix T transform an ellipsoid in to a sphere, as shown in Figure 1. This can be seen later in Section 3. Let $\hat{X}$ denote the set resulting from applying the above linear transformation to X, that is,

$$\hat{X} = \{Tx \mid x \in X\}.$$

Figure 1: Illustration of the transformation to an isotropic space: The ellipse is a level set of the directional curvature and the curvature is small along its semimajor axis. The isotropic transformation T "shortens" the triangle on the left along the directions in which f has small curvature

The following theorem [19] shows an error bound which depends on the radius of the smallest containement ball in the isotropic space.

THEOREM 1. *Let l be the affine function that interpolates the functions f over the simplex Δ. Then, for all $x \in \Delta$*

$$||f(x) - l(x)|| \leq \gamma_\Delta \frac{r_c^2(\hat{\Delta})}{2}.$$

where γ_Δ is the maximal curvature of f in Δ and $r_c(\hat{\Delta})$ is the radius of the smallest containement ball of the transformed simplex $\hat{\Delta}$.

The transformation of Δ to $\hat{\Delta}$ can shorten an edge of Δ by a factor of up to $\sqrt{\frac{\xi_{max}(C)}{\xi_{min}(C)}}$ if this edge is aligned with the eigenvector corresponding to ξ_{min}. In the worst case, its length remains unchanged if it is aligned with the eigenvector corresponding to ξ_{max}.

Using this theorem we can determine a simplicial domain that guarantees a desired error bound ε. Let P be the set of current states of the system, and $\hat{P}$ be the polytope resulting from applying the transformation T to P. The hybridization domain construction using an isotropic transformation is summarized in Algorithm 1.

Algorithm 1 Hybridization domain construction using isotropic transformation

---

Compute the transformation matrix T.

Determine the maximal radius r_c of the smallest containement ball corresponding to the desired error bound ε.

Let B be the ball of radius r_c the centroid of which coincides with that of the polytope $\widehat{P}$.

Construct in the isotropic space an equilateral simplex $\widehat{\Delta}$ that is circumscribed by the ball B. Such simplices have the largest volume for the given radius r_c of the smallest containement ball.

Use the inverse transformation T^{-1} to map the simplex $\widehat{\Delta}$ back to the original space. This results in a larger simplex Δ while guaranteeing the desired error bound.

Indeed, in this algorithm, the shape of the simplex is fixed while its orientation can be freely chosen. This offers a posibility of further optimization, such as reducing the frequency of domain constructions by orienting the simplex according to the dynamics of the system so that its trajectories stay in the domain for a longer time.

In the previous work [8], we also showed an effective application of this approach to quadratic systems where the Hessian matrices are constant and hence the maximal curvature γ_Δ can be straightforwardly computed. As mentioned earlier, in this paper we pursue this work and describe two new results.

The remainder of the paper is organized as follows. We first present the optimality property of the domain construction with isotropic transformation for quadratic systems. We then consider more general non-linear systems with nonconstant Hessian matrices and describe an algorithm for constructing domains using a curvature estimation. We also demonstrate the interest of the algorithm by means of some examples and report preliminary experimental results on the computational efficiency of the algorithm.

3. OPTIMAL DOMAINS FOR QUADRATIC FUNCTIONS

We show a class of quadratic functions f for which the domain construction based on equilateral simplices in an isotropic space is optimal. This optimality property is stated as follows: given an error tolerance ε, the computed simplex Δ has the largest volume and, in addition, the error between f and its linear interpolation over Δ is not greater than ε.

Let each quadratic function f^i be written as

$$f^i(x) = x^T H^i x + (m^i)^T x + p^i$$

where H^i is a real-valued matrix of size $n \times n$, $m^i \in \mathbb{R}^n$ and $p^i \in \mathbb{R}$. Note that we use the same notation H^i here because the Hessian matrix of f^i is exactly H^i. For every $i \in \{1, \ldots, n\}$, we define the interpolation error function as

$$e^i(x) = f^i(x) - l^i(x)$$

which is also a quadratic function. We now study this error

function and seek its maxima.

The error function can be expressed as:

$$e^i(x) = (w^i)^T x + q^i + x^T H^i x$$

where $w^i \in \mathbb{R}^n$ and $q^i \in \mathbb{R}$. Note that the level sets of this function form a family of conics with a common center, denoted by c^i. Indeed, they are ellipsoids if $det(H^i) > 0$ and hyperboloids if $det(H^i) < 0$. We now derive the error in a neighborhood of this common center. Let $\delta_x \in \mathbb{R}^n$ be a deviation from the common center c^i, then for every $i \in \{1, \ldots, n\}$

$$
\begin{aligned}
e^i(c^i + \delta_x) &= (w^i)^T(c^i + \delta_x) + q^i - (c^i + \delta_x)^T H^i(c^i + \delta_x) \\
&= [(w^i)^T c + q^i - (c^i)^T H_i c^i] + \\
&\quad (w^i)^T \delta_x - 2\delta_x^T H^i c^i - \delta_x^T H^i \delta_x \\
&= e^i(c^i) + (w^i)^T \delta_x - 2(c^i)^T H^i \delta_x - \delta_x^T H^i \delta_x.
\end{aligned}
$$

Since c is the common center of the family of conics corresponding to the error function, c satisfies the following

$$w^i - 2H^i c^i = 0.$$

Then,

$$
\begin{aligned}
(w^j)^T \delta_x &= 2(H^i c^i)^T \delta_x \\
&= 2(c^i)^T H^i \delta_x.
\end{aligned}
$$

It then follows from the above that

$$e^i(c^i + \delta_x) = e^i(c^i) - \delta_x^T H^i \delta_x.$$

We also observe that, the symmetric matrix H^i can be decomposed as

$$H^i = SW^T DW S^T$$

where D is a diagonal matrix with entries $\sigma_j \in \{-1, 0, +1\}$; W is a diagonal matrix whose entries on the diagonal are the square roots of the absolute values of the eigenvalues ξ_j of H^i; S is an orthonormal matrix containing the eigenvectors of H^i. We define a linear transformation

$$T^i = W^T S^T.$$

LEMMA 1. *Using the transformation $\widehat{\delta_x} = T^i \delta_x$, the term $\delta_x^T H^i \delta_x$ in the error $e^i(c^i + \delta_x)$ can be transformed into a quadratic form*

$$\delta_x^T H^i \delta_x = \sum_{j=1}^{n} \sigma_1^i \widehat{\delta}_{x_j}^2$$

where for all $j \in \{1, \ldots, n\}$ $\sigma_j^i \in \{-1, 0, +1\}$.

PROOF. Using $\delta_x = T^{-1}\widehat{\delta_x}$ and $H^i = SW^T DW S^T$, we obtain after some straightforward calculation:

$$
\begin{aligned}
\delta_x^T H^i \delta_x &= (T^{-1}\widehat{\delta_x})^T SW\, D\, W^T S^T T^{-1}\widehat{\delta_x} \\
&= (\widehat{\delta_x})^T D\widehat{\delta_x}.
\end{aligned}
$$

Therefore,

$$
\begin{aligned}
\delta_x^T H^i \delta_x &= \widehat{\delta_x}^T \begin{pmatrix} \sigma_1 & 0 & \ldots & 0 \\ 0 & \sigma_2 & \ldots & 0 \\ & & \ldots & \\ 0 & 0 & \ldots & \sigma_n \end{pmatrix} \widehat{\delta_x} \\
&= (\sigma_1 \widehat{\delta}_{x_1}^2 + \sigma_2 \widehat{\delta}_{x_2}^2 + \ldots + \sigma_n \widehat{\delta}_{x_n}^2).
\end{aligned}
$$

where $\forall j \in \{1, \ldots, n\} : \sigma_j \in \{-1, 0, +1\}$. In other words, using the linear transformation T^i we transforms the matrix H^i into a diagonal matrix D which has only entries $0, +1$ and -1 on the diagonal. ∎

Again, we can see that the interpolation error in the new space (resulting from the transformation T) is isotropic, that is it does not depend on the direction of $\widehat{\delta_x}$.

We identify a class of quadratic systems such that for every function f^i, σ_j^i are all equal to either $+1$ or -1. In the isotropic space the level sets of the error are spheres with a common center. The circumsphere of $\widehat{\Delta}$ is the level set of value zero (due to interpolation over the vertices). Hence, the maximal value of $|e^i(x)|$ is achieved at c^i and is directly related to the square of the radius of the circumsphere of $\widehat{\Delta}$.

Using the above reasoning, we can determine the maximal value of every error function $|e^i(x)|$ ($i \in \{1, \ldots, n\}$). Let f^i be the function that corresponds to the largest value. We then take the associated matrix T^i to be the isotropic transformation T for domain construction purposes. Note that in this case, the circumsphere radius is also the radius of the smallest containement ball of Δ in the Theorem 1. For a given fixed circumsphere radius (corresponding to an error tolerance), an equilateral simplex has the optimal shape because it has the largest volume.

The number of entries $+1$ of D is called the positive index of inertia of H^i, and the number of entry 1 is called the negative index of inertia. According to Sylvester's law of inertia, the positive and negative indices of H^i are also the number of positive and negative eigenvalues of H^i.

THEOREM 2. *For a class of quadratic functions f such that all the Hessian matrices have either only positive eigenvalues or only negative eigenvalues, the domain construction method based on equilateral simplices in the isotropic space is optimal.*

When this condition on the eigenvalues is not satisfied, the theorem no longer holds, that is starting from equilateral simplices in the isotropic space does not yield an optimal construction. For example, in 2 dimensions, in the case where the number of σ_j equal to $+1$ is equal to that of σ_j equal to -1 (which implies that the error is a harmonic function of $\widehat{\delta_x}$), the maximal error is not achieved at the common center but on the boundary of the simplex. Investigating the optimality conditions for the remaining cases is part of our undergoing work.

4. ESTIMATING CURVATURE TENSORS

We now present the second contribution of this paper, which involves an effective application of the domain construction algorithm to more general functions with non-constant Hessian matrices.

It can be seen from Theorem 1 that to obtain a good interpolation error bound, we need to know an accurate curvature tensor of f as defined in (6). We observe that, given a simplex Δ, by definition, for each $i \in \{1, \ldots, n\}$ the eigenvalues of the Hessian matrix H^i are inside the interval $[-\gamma_X, \gamma_X]$

where γ_X is the maximal curvature of f inside X. Hence, the error bound is determined by the maximal eigenvalue $\xi_{max}(C)$ of the matrix C. Note additonaly that $r_c(\widehat{\Delta})$ depends on $|det(T)|^{1/n}$ where $det(T)$ is the determinant of the transformation matrix defined in (7).

Therefore, we want to find a positive-definite matrix C that satisfies the condition (6) in the definition of curvature tensor and, in addition, makes $|\xi_{max}(C)||det(T)|^{2/n}$ as small as possible. To do so, we formulate this problem as solving the following constrained optimization problem:

$$\min |\xi_{max}(C)||det(T)|^{2/n}$$
$$\text{s.t. } \forall i \in \{1, \ldots, n\} \; \forall x \in X \; \forall d \in \mathbb{R}^n :$$
$$||d|| = 1 \; \wedge \; |\partial^2 f^i(x, d)| \leq |d^T C^i d|.$$

Again, we express C in its eigen-decomposition form. Let $\xi_1, \xi_2, \ldots, \xi_n$ be the eigenvalues in increasing order of C, that is $0 < \xi_1 \leq \xi_2 \leq \ldots \leq \xi_n$, and $v^1, v^2, \ldots, v^n$ be the corresponding eigenvectors. From now on we use superscripts to denote eigenvectors since subscripts will be used to denote their coordinates. Thus,

$$C = S \Xi S^T$$

where

$$\Xi = diag(\xi_1, \xi_2, \ldots, \xi_n)$$

Therefore, mimimizing over all possible matrices C satisfying (6) is equivalent to minimizing over all possible ξ_i and all possible orthogonal matrices S.

Notice that, by the definition of the matrix T,

$$|det(T)| = |det(C)|^{1/2} = |(\Pi_{j=1}^n \xi_j)|^{1/2}$$

The objective function can therefore be written as:

$$|\xi_{max}(C)||det(T)|^{2/n} = |\xi_n| \, |(\Pi_{j=1}^n \xi_j)|^{1/n}.$$

On the other hand, the constraint (6) can be written as:

$$\forall i \in \{1, \ldots, n\} \; \forall x \in X \; \forall d \in \mathbb{R}^n :$$
$$||d|| = 1 \; \wedge \; |\partial^2 f^i(x, d)| \leq |d^T S \Xi S^T d|.$$

This problem might not have a solution or it might have a solution with some eigenvalue equal to 0, which makes C singular. In the following we consider another approach, which involves approximating C by making the error bound as small as possible while respecting the constraint (6).

Since the error bound depends on the maximal eigenvalue of C and the product of the eigenvalues of C, we estimate C by determining successively its eigenvalues ξ_j and eigenvectors v^j such that each ξ_j is made as small as possible while satisfying the condition (6).

More precisely, in the first step we determine ξ_n such that

$$\forall x \in X \; \forall i \in \{1, \ldots, n\} \; d \in \mathbb{R}^n : ||d|| = 1 \wedge \xi_n \geq |\partial^2 f^i(x, d)|.$$

We can find ξ_n by solving the following n optimization problems

$$\xi_{n,i} = max_{x \in X \; \wedge \; ||d||=1} |\partial^2 f^i(x, d)|, i \in \{1, \ldots, n\}$$

Then we take the largest among the computed maximal values:

$$\xi_n = max_{i \in \{1, \ldots, n\}} \xi_{n,i}.$$

In other words, by (5), ξ_n is exactly the largest curvature of f in X. The corresponding eigenvector v^n is chosen as the unit vector d along which the second-order directional derivative of the corresponding f^i is the largest.

To determine the remaining $(n-1)$ eigenvalues, let $(n-1)$ vectors

$$v^1, \ldots, v^{n-1}$$

in $\mathbb{R}^n$ form an orthonormal basis for the orthogonal complement of $span\{v^n\}$ in $\mathbb{R}^n$. Then, any vector $d \in \mathbb{R}^n$ can be expressed as:

$$d = \sum_{j=1}^{n-1} \alpha_j v^j + \beta v^n \tag{8}$$

where $\alpha = (\alpha_1, \ldots, \alpha_{n-1}) \in \mathbb{R}^{n-1}$ and $\beta \in \mathbb{R}$.

Let

$$S^{n-1} = [v^1 v^2 \ldots v^{n-1}]$$

and Ξ^{n-1} be the diagonal matrix with $\xi_1, \xi_2, \ldots, \xi_{n-1}$ as its diagonal elements. Note that the superscripts here indicate that these matrices are used to compute the eigenvalue ξ_{n-1}. We define then

$$C^{n-1} = S^{n-1} \Xi^{n-1} (S^{n-1})^T.$$

Then, it is not hard to see that

$$d^T C d = \alpha^T C^{n-1} \alpha + \xi_n \beta^2.$$

The condition (6) becomes

$$\forall i \in \{1, \ldots, n\}, \; \forall \alpha \in \mathbb{R}^{n-1}, \; \forall \beta \in \mathbb{R}, \; \forall x \in X :$$
$$\alpha^T C^{n-1} \alpha \geq |\partial^2 f^i(x,d)| - \xi_n \beta^2. \tag{9}$$

We denote the right hand side as a function of α and β:

$$w^i(\alpha, \beta) = |\partial^2 f^i(x,d)| - \xi_n \beta^2,$$

and

$$\eta^i(\alpha) = max_{\beta \in \mathbb{R}} w^i(\alpha, \beta). \tag{10}$$

The condition (9) is then equivalent to

$$\forall i \in \{1, \ldots, n\}, \; \forall \alpha \in \mathbb{R}^{n-1}, \; \forall x \in X :$$
$$\alpha^T C^{n-1} \alpha \geq \eta^i(\alpha) \tag{11}$$

LEMMA 2. *If β maximizes $w^i(\alpha, \beta)$ in (10) then β satisfies*

$$\frac{\partial^2 f^i(x,d)}{|\partial^2 f^i(x,d)|} \partial g^i(x,d) = \xi_n \beta. \tag{12}$$

where

$$g^i(x) = \partial f^i(x, v^n)$$

Intuitively, $\partial g^i(x,d)$ is the first-order directional derivative of g^i with respect to the vector d, and $g^i(x) = \partial f^i(x, v^n)$ is the first-order directional derivative of f^i with respect to the vector v^n.

PROOF. To solve (10), we consider the critical points of $w^i(\alpha, \beta)$ with respect to β. To express the partial derivative of $w^i(\alpha, \beta)$, we first rewrite the second-order directional derivative of f^i as the following sum:

$$\partial^2 f^i(x,d) = \sum_{jk} H^i_{jk}(x) d_j d_k$$

where H^i_{jk} is the element of the jth line and the kth column of the Hessian matrix of the function f^i.

Then,

$$\begin{aligned}
\frac{\partial}{\partial \beta}(\partial^2 f^i(x,d)) &= \sum_{jk} H^i_{jk}(x)(d_j \frac{\partial d_k}{\partial \beta} + d_k \frac{\partial d_j}{\partial \beta}) \\
&= \sum_{jk} H^i_{jk}(x)(d_j v^n_k + d_k v^n_j) \quad \text{(from (8))} \\
&= 2 \sum_{jk} H^i_{jk}(x) d_j v^n_k \\
&= 2\partial(\partial f^i(x, v^n), d) \\
&= 2\partial g^i(x,d) \quad \text{(from (12))}
\end{aligned}$$

In addition,

$$\frac{\partial}{\partial \beta}(\xi_n \beta^2) = 2\xi_n \beta.$$

Note that the critical points that satisfy $\partial^2 f^i(x,d) = 0$ are not among the maxima of $w^i(\alpha, \beta)$. It then follows from the above that the critical points that are candidates to be among the maxima satisfy:

$$\frac{\partial^2 f^i(x,d)}{|\partial^2 f^i(x,d)|} \partial g^i(x,d) = \xi_n \beta.$$

This establishes the proof of the lemma. ∎

To determine β, we consider two cases:

- If $\partial^2 f^i(x,d) > 0$, the equation (12) can be rewritten as:

$$\sum_{j,k} H^i_{jk} v^n_j d_k = \xi_n \beta.$$

It then follows from (8) that

$$\sum_{j,k} H^i_{jk} v^n_j (\sum_{1}^{n-1} \alpha_j v^j_k + \beta v^n_k) = \xi_n \beta.$$

Note that ξ_n and v^n are now known, from the above we can determine β as a function of α and $v^1, v^2, \ldots, v^{n-1}$. From now on, for clarity, we denote by β^n the value of β satisfying the above, since this value corresponds to the eigenvalue ξ_n and the eigenvector v^n computed in the first step. Hence, if $\partial^2 f^i(x,d) > 0$

$$\beta^n = \frac{(v^n)^T H^i(\sum_{j=1}^{n-1} \alpha_j v^j)}{\xi_n - (v^n)^T H^i v^n}.$$

- If $\partial^2 f^i(x,d) < 0$, similarly we obtain

$$\beta^n = \frac{-(v^n)^T H^i(\sum_{j=1}^{n-1} \alpha_j v^j)}{\xi_n + (v^n)^T H^i v^n}.$$

With the above β^n, $\eta^i(\alpha)$ in (10) can be determined. Hence, the condition (9) becomes

$$\alpha^T C^{n-1} \alpha \geq u(\alpha) = |\partial^2 f^i(x,d)| - \xi_n(\beta^n)^2 \tag{13}$$

where $d = \sum_{1}^{n-1} \alpha_j v^j + \beta^n v^n$.

We now come to the same problem as the initial but with only $(n-1)$ eigenvalues to determine. We can repeat this procedure to determine all the eigenvalues. More precisely, we determine ξ_{n-1} by solving the following optimization problem over the variables $z = \sum_{j=1}^{n-1} \alpha_j v^j$ and x.

$$\xi_{n-1,i} = \max(|\partial^2 f^i(x, z + \beta^n v^n)| - \xi_n(\beta^n)^2)$$
$$\text{s.t. } x \in X \ \wedge$$
$$z \in \mathbb{R}^n \ \wedge$$
$$\|z + \beta^n v^n\| = 1.$$

Then, ξ_{n-1} is determined as the largest of all $\xi_{n-1,i}$.

To reduce further to the problem with $(n-2)$ eigenvalues, we write:

$$d^T C d = \alpha^T C^{n-2} \alpha + \xi_n(\beta^n)^2 + \xi_{n-1}(\beta^{n-1})^2.$$

where α is now a vector in $\mathbb{R}^{n-2}$.

Then, the sequence of optimization problems can be formulated as follows. For $k = n-1, n-2, \ldots, 1$

$$\xi_{n-k,i} = \max |\partial^2 f^i(x, z + \sum_{j=n-k+1}^{n} \beta^j v^j)| - \sum_{j=k+1}^{n} \xi_j(\beta^j)^2$$
$$\text{s.t. } x \in X \ \wedge$$
$$z \in \mathbb{R}^n \ \wedge$$
$$\|z + \sum_{j=k+1}^{n} \beta^j v^j\| = 1$$

(14)

Then, ξ_{n-k} is the largest value of all $\xi_{n-k,i}$.

In the above procedure, in order to proceed from the computation of the eigenvector ξ_{n-k} to that of ξ_{n-k-1} we need to compute the corresponding eigenvector v^{n-k}. As an example, in the step $k = n-1$, we obtain the solution $d = z + \beta^n v^n$ of the optimization problem. Let denote this d by d^{n-1} and we want to compute the corresponding eigenvector v^{n-1}. To this end, we use the following scheme. Indeed, at each step k, v^k is made orthogonal to the previous $v^{k+1}, \ldots, v^n$ by substracting the projection of d^k in the directions of $v^{k+1}, \ldots, v^n$.

$$u^k = d^k - \sum_{j=k+1}^{n} \frac{(v^j)^T d^k}{(v^j)^T v^j} v^j$$

Then we determine the eigenvector v^k as:

$$v^k = \frac{u^k}{\|u^k\|}.$$

Such n vectors v^k span the same subspace as n vectors d^k.

Finally, we construct the matrix C as follows:

$$C = S \begin{pmatrix} \xi_1 & 0 & \ldots & 0 \\ 0 & \xi_2 & \ldots & 0 \\ & & \ldots & \\ 0 & 0 & \ldots & \xi_n \end{pmatrix} S^T$$

where ξ_j with $j \in \{1, \ldots, n\}$ are the computed eigenvalues, and $S = [v^1 v^2 \ldots v^n]$ is an orthonormal matrix containing the computed eigenvectors.

5. REACHABILITY COMPUTATION USING HYBRIDIZATION

Once the curvature tensor matrix is estimated, we can compute from it an isotropic transformation T. This can then be used to create hybridization domains for reachability computation. The reachability computation accuracy depends on the precision of the curvature tensor approximation, since the latter is directly related to the error bound that is used to define the input set U. In the curvature estimation described in the previous section, the optimization problems are solved over all $x \in X$, that is the computed estimate is valid for the whole set X. The estimation precision can be improved by using a dynamical curvature estimation that is invoked each time a new hybridization domain needs to be created. In this case, the optimization domains can be chosen as a neighborhood of the current states of the system.

The reachability computation algorithm using hybridization is summarized by Algorithm 2 where P is a polytope containing all the initial states of the system. In each iteration, we first estimate the curvature tensor within a zone containing the current set P^k. This matrix is then used to construct the simplicial domain Δ. We perform the reachability computation from P^k under the approximate linear interpolating dynamics defined within Δ. This generates the polytope P^{k+1}. If this polytope contains points outside the current domain Δ we retrieve the previous polytope P^k and construct around P^k a new hybridization domain. Note that when the polytope P^k becomes too large to be included in Δ, splitting is required.

Algorithm 2 Reachability computation using hybridization

Input: Initial polytope P, interpolation error tolerance ε

$P^0 = P$, $k = 0$
$C = CurvatureEstimation(P^k)$
$\Delta = DomainConstruct(P^k, C, \varepsilon)$
while $k \leq k_{max}$ **do**
$\quad P^{k+1} = Reach_l(P^k, \Delta)$
$\quad$ **if** $P^{k+1} \cap \bar{\Delta} \neq \emptyset$ **then**
$\qquad\qquad\qquad$ /* $\bar{\Delta}$ is the complement of Δ */
$\quad\quad \Delta = DomainConstruct(P^k, \varepsilon)$
$\quad$ **else**
$\quad\quad k = k + 1$
$\quad$ **end if**
end while

It is worth mentioning that in this algorithm we use polytopes to represent reachable sets. However, the proposed hybridization and domain construction methods can be combined with the algorithms using other set representations (such as [15, 11]).

6. EXPERIMENTAL RESULTS

We implemented the domain construction algorithm and tested in on various examples. For non-linear optimization we use the publicly available NLopt library [13] which provides a common interface for a number of different optimization algorithms. For the computation of reachable sets of the approximate piecewise-affine systems, we used the algorithms implemented in the tool d/dt [2].

We first illustrate the interest of the algorithm by a number of experiments on a 2-dimensional system, the dynamics of which is described as follows:

$$\dot{x}_1 = x_2 - x_1^3 + x_1 x_2^2 \tag{15}$$
$$\dot{x}_2 = x_1^3 \tag{16}$$

This example is adapted from the one used in the study

of stabilization of systems with uncontrollable linearization in [18] (page 346). The initial set is a small box $[0.5, 0.51] \times [0.5, 0.51]$. The error tolerance is equal to 0.5.

Figure 2 shows the reachable set computed using a hybridization without isotropic transformation. The hybridization domains are chosen to be equilateral and oriented along the local evolution direction of the dynamics. We can see that without using an isotropic transformation, the domains are small and thus a lot of domains were created.

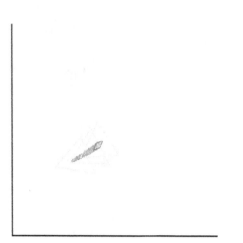

Figure 2: Domains constructed without isotropic transformation.

Figure 3 shows the reachable set computed using a hybridization with an isotropic transformation. In this experiment, the curvature estimation was done dynamically when each domain is created. We can see that the domains are larger for the same accuracy and less domains are needed.

Figure 3: Domains constructed with the curvature tensor dynamically estimated over large zones.

The reachability computation can be further improved by using smaller optimization domains around the reachable sets. This in general requires some rough approximation of

the reachable set within a number of next iterations. This is illustrated by the reachable set shown in Figure 4.

Figure 4: Domains constructed with the curvature tensor dynamically estimated over small zones.

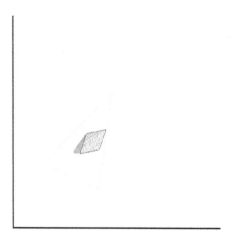

Figure 5: Reachability computation with a large error bound.

In order to illustrate the effect of error bounds, we fix the radius of the smallest containment ball in the isotropic space and perform two experiments: the first one with the curvature tensor estimated over a large zone and the second over a smaller zone. We observe that the hybridization domains in the two experiments are the same. However, a high curvature bound was computed in the first experiment and thus the corresponding error bound is large, which results in a large input set. This causes the system to expand fast, as one can see in Figure 6. On the other hand, with a better curvature estimate in the second experiment, the error bound is smaller and the reachable set computation is more accurate (see Figure 6).

In order to evaluate the performance of the algorithm, we performed a set of experiments on some polynomial systems (of degree 4) which are randomly generated. We report in the following the average computation times of 100 itera-

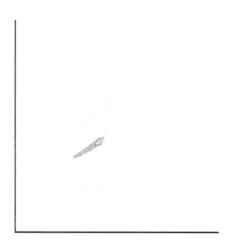

Figure 6: Reachability computation with a smaller error bound.

Dimension	Total time (s)	Optimisation time (s)
2	0.53	0.05
3	0.96	0.63
4	7.87	7.01
5	57.05	48.22
6	90.77	80.78
7	302.5	269.22

Figure 7: Computation times for polynomial systems.

tions for systems up to 7 dimensions. For each dimension, we tested 4 systems. In these experiments, for each system the curvature tensor matrix was estimated only once. The reason we did not go beyond 7 dimensions is that the optimisation took a lot of computation time (while the computation time for treating approximate piecewise affine systems is much less), as shown in Figure 7. Indeed, for a n-dimensional system, to estimate the curvature tensor matrix, we need to compute n eigenvalues, each of which requires solving n constrained optimization problems with $2n$ variables. Indeed, the curvature tensor estimation can be done a-priori over a large analysis zone and such a global estimate can be used for the whole reachability computation process. This alone can significantly improve the accuracy of the reachable set approximation, compared to the domain construction without isotropic transformation, as shown in the above 2-dimensional example. In order to include dynamic curvature estimation, we need more performant optimisation tools, such as those for specific classes of systems.

To sum up, our preliminary experiments demonstrated the interest of the proposed domain construction algorithm in terms of accuracy improvements. The practical efficiency of the algorithm is still limited by the required non-linear optimization. Beside exploiting more performant optimization algorithms, we plan to tackle this problem by exploring other methods for computing isotropic transformations without estimating the curvature tensor matrix.

7. CONCLUSION

In this paper we extended the curvature-based domain construction method to non-linear systems with non-constant Hessian matrices. In addition, we proved an optimality property of the domain construction for a class of quadratic systems. We demonstrated the effectiveness, in particular in terms of accuracy improvement, of the method on various examples. Future work directions include considering the optimal domain construction problem for larger classes of systems. Finding more efficient methods for computing curvature tensor matrices and isotropic transformations is also part of our future work.

8. REFERENCES

[1] M. Althoff, O. Stursberg, and M. Buss. Reachability Analysis of Nonlinear Systems with Uncertain Parameters using Conservative Linearization. *CDC'08*, 2008.

[2] E. Asarin, O. Bournez, T. Dang, and O. Maler. Approximate reachability analysis of piecewise linear dynamical systems. *HSCC'00*, LNCS 1790, 21-31, 2000.

[3] E. Asarin, T. Dang, and A. Girard. Reachability Analysis of Nonlinear Systems Using Conservative Approximation. *HSCC'03*, LNCS 2623, pp 20-35, Springer, 2003.

[4] E. Asarin, T. Dang, and A. Girard. Hybridization Methods for the Analysis of Nonlinear Systems. *Acta Informatica* **43**, 451-476, 2007.

[5] A. Chutinan and B.H. Krogh. Computational Techniques for Hybrid System Verification. *IEEE Trans. on Automatic Control* **48**, 64-75, 2003.

[6] T. Dang, C. Le Guernic and O. Maler. Computing Reachable States for Nonlinear Biological Models. *Computational Methods in Systems Biology CMSB'09*, LNCS 5688, pp 126-141, 2009.

[7] T. Dang. Approximate Reachability Computation for Polynomial Systems. *HSCC'06*, LNCS 3927, pp 138-152, 2006.

[8] T. Dang, O. Maler, and R. Testylier. Accurate hybridization of nonlinear systems. *HSCC'10*, ACM, pp 11-20, 2010.

[9] J. Della Dora, A. Maignan, M. Mirica-Ruse, and S. Yovine. Hybrid computation. Proceedings International Symposium on Symbolic and Algebraic Computation ISSAC, 2001.

[10] A. Girard. Reachability of Uncertain Linear Systems using Zonotopes. *HSCC'05*, LNCS 3414, 291-305, 2005.

[11] A. Girard, C. Le Guernic and O. Maler. Efficient Computation of Reachable Sets of Linear Time-invariant Systems with Inputs. *HSCC'06*, LNCS 3927, 257-271 2006.

[12] M.R. Greenstreet and I. Mitchell, Reachability Analysis Using Polygonal Projections. *HSCC'99* LNCS 1569, 103-116, 1999.

[13] S.G. Johnson. The NLopt nonlinear optimization package. http://ab-initio.mit.edu/nlopt

[14] M. Kloetzer and C. Belta, Reachability analysis of multi-affine systems. *HSCC Hybrid Systems: Computation and Control*, vol 3927 in LNCS, 348-362, Springer, 2006.

[15] A. Kurzhanskiy and P. Varaiya, Ellipsoidal Techniques for Reachability Analysis of Discrete-time Linear Systems, *IEEE Trans. Automatic Control* **52**, 26-38, 2007.

[16] M. Kvasnica, P. Grieder, M. Baotic and M. Morari. Multi-Parametric Toolbox (MPT) *HSCC'04*, LNCS 2993, pp 448-462, Springer, 2004.

[17] I. Mitchell and C. Tomlin. Level Set Methods for Computation in Hybrid Systems, *HSCC'00*, LNCS 1790, 310-323, 2000.

[18] S. Sastry. Nonlinear Systems: Analysis, Stability, and Control. Springer-Verlag, 1999.

[19] J.R. Shewchuk. What Is a Good Linear Element? Interpolation, Conditioning, and Quality Measures. Eleventh International Meshing Roundtable (Ithaca, New York), pp 115-126, Sandia National Laboratories, September 2002.

[20] A. Tiwari and G. Khanna. Nonlinear Systems: Approximating Reach Sets. *HSCC'04*, LNCS 2993, pp 600-614, 2004.

[21] P. Varaiya, Reach Set computation using Optimal Control, *KIT Workshop, Verimag, Grenoble*, 377-383, 1998.

A Dynamic Algorithm for Approximate Flow Computations

Pavithra Prabhakar
Department of Computer Science
University of Illinois at Urbana Champaign
pprabha2@illinois.edu

Mahesh Viswanathan
Department of Computer Science
University of Illinois at Urbana Champaign
vmahesh@illinois.edu

ABSTRACT

In this paper we consider the problem of approximating the set of states reachable within a time bound T in a linear dynamical system, to within a given error bound ϵ. Fixing a degree d, our algorithm divides the interval $[0, T]$ into sub-intervals of not necessarily equal size, such that a polynomial of degree d approximates the actual flow to within an error bound of ϵ, and approximates the reach set within each sub-interval by the polynomial tube. Our experimental evaluation of the algorithm when the degree d is fixed to be either 1 or 2, shows that the approach is promising, as it scales to large dimensional dynamical systems, and performs better than previous approaches that divided the interval $[0, T]$ evenly into sub-intervals.

Categories and Subject Descriptors

D.0 [**Software**]: General

General Terms

Verification

Keywords

Approximation, Linear Dynamical Systems, Post Computation

1. INTRODUCTION

Integral to the automatic verification of safety properties is the computation of the set of reachable states of a system. In the context of hybrid systems, the key challenge in reachability computation is to compute, for a given set of states X, all states that are reachable from X under the continuous dynamics, within (some) time T. While states reachable within a bounded time can be computed for some hybrid systems with simple dynamics [3, 1, 16, 19, 27], it must typically be approximated, since the problem of computing the reachable set is undecidable for most dynamical systems.

There are three principal techniques for computing an approximation to the reachable set of states. The first constructs an abstract transition system that *simulates* (in a formal sense) the dynamical system, and carries out the reachability computation on the abstract system [5, 2, 8, 7, 11, 17, 26]. In this method, the quality of the approximate solution cannot be measured, and so, often this is compensated by repeatedly refining the abstract transition system. The second approach, called *hybridization* [25, 4, 10], partitions the continuous state space, and approximates the continuous dynamics in each partition. Here one can explicitly bound the error between the reachable set of the hybrid system with simpler dynamics and the original dynamical system.

The third method is to directly compute the states reachable within time T. This has been carried out primarily for linear dynamical systems and a convex polytope as initial set X. For such systems, the algorithm proceeds as follows. First, the time interval $[0, T]$ is partitioned into equal intervals of size Δ. Then, the points reached at time Δ from the vertices of X are computed. The set of states reachable *within* time Δ is then approximated by the convex hull of the vertices of the set X and the points reached at time Δ from the vertices of X. Given Δ, T, and the dynamics, the error (or Hausdorff distance) between this convex hull and the actual set of states reachable within time Δ can be bounded. Based on this error bound, this convex hull is first "bloated" to contain all the reachable states and then approximated by a data structure of choice. Different data structures that have been considered and found useful include griddy polytopes [9], ellipsoids [18], level sets [21], polytopes [6], zonotopes [13, 14], and support functions [15]. After this, the computation for the time interval $[0, \Delta]$ is *translated* to obtain the reachable states for the time interval $[i\Delta, (i + 1)\Delta]$ — the states reached at time $i\Delta$ and $(i + 1)\Delta$ are obtained by translating the vertices of X and those at time Δ, respectively, and then the "bloated" convex hull of all of these points is approximated by the data structure of choice. This method has been found to scalable and successful, making the automated analysis of linear dynamical systems possible.

In this paper, we take a slightly different stance on the problem of directly approximating the reachable set. Instead of trying to bound the error of a reachability computation, we view the problem as one where given an error bound ϵ, one has to compute an over-approximation of the reachable set whose Hausdorff distance from the actual set of reachable states is bounded by ϵ. This subtle change in perspective, immediately suggests some natural changes to the basic algorithm outlined in the previous paragraph. First, the discretization of the time interval $[0, T]$ need not be in terms of equal-sized intervals. We could change interval sizes, as long as the error of approximating the reachable states within that interval can be bounded by ϵ. Second, in the "basic algorithm", the convex hull of the points of the initial set and those at time Δ is taken to be the approximation of the set of states reachable within time Δ. Instead, we view the approximation process as first approximating the *flow* in the interval $[0, \Delta]$ by a polynomial, and then taking the "polynomial tube" defined by this dynamics to be the approximation of the reachable states — when the polynomial is taken to be a linear function, then it corresponds to taking the convex hull of the points, as done in the basic algorithm. Combining these ideas, the algorithm we plan to study in this paper can be summarized as follows. Given an error bound ϵ and the degree d of the polynomials to be considered, we first find a time (hopefully, as large as possible) t such that dynamics in $[0, t]$ when approximated by degree d polynomials is within error bound ϵ of actual dynamics. The degree d polynomial approximating the actual dynamics can be found by constructing the appropriate Bernstein polynomial [20]. We approximate the set of reachable states in the interval $[0, t]$ by the degree d polynomial tube, and then repeat the process for the time interval $[t, T]$, each time *dynamically* figuring out the appropriate discretization.

Our algorithm, when compared with the basic algorithm previously studied, has both perceptible advantages and disadvantages. On first glance, the basic algorithm seems to be computationally simpler and potentially faster. For a system with linear dynamics, computing the set of states reached at time Δ involves computing (or rather approximating) matrix exponentials. This is a significant computational overhead. In the basic algorithm, this cost is minimized, as it is performed once for the first $[0, \Delta]$ interval, and for subsequent intervals it is obtained by translation rather than direct computation. In our algorithm, this must be computed afresh for each sub-interval. Moreover, in our algorithm, the intervals need to be determined dynamically, which is an additional overhead to the computation in each sub-interval. However, on the flip side, dynamically determining intervals is likely to give us larger sub-intervals in some places and therefore result in fewer intervals overall to consider. This could be a potentially significant advantage. This is because the eventual approximation of the reach set for the interval $[0, T]$ is given as the union of basic sets (represented by the chosen data structure) that are computed for each sub-interval; thus, the number of terms in the union is as large as the number of sub-intervals. Subsequent steps in verifying safety properties (or other properties of interest) involve taking intersections, and checking emptiness and membership of states in these sets. The complexity of these set-theoretic operations depends on how many terms there are in the union — this is true no matter what the chosen data structure is. Thus, the time (and memory) used in verification is directly influenced by how many steps the time interval $[0, T]$ is divided into. Note, that the "quality" of the solution computed by the basic algorithm is not better than the one computed by our algorithm, even if it uses smaller sub-intervals and more of them, because the overall quality is determined by the quality of the solution in the "worst" interval, which is then built into the bloating factor used by both algorithms.

In order to evaluate these competing claims, we implemented both the basic algorithm and our algorithm in Matlab. Observe that in both the basic algorithm and our algorithm, states reached at certain times must be computed, and then the convex hull or polynomial tube must be approximated by the data structure of choice. In our experimental evaluation, we choose to be agnostic about the relative merits of different data structures, and we make no claims about which data structure should be chosen. Therefore, we only compute the states reached at certain time steps, and not the data structure representing the reachable states. Once a data structure is chosen, the computational overhead in constructing the desired set will be the same whether the basic algorithm or our method is used (provided linear flows are used to approximate the actual flow in our method). Thus, our experimental setup is to evaluate under what conditions (types of matrices and time bound T) does unequal intervals plus associated computation costs beat uniform intervals with computation minimized by translation. We also try to understand, when it makes sense to use polynomials that are not linear to approximate the flow. Our results apply no matter what your favorite data structure is.

Our experimental results are surprising. We evaluated the two methods on both "natural" examples that have been studied before, and randomly generated matrices, and for different time intervals and error bounds. First we observed that our algorithm is scalable as it computes the points for both large matrices (we tried it on 100×100 matrices) and for many time steps (requiring thousands of iterations). Second, surprisingly, our algorithm, approximating flow by linear functions, *most of the time* outperforms the basic algorithm, sometimes by a few orders of magnitude. The gap in the performance between the two algorithms only widens considerably as the time bound T is increased. This can be explained by the fact that as the number of iterations increases the significant reduction in the number of intervals dominates the computation costs. Our algorithm not only uses significantly fewer number of intervals for the same precision (as would be expected) but the size of the *minimum* interval is also significantly larger than the size of the uniform interval chosen by the basic algorithm. This suggests that our algorithm reaps the benefits of dynamic computation of error bounds, over static determination of them. Next, we compare the potential benefits of using non-linear flows to approximate the actual flow. We consider polynomials of degree 2, as they are appropriate when considering ellipsoids as the data structure. Theoretically, the size of a dynamically determined sub-interval could be a factor of 2 larger when using polynomials of degree 2 when compared with linear functions. That, in turn, could translate to significant (exponential) savings in terms of the number of intervals. However, these theoretical possibilities were not observed in our experiments — the number of intervals for degree 2 polynomials were at most a factor of 2 smaller. This could be explained by our observation that most of the sub-intervals in the linear approximation tend to be small, and they are roughly of the same size. In such a scenario the theoretical benefits of using quadratic approximations don't translate to visible gains.

We would like to remark that the approximations that we compute depend on the machine precision, since the approximations are constructed by sampling the function at certain points, and are only as precise as that of the function values computed for the sample points. There are other approaches for reachability analysis [23, 22, 19, 27] which rely on the decidability of the satisfiability problem for the first order theory of reals, which in contrast are algebraic techniques with infinite precision computation. However, for reachability analysis for the class of linear dynamical system, the above techniques need to first compute a polynomial approximation of the dynamics.

The rest of the paper is organized as follows. We begin with some preliminaries in Section 2. Next, in Section 3, we outline our algorithm to compute post for general dynamical systems by approximating flows using Bernstein polynomials. In Section 4, we describe the specific algorithm for linear dynamical systems. We then give details of our experimental results (Section 5) before presenting our conclusions.

2. PRELIMINARIES

Let $\mathbb{N}$ denote the set of natural numbers and let $\mathbb{R}$ and $\mathbb{R}_{\geq 0}$ denote the set of real numbers and non-negative real numbers, respectively. Given $x \in \mathbb{R}^n$, let $(x)_i$ denote the projection of x onto the i-th component, that is, if $x = (x_1, \cdots, x_n)$, then $(x)_i = x_i$. Given a function $F : A \to \mathbb{R}^n$, let $F_i : A \to \mathbb{R}$ denote the function given by $F_i(a) = (F(a))_i$. Given a function $F : \mathbb{R}_{\geq 0} \to B$ and $[a, b] \subseteq \mathbb{R}_{\geq 0}$, let $F[a, b] : [0, b - a] \to B$ denote the function given by $F[a, b](c) = F(a + c)$.

We will use ∞-norms for measuring the distance between two vectors. Given $x, y \in \mathbb{R}^n$, let $\|x - y\|$ denote the distance between x and y in the ∞-norm, that is, $\|x - y\| = max_{1 \leq i \leq n} |(x)_i - (y)_i|$. Also, given two functions $G : A \to \mathbb{R}^n$ and $H : A \to \mathbb{R}^n$, the distance between the functions, denoted $\|G - H\|$, is given by $\|G - H\| = \inf_{x \in A} \|G(x) - H(x)\|$. Given two sets $A, B \subseteq \mathbb{R}^n$, the Hausdorff distance between the two sets, denoted $d_H(A, B)$, is defined as

$$d_H(A, B) = max(\sup_{x \in A} \inf_{y \in B} \|x - y\|, \sup_{x \in B} \inf_{y \in A} \|x - y\|)$$

A polynomial over a variable x of degree k, denoted $p(x)$, is a term of the form $a_0 + a_1 x^1 + \cdots + a_k x^k$, where $a_i \in \mathbb{N}$ for all $1 \leq i \leq k$ and $a_k \neq 0$. Given a $v \in \mathbb{R}$, let $p(v)$ denote the value obtained by substituting v for x in the expression $p(x)$ and evaluating the resulting expression. A function $P : [a, b] \to \mathbb{R}^n$, for some $a, b \in \mathbb{R}$, is a polynomial function if there exist polynomials, $p_1, \cdots, p_n$ over x, such that for all $v \in [a, b]$, $P_i(v) = p_i(v)$. The degree of the polynomial function P is the maximum degree of the polynomials representing it, that is, degree of P is maximum of the degrees of $p_1, \cdots, p_n$ (the degree is unique). Note that polynomial functions are continuous functions. A piecewise polynomial function is a function whose domain can be divided into finite number of intervals such that the function restricted to each of these intervals is a polynomial function. A piecewise polynomial function (PPF) is a continuous function $P : [a, b] \to \mathbb{R}^n$, where $a, b \in \mathbb{R}$, such that there exists a sequence $t_1, \cdots, t_k$ such that $a < t_1 < \cdots < t_k < b$ and $P[a, t_1], P[t_1, t_2], \cdots, P[t_k, b]$ are all polynomial functions.

3. POST COMPUTATION BY FLOW APPROXIMATION

In this section, we describe a general algorithm to approximate the flow of a dynamical system by a piecewise polynomial function of any fixed degree within any approximation error bound. The approximations are based on Bernstein Polynomials.

3.1 Bernstein Polynomial Approximations

Our general algorithm for approximating a flow within a given error bound is based on the well-known Weierstrass Approximation Theorem, which says that an arbitrary function over the reals with a compact domain can be approximated by a polynomial such that the distance between the two functions is within a given error bound.

THEOREM 1 (WEIERSTRASS). *Given a continuous function $F : \mathbb{R} \to \mathbb{R}$, a compact subset $[a, b]$ of $\mathbb{R}$ and an $\epsilon > 0$, there exists a polynomial function $P : \mathbb{R} \to \mathbb{R}$ such that*

$$|F(x) - P(x)| < \epsilon, \forall x \in [a, b].$$

We can use the polynomial function guaranteed by this theorem to obtain a polynomial approximation of the flow of a dynamical system. Application of this theorem for hybrid system verification by construction of ϵ-simulations can be found in [24].

The above theorem is an existential theorem, and by itself does not suggest a way to obtain the polynomials. However, there exist a class of polynomials called Bernstein Polynomials, which can be constructed for a given function F, and an $\epsilon > 0$, such that the distance between the function F and the corresponding polynomial is within ϵ. The approximate polynomial is constructed by evaluating the function F at some finite number of points and using these values to compute the coefficients of the polynomial. Let $F : [a, b] \to \mathbb{R}$ be a function. Then a Bernstein polynomial of degree n approximating F, denoted by $Bern_n(F)$ is given by:

$$(Bern_n(F))(x) =$$

$$\sum_{k=0}^{n} F(a + k(b - a)/n) * \binom{n}{k} *$$

$$((x - a)/(b - a))^k (1 - (x - a)/(b - a))^{n-k},$$

for all $a \leq x \leq b$.

The next two lemmas essentially show that the approximation error introduced by the polynomial approximation can be made arbitrarily small. In particular, given an $\epsilon > 0$, we can choose an n effectively such that the distance between the two functions is bounded by ϵ. Let us denote by F_{diff}, the absolute difference between the maximum and minimum values of F in its domain, i.e, $F_{diff} = \max_{x_1, x_2 \in [a,b]} |F(x_2) - F(x_1)|$. Then, we have the following from [20]:

LEMMA 1. *Let $F : [a, b] \to \mathbb{R}$ be a continuous function and $\epsilon > 0$. Let $\delta > 0$ be such that for all $x_1, x_2 \in [a, b]$, $|x_2 - x_1| \leq \delta$ implies $|F(x_2) - F(x_1)| \leq \epsilon$. Then $|F(x) - Bern_n(F)(x)| \leq \epsilon$ if $n > F_{diff}/(\epsilon \delta^2)$.*

LEMMA 2. *Let $F : [a, b] \to \mathbb{R}$ be a continuous function satisfying the Lipschitz condition $|F(x) - F(y)| < L|x - y|$ for $x, y \in [a, b]$. Then $|F(x) - Bern_n(F)(x)| < L/(2\sqrt{(n)})$.*

Note that both the lemmas give an n such that the error or distance between F and $Bern_n(F)$ is within ϵ. In particular, given an ϵ the first lemma tells us to choose an $n > F_{diff}/(\epsilon \delta^2)$, and the second lemma tells us that a choice of $n > (L/2\epsilon)^2$ would ensure that the error is within ϵ.

3.2 General Algorithm

Our aim is to approximate a flow F over a time interval $[0, T]$ by a polynomial of very low degree, such as a *linear* or a *quadratic* polynomial. Lemmas 1 and 2 give us a polynomial of a certain degree approximating the function over the interval $[0, T]$ ensuring the desired error bound. However, the degree of the polynomial can be large. Hence instead of approximating by a single polynomial of high degree, we present an algorithm which splits the interval $[0, T]$ into smaller intervals, and approximates the flow separately in each of the smaller intervals, thereby giving a piecewise continuous polynomial approximation of a fixed degree.

Consider the following dynamical system.

$$\dot{x} = f(x), x \in \mathbb{R}^n, x(0) \in X_0,$$

where X_0 is a set of initial vectors. We will assume that f is a 'nice' function (for example, Lipschitz continuous) such that it has a unique solution $\Phi : \mathbb{R}^n \times \mathbb{R}_{\geq 0} \to \mathbb{R}^n$, satisfying $d/dt(\Phi(x_0, t)) = f(\Phi(x_0, t))$ for all $x_0 \in X_0$ and $t \in \mathbb{R}_{\geq 0}$. Note that Φ is assumed to be continuous and differentiable.

Let us fix an initial vector $x_0 \in X_0$, and a time $T \in \mathbb{R}_{\geq 0}$. Let $F : [0, T] \to \mathbb{R}^n$ be the function $F(t) = \Phi(x_0, t)$ for all $0 \leq t \leq T$. We will approximate each F_i by a piecewise polynomial function P_i of degree $\leq m$ within an error bound of ϵ. Hence $\|F - P\| < \epsilon$. The general algorithm is outlined below.

Algorithm 1 Varying Time Step Algorithm

Input: $m \in \mathbb{N}, \epsilon \in \mathbb{R}_{\geq 0}, F : [0, T] \to \mathbb{R}$
Output: Sequence of polynomials

$t := 0$
while $t < T$ **do**
  Choose $0 < t_i < T$ s.t.
  $\|Bern_m(F[t, t + t_i]) - F[t, t + t_i]\| \leq \epsilon$
  Output $Bern_m(F[t, t + t_i])$
  $t := t + t_i$
end while

Starting at time $t = 0$, find a time $0 < t_1 \leq T$ such that $\|Bern_m(F_i[0, t_1]) - F_i[0, t_1]\| < \epsilon$. There always exists such a t_1, since continuity of F implies that $F_i[0, t_1]_{diff}$ can be made arbitrarily small by taking t_1 to be sufficiently small, and therefore one can satisfy the condition $F_i[0, t_1]_{diff}/\epsilon\delta^2 < m$ in Lemma 1. This reduces the problem to finding a piecewise polynomial approximation of the function $F_i[t_1, T]$, and we proceed in the same manner to compute $t_2, t_3, \cdots$. Since the function values of the Bernstein polynomial and the function it is approximating match at the end-point, the piecewise polynomial function P, in any interval $[0, \sum_{i=1}^{k} t_i]$ is continuous. To ensure that the number of iterations is finite, we need to ensure that the we make progress. This can be guaranteed by ensuring that in each step, the t_i chosen is at least Δ, for some $\Delta > 0$. Note that there always exists a Δ, which can be chosen at any step which satisfies the condition in Lemma 1. To see this, let $\gamma = m\epsilon\delta^2$, where m is the degree of the polynomials we are considering, ϵ is the desired bound on approximation error, and δ is the parameter in the definition of continuity for F_i corresponding to ϵ. Since F_i is continuous and bounded, there exists a $\Delta > 0$ such that for all $t, t' \in [0, T]$, $|t - t'| \leq \Delta$ implies $|F_i(t) - F_i(t')| \leq \gamma$. Hence choosing Δ at any step ensures

that we make progress. In order to materialize the above sketch of the algorithm, we need to be able to compute F_{diff} or some upper bound on it, which ensures progress. In the next section, we present two methods to compute the t_is for the class of linear dynamical systems.

4. APPROXIMATION OF LINEAR DYNAMICAL SYSTEMS

In this section, we consider linear dynamical systems and present our algorithm in detail. Consider the following system:

$$\dot{x} = Ax, x \in \mathbb{R}^n, x(0) \in X_0,$$

where $X_0 \subseteq \mathbb{R}^n$ is a bounded convex polyhedron. The solution of the above equation is given by:

$$\Phi(x_0, t) = e^{At}x_0, x_0 \in X_0, t \in \mathbb{R}_{\geq 0}.$$

Let us define $Post_\Phi(X, [0, T]) = \{\Phi(x, t) \mid x \in X, t \in [0, T]\}$. We consider the problem of computing an over approximation of $Post_\Phi(X_0, [0, T])$ such that the error in the approximation is within an ϵ. More precisely, we wish to find a set $\widehat{Post_\Phi}(X_0, [0, T])$ such that $\widehat{Post_\Phi}(X_0, [0, T])$ is an over approximation, that is, $Post_\Phi(X_0, [0, T]) \subseteq \widehat{Post_\Phi}(X_0, [0, T])$, and the Hausdorff distance between the two sets is bounded by ϵ, that is, $d_H(Post_\Phi(X_0, [0, T]), \widehat{Post_\Phi}(X_0, [0, T])) \leq \epsilon$.

First we show that the flow function for a linear system preserves convexity and hence it suffices to approximate only the flows starting from the vertices of X_0.

PROPOSITION 1. *Let* $x = \alpha_1 x_1 + \cdots + \alpha_k x_k$ *where* $x_i \in \mathbb{R}^n$ *and* $\sum_{i=1}^{k} \alpha_k = 1$. *Then* $\Phi(x, t) = \alpha_1 \Phi(x_1, t) + \cdots + \alpha_k \Phi(x_k, t)$.

Let $Vertices(X_0)$ denote the set of vertices of X_0. Let us fix a time T. Given a $v \in Vertices(X_0)$, let $F_v : [0, T] \to \mathbb{R}^n$ be the function $F_v(t) = \Phi(v, t)$ for all $t \in [0, T]$. For each $v \in Vertices(X_0)$, let $\hat{F}_v$ denote a function such that $\|\hat{F}_v - F_v\| \leq \epsilon$. Let $\hat{R} = \{\alpha_1 \hat{F}_{v_1}(t) + \cdots + \alpha_k \hat{F}_{v_k}(t) \mid \alpha_1 + \cdots + \alpha_k = 1, t \in [0, T]\}$. The next lemma says that the Hausdorff distance between the exact post set and $\hat{R}$ is bounded by ϵ.

LEMMA 3.

$$d_H(Post_\Phi(X_0, [0, T]), \hat{R}) \leq \epsilon.$$

The above proposition tells us that it suffices to approximate the flows starting at the vertices of the polyhedron. More precisely, if we approximate the flows at the vertices within an error bound of ϵ in an interval $[0, T]$, then at any time $t \in [0, T]$, the Hausdorff distance between the actual and approximate sets is with ϵ.

In the literature, various methods have been proposed to compute $\widehat{Post_\Phi}$. These methods can be seen as consisting of the following two steps.

- Depending on the ϵ, a time step Δ is chosen. Let $V_0 = Vertices(X_0)$ and $V_i = Post_\Phi(V_0, [i\Delta, i\Delta])$ for $i > 0$ be the set of points reached from the vertices of X_0 after i time steps of size Δ. First $V_1 = Post_\Phi(V_0, [\Delta, \Delta])$ is computed. Then the convex hull C_0 of V_0 and V_1 is bloated by ϵ, and the resulting set is enclosed by a data structure of a certain form to obtain an overapproximation C_0' of $Post_\Phi(X_0, [0, \Delta])$.

- Similarly, to obtain an overapproximation of $Post_\Phi(X_0, [i\Delta, (i+1)\Delta))$, the convex hull C_i of V_i and V_{i+1} is bloated by ϵ, and enclosed in a data structure. However, instead of computing V_i directly from V_0, it is computed iteratively from V_{i-1}, that is, V_i is computed from V_{i-1} by a linear transformation using the matrix e^Δ.

We think of the above algorithm as first computing an approximation of the flow function, which is the piecewise linear function obtained by joining the corresponding points in V_is, and then enclosing the reach set given by the approximated flow function by a set of a certain form. The above algorithms compute a piecewise linear approximation of the flow function by dividing the interval $[0, T]$ into equal intervals of size Δ. Our main contribution is a novel algorithm for computing an approximation of the flow function, which does not divide the interval uniformly, but dynamically computes the next time step. The obvious advantage is the reduction in the number of times steps, since a time step chosen by the dynamic algorithm is always larger than the constant time step Δ chosen by the uniform time step algorithm. This in turn implies that the size of the final representation of the post set would be smaller, and the size plays a crucial role in further analysis. However, there is a overhead involved with the dynamic algorithm, which is in computing the set of vertices V_is at various time points, since these V_is can no more be computed iteratively by multiplication using a fixed matrix. Since the timesteps Δ keep changing, there does not exist a fixed matrix e^Δ which can be used to obtain V_i from V_{i-1} for every i. So the new algorithm involves computing a new matrix exponential e^{Δ_i} at each step. However, as we will see in the next section, our experimental evaluations show that the overhead introduced due to computation of a new matrix exponential at each time step becomes negligible due to the huge decrease in the number of steps. In other words, the cost of doing the large number of matrix multiplications in the constant time step algorithm is greater than the cost of computing new matrix exponentials followed by matrix multiplications for a small number of timesteps in the dynamic algorithm.

To compute the approximation of the flow function for a linear dynamical system, we instantiate Algorithm 1 to obtain an effective algorithm for linear dynamical systems. As mentioned in the discussion of Algorithm 1, we need to present a method to compute the t_is in each step such that progress is ensured. Next we present two methods for computing t_is.

4.1 Computing t_i: First method

In this section we use Lemma 2 to compute the bound t_i. Let us fix an $x_0 \in \mathbb{R}^n$ and an $n \times n$ matrix A. Let $F : [0, T] \to \mathbb{R}^n$ be the function $F(t) = e^{At}x_0$. First we show that F satisfies the Lipschitz condition, and the Lipschitz constant can be bounded by a function of T.

LEMMA 4. *Let $F : [0, T] \to \mathbb{R}^n$ be as defined above. Then for each $1 \le i \le n$, F_i is Lipschitz continuous with the Lipschitz constant $L = \|A\|e^{\|A\|T}\|x_0\|$.*

The next lemma gives us a lower bound on the time step t_i that can be chosen at each step such that the approximate polynomial is within distance ϵ from the original function.

LEMMA 5. *Let $F : [0, T] \to \mathbb{R}^n$ be as defined above. For $t_1 = \log_e(2\sqrt{m}\epsilon/\|A\|\|x_0\|)/\|A\|$, $\|F[0, t_1] - B_m(F[0, t_1])\| \le \epsilon$.*

Using the t_i in the definition of Lemma 5 is desirable since it gives a closed form expression for computing the t_i. However, the problem with the above expression is that the expression being computed might not result in a positive number in which case we are in trouble. Next we present another method for computing lower bound for t_i, which always gives a positive answer.

4.2 Computing t_i: Second method

In this section, we use Lemma 1 to compute a bound on the t_is.

LEMMA 6. *Let $F : [0, T] \to \mathbb{R}^n$ be as defined above. Let t_1 be such that $e^{3\|A\|t_1}t_1 < m\epsilon^3/\|A\|^3\|x_0\|^3$. Then we have that $\|B_m(F[0, t_1]) - F[0, t_1]\| \le \epsilon$.*

There always exists a positive real number $t_1 \le t$ satisfying the inequality $e^{at_1}t_1 < b$ where $a = 3\|A\|$ and $b = m\epsilon^3/\|A\|^3\|x_0\|^3$, since the function $e^{ax}x \to 0$ as $x \to 0$. Computing a t_1 such that $e^{at_1}t_1 = b$ might not be possible, instead one can obtain an upper bound on this value. For example, we know that $t_1 \le t$. Hence we can consider $t_1 = b/e^{at}$. We use the following alternative bound. If $at_1 < 1$, then we can upper bound e^{at_1} by $1/(1-at_1)$. This gives us a bound $t_1 < b/(1+ab)$. Hence $t_1 < \min\{b/(1+ab), 1/2a_1\}$ is a positive bound for t_1.

The algorithm for computing a piecewise polynomial approximation of a linear dynamical system is given in Algorithm 2.

Algorithm 2 Post Computation Algorithm for Linear Dynamical Systems

---

Input: $m \in \mathbb{N}, \epsilon \in \mathbb{R}_{\ge 0}, V_0 \subseteq_{Fin} \mathbb{R}^n, A \in \mathbb{R}^{n \times n}, T > 0$
Output: Sequence of a set of n polynomials

Let $F^x : [0, T] \to \mathbb{R}^n$ be $F^x(t) = e^{At}x$
for all $v \in V_0$ **do**
  $t := 0$
  $x := v$
  while $t < T$ **do**
    Choose $\tau_1 > 0$ s.t
    $e^{3\|A\|\tau_1}\tau_1 < m\epsilon^3/\|A\|^3\|x\|^3$
    Let $\tau_2 = \log_e(2\sqrt{m}\epsilon/\|A\|\|x\|)/\|A\|$
    Let $t_i = \max \tau_1, \tau_2$
    Output $Bern_m(F_j^v[t, t+t_i])$, for each $1 \le j \le n$
    $t := t + t_i$
    $x := e^{At}v$
  end while
end for

4.3 Termination of the Algorithm

In each step, we take as the next time step the maximum of the values obtained by methods in Lemma 5 and Lemma 6. This time step is always going to be positive, since Method 2 always gives a positive answer. Next we show that the time step we choose in any iteration has a positive lower bound. Hence, the algorithm always terminates.

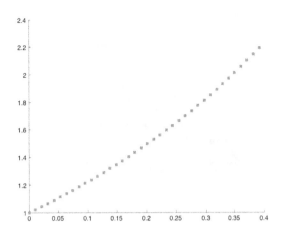

Figure 1: Constant time step Algorithm

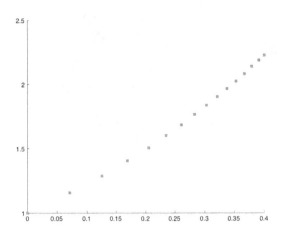

Figure 2: Varying time step Algorithm

Assume that in each step of method 2, we choose $t_i = \min\{b/(1+ab), 1/2a_1\}$.

$$b/(1+ab) = \frac{(m\epsilon^3/(\|A\|^3\|x_i\|^3))}{(1+(am\epsilon^3/(\|A\|^3\|x_i\|^3)))}$$

$$= \frac{(m\epsilon^3/(\|A\|^3))}{(\|x_i\|^3 + (am\epsilon^3/(\|A\|^3)))}$$

$$\geq \frac{(m\epsilon^3/(\|A\|^3))}{(e^{aT}\|x_0\|^3 + (am\epsilon^3/(\|A\|^3)))}$$

Therefore, the time steps t_i are lower bounded by a positive number.

Figure 1 and Figure 4.5 illustrate the difference between the constant step and varying step algorithms. For each algorithm, the points (t, y) are plotted, where t ranges over the times $\sum_{j=0}^{i} t_i$, where $t_1, t_2, \cdots$ is the sequence of time steps chosen by the algorithm, and y is given by $e^{At}x$. Observe that in the case of timestep varying algorithm, initially larger steps are chosen, and when the time approaches close to T, the timesteps taken by the constant time step algorithm and the varying time step algorithm become identical. Notice that the sampling rate of the varying time step algorithm depends on the value of the derivative of the function at various points, where as the constant time step algorithm makes no such distinction.

4.4 Function Evaluation Errors

Observe that we need to compute the value of the functions at several time points. In practice, these values can seldom be computed exactly. One can only hope to compute approximations within arbitrary error bounds. Let the function values be computed with in an error bound of γ. The approximate values of the samples can be thought of as the samples of a new function $\hat{F}$ such that $\|\hat{F} - F\| \leq \gamma$. Given an ϵ, let P be the piecewise polynomial approximation of F constructed using our algorithm, and let γ be a bound on the error in evaluating the function F. Then the error $\|P - F\| \leq \|P - \hat{F}\| + \|\hat{F} - F\| \leq \epsilon + \gamma$. Hence, by reducing γ, we can get as precise an approximation as desired.

4.5 Comparison with other polynomial approximations

In this paper, we considered Bernstein polynomials to approximate an arbitrary function by a polynomial function. Another popular technique to obtain polynomial approximations is to use Taylor's series expansion of the function and truncate the infinite sum after some points to obtain a polynomial. There are a few caveats in using Taylor's approximation in general. First it assumes that the function is smooth, and the derivatives can be computed. Secondly, it does not give a closed form expression for m, the number of terms in the Taylor's expansion that should be considered to obtain an ϵ bound on the approximation error.

5. EXPERIMENTAL EVALUATION

In this section, we explain our experimental set up for evaluating the performance of our algorithm and comparison with other methods. We implemented our algorithm is Matlab 7.4.0, and the experiments were conducted on Mac OS X Version 10.4.11, with a 2.16 GHz Intel Core 2 Duo processor and 1GB SDRAM. We performed our experiments on linear and quadratic approximations. We will report and explain our results for both the approximations in the following sections.

5.1 Linear approximation

Our experimental evaluation of the two algorithms chooses to be agnostic of the relative benefits of different data structures, and attempts to highlight the relative advantages of each algorithm, independent of the chosen data structure. No matter what the chosen data structure, the algorithms require computing the states reached at certain time steps. Once these points are computed, the data structure approximating a convex hull of the points needs to be computed. Thus, in our experimental evaluation we only compared the computational costs of finding the states reached at the required times using the two algorithms. In the basic algorithm, the interval size is fixed to be Δ, and then the states reached at time Δ are computed by multiplying initial set of states with matrix exponential e^Δ, but for subsequent times $i\Delta$ they are computed by translation that involves only multiplication with e^Δ which is evaluated only once.

Matrix	ϵ	T	n	m	RT_V	RT_C	t_{max}	t_{min}	Constant
2DR	4.93E-03	1	3.34E+02	8.53E+02	1.80E-01	2.79E-02	5.54E-03	1.78E-03	1.17E-03
2DR	6.78E-02	2	1.80E+01	1.60E+03	1.77E-02	1.15E-01	1.96E-01	7.81E-02	1.25E-03
2DR	3.60E-01	5	1.20E+01	4.27E+03	1.68E-02	4.99E-01	4.83E-01	4.83E-01	1.17E-03
2DR	1.85E+00	10	9.20E+01	8.47E+03	5.64E-02	1.51E+00	4.60E-01	9.42E-03	1.18E-03
5DR	4.92E-01	1	1.00E+01	5.65E+02	1.49E-02	4.69E-02	1.23E-01	1.07E-01	1.77E-03
5DR	3.14E+00	2	1.50E+01	1.14E+03	1.96E-02	1.15E-01	1.44E-01	1.44E-01	1.76E-03
5DR	2.11E+02	5	3.30E+01	2.99E+03	2.95E-02	4.97E-01	1.57E-01	1.57E-01	1.67E-03
5DR	3.04E+05	10	6.20E+01	6.15E+03	4.81E-02	1.62E+00	1.65E-01	1.65E-01	1.63E-03
100DR	2.01E-02	1	3.70E+02	9.16E+02	4.81E+00	5.33E-01	2.76E-03	2.61E-03	1.09E-03
100DR	2.56E-02	2	3.17E+02	1.84E+03	4.42E+00	2.93E+00	6.57E-03	5.93E-03	1.09E-03
100DR	6.13E-02	5	8.20E+01	4.59E+03	1.24E+00	2.14E+01	7.69E-02	4.97E-02	1.09E-03
100DR	2.65E-01	10	1.30E+01	9.17E+03	4.05E-01	9.23E+01	8.65E-01	8.22E-01	1.09E-03

Figure 3: Random matrices

Matrix	ϵ	T	n	m	RT_V	RT_C	t_{max}	t_{min}	Constant
Z2	1E-01	1	1.30E+02	7.47E+02	7.43E-02	2.55E-02	1.13E-02	5.96E-03	1.34E-03
Z2	1E-01	2	1.86E+02	6.69E+03	1.04E-01	9.95E-01	2.69E-02	5.96E-03	2.99E-04
Z2	1E-01	3	2.42E+02	4.49E+04	1.33E-01	2.78E+01	2.69E-02	5.96E-03	6.68E-05
Z5	1E-01	1	1.04E+02	5.61E+02	6.22E-02	4.00E-02	1.34E-02	7.91E-03	1.78E-03
Z5	1E-01	2	1.52E+02	5.02E+03	8.75E-02	1.04E+00	3.82E-02	7.91E-03	3.98E-04
Z5	1E-01	3	1.87E+02	3.38E+04	1.07E-01	4.33E+01	3.82E-02	7.91E-03	8.89E-05
Nav	1E+00	1	7.00E+00	5.00E+00	1.47E-02	2.30E-02	1.92E-01	1.92E-01	1.92E-01
Nav	1E+00	2	1.20E+01	9.30E+01	1.65E-02	2.30E-02	1.92E-01	1.92E-01	2.14E-02
Nav	1E+00	3	1.70E+01	6.37E+03	1.95E-02	1.45E+00	1.92E-01	1.92E-01	4.71E-04

Figure 4: Standard Examples for total time $T = 1, 2, 3$

In contrast, in our algorithm, the states reached at each of the designated time steps is computed from scratch using matrix exponentials, and the size of the next interval is found dynamically. Recall from Section 4, that when the initial set is a convex polyhedron, reach set computation involves computing this approximated flow for each vertex. Thus, in our experimental evaluation, we start from a single point, as this will faithfully reflect the costs of starting from a polyhedron. Finally, the error bound chosen for the varying time step algorithm was the one guaranteed by the fixed time step algorithm.

To determine feasibility of the algorithms, we first ran them on some randomly generated matrices. The entries of the matrices were random values in the interval $[-1, 1]$. The results of our experiments are shown in Figure 3. The rows labelled 2DR report results for 2×2 matrices, those labelled 5DR for 5×5 matrices, and finally those labelled 100DR for 100×100 matrices. The columns reported in the table are as follows: ϵ gives the error bound; T gives the time bound chosen for the experiment; m and n are the number of sub-intervals used by the constant timestep algorithm and varying timestep algorithms, respectively; RT_C and RT_V are the running times of the constant timestep and varying timestep algorithms, respectively; t_{max} and t_{min} are the largest and smallest time intervals considered by the varying timestep algorithm; "Constant" is the size of the interval used by the constant timestep algorithm. For these matrices, we chose a time step for the constant timestep algorithm to be of the order of 10^{-3}, and used the resulting error bound as ϵ. The results show that the varying timestep algorithm is scalable and has a running time comparable to the constant timestep algorithm; in many cases the varying timestep algorithm is faster by 2 orders of magnitude. The number of sub-intervals used by the varying timestep algorithm (n) is always less than that used by the constant timestep method (m) by either a magnitude or two orders of magnitude. The other surprising observation is that the smallest time interval used by the varying timestep method is in all cases larger than the interval used by the constant timestep algorithm.

We also experimented on benchmark examples considered in [13]. Nav is the navigation benchmark first suggested in [12], while Z2 and Z5 are the 2 dimensional and 5 dimensional examples from [13]. The matrices describing their dynamics is as follows.

$$Nav = \begin{bmatrix} 0 & 0 & 1 & 0 \\ 0 & 0 & 0 & 1 \\ 0 & 0 & -1.2 & 0.1 \\ 0 & 0 & 0.1 & -1.2 \end{bmatrix}, Z2 = \begin{bmatrix} -0.1 & -0.4 \\ 0.4 & -0.1 \end{bmatrix}$$

$$Z5 = \begin{bmatrix} -0.1 & -0.4 & 0 & 0 & 0 \\ 0.4 & -0.1 & 0 & 0 & 0 \\ 0 & 0 & -0.3 & 0.1 & 0 \\ 0 & 0 & -0.1 & -0.3 & 0 \\ 0 & 0 & 0 & 0 & -0.2 \end{bmatrix}$$

We tried to study the effect of increasing the time bound T on the running time of these algorithms and so we considered $T = 1, 2, 3$. Figure 4 shows our results for these benchmark examples and varying time. It shows that as T increases, the varying timestep algorithm's relative performance improves both in terms of the number of sub-intervals considered and

Matrix	t_{min}	t_{max}	RT_V	n	t^Q_{min}	t^Q_{max}	RT_Q	n_Q
$Z2$	5.96E-03	1.13E-02	9.88E-02	1.30E+02	7.21E-03	2.20E-02	4.74E-02	6.70E+01
$Z2$	5.96E-03	2.69E-02	1.28E-01	1.86E+02	1.26E-05	5.10E-02	5.98E-02	9.60E+01
$Z2$	1.00E-03	2.69E-02	1.60E-01	2.42E+02	1.18E-02	5.10E-02	7.49E-02	1.24E+02
$Z5$	2.48E-03	1.34E-02	1.21E-01	1.04E+02	1.56E-02	2.55E-02	4.10E-02	5.30E+01
$Z5$	1.60E-03	3.82E-02	1.12E-01	1.52E+02	1.56E-02	6.45E-02	5.17E-02	7.80E+01
$Z5$	7.91E-03	3.82E-02	1.31E-01	1.87E+02	1.56E-02	7.02E-02	6.25E-02	9.70E+01
Nav	1.92E-01	1.92E-01	3.54E-02	7.00E+00	1.92E-01	1.92E-01	1.10E-02	7.00E+00
Nav	1.92E-01	1.92E-01	3.84E-02	1.20E+01	1.92E-01	1.92E-01	1.44E-02	1.20E+01
Nav	1.92E-01	1.92E-01	4.09E-02	1.70E+01	1.92E-01	1.92E-01	1.57E-02	1.70E+01

Figure 5: Comparison of Quadratic and Linear Approximations

the running time, with the gap increasing to as much two orders of magnitude.

5.2 Quadratic approximation

We also implemented the varying time step algorithm for approximation by piecewise quadratic approximation. Theoretically, a single timestep of the quadratic approximation could be twice as much as that of the linear approximation. Interestingly, this could lead to a huge reduction in the number of total time steps. Consider an exponentially growing function, for which the time steps chosen by the linear approximation decrease by a constant factor in consecutive time steps. For example, consider the following sequence of timesteps $1, 1/2, 1/2^2, \cdots, 1/2^k$. A doubling of the time step in the quadratic approximation could lead to the skipping of k timesteps in the above example. However, we did not observe this phenomenon in our experiments which are tabulated in Figure 5. The columns t_{min}, t_{max}, RT_V, and n report the smallest time interval, largest time interval, running time, and number of intervals when using a linear approximations; the columns with superscript or subscript Q refer to the same quantities for the quadratic approximation. The improvement for the quadratic approximation was not as dramatic as we hoped it might, and in the best case was better by a factor of 2.

6. CONCLUSIONS

We presented a new algorithm for approximating the set of states reachable within time bound T in a linear dynamical system to within an arbitrary error bound ϵ. The main innovations in our algorithm over previous approaches is that the interval $[0, T]$ is dynamically subdivided into unequal sized intervals and then the flow in each interval is approximated by a polynomial of fixed degree. Our experimental evaluation of our algorithm reveals that the approach is scalable to high dimensional system, and is both faster and yields fewer sub-intervals than previous approaches that considered statically dividing the interval $[0, T]$ into equal sized sub-intervals.

7. ACKNOWLEDGEMENTS

We would like to thank Geir Dullerud for pointing us to the use of Bernstein Polynomials for approximations, and Parasara Sridhar Duggirala and Sayan Mitra for various helpful discussions. This research was support by grants NSF CCF 0448178, NSF CCF 1016989, and NSF CNS 1016791.

8. REFERENCES

[1] R. Alur, C. Courcoubetis, N. Halbwachs, T.A. Henzinger, P.-H. Ho, X. Nicollin, A. Olivero, J. Sifakis, and S. Yovine. The algorithmic analysis of hybrid systems. *Theoretical Computer Science*, 138(1):3–34, 1995.

[2] R. Alur, T. Dang, and F. Ivancic. Counter-Example Guided Predicate Abstraction of Hybrid Systems. In *Proceedings of the International Conference on Tools and Algorithms for the Construction and Analysis of Systems*, pages 208–223, 2003.

[3] R. Alur and D. Dill. A theory of timed automata. *Theoretical Computer Science*, 126:183–235, 1994.

[4] E. Asarin, T. Dang, and A. Girard. Hybridization methods for the analysis of nonlinear systems. *Acta Informatica*, 43(7):451–476, 2007.

[5] A. Chutinan and B.H. Krogh. Infinite state transition system verification using approximate quotient transition systems. *IEEE Transactions on Automatic Control*, 46:1401–1410, 2001.

[6] A. Chutinan and B.H. Krogh. Computational techniques for hybrid system verification. *IEEE Transactions on Automatic Control*, 48(1):64–75, 2003.

[7] E.M. Clarke, A. Fehnker, Z. Han, B. Krogh, J. Ouaknine, O. Stursberg, and M. Theobald. Abstraction and Counterexample-Guided Refinement in Model Checking of Hybrid Systems. *International Journal on Foundations of Computer Science*, 14(4):583–604, 2003.

[8] E.M. Clarke, A. Fehnker, Z. Han, B. Krogh, J. Ouaknine, O. Stursberg, and M. Theobald. Verification of Hybrid Systems Based on Counterexmple-Guided Abstraction Refinement. In *Proceedings of the International Conference on Tools and Algorithms for the Construction and Analysis of Systems*, pages 192–207, 2003.

[9] T. Dang and O. Maler. Reachability analysis via face lifting. In *Proceedings of the International Conference on Hybrid Systems: Computation and Control*, pages 96–109, 1998.

[10] T. Dang, O. Maler, and R. Testylier. Accurate hybridization of nonlinear systems. In *Proceedings of the International Conference on Hybrid Systems: Computation and Control*, pages 11–20, 2010.

[11] A. Fehnker, E.M. Clarke, S. Jha, and B. Krogh. Refining Abstractions of Hybrid Systems using Counterexample Fragments. In *Proceedings of the*

International Conference on Hybrid Systems: Computation and Control, pages 242–257, 2005.

[12] A. Fehnker and F. Ivancic. Benchmarks for hybrid systems verification. In Proceedings of the International Conference on Hybrid Systems: Computation and Control, pages 326–341, 2004.

[13] A. Girard. Reachability of uncertain linear systems using zonotopes. In Proceedings of the International Conference on Hybrid Systems: Computation and Control, pages 291–305, 2005.

[14] A. Girard and C.L. Guernic. Zonotope/Hyperplane intersecion for hybrid systems reachability analysis. In Proceedings of the International Conference on Hybrid Systems: Computation and Control, pages 215–228, 2008.

[15] C.L. Guernic and A. Girard. Reachability analysis of hybrid systems using support functions. In Proceedings of the International Conference on Computer Aided Verification, pages 540–554, 2009.

[16] T.A. Henzinger, P.W. Kopke, A. Puri, and P. Varaiya. What's decidable about hybrid automata? In Proceedings of the ACM Symposium on Theory of Computation, pages 373–382, 1995.

[17] S.K. Jha, B.H. Krogh, J.E. Weimer, and E.M. Clarke. Reachability for linear hybrid automata using iterative relaxation abstraction. In Proceedings of the International Conference on Hybrid Systems: Computation and Control, pages 287–300, 2007.

[18] A.B. Kurzhanski and P. Varaiya. Ellipsoidal techniques for reachability analysis. In Proceedings of the International Conference on Hybrid Systems: Computation and Control, pages 202–214, 2000.

[19] G. Lafferriere, G.J. Pappas, and S. Sastry. O-minimal Hybrid Systems. Mathematics of Control, Signals, and Systems, 13(1):1–21, 2000.

[20] G.G. Lorentz. Bernstein Polynomials. University of Toronto Press, 1953.

[21] I. Mitchell and C. Tomlin. Level set methods for computation in hybrid systems. In Proceedings of the International Conference on Hybrid Systems: Computation and Control, pages 310–323, 2000.

[22] Venkatesh Mysore, Carla Piazza, and Bud Mishra. Algorithmic Algebraic Model Checking II: Decidability of Semi-algebraic Model Checking and Its Applications to Systems Biology. In atva, pages 217–233, 2005.

[23] Carla Piazza, Marco Antoniotti, Venkatesh Mysore, Alberto Policriti, Franz Winkler, and Bud Mishra. Algorithmic Algebraic Model Checking I: Challenges from Systems Biology. In Proceedings of the International Conference on Computer Aided Verification, pages 5–19, 2005.

[24] P. Prabhakar, V. Vladimerou, M. Viswanathan, and G.E. Dullerud. Verifying tolerant systems using polynomial approximations. In Proceedings of the IEEE Real-Time Systems Symposium, pages 181–190, 2009.

[25] A. Puri, V.S. Borkar, and P. Varaiya. ε-approximation of differential inclusions. In Proceedings of the International Conference on Hybrid Systems: Computation and Control, pages 362–376, 1995.

[26] M. Segelken. Abstraction and Counterexample-guided Construction of Omega-Automata for Model Checking of Step-discrete linear Hybrid Models. In Proceedings of the International Conference on Computer Aided Verification, pages 433–448, 2007.

[27] V. Vladimerou, P. Prabhakar, M. Viswanathan, and G.E. Dullerud. STORMED hybrid systems. In Proceedings of the International Colloquium on Automata, Languages and Programming, pages 136–147, 2008.

9. APPENDIX

9.1 Proofs of the upper bounds

9.1.1 Proof of Lemma 4

PROOF. First let us recall the following identity. Given $n \times n$ matrices X and Y,

$$\|e^{X+Y} - e^X\| \leq \|Y\| e^{\|X\|} e^{\|Y\|}.$$

W.l.o.g assume $t_2 > t_1$.

$$\frac{\|e^{At_2} x_0 - e^{At_1} x_0\|}{\|t_2 - t_1\|}$$

$$= \frac{\|e^{At_1 + A(t_2 - t_1)} x_0 - e^{At_1} x_0\|}{|t_2 - t_1|}$$

$$\leq \frac{\|A(t_2 - t_1)\| e^{\|At_1\|} e^{\|A(t_2 - t_1)\|} \|x_0\|}{|t_2 - t_1|}$$

$$= \frac{\|A\| |t_2 - t_1| e^{\|A\| \|t_1\|} e^{\|A\| |t_2 - t_1|} \|x_0\|}{|t_2 - t_1|}$$

$$= \|A\| e^{\|A\| e (|t_1| + |t_2 - t_1|)} \|x_0\|$$

$$= \|A\| e^{\|A\| T} \|x_0\|$$

$$L \geq \max_{t_1, t_2} \frac{\|e^{At_2} x_0 - e^{At_1} x_0\|}{\|t_2 - t_1\|}$$

$$\geq \|A\| e^{\|A\| T} \|x_0\|$$

□

9.1.2 Proof of Lemma 5

PROOF. The Lipschitz constant L for the function $F[0, t_1]$ is given by $L = \|A\| e^{\|A\| t_1} \|x_0\|$ from Lemma 4.

$\|F[0, t_1] - B_m(F[0, t_1])\| < L/(2\sqrt{m})$ from Lemma 2.

$L/(2\sqrt{m}) \leq \epsilon$ implies $\|A\| e^{\|A\| t_1} \|x_0\| \leq 2\sqrt{m}\epsilon$.

Hence for $t_1 \leq \log_e(2\sqrt{m}\epsilon/\|A\|\|x_0\|)/\|A\|$,

$\|F[0, t_1] - B_n(F[0, t_1])\| \leq \epsilon$. □

9.1.3 Proof of Lemma 6

PROOF. From Lemma 1, we have that $\|B_n(F[0, t_1]) - F[0, t_1]\| \leq \epsilon$ if $n > F_{diff}/\epsilon\delta^2$.

We will find bounds on the values of F_{diff} and δ as a function of t_1.

$$F_{diff} = \max_{x,y \in [0,t_1]} \|F(x) - F(y)\|$$

$$= \max_{x,y \in [0,t_1]} \|e^{Ax}x_0 - e^{Ay}x_0\| \leq \|A\|e^{\|A\|t_1}\|x_0\|t_1$$

(See proof of Lemma 5.)

Next we need to find a lower bound on δ such that

$$\forall x, y \in [0, t_1], |x - y| \leq \delta \implies \|F(x) - F(y)\| \leq \epsilon.$$

Or equivalently

$$\max_{x,y \in [0,t_1], |x-y| \leq \delta} \|F(x) - F(y)\| \leq \epsilon.$$

However,

$$\max_{x,y \in [0,t_1], |x-y| \leq \delta} \|F(x) - F(y)\| \leq \|A\|e^{\|A\|t_1}\|x_0\|\delta$$

(again from the proof of Lemma 5).

Hence it suffices to choose a δ which ensures

$$\|A\|e^{\|A\|t_1}\|x_0\|\delta \leq \epsilon.$$

Hence we can choose

$$\delta = \epsilon/(\|A\|e^{\|A\|t_1}\|x_0\|).$$

We want to choose a t_1 so as to satisfy $m > F_{diff}/\epsilon\delta^2$. It suffices to satisfy

$$m > \frac{\|A\|e^{\|A\|t_1}\|x_0\|t_1)}{(\epsilon(\epsilon/(\|A\|e^{\|A\|t_1}\|x_0\|))^2)}.$$

Or,

$$m\epsilon^3 > \|A\|^3 e^{3\|A\|t_1}\|x_0\|^3 t_1.$$

For t_1 such that

$$e^{3\|A\|t_1}t_1 < n\epsilon^3/\|A\|^3\|x_0\|^3,$$

we have $\|F(x) - B_m(F(x))\| \leq \epsilon$ for all $0 \leq x \leq t_1$. □

9.1.4 Proof of Lemma 3

PROOF. Given $x \in Post_\Phi(X_0, [0, T])$ we will find an $x' \in \hat{R}$ such that $\|x - x'\| \leq \epsilon$ and vice versa.

Let $x \in Post_\Phi(X_0, [0, T])$.

Then $x = e^{At}x_0$ for some $x_0 \in X_0$ and $t \in T$.

Let $Vertices(X_0) = \{v_1, \cdots, v_k\}$.

Since X_0 is a bounded convex polyhedron,

$$x_0 = \alpha_1 v_1 + \cdots + \alpha_k v_k,$$

for some $\alpha_1 + \cdots + \alpha_k = 1$.

Then $x = e^{At}x_0 = e^{At}(\alpha_1 v_1 + \cdots + \alpha_k v_k)$

$$= (\alpha_1 e^{At}v_1 + \cdots + \alpha_k e^{At}v_k).$$

Let $x' = \alpha_1 \hat{F}_{v_1}(t) + \cdots + \alpha_k \hat{F}_{v_k}(t)$.

Then $|x - x'| =$

$$|(\alpha_1 e^{At}v_1 + \cdots + \alpha_k e^{At}v_k) - (\alpha_1 \hat{F}_{v_1}(t) + \cdots + \alpha_k \hat{F}_{v_k}(t))|$$

$$\leq \alpha_1 |e^{At}v_1 - \hat{F}_{v_1}(t)| + \cdots + \alpha_k |e^{At}v_k - hatF_{v_k}(t)|$$

$$\leq \alpha_1 \epsilon + \cdots + \alpha_k \epsilon = \epsilon.$$

Similarly given an $x \in \hat{R}$, we can find a $x' \in Post_\Phi(X_0, [0, T])$ such that $|x - x'| \leq \epsilon$. □

Automatic Abstraction of Non-Linear Systems Using Change of Bases Transformations

Sriram Sankaranarayanan [*]
University of Colorado, Boulder, CO, USA.
firstname.lastname@colorado.edu

ABSTRACT

We present abstraction techniques that transform a given non-linear dynamical system into a linear system, such that, invariant properties of the resulting linear abstraction can be used to infer invariants for the original system. The abstraction techniques rely on a change of bases transformation that associates each state variable of the abstract system with a function involving the state variables of the original system. We present conditions under which a given change of basis transformation for a non-linear system can define an abstraction.

Furthermore, we present a technique to discover, given a non-linear system, if a change of bases transformation involving degree-bounded polynomials yielding a linear system abstraction exists. If so, our technique yields the resulting abstract linear system, as well. This approach is further extended to search for a change of bases transformation that abstracts a given non-linear system into a system of linear differential inclusions. Our techniques enable the use of analysis techniques for linear systems to infer invariants for non-linear systems. We present preliminary evidence of the practical feasibility of our ideas using a prototype implementation.

Categories and Subject Descriptors: **F.3.1**(Specifying and Verifying and Reasoning about Programs):Invariants, **C.1.m**(Miscellaneous): Hybrid Systems.

Terms: Theory, Verification.

Keywords: Ordinary Differential Equations, Hybrid Systems, Algebraic Geometry, Invariants, Verification, Abstraction.

1. INTRODUCTION

> ... the purpose of abstracting is not to be vague, but to create a new semantic level in which one can be absolutely precise. – Edsger Dijkstra (ACM Turing Lecture, 1972).

[*]This material is based upon work supported by the US National Science Foundation (NSF) under Grant No. 0953941. Any opinions, findings, and conclusions or recommendations expressed in this material are those of the author(s) and do not necessarily reflect the views of the NSF.

In this paper, we present techniques to search for a *"linear system within"* a given non-linear system. Specifically, we wish to discover affine differential abstractions of continuous systems defined by non-linear differential equations. Given a system of non-linear differential equations, we seek a *change of bases* transformation, mapping the trajectories of the non-linear system into those of a linear abstraction. We present conditions under which a change of bases transformation defines an *abstraction*. Therefore, we can use the invariants for the abstraction to infer invariants for the original system. In this regard, an affine system abstraction is quite useful. Numerous techniques have been proposed to verify safety properties of affine systems efficiently, including *zonotopes* [10], *template polyhedra* [25] and *support functions* [31, 11]. These techniques have been implemented in tools such as HyTech [12], Phaver [9] and TimePass [25, 27]. However, these techniques are mostly restricted to systems with affine dynamics. Relying on these techniques to infer properties of non-linear systems is, therefore, a natural step forward. In this paper, we make the following contributions:

1. We first present techniques for discovering a *linearizing change of bases transformation* that results in a linear abstraction whose dynamics are described by affine ordinary differential equations (ODEs). We prove that a linearizing transformation for a non-linear system corresponds one-to-one to a finite dimensional vector space of functions that also contains the (Lie) derivatives of its elements. The basis functions of a vector space satisfying this closure property yields the desired change of bases transformation. This, in turn, yields the desired linear abstraction.

2. We extend our technique to discover transformations of non-linear ODEs into *differential inequalities*. We show that these transformations are closely related to finitely generated cones of functions that satisfy the property of *closure with a positive semi-definite residue*. We consider two approaches for discovering such cones: one based on the Sum-of-Squares (SOS) relaxation for finding positive semi-definite polynomials [18], and the other based on a simpler, heuristic approach using polyhedral cones over a finite set of posynomials. The result of this abstraction is a non-autonomous linear system with non-negative (disturbance) inputs.

We have implemented our approaches and present interesting preliminary results on finding abstractions for non-linear ODEs. Our implementation, the benchmarks used in the evaluation and the outputs are available on-line or upon request. We motivate our approach on a simple non-linear system.

Example 1.1 (Motivating Example). *Consider a continuous system over* $\{x, y\}$: $\dot{x} = xy + 2x$, $\dot{y} = -\frac{1}{2}y^2 + 7y + 1$, *with* *initial conditions given by the set* $x \in [0, 1]$, $y \in [0, 1]$. *Using the transformation* $\alpha : (x, y) \mapsto (w_1, w_2, w_3)$ *wherein* $\alpha_1(x, y) = x$, $\alpha_2(x, y) = xy$ *and* $\alpha_3(x, y) = xy^2$, *we find that the dynamics over* $\vec{w}$ *can be written as*

$$\dot{w}_1 = 2w_1 + w_2, \ \dot{w}_2 = w_1 + 9w_2 + \frac{1}{2}w_3, \ \dot{w}_3 = w_1 + 7w_2 + 2w_3$$

Its initial conditions are given by $w_1 \in [0, 1]$, $w_2 \in [0, 1]$, $w_3 \in [0, 1]$. *We analyze the system using the TimePass tool as presented in our previous work [27] to obtain polyhedral invariants:*

$$3w_1 - 8w_2 + 2w_3 \leq 5 \ \wedge \ 2w_1 + 13w_2 - 17w_3 \leq 15$$
$$w_1, w_2, w_3 \geq 0 \ \wedge \ 4w_2 - w_3 \geq -1 \ \wedge \ w_1 - 2w_2 \leq 1$$
$$w_1 - 2w_3 \leq 1 \ \wedge \ 7w_2 - 9w_3 \leq 7 \ \wedge \ 29w_1 - 4w_2 + w_3 \geq -4$$

Substituting back, we can infer polynomial inequality invariants on the original system including,

$$3x - 8xy + 2xy^2 \leq 5 \ \wedge \ 2x + 13xy - 17xy^2 \leq 15 \ \wedge$$
$$x \geq 0 \ \wedge \ xy \geq 0 \ \wedge \ xy^2 \geq 0 \ \wedge \ 4xy - xy^2 \geq -1 \ \wedge$$
$$x - 2xy \leq 1 \ \wedge \ x - 2xy^2 \leq 1 \ \wedge \ \cdots$$

Note that not every transformation yields a linear abstraction. In fact, most transformations will not define an abstraction. The conditions for an abstraction are discussed in Section 3.

1.1 Related Work

Many different types of *discrete abstractions* have been studied for hybrid systems [2] including predicate abstraction [28] and abstractions based on invariants [17]. The use of counter-example guided abstraction-refinement for iterative refinement has also been investigated in the past (Cf. Alur et al. [1] and Clarke et al. [5], for example). In this paper, we consider continuous abstractions for continuous systems specified as ODEs using a change of bases transformation. As noted above, not all transformations can be used for this purpose. Our abstractions bear similarities to the notion of topological semi-conjugacy between flows of dynamical systems [15].

Reasoning about the reachable set of states for flows of non-linear systems is an important primitive that is used repeatedly in the analysis of non-linear hybrid systems. This has been addressed using a wide variety of techniques in the past, including algebraic and semi-algebraic geometric techniques, interval analysis, constraint propagation and Bernstein polynomials [26, 29, 21, 16, 23, 19, 20, 14, 22, 8]. In particular, the hybridization of non-linear systems is an important approach for converting it into affine systems by subdividing the invariant region into numerous sub-regions and approximating the dynamics as a hybrid system by means of a linear differential inclusion in each region [12, 3, 7]. However, such a subdivision can be expensive as the number of dimensions increases and may not be feasible if the invariant region is unbounded. Our techniques can work on unbounded invariant regions. On the other hand, our abstraction search is *incomplete*. As a result, the techniques presented here may result in a trivial abstraction that does not yield useful information about the system. Nevertheless, we have been able to present some preliminary evidence of usefulness of our ideas over some complex non-linear system benchmarks.

Previous work on invariant generation for hybrid system by the author constructs invariants by assuming a desired template form (ansatz) with unknown parameters and applying the "consecution" conditions such as *strong consecution* and *constant scale* consecution [26]. Matringe et al. present generalizations of these conditions using morphisms [14]. Therein, they observe that strong and constant scale consecution conditions correspond to a linear

abstraction of the original non-linear system of a restrictive form. Specifically, the original system is abstracted by a system of the form $\frac{dx}{dt} = 0$ for strong consecution, and a system of the form $\frac{dx}{dt} = \lambda x$ for constant-scale consecution. This paper builds upon this observation by Matringe et al. using fixed-point computation techniques to search for a general linear abstraction that is related to the original system by a change of basis transformation. Moving from equality invariants to inequalities, our work is closely related to the technique of differential invariants proposed by Platzer et al. A key primitive used in this technique can be cast as a search for abstractions of the form $\frac{dx}{dt} \geq 0$ [20]. The approach presented here uses fixed-point computation over cones to search for generalized linear differential inequality abstractions.

Fixed point techniques for deriving invariants of differential equations have been proposed by the author in previous papers [27, 24] These techniques have addressed the derivation of polyhedral invariants for affine systems [27] and algebraic invariants for systems with polynomial right-hand sides [24]. In this technique, we employ the machinery of fixed-points. Our primary goal is not to derive invariants, per se, but to search for abstractions of non-linear systems into linear systems.

Finally, our approach for polynomials is closely related to *Carlemann embedding* that can be used to linearize a given differential equation with polynomial right-hand sides [13]. The standard Carlemann embedding technique creates an infinite dimensional linear system, wherein, each dimension corresponds to a monomial or a basis polynomial. In practice, it is possible to create a linear approximation with known error bounds by truncating the monomial terms beyond a degree cutoff. Our approach for differential equation abstractions can be seen as a search for a "finite submatrix" inside the infinite matrix created by the Carleman linearization. The rows and columns of this submatrix correspond to monomials such that the derivative of each monomial in the submatrix is a linear combination of monomials that belong the submatrix. Our approach of differential inequalities allows for a *residue* involving monomials outside the submatrix that is required to be positive semi-definite.

2. PRELIMINARIES

In this section, we briefly introduce some basic concepts behind multivariate polynomials and hybrid systems. Let $\mathbb{R}$ denote the field of real numbers. Let $x_1, \ldots, x_n$ denote a set of variables, collectively represented as $\vec{x}$. The $\mathbb{R}[\vec{x}]$ denotes the ring of multivariate polynomials over $\mathbb{R}$.

A *monomial* over $\vec{x}$ is of the form $x_1^{r_1} x_2^{r_2} \cdots x_n^{r_n}$, succinctly written as $\vec{x}^{\vec{r}}$, wherein each $r_i \in \mathbb{N}$. A *term* is of the form $c \cdot m$ where $c \in \mathbb{R}$, $c \neq 0$ and m is a monomial. The degree of a monomial $\vec{x}^{\vec{r}}$ is given by $\sum_{i=1}^{n} r_i = \vec{1} \cdot \vec{r}$. The degree of a multivariate polynomial p is the maximum over the degrees of all monomials m that *occur* in p with a non-zero coefficient.

Vector Fields: A *vector field* F over a manifold $X \subseteq \mathbb{R}^n$ is a map $F : X \mapsto \mathbb{R}^n$ from each $\vec{x} \in X$ to a vector $F(\vec{x}) \in \mathbb{R}^n$ ($F(\vec{x}) \in T_x(X)$, the tangent space of X at $\vec{x}$). A vector field F is continuous if the map F is continuous. A polynomial vector field $F : X \mapsto \mathbb{R}[\vec{x}]^n$ is specified by a map $F(\vec{x}) = \langle p_1(\vec{x}), p_2(\vec{x}), \ldots, p_n(\vec{x}) \rangle$, wherein $p_1, \ldots, p_n \in \mathbb{R}[\vec{x}]$. A system of ordinary differential equations $D, \frac{dx_1}{dt} = p_1(x_1, \ldots, x_n), \cdots, \frac{dx_n}{dt} = p_n(x_1, \ldots, x_n)$, specifies the evolution of variables $(x_1, \ldots, x_n) \in X$ over time t. The system defines a vector field $F(\vec{x}) : \langle p_1(\vec{x}), \ldots, p_n(\vec{x}) \rangle$.

Def. 2.1 (Lie Derivative). *For a vector field $F : \langle f_1, \ldots, f_m \rangle$, the* Lie derivative *of a smooth function $f(\vec{x})$ is given by*

$$\mathcal{L}_F(f) = (\nabla f) \cdot F(\vec{x}) = \sum_{i=1}^{n} \left(\frac{\partial f}{\partial x_i} \cdot f_i \right)$$

Henceforth, wherever the vector field F is clear from the context, we will drop subscripts and use $\mathcal{L}(p)$ to denote the Lie derivative of p w.r.t F.

We assume that all vector fields F considered in this paper are (locally) Lipschitz continuous over the domain X. In general, all polynomial vector fields are locally Lipschitz continuous, but not necessarily *globally* Lipschitz continuous over an unbounded domain X. The Lipschitz continuity of the vector field F, ensures that given $\vec{x} = \vec{x}_0$, there exists a time $T > 0$ and a unique time trajectory $\tau : [0, T] \mapsto \mathbb{R}^n$ such that $\tau(t) = \vec{x}_0$ [15].

Def. 2.2 (Continuous System). *A continuous system over variables $x_1, \ldots, x_n$ consists of a tuple $\mathcal{S} : \langle X_0, \mathcal{F}, X_I \rangle$ wherein $X_0 \subseteq \mathbb{R}^n$ is the set of initial states, $\mathcal{F}$ is a vector field over the domain $X_I \subseteq \mathbb{R}^n$.*

Note that in the context of hybrid systems, the set X_I is often referred to as the *state invariant* or the *domain*.

2.1 Hybrid Systems

Hybrid systems consists of continuous state variables and a finite set of discrete modes. The dynamics of the continuous state variables are a function of the system's current discrete mode. Furthermore, the system performs (instantaneous) mode changes upon encountering a switching condition (or a transition guard).

Def. 2.3 (Hybrid System). *An hybrid system is a tuple $\langle \mathcal{S}, \mathcal{T} \rangle$, wherein $\mathcal{S} = \{ S_1, \ldots, S_k \}$ consists of k discrete modes and $\mathcal{T}$ denotes discrete transitions between the modes. Each mode $S_i \in \mathcal{S}$ is a continuous sub-system $\langle X_{0,i}, F_i, X_i \rangle$, defined by the vector field $\mathcal{F}_i$, the initial conditions $X_{0,i}$ and the invariance condition X_i.*

Each transition $\tau : \langle S_i, S_j, P_{ij} \rangle \in \mathcal{T}$ consists of an edge $S_i \to S_j$ along with an transition relation $P_{ij}[\vec{x}, \vec{x}']$ specifying the next state $\vec{x}'$ in relation to the previous state $\vec{x}$. Note that the transition is guarded by the assertion $\exists \vec{x}' P_{ij}[\vec{x}, \vec{x}']$.

3. CHANGE OF BASES TRANSFORMATION

In this section, we will present change of bases (CoB) abstractions and some of their properties.

Consider a map $\alpha : \mathbb{R}^k \mapsto \mathbb{R}^l$. Given a set $S \subseteq \mathbb{R}^k$, let $\alpha(S)$ denote the set obtained by applying α to all the elements of S. Similarly, the inverse map over sets is $\alpha^{-1}(T) : \{ s \mid \alpha(s) \in T \}$. Let $\mathcal{S} : \langle X_0, \mathcal{F}, X_I \rangle$ be a continuous system over variables $\vec{x} : (x_1, \ldots, x_n)$ and $\mathcal{T} : \langle Y_0, \mathcal{G}, Y_I \rangle$ be a continuous system over variables $\vec{y} : (y_1, \ldots, y_m)$.

Def. 3.1 (Simulation). *We say that $\mathcal{T}$ simulates $\mathcal{S}$ iff there exists a smooth mapping $\alpha : \mathbb{R}^n \mapsto \mathbb{R}^m$ such that*

1. *$Y_0 \supseteq \alpha(X_0)$ and $Y_I \supseteq \alpha(X_I)$.*

2. *For any trajectory $\tau : [0, T) \mapsto X_I$ of $\mathcal{S}$, $\alpha \circ \tau$ is a trajectory of $\mathcal{T}$.*

A simulation relation implies that any time trajectory of $\mathcal{S}$ can be mapped to a trajectory of $\mathcal{T}$ through α. However, since α need not be invertible, the converse need not hold. I.e, $\mathcal{T}$ may exhibit time trajectories that are not mapped onto by any trajectory in $\mathcal{S}$.

Let $\mathcal{S}$ and $\mathcal{T}$ be defined by Lipschitz continuous vector fields. The following theorem enables us to check given $\mathcal{S}$ and $\mathcal{T}$, whether $\mathcal{T}$ simulates $\mathcal{S}$.

Theorem 3.1. *$\mathcal{T}$ simulates $\mathcal{S}$ if the following conditions hold:*

1. *$Y_0 \supseteq \alpha(X_0)$.*

2. *$Y_I \supseteq \alpha(X_I)$.*

3. *$\mathcal{G}(\alpha(\vec{x})) = J_\alpha . \mathcal{F}(\vec{x})$, wherein, J_α is the Jacobian*

$$J_\alpha(x_1, \ldots, x_n) = \begin{bmatrix} \frac{\partial \alpha_1}{\partial x_1} & \cdots & \frac{\partial \alpha_1}{\partial x_n} \\ \vdots & \ddots & \vdots \\ \frac{\partial \alpha_m}{\partial x_1} & \cdots & \frac{\partial \alpha_m}{\partial x_n} \end{bmatrix},$$

and $\alpha(\vec{x}) = (\alpha_1(\vec{x}), \cdots, \alpha_m(\vec{x}))$, $\alpha_i : \mathbb{R}^n \mapsto \mathbb{R}$.

PROOF. Let τ_x be a trajectory over $\vec{x}$ for system $\mathcal{S}$. Note that at any time instant $t \in [0, t)$, $\frac{d\tau_x}{dt} = \mathcal{F}(\tau(t))$.

We wish to show that $\tau_y(t) = \alpha(\tau_x(t))$ is a time trajectory for the system $\mathcal{T}$. Since, $\tau_x(0) \in X_0$, we conclude that $\tau_y(0) = \alpha(\tau_x(0)) \in Y_0$. Since $\tau_x(t) \in X_I$ for all $t \in [0, T)$, we have that $\tau_y(t) = \alpha(\tau_x(t)) \in Y_I$. Differentiating τ_y we get,

$$\begin{aligned} \frac{d\tau_y}{dt} &= \frac{d\alpha(\tau_x(t))}{dt} &= J_\alpha \cdot \frac{d\tau_x}{dt} &= J_\alpha \cdot \mathcal{F}(\tau_x(t)) \\ &= \mathcal{G}(\alpha(\tau_x(t))) &= \mathcal{G}(\tau_y(t)). \end{aligned}$$

Therefore $\tau_y = \alpha \circ \tau_x$ conforms to the dynamics of $\mathcal{T}$. By Lipschitz continuity of $\mathcal{G}$, we obtain that τ_y is the unique trajectory starting from $\alpha \circ \tau(0)$. $\square$

Note that, in general, a trajectory $\tau_y(t) = \alpha(\tau_x(t))$ may exist for a longer interval of time than the interval $[0, T)$ over which τ_x is assumed to be defined.

Theorem 3.2. *Let $\mathcal{T}$ simulate $\mathcal{S}$ through a map α. If $Y \subseteq Y_I$ is a positive invariant set for $\mathcal{T}$ then $\alpha^{-1}(Y) \cap X_I$ is a positive invariant set for $\mathcal{S}$.*

PROOF. Assuming otherwise, let τ_x be a time trajectory that starts from inside $\alpha^{-1}(Y) \cap X_I$ and has a time instant t such that $\tau_x(t) \notin \alpha^{-1}(Y) \cap X_I$. Since we defined time trajectories so that $\tau_x(t) \in X_I$, it follows that $\tau_x(t) \notin \alpha^{-1}(Y)$. As a result, $\alpha(\tau_x(t)) \notin Y$. Therefore, corresponding to τ_x, we define a new trajectory $\tau_y = \alpha \circ \tau_x$ which violates the positive invariance of Y. This leads to a contradiction. $\square$

An application of the Theorem above is illustrated in Example 1.1.

Example 3.1. *Consider a mechanical system $\mathcal{S}$ expressed in generalized position coordinates (q_1, q_2) and momenta (p_1, p_2) defined using the following vector field:*

$$F(p_1, p_2, q_1, q_2) : \left\langle -2q_1 q_2^2, -2q_1^2 q_2, 2p_1, 2p_2 \right\rangle$$

with the initial conditions: $(p_1, p_2) \in [-1, 1] \times [-1, 1] \wedge (q_1, q_2) : (2, 2)$. Using the transformation $\alpha(p_1, p_2, q_1, q_2) : p_1^2 + p_2^2 + q_1^2 q_2^2$, we see that $\mathcal{S}$ is simulated by a linear system $\mathcal{T}$ over y, with dynamics given by $\frac{dy}{dt} = 0$, $y(0) \in [16, 18]$.

Incidentally, the form of the system $\mathcal{T}$ above indicates that α is an expression for a conserved quantity (in this case, the Hamiltonian) of the system.

3.1 Linearizing CoB Transformations

In this section, we define the notion of a linearizing CoB transformation. An affine system $\mathcal{T}$ is described by an affine vector field $\frac{d\vec{y}}{dt} = A\vec{y} + \vec{b}$ for an $m \times m$ matrix A and an $m \times 1$ vector $\vec{b}$.

Def. 3.2 (Linearizing CoB Transformation). *Let $\mathcal{S}$ be a (nonlinear) system. We say that α is a linearizing CoB transformation if it maps each trajectory of $\mathcal{S}$ to that of an affine system $\mathcal{T}$. In other words, α ensures that $\mathcal{S}$ is simulated by an affine system $\mathcal{T}$.*

The above definition of a linearizing CoB seems useful, in practice, only if α and $\mathcal{T}$ are already known. We may then use known techniques for safely bounding the reachable set of an affine system, given some initial conditions, and transform the result back through α^{-1} to obtain a bound on the reachable set for $\mathcal{S}$.

We now present a technique that searches for a map α to obtain an affine system $\mathcal{T}$ that simulates a given system $\mathcal{S}$ through α. We ignore the initial condition and invariant, for the time being, and simply focus on the dynamics of $\mathcal{T}$. In other words, we will search for a map α that satisfies

$$J_\alpha(\vec{x}) \cdot \mathcal{F}(\vec{x}) = A\alpha(\vec{x}) + \vec{b}$$

for some constant matrices A, b. Having found such a map, we can always find appropriate initial and invariance conditions for the simulating system $\mathcal{T}$, whose dynamics will be given by $\mathcal{G}(\vec{y}) = A\vec{y} + \vec{b}$ so that Definition 3.1 holds.

We proceed by recasting a linearizing CoB transformation in terms of a vector space that is closed under the action of taking Lie-derivatives.

3.2 Vector Spaces Closed Under Lie Derivatives.

Recall the requirement for α serving as a linearizing change of variables transformation for a vector field $\mathcal{F}$:

$$J_\alpha(\vec{x}) \cdot \mathcal{F}(\vec{x}) = A\alpha(\vec{x}) + \vec{b}$$

for some constant matrices A, b.

Let $\alpha(\vec{x}) : (\alpha_1(\vec{x}), \dots, \alpha_m(\vec{x}))$ be a smooth mapping $\alpha : \mathbb{R}^n \mapsto \mathbb{R}^m$, wherein each $\alpha_i : \mathbb{R}^n \mapsto \mathbb{R}$. Recall that $\mathcal{L}_F(\alpha_i(\vec{x})) = (\nabla\alpha_i) \cdot \mathcal{F}(\vec{x})$ denotes the Lie derivative of the function $\alpha_i(\vec{x})$ w.r.t vector field $\mathcal{F}$.

Lemma 3.1. $J_\alpha \cdot \mathcal{F}(\vec{x}) = \begin{pmatrix} \mathcal{L}_F(\alpha_1(\vec{x})) \\ \mathcal{L}_F(\alpha_2(\vec{x})) \\ \vdots \\ \mathcal{L}_F(\alpha_m(\vec{x})) \end{pmatrix}.$

PROOF. Recall the definition of the Jacobian J_α:

$$J_\alpha(x_1, \dots, x_n) = \begin{bmatrix} \frac{\partial y_1}{\partial x_1} & \cdots & \frac{\partial y_1}{\partial x_n} \\ \vdots & \ddots & \vdots \\ \frac{\partial y_m}{\partial x_1} & \cdots & \frac{\partial y_m}{\partial x_n} \end{bmatrix} = \begin{bmatrix} \nabla\alpha_1 \\ \vdots \\ \nabla\alpha_m \end{bmatrix}.$$

It follows that,

$$J_\alpha.\mathcal{F} = \begin{pmatrix} (\nabla\alpha_1) \cdot (\mathcal{F}) \\ (\nabla\alpha_2) \cdot (\mathcal{F}) \\ \vdots \\ (\nabla\alpha_m) \cdot (\mathcal{F}) \end{pmatrix} = \begin{pmatrix} \mathcal{L}_F(\alpha_1(\vec{x})) \\ \mathcal{L}_F(\alpha_2(\vec{x})) \\ \vdots \\ \mathcal{L}_F(\alpha_m(\vec{x})) \end{pmatrix}.$$

$\square$

Given functions $\alpha_1, \dots, \alpha_m : \mathbb{R}^n \mapsto \mathbb{R}$, and the special constant function $\mathbf{1} : \mathbb{R}^n \mapsto \{1\}$, we consider the vector space generated by these functions:

$$Span\left(\mathbf{1}, \alpha_1, \dots, \alpha_m\right) = \left\{ c_0 \cdot \mathbf{1} + \sum_{i=1}^m c_i\alpha_i \mid c_0, \dots, c_m \in \mathbb{R} \right\}.$$

Theorem 3.3 (Vector Space Closure Theorem). A map α : $(\alpha_1, \dots, \alpha_m)$ represents a linearizing CoB transformation for a system iff the vector space $V : Span(\mathbf{1}, \alpha_1, \dots, \alpha_m)$ is closed under the operation of Lie-derivatives. I.e, $\forall g \in V$, $\mathcal{L}_F(g) \in V$.

PROOF. Let α be a linearizing CoB transformation mapping trajectories of $\mathcal{F}$ onto $\mathcal{G} : \frac{d\vec{y}}{dt} = A\vec{y} + \vec{b}$. Therefore, for each α_i,

$$\mathcal{L}_F(\alpha_i) = b_i + \sum_j A_{ij}\alpha_j \qquad (1)$$

Any element $\beta \in V$ can be written as $\beta = c_0 + \sum_k c_k\alpha_k$ (a linear combination of the bases of the vector space). Using (1), $\mathcal{L}_F(\beta)$ can be written, once again, as a linear combination of α_is and $\mathbf{1}$. Therefore $\mathcal{L}_F(\beta) \in V$.

Conversely, if V is closed under the action of a Lie-derivative, then its bases $\mathbf{1}, \alpha_1, \dots, \alpha_m$ satisfy the condition $\mathcal{L}_F(\alpha_i) = b_i\mathbf{1} + \sum_j a_{ij}\alpha_j$.

From Lemma 3.1, $J_\alpha\mathcal{F}(\vec{x}) = \begin{pmatrix} \mathcal{L}_F(\alpha_1(\vec{x})) \\ \mathcal{L}_F(\alpha_2(\vec{x})) \\ \vdots \\ \mathcal{L}_F(\alpha_m(\vec{x})) \end{pmatrix} = A\alpha(\vec{x}) + \vec{b}$,

wherein, $A = [a_{ij}]$ and $\vec{b} = (b_i)$. This proves that α is a linearizing CoB transformation. $\square$

Example 3.2. *Consider the ODE from Example 1.1 recalled below:*

$$\begin{array}{rcl} \frac{dx}{dt} & = & xy + 2x \\ \frac{dy}{dt} & = & -\frac{1}{2}y^2 + 7y + 1 \end{array}$$

We claim that the vector space V generated by the set of functions $\{\mathbf{1}, xy, xy^2, x\}$ is closed under the operation of computing Lie derivatives. To verify, we compute the Lie derivative of a function of the form $c_0 + c_1x + c_2xy + c_3xy^2$ to obtain

$$c_1(xy + 2x) + c_2(\frac{1}{2}xy^2 + 9xy + x) + c_3(9xy^2 + 2xy)$$

which is seen to belong to V. As a result the CoB abstraction $\alpha(x, y) : (x, xy, xy^2)$ linearizes the system.

4. SEARCHING FOR ABSTRACTIONS

In this section, we will present search strategies for finding a linearizing change of bases abstraction, if one exists. Following Theorem 3.3, our goal is to find a vector space generated by some functions $\alpha_1, \dots, \alpha_k$ that are closed under the action of taking Lie derivatives. Given a set of functions $B = \{f_1, \dots, f_k\}$, we write $Span(B)$ to denote the vector space spanned by the functions in B:

$$Span(B) = \left\{ \sum_{j=1}^k a_jf_j(\vec{x}) \mid a_j \in \mathbb{R} \right\}$$

We will proceed using a subspace iteration as follows:

1. Choose an initial vector space $V_0 = Span(\{\alpha_0, \dots, \alpha_N\})$. For instance, V_0 can be generated by all *monomial terms whose degrees are less than a cutoff*. In general, any *ansatz* of the form $\sum_i c_if_i$, for functions $f_i(\vec{x})$ and parameters c_i, can be written as a vector space $V_0 = Span(\{f_0, \dots, f_k\})$.

2. At each step, iteratively *refine* V_i for $i \geq 0$ to yield V_{i+1}, a subspace of V_i.

3. Stop when $V_{n+1} = V_n$. If V_n is a non-trivial vector space then, a non-trivial, linearizing change of bases transformation can be extracted along with the resulting system from the generators of V_n.

Initial Basis: For ODEs with polynomial right-hand sides, the initial basis can be generated by all monomial terms up to some degree bound d. However, our technique can be extended to handle other types of functions including trigonometric functions as long as these functions are continuous and differentiable.

Let $V_0 : Span\left(\{\mathbf{1}, \alpha_0, \ldots, \alpha_N\}\right)$ be the initial basis.

Refining the Basis: Let $\{\mathbf{1}, \alpha_1, \ldots, \alpha_k\}$ be the basis of the vector space V_i for the ith iteration. Our goal is to refine this basis to find a subspace $V_{i+1} \subseteq V_i$ such that

$$V_{i+1} = \mathcal{D}(V_i) = \{f \in V_i \mid \mathcal{L}_F(f) \in V_i\}$$

Note that by definition, V_{i+1} is a subset of V_i. It remains to show that V_{i+1} is a vector space.

Lemma 4.1. *If V_i is a vector space, then $V_{i+1} = \mathcal{D}(V_i)$ is a subspace of V_i.*

PROOF. Let V_i be generated by the basis $\{\mathbf{1}, \alpha_1, \ldots, \alpha_k\}$. That $V_{i+1} \subseteq V_i$ follows from its definition. It remains to show that V_{i+1} is a vector space. First $\mathbf{1} \in V_{i+1}$. Let $f_1, \ldots, f_l$ be functions in V_{i+1}. Therefore, by definition, $\mathcal{L}_F(f_1), \ldots, \mathcal{L}_F(f_l) \in V_i$. Consider their affine combination $f : a_0\mathbf{1} + \sum_{j=1}^{l} a_j f_j$ for some $a_0, a_1, \ldots, a_l \in \mathbb{R}$. Its Lie derivative is

$$\mathcal{L}_F\left(a_0 + \sum_{j=1}^{l} a_j f_j\right) = a_j \sum_{j=1}^{l} \mathcal{L}_F(f_j).$$

Therefore, $\mathcal{L}_F(f)$ can be written as a linear combination of the Lie derivatives of $f_1, \ldots, f_l$ which are themselves in V_i. Therefore, $\mathcal{L}_F(f) \in V_i$ and therefore $f \in V_{i+1}$. Thus, any linear combination of elements of V_{i+1} also belongs to V_{i+1}. $\square$

Theorem 4.1. *Given an initial vector space V_0 and vector field $\mathcal{F}$, the subspace iteration converges in finitely many steps to a subspace $V^* \subseteq V_0$. Let $\alpha_1, \ldots, \alpha_m$ be the basis functions that generate V^*.*

1. *The transformation $\alpha : (\alpha_1, \ldots, \alpha_m)$ generated by the basis functions of the final vector space is linearizing.*

2. *For every linearizing CoB transformation $\beta : (\beta_1, \ldots, \beta_k)$, wherein each $\beta_i \in V_0$, it follows that $\beta_i \in V^*$.*

PROOF. The convergence of the iteration follows from the observation that if $V_{i+1} \subset V_i$, the dimension of V_{i+1} is at least one less than that of V_i. Since V_0 is finite dimensional, the number of iterations is upper bounded by the number of basis functions in V_0.

The first statement follows directly from Theorem 3.3.

Finally, us assume that a linearizing transformation β exists such that $\beta_i \in V_0$. We note that the space U generated by $\mathbf{1}, \beta_1, \ldots, \beta_k$ is a subset of V_0. We can also prove that if $U \subseteq V_i$, then $U \subseteq V_{i+1}$. As a result, we prove by induction that $U \subseteq V^*$. $\square$

Example 4.1. *Consider the system $\mathcal{F}$ from Example 1.1 recalled below:*

$$\begin{array}{rcl} \frac{dx}{dt} &=& xy + 2x \\ \frac{dy}{dt} &=& -\frac{1}{2}y^2 + 7y + 1 \end{array}$$

The initial basis for V_0 can be chosen to be the set of all monomials whose degree is less than some limit d. For simplicity, let us choose the basis for V_0 to be: $\{\mathbf{1}, x, y, xy, x^2, y^2, x^2y, xy^2\}$. Any element of V_0 can be written as

$$f : c_0\mathbf{1} + c_1 x + c_2 y + c_3 xy + c_4 x^2 + c_5 y^2 + c_6 x^2 y + c_7 xy^2.$$

Its Lie derivative can be written as:

$$\mathcal{L}_F(f) : \begin{pmatrix} c_2 + (2c_1 + c_3)x + (7c_2 + 2c_5)y+ \\ (c_1 + 9c_3 + 2c_7)xy + (2c_4 + c_6)x^2 \\ -\frac{1}{2}(c_2 + 14c_5)y^2 + (2c_4 + 11c_6)x^2y+ \\ (\frac{1}{2}c_3 + 16c_7)xy^2 - c_5 y^3 + \frac{3c_6}{2}x^2y^2 \end{pmatrix}$$

We note that $\mathcal{L}_F(f) \in V_0$ iff $c_5 = 0$, $c_6 = 0$ *(so that terms corresponding to y^3, x^2y^2 vanish. This yields the bases for V_1:*

$$\{\mathbf{1}, x, y, xy, x^2, xy^2\}$$

Once again, computing the Lie derivative yields the constraint: $c_2 = 0, c_4 = 0$, *yielding the basis:*

$$\{\mathbf{1}, x, xy, xy^2\}$$

The iteration converges in two steps, yielding the linearizing transformation $\alpha : (x, xy, y^2)$, as expected.

Note that it is possible for the converged result V^* to be trivial. I.e, it is generated by the constant function $\mathbf{1}$.

Example 4.2. *Consider the van der Pol oscillator whose dynamics are given by*

$$\dot{x} = y, \; \dot{y} = \mu(y - \frac{1}{3}y^3 - x).$$

Our search for polynomials ($\mu = 1$) of degree up to 20 did not yield a non-trivial transformation.

For a trivial system, the resulting affine system $\mathcal{T}$ is $\frac{dy}{dt} = 0$ under the map $\alpha(\vec{x}) = 0$. Naturally, this situation is not quite interesting but will often result, depending on the system $\mathcal{S}$ and the initial basis chosen V_0. We now discuss common situations where the vector space V^* obtained as the result is guaranteed to be non-trivial. Section 5 presents techniques that can search for a broader class of affine differential inequations instead of just equations.

4.1 Strong and Constant Scale Consecution

The notion of "strong" consecution, "constant scale" consecution and "polynomial scale" consecution were defined for equality invariants of differential equations in our previous work [26] and subsequently expanded upon by Matringe et al. [14] using the notion of morphisms. We now show that the techniques presented in this section can capture these notions, ensuring that all the systems handled by the techniques presented in our previous work [26] can be handled by the techniques here (but not vice-versa).

Def. 4.1 **(Strong and Constant Scale Invariants).** *A function f satisfies the* strong scale *consecution requirement for a vector field $\mathcal{F}$ iff $\mathcal{L}_F(f) = 0$. In other words, f is a conserved quantity.*

Similarly, f satisfies the constant scale *consecution iff $\exists \lambda \in \mathbb{R}, \mathcal{L}_F(f) = \lambda f$.*

The following theorem is a corollary of Theorem 4.1 and shows that the ideas presented in this section can capture the notion of strong and constant scale consecution without requiring quantifier elimination, solving an eigenvalue problem [26] or finding roots of a univariate polynomial [14].

Theorem 4.2. *The result of the iteration V^* starting from an initial space V_0 contains all the strong and constant scale invariant functions in V_0.*

PROOF. This is a direct consequence of Theorem 4.1 by noting that for a constant scale consecuting function f, the subspace $U \subseteq V_0$ spanned by f is closed under Lie derivatives. $\square$

Furthermore, if such functions exist in V_0 the result after convergence V^* is guaranteed to be a non-trivial vector space (of positive dimension). Finally, constant scale and strong scale functions can be extracted by computing the affine equality invariants of the linear system $\mathcal{T}$ that can be extracted from V^*.

4.1.1 Stability

We briefly address the issue of deducing stability (or instability) of a system $\mathcal{S}$ using an abstraction to a system $\mathcal{T}$. Note that every equilibrium of $\mathcal{S}$ maps onto an equilibrium of $\mathcal{T}$, but not vice-versa. Furthermore, the map $\alpha(\vec{x}) = (\mathbf{0}, \ldots, \mathbf{0})$ is an abstraction from any non-linear system to one with an equilibrium at origin. Therefore, unless restrictions are placed on α, we are unable to draw conclusions on liveness properties for $\mathcal{S}$ based on $\mathcal{T}$. If α has a continuous inverse, then $\mathcal{T}$ is topologically diffeomorphic to $\mathcal{S}$ [15]. This allows us to correlate equilibria of $\mathcal{T}$ with those of $\mathcal{S}$. The preservation of stability under mappings of state variables has been studied by Vassilyev and Ul'yanov [32]. We are currently investigating restrictions that will allow us to draw conclusions about liveness properties of $\mathcal{S}$ from those of $\mathcal{T}$.

5. DIFFERENTIAL INEQUALITIES

In this section, we extend our results to search for transformations that result in an affine differential inequality rather than an equality. Affine differential inequalities represent a broader class of systems that include equalities. Therefore, we expect to find non-trivial affine differential inequality abstractions for a larger class of non-linear systems.

A function $f(\vec{x}) : \mathbb{R}^n \mapsto \mathbb{R}$ is *positive semi-definite* (psd) iff $f(\vec{x}) \geq 0$ for all $\vec{x} \in \mathsf{dom}(f)$. A function is positive definite iff it is positive semi-definite and non-zero everywhere.

Def. 5.1 (Affine Differential Inequality). *An affine differential inequality is a non-autonomous system of the form:*

$$\frac{d\vec{y}}{dt} = A\vec{y} + \vec{b} + \vec{u}(t) ,$$

for $m \times m$ matrix A, $m \times 1$ vector $\vec{b}$ and disturbance inputs $\vec{u}(t) : \mathbb{R}_{\geq 0} \mapsto \mathbb{R}^m$. The disturbance inputs are assumed to be integrable and positive semi-definite. Since the input $\vec{u}$ is psd, we write the differential inequality informally as: $\frac{d\vec{y}}{dt} \geq A\vec{y} + \vec{b}$.

In order to define a transformation that results in an abstract system of differential inequalities, we first define the notion of a set of functions that are closed under a Lie derivative with a *positive semi-definite residue*.

Def. 5.2 (Closure with PSD Residue). *A finite set of functions $S : \{\alpha_1, \ldots, \alpha_k\}$ is closed under Lie derivatives with a positive semi-definite residue iff for all $\alpha_i \in S$, the Lie derivative of α_i is of the form:*

$$\mathcal{L}_F(\alpha_i) = b_i \mathbf{1} + \sum_j a_{ij}\alpha_j + \rho_i(\vec{x}) ,$$

wherein, a_{ij}, b_i are real-valued constants, and ρ_i is a continuous, positive semi-definite function over $\vec{x}$.

Example 5.1 (Projectile with Energy Dissipation). *Consider the dynamics of a projectile moving "upwards" ($v_y \leq 0$) with dissipation of its kinetic energy due to friction. The variables (x, y) represent the position and (v_x, v_y) its velocity vector. We assume that the drag due to friction is a polynomial function of the velocity. The dynamics are given by $\dot{x} = v_x$, $\dot{y} = v_y$, $\dot{v_x} = -\frac{1}{10}v_x - \frac{1}{100}v_x^2$, $\dot{v_y} = -10 - \frac{1}{10}v_y - \frac{1}{100}v_y^2$. The system operates in the region $v_x \geq 0$ and $v_y \leq 0$.*

Consider the functions $\alpha_1 : -10v_y - y$ and $\alpha_2 : -10v_x - x$. We have $\mathcal{L}(\alpha_1) = 100 + .1v_y^2 = 100 + \rho_1$, wherein $\rho_1 = .1v_y^2$ is psd. Similarly, $\mathcal{L}(\alpha_2) = \rho_2$, for psd $\rho_2 = .1v_x^2$. Therefore, the set $S = \{\alpha_1, \alpha_2\}$ satisfies the conditions of Def. 5.2.

We now establish the connection between sets of functions that are closed under Lie derivatives with a psd residue and affine differential inequalities. Consider a Lipschitz continuous vector field $\mathcal{F}$ over X_I. Let $S : \{\alpha_1, \ldots, \alpha_k\}$ be a finite set of continuous and differentiable functions that are closed under Lie derivatives with a psd residue (Def. 5.2).

For a differential inequality $\frac{d\vec{y}}{dt} \geq A\vec{y} + \vec{b}$, we let $s_{y_0, u} : [0, T') \mapsto \mathbb{R}^m$ denote the unique time trajectory for the ODE $\frac{d\vec{y}}{dt} = A\vec{y} + \vec{b} + \vec{u}(t)$, with initial conditions $s(0) = \vec{y}_0$ and with a continuous input $\vec{u}(t)$ (such that $\vec{u}(t) \geq 0$ for all $t \geq 0$).

Theorem 5.1. *If a finite set $S : \{\alpha_1, \ldots, \alpha_m\}$ exists that is closed under Lie derivatives with a psd residue, then there exists an affine differential inequality $\frac{d\vec{y}}{dt} \geq A\vec{y} + \vec{b}$ over m variables $y_1, \ldots, y_m$, such that for each time trajectory $\tau_x : [0, T) \mapsto \mathbb{R}^n$ for $\mathcal{F}$, there exists a continuous, positive semi-definite input function $\vec{u}(t) : [0, T) \to \mathbb{R}^m$ such that*

$$\alpha(\tau_x(t)) = s_{\alpha(\vec{x}_0), u}(t), \ \forall \, t \in [0, T)$$

PROOF. Consider the Lie derivative for each α_i. It can be written as $\mathcal{L}_F(\alpha_i) = b_i + \sum_{j=1}^k a_{jk}\alpha_j + \rho_i$. Fixing some trajectory $\tau : [0, T) \mapsto \mathbb{R}^n$ for $\mathcal{F}$, let $\vec{u}(t) : (\rho_1(\tau(t)), \ldots, \rho_k(\tau(t)))$ be obtained by evaluating the residues over the trajectory τ. The continuity of τ and ρ_i imply the continuity and thus, the integrability of $\vec{u}$. As a result, the trajectory τ is mapped by $\alpha : (\alpha_1(\vec{x}), \ldots, \alpha_k(\vec{x}))$ to a trajectory of the ODE $\frac{d\vec{y}}{dt} = A\vec{y} + \vec{b} + \vec{u}(t)$, which is a trajectory of the inequality $\frac{d\vec{y}}{dt} \geq A\vec{y} + \vec{b}$. $\square$

A set Y is a positive invariant set for a differential inequality $\vec{y}' \geq A\vec{y} + \vec{b}$ iff starting from any point $\vec{y}_0 \in Y$ and for any input function $\vec{u}(t)$ that is integrable and positive semi-definite, the resulting trajectory $s_{y_0, u}$ lies entirely in Y.

Theorem 5.2. *Let Y be an invariant set for the differential inequality $\vec{y}' \geq A\vec{y} + \vec{b}$ that abstracts a system $\mathcal{S} : \langle X_0, \mathcal{F}, X_I \rangle$ through a map α. It follows that $\alpha^{-1}(Y) \cap X_I$ is a positive invariant set for $\mathcal{S}$.*

Example 5.2. *Continuing with the system in Ex. 5.1, we find that the differential inequality: $\frac{dy_1}{dt} \geq 100$, $\frac{dy_2}{dt} \geq 0$ abstracts the dynamics of the system.*

Example 5.3. *Consider the system:*

$$\frac{dx}{dt} = x + 2xy, \ \frac{dy}{dt} = 1 + 3y - y^2$$

A simple examination suggests that $\alpha_0(x, y) = -y$ has a positive semi-definite residue. However, consider the functions $\alpha_1(x, y) : x^2$ and $\alpha_2(x, y) : 2x^2y - x^2$. We find that $\mathcal{L}(\alpha_1) : 2x^2 + 4x^2y = 2\alpha_2 + 4\alpha_1$, $\mathcal{L}(\alpha_2) : 6x^2y + 6x^2y^2 = 3\alpha_2 + 3\alpha_1 + \rho$, wherein $\rho = 6x^2y^2$. Therefore, the system

$$\frac{dy_1}{dt} \geq -1 - 3y_1, \ \frac{dy_2}{dt} = 4y_2 + 2y_3, \ \frac{dy_3}{dt} \geq 3y_2 + 3y_3 ,$$

is obtained from the map α as an abstraction.

Having described CoB transformations for differential inequalities, we now focus on a best-effort algorithm for finding a CoB transformation. The difficulty in discovering an abstraction arises from the fact that (a) finding if a given polynomial is psd is NP-hard [1] and (b) the components of the mapping α do not form a structure such as vector spaces over which a terminating iteration can be readily defined.

---

[1] The problem is harder if trigonometric functions are involved. Therefore, we restrict our focus to polynomials.

5.1 Finding Polynomial Abstractions

For the remainder of this section, we focus on discovering inequality abstractions for ODEs with polynomial right-hand sides through maps α that involve polynomials. Note that the term *posynomial* refers to a positive semi-definite polynomial. Before proceeding, we recall the definition *finitely generated cones*.

Def. 5.3 (**Finitely Generated Cone**). *A finitely generated cone* $C = \mathsf{cone}(\alpha_0, \ldots, \alpha_k)$ *is the set of all functions obtained as conic combinations of its generators:* $C = \{f : \sum_{j=0}^{k} \lambda_j \alpha_j \mid \lambda_j \geq 0,\ \lambda_j \in \mathbb{R}\}.$

A finitely generated cone $C : \mathsf{cone}(\alpha_1, \ldots, \alpha_k)$ is closed under Lie derivatives w.r.t a vector field $\mathcal{F}$ *with a positive semidefinite (psd) residue* iff

$$\forall f \in C,\ \mathcal{L}_F(f) = a_0 + \sum_i a_i \alpha_i + \rho,\ \text{for positive semi-definite } \rho.$$

Lemma 5.1. *A set* $S = \{\alpha_1, \ldots, \alpha_k\}$ *is closed under Lie derivatives w.r.t* $\mathcal{F}$ *with a psd residue iff the* $\mathsf{cone}(S)$ *is also closed with a psd residue.*

5.1.1 Approach

Along the lines of our approach in Section 4, we adopt the following strategy:

1. Choose a *finitely generated cone* $C_0 : \mathsf{cone}(\mathbf{1}, \alpha_1, \ldots, \alpha_k)$ of functions. In practice, we form the initial cone by choosing all monomials m of degree at most d and adding the polynomials $+m$ and $-m$ to the generators of C_0.

2. We refine the cone C_i at step i, starting from $i = 0$, to obtain a cone $C_{i+1} \subseteq C_i$. If $C_{i+1} = C_i$, we stop and use the generators of the cone to extract a mapping.

3. Unlike vector spaces, the iteration over cones need not necessarily converge even for finitely generated (polyhedral) cones. Therefore, we will use a heuristic "widening" operator that will force convergence in finitely many steps [6].

Let us assume that a cone $C : \mathsf{cone}(\mathbf{1}, \alpha_1, \ldots, \alpha_k)$ fails to be closed (with psd residues). Our goal is to derive a sub-cone $D \subseteq C$ that is finitely generated and satisfies the closure conditions:

$$D \subseteq \{f \in C \mid \mathcal{L}_F(f) = c_0 + \sum_i c_i \alpha_i + \rho,\ c_i \in \mathbb{R},\ \rho \text{ is psd}\}.$$

A *naive* strategy for finding D is to drop those generators (α_is) in the basis that fail the closure condition. However, this strategy fails to find interesting cones in practice.

5.1.2 Using Sum-Of-Squares Programming

Sum-Of-Squares relaxation is a well known technique that allows us to relax nonlinear programs involving polynomial inequalities into semi-definite programs. Originally discovered by Shor, the SOS relaxation has been applied widely in control theory and verification [18, 21].

Let $C : \mathsf{cone}(\mathbf{1}, \alpha_1, \ldots, \alpha_k)$ be a finitely generated cone. Any element of the cone can be written as $f(\vec{\lambda}, \vec{x}) : \lambda_0 + \sum_{i=1}^{k} \lambda_i \alpha_i$, for multipliers $\vec{\lambda} : \lambda_0, \ldots, \lambda_k \geq 0$.

1. Consider the ansatz $f(\vec{\lambda}, \vec{x})$, with parameters $\vec{\lambda}$ representing the non-negative multipliers.

2. Compute its Lie derivative w.r.t $\mathcal{F}$, to obtain the polynomial $\mathcal{L}_F(f(\vec{\lambda}, \vec{x}))$.

3. Equate $\mathcal{L}_F(f)$ with the form $g = c_0 + \sum_{i=1}^{k} c_i \alpha_i + \rho$, wherein $c_0, \ldots, c_k$ are real-valued parameters (not necessarily non-negative) and ρ is an unknown generic polynomial template over parameters $d_1, \ldots, d_N$ that we require to be a posynomial.

4. We derive constraints by comparing monomial terms in $\mathcal{L}_F(f)$ and g. The positive semi-definiteness of ρ is encoded using the SOS relaxation. The resulting system of constraints has the following form:

$$A\vec{\lambda} + \vec{b} = P\vec{c} + Q\vec{d},\ \vec{\lambda} \geq 0,\ Z(\vec{d}) \succeq 0.$$

for matrices $A, P, Q, \vec{b}$ over reals, unknowns $\vec{\lambda}, \vec{c}, \vec{d}$ and matrix $Z(\vec{d})$ whose entries are linear expressions over $\vec{d}$. Z is constrained to be a positive semi-definite matrix.

Unfortunately, the set of values of $\vec{\lambda}$ for which the semi-definite program above is feasible need not form a finitely generated cone (the cone may have infinitely many generators). Therefore, we need to underapproximate the cone by extracting finitely many generators. This can be performed in many ways, including finding optimal solutions to the SDP for various randomly chosen values for the objective function.

5.1.3 Using Convex Polyhedral Cones

A weaker alternative to using SOS relaxation to characterize a finitely generated cone of positive semi-definite polynomials by starting from a finite set of known posynomials of bounded degree. Let $\mathrm{POS} = \{p_1, \ldots, p_m\}$ be a finite set of posynomials. Then any conic combination of polynomials in POS is also a posynomial, yielding us a finitely generated cone of posynomials.

A polynomial all of whose monomials have variables with even powers of the form $\Pi_{i=1}^{n} x_i^{2r_i}$, $r_i \in \mathbb{N}$, and all of whose coefficients are non-negative is a posynomial. Collecting all these monomials up to some degree and forming the cone generated by these monomials yields a polyhedral cone of posynomials.

Alternatively, we may derive a finite set by extracting finitely many feasible solutions to a SOS program that encodes that an unknown template polynomial is positive semi-definite.

Given a cone $C : \mathsf{cone}(\mathbf{1}, \alpha_1, \ldots, \alpha_k)$, we wish to refine the cone to obtain a new cone D that satisfies

$$D \subseteq \{f \in C \mid \mathcal{L}_F(f) = c_0 + \sum_i c_i \alpha_i + \rho,\ \rho \text{ is psd}\}.$$

Assuming a cone of posynomials generated by POS, we proceed as follows:

1. Create an ansatz $f(\vec{\lambda}, \vec{x}) : \lambda_0 + \sum_i \lambda_i \alpha_i$.

2. Compute the Lie derivative: $\mathcal{L}_F(f)$. This will be a polynomial over $\vec{\lambda}, \vec{x}$.

3. We equate $\mathcal{L}_F(f)$ with a template polynomial g,

$$g = c_0 + \sum_i c_i \alpha_i + \sum_j \gamma_j p_j,\ \text{where } \gamma_j \geq 0,$$

and each $p_j \in P$ is a known posynomial.

4. Finally, we obtain linear constraints of the form:

$$A\vec{\lambda} = P\vec{c} + Q\vec{\gamma},\ \vec{\gamma} \geq 0,\ \vec{\lambda} \geq 0.$$

5. Eliminating $\vec{c}, \vec{\gamma}$ yields linear inequality constraints over $\vec{\lambda}$ whose generators yield the new cone D.

The technique, as described above, requires a set POS of posynomials and uses polyhedral projection, which can be expensive in practice. The following example illustrates how this can be avoided in practice.

Example 5.4. *We recall the system in Example 5.3.*

$$\frac{dx}{dt} = x + 2xy, \quad \frac{dy}{dt} = 1 + 3y - y^2$$

Consider the cone C generated by

$$\{\mathbf{1}, \alpha_1 : x^2, \alpha_2 : y^2, \alpha_3 : x, \alpha_4 : -y, \alpha_5 : x^2 y\}.$$

A generic polynomial inside the cone can be written as

$$f : \lambda_0 + \lambda_1 x^2 + \lambda_2 y^2 + \lambda_3 x + \lambda_4 y + \lambda_5 x^2 y.$$

Its Lie derivative is given by:

$$-\lambda_4 + \lambda_3 x + (2\lambda_2 - 3\lambda_4)y + (2\lambda_1 + \lambda_5)x^2 + 2\lambda_3 xy + (5\lambda_5 + 4\lambda_1)x^2 y + 3\lambda_5 x^2 y^2 - 2\lambda_2 y^3$$

To avoid an expensive elimination, we heuristically split the lie derivative into two types of terms: (a) terms that need to be linear combinations of α_is and (b) terms that belong to the posynomial. Our heuristic collects all the monomial terms that are part of each polynomial α_i. The corresponding terms in the Lie derivative along with their coefficients are constrained to be a linear combination of α_is. The remainder is constrained to be a posynomial.

In this example, the posynomial part is $2\lambda_3 xy + 3\lambda_5 x^2 y^2 - 2\lambda_2 y^3$. We note that this is a posynomial if

$$\lambda_3 = 0, \lambda_2 = 0, \lambda_5 \geq 0.$$

We now add constraints that force the remainder of the Lie derivative to be a linear combination of the α_is:

$$-\lambda_4 + \lambda_3 x + (2\lambda_2 - 3\lambda_4)y + (2\lambda_1 + \lambda_5)x^2 + (5\lambda_5 + 4\lambda_1)x^2 y$$
$$=$$
$$c_0 + c_1 x^2 + c_2 y^2 + c_3 x + c_4 y + c_5$$

This yields linear equality constraints with $\vec{\lambda}$ and $\vec{c}$. The $\vec{c}$ variables can be eliminated by Gaussian elimination. In this instance, the resulting constraint after elimination of $\vec{c}$ variables is true. Combining, the overall constraint is $\lambda_3 = 0$, $\lambda_2 = 0$, $\vec{\lambda} \geq 0$. This yields the cone generated by $\{\mathbf{1}, \alpha_1 : x^2, \alpha_4 : -y, \alpha_5 : x^2 y\}$, as a result of the refinement operation.

5.1.4 Ensuring Termination

The iterative process of refinement starts from some initial cone C_0 that contains all monomial terms and their negations upto some degree cutoffs. At each step, we perform a refinement to compute a refined cone C_{i+1} from the current cone C_i. The refinement can be performed either by formulating a semidefinite program and extracting finitely many feasible solutions or by using polyhedral cones generated by a finite set of posynomials (eg., all monomials which are squares with non-negative coefficients).

In both cases, the refinement iteration is not guaranteed to converge in finitely many steps. To force convergence, we use a *widening* operator [6].

The widening operator ∇ applied to two successive iterates $C : C_i \nabla C_{i+1}$ produces a new finitely generated cone C satisfies $C \subseteq C_i$, $C \subseteq C_{i+1}$ and furthermore, for any sequence $C_0 \supseteq C_1 \supseteq C_2 \supseteq \cdots$, the *widened sequence*

$$C_0, C_1, D_1 : C_0 \nabla C_1, D_2 : D_1 \nabla C_2, \ldots, D_i : D_{i-1} \nabla C_i, \ldots$$

is always guaranteed to converge in finitely many steps.

Def. 5.4 **(Widening Over (Dual) Cones).** *Given two finitely generated cones C_1, C_2 such that $C_2 \subseteq C_1$, the standard widening $C_1 \nabla C_2$ is defined as $\mathsf{cone}(\{f \in \mathsf{Generators}(C_1) \mid f \in C_2\})$.*

In other words, the standard widening drops those generators in C_1 that do not belong to C_2. It is easy to see why a widened iteration using standard widening terminates in finitely many steps. At each widening step $D_{i+1} : D_i \nabla C_{i+1}$, we drop a generator from the cone D_i that do not belong to C_{i+1}. Since there are finitely many generators to begin with in D_1, the widened iteration terminates in finitely many steps.

5.1.5 Abstractions over Domains

Thus far, our techniques have considered the dynamics of the system $\mathcal{S} : \langle X_0, \mathcal{F}, X_I \rangle$ being abstracted, without using knowledge of its invariant set (or domain) X_I. Often, the abstraction being sought is over some domain $X_I \subseteq \mathbb{R}^n$. Our techniques for finding inequality abstractions can be readily modified to treat invariant domains that need not necessarily be bounded. We now briefly present the generalizations to the definitions and the iterative refinement technique to operate over domains X_I.

Def. 5.5 **(Closure over Domain).** *A finite set of functions $S : \{\alpha_1, \ldots, \alpha_k\}$ is closed under Lie derivatives with a positive semidefinite residue over a domain X_I iff for all $\alpha_i \in S$, the Lie derivative of α_i is of the form: $\forall \vec{x} \in X_I, \mathcal{L}_F(\alpha_i)(\vec{x}) = b_i \mathbf{1} + \sum_j a_{ij} \alpha_j + \rho_i(\vec{x})$, wherein, $a_{ij}, b_i \in \mathbb{R}$, and ρ_i is psd.*

In contrast to Def. 5.2, the Lie derivative of α_i equals a linear combination with positive residue over the domain X_I, as opposed to everywhere. The iterative refinement techniques incorporate the generalized definition. The refinement of the cone C has to guarantee that the refined cone D satisfies

$$D \subseteq \{f \in C \mid (\forall \vec{x} \in X_I), \mathcal{L}_F(f) = \sum_j a_j \alpha_j + \rho, \rho(\vec{x}) \geq 0\}.$$

The iteration using SOS relaxation is easily modified under the assumption that X_I is represented as the feasible region of a system of polynomial inequalities. The approach using a finitely generated cone of posynomials can be modified by allowing the cone to include monomials that are known to be positive over X_I (as opposed to everywhere).

If the domain X_I is bounded, our technique defaults to a *truncated* Carleman linearization involving all monomials upto a degree cutoff. Soundness of the truncation is ensured by obtaining interval bounds on the *residues* for each monomial. This approach can form the basis for a hybridization abstraction wherein accuracy can be improved by repeatedly subdividing the domain X_I. [7].

5.1.6 Application to Hybrid Systems

Thus far, we have presented our techniques for abstracting continuous system. The continuous system abstraction for the dynamics corresponding to a mode can be directly employed to compute reachable sets for a given initial condition. This can be used repeatedly as a primitive inside hybrid systems analysis tools.

It is possible to extend the notion of change of bases transformations to discrete transitions using closure under the weakest precondition operator as opposed to Lie derivatives. As a result, we can integrate the conditions for closure under Lie derivatives for continuous systems and the closure under pre-conditions for discrete transitions to yield an extension of our theories for hybrid systems. The full technique for hybrid systems will be described in an extended version of this paper.

Table 1: Summary of experimental results over benchmark systems. Legend: Var: number of variables, deg: degree bound, T: time taken (seconds), iter: number of iterations, eq: number of equalities, ineq: number of inequalities, †: transformations purely involving the system parameters were removed.

Sys	Vars	deg	T	iter	eq	ineq
proj-drag	4	3	.5	15	5	2
spr-mass2	5	3	.1	10	$1^\dagger$	1
spr-mass2	5	4	.6	15	$1^\dagger$	4
spr-mass2	5	7	61.7	28	$1^\dagger$	5
coll-avoid-2	14	2	.2	7	$14^\dagger$	0
coll-avoid-2	14	4	144	23	$160^\dagger$	0
bio-net	12	3	4.9	18	1	2
bio-net	12	4	112	19	1	2
bio-net-par	26	2	1.7	4	$16^\dagger$	1
bio-net-par	26	3	181	5	~ 100	~ 14

6. EXPERIMENTAL EVALUATION

In this section, we describe a prototype implementation of some of the ideas presented thus far. We also report on the experimental evaluation of some benchmark systems using our implementation. Our implementation, the benchmark systems and the outputs are available on-line for review [2].

6.1 Implementation

We have implemented the search for a change of bases transformation using vector space iteration as well as iteration over polyhedral cones using monomials of the form $\Pi x_i^{2r_i}$ as the generator of the cone of posynomials. Our implementation (in OCaml) reads in the description of a continuous system and a degree bound for the abstraction search. It then performs a vector space iteration followed by a polyhedral iteration. The polyhedral iteration uses the *Parma Polyhedral Library* [4]. However, polyhedral cone operations such as computing the generators is worst-case exponential in the number of variables. We implement an optimized version of the iteration over polyhedral cones that separates the linear equality and inequality constraints. The equality constraints are maintained in a triangular form so that we may minimize the number of variables involved in the inequalities. This enables us to handle systems with upwards of 5000 basis polynomials. To improve the quality of the result, we use a *delayed widening* strategy that starts applying the widening operator only after there are no more linear equality constraints to be added.

6.2 Experimental Evaluation

In this section, we describe the results of our technique on some benchmarks. Table 1 summarizes the results of our analysis and the performance of our implementation over various benchmarks assuming various degree bounds. We discuss some of these benchmarks below, briefly. The details of the invariants discovered are part of our release and will be discussed in an extended version.

Two Spring Mass System With Friction We model a mechanical system with two masses that are connected to each other and to a fixed end using springs with constants k_1, k_2 and masses m_1, m_2 adjusted so that $\frac{k_1}{m_1} = \frac{k_2}{m_2} = k$. Furthermore, we assume that $m_1 = 5m_2$. The variables $\vec{x} : (x_1, x_2, v_1, v_2, k)$ representing the displacements, velocities and the spring constant. We assume that the drag due to friction is proportional to the velocity. The

^{2}Cf. http://www.cs.colorado.edu/~srirams/code/nlsys-release.tar.gz.

dynamics are described by the vector field

$$(v_1, v_2, -kx_1 - \frac{k}{5}(x_1 - x_2) - .1v_1, k(x_1 - x_2) - .1v_2, 0).$$

The search for degree bound 3 yields the transformation below:

$$\alpha_0 : k, \; \alpha_1 : -10v_1v_2 + 5v_1^2 + x_2^2k - 12x_1x_2k + 11x_1^2k.$$

The Lie derivative of α_1 is v_2^2, a posynomial. The application of our technique to the conservative system without friction also discovers non-trivial transformations in the form of conserved quantities and inequality invariants.

Collision Avoidance We consider the algebraic abstraction of the collision avoidance system analyzed recently by Platzer et al. [20] and earlier by Tomlin et al. [30]. The two airplane collision avoidance system consists of the variables (x_1, x_2) denoting the position of the first aircraft, (y_1, y_2) for the second aircraft, (d_1, d_2) representing the velocity vector for aircraft 1 and (e_1, e_2) for aircraft 2. ω, θ abstract the trigonometric terms. In addition, the parameters a, b, r_1, r_2 are also represented as system variables. The dynamics are modeled by the following differential equations:

$$\begin{array}{llll}
x_1' = d_1 & x_2' = d_2 & d_1' = -\omega d_2 & d_2' = \omega d_1 \\
y_1' = e_1 & y_2' = e_2 & e_1' = -\theta e_2 & e_2' = \theta e_1 \\
a' = 0 & b' = 0 & r_1' = 0 & r_2' = 0
\end{array}$$

A search for transformations of degree 2 yields a closed vector space with 27 basis functions within 0.2 seconds. The basis functions include a, b, r_1, r_2 and all degree two terms involving these. Removing these from the basis, gives us 14 basis functions that yield a transformation to a 14 dimensional affine ODE.

Biochemical reaction network: Finally, we analyze a biochemical reaction network benchmark from Dang et al. [7]. The ODE along with the values are parameters in our model coincide with those used by Dang et al. The ODE consists of 12 variables. Our search for degree bound ≤ 2 discovers a transformation generated by three basis functions (in roughly .3 seconds). This leads to two differential equalities and one inequality. Note that our analysis does not assume any information about the invariant region. Searching for an abstraction over the invariant region may help us derive a finer abstraction of this ODE along the lines of Dang et al. [7]. Table 1 reports on the results of two versions of the system: with numerical values for the various parameters involved and by encoding these parameters as extra variables whose derivatives are zero. The ability to treat a 26 dimensional system (reasoning over vector-spaces and cones of dimension ~ 3600) is quite a promising result for our approach.

7. CONCLUSION

Thus far, we have presented some techniques for discovering linear abstractions through a change of bases transformation and an evaluation of our techniques using a prototype implementation. In the future, we wish to implement our techniques to search for abstraction over domains. The extension of our techniques to handle non-linear switched and hybrid systems is ongoing.

8. REFERENCES

[1] Rajeev Alur, Thao Dang, and Franjo Ivančić. Counter-example guided predicate abstraction of hybrid systems. In *TACAS*, volume 2619 of *LNCS*, pages 208–223. Springer, 2003.

[2] Rajeev Alur, Thomas A. Henzinger, G. Lafferriere, and George Pappas. Discrete abstractions of hybrid systems. *Proc. of IEEE*, 88(7):971–984, 2000.

[3] Eugene Asarin, Thao Dang, and Antoine Girard. Hybridization methods for the analysis of nonlinear systems. *Acta Informatica*, 43:451—476, 2007.

[4] R. Bagnara, P. M. Hill, and E. Zaffanella. The Parma Polyhedra Library: Toward a complete set of numerical abstractions for the analysis and verification of hardware and software systems. Quaderno 457, Dipartimento di Matematica, Università di Parma, Italy, 2006.

[5] Edmund M. Clarke, Orna Grumberg, Somesh Jha, Yuan Lu, and Helmut Veith. Counterexample-guided abstraction refinement for symbolic model checking. *J. ACM*, 50(5):752–794, 2003.

[6] Patrick Cousot and Rhadia Cousot. Abstract Interpretation: A unified lattice model for static analysis of programs by construction or approximation of fixpoints. In *ACM Principles of Programming Languages*, pages 238–252, 1977.

[7] Thao Dang, Oded Maler, and Romain Testylier. Accurate hybridization of nonlinear systems. In *HSCC '10*, pages 11–20. ACM, 2010.

[8] Thao Dang and David Salinas. Image computation for polynomial dynamical systems using the bernstein expansion. In *Computer Aided Verification*, volume 5643 of *Lecture Notes in Computer Science*, pages 219–232. Springer, 2009.

[9] Goran Frehse. PHAVer: Algorithmic verification of hybrid systems past HyTech. In *HSCC*, volume 2289 of *LNCS*, pages 258–273. Springer, 2005.

[10] Antoine Girard. Reachability of uncertain linear systems using zonotopes. In *HSCC*, volume 3414 of *LNCS*, pages 291–305. Springer, 2005.

[11] Colas Le Guernic and Antoine Girard. Reachability analysis of linear systems using support functions. *Nonlinear Analysis: Hybrid Systems*, 4(2):250 – 262, 2010.

[12] Thomas A. Henzinger, Pei-Hsin Ho, and Howard Wong-Toi. Algorithmic analysis of nonlinear hybrid systems. *IEEE Transactions on Automatic Control*, 43:540–554, 1998.

[13] K. Kowalski and W-H. Steeb. *Non-Linear Dynamical Systems and Carleman Linearization*. World Scientific, 1991.

[14] Nadir Matringe, Arnoldo Viera Moura, and Rachid Rebiha. Morphisms for non-trivial non-linear invariant generation for algebraic hybrid systems. In *HSCC*, volume 5469 of *LNCS*, pages 445–449, 2009.

[15] James D. Meiss. *Differential Dynamical Systems*. SIAM publishers, 2007.

[16] Venkatesh Mysore, Carla Piazza, and Bud Mishra. Algorithmic algebraic model checking II: Decidability of semi-algebraic model checking and its applications to systems biology. In *ATVA*, volume 3707 of *LNCS*, pages 217–233. Springer, 2005.

[17] Meeko Oishi, Ian Mitchell, Alexandre M. Bayen, and Claire J. Tomlin. Invariance-preserving abstractions of hybrid systems: Application to user interface design. *IEEE Trans. on Control Systems Technology*, 16(2), Mar 2008.

[18] Pablo A Parillo. Semidefinite programming relaxation for semialgebraic problems. *Mathematical Programming Ser. B*, 96(2):293–320, 2003.

[19] André Platzer. Differential dynamic logic for hybrid systems. *J. Autom. Reasoning*, 41(2):143–189, 2008.

[20] André Platzer and Edmund Clarke. Computing differential invariants of hybrid systems as fixedpoints. *Formal Methods in Systems Design*, 35(1):98–120, 2009.

[21] Stephen Prajna and Ali Jadbabaie. Safety verification using barrier certificates. In *HSCC*, volume 2993 of *LNCS*, pages 477–492. Springer, 2004.

[22] Stefan Ratschan and Zhikun She. Safety verification of hybrid systems by constraint propagation based abstraction refinement. In *HSCC*, volume 3414 of *LNCS*, pages 573–589. Springer, 2005.

[23] Enric Rodriguez-Carbonell and Ashish Tiwari. Generating polynomial invariants for hybrid systems. In *HSCC*, volume 3414 of *LNCS*, pages 590–605. Springer, 2005.

[24] Sriram Sankaranarayanan. Automatic invariant generation for hybrid systems using ideal fixed points. In *Hybrid Systems: Computation and Control*, pages 211–230. ACM Press, 2010.

[25] Sriram Sankaranarayanan, Thao Dang, and Franjo Ivančić. Symbolic model checking of hybrid systems using template polyhedra. In *TACAS*, volume 4963 of *LNCS*, pages 188–202. Springer, 2008.

[26] Sriram Sankaranarayanan, Henny Sipma, and Zohar Manna. Constructing invariants for hybrid systems. *Formal Methods in System Design*, 32(1):25–55, 2008.

[27] Sriram Sankaranarayanan, Henny B. Sipma, and Zohar Manna. Fixed point iteration for computing the time-elapse operator. In *HSCC*, LNCS. Springer, 2006.

[28] Ashish Tiwari. Abstractions for hybrid systems. *Formal Methods in Systems Design*, 32:57–83, 2008.

[29] Ashish Tiwari and Gaurav Khanna. Non-linear systems: Approximating reach sets. In *HSCC*, volume 2993 of *LNCS*, pages 477–492. Springer, 2004.

[30] Claire J. Tomlin, George J. Pappas, and Shankar Sastry. Conflict resolution for air traffic management: A study in multi-agent hybrid systems. *IEEE Trans. on Aut. Control*, 43(4):509–521, April 1998.

[31] Pravin Varaiya. Reach set computation using optimal control. In *Proc. KIT Workshop on Hybrid Systems*, 1998.

[32] S. Vassilyev and S. Ul'yanov. Preservation of stability of dynamical systems under homomorphisms. *Differential Equations*, 45:1709–1720, 2009.

Human-Data Based Cost of Bipedal Robotic Walking*

Aaron D. Ames
Mechanical Engineering
Texas A&M University
College Station, TX 77843
aames@tamu.edu

Ram Vasudevan
Electrical Engineering and
Computer Sciences
University of California,
Berkeley
Berkeley, CA 94720
ramv@eecs.berkeley.edu

Ruzena Bajcsy
Electrical Engineering and
Computer Sciences
University of California,
Berkeley
Berkeley, CA 94720
bajcsy@eecs.berkeley.edu

ABSTRACT

This paper proposes a cost function constructed from human data, the *human-based cost*, which is used to gauge the "human-like" nature of robotic walking. This cost function is constructed by utilizing motion capture data from a 9 subject straight line walking experiment. Employing a novel technique to process the data, we determine the times when the number of contact points change during the course of a step which automatically determines the ordering of discrete events or the *domain breakdown* along with the amount of time spent in each domain. The result is a weighted graph or *walking cycle*, associated with each of the subjects walking gaits. Finding a weighted cycle that minimizes the cut distance between this collection of graphs produces an *optimal* or *universal* domain graph for walking together with an *optimal walking cycle*. In essence, we find a single domain graph and the time spent in each domain that yields the most "natural" and "human-like" bipedal walking. The human-based cost is then defined as the cut distance from this optimal gait. The main findings of this paper are two-fold: (1) when the human-based cost is computed for subjects in the experiment it detects medical conditions that result in aberrations in their walking, and (2) when the human-based cost is computed for existing robotic models the more human-like walking gaits are correctly identified.

Categories and Subject Descriptors

I.6.8 [**Simulation and Modeling**]: Types of Simulation—*Continuous, Discrete Event*

General Terms

Theory

*This research is supported in part by NSF Awards CCR-0325274, CNS-0953823, ECCS-0931437, IIS-0703787, IIS-0724681, IIS-0840399 and NHARP Award 000512-0184-2009.

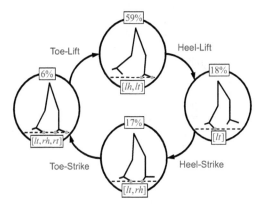

Figure 1: An example of a *domain breakdown*, **i.e., the discrete phases of a walking gait, based upon a specific temporal ordering. The red dots indicate the constraints enforced in each discrete phase (or domain).**

Keywords

Bipedal walking, Gait design, Human locomotion, Hybrid systems, Cost functions

1. INTRODUCTION

While constructing a bipedal walking robot, beyond the immediate goal of obtaining stable walking, a cost function is generally chosen to optimize certain system parameters. The choice of a cost function can have a dramatic impact on the resulting gait. In contrast to other robotic applications, the goal of bipedal walking is typically not to minimize the energy expended but rather to achieve the more nebulous goal of natural or human-like walking. The most popularly chosen cost function to obtain bipedal walking are torque squared [5, 11, 20] or the specific cost of transport [8, 12, 15]; however, no clear connection exists between minimizing these types of costs and achieving anthropomorphic walking. This lack of connection motivates the question: can a cost function be constructed that, when minimized, produces human-like gait?

This paper proposes a cost function based upon human-walking data: the *human-based cost*, built upon the idea of comparing the *temporal ordering of events* for humans and robots and, more specifically, the amount of time spent in each successive domain. One of the most important decisions made during the design of controllers for bipedal robots is the temporal ordering of events that occur during the walking gait, i.e., the discrete phases (or *domains*) of the walking termed a *domain breakdown*. This decision alone determines the constraints that are enforced at any given time during the walking gait, and thus determines the continuous dynamics (through holonomic constraints) on each phase and the discrete dynamics (impact equations) between each phase. Therefore, given the equations of motion for a bipedal robot, the temporal ordering of events completely determines the mathematical model for the biped. The temporal ordering of events not only determines the underlying mathematical model of a bipedal robot, but is essential during control design; specific controllers are often designed or constrained by the specific choice of domain breakdown. When controllers are obtained that yield a walking gait, this gait is again related back to the domain breakdown since one can consider the amount of time spent in each discrete domain. The end result is a weighted cycle associated to a walking gait, termed a *walking cycle* (see Fig. 1 for an example).

Given the domain breakdown's significance, this paper proposes employing walking cycles collected from human walking experiments to develop a cost function, the human-based cost, that judges the "human-like" nature of a specific gait. We begin by considering a 9 subject straight line walking experiment recorded using motion capture. From this data, we develop a novel and automatic method for determining the domain breakdown of each subject. The end result is a collection of walking cycles. We then find the cycle graph and associated weighting that is the minimum cut distance between all of the subjects, termed the *optimal walking cycle*, which gives a "universal" domain breakdown and associated walking cycle for walking. The human-based cost is the cut distance between a walking cycle and the optimal walking cycle. To demonstrate the usefulness of the proposed cost function we compute the human-based cost for both the subjects in the experiment along with existing robotic models. When the human-based cost is computed for the subjects in the experiment we are able to automatically find preexisting medical conditions without *a priori* knowledge of such conditions. Computing the human-based cost for existing bipedal robotic models confirms that robotic gaits that are more anthropomorphic in nature have a lower human-based cost.

The applications of the results of this paper have the potential to be far-reaching in the bipedal robotic community. There is currently a fractured landscape when one considers only domain breakdowns. Most models assume a single domain model [9, 18, 20] which we show in this paper, even under the best case scenario, results in an unnatural gait due to their high human-based cost. When bipedal models are extended beyond a single domain, there is no unity as to which domain breakdown should be used; temporal orderings have been chosen ranging from one discrete phase to five, e.g., [10] considers one, [1, 6, 7, 13, 19] considers two, [16, 19] considers three, [4, 11] considers four, and [17] considers five. This lack of consistency among models in the literature motivates the desire to determine if there does in fact exist a single "universal" domain breakdown that should be used when modeling bipedal robots, especially in the context of obtaining human-like bipedal walking. This is what the optimal walking cycles that are determined in this paper offer.

It is important to note that human gait has been studied at great length by the biomechanics community [2, 21, 22, 23]. This work has focused almost entirely on either understanding the kinematic nature of human gait or muscle coordination during human gait. In contrast to these classical biomechanics approaches, we use human data to determine the temporal ordering of events. Since a temporal ordering of events is crucial in determining the dynamical model of a biped, the discovery of a "universal" temporal ordering could dramatically aide in the development of a bipedal robot with an anthropomorphic gait.

2. FROM CONSTRAINTS TO MODELS

Bipedal robots display both discrete and continuous behavior, i.e., they are naturally modeled as *hybrid systems*. Bipeds evolve in a continuous fashion according to traditional equations of motion when a fixed number of points on the biped are in contact with the ground, e.g., when one foot is flat on the ground while the other swings forward. The discrete behavior in the system occurs when the number of contact points changes.

In this section, we formally introduce hybrid systems and discuss how the equations of motion of a robot together with a temporal ordering of constraints completely determines the hybrid model of the system. That is, when modeling bipedal robots, one needs only the Lagrangian of the robot and a domain breakdown.

2.1 Hybrid Systems.

Hybrid systems (or *systems with impulse effects*) have been studied extensively in a wide variety of contexts and have been utilized to model a wide range of bipedal robotic models. In this section, we introduce a definition of a hybrid system applicable to bipedal walking.

Graphs and Cycles. A graph is a tuple $\Gamma = (V, E)$, where V is the set of vertices and $E \subset V \times V$ is the set of edges; an edge $e \in E$ can be written as $e = (i, j)$, and the source of e is source$(e) = i$ and the target of e is target$(e) = j$. Since steady state bipedal walking is naturally periodic, we are interested in hybrid systems *on a cycle*; therefore, we are interested in graphs that contain cycles or are themselves cycles. A *directed cycle* (or just a cycle) is a graph $\ell = (V, E)$ such that the edges and vertices can be written as:

$$\begin{aligned} V &= \{v_0, v_1, \ldots, v_{p-1}\}, \\ E &= \{e_0 = (v_0, v_1), \ldots, e_{p-1} = (v_{p-1}, v_0)\}. \end{aligned} \tag{1}$$

Since in the case of a cycle, the edges are completely determined by the vertices, we sometimes simply denote a cycle by: $\ell : v_0 \to v_1 \to \cdots \to v_{p-1}$. In the case when a graph Γ is being considered with more than one cycle, we denote a cycle in the graph by $\ell \subset \Gamma$.

Example 1. *The domain graph pictured in Fig. 1 has an underlying graph that is a directed cycle: $\Gamma_u = (V_u, E_u)$. In particular, there are 4 vertices and edges, which results in the cycle:*

$$\ell_u : [lh, lt] \to [lt] \to [lt, rh] \to [lt, rh, rt].$$

With the notion of a directed cycle, we can introduce the formulation of a hybrid system that is of interest in this paper.

Definition 1. *A hybrid system in a cycle is a tuple*

$$\mathscr{HC} = (\ell, D, U, S, \Delta, FG),$$

where

- $\ell = (V, E)$ *is a directed cycle*

- $D = \{D_v\}_{v \in V}$ *is a set of domains, where $D_v \subseteq \mathbb{R}^{n_v} \times \mathbb{R}^{m_v}$ is a smooth submanifold of $\mathbb{R}^{n_v} \times \mathbb{R}^{m_v}$ (with $\mathbb{R}^{m_v}$ representing control inputs),*

- $U = \{U_v\}_{v \in V}$, *where $U_v \subset \mathbb{R}^{m_v}$ is a set of admissible controls,*

- $S = \{S_e\}_{e \in E}$ *is a set of guards, where $S_e \subseteq D_{\text{source}(e)}$,*

- $\Delta = \{\Delta_e\}_{e \in E}$ *is a set of reset maps, where $\Delta_e : \mathbb{R}^{n_{\text{source}(e)}} \to \mathbb{R}^{n_{\text{target}(e)}}$ is a smooth map,*

- $FG = \{(f_v, g_v)\}_{v \in E}$, *where (f_v, g_v) is a control system on D_v, i.e., $\dot{x} = f_v(x) + g_v(x)u$ for $x \in D_v$ and $u \in U_v$.*

2.2 Hybrid Systems from Constraints.

The remainder of this section illustrates how a Lagrangian for the biped, together with a domain breakdown (which determines the active constraints on each vertex of a directed cycle), allows one to explicitly construct a hybrid model of the system. Many details of this construction summarize the procedure presented in [11].

2.2.1 General Setup

We begin with a bipedal robot either in two or three dimensions—the discussion in this paper is applicable to either case. We first construct a Lagrangian for the biped when no assumptions on ground contact are made and then enforce the ground contact conditions through constraints as determined by the domain graph.

Lagrangians. Let R_0 be a fixed inertial or world frame, and R_b be a reference frame attached to the body of the biped with position $p_b \in \mathbb{R}^3$ and orientation $\phi_b \in SO(3)$. Consider a configuration space for the biped Q_r, i.e., a choice of (body or shape) coordinates for the robot where typically $q_r \in Q_r$ is a collection of (relative) angles between each successive link of the robot. The *generalized* coordinates of the robot are then given by $q = (p_b^T, \phi_b^T, q_r)^T \in Q = \mathbb{R}^3 \times SO(3) \times Q_r$, with Q the generalized configuration space.

The Lagrangian of a bipedal robot, $L : TQ \to \mathbb{R}$, can be stated in terms of kinetic and potential energies as:

$$L(q, \dot{q}) = K(q, \dot{q}) - V(q).$$

The Euler-Lagrange equations yield the equations of motion, which for robotic systems [14] are stated as:

$$D(q)\ddot{q} + H(q, \dot{q}) = B(q)u, \qquad (2)$$

where $D(q)$ is the inertia matrix, $B(q)$ is the torque distribution matrix (and only depends on q_r), $B(q)u$ is the vector of actuator torques and $H(q, \dot{q}) = C(q, \dot{q})\dot{q} + G(q) - \Gamma(q, \dot{q})$ contains the Coriolis, gravity terms and non-conservative forces grouped into a single vector.

Contact Points. The continuous dynamics of the system depend on which constraints are enforced at any given time, while the discrete dynamics depend only on the temporal ordering of constraints. Constraints and their enforcement are dictated by the number of contact points of the system

with the ground. Specifically, the *set of contact points* is the set $\mathcal{C} = \{c_1, c_2, \ldots, c_k\}$, where each c_i is a specific type of contact possible in the biped, either with the ground or in the biped itself (such as the knee locking).

If the knees do not lock, and assuming reasonable behavior by the feet, e.g., no standing on one corner of the foot, there are four contact points of interest given by:

$$\mathcal{C}_u = \{lh, lt, rh, rt\},$$

where lh and lt indicate the left heel and toe, and rh and rt indicate right heel and toe, respectively. If the knees lock, additional contact points for the left and right knee, lk and rk, must be considered. This yields a set of contact points:

$$\mathcal{C}_l = \{lh, lt, lk, rh, rt, rk\}.$$

Constraints. Contact points introduce a *holonomic constraint* on the system, η_c for $c \in \mathcal{C}$, which is a vector valued function $\eta_c : Q \to \mathbb{R}^{n_c}$, that must be held constant for the contact point to be maintained, i.e., $\eta_c(q) = \text{constant} \in \mathbb{R}^{n_c}$ fixes the contact point but allows rotation about this point if feasible. It is useful to express the collection of all holonomic constraints in a single matrix $\eta(q) \in \mathbb{R}^{n_c \times |\mathcal{C}|}$ as:

$$\eta(q) = \begin{bmatrix} \eta_{lh}(q) & 0 & 0 & 0 \\ 0 & \eta_{lt}(q) & 0 & 0 \\ 0 & 0 & \eta_{rh}(q) & 0 \\ 0 & 0 & 0 & \eta_{rt}(q) \end{bmatrix}$$

where $n = \sum_{c \in \mathcal{C}} n_c$. To determine the holonomic constraints for the contact points of interest, i.e., $\mathcal{C}_u$ and $\mathcal{C}_l$, two cases must be considered: one for foot contact points and another for knee contact.

In the case of foot contact, consider a reference frame R_c at the contact point $c \in \{lh, lt, rh, rt\}$ such that the axis of rotation about this point (either the heel or toe) is in the z direction. Then the rotation matrix between R_0 and R_c can be written as the product of three rotation matrices $Rot(x, \phi_c^z)Rot(y, \phi_c^z)Rot(z, \phi_c^z)$ and the position and orientation of R_c relative to R_0 is given as $\eta_c(q) = (p_c(q)^T, \phi_c^x, \phi_c^y)^T$, where $p_c(q)$ is the position of c, since ϕ_c^z is free to move while ϕ_c^x and ϕ_c^y must be held constant. The end result of this choice of coordinates is a holonomic constraint $\eta_c(q) = $ constant, which fixes the foot contact point to the ground but allows rotation about the heel or toe depending on the specific type of foot contact. In the case of knee contact, let $q_c, c \in \{lk, rk\}$, be the relative angle of the left or right knee. The holonomic constraint is then given by $\eta_c(q) = q_c$, and enforcing the constraint $\eta_c(q) = 0$ keeps the knee locked.

Another class of constraints that are important are *unilateral constraints*, h_c for $c \in \mathcal{C}$, which are scalar valued functions, $h_c : Q \to \mathbb{R}$, that dictate the set of admissible configurations of the system; that is $h_c(q) \geq 0$ implies that the configuration of the system is admissible for the contact point c. Again, there are two types of these constraints to consider depending on whether foot or knee contact is being considered. In the case of foot contact, assuming that the walking is on flat ground, these constraints require the height of a contact point above the ground be non-negative: $h_c(q) = p_c^y(q) \geq 0$. In the case of knee contact, these constraints require the angle of the knee be positive: $h_c(q) = q_c \geq 0$. These constraints can be put in the form of a matrix $h(q) \in \mathbb{R}^{|Q| \times |Q|}$ in the same manner as the holonomic constraints.

Domain Breakdowns. A domain breakdown is a directed cycle together with a specific choice of contact points on every vertex of that graph. To define this formally, we assign to each vertex a binary vector describing which contact points are active in that domain.

Definition 2. *Let $\mathcal{C} = \{c_1, c_2, \ldots, c_k\}$ be a set of contact points and $\ell = (V, E)$ be a cycle. A domain breakdown is a function $\mathcal{B} : \ell \to \mathbb{Z}_2^k$ such that $B(v)_i = 1$ if c_i is in contact on v and $B(v)_i = 0$ otherwise.*

Example 2. *In the case of the graph Γ_u given in Example 1 and set of contact points $\mathcal{C} = \{lh, lt, rh, rt\}$, for the domain breakdown given in Fig. 1, this domain breakdown is formally given by $\mathcal{B}_u : \ell_u \to \mathbb{Z}_2^4$ where $\mathcal{B}_u([lh, lt])$, $\mathcal{B}_u([lt])$, $\mathcal{B}_u([lt, rh])$ and $\mathcal{B}_u([lt, rh, rt])$ are given by:*

$$\mathcal{B}_u(\ell) : \begin{bmatrix} 1 \\ 1 \\ 0 \\ 0 \end{bmatrix} \to \begin{bmatrix} 0 \\ 1 \\ 0 \\ 0 \end{bmatrix} \to \begin{bmatrix} 0 \\ 1 \\ 1 \\ 0 \end{bmatrix} \to \begin{bmatrix} 0 \\ 1 \\ 1 \\ 1 \end{bmatrix}.$$

2.2.2 Hybrid System Construction

We now demonstrate that given a Lagrangian, a directed cycle, and a domain breakdown, a hybrid system can be explicitly constructed. Since the Lagrangian is intrinsic to a robot, this result proves that a domain breakdown, which is determined by the enforced contact points, alone dictates the mathematical model of a biped.

Continuous Dynamics. We explicitly construct the control system $\dot{x} = f_v(x) + g_v(x)u$ through the constraints imposed on each domain through the domain breakdown.

For the domain $v \in V$, the holonomic constraints that are imposed on that domain are given by:

$$\eta_v(q) = \eta(q)\mathcal{B}(v),$$

where the domain breakdown dictates which constraints are enforced. Differentiating the holonomic constraint yields a *kinematic constraint*:

$$J_v(q)\dot{q} = 0,$$

where $J_v(q) = \text{RowBasis}\left(\frac{\partial \eta_v(q)}{\partial q}\right)$ is a basis for the row space of the Jacobian (this removes any redundant constraints so that J_v has full row rank). The kinematic constraint yields the *constrained dynamics* on the domain:

$$D(q)\ddot{q} + H(q, \dot{q}) = B(q)u + J_v(q)^T F_v \tag{3}$$

which enforces the holonomic constraint; here D, H and B are as in Equation (2) and F_v is the *wrench* containing forces and moments expressed in the reference frame R_c [14]. To determine the wrench F_v, we differentiate the kinematic constraint:

$$J_v(q)\ddot{q} + \frac{\partial J_v(q)}{\partial q}\dot{q} = 0$$

and combine this equation with Equation (3) to obtain an expression for $F_v(q, \dot{q}, u)$ which is affine in u. Therefore, for $x = (q^T, \dot{q}^T)^T$, Equation (3) yields the affine control system $\dot{x} = f_v(x) + g_v(x)u$.

Discrete Dynamics. We now construct the domains, guards and reset maps for a hybrid system using the domain breakdown.

Given a vertex $v \in V$, the domain is the set of admissible configurations of the system factoring in both friction and a unilateral constraint. Specifically, from the wrench $F_v(q, \dot{q}, u)$, one can ensure that the foot does not slip by considering inequalities on the friction which can be stated in the form: $\mu_v(q)^T F_v(q, \dot{q}, u) \geq 0$, with $\mu_v(q)$ a matrix of friction parameters and constants defining the geometry of the foot (see [11] for more details). These are coupled with the unilateral constraint on this domain, $h_v(q) = h(q)\mathcal{B}(v)$, to yield the set of admissible configurations:

$$A_v(q, \dot{q}, u) = \begin{bmatrix} \mu_v(q)^T F_v(q, \dot{q}, u) \\ h_v(q) \end{bmatrix} \geq 0. \tag{4}$$

The domain is thus given by:

$$D_v = \{(q, \dot{q}, u) \in TQ \times \mathbb{R}^{m_v} : A_v(q, \dot{q}, u) \geq 0\}.$$

The guard is just the boundary of this domain with the additional assumption that set of admissible configurations is decreasing, i.e., the vector field is pointed outside of the domain, or for an edge $e = (v, v') \in E$,

$$G_e = \{(q, \dot{q}, u) \in TQ \times \mathbb{R}^{m_v} : A_v(q, \dot{q}, u) = 0$$
$$\text{and } \dot{A}_v(q, \dot{q}, u) \leq 0\}.$$

The impact equations are given by considering the constraints enforced on the subsequent domain. For an edge $e = (q, q') \in E$, the post-impact velocity $\dot{q}^+$ is given in terms of the pre-impact velocity $\dot{q}^-$ by:

$$\dot{q}^+ = P_e(q, \dot{q}^-) = (I - D^{-1}J_{q'}^T(J_{q'}D^{-1}J_{q'}^T)^{-1}J_{q'})\dot{q}^-$$

with I the identity matrix. This yields the reset map[1]:

$$R_e(q, \dot{q}) = \begin{bmatrix} q \\ P_e(q, \dot{q}) \end{bmatrix}$$

The end result is that given a domain breakdown and a bipedal robot, the hybrid model for the biped is completely determined.

3. DOMAIN BREAKDOWNS FROM HUMAN DATA

In this section, we determine the domain breakdown for 9 human subjects during walking. We begin by discussing the experiment and how the data was handled. We then present a method for extracting the times when the constraint for a given contact point is enforced through a method that fits the "simplest" function to the motion of the contact point when it is not enforced; the time intervals during a step when the constraints are enforced are simply the times when this function is not being followed. The end result of this procedure is a temporal ordering of events, which yields a domain breakdown. The domain breakdowns for all 9 subjects are presented in the case of no knee-lock and knee-lock (the motivation for considering both cases is discussed further in this section).

Walking Experiment. Data was collected on 9 subjects using the Phase Space System[2], which computes the 3D position of 19 LED sensors at 480 frames per second using 12 cameras at 1 millimeter level of accuracy. The cameras

---

[1]Note that in order to get periodic behavior in the walking, the "left" and "right" leg must be "swapped" at one of the transitions; this "trick" is common throughout the literature.
[2]http://www.phasespace.com/

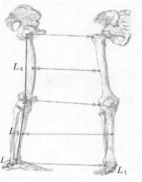

Figure 2: Illustrations of the experimental setup (left) and sensor placement on each foot (middle and right). Each LED sensor was placed at the joints as illustrated with the red dots on the right lateral (middle) and anterior aspects (right) of the each leg.

were calibrated prior to the experiment and were placed to achieve a 1 millimeter level of accuracy for a space of size 5 by 5 by 5 meters cubed. 8 LED sensors were placed on each leg at the joints on on the heel and toe, 1 LED sensor was placed on the sternum, 1 LED sensor was placed on the back behind the sternum, and 1 LED sensor was placed on the belly button. Each trial of the experiment required the subject to walk 3 meters along a line drawn on the floor. Each subject performed 12 trials, which constituted a single experiment. 3 female and 6 male subjects with ages ranging between 17 and 77, heights ranging between 161 and 189 centimeters, and weights ranging between 47.6 and 90.7 kilograms. Table 3 describes the measurements of each of the subjects. The data for each individual is then rotated so that the walking occurs in the x-direction and for each subject, the 12 walking trials are averaged (after appropriately shifting the data in time) which results in a single trajectory for each constraint for each subject for at least two steps (one step per leg); the resulting data can be seen in Fig. 4. Any interested researcher can perform analysis on the collected data[3].

---

[3]`http://www.eecs.berkeley.edu/~ramv/HybridWalker`

	Sex	Age	Weight	Height	L_1	L_2	L_3	L_4
1	M	30	90.7	184	14.5	8.50	43.0	44.0
2	F	19	53.5	164	15.0	8.00	41.0	44.0
3	M	17	83.9	189	16.5	8.00	45.5	55.5
4	M	22	90.7	170	14.5	9.00	43.0	39.0
5	M	30	68.9	170	15.0	8.00	43.0	43.0
6	M	29	59.8	161	14.0	8.50	37.0	40.0
7	M	26	58.9	164	14.0	9.00	39.0	41.0
8	F	77	63.5	163	14.0	8.00	40.0	42.0
9	F	23	47.6	165	15.0	8.00	45.0	43.0

Figure 3: Table describing each of the subjects. The subject number is in the left column and the L_1, L_2, L_3, L_4 measurements correspond to the lengths described in Fig. 2. The measurement in column 4 is in kilograms (and was self reported) and the measurements in columns 5-9 are in centimeters.

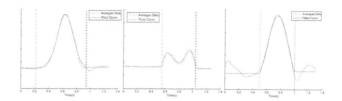

Figure 4: The data for the height of the heel (left), height of the toe (middle) and angle of the knee (right) together with the fittings of a constant, Gaussian, and constant for the heel (left), a constant, 4th order polynomial, and constant for the toe (middle) and a constant, Gaussian, and constant for the knee (right). The vertical lines indicate the transitions points between the fitting functions.

Function Fitting. In order to determine the domain breakdowns for the subjects in the walking experiment, it is necessary to determine the times when the number of contact points change, i.e., the *event times*. Rather than looking for when the contact point is constrained (via thresholding), we choose a "simple" function that the contact point follows when unconstrained. When this function is not being followed, the constraints are enforced and thus constant. Therefore, once this function has been identified, the rest of the process to determine the domain breakdown becomes automatic.

To formalize the idea of function fitting to determine event times, given a set of contact points, $\mathcal{C}$, let $s_c(n, a)$ be the "simplest" function that the contact point $c \in \mathcal{C}$ follows while unconstrained ($a \in R^k$ is a vector that parameterizes the function). Denote the contact point sensor data by $y_c(n)$ where $n \in \{1, \ldots, T\}$. When the contact point is constrained the sensor data appears constant, and when the contact point is unconstrained the sensor data follows $s_c(n, a)$. We therefore define the function:

$$f_c(n, \tau_l, \tau_s, a) = \begin{cases} s_c(\tau_l, a) & \text{if} & n \leq \tau_l \\ s_c(n, a) & \text{if} & \tau_l < t_n < \tau_s \\ s_c(\tau_s, a) & \text{if} & \tau_s \leq n \end{cases}$$

where $\tau_l, \tau_s \in \{1, \ldots, T\}$ are the event times indicating when the contact point becomes unconstrained, τ_l (lift), and constrained, τ_s(strike)[4]. To determine the event times that best fit the data, we solve the following optimization problem:

$$\min_{\tau_l, \tau_s \in \{1, \ldots, T\}} \min_{a \in \mathbb{R}^k} \frac{1}{T} \sum_{n=1}^{T} |f_c(n, \tau_l, \tau_s, a) - y_c(n)|.$$

To illustrate this procedure, consider the averaged data plotted against time for the heel and toe contact point sensors in Fig. 4. Looking at this data, the behavior of the heel appears to follow a constant, followed by a Gaussian, followed by a constant; therefore, we claim that the "simplest" function that the heel follows when unconstrained is a Gaussian. In a similar fashion, the averaged data for the toe appears to follow a constant, followed by a 4th order polynomial, followed by a constant. Using these observations, we fit the averaged heel and toe data to these functions using the described procedure. The results of this fitting are drawn in the same figure with the transition points τ_l and τ_s

---

[4]Here we assume without loss of generality that $\tau_l < \tau_s$.

indicated by vertical lines. The correlation coefficients for the illustrated heel and toe examples are 0.9968 and 0.9699, respectively.

Inspecting the data for the angle of the knee over time given in Fig. 4, one concludes that a human does not lock their knee through the course of walking, i.e., there is no period in which the knee is constant. In fact, there is clearly a period when the knee swings (the larger "bump") and a period when the knee goes through smaller oscillations. The smaller oscillations correspond to the period when the human has their weight on the knee, and the oscillations can be understood to be the natural spring and damping response of the muscles and tendons. Robotic bipedal walkers include a knee-lock domain since in practice it reduces energy consumption. Since we want to quantify how human-like such robotic walkers are, we do not want to disqualify the existence of knee-lock in the human data. It is for this reason that we consider both the case of knee-lock and no knee-lock.

One can view the period of smaller oscillations as the knee being "locked." Under this assumption, the "simplest" function that the knee follows is a constant, followed by a Gaussian, followed by a constant. The results of this fitting are drawn in Fig. 4 with the transition points τ_l and τ_s indicated by vertical lines.

Determining the Domain Breakdown. Given the data for a contact point $c \in \mathcal{C} = \{c_1, c_2, \ldots, c_k\}$, we determine the lift and strike times for the contact point, τ_l^c and τ_s^c, over the time interval of the averaged data using the aforementioned techniques. Since the data is over at least two steps (one step with each leg), there may be multiple lift and strike times over the period of the data. Using the aforementioned method we make no assumptions about simultaneous contact point enforcement. Denote by J_c the period where constraints associated with a contact point are enforced, i.e., $n \in J_c$ if $f_c(n) = $ constant with f_c the fitting function for the contact point $c \in \mathcal{C}$; these intervals are shown in blue in Fig. 5 over the course of one step (not the entire data period) in the case of $\mathcal{C}_u$ (or no knee-lock). Analogous to the definition of a domain breakdown (Def. 2), we can define a binary vector, $b(n) \in \mathbb{Z}_2^{|\mathcal{C}|}$, encoding which contact points are in force at any given time by letting $b(n)_i = 1$ if $n \in J_{c_i}$ and $b(n)_i = 0$ otherwise. This function only changes value a finite number of times and denotes these distinct values by $b(m), m = 0, 1 \ldots, M$.

To determine the domain breakdown associated with the walking, we begin by defining the directed cycle Γ (if it exists, which we do not assume). Looking at the data, we check to see if there exists an integer $p \in \mathbb{N}$ such that

$$b(m) = \begin{bmatrix} 0 & I \\ I & 0 \end{bmatrix} b(m+p)$$

for $0 \leq m \leq p$ with I the identity matrix and $0, I \in \mathbb{R}^{\frac{|\mathcal{C}|}{2} \times \frac{|\mathcal{C}|}{2}}$; the matrix that is multiplied by b serves the purpose of reordering the right and left leg. If this p can be found, periodic walking over the course of two steps exists with the left leg mirroring the behavior of the right leg. In this case, one constructs a directed cycle with p domains (as in Equation (1)) and this is the graph Γ. Finally, the domain breakdown $\mathcal{B}$ is given by $\mathcal{B}(v_m) = b(m)$. The application of this procedure to a single subject in the case of no knee-lock is illustrated in Fig. 5.

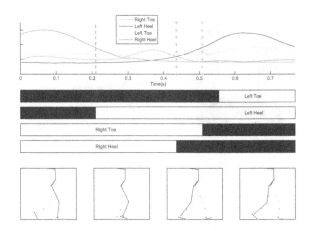

Figure 5: An overview of how the domain breakdown is achieved (with no knee-lock). The top row illustrates the height of the toe and heel of each leg over one step along with the lifting and strike time for each constraint (illustrated by vertical lines). The middle row illustrates which constraints are active based upon the fitting. The bottom row shows the resulting domain breakdown where enforced constraints are drawn with green circles.

Results. We perform the process outlined in this section on the set of contact points $\mathcal{C}_u = \{lh, lt, rh, rt\}$ and $\mathcal{C}_l = \{lh, lt, lk, rh, rt, rk\}$ on the 9 subjects that performed the walking experiment. The end result is that we find domain breakdowns for each subject, i.e., each subject had periodic walking. The domain breakdowns, along with the percentage of time spent in each domain in the case of no knee-lock and knee-lock for each individual, are illustrated in Fig. 6 and 8, respectively.

When knee-lock is not considered in the domain breakdown, observe that all subjects exhibit a *universal* domain breakdown shown in Fig. 1 in spite of great differences in age, height and weight. This is particularly surprising since we made no *a priori* assumptions about the ordering of contact point enforcement and did not demand simultaneous contact point enforcement. If we include knee-lock during the determination of the domain breakdown, there is not a single domain breakdown that is common to all subjects: out of the 9 subjects, there are 7 different domain breakdowns. This is probably due to the fact that humans do not actually lock their knees during walking. Nevertheless, during the design of a bipedal robot knee-lock can be an important domain to include since it can simplify the mechanical and controller development. In this instance, it is still useful to have a "universal" domain breakdown for the purposes of robotic design, and we construct such a "universal" domain breakdown in the next section by defining a distance metric on the space of domain breakdowns.

4. HUMAN-BASED COST OF WALKING

In this section, we construct a cost function that measures the anthropomorphic nature of robotic bipedal walking termed the *human-based cost*. We do this by first defining a metric on the space of weighted cycles, the *cut distance*, which allows us to compare different walking gaits and to construct an optimal walking cycle by minimizing the distance between the weighted cycles observed in the human walking data. Using the cut distance, we next define the *human-based cost* which allows us to compute the distance from a specific walking gait (either human or robotic) to the optimal walking cycle. The remainder of this section is devoted to using the human-based cost to determine the extent to which popular robotic models from the literature are anthropomorphic.

Distance Between Cycles. We employ the notion of *cut* (or *rectangular*) distance between two weighted graphs to compare different domain breakdowns (the definition in its general form can be found in [3]). Since we are only interested in the specific domains visited and the corresponding time spent in each of these domains, we define a notion of weighted cycle and a corresponding distance between weighted cycles that is pertinent to the application being considered.

Definition 3. *A walking cycle is a pair (α, ℓ) where $\ell = (V, E)$ is a cycle and $\alpha : \ell \to \mathbb{R}^{|V|}$ is a function such that $\alpha(v) \geq 0$ and $\sum_{v \in V} \alpha(v) = 1$. Denoting a cycle by $\ell : v_0 \to v_1 \to \cdots \to v_p$, we denote a walking cycle by:*

$$\alpha(\ell) : \alpha(v_0) \to \alpha(v_1) \to \cdots \to \alpha(v_p).$$

Example 3. *Each of the domain breakdowns presented in Fig. 6 gives us a distinct walking cycle. For example, Subject 1 has a walking cycle $S_1 = (\alpha_1, \ell_1)$ given by:*

$$
\begin{array}{ccccccc}
\ell_1 : & [lh, lt] & \to & [lt] & \to & [lt, rh] & \to & [lt, rh, rt] \\
\alpha_1(\ell_1) : & 26.5\% & \to & 49.7\% & \to & 21.4\% & \to & 2.4\%.
\end{array}
$$

Here, and throughout this paper, weightings are stated in percentages to indicate the percentage of time the human spends in a domain through the course of one step.

We now introduce a definition of cut distance that is a slight modification of the definition presented in [3]. The only differences are that we do not force the weighted graphs to have nodes with positive weights, and we require the weights to sum to one.

Definition 4. *Let (α_1, ℓ_1) and (α_2, ℓ_2) be two walking cycles. Viewing both α_1 and α_2 as functions on $V_1 \cup V_2$ by letting $\alpha_1(i) \equiv 0$ if $i \in V_2 \backslash V_1$ and $\alpha_2(j) \equiv 0$ if $j \in V_1 \backslash V_2$, the cut distance between two cycles is given by:*

$$
d(\alpha_1, \ell_1, \alpha_2, \ell_2) =
$$

$$
\max_{I, J \subset V_1 \cup V_2} \left| \sum_{i \in I, j \in J} (\alpha_1(i)\alpha_1(j)\beta_1(i,j) - \alpha_2(i)\alpha_2(j)\beta_2(i,j)) \right|
$$

$$
+ \sum_{k \in V_1 \cup V_2} |\alpha_1(k) - \alpha_2(k)|, \tag{5}
$$

where $\beta_1(i,j) = 1$ for all edges $(i,j) \in E_1$ and $\beta_2(i,j) = 1$ for all edges $(i,j) \in E_2$.

It is straightforward to check that the modified definition of the cut distance satisfies the requirements of a metric (i.e., non-negativity, identity of indiscernibles, symmetry and the triangle inequality). Intuitively, the cut distance compares just how different two walking cycles are when considering all possible "cuts" between the pair of cycles.

Human-Based Cost. The idea for developing a human-based cost of walking is that human walking data can be used to develop a cost function in which to judge other non-human walking. In the context of this paper, we develop a cost based upon the domain breakdown and resulting walking cycles associated with a subjects walking gait.

Definition 5. *Consider N subjects with associated domain breakdowns and walking cycles $S_i = (\alpha_i, \ell_i)$ for $i \in \{1, \ldots, N\}$. Letting $\mathcal{L} = \bigcup_{i=1}^{N} \ell_i$ be the graph obtained by combining all of the cycles ℓ_i, we define the optimal walking cycle by:*

$$(\alpha^*, \ell^*) = \operatorname*{argmin}_{(\alpha, \ell) \in \mathbb{R}^{|\ell|} \times \mathcal{L}} \frac{1}{N} \sum_{i=1}^{N} d(\alpha, \ell, \alpha_i, \ell_i). \tag{6}$$

The optimal walking cycle is just the walking cycle through the graph of all cycles obtained from the walking that best fits the data under the cut distance. The optimal walking cycle allows one to describe the extent to which a walking gait is human-like.

Definition 6. *Given a biped (either human or bipedal robot) with associated domain breakdown and walking cycle $R = (\alpha_r, \ell_r)$, the human-based cost (HBC) of walking is defined to be:*

$$\mathcal{H}(R) = d(\alpha_r, \ell_r, \alpha^*, \ell^*).$$

It is important to note that the optimal walking cycle may not be unique, and so there may be multiple HBCs of walking constructed from a single experiment, i.e., the HBC is not necessarily unique (in this paper, we found a unique HBC for the case of knee-lock, but multiple HBCs for the case of no knee-lock.) Unsurprisingly, multiple experiments might yield different optimal walking cycles and different HBCs, but if the experiments are carried out consistently they should be compatible and could be merged into a single HBC.

4.1 HBC without Knee-Lock

Using the walking cycles constructed from the human data illustrated in Fig. 6, we now compute the optimal walking cycle and compute the HBCs for the subjects and bipedal robots that have appeared in the literature. All subjects have the same universal cycle ℓ_u, so $S_i = (\alpha_i, \ell_u)$ for $i = 1, \ldots, 9$. The optimal walking cycle is given by (α^*, ℓ_u), where α^* is computed using Equation (6) yielding:

$$
\begin{array}{ccccccc}
\ell_u : & [lh, lt] & \to & [lt] & \to & [lt, rh] & \to & [lt, rh, rt] \\
\alpha^*(\ell_u) : & 59\% & \to & 18\% & \to & 17\% & \to & 6\%.
\end{array}
$$

If the objective of a robotic biped is to obtain anthropomorphic walking, this optimal walking cycle should be followed as closely as possible. To demonstrate this, we use the optimal walking cycle to compute the HBC in several cases.

Humans. Although all of the subjects have a universal domain breakdown, this does not imply that they have identical walking gaits. To quantify the differences in walking between the different subjects, we compute the HBC for each subject. The results of this computation are illustrated in Fig. 7; for this comparison, the optimal walking cycle is computed using the data from all of the subjects and then each subject is compared to the optimal walking cycle, α^*, via the HBC. From these computations, it is clear that Subject 1 has an unusually high cost when compared to the other subjects whose costs are fairly uniform. In fact,

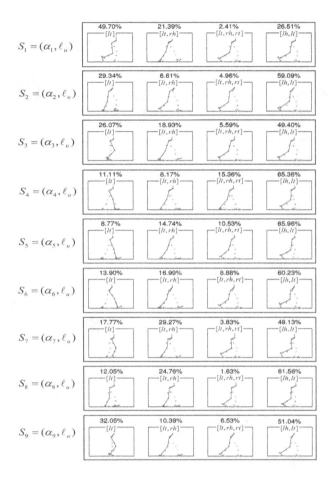

Figure 6: The domain breakdowns without knee lock for the 9 subjects in the order listed in Tab. 3 participating in the experiment, along with the corresponding walking cycle. Each illustration is a snapshot of the subject's configuration at the beginning of the domain. The green circles indicate the contact point that is enforced in a particular configuration.

Subject 1 has a history of back problems (herniated disks at L4-L5 and a pinched nerve at L5-S1) and was taking pain medication at the time of the experiment. Thus, even at the level of comparing human subjects, the HBC seems to identify less "natural" walking.

Due to the high HBC of Subject 1, this subject can be treated as an outlier in the data set. Therefore, it may be desirable to not include this subject in the calculation of the optimal walking cycle. For example, if the HBC were to be used for the detection of medical conditions, one could compile a HBC from healthy subjects. A subject with a suspected medical condition could then be compared to this HBC (much as existing robotic models are compared via the HBC below). To illustrate this idea, we compute the optimal walking cycle from the healthy subjects, or subjects 2-9. This results in the optimal walking cycle:

$$\alpha_2^*(\ell_u) \quad : \quad 60\% \quad \rightarrow \quad 17\% \quad \rightarrow \quad 17\% \quad \rightarrow \quad 6\%,$$

which has slightly different weightings than α^* due to the exclusion of Subject 1. The corresponding healthy HBC of Subject 1 is computed using α_2^* resulting in a slightly

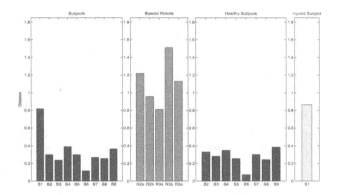

Figure 7: The HBC for the 9 subjects in the experiment and 5 bipedal robotic models that have appeared in the literature (the number of domains in each of the models is illustrated via a subscript) using α^*, and the HBC for the 8 healthy subjects and the "injured" subject using α_2^*.

higher cost than the HBC associated with α^* as expected (0.8628 as opposed to 0.8183). The healthy HBC for all of the subjects using α_2^* is illustrated in Fig. 7. Note that due to the exclusion of the "injured" subject, the costs among the healthy subjects are even more uniform.

Robots. We next use the HBC to compute the cost of walking for bipedal robots that have been considered in the literature with no knee-lock.

In [19], numerous bipedal modes with different numbers of domains (between 1 and 3) and walking gaits are considered. We focus on two with cycles: $\ell_2 : [lh, lt] \rightarrow [lt]$ and $\ell_3^a : [lh, lt] \rightarrow [lt] \rightarrow [lt, rh, rt]$. Associated with the walking found in that paper, there are three walking cycles: $R_{2a} = (\alpha_{2a}, \ell_2)$, $R_{2b} = (\alpha_{2b}, \ell_2)$ and $R_{3a} = (\alpha_{3a}, \ell_3^a)$, for which the HBC can be computed as shown in Fig. 7. From the results of the computed HBC, we conclude that the model R_{3a} produces the most anthropomorphic walking as it has a dramatically lower cost than the other two models. Interestingly, the authors of the paper state that this walking cycle appeared "the closest to human gait" just as the HBC discerns.

In [16], two bipedal walking gaits are obtained with a model consisting of a cycle: $\ell_3^b : [lh, lt] \rightarrow [lt] \rightarrow [rh]$, where $[rh]$ is a domain unseen in the human walking wherein the biped only has a single contact point at the right heel. Associated with this cycle are two walking cycles: $R_{3b} = (\alpha_{3b}, \ell_3^b)$ and $R_{3c} = (\alpha_{3c}, \ell_3^b)$ for which the HBC can be computed; the results are shown in Fig. 7. Interestingly, despite the fact that both of these models have three domains, they do not produce an HBC as low as walking cycle R_{2b} indicating that adding more domains does not necessarily result in more human-like walking.

4.2 HBC with Knee-Lock

For the 9 subjects that performed the walking experiment with associated walking cycles $S_i = (\alpha_i, \ell_i)$ illustrated in Fig. 8, we minimize Equation (6) to find the optimal walking cycle. In performing this minimization we find four optimal walking cycles: $L_i = (\alpha_i^*, \ell_i^*)$, $i = 1, 2, 3, 4$, i.e., four minima with essentially the same cost (3.69, 3.76, 3.79 and 3.94, respectively, with the cost of any other cycle being almost

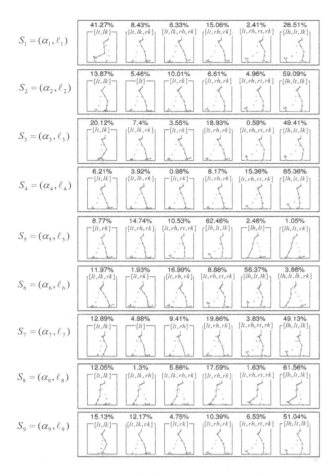

Figure 8: The domain breakdowns with knee lock for the 9 subjects in the order listed in Tab. 3 participating in the experiment, along with the corresponding walking cycle. In this case, there are 13 vertices traversed by the 9 subjects.

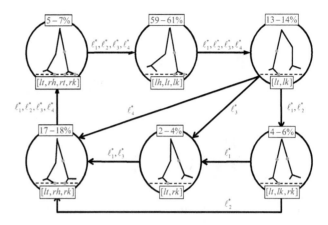

Figure 9: The "universal" domain graph for bipedal walking with knee-lock, consisting of the four optimal walking cycles, with the weight of each cycle on each domain taking values in the indicated interval and the path traversed by the cycle indicated by the edge labels.

twice as high). These optimal walking cycles are shown in Fig. 9.

To determine if one of these optimal walking cycles is preferable to the rest, we can take each of these optimal walking cycles and compute the HBC of the other optimal walking cycles with respect to it as illustrated in the right plot in Fig. 10. Notice that all of the HBCs are nearly uniform. We claim that the "universal" domain graph for bipedal walking with knees is illustrated in Fig. 9 in that any cycle taken in this graph is optimal. The interesting aspect of this domain graph is that, as pointed out earlier, walking with knee-lock is more suited to robots than humans. As such, the domain graph chosen is a design decision in the modeling of the robot so the specific cycle chosen in the "universal" domain graph can be guided by the robot being designed.

Humans. The human-based cost associated with each of the optimal walking cycles computed for each subject in the experiment is illustrated in Fig. 10. One can see that each optimal gives a very similar HBC, which is both consistent with the fact that they are all optimal and points to the fact that we do, in fact, get a "single" human-based cost for all intents and purposes. Looking at the computed human-based

costs, Subject 1 again has the largest cost, again consistent with the previously mentioned medical conditions. Interestingly, notice that Subject 8 has the lowest cost. This subject is the oldest, at 77 years of age. Since older individuals tend to walk with a "stiffer" gait, and because knee lock does not appear to naturally occur during human walking, we postulate that the low HBC is identifying this additional stiffness. This observation could have important health care ramifications.

Robots. We also can use the HBC to compare the anthropomorphic nature of different walking gaits for robots with knees-lock that have appeared in the literature consisting of a single domain, R_1 (corresponding to the straight-leg walking of the compass gait biped [9, 12]), two domains, R_2 (corresponding to bipeds with knees that are both unlocked and locked throughout the walking [1, 7, 13]) and both four and five domain models with both knees and feet, R_4 and R_5, respectively [11, 17]. We can compute the HBC for these different robotic walking gaits with respect to the 4 optimal walking gaits as illustrated in Fig. 10. From the computed cost it is clear that as the number of domains increases to better represent the universal domain breakdown in Fig. 9, the HBC decreases accordingly. In particular, R_4 and R_5 have substantially lower HBC than the other robotic walking cycles due to the additional domains, all of which can be found in the "universal" domain breakdown for knee-lock as shown in Fig. 9. In fact, the only domain found in these two walking cycles that is not found in the "universal" domain breakdown is $[lh, lt, lk, rk]$, and the time spent in this domain is small enough, while the time spent in the other domains (which is startlingly similar to the time spent in these domains for the optimal walking cycles) is large enough to deliver a low HBC.

5. CONCLUSION

This paper presented a "universal" domain breakdown and corresponding optimal walking cycle, both without and with knee-lock, obtained from human walking data. From this, the human-based cost was constructed. It was demonstrated

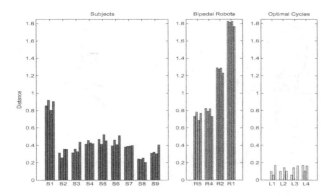

Figure 10: The HBC for the 9 subjects in the experiment computed with the 4 optimal walking cycles found (left), 4 bipedal robotic models that have appeared in the literature (middle) (the number of domains in each of the models is the subscript), and the HBC of each optimal walking cycle computed with the remaining 3 optimal walking cycles (right).

that when the HBC was computed for the human-subjects preexisting medical conditions were successfully identified. When the HBC was computed for existing bipedal walking robots, the robots with more "human-like" walking gaits were correctly identified. This points to the usefulness of the HBC both in identifying medical conditions in humans, and obtaining anthropomorphic walking in bipedal robots. The results of this paper are also applicable to future bipedal robot design. If the universal domain breakdown is used for the robotic model, and the parameters of the controller used to achieve walking are chosen so as to minimize the HBC, we claim that the end result promises to be natural and human-like walking.

6. REFERENCES

[1] A. D. Ames, R. W. Sinnet, and E. D. B. Wendel. Three-dimensional kneed bipedal walking: A hybrid geometric approach. In *Hybrid Systems: Computation and Control*, San Francisco, CA, 2009.

[2] T. Andriacchi and E. Alexander. Studies of human locomotion: past, present and future. *Journal of Biomechanics*, 33(10):1217–1224, 2000.

[3] C. Borgs, J. Chayes, L. Lovász, V. Sós, and K. Vesztergombi. Convergent sequences of dense graphs I: Subgraph frequencies, metric properties and testing. *Advances in Mathematics*, 219(6):1801–1851, 2008.

[4] D. J. Braun and M. Goldfarb. A control approach for actuated dynamic walking in bipedal robots. *IEEE Transactions on Robotics*, 25:1–12, 2009.

[5] C. Chevallereau, J. Grizzle, and C. Shih. Asymptotically stable walking of a five-link underactuated 3D bipedal robot. *IEEE Transactions on Robotics*, 25(1):37–50, 2009.

[6] J. H. Choi and J. W. Grizzle. Planar bipedal walking with foot rotation. In *American Control Conference*, 2005.

[7] S. H. Collins, M. Wisse, and A. Ruina. A 3-D passive dynamic walking robot with two legs and knees.

International Journal of Robotics Research, 20:607–615, 2001.

[8] M. Garcia, A. Chatterjee, and A. Ruina. Speed, efficiency, and stability of small-slope 2D passive dynamic bipedal walking. In *IEEE International Conference on Robotics and Automation*, 1998.

[9] A. Goswami, B. Thuilot, and B. Espiau. Compass-like biped robot part I : Stability and bifurcation of passive gaits. Rapport de recherche de l'INRIA, 1996.

[10] J. Grizzle, G. Abba, and F. Plestan. Asymptotically stable walking for biped robots: Analysis via systems with impulse effects. *IEEE Transactions on Automatic Control*, 46(1):51–64, 2001.

[11] J. W. Grizzle, C. Chevallereau, A. D. Ames, and R. W. Sinnet. 3d bipedal robotic walking: Models, feedback control, and open problems. In *NOLCOS*, Bologna, Italy, 2010.

[12] T. McGeer. Passive dynamic walking. *The International Journal of Robotics Research*, 9(2):62, 1990.

[13] T. McGeer. Passive walking with knees. In *IEEE International Conference on Robotics and Automation*, Cincinnati, OH, 1990.

[14] R. M. Murray, Z. Li, and S. S. Sastry. *A Mathematical Introduction to Robotic Manipulation*. Boca Raton, FL, 1993.

[15] K. Ono, T. Furuichi, and R. Takahashi. Self-excited walking of a biped mechanism with feet. *The International Journal of Robotics Research*, 23(1):55, 2004.

[16] T. Schaub, M. Scheint, M. Sobotka, W. Seiberl, and M. Buss. Effects of compliant ankles on bipedal locomotion. In *IROS*, St. Louis, Missouri, USA, 2009.

[17] R. W. Sinnet and A. D. Ames. 2D bipedal walking with knees and feet: A hybrid control approach. In *48th IEEE Conference on Decision and Control*, Shanghai, P.R. China, 2009.

[18] R. Tedrake, T. Zhang, and H. Seung. Learning to walk in 20 minutes. In *Proceedings of the Fourteenth Yale Workshop on Adaptive and Learning Systems*, New Haven, Connecticut, USA, 2005.

[19] D. Tlalolini, C. Chevallereau, and Y. Aoustin. Comparison of different gaits with rotation of the feet for planar biped. *Robotics and Autonomous Systems*, 57:371–383, 2009.

[20] E. R. Westervelt, J. W. Grizzle, C. Chevallereau, J. Choi, and B. Morris. *Feedback Control of Dynamic Bipedal Robot Locomotion*. Control and Automation. Boca Raton, FL, June 2007.

[21] D. A. Winter. *Biomechanics and Motor Control of Human Movement*. Hoboken, NJ, 2005.

[22] F. Zajac, R. Neptune, and S. Kautz. Biomechanics and muscle coordination of human walking-Part I: Introduction to concepts, power transfer, dynamics and simulations. *Gait and Posture*, 16(3):215–232, 2002.

[23] V. Zatsiorky, S. Werner, and M. Kaimin. Basic kinematics of walking: step length and step frequency: a review. *Journal of sports medicine and physical fitness*, 34(2):109–134, 1994.

Characterizing Knee-Bounce in Bipedal Robotic Walking: A Zeno Behavior Approach[*]

Aaron D. Ames
Mechanical Engineering
Texas A&M University
College Station, TX 77843
aames@tamu.edu

ABSTRACT

This paper studies the walking behavior of kneed bipedal robots with knee-lock and knee-bounce, formally demonstrating that if knee-locking results in stable bipedal walking, then small amounts of knee-bounce still will result in a walking gait for the robot. To achieve this result, hybrid system models of bipeds are considered wherein knee-bounce corresponds to Zeno behavior. Using results on Zeno stability, we propose a notion of *generalized completion* that allows solutions to be carried beyond the Zeno point, i.e., carried beyond knee-bounce. We assume that the completed hybrid system has a periodic orbit when the impacts are perfectly plastic—a *plastic periodic orbit*, or walking gait with knee-lock. The main result of this paper is that when the assumption of perfectly plastic impacts is relaxed, if the plastic periodic orbit is stable and the Zeno point is Zeno stable, then there exists a periodic orbit in the case of non-plastic impacts, i.e., a *Zeno periodic orbit* corresponding to walking with knee-bounce. This formal result is applied to a specific example of a bipedal robot with knees.

Categories and Subject Descriptors

G.1.0 [**Numerical Analysis**]: General—*Stability (and instability)*; I.6.8 [**Simulation and Modeling**]: Types of Simulation—*Continuous, Discrete Event*

General Terms

Theory

Keywords

Hybrid mechanical systems, Bipedal robotic walking, Zeno behavior, Stability, Periodic orbits.

[*]This research is supported in part by NSF Award CNS-0953823 and NHARP Award 000512-0184-2009.

1. INTRODUCTION

Mechanical knees are an important component of achieving natural and "human-like" walking in bipedal robots [6, 17]. Mechanical "knee-caps", i.e., mechanical stops, are typically added to these mechanical knees to prevent the leg from hyper-extending (see the figure on the right for a bipedal robot with mechanical knees that lock built by Cornell[1], replicating the passive biped with knees of McGeer [1, 17]). Yet with this benefit comes a cost: *knee-bounce*, which occurs when the shin bounces off the mechanical stop repeatedly as the leg attempts to lock. It has been shown that this behavior can destabilize the robot in certain situations; see [1] for a passive bipedal robot that falls due to knee-bounce. To address this problem, mechanical catches are often added to the knees to prevent bouncing behavior, i.e., *knee-lock* is enforced through mechanical means. It can be seen in [1] that adding knee-lock to the robot resulted in stable walking. Although the addition of knee-lock to a robot can result in walking where it would not be present with knee-bounce, it comes at the cost of additional mechanical complexity.

This motivates the question: *is it necessary to add mechanical catches that enforce knee-lock to bipedal robots to obtain walking?* This paper formally shows that mechanical catches are not always necessary to achieve bipedal walking, i.e., that if stable walking exists for a robot with knee-lock, then walking will also exist with knee-bounce as long as the knee-bounce is sufficiently small. To achieve this result, it is necessary to understand what knee-lock and knee-bounce correspond to in a formal setting.

Bipedal robots are naturally modeled by systems with discrete and continuous behavior: hybrid systems. The continuous component consists of dynamics dictated by Lagrangians modeling the robot with the number of contact points (such as foot and knee contact) enforced through holonomic constraints. The discrete behavior occurs when the number of contact points changes, e.g., the knee locks or unlocks or the foot strikes the ground, resulting in an instantaneous change in the velocity of the system. In the setting of hybrid systems, knee-bounce and knee-lock can be formally understood to be, respectively, a result of non-plastic

[1]Photo credit: Rudra Pratap,
http://ruina.tam.cornell.edu/hplab/pdw.html

and plastic impacts at the knee. In this setting, knee-bounce corresponds to *Zeno behavior*—when an infinite number of discrete transitions, or impacts, occur in a finite amount of time—where the point to which the Zeno solution converges, the *Zeno point*, corresponds to the leg being straight. The behavior of knee-bounce can thus be compared to the behavior of knee-lock, where the Zeno point is reached instantaneously due to the plastic impact.

Using the formalisms of hybrid systems and Zeno behavior, we can approach the problem of knee-bounce rigorously. In particular, we consider *Lagrangian hybrid systems* which model mechanical systems undergoing impacts as dictated by *unilateral constraints* on the configuration space; the amount of energy lost through impact is dictated by the *coefficient of restitution*. Hybrid systems of this form have a single discrete domain and can display Zeno behavior. Easily verifiable conditions on the existence of this behavior based upon the coefficient of restitution and the unilateral constraint function have been proven [13, 14, 15]. The first result of this paper is a bound on the distance between the *Zeno point*—the limit point of a Zeno solution—for plastic and non-plastic impacts. This result is essential in establishing the main result of this paper.

When there exists Zeno behavior, the hybrid system can be *completed* to allow for solutions to extend beyond the finite Zeno time. In particular, this paper presents a notion of *generalized completion*, extending previous notions of hybrid system completion [3, 4, 18, 22, 23] to a setting that will be applicable to bipedal robot models. Specifically, an additional domain is added to the hybrid system—the *post-Zeno* domain—which enforces the unilateral constraint as a holonomic constraint. A solution transitions to this domain when the Zeno point is reached, and transitions back when conditions in the post-Zeno domain are reached (generalizing the traditional transition back to the pre-Zeno domain based upon the Lagrange multipliers associated to the holonomic constraint). For a bipedal robot, the pre-Zeno domain is when the leg is bent, the post-Zeno domain is when the leg is straight, a transition to the post-Zeno domain occurs when the knee locks (in finite time for knee-bounce and instantaneously for knee-lock), and a transition to the pre-Zeno domain occurs when the foot strikes the ground.

The objective of this paper is to consider periodic orbits in completed hybrid systems and to show that the existence of periodic orbits in the case of plastic impacts—termed *plastic* periodic orbits—implies the existence of periodic orbits for non-plastic impacts—termed *Zeno* periodic orbits. Beginning with a plastic periodic orbit, the assumption of plastic impacts is relaxed, i.e., the coefficient of restitution is no longer assumed to be zero. In this case, the hybrid system will display non-trivial Zeno behavior. The main result is that if the plastic periodic orbit is stable and the Zeno point is Zeno stable, then for sufficiently small coefficients of restitution there exists a Zeno periodic orbit. In order to demonstrate the practical usefulness of the results of this paper, we apply them to a bipedal robot model with knees. In this setting plastic periodic orbits correspond to knee-lock and Zeno periodic orbits correspond to knee-bounce. We show numerically that the conditions of the main result are satisfied, and we confirm through simulation that the existence of a stable walking gait with knee-lock implies the existence of a stable walking gait with knee-bounce.

This paper, therefore, provides the first steps towards understanding Zeno behavior–and more general phenomena unique to hybrid systems—in complex hybrid mechanical systems. Due to the relationship between knee-bounce in robotic walkers and Zeno behavior shown in this paper, it is evident that understanding the abstract formalisms used to model physical mechanical systems can lead to important insights into the behavior of these systems; moreover, these insights can be used to aid in the design of these systems through the knowledge that design decisions can affect the behavior of the system. For example, the results of this paper imply that mechanical knee-locks are not necessarily when constructing physical bipedal robots as long as the knee bounce (or coefficient of restitution) is kept small.

2. HYBRID MECHANICAL SYSTEMS

Bipedal walkers are naturally modeled by hybrid systems. This section, therefore, introduces the basic terminology of hybrid systems in a general enough setting so as to formally describe both bipedal robotic models and completed hybrid systems.

Definition 1. *A hybrid system is a tuple*

$$\mathscr{H} = (\Gamma, D, G, R, F),$$

where

- $\Gamma = (V, E)$ *is an oriented graph, i.e., V and E are a set of vertices and edges, respectively, and there exists a source function* sor $: E \to V$ *and a target function* tar $: E \to V$ *which associates to an edge its source and target, respectively.*

- $D = \{D_v\}_{v \in V}$ *is a set of domains, where $D_v \subseteq \mathbb{R}^{n_v}$ is a smooth submanifold of $\mathbb{R}^{n_v}$,*

- $G = \{G_e\}_{e \in E}$ *is a set of guards, where $G_e \subseteq D_{\text{sor}(e)}$,*

- $R = \{R_e\}_{e \in E}$ *is a set of reset maps, where $R_e : G_e \to D_{\text{tar}(e)}$ is a smooth map,*

- $F = \{f_v\}_{v \in E}$, *where f_v is a smooth dynamical system on D_v, i.e., $\dot{x} = f_v(x)$ for $x \in D_v$.*

Definition 2. *An execution of a hybrid system $\mathscr{H}$ is a tuple $\chi = (\Lambda, \mathcal{I}, \rho, \mathcal{C})$, where*

- $\Lambda = \{0, 1, 2, \ldots\} \subseteq \mathbb{N}$ *is a finite or infinite indexing set,*

- $\mathcal{I} = \{I_i\}_{i \in \Lambda}$ *where for each $i \in \Lambda$, I_i is defined as follows: $I_i = [t_i, t_{i+1}]$ if $i, i+1 \in \Lambda$ and $I_{N-1} = [t_{N-1}, t_N]$ or $[t_{N-1}, t_N)$ or $[t_{N-1}, \infty)$ if $|\Lambda| = N$, N finite. Here, for all $i, i+1 \in \Lambda$, $t_i \le t_{i+1}$ with $t_i, t_{i+1} \in \mathbb{R}$, and $t_{N-1} \le t_N$ with $t_{N-1}, t_N \in \mathbb{R}$,*

- $\rho : \Lambda \to V$ *is a map such that for all $i, i+1 \in \Lambda$, $(\rho(i), \rho(i+1)) \in E$. This is the discrete component of the execution,*

- $\mathcal{C} = \{c_i\}_{i \in \Lambda}$ *is a set of continuous trajectories, and they must satisfy $\dot{c}_i(t) = f_{\rho(i)}(c_i(t))$ for $t \in I_i$.*

We require that when $i, i+1 \in \Lambda$,

$$
\begin{array}{lll}
\text{(i)} & c_i(t) \in D_{\rho(i)} \; \forall \; t \in I_i & \\
\text{(ii)} & c_i(t_{i+1}) \in G_{(\rho(i), \rho(i+1))} & (1) \\
\text{(iii)} & R_{(\rho(i), \rho(i+1))}(c_i(t_{i+1})) = c_{i+1}(t_{i+1}). &
\end{array}
$$

When $i = |\Lambda| - 1$, we still require that (i) holds. The initial condition for the hybrid execution is $c_0(t_0) \in D_{\rho(0)}$.

Simple hybrid systems. A simple hybrid system is a hybrid system with a single domain and edge. Systems of this form have been widely studied, especially with respect to Zeno behavior [3, 14, 22]. In addition, the model with nonplastic impacts to be considered later will be represented by a simple hybrid system.

Formally, a *simple hybrid system* is a hybrid system with $\Gamma = (\{v\}, \{e = (v, v)\})$. Since in this case there is only a single domain, guard, reset map and vector field, we write a simple hybrid system as a tuple:

$$\mathscr{SH} = (D, G, R, f),$$

where D is a domain (not a set of domains), G is a guard, R is a reset map and f is a vector field.

Consider an execution of a simple hybrid system $\chi^{\mathscr{SH}_{\mathbf{L}}} = (\Lambda, \mathcal{I}, \rho, \mathcal{C})$. Since there is only one domain, the only choice for the discrete component of the execution is $\rho(i) \equiv v$. Therefore, we can write an execution of a simple hybrid system as $\chi^{\mathscr{SH}_{\mathbf{L}}} = (\Lambda, \mathcal{I}, \mathcal{C})$.

We now consider simple hybrid systems modeling mechanical systems: *Lagrangian hybrid systems*. These systems are obtained from *hybrid Lagrangians* which consist of a configuration space, a Lagrangian and a unilateral constraint (systems of this form have been well-studied in the mechanics literature [4, 5, 18, 19, 27]).

Dynamical systems from Lagrangians. Let $q \in Q$ be the *configuration* of a mechanical system.[2] In this paper, we will consider Lagrangians, $L : TQ \to \mathbb{R}$, describing mechanical or robotic systems, which are of the form

$$L(q, \dot{q}) = \frac{1}{2} \dot{q}^T M(q) \dot{q} - V(q), \tag{2}$$

with $M(q)$ the (positive definite) inertial matrix, $\frac{1}{2} \dot{q}^T M(q) \dot{q}$ the kinetic energy and $V(q)$ the potential energy. Assume there is a feedback *control law* $\Upsilon(q, \dot{q})$, which is a given smooth function $\Upsilon : TQ \to Q$. In this case, the Euler-Lagrange equations yield the (controlled) equations of motion for the system given in coordinates by:

$$M(q)\ddot{q} + C(q, \dot{q}) + N(q) = \Upsilon(q, \dot{q}), \tag{3}$$

where $C(q, \dot{q})$ is the vector of centripetal and Coriolis terms (cf. [21]) and $N(q) = \frac{\partial V}{\partial q}(q)$. Defining the *state* of the system as $x = (q, \dot{q})$, the Lagrangian vector field, f_L, associated to L takes the familiar form:

$$\begin{aligned} \dot{x} &= f_L(x) \\ &= \begin{pmatrix} \dot{q} \\ M(q)^{-1}(-C(q, \dot{q}) - N(q) + \Upsilon(q, \dot{q})) \end{pmatrix}. \end{aligned} \tag{4}$$

Holonomic constraints. We now define the holonomically constrained dynamical system with a Lagrangian L and a *holonomic constraint* $\eta : Q \to \mathbb{R}$. For such systems, the constrained equations of motion can be obtained from the equations of motion for the unconstrained system (3), and are given by (cf. [21])

$$M(q)\ddot{q} + C(q, \dot{q})\dot{q} + N(q) = d\eta(q)^T \lambda + \Upsilon(q, \dot{q}), \tag{5}$$

where λ is the Lagrange multiplier which represents the contact force and $d\eta(q) = \left(\frac{\partial \eta}{\partial q}(q) \right)^T$.

---

[2]For simplicity, in the models considered, we assume that the configuration space is identical to $\mathbb{R}^n$

Differentiating the constraint equation $\eta(q) = 0$ twice with respect to time and substituting the solution for $\ddot{q}$ in (5), the solution for the constraint force λ is obtained (see [21]). From the constrained equations of motion (5) and (3), for $x = (q, \dot{q})$, we get the vector field

$$\dot{x} = f_L^{\eta}(x) = f_L(x) + \begin{pmatrix} 0 \\ M(q)^{-1} d\eta(q)^T \lambda(q, \dot{q}) \end{pmatrix}. \tag{6}$$

Unilateral Constraints. The domain, guard and reset map (for knee lock) will be obtained from *unilateral constraint* $h : Q \to \mathbb{R}$ which gives the set of admissible configurations of the system; we assume that the zero level set $h^{-1}(0)$ is a smooth manifold.

Define the domain and guard, respectively, as

$$\begin{aligned} D_h &= \{(q, \dot{q}) \in TQ : h(q) \geq 0\}, \tag{7} \\ G_h &= \{(q, \dot{q}) \in TQ : h(q) = 0 \text{ and } dh(q)\dot{q} \leq 0\}. \end{aligned}$$

The reset map associated to a unilateral constraint is obtained through impact equations of the form (see [5, 19]):

$$R_h(q, \dot{q}) = \tag{8}$$
$$\begin{pmatrix} q \\ \dot{q} - (1 + \varepsilon) \frac{dh(q)\dot{q}}{dh(q)M(q)^{-1}dh(q)^T} M(q)^{-1} dh(q)^T \end{pmatrix}.$$

Here $0 \leq \varepsilon \leq 1$ is the *coefficient of restitution*, which is a measure of the energy dissipated through impact; for a perfectly plastic impact $\varepsilon = 0$ and for a perfectly elastic impact $\varepsilon = 1$. This reset map corresponds to rigid-body collision under the assumption of *frictionless impact*. Examples of more complicated collision laws that account for friction can be found in [5] and [27].

Definition 3. *A simple hybrid Lagrangian is defined to be a tuple* $\mathbf{L} = (Q, L, h)$, *where*

- Q *is the configuration space (assumed to be* $\mathbb{R}^n$*),*

- $L : TQ \to \mathbb{R}$ *is a Lagrangian of the form* (2),

- $h : Q \to \mathbb{R}$ *is a unilateral constraint.*

Simple Lagrangian hybrid systems. For a given Lagrangian, there is an associated dynamical system. Similarly, given a hybrid Lagrangian $\mathbf{L} = (Q, L, h)$ the *simple Lagrangian hybrid system (SLHS)* $\mathscr{SH}_{\mathbf{L}}$, associated to $\mathbf{L}$ is the simple hybrid system: $\mathscr{SH}_{\mathbf{L}} = (D_h, G_h, R_h, f_L)$.

Remark 1. *We often will want to make clear the dependence of the reset map on the coefficient of restitution ε, in which case we will write R_h^{ε}. Therefore, in the case of perfectly plastic impacts, the reset map is given by R_h^0. In the case of SLHS's, we will use the same convention writing $\mathscr{SH}_{\mathbf{L}}^{\varepsilon}$.*

3. ZENO BEHAVIOR

We now introduce Zeno behavior and the corresponding notion of Zeno equilibria, and we consider the stability of these equilibria. The first result of the paper is then presented; this gives bounds in the distance between Zeno points for plastic and non-plastic impacts—a result that is vital in proving the main result of this paper.

Definition 4. *An execution $\chi^{\mathscr{H}}$ is Zeno if $\Lambda = \mathbb{N}$ and*

$$t_\infty := \lim_{i \to \infty} t_i - t_0 = \sum_{i=0}^\infty t_{i+1} - t_i < \infty.$$

Here t_∞ is called the Zeno time.

Zeno behavior in SLHS's. If $\chi^{\mathscr{SH}_{\mathbf{L}}}$ is a Zeno execution of a SLHS $\mathscr{SH}_{\mathbf{L}}$, then its *Zeno point* is defined to be

$$x_\infty = (q_\infty, \dot{q}_\infty) = \lim_{i \to \infty} c_i(t_i) = \lim_{i \to \infty} (q_i(t_i), \dot{q}_i(t_i)). \quad (9)$$

These limit points are intricately related to a type of equilibrium point that is unique to hybrid systems: Zeno equilibria.

Definition 5. *A Zeno equilibrium point of a simple hybrid system $\mathscr{SH}$ is a point $x^* \in G$ such that $R(x^*) = x^*$ and $f(x^*) \neq 0$.*

We also can consider the stability of Zeno equilibria (see [7, 8] for complementary notions of stability as it relates to Zeno behavior).

Definition 6. *A Zeno equilibrium x^* of a simple hybrid system $\mathscr{SH}$ is bounded-time locally Zeno stable if for every $U \subset D$ and every $\epsilon > 0$, there exists an open set $W \subset U$ with $x^* \in W$ such that for every $x_0 \in W$, there exists a unique[3] execution χ with $c_0(t_0) = x_0$ and $\Lambda = \mathbb{N}$. This execution is Zeno with $t_\infty < \epsilon$ and $c_i(t) \in U$ for all $i \in \mathbb{N}$ and $t \in I_i$.*

Zeno stability in simple Lagrangian hybrid systems. If $\mathscr{SH}_{\mathbf{L}}$ is a SLHS, then due to the special form of these systems we find that the point $(q^*, \dot{q}^*)$ is a Zeno equilibria iff $\dot{q}^* = P_h(q, \dot{q}^*)$, with P_h given in (8). In particular, the special form of P_h implies that this holds iff $dh(q^*)\dot{q}^* = 0$. Therefore the set of all Zeno equilibria for a SLHS is:

$$\mathcal{Z} = \{(q, \dot{q}) \in D_h : h(q) = 0 \text{ and } dh(q)\dot{q} = 0\}. \quad (10)$$

Note that if $\dim(Q) > 1$, the Zeno equilibria in Lagrangian hybrid systems are always non-isolated.

Let $\ddot{h}(q, \dot{q})$ be the acceleration of $h(q)$ along trajectories of the unconstrained dynamics (3), given by:

$$\ddot{h}(q, \dot{q}) = \dot{q}^T H(q)\dot{q} + dh(q)M(q)^{-1}(-C(q, \dot{q})\dot{q} - N(q)),$$

where $H(q)$ is the Hessian of h at q. The following theorem, which was presented in [13, 14, 15], provides sufficient conditions for existence of Zeno executions in the vicinity of a Zeno equilibrium point.

Theorem 1 ([13, 15]). *Let $\mathscr{SH}_{\mathbf{L}}$ be a simple Lagrangian hybrid system and let $x^* = (q^*, \dot{q}^*)$ be a Zeno equilibrium point of $\mathscr{SH}_{\mathbf{L}}$. If $0 \leq \varepsilon < 1$ and $\ddot{h}(q^*, \dot{q}^*) < 0$, then $(q^*, \dot{q}^*)$ is bounded-time locally Zeno stable.*

Note that Thm. 1 was proven in [13, 15] using "Lyapunov-like" functions which mapped hybrid systems to *first quadrant interval hybrid systems* which can be viewed as the "simplest Zeno stable systems." In particular, in the case of mechanical systems, the following function was considered:

$$\psi(q, \dot{q}) = \begin{pmatrix} \dot{h}(q, \dot{q}) + \sqrt{\dot{h}(q, \dot{q})^2 + 2h(q)} \\ -\dot{h}(q, \dot{q}) + \sqrt{\dot{h}(q, \dot{q})^2 + 2h(q)} \end{pmatrix},$$

---

[3]This is just the *maximal* execution with initial condition $c_0(t_0) = x_0$.

which has the following properties (as proven in [15]) for a Zeno execution χ with initial condition $c_0(t_0) = x_0 \in G_h$:

(P1) From the definition of the reset map R_h^ε

$$\psi(c_i(t_i))_1 = \varepsilon\psi(c_{i-1}(t_i))_2.$$

(P2) From the proof of Thm. 2 in [15] (from combining equations (4) and (5) and using (P1)), the Zeno time has an upper bound:

$$t_\infty < \frac{\varepsilon}{1 - \varepsilon}\psi(x_0)_2 \frac{1}{|\alpha|}$$

where $\alpha = \max_{x \in U} d\psi(x)_1 f_L(x)$.

(P3) Again from the proof of Thm. 2 in [15] (specifically by combining equations (5) and (6) with (P1))

$$\sum_{i=1}^\infty \psi(c_i(t_{i+1}))_2 \leq \frac{\varepsilon}{1 - \varepsilon}\psi(x_0)_2 \left|\frac{\beta}{\alpha}\right|$$

where $\beta = \max_{x \in U} d\psi(x)_2 f_L(x)$.

These properties will be utilized in proving the first result of this paper, but first some set-up is necessary.

Zeno points for plastic and non-plastic impacts. Let $\mathscr{SH}_{\mathbf{L}}^\varepsilon$ be a SLHS with a coefficient of restitution $\varepsilon > 0$. For $x_0 \in G_h$, if there is a Zeno execution χ with this point as its initial condition then it has a well-defined Zeno point $x_\infty^\varepsilon(x_0)$. We are interested in comparing this Zeno point with the point obtained by applying a perfectly plastic impact $x_\infty^0(x_0) = R_h^0(x_0)$, which is just the Zeno point of an execution of $\mathscr{SH}_{\mathbf{L}}^0$ with initial condition x_0. Necessarily, it follows that $x_\infty^0(x_0)$ is a Zeno equilibrium point.

We are interested in comparing $x_\infty^\varepsilon(x_0)$ and $x_\infty^0(x_0) = R_h^0(x_0)$. Intuitively, these two points should converge toward one another in a continuous fashion as $\varepsilon \to 0$. This is what the following proposition verifies.

Proposition 1. *Let $\mathscr{SH}_{\mathbf{L}}^\varepsilon$ be a simple Lagrangian hybrid system with a coefficient of restitution $\varepsilon > 0$. Let $x^* = (q^*, \dot{q}^*)$ be a Zeno equilibrium point of $\mathscr{SH}_{\mathbf{L}}^\varepsilon$ that is bounded-time locally Zeno stable, and let W, U and ϵ be as in Def. 6. If $x_0 = (q_0, \dot{q}_0) \in G_h \cap W$, then there exist positive constants A_1, A_2 and A_3 such that:*

$$\|x_\infty^\varepsilon(x_0) - x_\infty^0(x_0)\| \quad (11)$$
$$< \varepsilon\left(A_1 + \frac{1}{1-\varepsilon}A_2 + \frac{1+\varepsilon}{1-\varepsilon}A_3\right)|dh(q_0)\dot{q}_0|.$$

Note that the constants A_1, A_2 and A_3 in this lemma simply give bounds on the growth of the vector field and reset map over the region U. As a result, this lemma has a clear physical intuition. Basically the distance between the Zeno point for a plastic and non-plastic impact is determined by two main factors: the coefficient of restitution and the speed when the guard is reached, $|\dot{h}(q_0, \dot{q}_0)| = |dh(q_0)\dot{q}_0|$.

PROOF. Since $x_0 \in G_h$, it follows that $t_0 = t_1$ and $c_0(t_1) = c_0(t_0) = x_0$ and so by the definition of R_h^ε and ψ

$$\|c_1(t_1) - x_\infty^0(x_0)\| = \|R_h^\varepsilon(x_0) - R_h^0(x_0)\| \leq \varepsilon K\psi(x_0)_2$$

where

$$K = \max_{x = (q, \dot{q}) \in U} \frac{1}{2} \frac{\|M(q)^{-1}dh(q)^T\|}{dh(q)M(q)^{-1}dh(q)^T}.$$

More generally, for $x \in G_h$, again from the definition of R_h^ε and ψ,

$$||R_h(x) - x_\infty^0(x_0)|| \le ||x - x_\infty^0(x_0)|| + (1+\varepsilon)K\psi(x)_2.$$

Therefore, for $i \ge 2$,

$$||c_i(t_i) - x_\infty^0(x_0)|| \le$$
$$||c_{i-1}(t_i) - x_\infty^0(x_0)|| + (1+\varepsilon)K\psi(c_{i-1}(t_i))_2$$

and

$$||c_{i-1}(t_i) - x_\infty^0(x_0)||$$
$$= ||c_{i-1}(t_{i-1}) - x_\infty^0(x_0) + \int_{t_{i-1}}^{t_i} f_L(x(\tau))d\tau||$$
$$\le ||c_{i-1}(t_{i-1}) - x_\infty^0(x_0)|| + F(t_i - t_{i-1})$$

where $F = \max_{x \in U} ||f_L(x)||$. It follows that

$$||c_i(t_i) - x_\infty^0(x_0)|| \le ||c_{i-1}(t_{i-1}) - x_\infty^0(x_0)||$$
$$+ F(t_i - t_{i-1}) + (1+\varepsilon)K\psi(c_{i-1}(t_i))_2.$$

By induction, or through simple iteration, we thus have that

$$||c_i(t_i) - x_\infty^0(x_0)|| \le \varepsilon K\psi(x_0)_2$$
$$+ F\sum_{j=1}^{i-1}(t_{j+1} - t_j) + (1+\varepsilon)K\sum_{j=1}^{i-1}\psi(c_j(t_{j+1}))_2.$$

It follows from (P1), (P2) and (P3), together with the fact that $t_0 = t_1$, that

$$||x_\infty^\varepsilon(x_0) - x_\infty^0(x_0)|| = \lim_{i \to \infty} ||c_i(t_i) - x_\infty^0(x_0)||$$
$$\le \varepsilon K\psi(x_0)_2$$
$$+ F\sum_{j=1}^{\infty}(t_{j+1} - t_j) + (1+\varepsilon)K\sum_{j=1}^{\infty}\psi(c_j(t_{j+1}))_2$$
$$\le \varepsilon K\psi(x_0)_2$$
$$+ F\frac{\varepsilon}{1-\varepsilon}\psi(x_0)_2\frac{1}{|\alpha|} + (1+\varepsilon)K\frac{\varepsilon}{1-\varepsilon}\psi(x_0)_2\left|\frac{\beta}{\alpha}\right|$$

Since $\psi(x_0)_2 = 2|dh(q_0)\dot{q}_0|$ by definition, picking

$$A_1 = 2K, \qquad A_2 = 2F\frac{1}{|\alpha|}, \qquad A_3 = 2K\left|\frac{\beta}{\alpha}\right|$$

gives the desired result. $\square$

4. COMPLETED HYBRID SYSTEMS & ZENO PERIODIC ORBITS

Completed hybrid systems allow Zeno executions to be carried past the Zeno point. Let $\mathscr{SH}_\mathbf{L}$ be a SLHS then, as the execution converges toward the Zeno point, $h \to 0$. This implies that after the Zeno point is reached, there should be a switch to a holonomically constrained dynamical system with holonomic constraint $\eta = h$. Let $\mathscr{D}_h = (\mathcal{Z}, f_L^h)$ be the dynamical system obtained from this unilateral constraint as in (5) with $\mathcal{Z}$ the set in (10).

Traditionally, completed hybrid systems have been defined in the following manner [3, 4, 18, 22, 23, 24] (and are often termed complementary Lagrangian hybrid systems): if $\mathbf{L}$ is a simple hybrid Lagrangian and $\mathscr{SH}_\mathbf{L}$ the corresponding

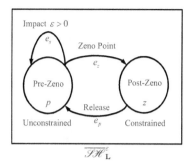

Figure 1: A graphical representation of a SHS and its associated completed hybrid system.

SLHS, the corresponding *completed Lagrangian hybrid system*[4] is:

$$\overline{\mathscr{SH}}_\mathbf{L} := \begin{cases} \mathscr{D}_h & \text{if} & h(q) = 0, \ dh(q)\dot{q} = 0, \\ & & \text{and } \lambda(q, \dot{q}) > 0 \\ \mathscr{SH}_\mathbf{L} & & \text{otherwise} \end{cases}$$

where λ is the Lagrange multiplier obtained from h. Systems of this form have been well-studied in the above references, and conditions have been given on how to practically simulate completed hybrid systems by truncating the executions in a formal manner (see [23, 24]).

While the notion of a completed hybrid system has proven very useful, we wish to extend it to include the possibility of unilateral constraints in the "post-Zeno" domain $\mathscr{D}_h$ that would cause the lagrange multiplier to switch sign (causing a switch back to the pre-Zeno domain). The second consideration is that we want this more general definition of a completed hybrid system to include as a special case the previous definition while simultaneously being able to model physical situations that occur with bipeds.

With this in mind, we consider the following definition of a *generalized completed hybrid system* (see Fig. 1).

Definition 7. *Let $\mathscr{SH}_\mathbf{L}$ be a SLHS associated to a hybrid Lagrangian $\mathbf{L} = (Q, L, h)$. A completed SLHS[5] is a tuple:* $\overline{\mathscr{SH}}_\mathbf{L}^\varepsilon := (\overline{\Gamma}, \overline{D}, \overline{G}, \overline{R}, \overline{F})$, *where*

- $\overline{\Gamma} = \{(p, z), e_s = (p, p), e_z = (p, z), e_p = (z, p)\}$,

- $\overline{D} = \{\overline{D}_p, \overline{D}_z\}$ where $\overline{D}_p = D_h$ and $\overline{D}_z \subset \mathcal{Z}$ satisfying:

$$\lambda(q, \dot{q}) \ge 0 \text{ if } (q, \dot{q}) \in \overline{D}_z, \tag{12}$$

- $\overline{G} = \{\overline{G}_{e_s}, \overline{G}_{e_z}, \overline{G}_{e_p}\}$ where $\overline{G}_{e_s} = G_h \backslash \mathcal{Z}$, $\overline{G}_{e_z} = \mathcal{Z}$ and $\overline{G}_{e_p} \subset \overline{D}_z$,

- $\overline{R} = \{\overline{R}_{e_s}, \overline{R}_{e_z}, \overline{R}_{e_p}\}$ where $\overline{R}_{e_s} = R_h$ (which depends on the coefficient of restitution ε), $\overline{R}_{e_z} = I$ and $\overline{R}_{e_p} : \overline{G}_{e_p} \to D_p$ satisfying:

$$\lambda(\overline{R}_{e_p}(q, \dot{q})) \le 0 \text{ for } (q, \dot{q}) \in \overline{G}_{e_p}, \tag{13}$$

- $\overline{F} = \{\overline{f}_p, \overline{f}_z\}$ where $\overline{f}_p = f_L$ and $\overline{f}_z = f_L^h$.

---

[4]As was originally pointed out in [3], this terminology (and notation) is borrowed from topology, where a metric space can be completed to ensure that "limits exist."

[5]We make the dependence of the system on the coefficient of restitution ε explicit since it is the main object of interest.

Note that the guard that forces a transition from the "pre-Zeno" domain $\overline{D}_p$ to the "post-Zeno" domain $\overline{D}_z$ is just the set of Zeno equilibria for the simple hybrid system $\mathscr{SH}_L$, i.e., a transition only occurs at the limit point of a Zeno execution. Also note that conditions (12) and (13) are consistency conditions that ensure that the constraint force has the right sign through the "post-Zeno" domain, and that when a transition back to the "pre-Zeno" domain occurs the constraint will no longer be enforced. Finally note that the traditional notion of completion is just a special case of Def. 7 with:

$$\overline{D}_z = \{(q, \dot{q}) \in \mathcal{Z} : \lambda(q, \dot{q}) \geq 0\},$$

(the largest domain satisfying (12)),

$$\overline{G}_{e_p} = \{(q, \dot{q}) \in \overline{D}_z : \lambda(q, \dot{q}) = 0\},$$

and $\overline{R}_{e_p} = I$ (which therefore satisfies (13)).

To better understand the motivation for the definition of a generalized completed hybrid system, and why the traditional definition must be extended, the specific application of bipedal robots must be considered. In this case, the traditional case where the solution switches back to the post-Zeno domain with an identity reset map would not be consistent with the bipedal robotic model. In particular, in the case of a biped the reset map $\overline{R}_{e_p}$ is the reset map associated with foot impact (see Fig. 4 for a graphical representation). The specifics of how a generalized completed hybrid system is obtained in bipedal walking will be presented in detail in Sec. 6, but before studying these systems in the context of walking, their properties must be studied.

Zeno points in completed systems. If $\overline{\mathscr{SH}}_L^\varepsilon$ is a completed SLHS, then for executions with initial conditions in the pre-Zeno domain we again can consider the Zeno point of these executions in the case when the coefficient of restitution $\varepsilon > 0$. Specifically, let $\overline{\chi} = (\Lambda, \mathcal{I}, \rho, \mathcal{C})$ be an execution of $\overline{\mathscr{SH}}_L^\varepsilon$ with $c_0(t_0) \in \overline{D}_p$. By the definition of the completed system and because $\varepsilon > 0$, the solution is Zeno and will never leave the pre-Zeno domain $\overline{D}_p$. Therefore, for this execution $\rho(i) \equiv p$. The Zeno point is thus given in (9) as in the case of simple hybrid systems.

In the case when $\varepsilon = 0$, the completed system $\overline{\mathscr{SH}}_L^0$ is "instantaneously Zeno." That is, the system displays the following behavior: when the guard $\overline{G}_{e_z}$ is reached, there is a perfectly plastic impact $\overline{R}_{e_z}$ which causes the execution to land directly on the guard $\overline{G}_{e_z}$ and thus there is an instantaneous transition to the post-Zeno domain $\overline{D}_z$. This will imply, for example, that periodic orbits for systems of this form are 3-periodic.

Periodic orbits of completed systems. Let $\overline{\mathscr{SH}}_L^\varepsilon$ be a completed SLHS.

In the special case of plastic impacts ($\varepsilon = 0$), a *plastic periodic orbit* is an execution $\overline{\chi}$ of $\overline{\mathscr{SH}}_L^0$ with initial condition $x^* = c_0(t_0) \in \overline{D}_z$ satisfying:

- $\Lambda = \mathbb{N}$,

- $\lim_{i \to \infty} t_i - t_0 = \infty$,

- $\rho(i) = \begin{cases} z & \text{if} \quad i = 0, 3, 6, 9, \ldots \\ p & \text{if} \quad i = 1, 2, 4, 5, \ldots \end{cases}$,

- $c_{3i}(t_{3i}) = c_{3(i+1)}(t_{3(i+1)})$.

The *period* of the orbit is $T = t_3 - t_0$.

For non-plastic impacts ($\varepsilon > 0$), a *Zeno periodic orbit* is an execution $\overline{\chi}$ of $\overline{\mathscr{SH}}_L^\varepsilon$ with initial condition $x^* = c_0(t_0) \in \overline{D}_z$ satisfying:

- $\Lambda = \mathbb{N}$,

- $\lim_{i \to \infty} t_i - t_0 = t_\infty < \infty$,

- $\rho(0) = z$ and $\rho(i) \equiv p$ for all $i \geq 1$,

- $x_\infty = \lim_{i \to \infty} c_i(t_i) = c_0(t_0) = x^*$.

The *period* of the orbit is $T = t_\infty$.

Stability of hybrid periodic orbits. We now define the stability of plastic periodic orbits. Note that we also could define the stability of Zeno periodic orbits, but as this definition is sufficiently more complicated and not necessary to the results presented here, we restrict our attention to the definition of the stability of plastic periodic orbits.

Definition 8. *A plastic periodic orbit $\overline{\chi}$ of $\overline{\mathscr{SH}}_L^0$ with initial condition $x^* \in \overline{D}_z \subset \mathcal{Z}$ is locally exponentially stable if there exists a neighborhood $X \subset \overline{D}_z$ of x^* and positive constants M and $\mu \in (0, 1)$ such that for any initial condition $x_0 = c_0(t_0) \in X$, the execution $\overline{\chi}'$ with this initial condition satisfies $\|c'_{3k}(t_{3k}) - x^*\| \leq M\|x_0 - x^*\|\mu^k$ for $k \in \mathbb{N}$.*

4.1 Main Result

The main result of this paper is that if there exists an exponentially stable plastic periodic orbit, then there exist Zeno periodic orbits for small coefficients of restitution. It is important to note that this result is in the spirit of [22] with three major differences: (1) it is more general in that we do *not* require $M \leq 1$ as was the case in [22] allowing it to be applied to bipedal robotic controllers which usually do not satisfy the assumption that $M \leq 1$, (2) the conditions of the theorem are easier to verify from a computational perspective (again, important for bipedal robots), and (3) the techniques used to prove the result are fundamentally different.

Theorem 2. *Let $\overline{\mathscr{SH}}_L^0$ have a plastic periodic orbit $\overline{\chi}$ with initial condition $x^* = (q^*, \dot{q}^*) \in \overline{D}_z \subset \mathcal{Z}$ that is locally exponentially stable. If $\ddot{h}(q^*, \dot{q}^*) < 0$, then there exists a positive constant r such that for any coefficient of restitution $0 < \varepsilon < r$ there exists a Zeno periodic orbit of $\overline{\mathscr{SH}}_L^\varepsilon$.*

The proof of this theorem will rely extensively on Poincaré maps associated to periodic orbits in hybrid systems (space constraints prevent a detailed introduction, but a complete definition can be found in [28]). For a completed hybrid system, let $\varphi_\tau^i(x) = \varphi^i(\tau^i(x), x)$, for $i = p, z$, be the flow associated to the vector field $\overline{f}_i$, where here $\tau^i : \overline{R}_{e_i}(\overline{G}_{e_i}) \to \mathbb{R}$ is the *time to impact function* which gives the time it takes to reach a guard from the image of another guard.

For a plastic periodic orbit, the Poincaré map is the partial function given by: $P : \mathcal{Z} = \overline{G}_{e_z} \to \mathcal{Z}$, where $P(x) = \overline{R}_{e_z}(\varphi_\tau^p(\overline{R}_{e_p}(\varphi_\tau^z(x))))$. The fixed point of the Poincareé map, $x^* = P(x^*)$ is just the point at which the periodic orbit intersects the surface $\mathcal{Z}$, i.e., it is simply the point x^* given in Def. 8. As with smooth dynamical systems [25], the (exponential) stability of a plastic periodic orbit (or periodic orbits in hybrid systems in general [20]) is equivalent to the (exponential) stability of the discrete-time dynamical system obtained through the Poincaré map, $x_{k+1} = P(x_k)$, at the equilibrium point x^*. This can be best understood by noting that in Def. 8 $c'_{3k}(t_{3k}) = P(c'_{3(k-1)}(t_{3(k-1)}))$.

PROOF. Since $x^* \in \overline{D}_z$, it is by definition a Zeno equilibrium point, and because $\ddot{h}(q^*, \dot{q}^*) < 0$ it is bounded-time locally Zeno stable by Thm. 1. Let W and U be the neighborhoods in G_h given in Def. 6. Note that here we implicitly assume (without loss of much generality) that both W and U do not intersect $\overline{G}_{e_p}$ (this condition can be guaranteed as long as x^* is not in $\overline{G}_{e_p}$ and the coefficient of restitution is picked to be sufficiently small); that is, we assume that the Zeno point is reached before foot-strike, or that knee-lock occurs before foot-strike. In addition, let X be the neighborhood of x^* in $\mathcal{Z}$ given in Def. 8 which exists due to the assumption of local exponential stability of the plastic periodic orbit. Note that we can suppose[6] that for any execution $\overline{\chi}^0$ or $\overline{\chi}^\varepsilon$ of $\overline{\mathscr{SH}}_{\mathbf{L}}^0$ or $\overline{\mathscr{SH}}_{\mathbf{L}}^\varepsilon$, respectively, with initial condition $x_0 \in X$, it follows that $c_1^0(t_2) = c_1^\varepsilon(t_2) \in W$. Finally, for the sake of notational simplicity, assume that $x^* = 0$.

Since $\overline{\chi}$ is a locally exponentially stable plastic periodic orbit, P is locally exponentially stable at 0, and because P is smooth (since it is given by composing smooth functions), it follows by the converse Lyapunov theorem for discrete-time dynamical systems (see [11]) that there exists a function $V : X \to \mathbb{R}$ satisfying

$$c_1\|x\|^2 \leq V(x) \leq c_2\|x\|^2 \tag{14}$$

$$\Delta V(x) = V(P(x)) - V(x) \leq -c_3\|x\|^2 \tag{15}$$

$$|V(x) - V(y)| \leq c_4\|x - y\|(\|x\| + \|y\|) \tag{16}$$

for all $x, y \in X$ and positive constants c_1, c_2, c_3 and c_4. Let $\alpha = \min_{x \in \partial X} V(x)$, and take $\beta \in (0, \alpha)$. Then the set $\Omega_\beta = \{x \in X : V(x) \leq \beta\}$ is in the interior of X and is invariant under P since $\Delta V(x) < 0$, i.e., if $x_0 \in \Omega_\beta$ then $P(x_0) \in \Omega_\beta$.

The goal is to show that if an execution of $\overline{\mathscr{SH}}_{\mathbf{L}}^\varepsilon$ has an initial condition $x_0 \in \Omega_\beta$ then the Zeno point $x_\infty^\varepsilon(x_0) \in \Omega_\beta$ for a sufficiently small coefficient of restitution. To see this, let $y_0 = c_1^0(t_2) = c_1^\varepsilon(t_2)$ wherein it follows that $x_\infty^\varepsilon(x_0) = x_\infty^\varepsilon(y_0)$ and $P(x_0) = x_\infty^0(y_0)$. From (15) and (16),

$$
\begin{aligned}
&V(x_\infty^\varepsilon(x_0)) - V(x_0) \\
&= V(x_\infty^\varepsilon(x_0)) - V(P(x_0)) + V(P(x_0)) - V(x_0) \\
&\leq |V(x_\infty^\varepsilon(x_0)) - V(P(x_0))| + V(P(x_0)) - V(x_0) \\
&\leq c_4\|x_\infty^\varepsilon(y_0) - x_\infty^0(y_0)\|(\|x_\infty^\varepsilon(y_0)\| + \|x_\infty^\varepsilon(y_0)\|) \\
&\qquad\qquad\qquad\qquad -c_3\|x_0\|^2.
\end{aligned}
$$

Now, by Prop. 1, it follows that $\|x_\infty^\varepsilon(y_0) - x_\infty^0(y_0)\|$ goes to zero continuously as $\varepsilon \to 0$. As a result, for all $0 < \varepsilon < r$ (with r sufficiently small), $V(x_\infty^\varepsilon(x_0)) - V(x_0) \leq 0$. Therefore, $x_0 \in \Omega_\beta$ implies that $V(x_0) \leq \beta$ and $V(x_\infty^\varepsilon(x_0)) \leq V(x_0) \leq \beta$ so $x_\infty^\varepsilon(x_0) \in \Omega_\beta$.

We have established that Ω_β is invariant under the continuous[7] map $x_\infty^\varepsilon : \Omega_\beta \to \Omega_\beta$ associating to a point its Zeno point. By the fixed point theorem [10], there exists a fixed point x_z such that $x_\infty^\varepsilon(x_z) = x_z$. By the definition of Zeno periodic orbits, this implies the existence of such an orbit, i.e., the execution of $\overline{\mathscr{SH}}_{\mathbf{L}}^\varepsilon$ with initial condition x_z. $\square$

---

[6]This supposition can be enforced by considering a subset of X if needed.

[7]Continuity is guaranteed as long as the neighborhood X does not intersect the guard $\overline{G}_{e_p}$; space constraints prevent a proof of this fact, but the reasoning is similar to the justification of continuity for hybrid systems with a single constraint [5].

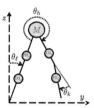 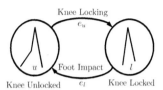

Figure 2: The configuration space of the 2D biped with knees (left) and a graphical representation of the domains of the hybrid system $\mathscr{H}^{\mathrm{B}}$ (right).

5. BIPEDAL MODEL WITH KNEE-LOCK

We begin by studying the case when a bipedal robot has knees that lock, i.e., knees in which the impact is perfectly plastic; this could be achieved physically by, for example, using knees with mechanical catches [17]. This section introduces the hybrid model for this system along with controllers that result in stable walking. The specific control laws that will be used are based upon the idea of *controlled symmetries* which mimic the gait of passive walkers walking down shallow slopes [16, 12, 6] by shaping the potential energy of the system [26]. The biped that will be considered has been studied in both 2D and 3D in [2].

The model of interest is a controlled bipedal robot with knees that walks on flat ground for which we will explicitly construct the hybrid system:

$$\mathscr{H}^{\mathrm{B}} = (\Gamma^{\mathrm{B}}, D^{\mathrm{B}}, G^{\mathrm{B}}, R^{\mathrm{B}}, F^{\mathrm{B}}).$$

We now introduce the elements of this hybrid system in a step-by-step fashion.

Discrete Structure The discrete structure for the model is given by $\Gamma^{\mathrm{B}} = (\{u, l\}, \{e_u = (u, l), e_l = (l, u)\})$. That is, there are two domains u, l and two edges e_u, e_l (see Fig. 2). In the first domain the biped's non-stance knee is unlocked and in the second domain the biped's knee is locked. Transitions occur from domain u to domain l when the knee locks, and from l to u when the foot strikes the ground. Note that the discrete structure of this model enforces temporal ordering to events (kneelock and footstrike) and this discrete structure implies that $D^{\mathrm{B}} = \{D_u^{\mathrm{B}}, D_l^{\mathrm{B}}\}$, $G^{\mathrm{B}} = \{G_{e_u}^{\mathrm{B}}, G_{e_l}^{\mathrm{B}}\}$, $R^{\mathrm{B}} = \{R_{e_u}^{\mathrm{B}}, R_{e_l}^{\mathrm{B}}\}$ and $F^{\mathrm{B}} = \{f_u^{\mathrm{B}}, f_l^{\mathrm{B}}\}$.

Configuration space and Lagrangian. Consider the configuration $Q^{\mathrm{B}} = \mathbb{R}^3$ with coordinates $q = (\theta_l, \theta_h, \theta_k)$, where θ_l is the angle of the leg from vertical, θ_h is the angle of the hip and θ_k is the angle of the knee (see Fig. 2). The Lagrangian for this system is of the form:

$$L^{\mathrm{B}}(q, \dot{q}) = \frac{1}{2}\dot{q}^T M^{\mathrm{B}}(q)\dot{q} - V^{\mathrm{B}}(q)$$

which is computed in the standard way.

Unilateral constraints. We will consider two unilateral constraints for this system. The first unilateral constraint enforces the knee being unlocked. It is therefore given by: $h_u^{\mathrm{B}}(q) = \theta_k$. The second unilateral constraint is the constraint that the foot is above the ground, and it is therefore given by:

$$h_l^{\mathrm{B}}(q) = 2\ell \cos\left(-\theta_l - \frac{1}{2}\theta_h\right)\cos\left(\frac{1}{2}\theta_h\right),$$

with ℓ the length of the leg.

Figure 3: A walking gait of the 2D biped with plastic impacts at the knee.

Domain 1 (knee unlocked). The domain D_u^B is obtained as in (7) from h_u^B. The vector field f_u^B is obtained as in (4) from the Lagrangian L^B and the feedback control law:

$$\Upsilon^B(q, \dot{q}) = \frac{\partial V^B}{\partial q}(q) - \frac{\partial V^B}{\partial q}(q + (\gamma, 0, 0)^T)$$

where γ is a control gain that can be viewed as the "slope" that would yield walking for passive biped walking down a slope of γ radians. Note that applying this control law implies that f_u^B is just the dynamical system associated with the Lagrangian:

$$L_\gamma^B(q, \dot{q}) = \frac{1}{2}\dot{q}^T M^B(q)\dot{q} - V^B(q + (\gamma, 0, 0)^T)$$

with no feedback control law. Also note that this control law uses full actuation at all joints, including the knee.

Domain 2 (knee locked). The domain D_l^B is obtained as in (7) from h_l^B. The vector field f_l^B is obtained by imposing the holonomic constraint $\eta = h_l^B$ in f_u^B (as outlined in (6)); this enforces the condition that the knee is locked in this domain. In particular, the control law in this domain is again the controlled symmetries control law enforced in Domain 1 since we use the vector field f_u^B (which includes this control law) to obtain the constrained dynamics on this domain. This implies that there is again a torque at the knee.

Edge 1 (knee locking). The guard $G_{e_u}^B$ is obtained as in (7) from h_u^B. The reset map $R_{e_u}^B$ is also obtained from h_u^B as in (8). Note that for this model we assume the knee impact is *perfectly plastic*, i.e., $\varepsilon = 0$. The rest of this paper will be devoted to understanding what happens if this assumption is not satisfied.

Edge 2 (Foot impact). The guard $G_{e_l}^B$ is obtained as in (7) from h_l^B. The reset map $R_{e_l}^B$ models a perfectly plastic impact at the foot which also relabels the stance and non-stance leg to account for the two legs switching roles. This is obtained through the same process outlined in [9], but space constraints prevent the inclusion of this equation.

Walking gait. For the model under consideration: $m_c = 0.05$kg, $m_t = 0.5$kg, $M_h = 0.5$kg, $\ell = 1$m, $r_c = 0.372$m, $r_t = 0.175$m. For a control gain of $\gamma = 0.0504$, the result is stable walking, i.e., there is a stable periodic orbit (see Fig. 5). For the purpose of illustration, the walking gait for these parameters and control gain can be seen in Fig. 3.

6. BIPEDAL MODEL WITH KNEE-BOUNCE

We now consider the case when the assumption of perfectly plastic impacts at the knee does not hold, i.e., the

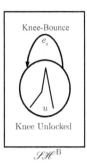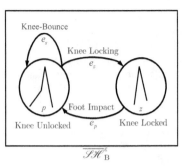

Figure 4: A graphical representation of (a) the simple hybrid system modeling knee-strike with a non-plastic impact, i.e., knee-bounce, and (b) the completed hybrid system in which the knee locks after the Zeno point is reached.

case where there is knee-bounce. Relaxing this assumption implies that the transition to domain 2 (knee locked) never formally takes place since this would involve an infinite number of discrete jumps, i.e., there is Zeno behavior. Therefore, relaxing this assumption results in a completely different hybrid system.

Non-plastic impacts at the Knee. We now relax the assumption that $\varepsilon = 0$ for the reset map obtained from the impact equations for knee impact.

Consider the hybrid Lagrangian $\mathbf{L}_u^B = (Q^B, L^B, h_u^B)$ with Q^B, L^B and h_u^B defined as in Sect. 5. From this hybrid Lagrangian we obtain a simple Lagrangian hybrid system

$$\mathscr{SH}^B = (D_u^B, G_{e_u}^B, R_{h_u^B}, f_u^B), \tag{17}$$

where D_u^B, $G_{e_u}^B$ and f_u^B are the same continuous domain, guard and vector field for Domain 1 and Edge 1 of the hybrid system $\mathscr{H}^B$ given in Sect. 5. The reset map $R_{h_u^B}$ is obtained from h_u^B as given in (8) where now $0 < \varepsilon < 1$, i.e., it is a non-plastic impact (that also is not allowed to be perfectly elastic). See Fig. 4(a) for a graphical representation of this model, where the discrete transition occurs at *knee strike* not *knee lock*, i.e., the knee never locks because the impacts are non-plastic.

Zeno Behavior. Due to the non-plastic impacts, the hybrid system $\mathscr{SH}^B$ is Zeno (by Thm. 1).

First, due the simple form of the unilateral constraint function on this domain ($h_u^B(q) = \theta_k$), the set of Zeno equilibra is $\mathcal{Z}^B = \{(q, \dot{q}) \in \mathbb{R}^6 : \theta_k = 0 \text{ and } \dot{\theta}_k = 0\}$. That is, the Zeno equilibria are the set of points such that the knee angle is zero with zero velocity, i.e., the set of points where the leg is straight.

To check for Zeno behavior, it is necessary to consider $\ddot{h}_u^B$, which in this case is given by: $\ddot{h}_u^B(q, \dot{q}) = (f_u^B(q, \dot{q}))_6$. Note that this is a complex function, and so it is not possible to give a simple characterization of the points $(q^*, \dot{q}^*) \in \mathcal{Z}^B$ such that $\ddot{h}_u^B(q^*, \dot{q}^*) < 0$.

Completion of bipedal model. From the hybrid system $\mathscr{SH}^B$ we obtain a completed hybrid system $\overline{\mathscr{SH}}_B^\varepsilon$, where ε is the coeficent of resolution. This is given by "combining" the hybrid system $\mathscr{H}^B$ given in Sect. 5 and the simple hybrid system $\mathscr{SH}^B$ (Fig. 4(b)). Let $\overline{\mathscr{SH}}_B^\varepsilon = (\overline{\Gamma}, \overline{D}, \overline{G}, \overline{R}, \overline{F})$, where $\overline{\Gamma}$ is given as in Def. 7 and

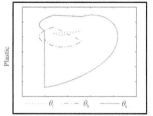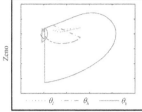

Figure 5: The periodic orbit associated to the walking gait for $\mathscr{H}^{\mathrm{B}}$, i.e., walking with knee-lock, which is equivalent to the plastic periodic orbit for $\overline{\mathscr{S\!H}}^{0}_{\mathrm{B}}$ (left). The Zeno periodic orbit for $\overline{\mathscr{S\!H}}^{\varepsilon}_{\mathrm{B}}$ with $\varepsilon = 0.25$, i.e., walking with knee-bounce (right).

- $\overline{D} = \{\overline{D}_p, \overline{D}_z\}$ where $\overline{D}_p = D_u^{\mathrm{B}}$ and $\overline{D}_z = D_l^{\mathrm{B}}$,

- $\overline{G} = \{\overline{G}_{e_s}, \overline{G}_{e_z}, \overline{G}_{e_p}\}$ where $\overline{G}_{e_s} = G_{e_u}^{\mathrm{B}} \setminus \mathcal{Z}^{\mathrm{B}}$, $\overline{G}_{e_z} = \mathcal{Z}^{\mathrm{B}}$ and $\overline{G}_{e_p} = G_{e_l}^{\mathrm{B}}$,

- $\overline{R} = \{\overline{R}_{e_s}, \overline{R}_{e_z}, \overline{R}_{e_p}\}$ where $\overline{R}_{e_s} = R_{h_u^{\mathrm{B}}}$ (which depends on ε), $\overline{R}_{e_z} = I$ and $\overline{R}_{e_p} = R_{e_l}^{\mathrm{B}}$,

- $\overline{F} = \{\overline{f}_p, \overline{f}_z\}$ where $\overline{f}_p = f_u^{\mathrm{B}}$ and $\overline{f}_z = f_l^{\mathrm{B}}$.

Note that $\overline{\mathscr{S\!H}}^{0}_{\mathrm{B}}$ and $\mathscr{H}^{\mathrm{B}}$ have the same qualitative behavior although they have slightly different structures. That is, the completion of $\mathscr{S\!H}^{\mathrm{B}}$ when $\varepsilon = 0$ is just the hybrid system $\mathscr{H}^{\mathrm{B}}$. We are, of course, interested in what happens when the assumption that $\varepsilon = 0$ is relaxed for the biped and its effect on walking gaits.

Application of Theorem 2 and Simulation Results.
We now apply Thm. 2 to the completed hybrid system modeling the biped with non-plastic impacts at the knee $\overline{\mathscr{S\!H}}^{\varepsilon}_{\mathrm{B}}$ to show that a plastic periodic orbit for $\varepsilon = 0$ implies the existence of a Zeno periodic orbit for $\varepsilon > 0$, i.e., that walking with knee-lock implies walking with knee-bounce when the knee-bounce is sufficiently small.

Due to the equivalence of $\overline{\mathscr{S\!H}}^{0}_{\mathrm{B}}$ and $\mathscr{H}^{\mathrm{B}}$, and since there was a periodic orbit for $\mathscr{H}^{\mathrm{B}}$, there is a plastic periodic orbit for $\overline{\mathscr{S\!H}}^{0}_{\mathrm{B}}$ as pictured in Fig. 5. The exponential stability of this control law can be verified by considering the Poincaré map; the exponential stability of this map implies the exponential stability of the plastic periodic orbit. Moreover, the exponential stability of the Poincaré map can be verified by considering the eigenvalues of its linearization and ensuring that none have magnitude greater than 1. In this case, the largest eigenvalue has magnitude 0.7329 indicating exponential stability of the plastic periodic orbit. Finally, the value of $\ddot{h}$ at the Zeno equilibria point is $\ddot{h}(x^*) = -50.135$. Therefore, the assumptions of Thm. 2 are satisfied.

As a result of Thm. 2, there exists a Zeno periodic orbit for $\overline{\mathscr{S\!H}}^{\varepsilon}_{\mathrm{B}}$ for a range of coefficients of restitution $0 < \varepsilon \leq r$. Of course, there is not an explicit value for r stated in the theorem, but we were able to find a rather large range of coefficients of restitution resulting in Zeno periodic orbits. One of these orbits can be seen in Fig. 5 for $\varepsilon = 0.25$. The effect of Zeno behavior on the biped can be clearly seen in this figure due to the "bouncing" behavior of θ_k, i.e., the Zeno periodic orbit clearly displays knee-bounce. The effect of knee-bounce on the behavior of the biped can be better seen when comparing the positions and velocities of the knee

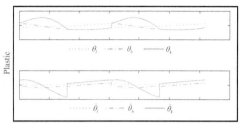
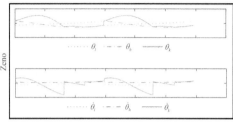

Figure 6: The positions and velocities over time for the walking gait in the case of plastic impacts (left) and non-plastic impacts (right) at the knee.

over time in the case of a zero and nonzero coefficient of restitution as seen in Fig. 6.

It is important to note that, although Thm. 2 implies the existence of a Zeno periodic orbit, it does not give any guarantees on the stability of this orbit. While a formal result of this nature would be interesting, for all practical purposes the stability of the Zeno periodic orbit can be checked the same way that the stability of the plastic periodic orbit was checked: by numerically computing the eigenvalues of the linearization of the Poincaré map. In the case of the Zeno periodic orbit in Fig. 5 we find that the largest eigenvalue has magnitude 0.2245, implying that we in fact get a stable walking gait in the case of knee-bounce.

Videos of the walking gait with both knee-lock and knee-bounce can be found at [1]. It is interesting to compare these walking gaits with the behavior of the passive dynamic walking with knees by McGeer (videos of this can also be found at [1]). In the case of knee lock, it can be seen that the behavior of the simulated and actual robotic walking are very similar. In the case of knee-bounce, we postulate that the McGeer biped falls due to there being a larger coefficient of restitution associated with knee-bounce than is the case with the simulated biped with knee-bounce, i.e., we found similar behavior in the simulated system when ε was taken to be larger than about 0.4. Yet, despite the differences in the coefficient of restitution between the physical and simulated system, the similarity between knee-bounce in simulation and reality is quite remarkable. This indicates that models with Zeno behavior can effectively simulate real physical systems in order to say useful things about the behavior of these systems.

7. CONCLUSION

Motivated by the issue of knee-bounce in bipedal robotic walking, this paper shows that knee-bounce may not always negatively affect the stability of bipedal walking as long as the bounce is kept sufficiently small. This is demonstrated through the observation that knee-bounce in walking is just an example of Zeno behavior in hybrid systems. Conditions on when Zeno behavior exists are used to characterize the difference between orbits in hybrid systems with plastic and non-plastic impacts. With this in hand, the notion of generalized completion of a hybrid system is introduced, extending the traditional notion of completion to a setting that allows bipedal robots to be modeled with this formalism. The main result of this paper is that if a plastic periodic orbit is stable (the biped has stable walking with knee-lock) then under easily verifiable conditions a Zeno periodic orbit exists (the biped has stable walking with knee-bounce as long as the knee-bounce is sufficiently small). These results are applied to a specific example of a bipedal robot with knees, and walking gaits are presented in the case of both knee-lock and knee-bounce.

Since this paper considered Zeno behavior that occurs at the knee, the natural question to ask is: what happens if Zeno behavior occurs at the foot, or the foot and knee simultaneously? Addressing this question will be surprisingly difficult due to the differences between the impact equations at the knee and foot. At the knee, the impacts are a result of unilateral constraints, and impacts related to these types of constraints have been well-studied; it was by building upon previous results from the author and other researchers that the main results of these paper were able to be shown. This preexisting work was non-trivial, taking years to establish. The first step in extending this work to more interesting types of impacts, such as those that occur at the feet, is to study Zeno behavior in the context of these impacts.

Acknoledgements. The author would like to thank Andy Lamperski and Yizhar Or for the many enlightening conversations on Zeno behavior, and the collaborations that laid the groundwork for this paper.

8. REFERENCES

[1] Videos of bipedal robotic walking with knee-bounce and knee-lock.
http://www.youtube.com/user/ProfAmes.

[2] A. D. Ames, R. W. Sinnet, and E. D. B. Wendel. Three-dimensional kneed bipedal walking: A hybrid geometric approach. In *HSCC*, LNCS, pages 16–30. Springer-Verlag, 2009.

[3] A. D. Ames, H. Zheng, R. D. Gregg, and S. Sastry. Is there life after Zeno? Taking executions past the breaking (Zeno) point. In *Proc. American Control Conference*, pages 2652 – 2657, 2006.

[4] J. M. Bourgeot and B. Brogliato. Asymptotic tracking of periodic trajectories for a simple mechanical systems subject to nonsmooth impacts. *IEEE Transactions on Automatic Control*, 46:1122–1126, 2001.

[5] B. Brogliato. *Nonsmooth Mechanics*. Springer-Verlag, 1999.

[6] S. H. Collins, M. Wisse, and A. Ruina. A 3-d passive dynamic walking robot with two legs and knees. *International Journal of Robotics Research*, 20:607–615, 2001.

[7] R. Goebel and A. R. Teel. Lyapunov characterization of Zeno behavior in hybrid systems. In *IEEE Conference on Decision and Control*, 2008.

[8] R. Goebel and A. R. Teel. Zeno behavior in homogeneous hybrid systems. In *IEEE Conference on Decision and Control*, 2008.

[9] J. Grizzle, G. Abba, and F. Plestan. Asymptotically stable walking for biped robots: Analysis via systems with impulse effects. *IEEE Transactions on Automatic Control*, 46(1):51–64, 2001.

[10] M. W. Hirsch. *Differential Topology*. Springer, 1980.

[11] H. K. Khalil. *Nonlinear Systems*. Prentice Hall, second edition, 1996.

[12] A. D. Kuo. Stabilization of lateral motion in passive dynamic walking. 18(9):917–930, 1999.

[13] A. Lamperski and A. D. Ames. On the existence of Zeno behavior in hybrid systems with non-isolated Zeno equilibria. In *IEEE Conference on Decision and Control*, 2008.

[14] A. Lamperski and A. D. Ames. Sufficient conditions for Zeno behavior in Lagrangian hybrid systems. In *HSCC*, volume 4981 of *LNCS*, pages 622–625. Springer Verlag, 2008.

[15] A. Lamperski and A. D. Ames. Local Zeno stability theory. Submitted for Publication, available upon request, 2010.

[16] T. McGeer. Passive dynamic walking. 9(2):62–82, 1990.

[17] T. McGeer. Passive walking with knees. In *IEEE International Conference on Robotics and Automation*, Cincinnati, OH, 1990.

[18] L. Menini and A. Tornambè. Tracking control of complementary Lagrangian hybrid systems. *International Journal of Bifurcation and Chaos*, 15(6):1839–1866, 2005.

[19] J. J. Moreau. Unilateral contact and dry friction in finite freedom dynamics. *Nonsmooth Mechanics and Applications, CISM Courses and Lectures*, 302, 1988.

[20] B. Morris and J. W. Grizzle. A restricted Poincaré map for determining exponentially stable periodic orbits in systems with impulse effects: Application to bipedal robots. In *44th IEEE Conference on Decision and Control and European Control Conference*, 2005.

[21] R. M. Murray, Z. Li, and S. S. Sastry. *A Mathematical Introduction to Robotic Manipulation*. Taylor & Francis/CRC, 1994.

[22] Y. Or and A. D. Ames. Existence of periodic orbits with Zeno behavior in completed Lagrangian hybrid systems. In *HSCC*, LNCS, pages 291–305. Springer-Verlag, 2009.

[23] Y. Or and A. D. Ames. Formal and practical completion of Lagrangian hybrid systems. In *ASME/IEEE American Control Conference*, 2009.

[24] Y. Or and A. D. Ames. Stability and completion of zeno equilibria in Lagrangian hybrid systems. In *Accepted for publication in IEEE Transactions on Automatic Control*, 2010.

[25] L. Perko. *Differential equations and dynamical systems*. Springer, 2001.

[26] M. W. Spong and F. Bullo. Controlled symmetries and passive walking. *IEEE Transactions on Automatic Control*, 50(7):1025–1031, 2005.

[27] W. J. Stronge. *Impact Mechanics*. Cambridge University Press, 2004.

[28] E. D. B. Wendel and A. D. Ames. Rank properties of Poincare maps for hybrid systems with applications to bipedal walking. In *HSCC*, 2010.

Impulsive Control for Nanopositioning: Stability and Performance

Tomas Tuma
IBM Research - Zurich
Säumerstrasse 4
8803 Rüschlikon, Switzerland
uma@zurich.ibm.com

Angeliki Pantazi
IBM Research - Zurich
Säumerstrasse 4
8803 Rüschlikon, Switzerland
agp@zurich.ibm.com

John Lygeros
Automatic Control Laboratory
Swiss Federal Institute of
Technology
8092 Zürich, Switzerland
lygeros@control.ee.ethz.ch

Abu Sebastian
IBM Research - Zurich
Säumerstrasse 4
8803 Rüschlikon, Switzerland
ase@zurich.ibm.com

ABSTRACT

In this paper, impulsive control is applied to a class of linear feedback systems and studied both theoretically and experimentally, with a particular focus on the usage in nanopositioning. By using impulsive control, improvements in tracking performance and tolerance to measurement noise can be achieved which are beyond the limits of conventional linear feedback. We derive sufficient conditions for bounded-input-bounded-state stability of the resulting hybrid systems, investigate their performance and unveil an inherent connection to the recently published signal transformation approach. It is demonstrated that for a triangular reference signal, impulsive control outperforms the signal transformation approach both in terms of performance and implementation complexity. Furthermore, we show that the measurement noise-induced positioning error of an impulsive feedback system is significantly lower than in case of a comparable, conventional linear feedback. The experimental results are obtained on a fast microelectromechanical nanopositioner.

Categories and Subject Descriptors

G.0 [**Mathematics of Computing**]: General; J.2 [**Computer applications**]: Physical sciences and engineering

General Terms

Design, Measurement, Performance, Theory

Keywords

hybrid systems, impulsive control, nanopositioning

1. INTRODUCTION

In practical applications, linear feedback control has been one of the most widely employed and understood control architectures. With the increasing performance requirements, however, the principal limitations of linear feedback may constrain the overall system. This is often the case in nanoscale motion control, such as in scanning probe microscopy [1] and nanoscale data storage [2], where the ultra-high precision sensors typically suffer from a significant level of measurement noise. When nanopositioning systems are controlled using simple linear feedback, the measurement noise affects the positioning accuracy increasingly with the feedback bandwidth. Consequently, the speed of operation has to be offset against the accuracy. Since both accuracy and operation speed are a critical design criterion, the incurred trade-off imposes a severe design limitation.

Some of the limitations in linear feedback control can be overcome by more advanced control strategies, such as two degree of freedom control [3], direct shaping of the controller noise transfer function [4] and various types of switching control [5] and nonlinear control. However, the price for their better performance is typically paid in high complexity, both at the design and implementation stage. In this paper, we focus on a structurally simple, non-linear approach that has recently appeared in the field of precise motion control. Impulsive control [6] is here used as a non-linear, discrete-time-based enhancement of linear continuous systems, and as such results in dynamics of inherently hybrid nature. Impulsive feedback systems feature a low complexity but require careful analysis of the dynamical behavior.

We analyze a particular type of impulsive control which can be used to improve tracking of piecewise affine signals, *impulsive state multiplication* (ISM). ISM multiplies the states of a controller by non-zero factors at discrete time instances. For example, if the tracked signal is a triangular waveform, the impulses are applied at the turnaround points of the triangle. In a more general setting, the timing and magnitude of the impulses correspond to the changes of slope or discontinuities in the reference signal. It was shown that with ISM, high bandwidth signals can be tracked without unnecessarily increasing the linear bandwidth of the feedback controller. This is possible because ISM decouples the controller from the rate of change in the reference signal.

This paper addresses some of the open questions in theory and practical application of impulsive control. In particular, we derive a bounded-input-bounded-state stability condition which can be used to choose a safe magnitude and frequency of impulses given the underlying linear system. Further, we explain the transient and steady-state behavior of an impulsive control loop. We reveal an inherent connection between ISM and the recently published signal transformation approach to nanopositioning (STA) [7] and show that for a triangular reference signal, ISM offers better transient performance and simpler implementation.

The paper is organized as follows. In Section 2, the concept of impulsive state multiplication is presented together with an impulsive control law for tracking of piecewise constant and piecewise affine signals. In Section 3, the relation between impulsive control and STA is derived. Section 4 contains experimental results and Section 5 concludes the paper.

2. IMPULSIVE CONTROL

In this section, we introduce ISM as a general concept and analyze its bounded-input-bounded-state stability. Next, impulsive control laws for tracking of piecewise constant and piecewise affine signals are presented. For the piecewise constant reference, a performance guarantee is proven for an ideal positioner.

2.1 Impulsive state multiplication

Impulsive state multiplication (ISM) is introduced for any linear, time invariant (LTI) system as follows:

DEFINITION 1 (IMPULSIVE STATE MULTIPLICATION). *Let* $\{t_i\}_{i=1}^{\infty}$ *be a sequence of time instants such that* $t_1 < t_2 <$ *... <* $t_i <$ *.... Let* $\{q_i\}_{i=1}^{\infty}$ *be a sequence of real coefficients, i.e.* $q_i \in \mathbb{R}$ *for* $i = 1, ..., n$. *Let* K *be a linear, time-invariant system. The following system is* K *with impulsive state multiplication,* $ISM(K, t, q)$:

$$\dot{x}_K(t) = A_K x_K(t) + B_K u(t) \text{ when } t \neq t_i, i = 1, 2, 3, ...$$

$$x_K(t_i) := q_i x_K(t_i^-) \text{ for } i = 1, 2, 3, ...$$

$$y(t) = C_K x_K(t)$$

where $t \in \mathbb{R}_+$, $x_K(t) \in \mathbb{R}^n$, $u(t) \in \mathbb{R}^m$, $y(t) \in \mathbb{R}^p$, $A_K \in \mathbb{R}^{n \times n}$, $B_K \in \mathbb{R}^{n \times m}$, $C_K \in \mathbb{R}^{p \times n}$.

In Definition 1, LTI system K is extended with an impulsive control law that is parametrized by the sequence of times and multiplication factors $\{t_i\}$ and $\{q_i\}$, respectively. The resulting hybrid system, $ISM(K, t, q)$, evolves in a continuous manner at all times except the discrete time instants $\{t_i\}$ when its state is multiplied by the corresponding factor from $\{q_i\}$. Extensions of this definition are possible, such as multiplication of a subset of the state, additional arithmetic operations, conditional application of the impulses etc.

In what follows, ISM will be analyzed as an extension of a linear, single-input-single-output feedback system. Fig. 1 depicts the resulting control architecture in which P is the positioner to be controlled and K is the feedback controller. The feedback controller generates the control signal u based on the tracking error e which is computed as the difference between the reference signal r and measured output y.

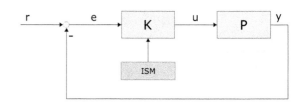

Figure 1: Feedback control loop with impulsive state multiplication of the controller states

2.2 Stability

Impulsive changes of the controller state can cause instability of a feedback loop which would be stable otherwise. A sufficient condition for bounded-input-bounded-state stability is presented.

THEOREM 1. *Let* $\{t_i\}_{i=1}^{\infty}$, $\{q_i\}_{i=1}^{\infty}$, *be sequences of time instants and corresponding state multiplication factors. Consider the system in Fig. 1 with input* $r(t)$ *and its state-space model*

$$\dot{x}(t) = Ax(t) + Br(t) \qquad when \ t \neq t_i$$

$$x(t_i) := \begin{bmatrix} q_i I & 0 \\ 0 & I \end{bmatrix} x(t_i^-)$$

$$y(t) = Cx(t)$$

where in $x(t) = \begin{bmatrix} x_K(t) & x_P(t) \end{bmatrix}^T$, x_K *is the state of controller* $ISM(K, t, q)$ *and* x_P *is the state of* P.

Assume that $0 < t_{i+1} - t_i < \Theta$ *and that reference* r *is bounded, i.e.* $\|r(t)\|_\infty < \infty$. *If there exists* $q < 1$ *such that*

$$\left\| e^{A(t_{i+1} - t_i)} \begin{bmatrix} q_i I & 0 \\ 0 & I \end{bmatrix} \right\| < q$$

for $i = 1, 2, 3...$ *then*

$$\|x(t)\|_\infty < \infty$$

PROOF. *Since* $t_{i+1} - t_i < \Theta$, *it is sufficient to show that* $x(t_i^-)$ *remains bounded for* $i \in \mathbb{N}$. *Observe that for* $i = 1, 2, 3...$

$$x(t_{i+1}^-) = e^{A(t_{i+1} - t_i)} \begin{bmatrix} q_i I & 0 \\ 0 & I \end{bmatrix} x(t_i^-) +$$

$$+ \int_{t_i}^{t_{i+1}} e^{A(t_{i+1} - \tau)} Br(\tau) d\tau$$

Let $\|x\|$ *denote any norm in Euclidean space and* $\|A\|$ *denote the corresponding induced matrix norm. Let* $M := \sup_{i \in \mathbb{N}} \int_{t_i}^{t_{i+1}} \|r(\tau)\| d\tau$ *and* $G := M e^{\|A\|\Theta} \|B\|$ *and note that, thanks to our assumptions on* $r(t)$ *and* $t_{i+1} - t_i$, M *and* G *are both finite. It follows from the properties of norm that for* $i = 1, 2, 3, ...$

$$\|x(t_{i+1}^-)\| \leq q\|x(t_i^-)\| + G$$

which gives for $i = 2, 3, 4, ...$

$$\|x(t_i^-)\| \leq q^{i-1} \|x(t_1^-)\| + \sum_{k=0}^{i-2} q^k G$$

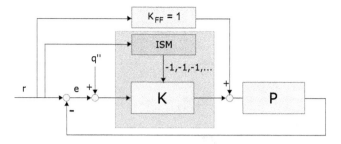

Figure 4: Impulsive control scheme equivalent to the signal transformation approach for tracking of triangular waveforms

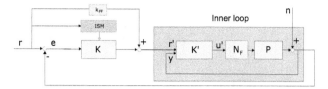

Figure 5: Control scheme for the demonstration of impulsive control. The inner feedback loop with controller K' compensates for the dominant resonances of microscanner P. Low bandwidth controller K is subject to impulsive state multiplication driven by the reference signal.

THEOREM 4. *Let r be a triangular waveform in range $[0, A]$ with period T. Consider the control schemes in Fig. 3 where K is a double integral controller, $K(s) = \frac{k_1 s + k_2}{s^2}$, and define the signal transformation functions as in [7]*

$$\Phi(x) = (-1)^{(i-1)} x(t) + 2A \lfloor \frac{i}{2} \rfloor$$

$$\Phi^{-1}(x) = (-1)^{(i-1)} x(t) + (-1)^i 2A \lfloor \frac{i}{2} \rfloor$$

where $i \in \mathbb{N}$ indexes the half-periods of the reference signal and denote $I(x) := (-1)^{(i-1)} x(t)$. Then the control schemes in Fig. 3 (a) and Fig. 3 (b) are equivalent in the sense that $u_1(t) = u_2(t)$ for all time. In Fig. 3 (b),

$$q''(t) = (-1)^{\lfloor \frac{2t}{T} \rfloor} \frac{2A}{k_1 T} e^{-\frac{k_2}{k_1} t}$$

is a bounded signal which exponentially decays to zero.

Theorem 4 decomposes the time-varying affine transformation functions into a feedforward connection and a pair of operators which perform sign alternations of the error signal and control effort. Given this structure, the controller and the adjacent operators can be merged according to Theorem 3. Consequently, we obtain

COROLLARY 1. *Consider an STA control loop designed for tracking of a triangular signal with frequency $1/T$ and amplitude A, with $K = \frac{k_1 s + k_2}{s^2}$ a double integral controller. Define the sequences $t_i := \frac{T}{2} i$ and $q_i := -1$ for $i \in \mathbb{N}$. Then the STA scheme of Fig. 3 (a) is equivalent to the scheme of Fig. 4 with controller $ISM(K, t, q)$.*

Corollary 1 unveils the relation between STA and impulsive control by converting STA into an impulsive form. This has several advantages. It follows from Corollary 1 that by removing the bounded signal q'', an improvement in the transient response of STA can be obtained. Corollary 1 also gives a direct and simple way to implement STA by merely multiplying the controller state by -1 at the turnaround points of the triangular reference signal.

Fig. 3 (b) also shows that a feed-forward element is implicitly present in STA. Using the ISM-based scheme of Fig. 4, it is possible to recognize this explicitly and track the triangular waveforms with only a single integrator K if the DC gain of P is known exactly. It also follows that appropriate linear control schemes for evaluation of STA performance should account for the feed-forward element and can be derived from the scheme in Fig. 3 (b).

Finally, the relation between STA and impulsive control provides alternative ways to examine the system stability and performance.

4. EXPERIMENTS

The performance of impulsive control is demonstrated experimentally on a microelectromechanical positioner. First, the mechanical resonances of the positioner are damped by means of feedback control. Second, two experiments are presented which demonstrate the tracking performance and reduction of measurement noise when using impulsive control.

4.1 Positioning system

The experiments were performed on a 2D nanopositioner with displacement range of approximately 120 μm in two orthogonal directions [11]. The positioner consists of a silicon wafer (scan table) which is actuated by voice-coil actuators on a spring support. The motion in each direction is sensed by a pair of thermal position sensors [12] which rely on a microheater that partially overlaps the scan table. A displacement of the scan table results in a change of the heat conduction between the scan table and the microheater and can be detected as a change in the electrical resistance of the microheater.

Precision control of the nanopositioner was studied in detail in [13] and [2]. For the purposes of control design in this section, the nanopositioner is modeled as a second order system with the transfer function

$$P = \frac{9.089 \cdot 10^6}{s^2 + 62.58s + 9.322 \cdot 10^5}$$

The transfer function captures the dominant resonant mode of the positioner which occurs at 153 Hz; further resonant modes are located at frequencies in excess of 3 kHz.

Fig. 5 depicts the control architecture used in the experiments. The vibrational modes of the positioner P were damped by means of an inner feedback loop in order to achieve a flat frequency response in the region of interest. The two experiments feature two different designs of the inner feedback controller K' which accentuate tracking performance and sensitivity to measurement noise, respectively. N_F denotes two Butterworth band-stop filters which were used to compensate for the higher order vibrational modes of P. The ISM- and STA-based control architectures were implemented on top of the inner feedback loop. Because of the robust positioner dynamics and accurate identification, a

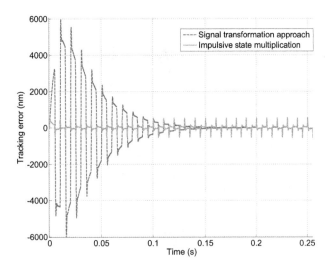

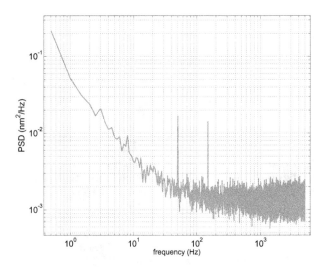

Figure 6: Comparison of transient tracking error of STA and ISM

Figure 7: Spectral density of the measured thermal sensor noise

single integrator K with a constant feed-forward term K_{FF} could be used to track the piecewise affine reference signals.

4.2 Tracking comparison with STA

In the experiment, the inner loop controller K' was chosen as

$$K' = \begin{bmatrix} K_D & -K_D \end{bmatrix}$$

The H_∞ control framework was used to design K_D, see e.g. [14]. The resulting fifth order controller achieved closed loop bandwidth of approximately 873 Hz.

The ISM-based control scheme shown in Fig. 2 and STA control scheme shown in Fig. 3 (a) were compared while running on top of the inner feedback loop. The reference signal was a 100 Hz triangular waveform in range 0 to 5 μm (linear velocity 1 mm/s). In the ISM-based control scheme, controller K was chosen as $K(s) = \frac{200}{s}$ and K_{FF} was set to the inverse of the DC gain of the inner loop, in this case to 1. The state of K was multiplied by -1 at every turnaround point of the triangular reference. In the STA control scheme, controller K was chosen as $K(s) = \frac{200s+5000}{s^2}$ and the signal transformation functions were defined as in Theorem 4 with $A = 5$ μm and $i = 1, 2, 3, ...$ incremented at every turnaround point of the triangular reference.

Fig. 6 shows the tracking error of the two schemes compared. While the steady state performance was excellent in both cases, impulsive control improved the transient tracking error by two orders of magnitude. Note also that only an extremely simple control law was used without the need to transform the input, control and measurement signals. This greatly simplified the digital implementation of the system.

4.3 Sensitivity to measurement noise

In high bandwidth nanopositioning systems, the presence of measurement noise can adversely affect the positioning accuracy. For instance, see Fig. 7 which shows the power spectral density of the thermal sensor noise as experimentally measured. The significant low frequency noise component may be detrimental to positioning accuracy if fast tracking is required.

The sensitivity to measurement noise was compared for a

high bandwidth linear feedback loop and a low bandwidth feedback loop with an ISM-based controller. In order to minimize the impact of the microscanner damping, the noise sensitivity transfer function of the inner loop controller K' was shaped [4] to attenuate noise in the frequency regions where damping was not necessary. The controller was chosen to be in the feedback path of the loop, i.e.

$$K' = \begin{bmatrix} -K_E & K_F \end{bmatrix}$$

where feed-forward term K_F was set to the inverse of the DC gain of the inner loop.

Controller K_E was synthesized using the H_∞ control framework, see e.g. [14]. The weighting functions were set such that the measurement noise was amplified only in the region around 300 Hz and attenuated elsewhere. A sixth order controller was obtained and the resulting -3 dB bandwidth of the complementary sensitivity function of the inner loop was 230 Hz.

The reference signal, a 20 Hz triangular waveform with amplitude of 5 μm, was tracked by a single integral feedback controller K. The bandwidth of the feedback loop without ISM was set to approximately 50 Hz in order to track the reference signal accurately. In the feedback loop with ISM, a slow controller was used with bandwidth of less than 1 Hz. According to the ISM control law, the state of the ISM-based controller was multiplied by a factor of -1 at every turnaround point of the reference signal.

Fig. 8 shows the tracking performance of the feedback loop with ISM as obtained experimentally. The 20 Hz reference signal was tracked precisely despite the low linear bandwidth of the feedback loop. In particular, the tracking lag which would appear as an unknown time shift of the signal was removed. Without ISM, the tracking lag would result in a tracking error with fundamental frequency around 40 Hz and would not be corrected by the slow linear controller.

Compensating for the high frequency tracking error by means of a low bandwidth controller enabled us to improve the tolerance to the measurement noise. To demonstrate this, we captured the measurement noise experimentally and performed a simulation to estimate the motion induced by the measurement noise in the aforementioned experiments.

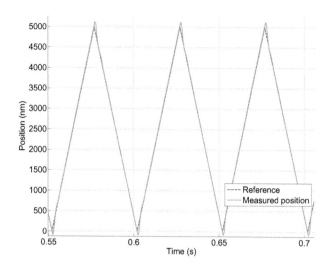

Figure 8: Tracking of a 20 Hz triangular waveform using a slow controller with ISM and an inner loop controller with a shaped noise sensitivity transfer function.

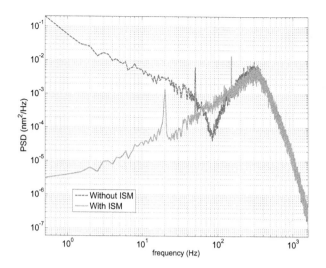

Figure 9: Estimate of the microscanner motion induced by measurement noise for an impulsive and non-impulsive controller of a comparable tracking performance.

Fig. 9 shows the spectra of the positioning error during the tracking operation. The positioning error in range 0 to 100 Hz decreased by using ISM from 0.5475 nm to 0.1774 nm while a comparable tracking performance was achieved. The peak at 20 Hz corresponds to the frequency of the signal tracked, the peaks at 50 Hz and 150 Hz are spurious and come from the line noise present in the noise measurements (see also Fig. 7).

5. CONCLUSION

A hybrid control architecture based on impulsive control was analyzed and experimentally studied. A bounded-input-bounded-state stability condition was established and the relation to the signal transformation approach (STA) was derived. The method was evaluated in simulation and experimentally using a microelectromechanical nanopositioner. In comparison to STA, a significant improvement in the transient tracking error was achieved. At the same time, the impulsive controller does not necessitate any signal transformations and greatly simplifies the implementation. Furthermore, it was shown that impulsive controllers have low sensitivity to measurement noise which makes them suitable for high bandwidth control loops.

6. ACKNOWLEDGMENTS

We thank Urs Egger for his valuable technical assistance. Special thanks go to Haris Pozidis for his support of this work.

7. REFERENCES

[1] S. M. Salapaka and M. V. Salapaka. Scanning probe microscopy. *IEEE Control Systems Magazine*, 28 (2):65 – 83, 2008.

[2] A. Sebastian, A. Pantazi, H. Pozidis, and E. Eleftheriou. Nanopositioning for probe-based data storage. *IEEE Control Systems Magazine*, 28 (4):26 – 35, 2008.

[3] C. Lee and S. M. Salapaka. Robust broadband nanopositioning: fundamental trade-offs, analysis, and design in a two-degree-of-freedom control framework. *Nanotechnology*, 20 (3):035501, 2009.

[4] A. Sebastian, A. Pantazi, S. O. R. Moheimani, H. Pozidis, and E. Eleftheriou. Achieving subnanometer precision in a MEMS-based storage device during self-servo write process. *IEEE Transactions on Nanotechnology*, 7 (5):586 – 595, 2008.

[5] D. Liberzon. *Switching in systems and control.* Systems & Control: Foundations & Applications. Birkhäuser Boston Inc., 2003.

[6] T. Tuma, A. Pantazi, J. Lygeros, and A. Sebastian. Tracking of high frequency piecewise affine signals using impulsive control. *Proceedings of the 5th IFAC Symposium on Mechatronic Systems*, pages 90 – 95, 2010.

[7] A. Sebastian and S. O. R. Moheimani. Signal transformation approach to fast nanopositioning. *Review of scientific instruments*, 80 (7):076101, 2009.

[8] D. D. Bainov and P. S. Simeonov. *Systems with impulse effect: Stability, theory and applications.* Ellis Horwood Ltd., 1989.

[9] S. O. R. Moheimani and B. J. G. Vautier. Resonant control of structural vibration using charge-driven piezoelectric actuators. *IEEE Transactions on Control Systems Technology*, 13(6):1021–1035, 2005.

[10] A. J. Fleming and S. O. R. Moheimani. Sensorless vibration suppression and scan compensation for piezoelectric tube nanopositioners. *IEEE Transactions on Control System Technology*, 14(1):33–44, January 2006.

[11] M. A. Lantz, H. E. Rothuizen, U. Drechsler, W. Häberle, and M. Despont. A vibration resistant nanopositioner for mobile parallel-probe storage applications. *Journal of Microelectromechanical Systems*, 16 (1):130 – 139, 2007.

[12] M. A. Lantz, G. K. Binnig, M. Despont, and U. Drechsler. A micromechanical thermal displacement sensor with nanometer resolution. *Nanotechnology*, 16:1089 – 1094, 2005.

[13] A. Pantazi, A. Sebastian, G. Cherubini, M. Lantz, H. Pozidis, H. Rothuizen, and E. Eleftheriou. Control of MEMS-based scanning-probe data-storage devices. *IEEE Transactions on Control Systems Technology*, 15:824 – 841, 2007.

[14] S. Skogestad and I. Postletwaithe. *Multivariable feedback control*. John Wiley and Sons, 1996.

A Predictive Control Solution for Driveline Oscillations Damping

C.F. Caruntu
"Gheorghe Asachi" Technical
University of Iasi
Romania
caruntuc@tuiasi.ro

A.E. Balau
"Gheorghe Asachi" Technical
University of Iasi
Romania
abalau@tuiasi.ro

M. Lazar
Eindhoven University of
Technology
The Netherlands
m.lazar@tue.nl

P.P.J. v.d. Bosch
Eindhoven University of
Technology
The Netherlands
p.p.j.v.d.bosch@tue.nl

S. Di Cairano
Ford Research and Adv.
Engineering
Dearbon, MI, USA
dicairano@ieee.org

ABSTRACT

This paper deals with the problem of damping driveline oscillations, which is crucial to improving driveability and passenger comfort. Recently, this problem has received an increased interest due to the introduction in several production vehicles of the *dual-clutch powershift* automatic transmission with dry clutches. This type of transmission improves fuel economy, but it results in a challenging control problem, due to driveline oscillations. These oscillations, also called "shuffles", occur during gear-shift, while traversing backlash or when tip-in and tip-out maneuvers are performed. The first contribution of this paper is the derivation of an accurate piecewise affine drivetrain model with three inertias. The second contribution is concerned with the design of a horizon-1 predictive controller based on flexible Lyapunov functions. Several simulations based on realistic scenarios show that the proposed control scheme can handle both the performance and physical constraints, and the strict limitations on the computational complexity.

Categories and Subject Descriptors

G.1.0 [**Numerical analysis**]: General—*Stability (and instability)*

General Terms

Design

Keywords

Automotive drivelines, Predictive control, Lyapunov methods

1. INTRODUCTION

Recent studies in automotive engineering explore various engine, transmission and chassis models and advanced control methods to increase overall vehicle performance, fuel economy, safety and comfort. Driveability, the ability to quickly and appropriately respond to driver commands and a high degree of driving comfort are expected in a modern vehicle. Due to elasticity of the driveline components, mechanical resonance may occur. This phenomenon is known as driveline oscillations or "shuffles". When driveline oscillations are induced, the driveability of the vehicle is reduced, as the oscillations are transmitted via the chassis to the driver who experiences longitudinal jerking. The control design objective is to increase the passenger comfort by reducing the oscillations that occur during gearshift, while traversing backlash or when tip-in and tip-out maneuvers are performed.

In recent years the driveline oscillation problem has received an increasing interest due to the introduction of dual-clutch powershift (or shortly, DPS) automatic transmissions. These dry clutch transmissions offer improved fuel economy, easier packaging and reduced weight with respect to the standard wet-clutch planetary gear transmissions. Also the torque converter, which provides a smooth hydrodynamic coupling between the engine and the transmission and which is present in standard automatic transmissions, can be removed. However, the absence of the torque converter makes the torque transfer path from the engine to the wheels entirely mechanical, which means that disturbances, including the inherent reciprocating behavior of the engine, have more impact on the driveline. In manual transmissions the driver avoids or reduces the shuffle mode by modulating the clutch pedal. In automatic DPS transmissions the engine torque is modulated by a control algorithm, since torque actuators (airflow and/or spark timing) attain a higher precision and higher bandwidth than clutch actuators. In this paper a specific design for this control set-up is proposed.

The automotive driveline is an essential part of the vehicle and its dynamics have been modeled differently, according to the driving necessities. The complexity of the numerous models reported in the literature varies [1], but the two masses models are more commonly used, and this fact is justified in [2], where it is shown that this model is able to capture the first torsional vibrational mode. There are also more complex three-masses models reported in different research papers, as it will be indicated next. In [3] a simple driveline model with two inertias, one for the engine and the transmission, and one for the wheel and the vehicle mass, was presented. A more complex two-masses model, including a nonlinearity introduced by the backlash, was presented in [4]. A mathematical model of a drivetrain was introduced in [5] and [6] in the form of a third order linear state-space model. A simple model with the pressure in the engine manifold and the engine speed as state variables and the throttle valve angle as control input was presented in [7]. A piecewise linear (PWL) two-masses model, with one inertia representing the engine and the other inertia representing the vehicle (including the clutch, main-shaft and the powertrain), was presented in [8]. [9] presents a piecewise affine (PWA) two masses model for a driveline including a backlash nonlinearity. In [10] a two masses model was presented with the driveline main flexibility represented by the driveshafts, as well as a three mass model to reproduce the behavior of a vehicle with a dual-mass flywheel. Linear and nonlinear three masses models, in which the clutch flexibility was also considered, were presented in [11]. A complex three masses model that includes certain nonlinear aspects of the clutch was presented in [12].

The first contribution of this paper is the development of a PWA model for an automotive driveline, which brings several improvements with respect to the above mentioned models, by considering the driving load given by the airdrag torque, gravity and rolling resistance, and four modes to describe the clutch dynamics. Taking into account all these factors yields a more accurate model of the driveline dynamics.

Concerning the control strategy, different approaches to damp driveline oscillations have also been proposed in literature. In [3] a linear quadratic regulator (LQR) design that damps driveline oscillations by compensating the driver's engine torque demand was presented. Other linear quadratic gaussian controllers designed with loop transfer recovery were presented in [2, 13–15]. Furthermore, [6] proposed the usage of a feedforward controller in combination with a LQR controller and considering the engine as an actuator to damp powertrain oscillations. A robust pole placement strategy was employed in [16, 17], an $\mathscr{H}_\infty$ optimization approach was presented in [18], while model predictive control (MPC) strategies were proposed in [5, 9, 19]. In [5], a model-based approach for anti-jerk control of passenger cars that minimizes driveline oscillations while retaining fast acceleration was introduced. The controller was designed with the help of the root locus method and an analogy to a classical PI-controller was drawn. In [9], a constraint was imposed on the difference between the motor speed and the load speed to minimize the driveline oscillations, while reducing the impact of forces between the mechanical parts.

MPC is increasingly seen as an attractive methodology for automotive applications due to its capability to directly handle various specifications including the optimization of a cost function while enforcing constrains on state and control variables. This fits the problem considered in this paper, i.e., damping driveline oscillations. However, due to the complexity of the model and the stringent time limitations for finding a solution on-line, standard "sufficiently long horizon" MPC approaches [20] do not represent a viable solution. As such, the second contribution of this paper is to make use of a recently introduced design method for horizon-1 MPC, based on flexible control Lyapunov functions [21], with the aim of reducing the driveline oscillations. The algorithm therein has the potential to satisfy the timing requirements, due to the short horizon, while it can still offer a non-conservative solution to stabilization due to the flexibility of the Lyapunov function. Several simulation scenarios defined in collaboration with Ford Research and Advanced Engineering, US validate the proposed approach and indicate that the proposed scheme, besides yielding a feasible algorithm, outperforms controllers obtained using typical approaches, such as PID control.

The remainder of the paper is structured as follows. In Section 2, a PWA model of the drivetrain is presented, which considers both driveshaft and clutch flexibilities. Then, in Section 3, the predictive

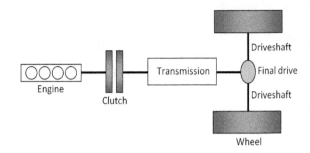

Figure 1: Schematic vehicle structure.

control strategy based on flexible Lyapunov functions is designed. Section 4 presents extensive simulation results, while conclusions are summarized in Section 5.

1.1 Notation and basic definitions

Let $\mathbb{R}$, $\mathbb{R}_+$, $\mathbb{Z}$ and $\mathbb{Z}_+$ denote the field of real numbers, the set of non-negative reals, the set of integer numbers and the set of non-negative integers, respectively. For every $c \in \mathbb{R}$ and $\Pi \subseteq \mathbb{R}$ define $\Pi_{\geq c} := \{k \in \Pi \mid k \geq c\}$ and similarly $\Pi_{\leq c}$, $\mathbb{R}_\Pi := \Pi$ and $\mathbb{Z}_\Pi := \mathbb{Z} \cap \Pi$. For a vector $x \in \mathbb{R}^n$ let $\|x\|$ denote an arbitrary p-norm and let $[x]_i, i \in \mathbb{Z}_{[1,n]}$, denote the i-th component of x. Let $\|x\|_\infty := \max_{i \in \mathbb{Z}_{[1,n]}} |[x]_i|$, where $|\cdot|$ denotes the absolute value. For a matrix $Z \in \mathbb{R}^{m \times n}$ let $\|Z\|_\infty := \sup_{x \neq 0} \frac{\|Zx\|}{\|x\|}$ denote its corresponding induced matrix norm. $I_n \in \mathbb{R}^{n \times n}$ denotes the identity matrix. A function $\varphi : \mathbb{R}_+ \to \mathbb{R}_+$ belongs to class $\mathscr{K}$ if it is continuous, strictly increasing and $\varphi(0) = 0$. A function $\varphi \in \mathscr{K}$ belongs to class $\mathscr{K}_\infty$ if $\lim_{s \to \infty} \varphi(s) = \infty$.

2. PWA DRIVELINE MODEL

The structure of a passenger car consists, in general, of the following parts: engine, clutch, transmission, final drive, driveshafts and wheels, as it can be seen in Fig. 1. To develop a controller, an accurate drivetrain model is required to predict the vehicle's response to a torque input. The model can then be used to design and simulate the closed-loop control system.

A PWA three inertias model, which takes into consideration the clutch flexibility together with the driveshaft flexibility, was derived from the laws of motion [10–12], in which the first inertia corresponds to the engine, the second one includes the inertia of the gearbox and the inertia of the final drive, and the last inertia corresponds to the wheel and vehicle mass, as it is represented in Fig. 2. The clutch torsional flexibility is a result of an arrangement of smaller springs in series with springs with much larger stiffness. The reason for this arrangement is vibration insulation. Unlike in the above mentioned previous modeling approaches, where only two or three working modes are considered for the clutch dynamics, in this paper four working modes are introduced, i.e., open, closing, closed and locked, which yields a more accurate model of the clutch.

In the *open* mode, there is no mechanical connection between the engine and the rest of the driveline, so no torque is transmitted towards the wheels. In the *closing* mode, the smaller springs in the clutch are compressed, which means that the engine torque is gradually transmitted to the driveline. In the *closed* mode, the stiffer springs in the clutch are compressed and the gradual transmission of torque to the driveline continues. The *locked* mode corresponds to the phase when the springs in the clutch cannot be compressed any further, i.e., the clutch hits a mechanical stop, and the maxi-

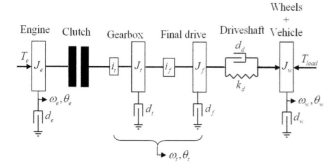

Figure 2: Three inertias drivetrain model including the clutch.

mum amount of torque is transmitted from the engine to the wheels through the driveline.

2.1 Continuous-time model

Considering as state variables the torsional angle between engine and transmission, the torsional angle between transmission and wheel, the angular speed of the engine, the angular speed of the transmission and the angular speed of the wheel, and as control input of the system, the engine torque, i.e.,

$$x_1 = \theta_e - \theta_t g_t, \quad x_2 = \frac{\theta_t}{g_f} - \theta_w,$$

$$x_3 = \omega_e, \quad x_4 = \omega_t, \quad x_5 = \omega_w, \quad u = T_e, \tag{1}$$

the following PWA state-space model is obtained:

$$\dot{x}(t) = A_{ci}x(t) + b_c u(t) + f_c \quad \text{if} \quad x(t) \in \Omega_i, \tag{2}$$

where $x := (x_1, \dots, x_5)^\top \in \mathbb{R}^5$ and $i \in \mathscr{I} := \mathbb{Z}_{[1,4]}$. Here i denotes the active mode at time $t \in \mathbb{R}_+$, $A_{ci} \in \mathbb{R}^{5 \times 5}$, $b_c \in \mathbb{R}^{5 \times 1}$ are the system matrices and $f_c \in \mathbb{R}^{5 \times 1}$ is the affine term. Notice that although there are 3 angles, only two states are introduced as only the angle difference is relevant. The collection of sets $\{\Omega_i \mid i \in \mathscr{I}\}$ defines a partition of the state-space $\mathbb{X} \subseteq \mathbb{R}^5$ as follows:

$$\begin{cases} \Omega_1 := \{x \in \mathbb{R}^5 \mid x_3 \leq \omega_e^{\text{closing}}\}, & \text{- open} \\ \Omega_2 := \{x \in \mathbb{R}^5 \mid x_3 > \omega_e^{\text{closing}} \ \& \ |x_1| \leq \theta_1\}, & \text{- closing} \\ \Omega_3 := \{x \in \mathbb{R}^5 \mid x_3 > \omega_e^{\text{closing}} \ \& \ \theta_1 < |x_1| \leq \theta_2\}, & \text{- closed} \\ \Omega_4 := \{x \in \mathbb{R}^5 \mid x_3 > \omega_e^{\text{closing}} \ \& \ \theta_2 < |x_1|\}, & \text{- locked} \end{cases} \tag{3}$$

where $\omega_e^{\text{closing}}$ is the engine closing speed and θ_1, θ_2 are threshold values for the torsional angle between the engine and the transmission, which determine the working mode of the clutch. Note that when a transition from the open mode to the closing mode occurs, the following reset condition must be imposed:

$$\forall t_1 \in \mathbb{R}_+, \forall t_2 \in \mathbb{R}_{>t_1}, \text{if } x(\tau) \in \Omega_1, \forall \tau \in \mathbb{R}_{[t_1, t_2)}$$
$$\text{and } x(t_2) \in \Omega_2, \text{ set } x_1(t_2) := 0. \tag{4}$$

As the engine angle θ_e tends to infinity in the open mode, so the state x_1 tends to infinity, synchronization of the engine angle and the transmission angle must be attained at the moment the clutch switches from the open mode to the closing mode. Notice that the reset condition does not apply to initial conditions, as the model cannot be initialized in the closing mode. The matrices A_{ci}, b_c and f_c can be obtained by deriving the whole dynamical model of the drivetrain using the generalized Newton's second law of motion. The equation of motion for the first rotational mass yields

$$J_e \dot{\omega}_e = T_e - T_c - d_e \omega_e, \tag{5}$$

where the engine is described as an ideal torque source T_e with a mass moment of inertia J_e and a viscous friction coefficient d_e; the engine speed is represented by ω_e. Although this will not be pursued here, a dynamical model of torque production (i.e., manifold dynamics and combustion delay) can also be included. On the other hand, instantaneous torque production can be achieved by creating and maintaining a torque reserve via airflow control. The torque reserve is then exploited by actuating the spark timing which provides fast torque production changes [22]. The clutch torque T_c can be modeled as

$$T_c = k_{ci}(\theta_e - \theta_t g_t) + d_{ci}(\omega_e - \omega_t g_t), \tag{6}$$

where the clutch stiffness k_{ci} and damping d_{ci} have different values for the four modes of operation. In (6), g_t is the transmission ratio, while ω_t stands for the transmission speed. The corresponding angles of the engine and the transmission are given by θ_e and θ_t.

The equation of motion of the second body, the transmission and the shafts, can be derived as

$$J_2 \dot{\omega}_t = g_t T_c - d_2 \omega_t - \frac{1}{g_f} T_d, \tag{7}$$

where $J_2 = J_t + \frac{J_f}{g_f^2}$ is the second inertia with damping $d_2 = d_t + \frac{d_f}{g_f^2}$ composed by the transmission and final drive damping and g_f is the final reduction gear ratio. The torque in the driveshafts can be expressed as

$$T_d = k_d \left(\frac{\theta_t}{g_f} - \theta_w \right) + d_d \left(\frac{\omega_t}{g_f} - \omega_w \right), \tag{8}$$

where ω_w is the wheel speed, θ_w is the wheel angle, and k_d, d_d are the driveshaft stiffness and damping, respectively.

Remark 2.1 In this paper, a parallel structure that is equivalent to the series structure adopted in [1] is used to model the stiffness and damping of the driveshafts. Testing how the series structure would affect the performance of the designed control strategy makes the object of future research. □

The final law of motion corresponds to the wheels and vehicle body and can be written as

$$J_3 \dot{\omega}_w = T_d - d_w \omega_w - T_{\text{load}}, \tag{9}$$

where $J_3 = J_w + m_{\text{COG}} r_w^2$ is the wheel and vehicle inertia with damping d_w, m_{COG} is the vehicle mass and r_w is the wheel radius. The load torque is modeled as

$$T_{\text{load}} = T_{\text{roll}} + T_{\text{grade}} + T_{\text{airdrag}}, \tag{10}$$

where T_{roll} is the rolling resistance torque, T_{grade} is the torque due to the road slope and T_{airdrag} is the aerodynamic drag torque of the vehicle body,

$$\begin{aligned} T_{\text{roll}} &= c_{r1} m_{\text{COG}} g \cos(\chi_{\text{road}}) r_w, \\ T_{\text{grade}} &= m_{\text{COG}} g \sin(\chi_{\text{road}}) r_w, \\ T_{\text{airdrag}} &= 0.5 \rho_{\text{air}} A_f c_d v_v^2 r_w, \end{aligned} \tag{11}$$

where c_{r1} is the rolling resistance coefficient, g is the gravitational acceleration, χ_{road} is the road slope gradient, ρ_{air} is the air mass density, A_f is the frontal area of the vehicle, c_d is the airdrag coefficient and $v_v = r_w \omega_w$ is the vehicle velocity. To avoid unnecessary complications, only the terms given by the rolling torque T_{roll} and the aerodynamic drag torque T_{airdrag} are considered, while the road slope gradient is assumed to be zero. Furthermore, instead of the nonlinear function that describes the aerodynamic drag torque given in (11), a linear approximation will be used with c_{r2} as an approximation parameter, i.e.,

$$T_{\text{airdrag}} = c_{r2}\omega_w. \tag{12}$$

Rewriting the equations in a matrix form, yields the model matrices

$$A_{ci} = \begin{pmatrix} 0 & 0 & 1 & -g_t & 0 \\ 0 & 0 & 0 & \frac{1}{g_f} & -1 \\ -\frac{k_{ci}}{J_1} & 0 & -\frac{D_{\text{sum1}}}{J_1} & \frac{d_{ci}g_t}{J_1} & 0 \\ \frac{k_{ci}g_t}{J_2} & -\frac{k_d}{g_f J_2} & \frac{d_{ci}g_t}{J_2} & -\frac{D_{\text{sum2}}}{J_2} & \frac{d_d}{g_f J_2} \\ 0 & \frac{k_d}{J_3} & 0 & \frac{d_d}{g_f J_3} & -\frac{d_{\text{wheel}}}{J_3} \end{pmatrix}, \tag{13}$$

with $D_{\text{sum1}} = d_{ci} + d_e$, $D_{\text{sum2}} = d_{ci}g_t{}^2 + d_2 + \frac{d_d}{g_f{}^2}$, $d_{\text{wheel}} = d_w + d_d + c_{r2}$ and

$$b_c = \begin{pmatrix} 0 & 0 & \frac{1}{J_1} & 0 & 0 \end{pmatrix}^\top, \quad f_c = \begin{pmatrix} 0 & 0 & 0 & 0 & -\frac{T_{\text{roll}}}{J_3} \end{pmatrix}^\top. \tag{14}$$

The engine torque (i.e., the control input) is restricted by lower and upper bounds and by a torque rate constraint as follows:

$$0 \leq u(t) \leq T_e^{\max}, \quad \forall t \in \mathbb{R}_+,$$
$$T_e^{\text{m}} \leq \dot{u}(t) \leq T_e^{\text{M}}, \quad \forall t \in \mathbb{R}_+, \tag{15}$$

where $T_e^{\max}$ is the maximum torque that can be generated by the internal combustion engine and T_e^{m}, T_e^{M} are torque rate bounds. Furthermore, the engine and wheel speeds are bounded, i.e.,

$$\omega_e^{\min} \leq x_3(t) \leq \omega_e^{\max}, \quad \forall t \in \mathbb{R}_+, \tag{16}$$

$$\omega_w^{\min} \leq x_5(t) \leq \omega_w^{\max}, \quad \forall t \in \mathbb{R}_+, \tag{17}$$

where $\omega_e^{\min}$ and $\omega_e^{\max}$ are the idle speed and the engine limit speed, respectively, and $\omega_w^{\min}$ and $\omega_w^{\max}$ are the minimum and the maximum speed of the wheels.

2.2 Discrete-time model

To obtain a discrete-time PWA model, each affine subsystem in (2) is discretized with sampling period T_s using the Tustin transform, which yields

$$x_{k+1}^m = A_{di}^m x_k^m + b_d^m u_k^m + f_d^m \quad \text{if} \quad x_k \in \Omega_i, \tag{18}$$

for all $k \in \mathbb{Z}_+$, where A_{di}^m and b_d^m are the corresponding discretized system matrices, f_d^m is the discretized affine term and x_k^m, u_k^m are the state and input of the system at time instant $k \in \mathbb{Z}_+$. The active mode i is selected for the discrete-time PWA system using (3), the same as done for the continuous-time PWA system. Letting I_5^1 denote I_5 with the first element on the diagonal equal to zero, the reset condition (4) now becomes

$$\forall k \in \mathbb{Z}_{\geq 1}, \text{ if } (x_{k-1}^m, x_k^m) \in \Omega_1 \times \Omega_2, \text{ set } x_k^m := I_5^1 x_k^m. \tag{19}$$

Note that the utilized discretization method is more or less standard for PWA systems; for example, it is also employed in the previous works on predictive control for driveline oscillations damping [7,9,12]. A simulation comparison, see Fig. 3, indicated that the so-obtained discretized model produces a similar response as the original continuous-time model, for a sampling period of 5ms. The more frequent transitions between the closing and the closed mode, and between the closed and the locked mode, for the discrete-time model, which can be observed in Fig. 3, are due to the fact that within one sampling period the state exceeds the corresponding mode thresholds.

The engine torque rate constraint now becomes

$$-T_e^\Delta \leq \Delta u_k^m \leq T_e^\Delta, \quad \forall k \in \mathbb{Z}_{\geq 1}, \tag{20}$$

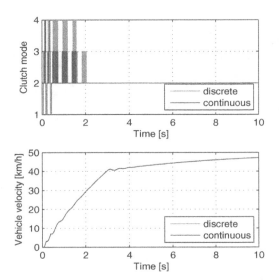

Figure 3: Clutch active mode and velocity histories.

where $\Delta u_k^m := u_k^m - u_{k-1}^m$ and T_e^Δ is the maximum allowed increase (decrease) in torque at each sampling instant. Torque rate constraints are important to allow full usage of the airflow to maintain the torque reserve, so that torque variations can be actuated instantaneously.

The desired objective is to reach a desired value of the wheel speed, i.e., x_5^{ss}, as fast as possible and with minimum overshoot, while damping driveline oscillations. Note that for a desired wheel speed value x_5^{ss}, one can obtain the corresponding steady-state values of the system states and input, i.e., x_1^{ss}, x_2^{ss}, x_3^{ss}, x_4^{ss} and u^{ss}. As such, the above objective can be formulated as asymptotic stabilization of the desired steady-state point while satisfying the required constraints.

In what follows, for simplicity of exposition, a coordinate transformation is performed in (18) to translate the problem into stabilization of the origin, i.e.,

$$x_k = x_k^m - x^{ss}, \quad u_k = u_k^m - u^{ss}, \tag{21}$$

where $x^{ss} = \begin{pmatrix} x_1^{ss} & x_2^{ss} & x_3^{ss} & x_4^{ss} & x_5^{ss} \end{pmatrix}^\top$, which yields the following system description:

$$x_{k+1} = A_{di}x_k + b_{di}u_k + f_{di} \quad \text{if} \quad x_k \in \Omega_i, \tag{22}$$

along with the corresponding reset condition (19). Here A_{di} and b_{di} are the discretized and transformed system matrices and f_{di} are the discretized and transformed affine terms. Notice that the transformed PWA model (21) has zero as an equilibrium within region Ω_2, i.e., $f_{d2} = 0$. Also, observe that u^{ss} can be interpreted as the feedforward component of the control action.

3. HORIZON-1 MPC SCHEME

Traditional control design methods such as PID or LQR cannot explicitly take into account hard constraints. In contrast, a MPC algorithm solves a finite-horizon open-loop optimization problem on-line, at each sampling instant, while explicitly taking input and state constraints into account. However, standard MPC techniques require a sufficiently long prediction horizon to guarantee stability, which makes the corresponding optimization problem too complex. Recently, a relaxation of the conventional notion of a Lyapunov function was proposed in [21], which resulted in a so-called flexible Lyapunov function. A first application of flexible Lyapunov functions in automotive control problems was presented in [23]. Therein it was indicated that flexible Lyapunov functions can be used to design stabilizing MPC schemes with a unitary horizon, without introducing conservatism. In this section we demonstrate how the theory introduced in [21] can be employed to design a horizon-1 MPC controller for the considered application. To this end, a generic system class and standard Lyapunov notions will be introduced.

Consider the discrete-time constrained nonlinear system

$$x_{k+1} = \phi(x_k, u_k), \quad k \in \mathbb{Z}_+, \tag{23}$$

where $x_k \in \mathbb{X} \subseteq \mathbb{R}^n$ is the state and $u_k \in \mathbb{U} \subseteq \mathbb{R}^m$ is the control input at the discrete-time instant k. $\phi : \mathbb{R}^n \times \mathbb{R}^m \to \mathbb{R}^n$ is an arbitrary nonlinear, possibly discontinuous, function with $\phi(0,0) = 0$. Notice that PWA systems of the form (22) are a subclass of (23). It is assumed that $\mathbb{X}$ and $\mathbb{U}$ are bounded sets with $0 \in \text{int}(\mathbb{X})$ and $0 \in \text{int}(\mathbb{U})$. Next, let $\alpha_1, \alpha_2 \in \mathcal{K}_\infty$ and let $\rho \in \mathbb{R}_{[0,1)}$.

Definition 3.1 A function $V : \mathbb{R}^n \to \mathbb{R}_+$ that satisfies

$$\alpha_1(\|x\|) \leq V(x) \leq \alpha_2(\|x\|), \quad \forall x \in \mathbb{R}^n \tag{24}$$

and for which there exists a, possibly set-valued, control law $\pi : \mathbb{R}^n \rightrightarrows \mathbb{U}$ such that

$$V(\phi(x,u)) \leq \rho V(x), \quad \forall x \in \mathbb{X}, \forall u \in \pi(x) \tag{25}$$

is called a control Lyapunov function (CLF) in $\mathbb{X}$ for system (23).

Consider the following inequality corresponding to (25):

$$V(x_{k+1}) \leq \rho V(x_k) + \lambda_k, \quad \forall k \in \mathbb{Z}_+, \tag{26}$$

where λ_k is an additional decision variable which allows the radius of the sublevel set $\{z \in \mathbb{X} \mid V(z) \leq \rho V(x_k) + \lambda_k\}$ to be flexible, i.e., it can increase if (25) is too conservative. Based on inequality (26) we can formulate the following optimization problem. Let $\alpha_3, \alpha_4 \in \mathcal{K}_\infty$ and $J : \mathbb{R} \to \mathbb{R}_+$ be a function such that $\alpha_3(|\lambda|) \leq J(\lambda) \leq \alpha_4(|\lambda|)$ for all $\lambda \in \mathbb{R}$ and let $\mu \in \mathbb{R}_{[0,1)}$. Let $\Omega \subseteq \mathbb{X}$ with the origin in its interior be a set where $V(\cdot)$ is a CLF for system (23). Such a region can be obtained for the desired application as the region of validity of an explicit PWA stabilizing state feedback controller obtained for the *unconstrained* model. More details on how to obtain a local CLF with corresponding PWA state-feedback law for model (22) are given in the next section.

Problem 3.2 Choose the CLF candidate V and the constants $\rho \in \mathbb{R}_{[0,1)}$, $\Delta \in \mathbb{R}_+$ and $M \in \mathbb{Z}_{>0}$ off-line. At time $k \in \mathbb{Z}_+$ measure x_k and minimize the cost $J(\lambda_k)$ over u_k, λ_k subject to the constraints

$$u_k \in \mathbb{U}, \ \phi(x_k, u_k) \in \mathbb{X}, \ \lambda_k \geq 0, \tag{27a}$$

$$V(\phi(x_k, u_k)) \leq \rho V(x_k) + \lambda_k, \tag{27b}$$

$$\lambda_k \leq \rho^{\frac{1}{M}}(\lambda_{k-1}^* + \rho^{\frac{k-1}{M}}\Delta), \quad \forall k \in \mathbb{Z}_{\geq 1}. \tag{27c}$$

Above λ_k^* denotes the optimum at time $k \in \mathbb{Z}_+$. Let $\pi(x_k) := \{u_k \in \mathbb{R}^m \mid \exists \lambda_k \in \mathbb{R} \text{ s.t. (27) holds}\}$ and let

$$\phi_{cl}(x, \pi(x)) := \{\phi(x,u) \mid u \in \pi(x)\}.$$

Theorem 3.3 Let a CLF V in Ω be known for system (23). Suppose that Problem 3.2 is feasible for all states x in $\mathbb{X}$. Then the difference inclusion

$$x_{k+1} \in \phi_{cl}(x_k, \pi(x_k)), \quad k \in \mathbb{Z}_+, \tag{28}$$

is asymptotically stable in $\mathbb{X}$.

The proof of Theorem 3.3 starts from the fact that (27c) implies $\lim_{k \to \infty} \lambda_k^* = 0$ and then employs standard arguments for proving input-to-state stability and Lyapunov stability. For brevity a complete proof is omitted here and the interested reader is referred to [21] for more details. However, in [21] a more conservative condition than (27c) was used, which corresponds to setting $\Delta = 0$ and $M = 1$. As such, it is necessary to prove that (27c) actually implies $\lim_{k \to \infty} \lambda_k^* = 0$, which is accomplished in the next lemma.

Lemma 3.4 Let $\Delta \in \mathbb{R}_+$ be a fixed constant to be chosen a priori and let $\rho \in \mathbb{R}_{[0,1)}$ and $M \in \mathbb{Z}_{>0}$. If

$$0 \leq \lambda_k \leq \rho^{\frac{1}{M}}(\lambda_{k-1}^* + \rho^{\frac{k-1}{M}}\Delta), \quad \forall k \in \mathbb{Z}_{\geq 1}, \tag{29}$$

then $\lim_{k \to \infty} \lambda_k = 0$.

PROOF. Since λ_k^* is bounded for all $k \in \mathbb{Z}_+$ [21], there exists an $N_\lambda \in \mathbb{R}_+$ such that $\lambda_0^* \leq N_\lambda$. Applying recursively (29), the following inequality is obtained:

$$\lambda_k \leq \rho^{\frac{k}{M}}(N_\lambda + k\Delta), \quad \forall k \in \mathbb{Z}_{\geq 1}. \tag{30}$$

Since $\rho \in \mathbb{R}_{[0,1)}$, $\lim_{k \to \infty} \rho^{\frac{k}{M}} = 0$. Moreover,

$$\lim_{k \to \infty} k\rho^{\frac{k}{M}} = \lim_{k \to \infty} k(\rho^{-1})^{-\frac{k}{M}} = \lim_{k \to \infty} \frac{k}{(\rho^{-1})^{\frac{k}{M}}} \tag{31}$$

and applying l'Hôpital's rule [24] it follows that

$$\lim_{k \to \infty} \frac{k}{(\rho^{-1})^{\frac{k}{M}}} = \lim_{k \to \infty} \frac{\frac{d}{dk}k}{\frac{d}{dk}(\rho^{-1})^{\frac{k}{M}}} = \lim_{k \to \infty} \frac{1}{(\rho^{-1})^{\frac{k}{M}}\frac{\ln(\rho^{-1})}{M}} = 0, \tag{32}$$

which completes the proof. $\square$

Next, consider the following cost function to be minimized

$$\begin{aligned} J_1(x_k, u_k, \lambda_k) &:= J_{\text{MPC}}(x_k, u_k) + J(\lambda_k) \\ &:= \|P_x x_{k+1}\|_\infty + \|R u_k\|_\infty + \|G\lambda_k\|_\infty, \end{aligned} \tag{33}$$

subject to constraints:

$$0 - u^{ss} \leq u_k \leq T_e^{\max} - u^{ss},$$
$$-T_e^\Delta \leq \Delta u_k \leq T_e^\Delta, \tag{34}$$
$$x^{\min} \leq H(A_{d_i}x_k + b_{d_i}u_k + f_{d_i}) \leq x^{\max},$$

with $x^{\min} := \begin{pmatrix} \omega_e^{\min} - x_3^{ss} \\ \omega_w^{\min} - x_5^{ss} \end{pmatrix}$, $x^{\max} := \begin{pmatrix} \omega_e^{\max} - x_3^{ss} \\ \omega_w^{\max} - x_5^{ss} \end{pmatrix}$ and $H := \begin{pmatrix} 0 & 0 & 1 & 0 & 0 \\ 0 & 0 & 0 & 0 & 1 \end{pmatrix}$. The cost $J(\cdot)$ is chosen as required in Problem 3.2 and the matrices $P_x \in \mathbb{R}^{p_x \times n}$ and $R \in \mathbb{R}^{r \times n}$ are chosen as full-column rank matrices of appropriate dimensions. Consider the following infinity-norm based CLF

$$V(x) = \|Px\|_\infty, \tag{35}$$

where $P \in \mathbb{R}^{p \times n}$ is a full-column rank matrix to be determined, e.g., using techniques from [25]. This function satisfies (24) with $\alpha_1(s) = \frac{\sigma}{\sqrt{p}}s$, where $\underline{\sigma}$ is the smallest singular value of P, and $\alpha_2(s) = \|P\|_\infty s$. For $x_k \in \Omega_i$, substituting (22) and (35) in (27b) yields

$$\|P(A_{d_i}x_k + b_{d_i}u_k + f_{d_i})\|_\infty \leq \rho\|Px_k\|_\infty + \lambda_k \tag{36}$$

where x_k, P and $\rho \in \mathbb{R}_{[0,1)}$ are known at $k \in \mathbb{Z}_+$. In what follows it is shown that for a unitary horizon, the above MPC optimization problem can be formulated as a linear program (LP) via a particular set of equivalent linear inequalities, despite switching dynamics, while for any other larger horizon it would lead to a mixed integer linear programming (MILP) problem. By definition of the infinity norm, for $\|x\|_\infty \leq c$ to be satisfied, it is necessary and sufficient to require that $\pm[x]_j \leq c$ for all $j \in \mathbb{Z}_{[1,n]}$. So, for (36) to be satisfied, it is necessary and sufficient to require

$$\pm[P(A_{di}x_k + b_{di}u_k + f_{di})]_j \leq \rho\|Px_k\|_\infty + \lambda_k \tag{37}$$

for all $j \in \mathbb{Z}_{[1,p]}$. As such, solving Problem 3.2, which includes minimizing the cost function (33), can be reformulated as the following problem.

Problem 3.5 Measure x_k, determine the active mode i and

$$\min_{u_k, \lambda_k} \varepsilon_k^1 + \varepsilon_k^2 + \varepsilon_k^3 \tag{38}$$

subject to (27c), (34), (37) and

$$\pm[P_x(A_{di}x_k + b_{di}u_k + f_{di})]_j \leq \varepsilon_k^1, \ \forall j \in \mathbb{Z}_{[1,p_x]}, \tag{39a}$$

$$\pm R u_k \leq \varepsilon_k^2, \tag{39b}$$

$$G\lambda_k \leq \varepsilon_k^3. \tag{39c}$$

Problem 3.5 is a linear program, since x_k and λ_{k-1}^* are known at time $k \in \mathbb{Z}_{\geq 1}$ and thus, all constraints are linear in u_k, λ_k and ε_k^l, $l \in \mathbb{Z}_{[1,3]}$. The horizon-1 MPC algorithm is stated next.

Algorithm 3.6
At each sampling instant $k \in \mathbb{Z}_+$:
Step 1: Measure the current state x_k and obtain the active mode i;
Step 2: Solve the LP Problem 3.5 and pick any feasible control action, i.e., $u^f(x_k)$;
Step 3: Implement $u_k := u^f(x_k)$ as control action. □

The fact that only a feasible, rather than optimal, solution of Problem 3.5 is required in Algorithm 3.6, can reduce the execution time.

4. CASE STUDY VALIDATION

The continuous-time PWA model (2)-(4) was implemented in Matlab/Simulink and two different control strategies were applied to damp driveline oscillations, i.e., the horizon-1 predictive controller proposed in this paper and a PID controller. The sampling period of the system was chosen to be $T_s = 5$ms. The values of the parameters used in simulations, which relate to a medium size passenger car, were obtained with the help of Ford Research and Advanced Engineering, US, and they are given in Table 1 in the Appendix. The control objective is to reach a desired speed reference in a short time, but, at the same time, to increase the passenger comfort by reducing the oscillations that appear in the driveline. The axle wrap is calculated as the difference between the engine speed (divided by the total transmission ratio) and the wheel speed, and it is used as a measure of the driveline oscillations.

A PID controller was designed based on [26] and it was tuned to have a fast response, which yielded the proportional, integral and derivative terms $K_R = 30$, $T_i = 10^{-3}$ and $T_d = 9 \cdot 10^{-5}$, respectively.

Remark 4.1 The more common approach to the design of an explicit MPC controller was also applied. For the considered discrete-time PWA model and operating constraints, using the Multi Parametric Toolbox for Matlab, a feasible solution to the corresponding mpMILP problem was only obtained for the prediction horizon equal to 1, but the resulting performance was substandard. For a prediction horizon larger than 1, despite using a powerful working station and several robust mpMILP solvers, a solution could not

be obtained. This indicates the non-trivial nature of the considered case study. □

The horizon-1 predictive controller proposed in this paper uses the following weight matrices of the cost (33): $P_x = 0.71 \cdot I_5$, $R = 0.021$ and $G = 1$. The technique presented in [25] was used for the off-line computation of the infinity norm based local CLF $V(x) = \|Px\|_\infty$ for $\rho = 0.99$ and the PWA model of the drivetrain in closed-loop with $u_k := K_i x_k$ if $x_k \in \Omega_i$, $i \in \mathbb{Z}_{[1,4]}$. The following matrices were obtained

$$P = \begin{pmatrix} 32.63 & -12.80 & 0.15 & -0.45 & -2.01 \\ -42.24 & -10.67 & 0.16 & -0.32 & 10.34 \\ 29.85 & 439.64 & 0.34 & -0.04 & -52.76 \\ -151.96 & 22.31 & 7.78 & -0.63 & 3.96 \\ 33.98 & 360.39 & -0.16 & 0.06 & 64.13 \end{pmatrix},$$

$$K_1 = \begin{pmatrix} 45.54 & -17.58 & -1.69 & -8.42 & 46.33 \end{pmatrix},$$
$$K_2 = \begin{pmatrix} 13.27 & -30.15 & -5.35 & -6.83 & 94.15 \end{pmatrix},$$
$$K_3 = \begin{pmatrix} 17.84 & -26.83 & -7.02 & -6.78 & 88.49 \end{pmatrix},$$
$$K_4 = \begin{pmatrix} 23.40 & -30.89 & -6.52 & -7.21 & 31.03 \end{pmatrix}.$$
$$\tag{40}$$

The above control law was only employed off-line, to calculate the weight matrix P of the local CLF $V(\cdot)$, and it was never used for controlling the system.

The following paragraph is dedicated to the stability of the resulting closed-loop system for each technique. Clearly, no stability guarantee can be obtained for the PWA system in closed-loop with the PID controller. For the horizon-1 MPC scheme developed in this paper, recursive feasibility implies asymptotic stability. However, recursive feasibility is not a priori guaranteed and hinges mainly on the constraint (27c) on the future evolution of λ_k^*. For all simulation scenarios case studies, the values $\Delta = 500$ and $M = 5$ proved to be large enough to guarantee recursive feasibility for the desired operating scenarios.

The worst case time (over numerous simulation scenarios) needed for computation of the control input for the proposed horizon-1 predictive controller was less than 2ms, which meets the timing constraints.

Different simulations were conducted, to evaluate the vehicle behavior in response to acceleration, deceleration, tip-in and tip-out maneuvers and a stress test, which are presented in the following subsections. Note that, although the PID controller does not enforce constraints on control command, its output was saturated in order to enforce the engine limitations, i.e., the torque limit $T_e^{\max}$.

In all figures (Fig. 4 - Fig. 8), the thick black dashed-line represents the vehicle velocity reference, the light blue line represents the PID results and the dark red line represents the horizon-1 predictive controller results. Also, in the top left sub-figure of each figure, the green dashed-line represents the upper bound for the variable λ_k given by (27c) and the black line represents the optimal value λ_k^*.

4.1 Scenario 1: Acceleration test

A first simulation test is performed on an acceleration scenario where the vehicle has to accelerate from 0 km/h to 30 km/h. In what follows the performance of the resulting closed-loop systems for the acceleration scenario is analyzed using the trajectories plotted in Fig. 4. The amplitude of the axle wrap is represented in Fig. 4, bottom left. The evolution of the CLF relaxation variable λ_k^* and the corresponding upper bound defined by (27c) for $\rho = 0.99$, $\Delta = 500$ and $M = 5$ is shown in Fig. 4, top left. It can be observed that λ_k^* may decrease or even go to 0, after which it is allowed to increase again, as long as this does not violate the upper

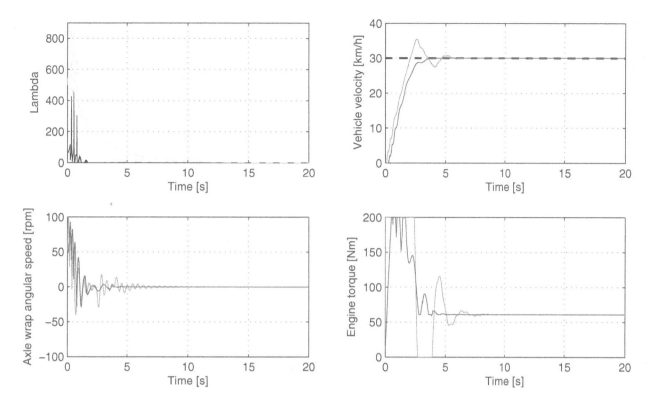

Figure 4: Scenario 1: Acceleration test.

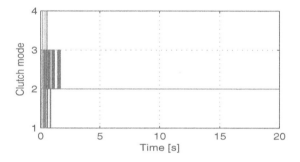

Figure 5: Scenario 1: Clutch mode of operation.

bound. However as $k \to \infty$, λ_k^* is forced to converge to 0. In Fig. 5 the clutch mode history was represented for the PID controller and for the horizon-1 predictive controller, to show that in the transient the closed-loop system frequently switches between the operating modes.

4.2 Scenario 2: Deceleration test

The second simulation scenario consist of decelerating the vehicle from 30 km/h to 10 km/h and the results obtained are illustrated in Fig. 6. Although both controllers obtained almost the same settling time, the PID controller produces some undesired axle wrap oscillations, indicating that the horizon-1 predictive controller has a superior behavior in terms of damping the drivetrain oscillations. The evolution of the CLF relaxation variable λ_k^* and the corresponding upper bound is illustrated in Fig. 6, top left. Although the upper bound starts from 500, due to Δ, only the values below

50 were plotted. This makes it possible to observe the evolution of λ_k^*.

4.3 Scenario 3: Tip-in tip-out maneuvers

The results of a tip-in, tip-out maneuver simulation, in which the reference vehicle velocity goes from 30 km/h to 10 km/h and back to 30 km/h, are presented in Fig. 7. It can be seen that the horizon-1 predictive controller has a slightly faster response with no overshoot when it approaches the reference velocity. Moreover, the oscillations of the axle wrap are damped much faster in the acceleration phase. In the deceleration phase, again, the PID controller produces undesired oscillations of the axle wrap. Note that the controller performance during deceleration is limited by the actuator authority. For this experiment the evolution of the CLF relaxation variable λ_k^* and the corresponding upper bound defined by (27c) are shown in Fig. 7, top left. Due to changing the reference vehicle velocity, the upper bound of the CLF relaxation variable defined by (27c) may become unfeasible, so whenever a change in the reference vehicle velocity occurs, the value of the upper bound was re-initialized.

4.4 Scenario 4: Stress test

The results of a stress test, in which the reference velocity is a square wave that changes rapidly between 30 km/h and 20 km/h, are presented in Fig. 8. The purpose is to check what happens to the axle wrap speed if it does not have enough time to settle between two set-point changes, which means that continuous perturbations may occur. The results illustrate how the horizon-1 predictive controller has again a smaller amplitude for the axle wrap angular speed, while the PID barely manages to cope with this kind of maneuver. Whenever a change in the reference vehicle velocity occurs, the value of the upper bound was re-initialized as done in

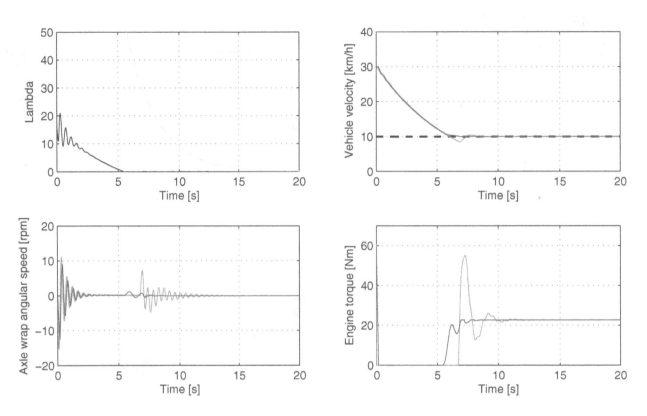

Figure 6: Scenario 2: Deceleration test.

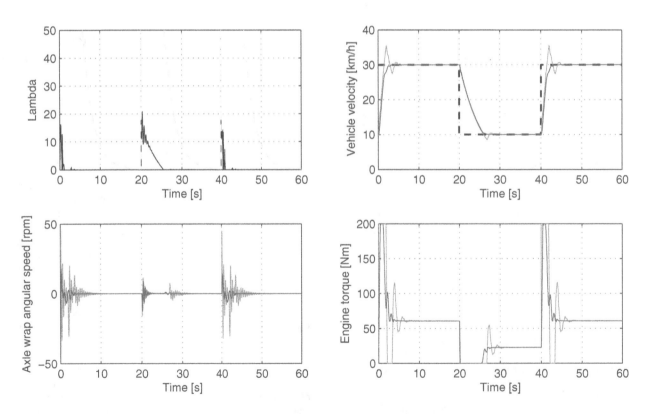

Figure 7: Scenario 3: Tip-in, tip-out maneuver simulation.

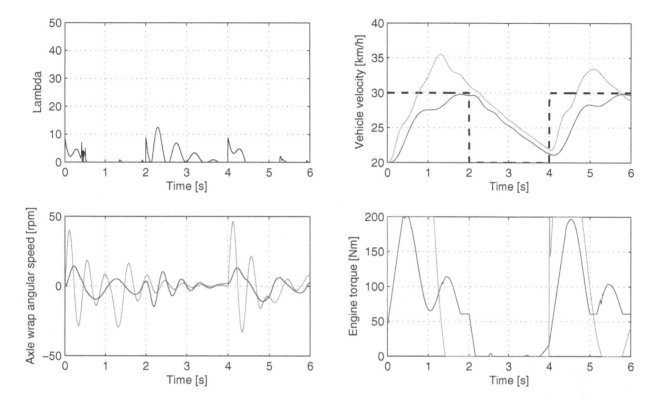

Figure 8: Scenario 4: Stress test.

the previous scenario. This is not visible in Fig. 8, top left, because the upper bound starts again from 500 and it does not reach values below 50 in such a short amount of time.

5. CONCLUSIONS

This paper considered the problem of damping driveline oscillations to improve passenger comfort. Recently, this problem has received an increased interest due to the introduction in several production vehicles of *dual-clutch powershift* automatic transmissions with dry clutches. As such, firstly, a novel, more accurate state-space piecewise affine drivetrain model with three inertias was derived, in which both driveshafts and clutch flexibilities were considered. Secondly, a horizon-1 predictive controller based on flexible Lyapunov functions was designed and tested in Matlab/Simulink for the developed drivetrain model. The simulations showed that the proposed control scheme can handle both the performance, physical constraints and the strict limitations on the computational complexity corresponding to vehicle drivetrain oscillations damping.

6. ACKNOWLEDGMENTS

The first two authors acknowledge the support of the National Center for Programs Management of Romania, under the research grant SICONA - 12100/2008.

The third author acknowledges the support of the Veni grant "Flexible Lyapunov Functions for Real-time Control", grant number 10230, warded by STW (Dutch Technology Foundation) and NWO (The Netherlands Organisation for Scientific Research).

The authors are grateful to *IEEE Fellow* Dr. Davor Hrovat and Prof. Dr. Corneliu Lazar for many useful discussions and helpful comments.

7. REFERENCES

[1] D. Hrovat, J. Asgari, and M. Fodor, "Automotive mechatronic systems," in *Mechatronic systems techniques and applications (vol. 2)*. Gordon and Breach Science Publishers, Inc., 2000, pp. 1–98.

[2] M. Pettersson, "Driveline modeling and control," Ph.D. dissertation, Linkoping University, 1997.

[3] P. Templin and B. Egardt, "An LQR torque compensator for driveline oscillation damping," in *IEEE Control Applications (CCA) & Intelligent Control (ISIC)*, St. Petersburg, Rusia, 2009, pp. 352–356.

[4] P. Templin, "Simultaneous estimation of driveline dynamics and backlash size for control design," in *IEEE International Conference on Control Applications (CCA)*, San Antonio, TX, USA, 2008, pp. 13–18.

[5] J. Baumann, D. D. Torkzadeh, A. Ramstein, U. Kiencke, and T. Schlegl, "Model-based predictive anti-jerk control," *Control Engineering Practice*, vol. 14, pp. 259–266, 2006.

[6] M. Bruce, B. Egardt, and S. Petterson, "On powertrain oscillation damping using feedforward and LQ feedback control," in *IEEE Conference on Control Applications (CCA)*, Toronto, Canada, 2005, pp. 1415–1420.

[7] B. Saerens, M. Diehl, J. Swevers, and E. Van den Bulck, "Model predictive control of automotive powertrains: first experimental results," in *47th IEEE Conference on Decision and Control*, Cancun, Mexico, 2008, pp. 5692–5697.

[8] A. Bemporad, F. Borrelli, L. Glielmo, and F. Vasca, "Optimal piecewise-linear control of dry clutch engagement," in *IFAC Workshop: Advances in Automotive Control*, Karlsruhe, Germany, 2001, pp. 33–38.

[9] P. Rostalski, T. Besselmann, M. Baric, F. Van Belzen, and

M. Morari, "A hybrid approach to modelling, control and state estimation of mechanical systems with backlash," *International Journal of Control*, vol. 80, pp. 1729–1740, 2007.

[10] M. Grotjahn, L. Quernheim, and S. Zemke, "Modelling and identification of car driveline dynamics for anti-jerk controller design," in *IEEE International Conference on Mechatronics*, Budapest, Hungary, 2006, pp. 131–136.

[11] U. Kiencke and L. Nielsen, *Automotive control systems: for engine, driveline, and vehicle.* Springer Verlag, 2005, vol. 290.

[12] A. C. Van Der Heijden, A. F. A. Serrarens, M. K. Camlibel, and H. Nijmeijer, "Hybrid optimal control of dry clutch engagement," *International Journal of Control*, vol. 80, pp. 1717–1728, 2007.

[13] J. Fredriksson, H. Weiefors, and B. Egardt, "Powertrain control for active damping of driveline oscillations," *Vehicle System dynamics*, vol. 37, pp. 359–376, 2002.

[14] M. Berriri, P. Chevrel, D. Lefebvre, and M. Yagoubi, "Active damping of automotive powertrain oscillations by a partial torque compensator," in *American Control Conference*, New York City, USA, 2007, pp. 5718–5723.

[15] M. Berriri, P. Chevrel, and D. Lefebvre, "Active damping of automotive powertrain oscillations by a partial torque compensator," *Control Engineering Practice*, vol. 16, pp. 874–883, 2008.

[16] S. Richard, P. Chevrel, and B. Maillard, "Active control of future vehicle drivelines," in *38th IEEE Conference on Decision and Control*, Phoenix, Arizona, 1999, pp. 3752–3757.

[17] P. Stewart, J. C. Zavala, and P. J. Fleming, "Automotive drive by wire controller design by multi-objectives techniques," *Control Engineering Practice*, vol. 13, pp. 257–264, 2005.

[18] D. Lefebvre, P. Chevrel, and S. Richard, "An H-infinity-based control design methodology dedicated to active control of vehicle longitudinal oscillations," *IEEE Transactions on Control Systems Technology*, vol. 11, pp. 948–956, 2003.

[19] A. Lagerberg and B. Egardt, "Model predictive control of automotive powertrains with backlash," in *16th IFAC World Congress*, Prague, Czech Republic, 2005.

[20] J. M. Maciejowski, *Predictive control with constraints.* Prentice Hall, Harlow, 2002.

[21] M. Lazar, "Flexible control Lyapunov functions," in *American Control Conference*, St. Louis, MO, USA, 2009, pp. 102–107.

[22] S. Di Cairano, D. Yanakiev, A. Bemporad, I. V. Kolmanovsky, and D. Hrovat, "Model predictive powertrain control: an application to idle speed regulation," ser. Lecture Notes in Control and Information Sciences. Springer, 2010, vol. 402, pp. 183–194.

[23] R. M. Hermans, M. Lazar, S. Di Cairano, and I. V. Kolmanovsky, "Low complexity model predictive control of electromagnetic actuators with a stability guarantee," in *28th American Control Conference*, St. Louis, Missouri, 2009, pp. 2708–2713.

[24] J. Stewart, *Calculus, early transcendentals.* Thomson Brooks/Cole, 2003.

[25] M. Lazar, "Model predictive control of hybrid systems: Stability and robustness," Ph.D. dissertation, Eindhoven University of Technology, The Netherlands, 2006.

[26] A. O'Dwyer, *Handbook of PI and PID controller tuning rules.* Imperial College Press, 2006, vol. 2.

APPENDIX

Table 1: Simulation vehicle parameter values

Symbol	Value	Measure Unit	Description
J_e	0.17	[kg m^2]	Engine inertia
J_t	0.014	[kg m^2]	Transmission inertia
J_f	0.031	[kg m^2]	Final drive inertia
J_w	1	[kg m^2]	Wheel inertia
d_d	65	[Nms/rad]	Flexible driveshaft damping
k_d	5000	[Nm/rad]	Flexible driveshaft stiffness
d_e	0.159	[Nms/rad]	Engine damping
d_t	0.1	[Nms/rad]	Transmission damping
d_f	0.1	[Nms/rad]	Final drive damping
d_w	0.1	[Nms/rad]	Wheel damping
g_t	3.5		Gearbox ratio (1st gear)
g_f	3.7		Final drive ratio
m_{COG}	1400	[kg]	Vehicle mass
r_w	0.32	[m]	Wheel radius
c_{r1}	0.01	[Nm/kg]	Rolling coefficient
c_{r2}	0.36	[Nms/rad]	Approximation coefficient
c_d	0.3	[rad^{-2}]	Airdrag coefficient
ρ_{air}	1.2	[kg/m^3]	Air density
A_f	2.7	[m^2]	Frontal area of the vehicle
g	9.8	[m/s^2]	Gravitational acceleration
χ_{road}	0	[rad]	Road slope
θ_1	0.1745	[rad]	Clutch switching boundary
θ_2	0.2094	[rad]	Clutch switching boundary
d_{c1}	0	[Nms/rad]	Clutch damping (open)
d_{c2}	3	[Nms/rad]	Clutch damping (closing)
d_{c3}	6	[Nms/rad]	Clutch damping (closed)
d_{c4}	10	[Nms/rad]	Clutch damping (locked)
k_{c1}	0	[Nm/rad]	Clutch stiffness (open)
k_{c2}	800	[Nm/rad]	Clutch stiffness (closing)
k_{c3}	1600	[Nm/rad]	Clutch stiffness (closed)
k_{c4}	3200	[Nm/rad]	Clutch stiffness (locked)
T_e^{max}	200	[Nm]	Maximum engine torque
T_e^{Δ}	6	[Nm]	Maximum engine torque increase/decrease
ω_e^{min}	62.83	[rad/s]	Engine idle speed
$\omega_e^{closing}$	83.76	[rad/s]	Engine closing speed
ω_e^{max}	628.3	[rad/s]	Maximum engine speed
ω_w^{min}	0	[rad/s]	Minimum wheel speed
ω_w^{max}	250	[rad/s]	Maximum wheel speed

Synthesis of Switching Controllers using Approximately Bisimilar Multiscale Abstractions[*]

Javier Cámara
POP ART project
INRIA Grenoble Rhône-Alpes
38334 Saint-Ismier cedex,
France

Antoine Girard
Laboratoire Jean Kuntzmann
Université de Grenoble
B.P. 53, 38041 Grenoble,
France

Gregor Gössler
POP ART project
INRIA Grenoble Rhône-Alpes
38334 Saint-Ismier cedex,
France

ABSTRACT

When available, discrete abstractions provide an appealing approach to controller synthesis. Recently, an approach for computing discrete abstractions of incrementally stable switched systems has been proposed, using the notion of approximate bisimulation. This approach is based on sampling of time and space where the sampling parameters must satisfy some relation in order to achieve a certain precision. Particularly, the smaller the sampling period, the finer the lattice approximating the state-space and the larger the number of states in the abstraction. This renders the use of these abstractions for synthesis of fast switching controllers computationally prohibitive. In this paper, we present a novel class of multiscale discrete abstractions for switched systems that allows us to deal with fast switching while keeping the number of states in the abstraction at a reasonable level. The transitions of our abstractions have various durations: for transitions of longer duration, it is sufficient to consider abstract states on a coarse lattice; for transitions of shorter duration, it becomes necessary to use finer lattices. These finer lattices are effectively used only on a restricted area of the state-space where the fast switching occurs. We show how to use these abstractions for multiscale synthesis of self-triggered switching controllers for reachability specifications under time optimization. We illustrate the merits of our approach by applying it to the boost DC-DC converter.

Categories and Subject Descriptors

I.2.8 [**Artificial Intelligence**]: Problem Solving, Control Methods and Search—*Control theory*

General Terms

Design, Performance

[*]This work was supported by the Agence Nationale de la Recherche (VEDECY project - ANR 2009 SEGI 015 01).

Keywords

Switched systems, Multiscale abstractions, Approximate bisimulation, Optimal control, Self-triggered controllers.

1. INTRODUCTION

The use of discrete abstractions for continuous dynamics has become standard in hybrid systems design (see e.g. [15] and the references therein). The main advantage of this approach is that it offers the possibility to leverage controller synthesis techniques developed in the areas of supervisory control of discrete-event systems [13] or algorithmic game theory [3]. Historically, the first attempts to compute discrete abstractions for hybrid systems were based on traditional systems behavioral relationships such as simulation or bisimulation [10], initially proposed for discrete systems most notably in the area of formal methods. These notions require inclusion or equivalence of observed behaviors which is often too restrictive when dealing with systems observed over metric spaces. For such systems, a more natural abstraction requirement is to ask for closeness of observed behaviors. This leads to the notions of approximate simulation and bisimulation introduced in [7].

These notions enabled the computation of approximately equivalent discrete abstractions for several classes of dynamical systems, including nonlinear control systems with or without disturbances (see [12] and [11], respectively) and switched systems [8]. These approaches are based on sampling of time and space where the sampling parameters must satisfy some relation in order to obtain abstractions of a prescribed precision. Particularly, it should be noticed that the smaller the time sampling parameter, the finer the lattice used for approximating the state-space; this may result in abstractions with a very large number of states when the sampling period is small. However, there are a number of applications where sampling has to be fast; though this is generally necessary only on a small part of the state-space. For instance, in sliding mode control of switched systems (see e.g. [14]), the sampling has to be fast only in the neighborhood of a sliding surface.

In this paper, we present a novel class of multiscale discrete abstractions for incrementally stable switched systems that allows us to deal with fast switching while keeping the number of states in the abstraction at a reasonable level. Following the self-triggered control paradigm [17, 18, 2], we assume that the controller of the switched system has to decide the control input and the time period during which it will be applied before the controller executes again. In this context, it is natural to consider abstractions where transitions have various durations. For transitions of longer duration, it is sufficient to consider abstract states on a coarse lattice. For transitions of shorter duration, it becomes necessary to use finer lattices. These finer lattices are effectively used only on a restricted area of the state-space where the fast switching occurs.

These abstractions allow us to use multiscale iterative approaches for controller synthesis as follows. An initial controller is synthesized based on the dynamics of the abstraction at the coarsest scale where only transitions of longer duration are enabled. An analysis of this initial controller allows us to identify regions of the state-space where transitions of shorter duration may be useful (e.g. to improve the performance of the controller). Then, the controller is refined by enabling transitions of shorter duration in the identified regions. The last two steps can be repeated until we are satisfied with the obtained controller.

The concept of approximately bisimilar multiscale discrete abstractions has already been explored in [16]. In this work, the multiscale feature was used for accomodating locally the precision of the abstraction while the time sampling period remained constant. On the contrary, our approach seeks for a uniform precision but varying time sampling periods.

The paper is organized as follows. In section 2, we present previous results on switched systems and approximate bisimulation. In section 3, we introduce our multiscale discrete abstractions for a class of incrementally stable switched systems. In section 4, we show how to use them for multiscale synthesis of controllers for reachability specifications under time optimization. Finally, we illustrate the merits of our approach by applying it to the boost DC-DC converter.

2. PRELIMINARIES

2.1 Incrementally stable switched systems

In this paper, we shall consider the class of switched systems formalized in the following definition.

DEFINITION 2.1. *A switched system is a quadruple* $\Sigma = (\mathbb{R}^n, P, \mathcal{P}, F)$, *where:*

- $\mathbb{R}^n$ *is the state space;*

- $P = \{1, \ldots, m\}$ *is the finite set of modes;*

- $\mathcal{P}$ *is the set of piecewise constant functions from $\mathbb{R}^+$ to P, continuous from the right and with a finite number of discontinuities on every bounded interval of $\mathbb{R}^+$;*

- $F = \{f_1, \ldots, f_m\}$ *is a collection of smooth vector fields indexed by P.*

A *switching signal* of Σ is a function $\mathbf{p} \in \mathcal{P}$, the discontinuities of $\mathbf{p}$ are called *switching times*. A piecewise $\mathcal{C}^1$ function $\mathbf{x} : \mathbb{R}^+ \to \mathbb{R}^n$ is said to be a *trajectory* of Σ if it is continuous and there exists a switching signal $\mathbf{p} \in \mathcal{P}$ such that, at each $t \in \mathbb{R}^+$ where the function $\mathbf{p}$ is continuous, $\mathbf{x}$ is continuously differentiable and satisfies:

$$\dot{\mathbf{x}}(t) = f_{\mathbf{p}(t)}(\mathbf{x}(t)).$$

We assume that the vector fields $f_1, \ldots, f_m$ are such that for all initial conditions and switching signals, there is existence and uniqueness of the trajectory of Σ. We will denote $\mathbf{x}(t, x, \mathbf{p})$ the point reached at time $t \in \mathbb{R}^+$ from the

initial condition x under the switching signal $\mathbf{p}$. We will denote $\mathbf{x}(t, x, p)$ the point reached by Σ at time $t \in \mathbb{R}^+$ from the initial condition x under the constant switching signal $\mathbf{p}(t) = p$, for all $t \in \mathbb{R}^+$.

The results presented in this paper rely on some notion of incremental stability (i.e. δ-GUAS [1, 8]). Essentially, incremental stability of a switched system means that all the trajectories associated with the same switching signal converge asymptotically to the same reference trajectory independently of their initial condition. Let us remark that, unless all vector fields share a common equilibrium, this does not imply stability. In the following, $\|.\|$ denotes the usual Euclidean norm over $\mathbb{R}^n$. Incremental stability of a switched system can be characterized using Lyapunov functions:

DEFINITION 2.2. *A smooth function $V : \mathbb{R}^n \times \mathbb{R}^n \to \mathbb{R}^+$ is a common δ-GUAS Lyapunov function for Σ if there exist $\mathcal{K}_\infty$ functions[1] $\underline{\alpha}$, $\overline{\alpha}$ and $\kappa > 0$ such that for all $x, y \in \mathbb{R}^n$, for all $p \in P$:*

$$\underline{\alpha}(\|x - y\|) \leq V(x, y) \leq \overline{\alpha}(\|x - y\|); \tag{1}$$
$$\frac{\partial V}{\partial x}(x, y) f_p(x) + \frac{\partial V}{\partial y}(x, y) f_p(y) \leq -\kappa V(x, y). \tag{2}$$

The computation of a δ-GUAS Lyapunov function is generally hard and out of the scope of the paper. However, if all vector fields are affine one can look for a quadratic δ-GUAS Lyapunov function by solving a set of linear matrix inequalities. As in [8], we will make in the following the supplementary assumption on the δ-GUAS Lyapunov function that there exists a $\mathcal{K}_\infty$ function γ such that

$$\forall x, y, z \in \mathbb{R}^n, \ |V(x, y) - V(x, z)| \leq \gamma(\|y - z\|). \tag{3}$$

This assumption was shown to be not restrictive provided V is smooth and we are interested in the dynamics of Σ on a compact subset of $\mathbb{R}^n$, which is often the case in practice.

In [8], it was shown that under the existence of common δ-GUAS Lyapunov function and equation (3) it is possible to compute discrete abstractions that are approximately equivalent to a switched system.

2.2 Approximate bisimulation

In this section, we present a notion of approximate equivalence which will relate a switched system to the discrete systems that we construct. We start by introducing the class of transition systems which allows us to model switched and discrete systems in a common framework.

DEFINITION 2.3. *A transition system is a tuple*

$$T = (Q, L, \longrightarrow, O, H, I)$$

consisting of:

- *a set of states Q;*

- *a set of labels L;*

- *a transition relation $\longrightarrow \subseteq Q \times L \times Q$;*

- *an output set O;*

- *an output function $H : Q \to O$;*

- *a set of initial states $I \subseteq Q$.*

---

[1]A continuous function $\gamma : \mathbb{R}^+ \to \mathbb{R}^+$ is said to belong to class $\mathcal{K}_\infty$ if it is strictly increasing, $\gamma(0) = 0$ and $\gamma(r) \to \infty$ when $r \to \infty$.

T is said to be *metric* if the output set O is equipped with a metric d, *discrete* if Q and L are finite or countable sets.

The transition $(q, l, q') \in \longrightarrow$ will be denoted $q \xrightarrow{l} q'$, or alternatively $q' \in \text{Succ}(q, l)$; this means that the system can evolve from state q to state q' under the action l. An action $l \in L$ belongs to the set of *enabled actions* at state q, denoted $\text{Enab}(q)$, if $\text{Succ}(q, l) \neq \emptyset$. If $\text{Enab}(q) = \emptyset$, then q is said to be a *blocking* state; otherwise it is said to be *non-blocking*. If all states are non-blocking, we say that the transition system T is non-blocking. The transition system is said to be *deterministic* if for all $q \in Q$ and $l \in \text{Enab}(q)$, $\text{Succ}(q, l)$ has only one element. A *trajectory* of the transition system is a finite sequence of transitions

$$\sigma = q_0 \xrightarrow{l_0} q_1 \xrightarrow{l_1} q_2 \xrightarrow{l_2} \ldots \xrightarrow{l_{N-1}} q_N.$$

$N \in \mathbb{N}$ is referred to as the *length* of the trajectory. The associated *observed behavior* is the finite sequence of outputs $o_0 o_1 o_2 \ldots o_N$ where $o_i = H(q_i)$, for all $i \in \{0, \ldots, N\}$.

Transition systems can serve as models for switched systems. Given a switched system $\Sigma = (\mathbb{R}^n, P, \mathcal{P}, F)$, we define the transition system $T(\Sigma) = (Q, L, \longrightarrow, O, H, I)$, where the set of states is $Q = \mathbb{R}^n$; the set of labels is $L = P \times \mathbb{R}^+$; the transition relation is given by

$$x \xrightarrow{p, \tau} x' \text{ iff } \mathbf{x}(\tau, x, p) = x',$$

i.e. the switched system Σ goes from state x to state x' by applying the constant mode p for a duration τ; the set of outputs is $O = \mathbb{R}^n$; the observation map H is the identity map over $\mathbb{R}^n$; the set of initial states is $I = \mathbb{R}^n$. The transition system $T(\Sigma)$ is non-blocking and deterministc, it is metric when the set of outputs $O = \mathbb{R}^n$ is equipped with the metric $d(x, x') = \|x - x'\|$. Note that the state space of $T(\Sigma)$ is uncountable.

Traditional equivalence relationships for transition systems rely on equality of observed behaviors. One of the most common notions is that of bisimulation equivalence [10, 15]. For metric transition systems, requiring strict equivalence of observed behaviors is often quite restrictive. A natural relaxation is to ask for closeness of observed behaviors where closeness is measured with respect to the metric on the output set. This leads to the notion of approximate bisimulation introduced in [7].

DEFINITION 2.4. *Let* $T_i = (Q_i, L, \xrightarrow{i}, O, H_i, I_i)$, *with* $i = 1, 2$ *be metric transition systems with the same sets of labels* L *and outputs* O *equipped with the metric* d. *Let* $\varepsilon \geq 0$ *be a given precision. A relation* $R \subseteq Q_1 \times Q_2$ *is said to be an* ε-*approximate bisimulation relation between* T_1 *and* T_2 *if for all* $(q_1, q_2) \in R$:

- $d(H_1(q_1), H_2(q_2)) \leq \varepsilon$;

- $\forall q_1 \xrightarrow{l}_1 q_1', \exists q_2 \xrightarrow{l}_2 q_2'$, *such that* $(q_1', q_2') \in R$;

- $\forall q_2 \xrightarrow{l}_2 q_2', \exists q_1 \xrightarrow{l}_1 q_1'$, *such that* $(q_1', q_2') \in R$.

The transition systems T_1 *and* T_2 *are said to be approximately bisimilar with precision* ε, *denoted* $T_1 \sim_\varepsilon T_2$, *if:*

- $\forall q_1 \in I_1, \exists q_2 \in I_2$, *such that* $(q_1, q_2) \in R$;

- $\forall q_2 \in I_2, \exists q_1 \in I_1$, *such that* $(q_1, q_2) \in R$.

If T_1 is a system we want to control and T_2 is a simpler system that we want to use for controller synthesis, then T_2 is called an *approximately bisimilar abstraction* of T_1.

3. MULTISCALE ABSTRACTIONS FOR SWITCHED SYSTEMS

Let Σ be a switched system and $\tau > 0$ a time-sampling parameter, $T_\tau(\Sigma)$ is the sub-transition system of $T(\Sigma)$ obtained by selecting the transitions of $T(\Sigma)$ that describe trajectories of duration τ. This is a natural assumption when the switching in Σ is determined by a time-triggered controller with period τ. In [8], an approach to compute approximately bisimilar abstractions of $T_\tau(\Sigma)$ was presented. It is based on a quantization of the state-space $\mathbb{R}^n$ which is approximated by the lattice:

$$[\mathbb{R}^n]_\eta = \left\{ q \in \mathbb{R}^n \mid q[i] = k_i \frac{2\eta}{\sqrt{n}},\ k_i \in \mathbb{Z},\ i = 1, \ldots, n \right\}$$

where $q[i]$ is the i-th coordinate of q and $\eta > 0$ is a state space discretization parameter. The resulting abstraction $T_{\tau, \eta}(\Sigma)$ is discrete since its set of states $[\mathbb{R}^n]_\eta$ and its set of actions P are respectively countable and finite. It is shown in [8] that under the existence of a common δ-GUAS Lyapunov function and equation (3), $T_\tau(\Sigma)$ and $T_{\tau, \eta}(\Sigma)$ are approximately bisimilar:

THEOREM 3.1. *[8] Consider a switched system* Σ, *time and state space sampling parameters* $\tau, \eta > 0$ *and a desired precision* $\varepsilon > 0$. *If there exists a common* δ-*GUAS Lyapunov function* V *for* Σ *such that equation (3) holds and*

$$\eta \leq \min \left\{ \gamma^{-1} \left((1 - e^{-\kappa \tau}) \underline{\alpha}(\varepsilon) \right), \overline{\alpha}^{-1} \left(\underline{\alpha}(\varepsilon) \right) \right\} \quad (4)$$

then, $T_\tau(\Sigma) \sim_\varepsilon T_{\tau, \eta}(\Sigma)$.

Particularly, it should be noted that given a time sampling parameter $\tau > 0$ and a desired precision $\varepsilon > 0$, it is always possible to choose $\eta > 0$ such that equation (4) holds. This essentially means that approximately bisimilar discrete abstractions of arbitrary precision can be computed for $T_\tau(\Sigma)$. However, the smaller τ or ε, the smaller η must be to satisfy equation (4). In practice, for a small time sampling parameter τ, the ratio ε/η can be very large and discrete abstractions with an acceptable precision may have a very large number of states (see e.g. [8]).

Unfortunately, there are a number of applications where the switching has to be fast; though this fast switching is generally necessary only on a restricted part of the state space. For instance, in sliding mode control (see e.g. [14]), the switching is fast only in the neighborhood of a sliding surface. In order to enable fast switching while dealing with abstractions with a reasonable number of states, one may consider discrete abstractions enabling transitions of different durations. For transitions of long duration, it is sufficient to consider abstract states on a coarse lattice to meet the desired precision ε. As we consider transitions of shorter durations, it becomes necessary to use finer lattices for the abstract state-space. These finer lattices are effectively used only on a restricted area of the state space, where the fast switching occurs. This makes it possible to keep the number of states in the abstraction at a reasonable level. This results naturally in a notion of multiscale discrete abstraction presented next.

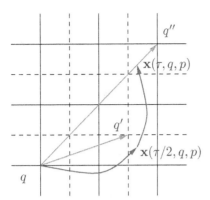

Figure 1: Principle for the computation of the discrete abstraction: $q' = \mathrm{Succ}_2(q, (p, \tau/2))$ where $q' = \arg\min_{r \in Q_2^1}(\|\mathbf{x}(\tau/2, q, p) - r\|)$ **and** $q'' = \mathrm{Succ}_2(q, (p, \tau))$ **where** $q'' = \arg\min_{r \in Q_2^0}(\|\mathbf{x}(\tau, q, p) - r\|)$.

We work with a sub-transition system of $T(\Sigma)$ obtained by selecting the transitions of $T(\Sigma)$ that describe trajectories of duration that are dyadic fractions of a time sampling parameter $\tau > 0$. The duration of the trajectories are elements of the finite set $\Theta_\tau^N = \{2^{-s}\tau \mid s = 0, \ldots, N\}$ for some scale parameter $N \in \mathbb{N}$. Intuitively, one can think that the switched system Σ has to be controlled by a controller with time period $2^{-N}\tau$. The controller chooses to apply a selected mode for a given duration in the set Θ_τ^N: the controller may not need to take a decision at each period $2^{-N}\tau$. This enables the design of self-triggered controllers [17, 18, 2], where the controller not only chooses the control action but also the next instant where it should be executed.

Given a switched system $\Sigma = (\mathbb{R}^n, P, \mathcal{P}, F)$, a time sampling parameter $\tau > 0$, and a scale parameter $N \in \mathbb{N}$, we define the transition system $T_\tau^N(\Sigma) = (Q_1, L, \xrightarrow{\;\;}_1, O, H_1, I_1)$ where the set of states is $Q_1 = \mathbb{R}^n$; the set of labels is $L = P \times \Theta_\tau^N$; the transition relation is given by

$$x \xrightarrow{p, 2^{-s}\tau}_1 x' \text{ iff } \mathbf{x}(2^{-s}\tau, x, p) = x';$$

the set of outputs is $O = \mathbb{R}^n$; the observation map H_1 is the identity map over $\mathbb{R}^n$; the set of initial states is $I_1 = \mathbb{R}^n$.

The computation of a discrete abstraction of $T_\tau^N(\Sigma)$ can be done by the following approach. We approximate the set of states $Q_1 = \mathbb{R}^n$ by a sequence of embedded lattices: for $s = 0, \ldots, N$, let $Q_2^s = [\mathbb{R}^n]_{2^{-s}\eta}$, i.e.

$$Q_2^s = \left\{ q \in \mathbb{R}^n \;\middle|\; q[i] = k_i \frac{2^{-s+1}\eta}{\sqrt{n}}, \; k_i \in \mathbb{Z}, \; i = 1, \ldots, n \right\}.$$

where $\eta > 0$ is a state space discretization parameter. Let us remark that we have the inclusions $Q_2^0 \subseteq Q_2^1 \subseteq \cdots \subseteq Q_2^N$. By simple geometrical considerations, we can check that for all $x \in \mathbb{R}^n$, for all $s = 0, \ldots, N$, there exists $q \in Q_2^s$ such that $\|x - q\| \leq 2^{-s}\eta$.

We now define the abstraction of $T_\tau^N(\Sigma)$ as the transition system $T_{\tau,\eta}^N(\Sigma) = (Q_2, L, \xrightarrow{\;\;}_2, O, H_2, I_2)$, where the set of states is $Q_2 = Q_2^N$; the set of actions remains the same $L = P \times \Theta_\tau^N$; the transition relation is given by

$$q \xrightarrow{p, 2^{-s}\tau}_2 q' \text{ iff } q' = \arg\min_{r \in Q_2^s}(\|\mathbf{x}(2^{-s}\tau, q, p) - r\|).$$

If the minimizer $r \in Q_2^s$ is not unique, then one can choose arbitrarily one of them. By definition of the set $Q_2^s = [\mathbb{R}^n]_{2^{-s}\eta}$, if $q \xrightarrow{p, 2^{-s}\tau}_2 q'$ then we have $\|\mathbf{x}(2^{-s}\tau, q, p) - q'\| \leq 2^{-s}\eta$. The approximation principle is illustrated in Figure 1. The set of outputs remains the same $O = \mathbb{R}^n$; the observation map H_2 is the natural inclusion map from Q_2^N to $\mathbb{R}^n$, i.e. $H_2(q) = q$; the set of initial states is $I_2 = Q_2^0$.

It is important to note that all the transitions of duration $2^{-s}\tau$ end in states belonging to Q_2^s. This means that the states on the finer lattices are only accessible by transitions of shorter duration. Note that the transition system $T_{\tau,\eta}^N(\Sigma)$ is discrete since its sets of states and actions are respectively countable and finite. Also, if we only consider transitions of duration τ, the dynamics of $T_{\tau,\eta}^N(\Sigma)$ coincides with that of the transition $T_{\tau,\eta}(\Sigma)$ defined in [8]. Both transition systems $T_\tau^N(\Sigma)$ and $T_{\tau,\eta}^N(\Sigma)$ are non-blocking and deterministic.

THEOREM 3.2. *Consider a switched system Σ, time and state space sampling parameters $\tau, \eta > 0$, scale parameter $N \in \mathbb{N}$, and a desired precision $\varepsilon > 0$. Let us assume that there exists a common δ-GUAS Lyapunov function V for Σ such that equation (3) holds. If*

$$\eta \leq \min\left\{ \min_{s=0\ldots N} \left[2^s \gamma^{-1}\left((1 - e^{-\kappa 2^{-s}\tau})\underline{\alpha}(\varepsilon)\right) \right], \overline{\alpha}^{-1}\left(\underline{\alpha}(\varepsilon)\right) \right\} \tag{5}$$

then $R = \{(x, q) \in Q_1 \times Q_2 \mid V(x, q) \leq \underline{\alpha}(\varepsilon)\}$ is an ε-approximate bisimulation relation between $T_\tau^N(\Sigma)$ and $T_{\tau,\eta}^N(\Sigma)$. Moreover, $T_\tau^N(\Sigma) \sim_\varepsilon T_{\tau,\eta}^N(\Sigma)$.

PROOF. We start by showing that the relation R is an ε-approximate bisimulation relation. Let $(x, q) \in R$, then $V(x, q) \leq \underline{\alpha}(\varepsilon)$ and we have from equation (1) that

$$\|x - q\| \leq \underline{\alpha}^{-1}(V(x, q)) \leq \varepsilon.$$

Thus, the first condition of Definition 2.4 holds.

Let $x \xrightarrow{p, 2^{-s}\tau}_1 x'$, then $x' = \mathbf{x}(2^{-s}\tau, x, p)$. Let

$$q' = \arg\min_{r \in Q_2^s}(\|\mathbf{x}(2^{-s}\tau, q, p) - r\|),$$

then $q \xrightarrow{p, 2^{-s}\tau}_2 q'$ and $\|\mathbf{x}(2^{-s}\tau, q, p) - q'\| \leq 2^{-s}\eta$. Let us check that $(x', q') \in R$. From equation (3),

$$
\begin{aligned}
|V(x', q') - V(x', \mathbf{x}(2^{-s}\tau, q, p))| &\leq \gamma(\|q' - \mathbf{x}(2^{-s}\tau, q, p)\|) \\
&\leq \gamma(2^{-s}\eta).
\end{aligned}
$$

It follows that

$$
\begin{aligned}
V(x', q') &\leq V(x', \mathbf{x}(2^{-s}\tau, q, p)) + \gamma(2^{-s}\eta) \\
&\leq V(\mathbf{x}(2^{-s}\tau, x, p), \mathbf{x}(2^{-s}\tau, q, p)) + \gamma(2^{-s}\eta) \\
&\leq e^{-\kappa 2^{-s}\tau} V(x, q) + \gamma(2^{-s}\eta)
\end{aligned}
$$

because V is a δ-GUAS Lyapunov function for Σ. Then,

$$V(x', q') \leq e^{-\kappa 2^{-s}\tau}\underline{\alpha}(\varepsilon) + \gamma(2^{-s}\eta) \leq \underline{\alpha}(\varepsilon)$$

because of equation (5) and γ is a $\mathcal{K}_\infty$ function, Hence, $(x', q') \in R$. In a similar way, we can prove that, for all $q \xrightarrow{p, 2^{-s}\tau}_2 q'$, there is $x \xrightarrow{p, 2^{-s}\tau}_1 x'$ such that $(x', q') \in R$. Hence R is an ε-approximate bisimulation relation between $T_\tau^N(\Sigma)$ and $T_{\tau,\eta}^N(\Sigma)$.

By definition of $I_2 = Q_{2,0} = [\mathbb{R}^n]_\eta$, for all $x \in I_1 = \mathbb{R}^n$, there exists $q \in I_2$ such that $\|x - q\| \leq \eta$. Then,

$$V(x, q) \leq \overline{\alpha}(\|x - q\|) \leq \overline{\alpha}(\eta) \leq \underline{\alpha}(\varepsilon)$$

because of equation (5) and $\overline{\alpha}$ is a $\mathcal{K}_\infty$ function. Hence, $(x,q) \in R$. Conversely, for all $q \in I_2$, $x = q \in \mathbb{R}^n = I_1$, then $V(x,q) = 0$ and $(x,q) \in R$. Therefore, $T_\tau^N(\Sigma)$ and $T_{\tau,\eta}^N(\Sigma)$ are approximately bisimilar with precision ε. $\square$

It is interesting to note that given a time sampling parameter $\tau > 0$ and a scale parameter $N \in \mathbb{N}$, for all desired precisions $\varepsilon > 0$, there exists $\eta > 0$ such that equation (4) holds. This essentially means that approximately bisimilar multiscale abstractions of arbitrary precision can be computed for $T_\tau^N(\Sigma)$.

4. CONTROLLER SYNTHESIS USING MULTISCALE ABSTRACTIONS

We illustrate the use of multiscale abstractions for synthesizing sub-optimal reachability controllers. This problem was considered in [9, 6] based on the use of uniform discrete abstractions. We extend the synthesis algorithm to multiscale abstractions that are computed on-the-fly, so as to provide a scalable trade-off between precision and cost, while guaranteeing a lower bound on the performance of the closed-loop system.

4.1 Problem formulation

Let us consider a system $T = (Q, L, \longrightarrow, O, H, I)$. For simplicity, we assume that T is deterministic. This is satisfied by the transition systems $T_\tau^N(\Sigma)$ and $T_{\tau,\eta}^N(\Sigma)$ defined in the previous section. In the following, we consider deterministic static state-feedback controllers. However, we just use the term controller for brevity.

DEFINITION 4.1. *A controller for T is a map $\mathcal{S} : Q \to L \cup \{\emptyset\}$ where $\emptyset$ is the dummy symbol. It is well-defined if for all $q \in Q$, $\mathcal{S}(q) \in \mathrm{Enab}(q) \cup \{\emptyset\}$. The dynamics of the controlled system is described by the transition system $T_\mathcal{S} = (Q, L, \xrightarrow{}_\mathcal{S}, O, H, I)$ where the transition relation is given by*

$$q \xrightarrow{l}_\mathcal{S} q' \iff \left[(l = \mathcal{S}(q)) \wedge (q \xrightarrow{l} q') \right].$$

$\mathcal{S}(q) = \emptyset$ essentially means that the controller is not defined at q. q is a blocking state of $T_\mathcal{S}$ if and only if $\mathcal{S}(q) = \emptyset$. Since we assumed that T is deterministic, the system $T_\mathcal{S}$ is deterministic as well. Moreover, since $\mathcal{S}$ enables at most one action at each state, it follows that for any non-blocking state $q \in Q$ there exists a unique transition starting from q. We assume that for all transitions $q \xrightarrow{l} q'$, the time needed by T for moving from state q to state q' is given by $\delta(l)$ where $\delta : L \to \mathbb{R}^+$. Then, for all trajectories of T,

$$\sigma = q_0 \xrightarrow{l_0} q_1 \xrightarrow{l_1} \ldots \xrightarrow{l_{N-1}} q_N$$

we define its *duration* as $\Delta(\sigma) = \delta(l_0) + \delta(l_1) + \cdots + \delta(l_{N-1})$. For instance, for the transition systems $T_\tau^N(\Sigma)$ and $T_{\tau,\eta}^N(\Sigma)$, we have that $L = P \times \Theta_{\tau,N}$ and for all $l = (p, 2^{-s}\tau) \in L$, $\delta(l) = 2^{-s}\tau$. The time-optimal reachability problem consists in steering the state of a system to a desired target and keeping the system safe along the way while minimizing the duration of the trajectory.

DEFINITION 4.2. *Let T be a transition system and $\mathcal{S}$ a controller. Let $O_S \subseteq O$ and $O_T \subseteq O_S$ be sets of outputs associated with safe states and target states, respectively. The*

entry time of $T_\mathcal{S}$ from $q_0 \in Q$ for specification (O_S, O_T) is the smallest $\theta \in \mathbb{R}^+$ such that there exists a trajectory of $T_\mathcal{S}$ starting from q_0, $\sigma = q_0 \xrightarrow{l_0}_\mathcal{S} \ldots \xrightarrow{l_{N-1}}_\mathcal{S} q_N$ with $\Delta(\sigma) = \theta$, and such that:

$$\forall i \in \{0, \ldots, N\},\ H(q_i) \in O_S \ \ and \ \ H(q_N) \in O_T.$$

The entry time is denoted by $J(T_\mathcal{S}, O_S, O_T, q_0)$. If such a θ does not exist, then we define $J(T_\mathcal{S}, O_S, O_T, q_0) = +\infty$.

We now define the notion of time-optimal controller:

DEFINITION 4.3. *We say that a controller $\mathcal{S}^*$ for T is time-optimal for specification (O_S, O_T) if for all controllers $\mathcal{S}$ the following holds:*

$$\forall q \in I,\ J(T_{\mathcal{S}^*}, O_S, O_T, q) \leq J(T_\mathcal{S}, O_S, O_T, q).$$

4.2 Hierarchical synthesis using abstractions

Let us assume that we want to compute a controller for the switched system $T_\tau^N(\Sigma)$ for a reachability specification (O_S, O_T). An approach to compute a sub-optimal controller using an approximately bisimilar abstraction was proposed in [6]. It consists in computing a controller for the discrete abstraction for a modified specification where the safe and target sets are given by contractions of O_S and O_T.

Let $O' \subseteq O$ and $\varepsilon \geq 0$, the ε-contraction of O' is the subset of O defined as follows

$$C_\varepsilon(O') = \left\{ o' \in O' \mid \forall o \in O, d(o, o') \leq \varepsilon \implies o \in O' \right\}.$$

We start by synthesizing a controller $\tilde{\mathcal{S}}$ for the discrete abstraction $T_{\tau,\eta}^N(\Sigma)$ for the modified reachability specification $(C_\varepsilon(O_S), C_\varepsilon(O_T))$ where parameters ε, τ, η and N are related according to Theorem 3.2. The synthesis of the controller $\tilde{\mathcal{S}}$ will be the main focus of the following section.

The use of ε-contractions of the safe and target sets ensures that the controller $\tilde{\mathcal{S}}$ for the abstraction $T_{\tau,\eta}^N(\Sigma)$ can be used to derive a controller $\mathcal{S}$ for $T_\tau^N(\Sigma)$ that meets the specification (O_S, O_T). Such a controller $\mathcal{S}$, with guaranteed performance, can be synthesized using the following result that can be easily adapted from [6]:

PROPOSITION 4.4. *Let R denote the ε-approximate bisimulation relation between $T_\tau^N(\Sigma)$ and $T_{\tau,\eta}^N(\Sigma)$ given by Theorem 3.2. Let $\tilde{\mathcal{S}}$ be a controller for abstraction $T_{\tau,\eta}^N(\Sigma)$ for specification $(C_\varepsilon(O_S), C_\varepsilon(O_T))$. Let us define the controller for the switched system $T_\tau^N(\Sigma)$*

$$\mathcal{S}(x) = \tilde{\mathcal{S}}\left(\arg \min_{q \in R(x)} J(T_{\tau,\eta,\tilde{\mathcal{S}}}^N(\Sigma), C_\varepsilon(O_S), C_\varepsilon(O_T), q) \right)$$

where $q \in R(x)$ stands for $(x,q) \in R$. Then, for all $x \in \mathbb{R}^n$, the entry time of $T_{\tau,\mathcal{S}}^N(\Sigma)$ for specification (O_S, O_T) satisfies:

$$J(T_{\tau,\mathcal{S}}^N(\Sigma), O_S, O_T, x) \leq$$
$$\min_{q \in R(x)} J(T_{\tau,\eta,\tilde{\mathcal{S}}}^N(\Sigma), C_\varepsilon(O_S), C_\varepsilon(O_T), q).$$

4.3 Discrete controller synthesis for multiscale abstractions

In principle, computing the optimal controller for the discrete abstraction $T_{\tau,\eta}^N(\Sigma) = (Q, L, \longrightarrow, O, H, I)$ for specification $(C_\varepsilon(O_S), C_\varepsilon(O_T))$ can be done using dynamic programming (see e.g. [5]). Termination of the dynamic programming algorithm is guaranteed provided $H^{-1}(C_\varepsilon(O_S))$

contains a finite number of states. Though, in practice, this number can be very large and the convergence of the algorithm very slow. For this reason, it may be desirable to compute only a sub-optimal controller $\tilde{S}$ but in a more reasonable time. We propose to exploit the multiscale structure of our abstraction for that purpose. The main idea is to compute a sequence of sub-optimal controllers $\tilde{S}_k$ for $T^N_{\tau,\eta}(\Sigma)$ with increasing performance. The sub-optimal controllers are obtained by restraining the sets of enabled transitions. Having less potential transitions to explore results in faster controller synthesis. Essentially, our approach (see Figure 2, detailed in Algorithm 1), can be summarized as follows.

Let $k = 0$, initially, we only enable the transitions of longer duration τ between states on the coarsest lattice. More precisely, let us define the initial map of enabled transitions Enab_0 such that for all $q \in Q$,

$$\text{Enab}_0(q) = \begin{cases} P \times \{\tau\} & \text{if } q \in [\mathbb{R}^n]_\eta \\ \emptyset & \text{otherwise} . \end{cases}$$

Then, we repeat the following procedure:

1. **Controller synthesis**: We compute the optimal controller $\tilde{S}_k$ for transition system $T^N_{\tau,\eta}(\Sigma)$ and specification $(C_\varepsilon(O_S), C_\varepsilon(O_T))$ with the additional constraint that for all $q \in Q$, $\tilde{S}_k(q) \in \text{Enab}_k(q) \cup \{\emptyset\}$.

2. **Refinement area identification**: We analyze the controller $\tilde{S}_k$ to identify a region of the state-space O_{Bk} where the performance of the controller should be improved by enabling transitions of shorter duration.

3. **Abstraction refinement**: We define a new map of enabled transitions Enab_{k+1} by extending Enab_k with transitions of duration $2^{-(k+1)}\tau$ within the identified region. We stop when $k = N$ or the performance of the controller $\tilde{S}_k$ stops improving beyond some threshold.

Algorithm 1 comp_controller_ms. Iterative computation of controller $\tilde{S}$ for multiscale abstraction $T^N_{\tau,\eta}(\Sigma)$.

inputs abstraction $T^N_{\tau,\eta}(\Sigma)$, specification $(C_\varepsilon(O_S), C_\varepsilon(O_T))$, initial map of enabled transitions Enab_0, initial refinement area detection parameters μ, ζ
output Controller $\tilde{S}$

1: $(\mu_0, \zeta_0) := (\mu, \zeta)$
2: $\tilde{S}_0 := \text{comp\_controller}(T^N_{\tau,\eta}(\Sigma), C_\varepsilon(O_S), C_\varepsilon(O_T), \text{Enab}_0)$
3: $k := 0$
4: **while** $\neg\text{stop\_criterion}$ **do**
5: $\quad O_{Bk} := \text{detect\_area}(\tilde{S}_k, \mu_k, \zeta_k)$
6: $\quad \text{Enab}_{k+1} := \text{refine\_area}(\tilde{S}_k, \mu_k, \zeta_k, \text{Enab}_k, O_{Bk})$
7: $\quad \tilde{S}_{k+1} := \text{comp\_controller}(T^N_{\tau,\eta}(\Sigma), C_\varepsilon(O_S), C_\varepsilon(O_T), \text{Enab}_{k+1})$
8: $\quad (\mu_{k+1}, \zeta_{k+1}) := \text{update\_params}(\mu_k, \zeta_k)$
9: $\quad k := k + 1$
10: **end while**
11: $\tilde{S} := \tilde{S}_k$
12: **return** $\tilde{S}$

In the following we will detail each of the steps of this iterative procedure. Let us remark that initially it is sufficient

Figure 2: Iterative controller synthesis overview.

to generate the abstraction only the coarsest level. The dynamics at the finer scales is computed only locally and on the fly during the abstraction refinement step.

4.3.1 Controller synthesis

The optimal controller $\tilde{S}_k$ for $T^N_{\tau,\eta}(\Sigma)$ and specification $(C_\varepsilon(O_S), C_\varepsilon(O_T))$ with the additional constraint that for all $q \in Q$, $\tilde{S}_k(q) \in \text{Enab}_k(q) \cup \{\emptyset\}$ can be computed using the following standard dynamic programming algorithm [5]:

Algorithm 2 comp_controller. Computation of controller $\tilde{S}_k$

inputs abstraction $T^N_{\tau,\eta}(\Sigma)$, specification $(C_\varepsilon(O_S), C_\varepsilon(O_T))$, map of enabled transitions Enab_k
output Controller $\tilde{S}_k$

1: $Q_S := H^{-1}(C_\varepsilon(O_S))$;
2: $Q_T := H^{-1}(C_\varepsilon(O_T))$;
3: $Q_k := \{q \in Q \mid \text{Enab}_k(q) \neq \emptyset\}$;
4: $\forall q \in Q_T, \ J^0_k(q) := 0$;
5: $\forall q \in Q \setminus Q_T, \ J^0_k(q) := +\infty$;
6: $i := 0$;
7: **repeat**
8: $\quad \forall q \in Q_T, \ J^{i+1}_k(q) := 0$;
9: $\quad \forall q \in Q \setminus (Q_S \cap Q_k), \ J^{i+1}_k(q) := +\infty$;
10: $\quad \forall q \in Q_S \cap Q_k$,

$$J^{i+1}_k(q) := \min_{l \in \text{Enab}_k(q)} \left(\delta(l) + J^i_k(\text{Succ}(q, l)) \right);$$

11: $\quad i := i + 1$;
12: **until** $\forall q \in Q, \ J^i_k(q) = J^{i-1}_k(q)$
13: $\forall q \in Q, \ J^*_k(q) := J^i_k(q)$;
14: $\forall q \in Q$, if $J^*_k(q) = +\infty, \ \tilde{S}_k(q) := \emptyset$;
15: $\forall q \in Q$, if $J^*_k(q) \neq +\infty$,

$$\tilde{S}_k(q) := \arg \min_{l \in \text{Enab}_k(q)} \left(\delta(l) + J^*_k(\text{Succ}(q, l)) \right);$$

Algorithm 2 is guaranteed to terminate in a finite number of steps if $H^{-1}(C_\varepsilon(O_S))$ contains a finite number of states which is the case if O_S is a bounded subset of $\mathbb{R}^n$. It is simple to verify that for all $q \in Q$,

$$J(T^N_{\tau,\eta,\tilde{S}_k}(\Sigma), C_\varepsilon(O_S), C_\varepsilon(O_T), q) = J^*_k(q).$$

It is important to note that it is not necessary to compute the full abstraction $T^N_{\tau,\eta}(\Sigma)$ for synthesizing the controller $\tilde{S}_k$. It is sufficient to compute the transitions that are enabled by the map Enab_k. This actually renders the synthesis of $\tilde{S}_k$ quite efficient.

REMARK 4.5. *For $k = 0$, the controller $\tilde{S}_0$ we obtain is actually the time optimal controller for the uniform abstraction $T_{\tau,\eta}(\Sigma)$ presented in section 3.1.*

REMARK 4.6. *If for all $q \in Q$, $\text{Enab}_k(q) \subseteq \text{Enab}_{k+1}(q)$, then it follows that for all $q \in Q$, $J^*_k(q) \geq J^*_{k+1}(q)$. This provides a simple criterium to be satisfied when choosing the map of enabled transitions Enab_{k+1} from Enab_k to ensure that the successive controllers we compute will have increasing performance.*

4.3.2 Refinement area identification

The performance of the controller $\tilde{S}_k$ is constrained by the fact that only transitions of duration greater than $2^{-k}\tau$ are enabled. Coarse-grained resolutions (i.e. small values of k)

tend to show higher discrepancies between the ideal optimal trajectories that the controlled system $T_{\tau,\eta}^N(\Sigma)$ should follow, and those yielded by the application of the controller $\tilde{S}_k$. This situation can be mitigated by enabling transitions of shorter duration $2^{-(k+1)}\tau$. However, enabling these transitions at all states of $T_{\tau,\eta}^N(\Sigma)$ results in an exponential growth in the number of transitions to be considered, making the computation of a controller very expensive, or even impossible, depending on the case. It is therefore more efficient to determine a subset of states where enabling transitions of duration $2^{-(k+1)}\tau$ will help improve the performance of the controller.

In time-optimal control, the states where the mode of consecutive transitions does not change represent little opportunity for performance improvement, since they tend to better approximate the optimal trajectories. In contrast, a good strategy to detect parts of the state space where we can gain some performance improvement is checking for trajectories where the number of mode changes exceeds a threshold ζ during a time lapse of duration μ. In order to detect those parts of the state space, we define the function MSC (Mode Switch Count) that obtains the number of mode changes within a trajectory in $T_{\tau,\eta,\tilde{S}_k}^N(\Sigma)$:

$$\mathrm{MSC}(q_0 \xrightarrow[\tilde{S}_k]{l_0} \ldots \xrightarrow[\tilde{S}_k]{l_{K-1}} q_K) = |\{i \mid l_i \neq l_{i+1}\}|$$

Moreover, we define function $\mathrm{Sw}(\tilde{S}_k,\mu,\zeta)$, which computes all the states that are included in a trajectory of $T_{\tau,\eta,\tilde{S}_k}^N(\Sigma)$ of duration μ and that contains more than ζ mode changes. Both μ and ζ are user-defined parameters. In particular, the length of the trajectories of $T_{\tau,\eta,\tilde{S}_k}^N(\Sigma)$ that are analyzed is the maximum $K \in \mathbb{N}$, such that $\delta(l_0) + \delta(l_1) + \cdots + \delta(l_{K-1}) \leqslant \mu$.

$$\mathrm{Sw}(\tilde{S}_k,\mu,\zeta) = \{ q \in \{q_0 \ldots q_K\} | \exists \sigma = q_0 \xrightarrow[\tilde{S}_k]{l_0} \ldots \xrightarrow[\tilde{S}_k]{l_{K-1}} q_K :$$
$$\Delta(\sigma) \leqslant \mu \wedge \mathrm{MSC}(\sigma) \geqslant \zeta\}$$

Using this measure, Algorithm 3 determines the region of the state-space that needs to be refined for the computation of $\tilde{S}_{k+1}$. Specifically, it determines the range of each of the variables in $\mathbb{R}^n$ for the refinement area (i.e., the bounding hyperrectangle of states in $\mathrm{Sw}(\tilde{S}_k,\mu,\zeta)$), returning them as a list of couples with the minimum and maximum values for each variable.

Algorithm 3 detect_area. Controller refinement area detection.

---

inputs controller $\tilde{S}_k$, refinement area detection parameters μ,ζ
output refinement area O_{Bk}

1: $Q' := \mathrm{Sw}(\tilde{S}_k,\mu,\zeta)$
2: **for all** $i \in \{1,\ldots,n\}$ **do**
3: $\quad O_{Bk}[i] := (\min_{q \in Q'} q[i], \max_{q \in Q'} q[i])$
4: **end for**
5: **return** O_{Bk}

4.3.3 Abstraction refinement

Once the refinement area of the state-space has been detected, we proceed to extend the map of enabled transitions Enab_k with transitions of duration $2^{-(k+1)}\tau$. Let us note as O_{Bk} the set of outputs associated to the refinement area

detected by Algorithm 3 for the controller $\tilde{S}_k$. Algorithm 4 computes the map Enab_{k+1} by: (i) computing the set of states Q' included in trajectories of $T_{\tau,\eta,\tilde{S}_k}^N(\Sigma)$ of duration $\leqslant \mu$ and with ζ or more mode switches, (ii) enabling transitions of duration $2^{-(k+1)}\tau$ for those states. The algorithm keeps on enabling transitions of the same duration for all the subsequently computed successor states within the refinement area O_{Bk}.

Algorithm 4 refine_area. Abstraction area refinement.

---

inputs controller $\tilde{S}_k$, map of enabled transitions Enab_k, refinement area detection parameters μ,ζ, detected refinement area O_{Bk}
output extended map of enabled transitions Enab_{k+1}

1: $\mathrm{Enab}_{k+1} := \mathrm{Enab}_k$
2: $Q' := \mathrm{Sw}(\tilde{S}_k,\mu,\zeta)$
3: $V := \emptyset$
4: **while** $\exists q \in Q'$ **do**
5: $\quad Q' := Q'\backslash\{q\}$
6: $\quad V := V \cup \{q\}$
7: $\quad$ **for all** $l \in (P \times \{2^{-(k+1)}\tau\})$ **do**
8: $\quad\quad Q_{\mathrm{Succ}} := \mathrm{Succ}(q,l) \cap H^{-1}(O_{Bk})$
9: $\quad\quad$ **if** $Q_{\mathrm{Succ}} \neq \emptyset$ **then**
10: $\quad\quad\quad \mathrm{Enab}_{k+1}(q) := \mathrm{Enab}_{k+1}(q) \cup \{l\}$
11: $\quad\quad\quad Q' := Q' \cup (Q_{\mathrm{Succ}}\backslash V)$
12: $\quad\quad$ **end if**
13: $\quad$ **end for**
14: **end while**
15: **return** Enab_{k+1}

Algorithm 4 is guaranteed to terminate in a finite number of steps, since the number of successor states to refine is finite (bounded by refinement area O_{Bk}), and the algorithm processes each state only once, keeping track of already traversed states. In the worst case, the algorithm traverses once all the states in $H^{-1}(O_{Bk})$.

As an alternative to Algorithm 4, we may directly compute Enab_{k+1} by refining the full set of states in $H^{-1}(O_{Bk})$:

$$\forall(q \in [\mathbb{R}^n]_{2^{-k}\eta} \cap H^{-1}(O_{Bk}), l \in P \times \{2^{-(k+1)}\tau\})$$

$$\mathrm{Enab}_{k+1}(q) = \begin{cases} \mathrm{Enab}_k(q) \cup \{l\} & \text{if } \mathrm{Succ}(q,l) \cap H^{-1}(O_{Bk}) \neq \emptyset \\ \mathrm{Enab}_k(q) & \text{otherwise .} \end{cases}$$

This alternative (referred to as full-area refinement in the following), is aimed at optimizing the performance of the resulting controller, although at the cost of a partial reduction of the efficiency in the overall controller synthesis process (we provide experimental results for both alternatives in Section 5).

5. EXPERIMENTAL RESULTS

We show an application of our multiscale synthesis approach for a concrete switched system: the boost DC-DC converter (see Figure 3). The boost converter has two operation modes depending on the position of the switch. The state of the system is $x(t) = [i_l(t)\ v_c(t)]^T$ where $i_l(t)$ is the inductor current and $v_c(t)$ the capacitor voltage. The dynamics associated with both modes are affine of the form $\dot{x}(t) = A_p x(t) + b$ ($p = 1,2$) with

$$A_1 = \begin{bmatrix} -\frac{r_l}{x_l} & 0 \\ 0 & -\frac{1}{x_c}\frac{1}{r_0+r_c} \end{bmatrix}, b = \begin{bmatrix} \frac{v_s}{x_l} \\ 0 \end{bmatrix},$$

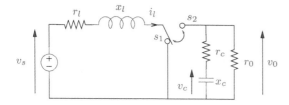

Figure 3: Boost DC-DC converter.

$$A_2 = \begin{bmatrix} -\frac{1}{x_l}(r_l + \frac{r_0 r_c}{r_0 + r_c}) & -\frac{1}{x_l}\frac{r_0}{r_0 + r_c} \\ \frac{1}{x_c}\frac{r_0}{r_0 + r_c} & -\frac{1}{x_c}\frac{1}{r_0 + r_c} \end{bmatrix}.$$

It is clear that the boost DC-DC converter is an example of a switched system. In the following, we use the numerical values from [4], that are, in the per unit system, $x_c = 70$ p.u., $x_l = 3$ p.u., $r_c = 0.005$ p.u., $r_l = 0.05$ p.u., $r_0 = 1$ p.u. and $v_s = 1$ p.u.. We consider the problem of steering in minimal time the state of the system in a desired region of operation while respecting some safety constraints. This is a time-optimal control problem.

For a better numerical conditioning, we rescaled the second variable of the system (i.e. the state of the system becomes $x(t) = [i_l(t)\ 5v_c(t)]^T$; the matrices A_1, A_2 and vector b are modified accordingly). It can be shown that the switched system has a common δ-GUAS Lyapunov function of the form $V(x, y) = \sqrt{(x-y)^T M(x-y)}$, with

$$M = \begin{bmatrix} 1.0224 & 0.0084 \\ 0.0084 & 1.0031 \end{bmatrix}.$$

This δ-GUAS Lyapunov function has the following characteristics: $\underline{\alpha}(s) = s$, $\overline{\alpha}(s) = 1.0127s$, $\kappa = 0.014$. Let us remark that (3) holds as well with $\gamma(s) = 1.0127s$. The specification of the time-optimal control problem is given by the safe set $O_S = [0.65, 1.65] \times [4.95, 5.95]$ and the target $O_T = [1.1, 1.6] \times [5.4, 5.9]$.

In the following, we use approximately bisimilar abstractions to synthesize sub-optimal switching controllers. We set the desired precision of abstractions to 0.1. Then, the contracted safe set and target are $C_\varepsilon(O_S) = [0.75, 1.55] \times [5.05, 5.85]$ and $C_\varepsilon(O_T) = [1.2, 1.5] \times [5.5, 5.8]$. For the sake of comparison, we choose to work both with uniform and multiscale abstractions.

The uniform abstractions $T_{\tau_i, \eta_i}(\Sigma)$ are computed according to [8] for time sampling parameters $\tau_0 = 1$, $\tau_1 = 0.5$ and $\tau_2 = 0.25$. The state-space sampling parameters are chosen according to Theorem 3.1, that is $\eta_0 = 9.7 \times 10^{-4}\sqrt{2}$, $\eta_1 = 4.85 \times 10^{-4}\sqrt{2}$ and $\eta_2 = 2.425 \times 10^{-4}\sqrt{2}$ respectively.

We also use multiscale abstractions $T_{\tau, \eta}^N$ for parameters $\tau = 1$, $\eta = 9.7 \times 10^{-4}\sqrt{2}$ and $N \in \{1, 2\}$. This corresponds to transitions of possible duration $\Theta_\tau^1 = \{1, 0.5\}$ and $\Theta_\tau^2 = \{1, 0.5, 0.25\}$. Hence, the controllers synthesized using $T_{\tau, \eta}^1$ and $T_{\tau, \eta}^2$ are to be compared with those of T_{τ_1, η_1} and T_{τ_2, η_2} respectively. For the detection of refinement regions in the multiscale controllers, we set as initial parameters $\mu_0 = 8$ and $\zeta_0 = 4$. For each subsequent refinement step, we set $\mu_{k+1} = \mu_k/2$, whereas we kept ζ constant. This essentially means that at scale k we refine the trajectories that have at least 4 switches in less that 8×2^{-k} time units.

The initial controller $\tilde{S}_0$ for multiscale abstractions $T_{\tau, \eta}^1$ and $T_{\tau, \eta}^2$ computed in Algorithm 1 is shown in Figure 4. This corresponds to the time-optimal controller for $T_{\tau_0, \eta_0}(\Sigma)$. In particular, we can observe that the path from an arbitrarily chosen state to $C_\varepsilon(O_T)$ (black rectangle) switches many

times in a zigzagging pattern within the region enclosed in the green rectangle, which is the region of the state-space detected by the application of Algorithm 3 for a subsequent refinement ($O_{B0} = [1.12, 1.54] \times [5.04, 5.78]$). This is clearly a region where faster switching would reduce the time to reach the target state. Figure 5 depicts the optimal controllers for T_{τ_1, η_1} and T_{τ_2, η_2} and the sub-optimal controllers computed by Algorithm 1 for $T_{\tau, \eta}^1$ and $T_{\tau, \eta}^2$, using point-wise (Algorithm 4) or full area refinement.

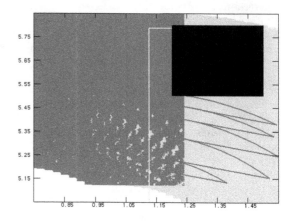

Figure 4: Initial controller $\tilde{S}_0$ for multiscale abstractions $T_{\tau, \eta}^1$ and $T_{\tau, \eta}^2$ (dark gray: mode 1, light gray: mode 2). The area inside the green box is the region detected for subsequent refinement.

Table 1 details the experimental results obtained for the synthesis of the aforementioned set of controllers. In particular, we summarize the results of multiscale controllers for one and two refinement steps, both using point-wise refinement (Algorithm 4), as well as full-area refinement. Moreover, we compare them with the controllers obtained from uniform abstractions with time and state sampling parameters similar to those present at the highest level of refinement in their multiscale counterparts.

Looking at the results, it is worth emphasizing that in general, there is a remarkable reduction in the overall time used to compute the controller using multiscale abstractions with respect to the use of uniform ones (up to a 79% improvement for two refinement steps using Algorithm 4). However, this reduction in computation time is obtained in all cases during the second iteration of the process (in the range of 70-80% versus negligible variations for the first iteration). This is due to the fact that the size of uniform abstractions grows exponentially with higher resolutions, whereas the refinement process that we use with multiscale abstractions bounds this growth by progressively reducing the region of the state-space to refine. In particular, we may observe in Figure 5 (center and right) how the refinement region is reduced in the second iteration (bottom), compared to the original refinement area in the first iteration (top).

Regarding controller performance, in order to minimize the bias in our measures of worst and average entry times, we have only considered the states at the coarsest level of resolution in the multiscale controllers, as well as the subsets of these same states on their uniform counterparts in order to compute our measures. In this respect, despite the aforedescribed reduction in computation time, the general levels of performance of the multiscale controllers show only small

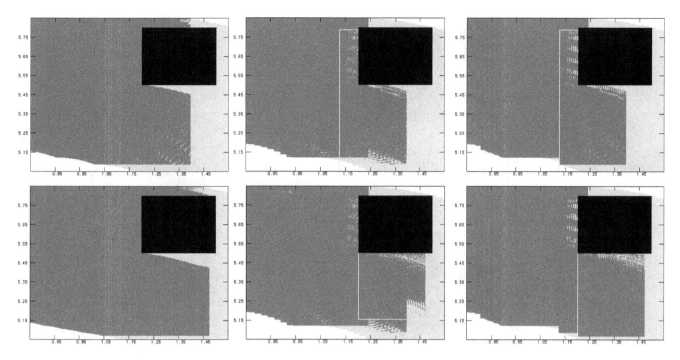

Figure 5: Left: optimal controllers for uniform abstractions $T_{\tau_1,\eta_1}(\Sigma)$ (top) and $T_{\tau_2,\eta_2}(\Sigma)$ (bottom). Center: sub-optimal controllers for multiscale abstractions $T^1_{\tau,\eta}(\Sigma)$ (top) and $T^2_{\tau,\eta}(\Sigma)$ (bottom) computed using point-wise refinement (Algorithm 4). Right: sub-optimal controllers for multiscale abstractions $T^1_{\tau,\eta}(\Sigma)$ (top) and $T^2_{\tau,\eta}(\Sigma)$ (bottom) computed using full-area refinement.

	Uniform Abstractions - $T_{\tau,\eta}(\Sigma)$			Multiscale Abstractions - $T^N_{\tau,\eta}(\Sigma)$	
				(point-wise / full-area refinement)	
	$\tau = 1s$ $\eta = 13.7 \times 10^{-4}$	$\tau = 0.5s$ $\eta = 6.85 \times 10^{-4}$	$\tau = 0.25s$ $\eta = 3.42 \times 10^{-4}$	$N = 1, \tau = 1s$ $\eta = 13.7 \times 10^{-4}$	$N = 2, \tau = 1s$ $\eta = 13.7 \times 10^{-4}$
Computation Time (s)	60.99	436.01	4167.23	425.75/437.37	846.31/1138.2
Average Entry Time (s)	16.52	10.58	9.21	10.73/10.71	9.69/9.43
Worst Entry Time (s)	54	34.5	29.25	40/35.5	38/31.5
Abstraction Size (# States)	170569	680625	2719201	486606/520668	1250205/1511404
Worst Entry Time Imp. (%)	-	36.11	45.83	25.92/34.25	29.62/41.66
Average Entry Time Imp. (%)	-	35.95	44.24	35.04/35.16	41.34/42.91
Rel. Computation Time Imp. (%)	-	-	-	2.35/-0.31	79.69/72.68
Rel. Worst Entry Time Imp. (%)	-	-	-	-10.18/-1.85	-16.2/-4.16
Rel. Average Entry Time Imp. (%)	-	-	-	-0.9/-0.78	-2.9/-1.33
Rel. Abstraction Size Imp. (%)	-	-	-	28.6/23.6	54.1/44.5

Table 1: Experimental results comparing controller synthesis for time-optimal control of the Boost DC-DC Converter using both uniform and multiscale abstractions.

variations with respect to the uniform controllers with similar resolutions. In the particular case of Algorithm 4, the average entry time shows a slight degradation (-0.9%) with respect to the equivalent uniform controller ($\tau = 0.5$) for 1 iteration. For 2 iterations, the performance further degrades up to -2.9%. However, this does not constitute a remarkable performance loss, especially if we consider that the reduction in the time used to compute the controller is more than 79% in this case. For full-area refinement, the performance of the controller slightly improves, with a degradation on average entry time of up to -1.33% in the controller obtained after 2 iterations of the process, with a computation time reduction still above 70%.

In the case of worst entry time, there is a stronger perfor-

mance reduction on the multiscale controllers, which yield worst entry times between 1.85% and 16.2% longer. In any case, it is worth considering that the set of controllable regions of the state-space in the controllers obtained from multiscale abstractions does not exactly match with those in the uniform versions, which are slightly more extensive. In particular, we have observed that the highest entry times in the uniform controllers always correspond to the additional set of controllable states in regions of the state-space not covered by the multiscale controllers. However, these additional states are not considered in our measures to provide an even comparison with the multiscale controllers. Hence, this measure provides a less representative indicator of the real controller's performance compared to average entry time.

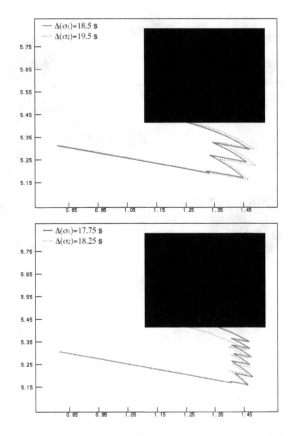

Figure 6: Trajectories of the switched system driven by controllers derived using Proposition 4.4: (top) based on controller of uniform abstraction T_{τ_1,η_1} (solid line) and multiscale abstraction $T_{\tau,\eta}^1$ (dashed line); (bottom) based on controller of uniform abstraction T_{τ_2,η_2} (solid line) and multiscale abstraction $T_{\tau,\eta}^2$ (dashed line).

In relation with the size of abstractions required to compute multiscale controllers, we can observe that there is a reduction with respect to uniform abstractions in the range 23-29% and 44-54% for 1 and 2 iterations, respectively.

Finally, Figure 6 shows trajectories of the actual switched system driven by controllers derived using Proposition 4.4 based on controllers of uniform abstractions T_{τ_1,η_1}, T_{τ_2,η_2} and multiscale abstractions $T_{\tau,\eta}^1$, $T_{\tau,\eta}^2$. First, we can observe that the specification is met: all trajectories reach the target without leaving the safe set. Regarding the performance of the controllers, the ones synthesized using uniform abstractions are more efficient than those synthesized using multiscale abstractions. This was expected, although it should be noticed that their performance levels are still comparable and tend to become closer in finer scales (18.5 vs 19.5 at the first scale, 17.75 vs 18.25 at the second).

6. CONCLUSION

In this paper, we have proposed the use of multiscale, approximately bisimilar discrete abstractions for the computation of controllers, applying them to the specific case of time-optimal control problems. In particular, our experimental results have shown that we can achieve a remarkable reduction in the computation time of such controllers in comparison with the use of uniform abstractions, while preserving similar levels of performance. Future work will deal with the application of multiscale abstractions to other kinds of control problems.

7. REFERENCES

[1] D. Angeli. A Lyapunov approach to incremental stability properties. *IEEE Trans. on Automatic Control*, 47(3):410–421, March 2002.

[2] A. Anta and P. Tabuada. To sample or not to sample: Self-triggered control for nonlinear systems. *IEEE Transactions on Automatic Control*, 2010. To appear.

[3] A. Arnold, A. Vincent, and I. Walukiewicz. Games for synthesis of controllers with partial observation. *Theoretical Computer Science*, 28(1):7–34, 2003.

[4] A. Beccuti, G. Papafotiou, and M. Morari. Optimal control of the boost dc-dc converter. In *IEEE Conf. on Decision and Control*, pages 4457–4462, 2005.

[5] D. P. Bertsekas. *Dynamic Programming and Optimal Control*. Athena Scientific, 2000.

[6] A. Girard. Synthesis using approximately bisimilar abstractions: time-optimal control problems. In *IEEE Conf. on Decision and Control*, 2010.

[7] A. Girard and G. Pappas. Approximation metrics for discrete and continuous systems. *IEEE Trans. on Automatic Control*, 52(5):782–798, 2007.

[8] A. Girard, G. Pola, and P. Tabuada. Approximately bisimilar symbolic models for incrementally stable switched systems. *IEEE Transactions on Automatic Control*, 55(1):116–126, 2010.

[9] M. Mazo Jr. and P. Tabuada. Approximate time-optimal control via approximate alternating simulations. In *ACC*, pages 10201–10206, 2010.

[10] R. Milner. *Communication and Concurrency*. Prentice Hall, 1989.

[11] G. Pola, A. Girard, and P. Tabuada. Approximately bisimilar symbolic models for nonlinear control systems. *Automatica*, 44(10):2508–2516, 2008.

[12] G. Pola and P. Tabuada. Symbolic models for nonlinear control systems: Alternating approximate bisimulations. *SIAM J. on Con. and Opt.*, 48(2):719–733, 2009.

[13] P. Ramadge and W. Wonham. Supervisory control of a class of discrete event systems. *SIAM Journal on Control and Optimization*, 25(1):206–230, 1987.

[14] P. Richard, H. Cormerais, and J. Buisson. A generic design methodology for sliding mode control of switched systems. *Nonlinear Analysis: Hybrid Systems and Applications*, 65(9):1751–1772, 2006.

[15] P. Tabuada. *Verification and Control of Hybrid Systems - A Symbolic Approach*. Springer, 2009.

[16] Y. Tazaki and J. Imura. Approximately bisimilar discrete abstractions of nonlinear systems using variable-resolution quantizers. In *America Control Conference*, pages 1015–1020, 2010.

[17] M. Velasco, J. Fuertes, and P. Marti. The self triggered task model for real-time control systems. In *24th IEEE Real-Time Systems Symposium*, pages 67–70, 2003.

[18] X. Wang and M. Lemmon. State based self-triggered feedback control systems with l2 stability. In *17th IFAC world congress*, 2008.

Joint Synthesis of Switching and Feedback for Linear Systems in Discrete Time

Ji-Woong Lee
Pennsylvania State University
Dept. of Electrical Engineering
University Park, PA 16802
jiwoong@psu.edu

Geir E. Dullerud
University of Illinois at Urbana-Champaign
Dept. of Mechanical Science and Engineering
Urbana, IL 61801
dullerud@illinois.edu

ABSTRACT

The paper is concerned with control of switched linear systems, and proves two new fundamental results (equivalence and separation) about feedback policies for these systems. Addressed is the problem of jointly synthesizing a switching rule and a feedback controller for guaranteed stability and performance levels subject to well-posedness-like constraints. Two main results of the paper are the equivalence of open-loop and closed-loop switching, and the separation between switching and feedback. That is, open-loop switching rules are non-conservative for stabilization and performance optimization, and in output feedback control they can be obtained separately from the feedback controller. The synthesis conditions for switching and feedback are convex and expressed in terms of linear matrix inequalities.

Categories and Subject Descriptors

J.2 [**Physical Sciences and Engineering**]: Engineering; G.1.6 [**Optimization**]: Convex programming

General Terms

Theory

Keywords

Stability, feedback control, hybrid systems, small gain property, linear matrix inequality

1. INTRODUCTION

The problem of determining optimal stabilizing switching rules typically arises in the context of supervisory control [21] and measurement scheduling [20]. In the problem of supervisory control, one is given a set of controllers for a single plant. At each time instant, the supervisor chooses a controller among this set based on the past and present state measurements, and uses the chosen controller to close the feedback loop. Due to the switching among different

controllers, the overall closed-loop system is time-varying but can potentially exhibit better stability and performance properties than when the feedback loop is closed using any single controller among the given set [21, 18]. On the other hand, in the problem of measurement scheduling, one is given a set of sensors for a single plant. At each time instant, the scheduler chooses a sensor among this set based on the past state measurements it has made, and samples a new state measurement using the chosen sensor. Again, due to the switching among multiple sensors, the overall system is time-varying and yet, once feedback-interconnected with a suitable controller, it can potentially yield better stability and performance properties than when feedback-controlled using any single sensor among the given set [20].

We will restrict our attention to linear systems in the discrete time domain and consider joint synthesis of a switching rule and a feedback controller for uniform exponential stability and guaranteed $\mathcal{H}_2$ and $\mathcal{H}_\infty$ type performance levels. A switched linear system under a general switching rule exhibits nonlinear behavior. To ensure such a nonlinear system possesses the kind of robustness that uniformly exponentially stable standard linear systems exhibit against state perturbations [4], we impose a separation of the total response of the system into a zero-input response–like term and a zero-state response–like term. Imposing this requirement guarantees that one can speak of "stability under zero input" and "performance under zero initial state" at the same time as in the standard $\mathcal{H}_2$ and $\mathcal{H}_\infty$ control problems. Similarly, to ensure the overall nonlinear system possesses the kind of small gain property and its converse that standard linear systems exhibit against dynamic uncertainty [5], we impose a separation of the response of the uncertain system into the dynamics associated with the plant and that with the uncertainty block. This requirement is analogous to the stability requirement: While the stability definition requires robustness against "static" state perturbations, the performance definition requires robustness against "dynamic" feedback uncertainties.

Under these stability and performance requirements, we propose and prove the following theses:

- *Equivalence:* Whenever a closed-loop switching rule is stabilizing and guarantees a performance level, there exists an open-loop switching rule that does so and results in a periodic switching sequence.

- *Separation:* The joint synthesis problem for a switching rule and a feedback controller is separated, and one can obtain a switching rule separately from the feedback controller.

Thus, we justify the standard problem of determining a single switching sequence [17] to guarantee stability and desired performance levels, and generalize the recent results on switching stabilization [14] and infinite-horizon joint LQG switching-feedback synthesis [11]. Our results are in contrast to the fact that there are cases where open-loop switching rules are not sufficient when mere asymptotic stability is required [25]. However, they do not contradict this fact because of the additional robustness requirements that we impose. In particular, our stability condition is stronger than the requirement that the radius of weak convergence be greater than one [7].

Aside from the robustness requirement on asymptotic stability, we also require that the feedback controller should be able to recall past switching paths as well as past state measurements. That is, following the lead taken in [13, 12, 15, 16], we require that the feedback controller be not only dynamic (with the information about past measurements stored in its state) but also switching path–dependent (with the information about past switching paths encoded in its coefficients). It turns out that, as is known in the context of linear time-varying systems [10], only a finite amount of memory, in a sense, is needed to store information about past state measurements and past switching paths. More specifically, the order of the controller can be no more than that of the plant, and the length of the past switching paths encoded in the controller coefficients can be no more than the period of the open-loop switching sequence. Such a requirement is crucial in guaranteeing nonconservative synthesis of switching rules and feedback controllers. In particular, our conditions for joint synthesis of switching and feedback are expressed in terms of linear matrix inequalities [2]. Moreover, these conditions give rise to switching path–dependent Lyapunov functions and cover the common Lyapunov function and multiple Lyapunov functions approaches [3] as special cases of path length zero and one, respectively. An optimal switching sequence can, in principle, be determined by solving an increasing sequence of semidefinite programs indexed by the length of past switching paths that the feedback controller should recall, and then the associated controller coefficients are determined in a straightforward manner using the linear matrix inequality embedding technique [22, 6] or change-of-variables approaches [23, 19].

In Section 2, we present joint synthesis of switching and feedback for uniform exponential stability subject to robustness against state perturbations. Then, in Section 3, we show the results carry over to optimizing $\mathcal{H}_2$ and $\mathcal{H}_\infty$ type performance levels as well, where the $\mathcal{H}_\infty$ type performance level is required to exhibit a small gain property against dynamic uncertainties. Concluding remarks are made in Section 4.

The set of real numbers is denoted by $\mathbb{R}$, and the set of positive (resp. nonnegative) integers by $\mathbb{N}$ (resp. $\mathbb{N}_0$). For $x \in \mathbb{R}^n$ and $\mathbf{A} \in \mathbb{R}^{m \times n}$, denoted by $\|x\|$ and $\|\mathbf{A}\|$ are the Euclidean norm of x and the spectral norm of $\mathbf{A}$, respectively. If $\mathbf{X}$, $\mathbf{Y} \in \mathbb{R}^{n \times n}$ are symmetric (i.e., $\mathbf{X} = \mathbf{X}^{\mathrm{T}}$ and $\mathbf{Y} = \mathbf{Y}^{\mathrm{T}}$), then we write $\mathbf{X} < \mathbf{Y}$ or $\mathbf{Y} - \mathbf{X} > \mathbf{0}$ to mean that $\mathbf{Y} - \mathbf{X}$ (resp. $\mathbf{X} - \mathbf{Y}$) is positive definite (resp. negative definite). The identity matrices, with their dimensions understood, are denoted $\mathbf{I}$; similarly, the zero matrices are denoted $\mathbf{0}$.

2. SWITCHING AND FEEDBACK FOR UNIFORM EXPONENTIAL STABILITY

2.1 Problem Statement

Let

$$\mathcal{S} = \{(\mathbf{A}_i, \mathbf{C}_i) : i = 1, \dots, N\} \tag{1}$$

be an indexed set of N matrix pairs, where $\mathbf{A}_i \in \mathbb{R}^{n \times n}$ and $\mathbf{C}_i \in \mathbb{R}^{l \times n}$ are given matrices for all $i = 1, \dots, N$. The set $\mathcal{S}$ shall be called a *switched linear system*, and it defines the family of linear time-varying models of the form

$$x(t + 1) = \mathbf{A}_{\theta(t)} x(t), \quad t \in \mathbb{N}_0; \tag{2a}$$
$$y(t) = \mathbf{C}_{\theta(t)} x(t), \quad t \in \mathbb{N}_0, \tag{2b}$$

over all *switching sequences*

$$\theta = (\theta(0), \theta(1), \dots) \in \{1, \dots, N\}^\infty.$$

If $\theta(t) = i$ for some $t \in \mathbb{N}_0$ and $i \in \{1, \dots, N\}$, then the switched linear system is said to be in *mode i* at time t. Once a switching sequence $\theta \in \{1, \dots, N\}^\infty$ and an initial state $x(0) \in \mathbb{R}^n$ are specified, the state-space model (2) generates a state sequence $x = (x(0), x(1), \dots)$ and a measurement sequence $y = (y(0), y(1), \dots)$. The following examples show that our definition covers the switched systems arising from supervisory control [21] and measurement scheduling [20].

EXAMPLE 1. *Consider the problem of stabilizing a linear plant*

$$x(t + 1) = \mathbf{A}x(t) + \mathbf{B}u(t), \quad t \in \mathbb{N}_0,$$

via state feedback of the form $u(t) = \mathbf{K}(t)x(t)$. *There are N feedback gains* $\mathbf{K}_1, \dots, \mathbf{K}_n$ *to choose from at each time instant. If the supervisor chooses the ith controller (based on the perfectly observed state $x(t)$), so that $\mathbf{K}(t) = \mathbf{K}_i$ at time t, then closing the feedback loop yields*

$$x(t + 1) = (\mathbf{A} + \mathbf{B}\mathbf{K}_{\theta(t)})x(t),$$

where $\theta(t) = i$. *Thus the problem of determining a stabilizing supervisory law is reduced to that of finding a stabilizing switching rule for the switched linear system*

$$\mathcal{S} = \{(\mathbf{A} + \mathbf{B}\mathbf{K}_i, \mathbf{I}) : i = 1, \dots, N\}.$$

EXAMPLE 2. *Consider the problem of guaranteeing the asymptotic stability of a state observer of the form*

$$x_K(t + 1) = \mathbf{A}x_K(t) + \mathbf{L}(\mathbf{C}(t)x_K(t) - y(t))$$

for a linear system that evolves according to

$$x(t + 1) = \mathbf{A}x(t), \quad t \in \mathbb{N}_0;$$
$$y(t) = \mathbf{C}(t)x(t), \quad t \in \mathbb{N}_0.$$

There are N sensors to choose from at each time instant. If the measurement scheduler chooses the ith sensor (based on the measurements $y(0), \dots, y(t-1)$), so that $\mathbf{C}(t) = \mathbf{C}_i$ at time t, then the state prediction error $e(t) = x(t) - x_K(t)$ is updated according to

$$e(t + 1) = (\mathbf{A} + \mathbf{L}\mathbf{C}_{\theta(t)})e(t),$$

where $\theta(t) = i$. *Thus the problem of determining a stabilizing measurement scheduling law is reduced to that of finding a stabilizing switching rule for the switched linear system*

$$\mathcal{S} = \{(\mathbf{A} + \mathbf{L}\mathbf{C}_i, \mathbf{C}_i) : i = 1, \dots, N\}.$$

In Example 1, the mode $\theta(t)$ is computed after sampling the state $x(t)$. However, in Example 2, the mode $\theta(t)$ is calculated before the measurement $y(t)$ is sampled. Otherwise, the two examples both give a switched system whose state-space model is of the form (2). Obviously, there are other, more complicated, situations that give rise to such state-space models.

In this section, we will obtain an exact condition for the existence of a stabilizing switching rule for $\mathcal{S}$. This condition will turn out to be equivalent to an exact condition for the existence of a single stabilizing switching sequence θ, which works independently of the state measurements. Furthermore, we will extend the condition to joint synthesis of switching and feedback for stabilizing $\mathcal{S}$.

2.2 Definitions

Let $\mathcal{S}$ be as in (1). Let

$$g = \{g_t : t \in \mathbb{N}_0\} \tag{3a}$$

be an indexed family of functions such that

$$\theta(t) = g_t(y^t) \in \{1, \ldots, N\}, \quad t \in \mathbb{N}_0, \tag{3b}$$

where $y^t = (y(0), \ldots, y(t))$. Then the family g defines a *switching rule* that generates a switching sequence θ for the switched linear system $\mathcal{S}$ based on the measurement sequence y. As in the measurement scheduling case, one may also consider switching rules that generate $\theta(t)$ based on past measurements y^{t-1} only. The results of this section and subsequent sections carry over to such switching rules as well. However, in order to maintain notational simplicity, we will not explicitly cover these rules.

A switching rule g is said to be asymptotically stabilizing for $\mathcal{S}$ if the state equation (2a) satisfies $x(t) \to 0$ as $t \to \infty$ under g. Similarly, a switching sequence θ is said to be asymptotically stabilizing for $\mathcal{S}$ if $x(t) \to 0$ as $t \to \infty$ over all initial states $x(0)$. Our stability notions are stronger than these and require that the state $x(t) \to 0$ with a uniform exponential decay rate over all initial state and over all state perturbations. Consider what we call the *internal switched linear system*:

$$x(t+1) = \mathbf{A}_{\theta(t)} x(t) + r(t), \quad t \in \mathbb{N}_0. \tag{4}$$

DEFINITION 3. *A switching rule g as in (3) is said to be* uniformly exponentially stabilizing *for the switched linear system $\mathcal{S}$ if there exist $c > 0$ and $\lambda \in (0, 1)$ such that the internal switched linear system (4), with $\theta(t) = g_t(y^t)$ for all $t \in \mathbb{N}_0$, satisfies*

$$\|x(t)\| \leq c\lambda^t \|x(0)\| \tag{5a}$$

for all $t \in \mathbb{N}_0$ and $x(0) \in \mathbb{R}^n$ whenever $r(0) = r(1) = \cdots = 0$, and

$$\|x(t)\| \leq c\lambda^{t-t_0} \|r(t_0)\| \tag{5b}$$

for all $t_0, t \in \mathbb{N}_0$ with $t \geq t_0$ and $r(t_0) \in \mathbb{R}^n$ whenever $x(0) = r(0) = \cdots = r(t_0 - 1) = r(t_0 + 1) = \cdots = 0$.

DEFINITION 4. *A switching sequence $\theta \in \{1, \ldots, N\}^\infty$ is said to be* uniformly exponentially stabilizing *for the switched linear system $\mathcal{S}$ if there exist $c > 0$ and $\lambda \in (0, 1)$ such that*

$$\|\mathbf{\Phi}_\theta(t, t_0)\| \leq c\lambda^{t-t_0}$$

for all $t_0, t \in \mathbb{N}_0$ with $t \geq t_0$, where

$$\mathbf{\Phi}_\theta(t, t_0) = \begin{cases} \mathbf{A}_{\theta(t-1)} \cdots \mathbf{A}_{\theta(t_0)}, & t > t_0; \\ \mathbf{I}, & t = t_0. \end{cases}$$

It is readily seen that the existence of a uniformly exponentially stabilizing switching sequence implies that of a uniformly exponentially stabilizing switching rule. We will adopt some of the notions introduced in [14] to show the converse also holds true and to establish an exact, convex synthesis condition. For each $M \in \mathbb{N}_0$, the elements of $\{1, \ldots, N\}^{M+1}$ will be called *switching paths* of length M. A nonempty set $\mathcal{N}$ of switching paths of length M is said to be *admissible* if, for each $(i_0, \ldots, i_M) \in \mathcal{N}$, there exist a $K \in \mathbb{N}$, with $K > M$, and a switching path $(i_{M+1}, \ldots, i_K)$ such that $(i_{K-M}, \ldots, i_K) = (i_0, \ldots, i_M)$ and $(i_t, \ldots, i_{t+M}) \in \mathcal{N}$ for all $t \in \{0, \ldots, K-M\}$. An admissible set $\mathcal{N}$ of switching paths is called *minimal* if none of the proper subsets of $\mathcal{N}$ is admissible. It is readily seen that, if $\mathcal{N}$ is a minimal set of switching paths of length M, then there exists a switching sequence θ such that

$$\mathcal{N} = \{(i_0, \ldots, i_M):$$
$$(i_0, \ldots, i_M) = (\theta(t), \ldots, \theta(t+M)), \ t \in \mathbb{N}_0\}. \tag{6}$$

Such a switching sequence is periodic and is unique up to a time shift [14].

2.3 Synthesis of Switching Rules

The following theorem characterizes the existence of uniformly exponentially stabilizing switching laws.

THEOREM 5. *Let $\mathcal{S}$ be as in (1). The following are equivalent:*

(a) *There is a uniformly exponentially stabilizing switching rule for $\mathcal{S}$.*

(b) *There is a uniformly exponentially stabilizing switching sequence for $\mathcal{S}$.*

(c) *There exist a path length $M \in \mathbb{N}_0$, a minimal set $\mathcal{N}$ of switching paths of length M, and matrices $\mathbf{Y}_{(i_0, \ldots, i_{M-1})} \in \mathbb{R}^{n \times n}$ such that*

$$\mathbf{Y}_{(i_0, \ldots, i_{M-1})} > \mathbf{0},$$
$$\mathbf{A}_{i_M} \mathbf{Y}_{(i_0, \ldots, i_{M-1})} \mathbf{A}_{i_M}^{\mathrm{T}} - \mathbf{Y}_{(i_1, \ldots, i_M)} < \mathbf{0}$$

for all $(i_0, \ldots, i_M) \in \mathcal{N}$.

Moreover, if condition (c) holds, then any periodic switching sequence θ satisfying (6) is uniformly exponentially stabilizing for $\mathcal{S}$.

PROOF. Suppose there exists a switching rule g which is uniformly exponentially stabilizing for $\mathcal{S}$, so that the internal switched linear system (4) satisfies (5a) for all t and $x(0)$ whenever r is identically zero, and (5b) for all $t \geq t_0$ and for all $r(t_0)$ whenever $x(0) = 0$ and $r(t) = 0$ for $t \neq t_0$. Let e_k be the kth standard basis vector for $\mathbb{R}^n$ for $k = 1, \ldots, n$, so that $\mathbf{I} = [e_1 \cdots e_n] \in \mathbb{R}^{n \times n}$. To show (a) implies (b), choose $x(0)$ and r as follows:

1. Set $x(0) = 0$ and obtain $\theta(0)$ for the internal system under the switching rule g;

2. Apply $r(0) = \mathbf{A}_{\theta(0)} e_1$ to the internal system at time $t = 0$ and obtain $\theta(1)$ under g;

3. Apply $r(1) = \mathbf{A}_{\theta(1)} \mathbf{A}_{\theta(0)} e_2$ to the internal system at time $t = 1$ and obtain $\theta(2)$ under g;

4. Proceed in this manner to obtain $r(0), \ldots, r(n-1)$ and $\theta(0), \ldots, \theta(n-1)$;

5. Apply $r(t) = 0$ to the internal system at all $t \geq n$ to obtain $\theta(n), \theta(n+1), \ldots$.

Next, choose any $x_0 = [\alpha_1 \cdots \alpha_n]^{\mathrm{T}} \in \mathbb{R}^n$ with $\|x_0\| = 1$. Then, for $T \in \mathbb{N}_0$ and $M \geq 1$ such that

$$\max_i \|\mathbf{A}_i\| \leq M \quad \text{and} \quad \sqrt{n} c \lambda^{T-n} M^n < 1,$$

we have

$$\|\mathbf{\Phi}_\theta(T, 0) x_0\| \leq \sum_{k=1}^n |\alpha_k| \|\mathbf{\Phi}_\theta(T, k) \mathbf{\Phi}_\theta(k, 0) e_k\|$$

$$\leq \sum_{k=1}^n |\alpha_k| c \lambda^{T-k} \|r(k-1)\|$$

$$= \sum_{k=1}^n |\alpha_k| c \lambda^{T-k} \|\mathbf{\Phi}(k, 0) e_k\|$$

$$\leq \sqrt{n} c \lambda^{T-n} M^n \|x_0\|,$$

which implies $\|\mathbf{\Phi}_\theta(T, 0)\| < 1$. Thus, the periodic switching sequence, where the switching path $(\theta(0), \ldots, \theta(T-1))$ repeats, is uniformly exponentially stabilizing. This establishes that (a) implies (b). Thus (a) and (b) are equivalent.

It remains to show that (b) is equivalent to (c). By [14, Theorem 2], condition (b) is equivalent to the existence of a path length $M \in \mathbb{N}_0$, an admissible set $\tilde{\mathcal{N}}$ of switching paths of length M, and matrices $\mathbf{X}_{(i_0, \ldots, i_{M-1})} > 0$ such that

$$\mathbf{A}_{i_M}^{\mathrm{T}} \mathbf{X}_{(i_1, \ldots, i_M)} \mathbf{A}_{i_M} - \mathbf{X}_{(i_0, \ldots, i_{M-1})} < \mathbf{0}$$

for all $(i_0, \ldots, i_M) \in \tilde{\mathcal{N}}$. A Schur complement argument yields that this inequality is equivalent to

$$\mathbf{A}_{i_M} \mathbf{X}_{(i_0, \ldots, i_{M-1})}^{-1} \mathbf{A}_{i_M}^{\mathrm{T}} - \mathbf{X}_{(i_1, \ldots, i_M)}^{-1} < \mathbf{0}$$

for $(i_0, \ldots, i_M) \in \tilde{\mathcal{N}}$. Choose a minimal $\mathcal{N} \subset \tilde{\mathcal{N}}$ and put

$$\mathbf{Y}_{(i_0, \ldots, i_{M-1})} = \mathbf{X}_{(i_0, \ldots, i_{M-1})}^{-1}$$

for each $(i_0, \ldots, i_M) \in \mathcal{N}$. Since $\mathcal{N}$ is admissible and since $\mathbf{Y}_{(i_0, \ldots, i_{M-1})} > \mathbf{0}$ for all $(i_0, \ldots, i_M) \in \mathcal{N}$, conditions (b) and (c) are equivalent by [14, Theorem 2]. Because $\mathcal{N}$ is minimal, there exists a periodic switching sequence θ satisfying (6). Such a θ is uniformly exponentially stabilizing for $\mathcal{S}$. This completes the proof. $\square$

Generation of all minimal sets of switching paths of length M amounts to identifying all elementary cycles in a directed graph whose nodes are switching paths of length $M-1$ [14, 9]. A well-known algorithm for finding all elementary cycles in a directed graph is presented in [8]. A counterexample in [25] shows that the existence of a switching rule that guarantees asymptotic stability does not necessarily imply the existence of an asymptotically stabilizing switching sequence. However, Theorem 5 says that adding a uniformity/robustness requirement to the stability notion makes the situation quite different.

EXAMPLE 6. *Suppose $N = 2$ and $\mathcal{S}$ has*

$$\mathbf{A}_1 = \begin{bmatrix} \frac{1}{2} & 0 \\ 0 & 2 \end{bmatrix}, \quad \mathbf{A}_2 = \begin{bmatrix} \frac{\sqrt{3}}{2} & \frac{1}{2} \\ -\frac{1}{2} & \frac{\sqrt{3}}{2} \end{bmatrix},$$

$$\mathbf{C}_1 = \mathbf{C}_2 = \begin{bmatrix} 1 & 0 \\ 0 & 1 \end{bmatrix}.$$

As is shown in [24], there exist asymptotically stabilizing switching rules for this $\mathcal{S}$. Such a switching rule $g = \{g_t\}$ is given by

$$\theta(t) = g_t(x^t)$$

$$= \begin{cases} 1 & \text{if } x(t) = [x_1 \ x_2]^{\mathrm{T}} \text{ with } |x_1| > 2|x_2|; \\ 2 & \text{otherwise.} \end{cases}$$

However, it is shown in [25] that there is no asymptotically stabilizing switching sequence for this case. Therefore, Theorem 5 suggests that no switching rule is uniformly exponentially stabilizing for $\mathcal{S}$. Indeed, if

$$x(0) = \begin{bmatrix} 2 \\ 2 \end{bmatrix}, \quad r(0) = \begin{bmatrix} 0 \\ 1 - \sqrt{3} \end{bmatrix},$$

and

$$r(t) = \begin{bmatrix} 0 \\ 0 \end{bmatrix} \quad \text{for } t \geq 1,$$

then the internal system (4) under g generates the switching sequence $\theta = (2, 1, 1, \ldots)$; letting $x_0(t)$ (resp. $x_r(t)$) be the response of the internal system to $x(0)$ (resp. $r(0)$), we have that both $x_0(t)$ and $x_r(t)$ diverge even though $x(t) = x_0(t) + x_r(t) \to 0$.

EXAMPLE 7. *Let us replace $\mathbf{A}_1$ in Example 6 with*

$$\mathbf{A}_1 = \begin{bmatrix} \frac{1}{2} & 0 \\ 0 & 1 \end{bmatrix}.$$

Then the spectral radius of $\mathbf{A}_2 \mathbf{A}_1$ is less than one, and hence the periodic switching sequence

$$\theta = (1, 2, 1, 2, \ldots),$$

where modes 1 and 2 alternate, is uniformly exponentially stabilizing. This switching sequence is generated by the following open-loop switching rule:

$$\theta(t) = g_t(x^t) = \begin{cases} 1 & \text{if } t \text{ is even}; \\ 2 & \text{if } t \text{ is odd.} \end{cases}$$

2.4 Joint Synthesis of Switching and Feedback

Let $\mathbf{A}_1, \ldots, \mathbf{A}_N \in \mathbb{R}^{n \times n}$, $\mathbf{B}_1, \ldots, \mathbf{B}_N \in \mathbb{R}^{n \times m}$, and $\mathbf{C}_1, \ldots, \mathbf{C}_N \in \mathbb{R}^{l \times n}$ be given. Consider the controlled plant model of the form

$$x(t+1) = \mathbf{A}_{\theta(t)} x(t) + \mathbf{B}_{\theta(t)} u(t), \quad t \in \mathbb{N}_0; \quad (8a)$$

$$y(t) = \mathbf{C}_{\theta(t)} x(t), \quad t \in \mathbb{N}_0. \quad (8b)$$

Assuming that the mode $\theta(t)$ is perfectly observed by the feedback controller at each time instant $t \in \mathbb{N}_0$, and that the controller is able to recall L most recent past modes, consider dynamic output feedback controllers of the form

$$x_K(t+1) = \mathbf{A}_{K, (\theta(t-L), \ldots, \theta(t))} x_K(t)$$
$$\qquad + \mathbf{B}_{K, (\theta(t-L), \ldots, \theta(t))} y(t), \quad t \in \mathbb{N}_0;$$

$$u(t) = \mathbf{C}_{K, (\theta(t-L), \ldots, \theta(t))} x_K(t)$$
$$\qquad + \mathbf{D}_{K, (\theta(t-L), \ldots, \theta(t))} y(t), \quad t \in \mathbb{N}_0.$$

Closing the feedback loop via such a controller yields the closed-loop state $\tilde{x}(t) = [x(t)^{\mathrm{T}} \ x_K(t)]^{\mathrm{T}}$ and the corresponding closed-loop system

$$\tilde{x}(t+1) = \left(\begin{bmatrix} \mathbf{A}_{\theta(t)} & \mathbf{0} \\ \mathbf{0} & \mathbf{0} \end{bmatrix} + \begin{bmatrix} \mathbf{0} & \mathbf{B}_{\theta(t)} \\ \mathbf{I} & \mathbf{0} \end{bmatrix} \right.$$
$$\left. \times \mathbf{K}_{(\theta(t-L),\dots,\theta(t))} \begin{bmatrix} \mathbf{0} & \mathbf{I} \\ \mathbf{C}_{\theta(t)} & \mathbf{0} \end{bmatrix} \right) \tilde{x}(t), \quad (9)$$

where

$$\mathbf{K}_{(\theta(t-L),\dots,\theta(t))} = \begin{bmatrix} \mathbf{A}_{K,(\theta(t-L),\dots,\theta(t))} & \mathbf{B}_{K,(\theta(t-L),\dots,\theta(t))} \\ \mathbf{C}_{K,(\theta(t-L),\dots,\theta(t))} & \mathbf{D}_{K,(\theta(t-L),\dots,\theta(t))} \end{bmatrix}$$
$$(10)$$

for $t \in \mathbb{N}_0$.

The following result is immediate from Theorem 5 and [13, Theorem 20], and enables us to determine in a nonconservative manner a switching rule and a feedback controller such that the closed-loop system (9) is uniformly exponentially stable.

COROLLARY 8. *Let the controlled plant be given by (8). The following are equivalent:*

(a) *There exist a switching rule g and a set of controller coefficients (10), where $\theta(t) = g_t(y^t)$, such that the closed-loop system (9) is uniformly exponentially stable.*

(b) *There exist a switching sequence θ and a set of controller coefficients (10) such that the closed-loop system (9) is uniformly exponentially stable.*

(c) *There exist a path length $M \in \mathbb{N}_0$, a minimal set $\mathcal{N}$ of switching paths of length M, and matrices $\mathbf{R}_{(i_0,\dots,i_{M-1})}$, $\mathbf{S}_{(i_0,\dots,i_{M-1})} \in \mathbb{R}^{n \times n}$ such that*

$$N(\mathbf{B}_{i_M}^{\mathrm{T}})^{\mathrm{T}} \big(\mathbf{A}_{i_M} \mathbf{R}_{(i_0,\dots,i_{M-1})} \mathbf{A}_{i_M}^{\mathrm{T}}$$
$$- \mathbf{R}_{(i_1,\dots,i_M)} \big) N(\mathbf{B}_{i_M}^{\mathrm{T}}) < \mathbf{0}, \quad (11a)$$

$$N(\mathbf{C}_{j_M})^{\mathrm{T}} \big(\mathbf{A}_{i_M}^{\mathrm{T}} \mathbf{S}_{(i_1,\dots,i_M)} \mathbf{A}_{i_M}$$
$$- \mathbf{S}_{(i_0,\dots,i_{M-1})} \big) N(\mathbf{C}_{j_M}) < \mathbf{0}, \quad (11b)$$

$$\begin{bmatrix} \mathbf{R}_{(i_1,\dots,i_M)} & \mathbf{I} \\ \mathbf{I} & \mathbf{S}_{(i_1,\dots,i_M)} \end{bmatrix} \geq \mathbf{0}. \quad (11c)$$

for all $(i_0,\dots,i_M) \in \mathcal{N}$, where $N(\mathbf{M})$ denotes any full-column-rank matrix whose columns span the null space of $\mathbf{M}$.

Moreover, if condition (c) holds, then the closed-loop system (9) is uniformly exponentially stable under a periodic switching sequence θ satisfying (6) and a set of feedback controller coefficients (10) with the number of recalled past modes $L = M$.

Once the linear matrix inequalities (11) have been solved for some path length M and some minimal set $\mathcal{N}$, a uniformly exponentially stabilizing set of controller coefficients (10) can be obtained based on the well-known linear matrix inequality embedding technique [22, 6], which combines projection and matrix completion arguments. This technique is guaranteed to produce a nonconservative controller synthesis. An algorithm is provided in [13, Algorithm 1], and so will not be repeated here.

3. SWITCHING AND FEEDBACK FOR GUARANTEED PERFORMANCE

3.1 Problem Statement

Let

$$\mathcal{T} = \{ (\mathbf{A}_i, \mathbf{B}_i, \mathbf{C}_{1,i}, \mathbf{D}_{1,i}, \mathbf{C}_{2,i}, \mathbf{D}_{2,i}) : i = 1,\dots,N \} \quad (12)$$

be an indexed set of matrix tuples, where $\mathbf{A}_i \in \mathbb{R}^{n \times n}$, $\mathbf{B}_i \in \mathbb{R}^{n \times m}$, $\mathbf{C}_{1,i} \in \mathbb{R}^{l_1 \times n}$, $\mathbf{D}_{1,i} \in \mathbb{R}^{l_1 \times m}$, $\mathbf{C}_{2,i} \in \mathbb{R}^{l_2 \times n}$, and $\mathbf{D}_{2,i} \in \mathbb{R}^{l_2 \times m}$, $i = 1, \dots, N$, are given matrices. The set $\mathcal{T}$ defines the family of linear time-varying state-space equations

$$x(t+1) = \mathbf{A}_{\theta(t)} x(t) + \mathbf{B}_{\theta(t)} w(t), \quad t \in \mathbb{N}_0; \quad (13a)$$
$$z(t) = \mathbf{C}_{1,\theta(t)} x(t) + \mathbf{D}_{1,\theta(t)} w(t), \quad t \in \mathbb{N}_0; \quad (13b)$$
$$y(t) = \mathbf{C}_{2,\theta(t)} x(t) + \mathbf{D}_{2,\theta(t)} w(t), \quad t \in \mathbb{N}_0, \quad (13c)$$

over all switching sequences $\theta \in \{1,\dots,N\}^\infty$. Given a switching sequence θ, a disturbance input sequence $w = (w(0), w(1), \dots)$, and an initial state $x(0) \in \mathbb{R}^n$, the linear time-varying system (13) generates an output sequence $z = (z(0), z(1), \dots)$ in addition to the state sequence x and measurement sequence y.

Our $\mathcal{H}_2$-type performance measure gives the square root of the average output variance per unit time of the state-space model (13) under white Gaussian disturbance input sequence w, and it indicates how well the system output is regulated under random disturbances. On the other hand, our $\mathcal{H}_\infty$-type performance measure requires that the switching rule g satisfy the following small gain property: Whenever a disturbance sequence w generates a switching sequence θ for (12), the state-space model (13) is robustly well-connected with all sufficiently small dynamic uncertainties.

DEFINITION 9. *A switching rule g as in (3) is said to achieve output regulation level $\gamma > 0$ for the switched linear system $\mathcal{T}$ if it is uniformly exponentially stabilizing for $\mathcal{T}$ whenever $w = 0$ and if there exists a $\tilde{\gamma} \in (0, \gamma)$ such that, whenever $x(0) = 0$, the state-space equations (13) satisfy*

$$\lim_{T \to \infty} \frac{1}{T+1} \sum_{t=0}^{T} \mathrm{E} \, \|z(t)\|^2 \leq \tilde{\gamma}^2,$$

where $\mathrm{E}(\cdot)$ denotes the expectation with respect to disturbance input sequences w with

$$\mathrm{E} \, w(t) = 0, \quad \mathrm{E} \, w(t) w(s)^{\mathrm{T}} = \begin{cases} \mathbf{I}, & t = s; \\ \mathbf{0}, & t \neq s. \end{cases}$$

DEFINITION 10. *A switching sequence $\theta \in \{1,\dots,N\}^\infty$ is said to achieve output regulation level $\gamma > 0$ for the switched linear system $\mathcal{T}$ if it is uniformly exponentially stabilizing for $\mathcal{T}$ whenever $w = 0$ and if*

$$\lim_{t \to \infty} \frac{1}{T+1} \sum_{t=0}^{T} \mathrm{tr} \, \big(\mathbf{C}_{1,\theta(t)} \mathbf{Y}_\theta(t_0, t) \mathbf{C}_{1,\theta(t)}^{\mathrm{T}}$$
$$+ \mathbf{D}_{1,\theta(t)} \mathbf{D}_{1,\theta(t)}^{\mathrm{T}} \big) \leq \tilde{\gamma},$$

where

$$\mathbf{Y}_\theta(t_0, t) = \begin{cases} \sum_{s=t_0}^{t-1} \mathbf{\Phi}_\theta(t, s+1)\mathbf{B}_{\theta(s)}\mathbf{B}_{\theta(s)}^{\mathrm{T}}\mathbf{\Phi}_\theta(t, s+1)^{\mathrm{T}}, \\ \qquad\qquad\qquad\qquad\qquad t > t_0; \\ \mathbf{0}, \qquad\qquad\qquad\qquad\quad t = t_0. \end{cases}$$
(14)

DEFINITION 11. *A switching rule g as in (3) is said to achieve disturbance attenuation level $\gamma > 0$ for the switched linear system $\mathcal{T}$ if it is uniformly exponentially stabilizing for $\mathcal{T}$ whenever $w = 0$ and if there exists a $\tilde{\gamma} \in (0, \gamma)$ such that, whenever $x(0) = 0$, the state-space model (13) feedback-interconnected with $w = r + \Delta z$ has $z = 0$ as the unique response to $w = 0$ over all linear operators Δ satisfying*

$$\sum_{t=0}^{\infty} \|w(t)\|^2 < \tilde{\gamma}^{-2} \sum_{t=0}^{\infty} \|z(t)\|^2$$
(15)

for all z with $\sum_{t=0}^{\infty} \|z(t)\|^2 < \infty$.

DEFINITION 12. *A switching sequence $\theta \in \{1, \ldots, N\}^\infty$ is said to achieve disturbance attenuation level $\gamma > 0$ for the switched linear system $\mathcal{T}$ if it is uniformly exponentially stabilizing for $\mathcal{T}$ whenever $w = 0$ and if there exists a $\tilde{\gamma} \in (0, \gamma)$ such that*

$$\|\mathbf{M}_\theta(t, t_0)\| \leq \tilde{\gamma}$$

for all t_0, $t \in \mathbb{N}_0$ with $t_0 \leq t$, where

$$\mathbf{M}_\theta(t, t_0) = \begin{bmatrix} \mathbf{D}_{1,\theta(t_0)} \\ \mathbf{C}_{1,\theta(t_0+1)}\mathbf{B}_{\theta(t_0)} \\ \vdots \\ \mathbf{C}_{1,\theta(t)}\mathbf{\Phi}_\theta(t, t_0+1)\mathbf{B}_{\theta(t_0)} \end{bmatrix}$$
$$\begin{matrix} \mathbf{0} & \cdots & \mathbf{0} \\ \mathbf{D}_{1,\theta(t_0+1)} & \cdots & \mathbf{0} \\ \vdots & \ddots & \vdots \\ \mathbf{C}_{1,\theta(t)}\mathbf{\Phi}_\theta(t, t_0+2)\mathbf{B}_{\theta(t_0+1)} & \cdots & \mathbf{D}_{1,\theta(t)} \end{matrix}.$$

Note that if θ is a switching sequence, and if $x(0) = 0$, then we have $z(0) = \mathbf{D}_{1,\theta(0)}w(0)$ and

$$z(t) = \sum_{s=0}^{t-1} \mathbf{C}_{1,\theta(t)}\mathbf{\Phi}_\theta(t, s+1)\mathbf{B}_{\theta(s)}w(s) + \mathbf{D}_{1,\theta(t)}w(t)$$

for all $t \in \mathbb{N}$. Thus it is readily seen that, if a switching sequence θ achieves an output regulation level, or a disturbance attenuation level, then there exists a switching rule g that achieves the same performance level.

We will show that the converse also holds and establish exact, convex synthesis conditions for such a switching rule. We will also extend these conditions to joint synthesis of switching and feedback.

3.2 Synthesis of Switching Rules

The following theorems characterize the existence of switching rules that achieve guaranteed performance levels. They also give convex synthesis conditions for obtaining these rules.

THEOREM 13. *Let $\mathcal{T}$ be as in (12); let $\gamma > 0$. The following are equivalent:*

(a) *There exists a switching rule that achieves output regulation level γ for $\mathcal{T}$.*

(b) *There exists a switching sequence that achieves output regulation level γ for $\mathcal{T}$.*

(c) *There exist a path length $M \in \mathbb{N}_0$, a minimal set $\mathcal{N}$ of switching paths of length M, and matrices $\mathbf{Y}_{(i_0, \ldots, i_{M-1})} \in \mathbb{R}^{n \times n}$ such that*

$$\mathbf{Y}_{(i_0, \ldots, i_{M-1})} > \mathbf{0},$$
(16a)

$$\mathbf{A}_{i_M}\mathbf{Y}_{(i_0, \ldots, i_{M-1})}\mathbf{A}_{i_M}^{\mathrm{T}} - \mathbf{Y}_{(i_1, \ldots, i_M)} < \mathbf{B}_{i_M}\mathbf{B}_{i_M}^{\mathrm{T}}$$
(16b)

for all $(i_0, \ldots, i_M) \in \mathcal{N}$, and such that

$$\frac{1}{|\mathcal{N}|} \sum_{(i_0, \ldots, i_M) \in \mathcal{N}} \mathrm{tr}\left(\mathbf{C}_{1,i_M}\mathbf{Y}_{(i_0, \ldots, i_{M-1})}\mathbf{C}_{1,i_M}^{\mathrm{T}}\right.$$
$$\left. + \mathbf{D}_{1,i_M}\mathbf{D}_{1,i_M}^{\mathrm{T}}\right) < \gamma^2, \quad (16c)$$

where $|\mathcal{N}|$ denotes the cardinality of $\mathcal{N}$.

Moreover, if condition (c) holds, then any periodic switching sequence θ satisfying (6) achieves output regulation level $\tilde{\gamma}$ for $\mathcal{T}$.

PROOF. The proof is similar to those of [15, Theorem 3.7] and [11, Theorem 1], so we will only sketch it. Suppose a switching rule g achieves an output regulation level $\gamma > 0$ for $\mathcal{T}$. Let θ be any realization of the random switching sequence generated under g. For $\varepsilon > 0$, let $\mathbf{Y}_t^{(\varepsilon)}$ be perturbed versions of $\mathbf{Y}_\theta(0, t)$ defined in (14), so that

$$\mathbf{Y}_{t_0}^{(\varepsilon)} = \mathbf{0},$$
$$\mathbf{A}_{\theta(t)}\mathbf{Y}_t^{(\varepsilon)}\mathbf{A}_{\theta(t)}^{\mathrm{T}} - \mathbf{Y}_{t+1}^{(\varepsilon)} = -\mathbf{B}_{\theta(t)}\mathbf{B}_{\theta(t)}^{\mathrm{T}} - \varepsilon\mathbf{I}$$

for $t \in \mathbb{N}_0$. Then there exists a sufficiently small ε and an $\eta \in (0, \gamma)$ such that

$$\mathrm{tr}\left(\mathbf{C}_{1,\theta(t)}\mathbf{Y}_t^{(\varepsilon)}\mathbf{C}_{1,\theta(t)}^{\mathrm{T}} + \mathbf{D}_{1,\theta(t)}\mathbf{D}_{1,\theta(t)}^{\mathrm{T}}\right) \leq \eta^2$$

for $t \in \mathbb{N}_0$. Choosing a sufficiently small $\mathbf{Y} > 0$ and putting

$$\mathbf{Y}_0 = \mathbf{Y}; \quad \mathbf{Y}_t = \mathbf{Y}_t^{(\varepsilon)}, \quad t \in \mathbb{N},$$

we obtain that there are α, $\beta > 0$, $\tilde{\gamma} \in (0, \gamma)$, and $\mathbf{Y}_t \in \mathbb{R}^{n \times n}$, $t \in \mathbb{N}_0$, such that

$$\alpha\mathbf{I} \leq \mathbf{Y}_t \leq \beta\mathbf{I},$$
(17a)

$$\mathbf{A}_{\theta(t)}\mathbf{Y}_t\mathbf{A}_{\theta(t)}^{\mathrm{T}} - \mathbf{Y}_{t+1} \leq -\mathbf{B}_{\theta(t)}\mathbf{B}_{\theta(t)}^{\mathrm{T}} - \alpha\mathbf{I},$$
(17b)

$$\mathrm{tr}\left(\mathbf{C}_{1,\theta(t)}\mathbf{Y}_t\mathbf{C}_{1,\theta(t)}^{\mathrm{T}} + \mathbf{D}_{1,\theta(t)}\mathbf{D}_{1,\theta(t)}^{\mathrm{T}}\right) \leq \tilde{\gamma}^2$$
(17c)

hold for $t \in \mathbb{N}_0$. Due to [15, Lemma 2.3], however, such a switching sequence θ achieves the disturbance attenuation level γ for $\mathcal{T}$, and hence conditions (a) and (b) are equivalent.

To show (b) is equivalent to (c), suppose there exists a switching sequence that achieves a disturbance attenuation level $\gamma > 0$. Then it follows from [11, Theorem 1] that there are a path length $M \in \mathbb{N}_0$ and a minimal set $\mathcal{N}$ of switching paths of length M such that inequalities (16) hold for all $(i_0, \ldots, i_M) \in \mathcal{N}$. Conversely, if inequalities (16) hold for all $(i_0, \ldots, i_M) \in \mathcal{N}$, then we recover (17) from (16) by choosing a periodic switching sequence θ that satisfies (6). $\square$

THEOREM 14. *Let* $\mathcal{T}$ *be as in* (12); *let* $\gamma > 0$. *The following are equivalent:*

(a) *There exists a switching rule that achieves disturbance attenuation level* γ *for* $\mathcal{T}$.

(b) *There exists a switching sequence that achieves disturbance attenuation level* γ *for* $\mathcal{T}$.

(c) *There exist a path length* $M \in \mathbb{N}_0$, *a minimal set* $\mathcal{N}$ *of switching paths of length* M, *and matrices* $\mathbf{Y}_{(i_0,\dots,i_{M-1})} \in \mathbb{R}^{n \times n}$ *such that*

$$\mathbf{Y}_{(i_0,\dots,i_{M-1})} > \mathbf{0} \tag{18a}$$

and

$$\begin{bmatrix} \mathbf{A}_{i_M} & \mathbf{B}_{i_M} \\ \mathbf{C}_{1,i_M} & \mathbf{D}_{1,i_M} \end{bmatrix} \begin{bmatrix} \mathbf{Y}_{(i_0,\dots,i_{M-1})} & \mathbf{0} \\ \mathbf{0} & \mathbf{I} \end{bmatrix}$$
$$\times \begin{bmatrix} \mathbf{A}_{i_M} & \mathbf{B}_{i_M} \\ \mathbf{C}_{1,i_M} & \mathbf{D}_{1,i_M} \end{bmatrix}^{\mathrm{T}} - \begin{bmatrix} \mathbf{Y}_{(i_1,\dots,i_M)} & \mathbf{0} \\ \mathbf{0} & \gamma^2 \mathbf{I} \end{bmatrix} < \mathbf{0} \tag{18b}$$

for all $(i_0,\dots,i_M) \in \mathcal{N}$.

Moreover, if condition (c) *holds, then any periodic switching sequence* θ *satisfying* (6) *achieves disturbance attenuation level* γ *for* $\mathcal{T}$.

PROOF. Suppose there is a switching rule g that achieves a disturbance attenuation level $\gamma > 0$ for $\mathcal{T}$. Set $x(0) = 0$ and choose a w to generate a switching sequence θ for $\mathcal{T}$ under g. To show by contradiction that this θ achieves the disturbance attenuation level γ, suppose $\|\mathbf{M}_\theta(t, t_0)\| \geq \gamma$ for some t, $t_0 \in \mathbb{N}_0$ with $t \geq t_0$. Let an uncertain linear operator $\boldsymbol{\Delta}$ be such that $w(s)$ is identically zero whenever $s < t_0$ or $s > t$ and such that

$$\begin{bmatrix} w(t_0) \\ \vdots \\ w(t) \end{bmatrix} = \frac{1}{\|\mathbf{M}_\theta(t, t_0)\|^2} \mathbf{M}_\theta(t, t_0)^{\mathrm{T}} \begin{bmatrix} z(t_0) \\ \vdots \\ z(t) \end{bmatrix}.$$

This $\boldsymbol{\Delta}$ satisfies (15) for any $\tilde{\gamma} \in (0, \gamma)$. However, since the matrix

$$\mathbf{I} - \mathbf{M}_\theta(t, t_0) \mathbf{M}_\theta(t, t_0)^{\mathrm{T}} / \|\mathbf{M}_\theta(t, t_0)\|^2$$

is singular, $z = 0$ is not the unique response of the feedback-interconnection of (13) and $w = r + \boldsymbol{\Delta}z$ to $r = 0$. This contradicts the assumption that g achieves the disturbance attenuation level γ, and hence (a) and (b) are equivalent.

To show (b) is equivalent to (c), suppose there exists a switching sequence θ that achieves a disturbance attenuation level $\gamma > 0$. Then [4, Theorem 11] gives that there exist α, $\beta > 0$, $\eta \in (0, \gamma)$, and $\mathbf{X}_t \in \mathbb{R}^{n \times n}$ satisfying

$$\alpha \mathbf{I} \leq \mathbf{X}_t \leq \beta \mathbf{I}, \tag{19a}$$

$$\begin{bmatrix} \mathbf{A}_{\theta(t)} & \mathbf{B}_{\theta(t)} \\ \mathbf{C}_{1,\theta(t)} & \mathbf{D}_{1,\theta(t)} \end{bmatrix}^{\mathrm{T}} \begin{bmatrix} \mathbf{X}_{t+1} & \mathbf{0} \\ \mathbf{0} & \eta^{-1}\mathbf{I} \end{bmatrix}$$
$$\times \begin{bmatrix} \mathbf{A}_{\theta(t)} & \mathbf{B}_{\theta(t)} \\ \mathbf{C}_{1,\theta(t)} & \mathbf{D}_{1,\theta(t)} \end{bmatrix} - \begin{bmatrix} \mathbf{X}_t & \mathbf{0} \\ \mathbf{0} & \eta\mathbf{I} \end{bmatrix} \leq -\alpha\mathbf{I} \tag{19b}$$

for all $t \in \mathbb{N}_0$. For $M \in \mathbb{N}_0$, define

$$\mathcal{W}_M(\theta) = \{(i_0,\dots,i_M):$$
$$(i_0,\dots,i_M) = (\theta(t),\dots,\theta(t+M)), \ t \in \mathbb{N}_0\}.$$

Then, by [12, Theorem 3.3], there are some $M \in \mathbb{N}_0$, $\tilde{\gamma} \in (\eta, \gamma)$, and $\mathbf{X}_{(i_0,\dots,i_{M-1})} \in \mathbb{R}^{n \times n}$ such that $\mathbf{X}_{(i_0,\dots,i_{M-1})} > \mathbf{0}$ and

$$\begin{bmatrix} \mathbf{A}_{i_M} & \mathbf{B}_{i_M} \\ \mathbf{C}_{1,i_M} & \mathbf{D}_{1,i_M} \end{bmatrix}^{\mathrm{T}} \begin{bmatrix} \mathbf{X}_{(i_1,\dots,i_M)} & \mathbf{0} \\ \mathbf{0} & \tilde{\gamma}^{-1}\mathbf{I} \end{bmatrix}$$
$$\times \begin{bmatrix} \mathbf{A}_{i_M} & \mathbf{B}_{i_M} \\ \mathbf{C}_{1,i_M} & \mathbf{D}_{1,i_M} \end{bmatrix} - \begin{bmatrix} \mathbf{X}_{(i_0,\dots,i_{M-1})} & \mathbf{0} \\ \mathbf{0} & \tilde{\gamma}\mathbf{I} \end{bmatrix} < \mathbf{0}$$

for all $(i_0,\dots,i_M) \in \mathcal{W}_M(\theta)$. Since the number of modes N is finite and the sequence θ is infinite, there exists at least one minimal set $\mathcal{N}$ of switching paths of length M such that $\mathcal{N} \subset \mathcal{W}_M(\theta)$, so that these inequalities are feasible over all switching paths $(i_0,\dots,i_M) \in \mathcal{N}$. Using the Schur complement formula on the last inequality, we obtain

$$\begin{bmatrix} \mathbf{A}_{i_M} & \mathbf{B}_{i_M} \\ \mathbf{C}_{1,i_M} & \mathbf{D}_{1,i_M} \end{bmatrix} \begin{bmatrix} \mathbf{X}^{-1}_{(i_0,\dots,i_{M-1})} & \mathbf{0} \\ \mathbf{0} & \tilde{\gamma}^{-1}\mathbf{I} \end{bmatrix}$$
$$\times \begin{bmatrix} \mathbf{A}_{i_M} & \mathbf{B}_{i_M} \\ \mathbf{C}_{1,i_M} & \mathbf{D}_{1,i_M} \end{bmatrix}^{\mathrm{T}} - \begin{bmatrix} \mathbf{X}^{-1}_{(i_1,\dots,i_M)} & \mathbf{0} \\ \mathbf{0} & \tilde{\gamma}\mathbf{I} \end{bmatrix} < \mathbf{0}.$$

Putting

$$\mathbf{Y}_{(i_0,\dots,i_{M-1})} = \tilde{\gamma} \mathbf{X}^{-1}_{(i_0,\dots,i_{M-1})}$$

for each $(i_0,\dots,i_{M-1})$ yields (18), and thus (b) implies (c). Conversely, given (c), reversing the above argument gives that (19) holds for all $t \in \mathbb{N}_0$, where $\eta \in (0, \gamma)$ is sufficiently close to γ and θ satisfies (6). Therefore, (b) and (c) are equivalent. $\square$

EXAMPLE 15. *Suppose* $N = 2$ *and* $\mathcal{T}$ *has*

$$\mathbf{A}_1 = \begin{bmatrix} \frac{1}{2} & 0 \\ 0 & 1 \end{bmatrix}, \quad \mathbf{A}_2 = \begin{bmatrix} \frac{\sqrt{3}}{2} & \frac{1}{2} \\ -\frac{1}{2} & \frac{\sqrt{3}}{2} \end{bmatrix}, \quad \mathbf{B}_1 = \mathbf{B}_2 = \begin{bmatrix} 0 \\ 1 \end{bmatrix},$$

$$\mathbf{C}_{1,1} = \mathbf{C}_{1,2} = \begin{bmatrix} 1 & 0 \\ 0 & 1 \end{bmatrix}, \quad \mathbf{C}_{2,1} = \mathbf{C}_{2,2} = \begin{bmatrix} 1 & 0 \\ 0 & 1 \end{bmatrix},$$

$$\mathbf{D}_{1,1} = \mathbf{D}_{1,2} = \begin{bmatrix} 0 \\ 0 \end{bmatrix}, \quad \mathbf{D}_{2,1} = \mathbf{D}_{2,2} = \begin{bmatrix} 1 \\ 0 \end{bmatrix}.$$

We will use condition (c) *of Theorem 14 to obtain best switching sequences in terms of their disturbance attenuation performance. There are two minimal sets of switching paths of length zero; namely,* $\{1\}$ *and* $\{2\}$. *Since the spectral radii of* $\mathbf{A}_1$ *and* $\mathbf{A}_2$ *are not less than 1, inequalities* (18) *are infeasible for any* $\gamma > 0$ *if* $M = 0$. *However, among three minimal sets of switching paths of length one (namely,* $\{(1,1)\}$, $\{(1,2),(2,1)\}$, *and* $\{(2,2)\}$),

$$\mathcal{N} = \{(1,2),(2,1)\}$$

renders (18) *feasible if and only if* $M = 1$ *and* $\gamma \geq 6.845$ *(up to four significant digits). Thus the periodic switching sequence*

$$\theta = (1,2,1,2,\dots)$$

achieves all disturbance attenuation levels $\gamma \geq 6.845$ *but does not achieve any disturbance attenuation level* $\gamma \leq 6.844$. *Incrementing the path length to* $M = 2$, *we obtain six minimal sets of switching paths of length two. Among them,*

$$\mathcal{N} = \{(1,1,2),(1,2,2),(2,2,1),(2,1,1)\}$$

renders (18) *feasible if and only if* $M = 2$ *and* $\gamma \geq 6.484$. *Thus the periodic switching sequence*

$$\theta = (1,1,2,2,1,1,2,2,\dots)$$

achieves all disturbance attenuation levels $\gamma \geq 6.484$ but does not achieve any disturbance attenuation level $\gamma \leq 6.483$. One can find a better switching switching sequence by increasing the path length further. For path length $M = 3$, the minimal set

$$\mathcal{N} = \{(1,1,1,2),(1,1,2,2),(1,2,2,2),$$
$$(2,2,2,1),(2,2,1,1),(2,1,1,1)\}$$

is the best one, and we conclude that the periodic switching sequence

$$\theta = (1,1,1,2,2,2,1,1,1,2,2,2,\dots),$$

where the switching sequence $(1,1,1,2,2,2)$ repeats, achieves all disturbance attenuation levels $\gamma \geq 6.466$ and does not achieve any disturbance attenuation level $\gamma \leq 6.465$. For this particular example, incrementing the path length even further to $M = 4$ does not improve the performance level.

3.3 Joint Synthesis of Switching and Feedback

Joint synthesis of a switching rule and a feedback controller is also possible for achieving guaranteed disturbance attenuation performance in view of Theorem 13 and [12, Theorem 3.7] and for achieving guaranteed output regulation performance in view of Theorem 14 and [15, Theorem 4.2]. The technical development is similar to that of Section 2.4, so the details are omitted. In particular, since we have proved that it suffices to look for open-loop switching rules, the joint synthesis condition for guaranteed output regulation is the same as that in [11, Theorem 1].

4. CONCLUSIONS

We proposed and proved the theses that, as far as uniform exponential stability and $\mathcal{H}_2$ and $\mathcal{H}_\infty$ type performance levels for discrete-time switched linear systems are concerned, it suffices to look for open-loop periodic switching rules for joint synthesis of measurement-dependent switching rules and switching path-dependent dynamic output feedback laws, and that a switching rule and a feedback controller can be designed separately. These results apply to both supervisory control and measurement scheduling problems.

Determining the best switching-feedback pairs requires solving an increasing sequence of semidefinite programs. As one moves from one such program to the next, the computational complexity is likely to grow exponentially. Although this is due to the "semi-decidable" problem nature [1], the presented results are nonconservative and expected to be useful for small problems. In particular, the results cover the well-known approaches based on common and multiple Lyapunov functions as special cases.

5. REFERENCES

[1] P.-A. Bliman and G. Ferrari-Trecate. Stability analysis of discrete-time switched systems through Lyapunov functions with nonminimal state. In *Proceedings of IFAC Conference on the Analysis and Design of Hybrid Systems*, pages 325–330, 2003.

[2] S. Boyd, L. El Ghaoui, E. Feron, and V. Balakrishnan. *Linear Matrix Inequalities in System and Control Theory*. SIAM, Philadelphia, PA, 1994.

[3] M. S. Branicky. Multiple Lyapunov functions and other analysis tools for switched and hybrid systems.

IEEE Transactions on Automatic Control, 43(4):475–482, 1998.

[4] G. E. Dullerud and S. Lall. A new approach for analysis and synthesis of time-varying systems. *IEEE Transactions on Automatic Control*, 44(8):1486–1497, 1999.

[5] G. E. Dullerud and F. Paganini. *A Course in Robust Control Theory: A Convex Approach*. Springer-Verlag, New York, NY, 2000.

[6] P. Gahinet and P. Apkarian. A linear matrix inequality approach to H_∞ control. *International Journal of Robust and Nonlinear Control*, 4(4):421–448, 1994.

[7] J. Hu, J. Shen, and W. Zhang. A generating function approach to the stability of discrete-time switched linear systems. In *Proceedings of the 13th ACM International Conference on Hybrid Systems: Computation and Control*, pages 273–282, 2010.

[8] D. B. Johnson. Finding all the elementary circuits of a directed graph. *SIAM Journal on Computing*, 4(1):77–84, 1975.

[9] S. A. Krishnamurthy and J.-W. Lee. A computational stability analysis of discrete-time piecewise linear systems. In *Proceedings of the 48th IEEE Conference on Decision and Control*, pages 1106–1111, 2009.

[10] J.-W. Lee. Inequality-based properties of detectability and stabilizability of linear time-varying systems in discrete time. *IEEE Transactions on Automatic Control*, 54(3):634–641, 2009.

[11] J.-W. Lee. Infinite-horizon joint LQG synthesis of switching and feedback in discrete time. *IEEE Transactions on Automatic Control*, 54(8):1945–1951, 2009.

[12] J.-W. Lee and G. E. Dullerud. Optimal disturbance attenuation for discrete-time switched and Markovian jump linear systems. *SIAM Journal on Control and Optimization*, 45(4):1329–1358, 2006.

[13] J.-W. Lee and G. E. Dullerud. Uniform stabilization of discrete-time switched and Markovian jump linear systems. *Automatica*, 42(2):205–218, 2006.

[14] J.-W. Lee and G. E. Dullerud. Uniformly stabilizing sets of switching sequences for switched linear systems. *IEEE Transactions on Automatic Control*, 52(5):868–874, 2007.

[15] J.-W. Lee and P. P. Khargonekar. Optimal output regulation for discrete-time switched and Markovian jump linear systems. *SIAM Journal on Control and Optimization*, 47(1):40–72, 2008.

[16] J.-W. Lee and P. P. Khargonekar. Detectability and stabilizability of discrete-time switched linear systems. *IEEE Transactions on Automatic Control*, 54(3):424–437, 2009.

[17] D. Liberzon and A. S. Morse. Basic problems in stability and design of switched systems. *Control Systems Magazine*, 19(5):59–70, 1999.

[18] B. Lincoln and B. Bernhardsson. LQR optimization of linear system switching. *IEEE Transactions on Automatic Control*, 47(10):1701–1705, 2002.

[19] I. Masubuchi, A. Ohara, and N. Suda. LMI-based controller synthesis: A unified formulation and solution. *International Journal of Robust and Nonlinear Control*, 8(8):669–686, 1998.

[20] L. Meier, III, J. Peschon, and R. M. Dressler. Optimal control of measurement subsystems. *IEEE Transactions on Automatic Control*, 12(5):528–536, 1967.

[21] A. S. Morse. Supervisory control of families of linear set-point controllers—part 1: Exact matching. *IEEE Transactions on Automatic Control*, 41(10):1413–1431, 1996.

[22] A. Packard. Gain scheduling via linear fractional transformations. *Systems & Control Letters*, 22(2):79–92, 1994.

[23] C. Scherer, P. Gahinet, and M. Chilali. Multiobjective output-feedback control via LMI optimization. *IEEE Transactions on Automatic Control*, 42(7):896–911, 1997.

[24] D. P. Stanford. Stability for a multi-rate sampled-data system. *SIAM Journal on Control and Optimization*, 17(3):390–399, 1979.

[25] D. P. Stanford and J. M. Urbano. Some convergence properties of matrix sets. *SIAM Journal on Matrix Analysis and Applications*, 15(4):1132–1140, 1994.

Robust Discrete Synthesis against Unspecified Disturbances[*]

Rupak Majumdar[1,2]   Elaine Render[2]   Paulo Tabuada[2]

[1]MPI-SWS   [2]UC Los Angeles

rupak@mpi-sws.org, elaine@cs.ucla.edu, tabuada@ee.ucla.edu

ABSTRACT

Systems working in uncertain environments should possess a *robustness* property, which ensures that the behaviours of the system remain close to the original behaviours under the influence of unmodeled, but bounded, disturbances. We present a theory and algorithmic tools for the design of robust discrete controllers for ω-regular properties on discrete transition systems. Formally, we define *metric automata* —automata equipped with a metric on states— and strategies on metric automata which guarantee robustness for ω-regular properties. We present graph-theoretic algorithms to construct such strategies in polynomial time. In contrast to strategies computed by classical automata-theoretic algorithms, the strategies computed by our algorithm ensure that the behaviours of the controlled system under disturbances satisfy a related property which depends on the magnitude of the disturbance. We show an application of our theory to the design of controllers that tolerate infinitely many transient errors provided they occur infrequently enough.

Categories and Subject Descriptors: D.2.4 [Software Engineering]: Software/Program Verification.

General Terms: Verification, Reliability.

Keywords: Cyber-physical systems, Robustness, Linear temporal logic

1. INTRODUCTION

High confidence design and implementation of cyber-physical systems (CPSs) is widely recognized as the enabler for a large number of exciting applications —in healthcare, energy-distribution, and industrial automation— with enormous societal impact. It is equally recognized that the current design and verification methodologies fall short of what is required to design these systems in a robust, yet cost-effective manner.

Current system design and verification techniques take a 0-1 view of the world: a property either holds or it doesn't, and a synthesized system provides no guarantees if the environment deviates

in any way from the model. In practice, this view can be overly restrictive. First, CPSs need to operate for extended periods of time in environments that are either unknown or difficult to describe and predict at design time. For example, sensors and actuators may have noise, there can be mismatches between the dynamics of the physical world and its model, software scheduling strategies can change dynamically. Thus, asking for a model that encompasses all possible scenarios places an undue burden on the programmer, and the detailed book-keeping of every deviation from nominal behaviour renders the specifications difficult to understand and maintain. Second, even when certain assumptions are violated at run-time, we would expect the system to behave in a *robust* way: either by continuing to guarantee correct behaviour or by ensuring that the resulting behaviour only deviates modestly from the desired behaviour upon the influence of small perturbations. Unfortunately, current design methodologies for CPSs fall short in this respect: the Boolean view cannot specify or guarantee that small changes in the physical world, in the software world, or in their interaction, still results in acceptable behaviour.

In this paper, we present a theory of robustness for cyber-physical systems. Our starting point is the observation that a notion of robustness and associated design tools have been successfully developed in continuous control theory. There, the control designer designs the system for the nominal case, while bounding the effects of uncertainties and errors on the performance of the system. Our goal is to provide a similar theory and algorithmic tools in the presence of both discrete and continuous changes on the one hand, and in the presence of more complex temporal specifications —given, for example, in linear temporal logic (LTL) or as ω-automata— on the other hand. We do this in three steps.

First, robustness is a topological concept. In order to define it, we need to give meaning to the words "closeness" or "distance." For this, we define a metric on the system states. Second, instead of directly modeling the effect of every disturbance, we model a *nominal* system (the case with no disturbance), together with a set of (unmodeled) disturbances whose effect can be bounded in the metric. That is, while making no assumption on the nature or origin of disturbances, we assume that the disturbances can only push a state to another one within a distance γ. Third, under these three assumptions, we show how we can derive strategies for ω-regular objectives that are robust in that the deviation from nominal behaviour can be bounded as a function of the disturbance and the parameters of the system.

To illustrate this last point, consider *reachability properties* $\Diamond F$, where the system tries to reach a given set of states F. We give algorithms to compute strategies that ensure F is reached in the nominal case, and additionally, when disturbances are present, guarantees that the system reaches a set F' which contains states at most

---

[*]This research was funded in part by the NSF awards 0834771, 0953994, and 1035916.

distance $\Delta(\gamma)$ from F, where Δ is an increasing function of the magnitude γ of the disturbance. More importantly, we show that an arbitrary strategy obtained through classical automata-theoretic constructions (e.g., [17, 29]) may only provide trivial robustness guarantees (e.g., a bounded disturbance can force the system to reach any arbitrary state). We show how similar arguments can be made to bound the system dynamics for Büchi and parity (and thus, for all LTL) specifications under the presence of disturbances.

Technically, our constructions lift arguments similar to arguments in robust control based on control Lyapunov functions to the setting of ω-regular properties. For reachability, the correspondence is simple: we require that the strategy decrements a "rank function" at a rate that depends on the distance to the target. For parity, the argument is more technical, and uses progress measures for parity games [14, 19]. Finally, we show algorithms to compute strategies with these properties based on shortest path algorithms on graphs.

We provide a simple application of our theory to the synthesis problem in the presence of transient faults. Transient faults, such as single-event upsets, are unpredictable disturbances in electronic systems that can cause bits in an electronic circuit to flip. They are becoming more relevant in electronic systems design due to reductions in feature sizes [6, 16, 21]. We show how, using our methodology, we can algorithmically synthesize controllers for LTL objectives which provide a time-space tradeoff: given that transient errors can cause a deviation of γ in the state, we show that if transient errors are infrequent, that is, occur at most once in $N(\gamma)$ cycles for a monotonic function N depending on the magnitude γ of the deviation (as well as the property), then our robust controller can guarantee that the property still holds for the system. However, a strategy computed by classical automata-theoretic techniques may not be able to provide this guarantee without first modeling all the parameters of the transient fault in detail.

Although we focus on robustness for discrete transition systems, the natural next step is to combine these results with the existing theory for continuous control systems in order to handle robustness in CPSs. We outline our plans for this unification in Section 7.

Related work. Our starting point is the theory of robust continuous control [25, 28] and the theory of infinite games with ω-regular objectives on discrete graphs [10, 24, 29]. There has not been much previous work combining robustness with automata theory. In discrete control systems, tolerance to errors is achieved by explicitly modeling faults and then solving a game assuming that the adversary determines when faults occur [11]. As mentioned earlier, the enumeration of possible faults can be tedious, if not impossible, at design time. Topologies for hybrid systems [7, 20] have been studied before, but the interaction with ω-regular specifications have not been.

Qualitative notions of fault tolerance have been studied in distributed systems, for example, by designing algorithms to be "self-stabilizing" on perturbations [9], or by requiring that an invariant is eventually restored after an error ("convergence") or that the system satisfied a more liberal invariant on error ("closure") [1]. However, quantitative notions, relevant to CPSs, have not been studied. Our synthesis procedure produces strategies that satisfy quantitative notions of closure and (under some assumptions on the rate of faults) convergence.

In a series of papers [3, 4, 5, 8], robustness measures were developed by comparing the number of environmental errors and the number of resulting system errors using cost functions. Bloem et al. [5] define k-robustness: roughly speaking, a system is k-robust if the ratio of system to environment errors is k. In general terms, the synthesis approach presented here results in K-robust strategies, where K is a constant associated with the rank function which serves as the basis for the strategy. The magnitude of the constant K can be influenced by the choice of cost function used in the synthesis algorithm. Moreover, we work in a simpler model, where the only adversarial action is the bounded disturbance, while the work of Bloem et al. considers an explicit adversary. Our framework has the advantage of leading to simple polynomial-time algorithms for synthesis, but may provide more conservative results than the game-solving algorithms from [3] when robustness is sought in the presence of explicit adversaries. A more detailed technical comparison with their work is provided in Section 3.

Tarraf et al. [23] develop a framework for quantifying robust stability in finite Mealy machines with finite input and output alphabets. Their robustness notions are generalizations of classical notions of gain stability in control theory. The focus is upon input-output stability and, although we adopt a state-space approach, the results derived in this paper for reachability are similar to those in [23]. However, we go beyond simple reachability properties and consider also Büchi and parity requirements. A more technical comparison appears in Section 3.

Measures of robustness against transient fault models have been studied in the context of combinational circuits and FPGAs [12, 13, 15, 18], but extensions to temporal behaviours have not been considered.

2. PRELIMINARIES

Let Q be a (finite or infinite) set. A function $d : Q \times Q \to \mathbb{R}_0^+$ is called a *metric* or *distance function* for Q if for all $p, q, r \in Q$, we have $d(p, q) = 0$ if and only if $p = q$; $d(p, q) = d(q, p)$ (*symmetry*); and $d(p, r) \leq d(p, q) + d(q, r)$ (*triangle inequality*). The pair (Q, d) where Q is a set and d is a metric for Q is called a *metric space*. For $q \in Q$ and a subset $Q' \subseteq Q$, we define $d(q, Q') = \inf_{q' \in Q'} d(q, q')$.

A function $R : Q \to \mathbb{R}$ is *Lipschitz continuous* if there exists some constant $K > 0$ such that for any two states $q, q' \in Q$:

$$|R(q) - R(q')| \leq K d(q, q').$$

The constant K is called the *Lipschitz constant* of the function R with respect to the distance d. Note that if the set Q is finite then every function has this property. Lipschitz continuity is a standard assumption in continuous robust control regardless of the cardinality of the state set and hence is reasonable here.

We model discrete control systems using *automata*. Intuitively, we consider a "nominal" automaton modeling the undisturbed dynamics of the system, and add a set of disturbance actions which can perturb the nominal behaviour. We consider a very general model for the disturbances by simply requiring their effects to be bounded but otherwise arbitrary.

For a set Σ, let Σ^* represent the set of finite strings of symbols from Σ, and let Σ^ω denote the set of infinite strings over Σ. The notation $|\Sigma|$ denotes the cardinality of the set Σ and Σ^+ represents the set of non-empty finite strings over Σ. A (metric) *automaton* is a tuple $A = ((Q, d), q_0, \Sigma, \delta, X, \gamma)$, where Q is a set of *states* and (Q, d) is a metric space, $q_0 \in Q$ is the unique *initial state*, Σ is a set of *input letters*, X is a set of *disturbance indices* including a special symbol ϵ signifying "no disturbance", $\delta : Q \times \Sigma \times X \to Q$ is the *transition function* specifying the next state given the current state, the input letter chosen by the system, and some member of X chosen by the environment, and finally, $\gamma \geq 0$ is a real-valued parameter such that for each $q \in Q$ and $a \in \Sigma$, we have

$$d(\delta(q, a, \epsilon), \delta(q, a, x)) \leq \gamma \text{ for every } x \in X.$$

An automaton is *finite* if Q, Σ, and X are all finite sets. For an automaton A, we define the *undisturbed automaton*, written A_ϵ, as the automaton resulting from restricting the set of disturbance indices X to the singleton $\{\epsilon\}$. For $q \in Q$, $a \in \Sigma$ and $x \in X$ we use the shorthand q^{ax} to denote the state $\delta(q, a, x)$.

Intuitively, the undisturbed automaton models a "nominal" design of a discrete transition system, and the set of disturbance indices X models possible environmental disturbances to the nominal design (the symbol $\epsilon \in X$ thus represents the case where there is no disturbance). The parameter γ limits the effects of the disturbances with respect to the nominal behaviour: when an action a is chosen at a state q, the disturbances can cause a state at most distance γ away from the nominal state to be reached instead. As a special case, if $\gamma = 0$, then the disturbances have no effect on the nominal behaviour (i.e., $q^{ax} = q^{a\epsilon}$ for each $q \in Q$, $a \in \Sigma$, and $x \in X$).

A *trace* $\sigma \in Q^* \cup Q^\omega$ of the automaton A is a (finite or infinite) sequence of states $\sigma = q_0 q_1 q_2 \ldots$ from Q such that q_0 is the initial state of the automaton, and there exist inputs $a_0, a_1, a_2, \ldots$ and disturbances $x_0, x_1, x_2, \ldots$ with $\delta(q_i, a_i, x_i) = q_{i+1}$ for $i \geq 0$. A *nominal trace* is a trace in the nominal automaton A_ϵ, that is, σ is such that $x_0 = x_1 = x_2 = \ldots = \epsilon$. For a finite trace $\sigma = q_0 q_1 \ldots q_n \in Q^*$, we define $|\sigma| = n + 1$.

The proposed model for the disturbances encompasses a wide range of concrete applications ranging from the discrete to the continuous world, as illustrated by the next two examples.

EXAMPLE 2.1. **[Digital Design with Single Bit-Flips]** *Consider an automaton A modeling a state machine whose states are encoded using a binary Gray code [26]. Each state of A is a sequence of n bits, and neighbouring states differ in only one bit. Disturbances occur as single-event upsets which can cause a single bit in the state to flip. The distance function for the automaton is defined to be the Hamming distance between n-bit strings. The set of disturbance actions $X \subset \{0,1\}^n$ contains all binary strings of length n with at most one non-zero digit. Under this definition, ϵ is equal to the binary string of length n consisting entirely of zeros. The transition function δ for A is defined from the transition function δ_ϵ of A_ϵ by $\delta(q, a, x) = \delta_\epsilon(q, a) + x$ for any $q \in \{0,1\}^n$ where $+$ is the XOR function. Hence, the potential effect of the disturbance is bounded by $\gamma = 1$.*

EXAMPLE 2.2. **[Robust Control]** *Consider a continuous control system in discrete time which can be viewed as an infinite-state automaton with transition function $\delta : \mathbb{R}^n \times \mathbb{R}^m \times \mathbb{R}^p \to \mathbb{R}^n$. The state set is $\mathbb{R}^n$, the input alphabet is $\mathbb{R}^m$ and $\mathbb{R}^p$ is the set of environmental disturbances. Disturbance signals $x : \mathbb{N} \to \mathbb{R}^p$ are often used as a lumped representation for several sources of uncertainly such as measurement errors or errors in the model of the transition function. Hence, the disturbance signals are assumed to be arbitrary but of bounded amplitude, that is, $\|x(k)\| \leq \gamma'$ for some constant $\gamma' \in \mathbb{R}_0^+$, some norm $\| \cdot \|$ on $\mathbb{R}^p$, and every $k \in \mathbb{N}$. A further typical assumption is Lipschitz continuity of δ. It then follows from these two assumptions that*

$$\|\delta(q, a, x) - \delta(q, a, 0)\| \leq K' \|x - 0\| \leq K' \gamma'$$

where K' is the Lipschitz constant. Therefore by defining the distance function d as $d(y, z) = \|y - z\|$ we conclude that $\gamma \in \mathbb{R}_0^+$ is equal to $K' \gamma'$ for this example.

We associate *acceptance conditions* with automata to distinguish between "good" and "bad" behaviours. A *reachability condition* is a set $F \subseteq Q$ of *terminal* states. A *reachability automaton* (A, F) consists of an automaton A together with a reachability condition

F. A finite trace of the automaton A satisfies the reachability condition F if and only if it ends at some state in the set F.

A *Büchi automaton* (A, F) is an automaton A together with a *Büchi acceptance condition* $F \subseteq Q$. For an infinite trace $\sigma = q_0 q_1 \ldots \in Q^\omega$ let $\zeta(\sigma) = \{q \in Q \mid \forall i \geq 0 \exists j > i, q_j = q\}$ denote the set of states appearing infinitely often on σ. A trace $\sigma \in Q^\omega$ satisfies the Büchi acceptance condition F if and only if $\zeta(\sigma) \cap F \neq \emptyset$. In other words, there exists at least one state in the set F which features infinitely often on the trace.

A *generalized Büchi acceptance condition* is a set of the form $\mathscr{F} = \{F_0, \ldots F_{n-1}\}$ and consists of a finite number of subsets of the state set Q. An automaton A paired with such an acceptance condition is called a *generalized Büchi automaton*. An infinite trace $\sigma \in Q^\omega$ of A satisfies the acceptance condition $\mathscr{F}$ if and only if $\zeta(\sigma) \cap F_i \neq \emptyset$ for all $i = 0, \ldots, n - 1$.

Finally a *parity automaton* $(A, \mathscr{F})$ is an automaton A together with a *parity acceptance condition* consisting of a finite number of pairwise disjoint subsets of the state set Q: $\mathscr{F} = \{F_1, \ldots, F_{2n+1}\}$ with $F_i \cap F_j = \emptyset$ for $i \neq j$.

The *parity* of a state $q \in Q$ is the index i of the unique set F_i containing q, if any, and undefined if there exists no such i. A trace $\sigma \in Q^\omega$ of A satisfies the acceptance condition $\mathscr{F}$ if and only if the least parity amongst the states in the set $\zeta(\sigma)$ is even. Note that we assume that the set of states may be partially colored [29]: the set $\bigcup_{i=1}^{2n+1} F_i$ does not necessarily cover the set Q. We make the extra assumption that for any state $q \in Q$, there exists a finite nominal trace connecting q to some state of even parity, and if q has odd parity, a state of lower even parity is assumed to be nominally reachable from q.

A *strategy* for an automaton A is a function $S : Q^+ \to \Sigma$ specifying an input choice for each finite trace. Given a strategy S, the set of *outcomes* is the set of traces $q_0 q_1 \ldots$ on which q_0 is the initial state of the automaton, and for each $i \geq 0$, we have $q_{i+1} = \delta(q_i, S(q_0 \ldots q_i), x)$ for some $x \in X$. The (unique) *nominal outcome* of a strategy is the trace $q_0 q_1 q_2 \ldots$ where q_0 is the initial state and for each $i \geq 0$ we have $q_{i+1} = \delta(q_i, S(q_0 \ldots q_i), \epsilon)$. A strategy S is *memoryless* if $S(w \cdot q) = S(w' \cdot q)$ for all $w, w' \in Q^*$ and $q \in Q$, that is, if it depends only on the last state on the trace. In this case, we omit the (irrelevant) prefix, and consider a strategy to be a function from Q to Σ.

A *disturbance strategy* is a function from $Q^+ \times \Sigma$ to X. Let $S : Q^+ \to \Sigma$ be a strategy and $T : Q^+ \times \Sigma \to X$ a disturbance strategy. The (unique) outcome $q_0 q_1 \ldots$ of S and T is the trace on which q_0 is the initial state of the automaton and for each $i \geq 0$ we have $q_{i+1} = \delta(q_i, S(q_0 \ldots q_i), T(q_0 \ldots q_i, S(q_0 \ldots q_i)))$.

Let (A, F) be an automaton together with an acceptance condition. A strategy is *nominally winning* in A if the nominal outcome satisfies F. It is known that reachability, Büchi, and parity conditions admit memoryless nominally winning strategies [10]. A strategy S is *winning* if for every disturbance strategy T, the unique outcome of S and T satisfies F.

EXAMPLE 2.3. *We show a simple example illustrating our definitions. Consider the reachability automaton (A, F) with $Q = \{q_0, \ldots, q_6\}$, $\Sigma = \{a, b\}$, $F = \{q_6\}$, $\gamma = 1$ and nominal behaviour defined as shown in Figure 1. The automaton A is equipped with a distance function $d : Q \times Q \to \mathbb{R}_0^+$. The relative distances of the states in Q are presented in Table 1 and are approximated by the layout of the automaton in Figure 1.*

Let $S_b : Q \to \Sigma$ be the memoryless strategy which chooses $b \in \Sigma$ for all $q \in Q$, and let $S_a : Q \to \Sigma$ be the memoryless strategy which chooses $a \in \Sigma$ for all $q \in Q$.

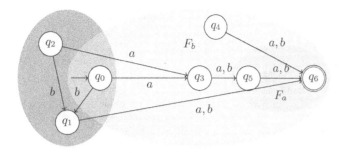

Figure 1: The undisturbed automaton A_ϵ

	q_0	q_1	q_2	q_3	q_4	q_5	q_6
q_0	0	1	2	4	4	4	5
q_1		0	1	5	5	5	6
q_2			0	6	6	7	8
q_3				0	1	3	3
q_4					0	3	3
q_5						0	1
q_6							0

Table 1: The distance between states in the automaton A of Figure 1.

Clearly both S_b and S_a are nominally winning for the reachability condition; they are both equally good strategies in classical automata theory.

Now consider the result of applying the strategies S_b and S_a in the disturbed automaton A. First note that the nominal trace connecting the initial state q_0 to the terminal state q_6 resulting from applying S_b is $q_0 q_1 q_6$. Inputting b at state q_0 could result in reaching any of the states in the left-hand ellipse, and hence a trace implementing S_b is only guaranteed to reach a state at distance 5 or less from the terminal state q_6. Therefore, in the disturbed automaton, strategy S_b is winning with respect to the inflated acceptance condition $\{q \in Q \mid d(q, F) \leq 5\}$ as shown in Figure 1. Now consider the strategy S_a. The nominal trace connecting q_0 to q_6 for this strategy is $q_0 q_3 q_5 q_6$. Note that $d(q_0, q_3)$ and $d(q_3, q_5)$ are both greater than the power of the disturbance $\gamma = 1$. Therefore in the disturbed automaton progress is still being made towards F until we reach q_5 which is at a distance of 1 from F. Hence the strategy S_a is winning with respect to the inflated reachability condition $\{q \in Q \mid d(q, F) \leq 1\}$ as shown in Figure 1.

In classical automata and game theory (e.g., [29]), the outcomes of the two strategies are indistinguishable: both strategies reach the set $F = \{q_6\}$ in the nominal case, and can result in traces which do not reach F in the disturbed case. However, the distance function in this model gives an extra method of comparison for the two strategies: the distance from F as a function of the bound on the disturbance γ. With this in mind it is obvious that the strategy S_a is a better choice for the automaton A.

We discuss the construction of the two strategies S_b and S_a in Section 3.

3. REACHABILITY

Let $(A, F) = ((Q, d), q_0, \Sigma, \delta, X, \gamma, F)$ be a reachability automaton satisfying a finite reachability assumption: for every state $q \in Q$ there exists a finite trace connecting q to some state in F. A *(reachability) rank function* with respect to the reachability con-

dition F is a function $R_F : Q \to \mathbb{R}_0^+$ where $R_F(q) = 0$ if and only if $q \in F$ and for some monotonically increasing function $\alpha : \mathbb{R}_0^+ \to \mathbb{R}_0^+$ satisfying $\alpha(0) = 0$ we have

$$\alpha(d(q, F)) \leq R_F(q) \; \forall q \in Q. \tag{1}$$

A rank function R_F is said to be a *control Lyapunov function* if there exists a monotonically increasing function $f : \mathbb{R}_0^+ \to \mathbb{R}_0^+$ satisfying $f(0) = 0$ and such that for each $q \in Q \backslash F$ there exists some $a \in \Sigma$ with

$$R_F(q^{a\epsilon}) - R_F(q) \leq -f(d(q, F)). \tag{2}$$

A control Lyapunov function R_F induces one or more memoryless strategy functions S defined by mapping a state $q \in Q$ to some input $S(q) = a \in \Sigma$ satisfying inequality (2). Such strategies are nominally winning. To see why this is the case, consider the nominal outcome $\sigma = q_0 q_1 \ldots$ under strategy S. Since (2) holds for every q in the trace σ, the function R_F decreases along this trace. Moreover, R_F being non-negative and decreasing along σ, necessarily reaches zero in finitely many steps. It then follows from (1) that $d(\sigma, F)$ also reaches zero in finitely many steps since $d(\sigma, F) \leq \alpha^{-1}(R_F(\sigma))$ where the inverse α^{-1} is also a monotonically increasing function vanishing at zero.

The existence of a control Lyapunov function implies that for every state $q \in Q$ there exists some finite nominal trace connecting q with some state in the terminal set F. This is a natural assumption in certain applications, such as in the control of physical systems, but it is typically not satisfied when considering software synthesis. Models in which this assumption is not present will be dealt with in future work.

The following proposition describes the "graceful degradation" or robustness properties possessed by strategies induced by control Lyapunov functions. When disturbances are present, the nominal outcome is not guaranteed but no catastrophic failure will occur. Instead, the deviation from the nominal outcome is linearly bounded by the power of the disturbance as measured by γ.

PROPOSITION 3.1. *Let R_F be a Lipschitz continuous control Lyapunov function for the reachability automaton (A, F) and let S be a nominally winning strategy induced from R_F. Then S is a winning strategy for the reachability automaton (A, F') with*

$$F' = \{q \in Q \mid f(d(q, F)) \leq K\gamma\}$$

for some monotonically increasing function f mapping zero to zero and depending upon R_F, K the Lipschitz constant of R_F and γ the disturbance bound of A.

PROOF. Let T be a disturbance strategy and consider the outcome $\sigma = q_0 q_1 \ldots$ of S and T. We first establish the inequality:

$$R_F(q^{ax}) - R_F(q) \leq K\gamma - f(d(q, F))$$

for any q appearing in σ:

$$
\begin{aligned}
R_F(q^{ax}) - R_F(q) &= R_F(q^{ax}) - R_F(q^{a\epsilon}) \\
&\quad + R_F(q^{a\epsilon}) - R_F(q) \\
&\leq |R_F(q^{ax}) - R_F(q^{a\epsilon})| \\
&\quad - f(d(q, F)) \\
&\leq K d(q^{ax}, q^{a\epsilon}) - f(d(q, F)) \\
&= K\gamma - f(d(q, F)).
\end{aligned}
$$

Note that as long as q is sufficiently far from F, $-f(d(q, F))$ is sufficiently negative, and the sum $K\gamma - f(d(q, F))$ remains negative. Hence, R_F continues to decrease along σ. The situation changes when we reach a state q' satisfying $K\gamma > f(d(q', F))$.

q	$d(q, q_6)$	$R_b(q)$	$R_a(q)$
q_0	5	2	18
q_1	6	1	12
q_2	8	2	24
q_3	3	2	8
q_4	3	1	6
q_5	1	1	1
q_6	0	0	0

Table 2: The rank functions R_b and R_a for the automaton A.

So we conclude that the outcome of S and T must reach the set F' in finitely many steps and that S is a winning strategy for the acceptance condition F'. $\square$

Returning to Example 2.3, we discuss the two rank functions $R_b : Q \to \mathbb{R}_0^+$ and $R_a : Q \to \mathbb{R}_0^+$ from which the strategies S_b and S_a are induced. Table 2 lists the distance from each state to the terminal state q_6 and the value of the two rank functions R_b and R_a.

The function $R_b : Q \to \mathbb{R}_0^+$ is the result of a classical graph theoretic shortest path approach - each state $q \in Q$ is mapped to the length of the shortest path connecting q to some state in F.

Let $\eta : \mathbb{R}_0^+ \to \mathbb{R}_0^+$ be the monotonically increasing function defined by $x \mapsto 2x$ for all $x \in \mathbb{R}_0^+$. Then R_a is a control Lyapunov function since for all $q \in Q$

$$R_a(q^{a\epsilon}) - R_a(q) \leq -\eta(d(q, q_6)).$$

Comparison with existing work. At this point it is convenient to compare our framework with the frameworks of Bloem et. al. [5] and of Tarraf et. al. [23]. Both of these references adopt an input-output perspective by relating environment errors (inputs) to system errors (outputs). In contrast, we adopt a state-space approach by endowing the set of states with a metric and placing no assumptions on the environment other than having bounded power.

In [5] the authors define the notion of *k-robustness* for automata. For a reachability automaton (A, F) two monotonically increasing functions which map zero to zero are defined: an *environmental error function* $e : \Sigma^* \to \mathbb{N} \cup \{\infty\}$ and a *system error function* $s : \Sigma^* \to \mathbb{N} \cup \{\infty\}$. A pair (e, s) of error functions for a given automaton is called an *error specification* for A. Then a strategy $S : Q \to \Sigma$ for A is κ-robust with respect to the error specification (e, s) if there exists $\beta \in \mathbb{N}$ such that for all $w \in \Sigma^*$ which label traces which are outcomes of S,

$$s(w) \leq \kappa e(w) + \beta.$$

In order to compare Bloem's and Tarraf's results with ours, we resort to some key ideas from robust control [25, 28]. First, we define an environment error signal $\mathsf{e} = \mathsf{e}_1 \mathsf{e}_2 \ldots \mathsf{e}_n \in \mathbb{R}^*$ and a system error signal $\mathsf{s} = \mathsf{s}_1 \mathsf{s}_2 \ldots \mathsf{s}_n \in \mathbb{R}^*$. The only assumption we place on e and s is that an absence of environment errors at time $k \in \mathbb{N}$ corresponds to $\mathsf{e}_k = 0$ and the absence of system errors at time $k \in \mathbb{N}$ corresponds to $\mathsf{s}_k = 0$. The error functions e and s in Bloem's framework can be seen as the cumulative versions of e and s, for example:

$$e(k) = \sum_{i=0}^{n} \mathsf{e}_i, \qquad s(k) = \sum_{i=0}^{n} \mathsf{s}_i.$$

In Tarraf's framework the role of e_i is played by $\rho(u(i))$ and the role of s_i is played by $\mu(y(i))$. We now regard an automaton as defining a transformation $F : \mathbb{R}^* \to \mathbb{R}^*$ from environment error

signals to system error signals $F(\mathsf{e}) = \mathsf{s}$. In general F will be a set valued function, but we assume it to be single valued to simplify the discussion.

The notion of finite-gain stability from robust control can now be introduced as follows:

A map $F : \mathbb{R}^* \to \mathbb{R}^*$ is said to be finite-gain stable with gain κ and bias β if the following inequality holds:

$$\sum_{i=0}^{n} F(\mathsf{e}) \leq \kappa \sum_{i=0}^{n} \mathsf{e} + \beta \qquad (3)$$

for every $\mathsf{e} \in \mathbb{R}^*$. A more condensed version of (3) is:

$$s \leq \kappa e + \beta$$

which is Bloem's notion of κ-robustness and Tarraf's notion of ρ/μ gain stability. It is well known in robust control and dissipative systems theory that the existence of a certain type of Lyapunov function (a storage function) implies finite-gain stability. In the context of reachability automata, we define e to be the effect of the environment actions on the state:

$$\mathsf{e}_k = d(q_k^{ax}, q_k^{a\epsilon}).$$

When $x = \epsilon$, $q_k^{ax} = q_k^{\epsilon}$ and $\mathsf{e}_k = 0$ since the behaviour coincides with the nominal behaviour under no environment disturbances. For problems of the form $\Diamond F$ we regard F as the set of states describing the desired operation for the system. Hence, any deviation from F is regarded as a system error. This formulation differs subtly from that presented at the start of this section - here the specification may be thought of as "always F" instead of "eventually F". By defining control Lyapunov functions for the full state set of a system instead of only those which are non terminal, the results above immediately apply in the "always F" case. The decision to present the results in this way will become clearer in the following sections on ω-regular acceptance conditions.

Returning to our point, the system error signal is defined as:

$$\mathsf{s}_k = f(d(q_k, F))$$

for any monotonically increasing function $f : \mathbb{R}_0^+ \to \mathbb{R}_0^+$ satisfying $f(0) = 0$. Standard arguments in dissipative systems theory [25] would then show that:

$$s \leq Ke + R_F(q_0)$$

with K being the Lipschitz constant of R_F. It is also known that finite-gain stability does not imply the notion of stability considered in this paper unless certain controllability/observability assumptions hold. This follows from the fact that it may not be possible to infer the decrease of R_F at every state only from the knowledge of e and s when not every state can be reached from q_0 or when s does not provide enough information about the state.

4. OMEGA-REGULAR OBJECTIVES

We now extend the results to more general ω-regular acceptance conditions. We do this in two steps. First, we provide a simple generalization to Büchi acceptance conditions. Then, we show how ideas based on *progress measures* [14, 19] can be used to provide robustness results for parity acceptance conditions. In every case we repeat the finite reachability assumption given at the start of the previous section.

4.1 Büchi acceptance conditions

Let $(A, F) = ((Q, d), q_0, \Sigma, \delta, X, \gamma, F)$ be a Büchi automaton with acceptance condition $F \subset Q$. First note that a Büchi acceptance condition asks only that for a trace $\sigma \in Q^\omega$, the intersection

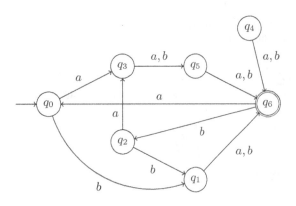

Figure 2: The undisturbed Büchi automaton A_e

of the set $\zeta(\sigma)$ with the set of terminal states is non-empty. Hence by viewing the Büchi condition as an infinite series of reachability conditions for the set F, the definitions and results for reachability also apply in the case of Büchi automata.

In particular, note that the definition of a control Lyapunov function asks only that inequality (2) holds for states outside of the set F. As before, a control Lyapunov function R_F induces a memoryless strategy $S : Q \to \Sigma$ which chooses an action a satisfying (2) for any state in $Q \setminus F$ and plays an arbitrary action for states in F. The strategy S is nominally winning: the argument that F is reached is identical to the reachability case, and our assumption that F can be reached from any state ensures that an arbitrary action from F will not prevent F from being visited again.

COROLLARY 4.1. *Let R_F be a Lipschitz continuous control Lyapunov function for the Büchi automaton (A, F) and let S be a nominally winning strategy induced from R_F. Then S is a winning strategy for the Büchi automaton (A, F') where*

$$F' = \{q \in Q \mid f(d(q, F)) \le K\gamma\}.$$

Consider the Büchi automaton (A, F) with $F = \{q_6\}$ and whose nominal behaviour is defined as shown in Figure 2. Note that this automaton is identical to the reachability automaton presented in Figure 1 (example 2.3) with the addition of two outgoing edges from the terminal state q_6. The distances between the respective states and the rank functions $R_b : Q \to \mathbb{R}_0^+$ and $R_a : Q \to \mathbb{R}_0^+$ are as before, and hence their values may be found in Tables 1 and 2. The two strategies $S_b : Q \to \Sigma$ and $S_a : Q \to \Sigma$ are induced as before for states in $Q \setminus F$. Observe that a control Lyapunov function for a Büchi automaton does not specify the value of the induced strategy for terminal states. There are of course two options, namely a and b, leading to the states q_0 and q_2 respectively.

For S_a we note that $R_a(q_0) < R_a(q_2)$ and so we set $S_a(q_6)$ equal to a. For the strategy S_b, note that $R_b(q_0) = R_b(q_2)$. For consistency we set $S_b(q_6) = b$. Then the strategy S_b is winning for the Büchi automaton (A, F_b) where $F_b = \{q \in Q \mid d(q, F) \le 5\}$ and S_a is winning for the Büchi automaton (A, F_a) with

$$F_a = \{q \in Q \mid d(q, F) \le 1\}.$$

4.2 Generalized Büchi conditions

We want to generalize the construction of rank functions to parity acceptance conditions. As a warm-up, we first describe the case for generalized Büchi acceptance conditions. It is a standard argument in automata theory to reduce a generalized Büchi automaton to a

Büchi automaton: the resulting Büchi automaton will have state set $Q \times \{0, \ldots, n-1\}$ where $|F| = n$. Hence for a system presented as a generalized Büchi automaton, Corollary 4.1 can be applied to an expanded state space, and winning strategies may be constructed. However, we give an alternate "direct" rank function construction based on progress measures that will introduce techniques useful in the parity case.

Let $(A, \mathscr{F}) = ((Q, d), q_0, \Sigma, \delta, X, \gamma, \mathscr{F})$ be a generalized Büchi automaton with $\mathscr{F} = \{F_0, \ldots, F_{n-1}\}$. For each $i = 0, 1, \ldots, n - 1$ define a (reachability) rank function $R_i : Q \to \mathbb{R}_0^+$ with respect to the set F_i as before. Then a *(generalized Büchi) rank function* $R : Q \to (\mathbb{R}_0^+)^n$ is defined for each $q \in Q$ by $R(q) = (R_0(q), R_1(q), \ldots, R_{n-1}(q))$.

The notion of Lipschitz continuity for functions is extended in the following way: a function $R : Q \to (\mathbb{R}_0^+)^n$ is Lipschitz continuous if there is a $K > 0$ such that for each $i \in \{0, \ldots, n-1\}$ and for all $q, q' \in Q$, we have $|R_i(q) - R_i(q')| \le Kd(q, q')$.

A relation and ordering on n-tuples of positive reals is defined as follows. For every $i \in \{0, 1, \ldots, n - 1\}$ we define the preorder $>^i$ on $(\mathbb{R}_0^+)^n$: let $a, b \in (\mathbb{R}_0^+)^n$ with $a = (a_0, \ldots, a_{n-1})$ and $b = (b_0, \ldots, b_{n-1})$. Then $a >^i b$ if and only if $a_i > b_i$. Similarly we define $a \ge^i b$ if and only if $a_i \ge b_i$. Based on $>^i$ we introduce another relation on $(\mathbb{R}_0^+)^n$, denoted by $\blacktriangleright^i$ and defined by $a \blacktriangleright^i b$ if and only one of the following two conditions holds:

$$a >^i b \quad \text{or} \quad a_{(i-1) \mod n} = 0.$$

Observe that, since the labeling of the sets in F begins at 0 instead of 1, the relation $>^0$ corresponds with the 1st index of the n-tuple, $>^1$ corresponds with the 2nd index, and so on.

PROPOSITION 4.2. *Let A be a generalized Büchi automaton with acceptance condition $\mathscr{F} = \{F_0, \ldots, F_{n-1}\}$. If a trace $\sigma = q_0 q_1 q_2 \ldots$ is such that*

$$R(q_0) \blacktriangleright^0 R(q_1) \blacktriangleright^0 R(q_2) \ldots R(q_{i_0}) \blacktriangleright^1 R(q_{i_0+1}) \blacktriangleright^1 \ldots$$
$$\ldots \blacktriangleright^1 R(q_{i_1}) \blacktriangleright^2 R(q_{i_1+1}) \blacktriangleright^2 \ldots$$
$$\ldots \blacktriangleright^{n-1} R(q_{i_{n-1}}) \blacktriangleright^0 R(q_{i_{n-1}+1}) \ldots \quad (4)$$

then σ satisfies the generalized Büchi acceptance condition $\mathscr{F}$.

PROOF. Let σ be a trace of the form given above. The definition of $\blacktriangleright$ implies that if two consecutive relations in (4) have different indices (say k and $k + 1$) then the state appearing between them must be contained in the set F_k. Hence

$$q_{i_0} \in F_0, q_{i_1} \in F_1, \ldots, q_{i_{n-1}} \in F_{n-1}, \ldots,$$

and σ features infinitely often a state in each of the sets in $\mathscr{F}$. $\square$

Intuitively, a trace of this form is initially moving towards the set F_0, and hence the relation is fixed as $\blacktriangleright^0$. Once a state in the set F_0 is reached, the second part of the definition of $\blacktriangleright$ comes into effect and the relation $\blacktriangleright^1$ is applied until a state in the set F_1 is reached. On reaching a state in the set F_{n-1}, the relation returns to $\blacktriangleright^0$, and so on.

Note that the other direction does not necessarily hold: an accepting trace will not necessarily have the above form. For example, the trace may visit the sets in a non-sequential order, or may visit multiple states from each set on each pass through the sets.

For brevity, we introduce some more notation. Let $d(q, \mathscr{F})$ denote the vector valued distance

$$d(q, \mathscr{F}) = (d(q, F_0), d(q, F_1), d(q, F_2), \ldots, d(q, F_{n-1})).$$

A generalized Büchi rank function R is said to be a *control Lyapunov function* if there exists a monotonically increasing function

$f : \mathbb{R}_0^+ \to \mathbb{R}_0^+$ with $f(0) = 0$ such that for every $q \in Q$ and every $i \in \{0, 1, \ldots, n-1\}$ there exists $a \in \Sigma$ with

$$R(q^{ae}) - R(q) \preceq^i -f(d(q, \mathscr{F})) \text{ or } R_i(q) = 0.$$

For a fixed i, the function R_i is a reachability control Lyapunov function with respect to the set F_i. This definition of control Lyapunov function asks that the set F_i is reachable from every state $q \in Q$ for all $i \in \{0, \ldots, n-1\}$. To see that this is necessary, consider for example a state $q \in Q$ from which the set F_i is not reachable for some $i > 0$. Then any state from which q may be reached, and any state reachable from q, may not feature on an accepting trace. Hence all such states are redundant (including q). Since we are implicitly assuming that there is no redundancy in the state set throughout this paper, we see that this definition is the natural extension of the control Lyapunov function definition to the generalized Büchi case.

Generalized Büchi automata do not admit memoryless strategies, where the strategy keeps track of the index $i + 1$ where i is the index of the last terminal set F_i which was visited on the trace. So a strategy for a generalized Büchi automaton $(A, \mathscr{F})$ is a function $S : Q \times \{0, \ldots, n-1\} \to \Sigma$ where for every $i \in \{0, \ldots, n-1\}$, the restriction $S(\cdot, i)$ is a memoryless reachability strategy. As in the case of Büchi acceptance conditions, for states $q \in F_i$ for some $i \in \{0, \ldots, n-1\}$ the strategy $S(q, i)$ prescribes any input, since every set in the acceptance condition can be reached via a finite trace from every state in Q.

PROPOSITION 4.3. *Let $R = (R_0, \ldots, R_{n-1})$ be a generalized Büchi control Lyapunov function constructed of n Lipschitz continuous (reachability) control Lyapunov functions R_i for the generalized Büchi automaton $(A, \mathscr{F})$. Let $S : Q \times \{0, \ldots, n-1\} \to \Sigma$ be a nominally winning strategy induced from R. Then S is a winning strategy for the generalized Büchi automaton $(A, \mathscr{F}')$ where $\mathscr{F}' = \{F_0', \ldots, F_{n-1}'\}$ with $F_i' = \{q \in Q \mid f(d(q, F_i)) < K\gamma\}$ and K is the Lipschitz constant of the function R with respect to d.*

PROOF. Restricting the control Lyapunov function to a fixed $i \in \{1, \ldots, n-1\}$, Corollary 4.1 may be applied to conclude that

$$R(q^{ax}) - R(q) \leq^i K\gamma - f(d(q, \mathscr{F})) \text{ or } R_i(q) < K\gamma$$

for K the Lipschitz constant of R with respect to d and γ the environmental error bound of A. Hence the strategy induced from the generalized Büchi rank function constructed from the functions R_i is winning for the inflated acceptance condition $\mathscr{F}'$. $\square$

4.3 Parity conditions

The simple notion of rank defined previously is insufficient for parity acceptance conditions. Instead we generalize progress measures for parity games [14, 19].

Let $(A, \mathscr{F}) = ((Q, d), q_0, \Sigma, \delta, X, \gamma, \mathscr{F})$ be a parity automaton with $\mathscr{F} = \{F_0, F_1, \ldots, F_{2n+1}\}$. Denote by $d(q, \mathscr{F})$ the vector valued distance

$$d(q, \mathscr{F}) = (d(q, F_0), d(q, F_2), d(q, F_4), \ldots, d(q, F_{2n}))$$

and let $\overline{\mathbb{R}}_0^+ = \mathbb{R}_0^+ \cup \{\infty\}$, the extended positive reals.

A (parity) rank function $R_{\mathscr{F}} : Q \to (\overline{\mathbb{R}}_0^+)^{n+1}$ is a function with $R_{\mathscr{F}}^i(q) = 0$ if and only if $q \in F_{2i}$ (where the notation $R_{\mathscr{F}}^i(q)$ denotes the ith component of the image of q under $R_{\mathscr{F}}$) and there exists a monotonically increasing function $\alpha : \overline{\mathbb{R}}_0^+ \to \overline{\mathbb{R}}_0^+$ such that $\alpha(0) = 0$ and

$$\alpha(d(q, F_i)) \leq R_{\mathscr{F}}^i(q)$$

for all $q \in Q$, $i \in \{0, \ldots, n\}$. Hence a parity rank function consists of $n + 1$ reachability rank functions defined upon the extended positive real numbers.

Let $\succ$ denote the lexicographic ordering on the $n + 1$ tuples $R_{\mathscr{F}}(q)$ for $q \in Q$, and let $\succ^i$ denote the lexicographic ordering restricted to the first i components. We define $\succeq^i$ in the obvious way: $a \succeq^i b$ if a is either less than b in the lexicographic ordering or equal to b. For $a, b \in (\overline{\mathbb{R}}_0^+)^{n+1}$ define $a \rhd^q b$ if and only if there exists $i \in \{0, 1, \ldots, n\}$ such that either

(i) $q \in F_{2i+1}$ and $a \succ^i b$ or

(ii) $q \in F_{2i}$ and $a \succeq^i b$ or

(iii) $q \notin \bigcup_{j \in \{0, \ldots, 2n+1\}} F_j$ and $a \succ b$.

Then a trace $\sigma = q_0 q_1 q_2 \ldots$ through the automaton $\mathscr{A}$ satisfies the parity condition $\mathscr{F}$ if

$$R_{\mathscr{F}}(q_0) \rhd^{q_0} R_{\mathscr{F}}(q_1) \rhd^{q_1} R_{\mathscr{F}}(q_2) \ldots.$$

Let $\bar{Q} \subseteq (F_0 \cup F_2 \cup \ldots \cup F_{2n})$ denote the set of states of even parity from which a state of lower or equal even parity cannot be reached. That is, the set contains all states $q \in F_{2i}$ for some $i \in \{0, \ldots, n\}$ such that there does not exist $k \leq i$ with some state $q' \in F_{2k}$ reachable from q.

A rank function $R_{\mathscr{F}}$ for a parity automaton $(A, \mathscr{F})$ is a *control Lyapunov function* if there exists some monotonically increasing function $f : (\overline{\mathbb{R}}_0^+)^{n+1} \to (\overline{\mathbb{R}}_0^+)^{n+1}$ satisfying $f(0^{n+1}) = 0^{n+1}$ such that for every $j \in \{1, \ldots, 2n+1\}$ and every $q \in F_j \setminus \bar{Q}$ there exists $a \in \Sigma$ with

$$R_{\mathscr{F}}(q^{ae}) - R_{\mathscr{F}}(q) \preceq^i -f(d(q, \mathscr{F})) \tag{5}$$

for some $2i \leq j$.

PROPOSITION 4.4. *Let $\sigma = q_0 q_1 q_2 \ldots \in Q^\omega$ be an infinite trace of the parity automaton $(A, \mathscr{F})$. Then if*

$$R_{\mathscr{F}}(q_0) \rhd^{q_0} R_{\mathscr{F}}(q_1) \rhd^{q_1} R_{\mathscr{F}}(q_2) \ldots, \tag{6}$$

σ satisfies $\mathscr{F}$. Moreover, if there exist a finite number of indices $i \in \mathbb{N}$ such that $R_{\mathscr{F}}(q_i) \rhd^{q_i} R_{\mathscr{F}}(q_{i+1})$ does not hold, the trace σ will still satisfy the parity condition $\mathscr{F}$.

Since an infinite trace of a finite state parity automaton necessarily is mostly comprised of a series of repeated loops, it is straightforward to argue that the least parity appearing on any loop must be even. Continuing on this line of thinking one observes that any such repeated loop comprising part of an infinite trace satisfying (6) above must consist entirely of even states. Hence a trace of this form will feature odd states only finitely often.

PROPOSITION 4.5. *Let $\sigma = q_0 q_1 q_2 \ldots \in Q^\omega$ be an infinite trace of the parity automaton $(A, \mathscr{F})$. If*

$$R_{\mathscr{F}}(q_{k+1}) - R_{\mathscr{F}}(q_k) \preceq^i -f(d(q_k, \mathscr{F})) \tag{7}$$

for all $q_k \in Q \setminus \bar{Q}$ appearing on σ and $\bar{Q}$ is finite then σ satisfies $\mathscr{F}$.

PROOF. Let $q_k \in Q \setminus \bar{Q}$. If $q_k \in F_{2i}$ for some i then (7) implies that $R_{\mathscr{F}}(q_k) \succeq^i R_{\mathscr{F}}(q_{k+1})$ and $q_k \rhd^{q_k} q_{k+1}$ as required.

Instead assume that $q_k \in F_j$ for some j odd. The function $f(d(q_k, \mathscr{F}))$ restricted to any i in $\{0, \ldots, n\}$ is non-zero, and so $R_{\mathscr{F}}(q_{k+1}) \prec^i R_{\mathscr{F}}(q_k)$ for some $2i < j$ and hence for all l satisfying $i \leq l < j$. Therefore we may again conclude that $R_{\mathscr{F}}(q_k) \rhd^{q_k} R_{\mathscr{F}}(q_{k+1})$.

Finally let $q_k \in \bar{Q}$ and $q_k \in F_{2i}$. Then q_k and q_{k+1} need not satisfy (7) and so may not satisfy the parity measure $\rhd$. Since $q \in \bar{Q}$ there exists no $l > k$ such that $q_l = q_k$. Indeed, if this were the case, it would contradict our assumption that a finite trace connecting q_k to a state of lower or equal parity does not exist. Since the cardinality of $\bar{Q}$ is finite there exist only a finite number of indices $l \in \mathbb{N}$ such that $R_{\mathscr{F}}(q_l) \rhd^{q_l} R_{\mathscr{F}}(q_{l+1})$ does not hold and applying Proposition 4.4 yields the result. $\square$

Given a control Lyapunov function $R_{\mathscr{F}}$ for a parity automaton $(A, \mathscr{F})$ the strategy $S : Q \to \Sigma$ induced from $R_{\mathscr{F}}$ is defined as follows. Let $q \in Q$.

(i) If $q \in Q \setminus \bar{Q}$ choose $S(q) = a$ such that $R_{\mathscr{F}}(q^{a\epsilon})$ satisfies the inequality (5) and is minimal with respect to the lexicographic ordering.

(ii) If $q \in \bar{Q}$ set $S(q) = a$ for any $a \in \Sigma$.

That such a strategy is winning follows from Proposition 4.5.

The following result takes advantage of the extra flexibility resulting from the possibility of only partially colouring the state set. If each set F_{2i} for $i = 0, \ldots, n$ has only non-parity states in its immediate neighbourhood, the sets may be inflated without overlap to ensure that a strategy induced from a control Lyapunov function is winning for an inflated acceptance condition $\mathscr{F}'$ which is defined below.

PROPOSITION 4.6. *Let $R_{\mathscr{F}}$ be a Lipschitz continuous control Lyapunov function for the parity automaton $(A, \mathscr{F})$ and let S be a nominally winning strategy induced from $R_{\mathscr{F}}$. Define the inflated parity acceptance condition $\mathscr{F}' = \{F_0', F_1, F_2', F_3, \ldots, F_{2n}', F_{2n+1}\}$ with $F_{2i}' = \{q \in Q \mid f(d(q, F_{2i})) \leq K\gamma\}$ for $i = 0, \ldots, n$. If $F = \bigcup_{i=0}^{n} F_{2i}$ is strictly contained in Q and further if $q \notin F_{2i}$ and $f(d(q, F_{2i})) \leq K\gamma$ imply that $q \notin F$ for $i = 0, \ldots, n$ then S is a winning strategy for $(A, \mathscr{F}')$.*

PROOF. An argument similar to the proof of Proposition 3.1 may be used to demonstrate that the following inequality holds

$$R_{\mathscr{F}}(q^{ax}) - R_{\mathscr{F}}(q) \preceq^i (K\gamma, \ldots, K\gamma) - f(d(q, \mathscr{F}))$$

for each $q \in Q$, some $a \in \Sigma$ and all $x \in X$.

Let $2i$ be the least colour appearing infinitely often on the nominal outcome of S in A. The above inequality implies that any outcome of S in A must feature infinitely often states in the set F_{2i}'. Since, by assumption, states in $F_{2i}' \setminus F_{2i}$ are not contained in F, the inflation from F_{2i} to F_{2i}' will not cause any state to have multiple parities, and by Proposition 4.5 we conclude that the strategy S is winning for the inflated acceptance condition $\mathscr{F}'$. $\square$

5. SYNTHESIS ALGORITHMS

The objective of synthesis is to construct a control Lyapunov function for a given automaton and acceptance condition. In this section we discuss one possible approach to this based on dynamic programming [2]. We focus on *finite* automata in this section.

PROPOSITION 5.1. *There is a polynomial-time algorithm to construct a reachability rank function which is also a control Lyapunov function for any finite reachability automaton.*

PROOF. Let (A, F) be a reachability automaton and let $\eta : \mathbb{R}_0^+ \to \mathbb{R}_0^+$ be a monotonically increasing function satisfying $\eta(0) = 0$.

From the nominal automaton A_ϵ we construct a weighted digraph $G_A = (Q, E)$ where $E \subseteq Q \times \mathbb{R}_0^+ \times Q$ is the set of directed edges with weights taken from $\mathbb{R}_0^+$, and defined as follows. For $q, q' \in Q$, there exists an edge $(q, \eta(d(q, F)), q') \in E$ if and only if there exists $a \in \Sigma$ such that $q' = \delta(q, a, \epsilon)$.

Next a modified shortest path algorithm [2] is applied to determine the length of the shortest path between each $q \in Q$ and the set F in G_A.

For each $q \in Q$ define $R(q)$ to be the length of the shortest possible trace connecting q to F in G_A and let $\sigma_F(q) \in Q^*$ denote one such trace.

It remains to show that the resulting function $R : Q \to \mathbb{R}_0^+$ is a control Lyapunov function. First note that since F is reachable from every state by assumption, $R(q)$ is finite for every $q \in Q$. Hence, for each $q \in Q$ there exists some $a \in \Sigma$ such that $q^{a\epsilon}$ is the next state on the trace $\sigma_F(q)$.

By definition, $R(q^{a\epsilon}) + \eta(d(q, F)) = R(q)$ and

$$R(q^{a\epsilon}) - R(q) \leq -\eta(d(q, F)).$$

Therefore for every state $q \in Q$ there exists some state $q^{a\epsilon}$ such that the above inequality is satisfied, and R is a control Lyapunov function as required.

The running time of the algorithm is dominated by the running time of the shortest path algorithm, which is quadratic in the number of states (assuming $\eta()$ is computed in a unit step). $\square$

Returning again to the example in Section 2, note that the rank function $R_2 : Q \to \mathbb{R}_0^+$ was derived via the method described in the proof of Proposition 5.1 with the weight function $\eta : \mathbb{R}_0^+ \to \mathbb{R}_0^+$ defined as $x \mapsto 2x$ for all $x \in \mathbb{R}_0^+$.

Control Lyapunov functions for Büchi and generalized Büchi acceptance conditions are based on control Lyapunov functions for reachability. Hence, the algorithm defined above applies to Büchi, and generalized Büchi automata.

COROLLARY 5.2. *There is a polynomial-time algorithm to construct a control Lyapunov function for any finite Büchi or generalized Büchi automaton.*

For parity acceptance conditions, the situation is a little more involved. Let A be an automaton with parity acceptance condition $\mathscr{F} = \{F_0, F_1, \ldots, F_{2n+1}\}$ for some $n > 0$. Recall the notation for $q \in Q$:

$$d(q, \mathscr{F}) = (d(q, F_0), d(q, F_2), \ldots, d(q, F_{2n})) \in (\overline{\mathbb{R}_0^+})^{n+1}.$$

PROPOSITION 5.3. *Let $(A, \mathscr{F})$ be a finite parity automaton satisfying the reachability assumptions outlined in Section 2. Then there exists a polynomial-time algorithm to construct a control Lyapunov function $R : Q \to (\mathbb{R}_0^+)^{n+1}$ for $(A, \mathscr{F})$.*

PROOF. Let $A = ((Q, d), q_0, \Sigma, \delta, X, \gamma)$ and let $\eta : \mathbb{R}_0^+ \to \mathbb{R}_0^+$ be a monotonically increasing function.

The state set Q is partitioned into $n + 2$ pieces,

$$Q = F_0 \cup \overline{F_0} \cup \overline{F_2} \cup \ldots \cup \overline{F_{2n}}$$

where the sets $\overline{F_{2i}}$ for $i = 0, \ldots, n$ are defined as follows.

- $q \in \overline{F_0}$ if and only if $0 < d(q, F_0) < \infty$;

- $q \in \overline{F_{2i}}$ if and only if $q \notin \overline{F_{2j}}$ for $j < i$ and $d(q, F_{2i}) < \infty$.

It is straightforward to see that the sets defined in this way indeed form a partition of the set Q. Note that for $i = 1, \ldots, n$, if $\overline{F_{2i}}$ is non-empty then necessarily $F_{2i} \cap \overline{F_{2i}} \neq \emptyset$.

We construct from $(A, \mathscr{F})$ the weighted digraph $G_A = (Q, E)$ with edge set E defined as follows:

$$(q, \eta(d(q, \mathscr{F})), q') \in E \iff \exists a \in \Sigma : q' = \delta(q, a, \epsilon).$$

Hence the edge weights in the graph G_A are taken from the set $(\overline{\mathbb{R}}_0^+)^{n+1}$.

For $i = 0, \ldots, n$ a shortest path algorithm is applied in G_A to states in the set $\overline{F}_{2i}$ with respect to the set F_{2i}. The ordering on the $(n+1)$-tuples used in the algorithm is the lexicographic ordering $\prec^i$.

For each $q \in \overline{F}_{2i}$ define $R(q)$ to be the length of the shortest possible trace connecting q to F_{2i} in G_A and let $\sigma_i(q) \in Q^*$ denote one such trace. Finally for $q \in F_0$ set $R(q) = 0$.

For each $q \in Q \setminus F_0$ there exists some $q^{a\epsilon}$ which is the next state in the shortest path from q to the set F_{2i} in the graph G_A for some $i \in \{0, \ldots, n\}$. By definition, $R(q^{a\epsilon}) + \eta(d(q, \mathscr{F})) = R(q)$. Since $d(q, F_{2i}) \geq d(q^{a\epsilon}, F_{2i})$ the lexicographic ordering implies that

$$R(q^{a\epsilon}) - R(q) \preceq^i -(\eta(d(q, F_0)), \ldots, \eta(d(q, F_{2n}))).$$

Hence for every state $q \in Q \setminus F_0$ there exists some state $q^{a\epsilon}$ such that the above inequality is satisfied, and R is a control Lyapunov function as required. $\square$

6. APPLICATION: TRANSIENT FAULTS

We now show an application of the robustness results of the preceding sections to the design of strategies under *transient faults*. We show that strategies synthesized using control Lyapunov functions are robust to infinitely occurring transient faults provided they occur infrequently enough.

Let $N \in \mathbb{N}$. A disturbance strategy $T : Q^+ \times \Sigma \to X$ is N-*bounded* if, whenever $T(\sigma, a) \neq \epsilon$ and $T(\sigma', b) \neq \epsilon$ for traces $\sigma, \sigma' \in Q^*$ with σ a prefix of σ' and $a, b \in \Sigma$, we have

$$|\sigma'| - |\sigma| \geq N.$$

Intuitively, disturbance strategies are N-bounded if any two occurrences of (non-trivial) disturbances are separated by at least N steps.

Let A be an automaton and F a (Büchi or parity) acceptance condition. Our main result is that for sufficiently large (but finite) N, a winning nominal strategy induced from a control Lyapunov function is winning against N-bounded disturbance strategies.

For a state $q \in Q$ of a Büchi automaton with acceptance set F and strategy S, let $\sigma^S(q) \in \mathbb{N}_0$ denote the length of the unique nominal trace connecting q to an element of the acceptance set F. For a state $q \in Q$ in a parity automaton with acceptance condition $\mathscr{F} = \{F_0, \ldots, F_{2n+1}\}$ and a strategy S, let $\sigma_i^S(q) \in \mathbb{N}_0$ denote the length of the unique nominal trace connecting q to some state in the set F_{2i}.

PROPOSITION 6.1. *Let A be an automaton, F a Büchi acceptance condition and $\mathscr{F}$ a parity acceptance condition.*

(i) *Let R_F be a control Lyapunov function for the Büchi automaton (A, F) and let S be a strategy induced from R_F. Let K be the Lipschitz constant of R_F with respect to d. Then S is winning against every N-bounded disturbance strategy with $N \geq \max_{q \in F'} |\sigma^S(q)|$ where*

$$F' = \{q \in Q \mid f(d(q, F)) \leq K\gamma\}.$$

(ii) *Let $R_{\mathscr{F}}$ be a control Lyapunov function for the parity automaton $(A, \mathscr{F})$ and let S be a strategy induced from R_F. Let K be the Lipschitz constant of $R_{\mathscr{F}}$ with respect to d. Then S is winning against every N-bounded disturbance strategy, where*

$$\infty > N \geq \max_{i=0,\ldots,n} \left(\max \left(\{|\sigma_i^S(q)| : q \in F_{2i}'\} \cap \mathbb{R} \right) \right)$$

for $F_{2i}' = \{q \in Q \mid f(d(q, F_{2i})) \leq K\gamma\}$.

In (ii) it is important that the value of N is finite. Indeed, that N is not finite is a possibility since there may exist sets of even parity which are not reachable from a given state $q \in Q$.

PROOF. For (i), we show that for any $q \in Q$ there exists a finite trace in A connecting q to F resulting from applying S.

First let $q \in Q$ be such that $f(d(q, F)) > K\gamma$. Then by Proposition 4.1 applying S beginning at q results in a finite trace connecting q to some state $q' \in Q$ such that $f(d(q', F)) \leq K\gamma$, regardless of how frequently the fault occurs.

Now assume $q \in Q$ is such that $f(d(q, F)) \leq K\gamma$. By assumption if the trace resulting from S connecting q_0 to q is such that $q = p^{ax}$ for some $x \neq \epsilon$, that is, the state q was reached due to the effects of a fault, the next N transitions on the trace will be nominal, that is, $x = \epsilon$. By definition of N and S the resulting subtrace of length N will visit a state in the set F.

Assume instead that $q = p^{a\epsilon} \notin F$. If the next state $q^{a'x}$ on the trace resulting from S is such that $x = \epsilon$ then $q^{a'x}$ will satisfy $f(d(q^{a'x})) \leq K\gamma$ and the same argument may be applied. Therefore if no fault occurs for the next N transitions some state in the set F will be reached. If a fault occurs, a state in the set F will be reached in the N transitions following the fault.

If $x \neq \epsilon$ then either $f(d(q^{ax}, F)) \leq K\gamma$ in which case the above argument applies, or $f(d(q^{ax}, F)) > K\gamma$ and the first argument applies. So we conclude that the strategy S is winning in the automaton A against an N-bounded disturbance strategy.

For (ii) the argument is similar. If q has even parity then the result follows. Assume instead that $q \in F_j$ with j odd. If for $i \in \{0, \ldots, n\}$, $\infty > f(d(q, \mathscr{F}_{2i})) > K\gamma$ with $2i < j$ then Proposition 4.6 applies: there exists a finite trace resulting from S connecting q to some state q' satisfying $f(d(q', F_{2k})) < K\gamma$ for some $k \in \{0, \ldots, n\}$ regardless of how frequently the fault occurs.

Now assume q and $i \in \{0, \ldots, n\}$ are such that $2i < j$ and $f(d(q, F_{2i})) \leq K\gamma$. By assumption if the trace resulting from S connecting q_0 to q is such that $q = p^{ax}$ for some $p \in Q, a \in \Sigma$ and $x \in X \setminus \{\epsilon\}$ the next N transitions will be nominal and the resulting subtrace will feature a state in the set F_{2i}. If instead $q = p^{a\epsilon}$ then either

(i) the next state on the trace is contained in a set F_{2i} for some $2i < j$ and we are done;

(ii) $q^{ax} \notin F_{2i}$ for some i and $x \neq \epsilon$ and a state in a set of lower even parity will be reached in the next N steps or

(iii) the next state q^{ax} is such that $x = \epsilon$. Then the argument is repeated: if a fault does not occur for the next N transitions then a state in a set of lower even parity will be visited. If a fault occurs, a state of even parity will be visited in the next N transitions following the fault.

Therefore a strategy S induced from a Lipschitz continuous control Lyapunov function is winning for the parity automaton $(A, \mathscr{F})$ against an N-bounded disturbance strategy. $\square$

Compare the above result to the equivalent bound one might establish for a strategy induced from a classical shortest path rank function in a Büchi automaton: in this case the value of N must be greater than the length of the longest simple path connecting a state in Q to a state in F. In our result N is defined with respect to a potentially much smaller subset of Q. Since the bound N is a monotonically increasing function of the environmental error γ this result provides a bridge between the state based view of faults and the running time of the system: a less powerful fault may occur more frequently than a more powerful one without disrupting a well designed strategy.

7. DISCUSSION

We have presented a theory of robustness for ω-regular properties of discrete transition systems. There are two natural extensions to our work. First, in our model, bounded disturbances are the only source of adversarial interaction. The presence of additional adversaries leads to (more complex) algorithms for solving two-player games [3, 29]. We believe our simpler model is already applicable in many settings —we are inspired by similar models in continuous control— and our polynomial-time algorithms render our results applicable in practice.

Second, how can we combine our results on discrete transition systems with the existing theory of robust control for continuous systems? Towards this objective, we will leverage the recent results reported in [22, 27] to guarantee the existence of discrete transition system abstractions of continuous control systems. These abstractions are related to the differential equation models by relationships that guarantee the following properties:

- any strategy synthesized for the abstraction can be refined to a strategy for the differential equation model, enforcing the same specification up to an error of ε. Here, ε is a design parameter that can be made arbitrarily small;

- the metric properties of differential equation models (the set of states is typically $\mathbb{R}^n$ and thus equipped with the Euclidean distance) can be naturally lifted to the discrete abstractions. This means that we can explicitly reason and enforce robustness for the physical component by working with its abstraction. Moreover, by combining these discrete abstractions with discrete transition system models of cyber components we can expect to obtain a comprehensive robustness theory for CPSs.

8. REFERENCES

[1] A. Arora and M. G. Gouda. Closure and convergence: a foundation of fault tolerant computing. *IEEE Transactions on Software Engineering*, 19(11):1015–1027, 1993.

[2] R. Bellman. *Dynamic Programming*. Princeton University Press, 1957.

[3] R. Bloem, K. Chatterjee, K. Greimel, T.A. Henzinger, and B. Jobstmann. Robustness in the presence of liveness. In *CAV 2010*, volume 6174 of *Lecture Notes in Computer Science*, pages 410–424. Springer, 2010.

[4] R. Bloem, K. Chatterjee, T. A. Henzinger, and B. Jobstmann. Better quality in synthesis through quantitative objectives. In *CAV 2009: Computer-Aided Verification*, volume 5643 of *Lecture Notes in Computer Science*, pages 140–156. Springer-Verlag, 2009.

[5] R. Bloem, K. Greimel, T.A. Henzinger, and B. Jobstmann. Synthesizing robust systems. In *FMCAD 09: Formal Methods in Computer-Aided Design*, pages 85–92. IEEE, 2009.

[6] S. Borkar. Electronics Beyond Nano-scale CMOS. In *DAC 06: Design Automation Conference*. ACM, 2006.

[7] M.S. Branicky. Topology of hybrid systems. In *CDC 93: Conference on Decision and Control*, pages 2309–2314. IEEE, 1993.

[8] P. Černý, T. A. Henzinger, and A. Radhakrishna. Simulation distances. In *CONCUR 2010 - Concurrency Theory*, volume 6269 of *Lecture Notes in Computer Science*, pages 253–268. Springer-Verlag, 2010.

[9] E. W. Dijkstra. Self-stabilizing systems in spite of distributed control. *Communications of the ACM*, 17(11):643–644, 1974.

[10] E.A. Emerson and C. Jutla. Tree automata, mu-calculus and determinacy. In *FOCS 91: Foundations of Computer Science*, pages 368–377. IEEE, 1991.

[11] A. Girault and E. Rutten. Automating the addition of fault tolerance with discrete controller synthesis. *Formal Methods in System Design*, 35(2):190–225, 2009.

[12] S. Golshan and E. Bozorgzadeh. Single-Event-Upset (SEU) Awareness in FPGA Routing. In *DAC 07: Design Automation Conference*. ACM, 2007.

[13] Y. Hu, Z .Feng, L. He, and R. Majumdar. Robust FPGA resynthesis based on fault-tolerant boolean matching. In *ICCAD 08: International Conference on Computer-Aided Design*. ACM, 2008.

[14] N. Klarlund. *Progress measures and finite arguments for infinite computations*. PhD thesis, Cornell University, 1990.

[15] S. Krishnaswamy, S. Plaza, I. Markov, and J. Hayes. Signature-based SER analysis and design of logic circuits. *Trans. CAD*. ACM, 2009.

[16] A. Lesea, S. Drimer, J.J. Fabula, C. Carmichael, and P. Alfke. The Rosetta experiment: atmospheric soft error rate testing in differing technology FPGAs. *IEEE Transactions on Device and Materials Reliability*, 5(3):317–328, 2005.

[17] R. McNaughton. Infinite gam,es played on finite graphs. *Annals of Pure and Applied Logic*, 65(2):149–184, 1993.

[18] N. Miskov-Zivanov and D. Marculescu. Formal modeling and reasoning for reliability analysis. In *DAC 10: Design Automation Conference*, pages 531–536. ACM, 2010.

[19] K. Namjoshi. Certifying model checkers. In *CAV 01: Computer Aided Verification*, volume 2102 of *Lecture Notes in Computer Science*, pages 2–13. Springer-Verlag, 2001.

[20] A. Nerode and W. Kohn. Models for hybrid systems: Automata, topologies, controllability, observability. In *Hybrid Systems*, volume 736 of *Lecture Notes in Computer Science*, pages 297–316. Springer-Verlag, 1993.

[21] E. Normand. Single event upset at ground level. *IEEE Transactions on Nuclear Science*, 43(6):2742–2750, 1996.

[22] G. Pola, A. Girard, and P. Tabuada. Approximately bisimilar symbolic models for nonlinear control systems. *Automatica*, 44(10):2508–2516, 2008.

[23] D.C. Tarraf, A. Megretski, and M.A. Dahleh. A framework for robust stability of systems over finite alphabets. *IEEE Transactions on Automatic Control*, 53(5):1133–1146, 2008.

[24] W. Thomas. On the synthesis of strategies in infinite games. In *STACS 95: Theoretical Aspects of Computer Science*, volume 900 of *Lecture Notes in Computer Science*, pages 1–13. Springer-Verlag, 1995.

[25] A.J. van der Schaft. *L2-Gain and Passivity Techniques in Nonlinear Control*, volume 218 of *Lecture Notes in Control and Information Sciences*. Springer-Verlag, 2000.

[26] J.F. Wakerly. *Digital Design Principles and Practices*. Prentice Hall, 1994.

[27] M. Zamani, G. Pola, and Paulo Tabuada. Symbolic models for unstable nonlinear control systems. In *Proceedings of the 2010 American Control Conference*, 2010.

[28] K. Zhou, J. Doyle, and K. Glover. *Robust and Optimal Control*. Prentice Hall, 1996.

[29] W. Zielonka. Infinite games on finitely coloured graphs with applications to automata on infinite trees. *Theor. Comput. Sci.*, 200(1-2):135–183, 1998.

Synthesis of Memory-Efficient Real-Time Controllers for Safety Objectives*

Krishnendu Chatterjee
Institute of Science and Technology (IST) Austria
Krishnendu.Chatterjee@ist.ac.at

Vinayak S. Prabhu
University of Porto
vinayak@eecs.berkeley.edu

ABSTRACT

We study synthesis of controllers for real-time systems, where the objective is to stay in a given safe set. The problem is solved by obtaining winning strategies in the setting of concurrent two-player *timed automaton games* with safety objectives. To prevent a player from winning by blocking time, we restrict each player to strategies that ensure that the player cannot be responsible for causing a zeno run. We construct winning strategies for the controller which require access only to (1) the system clocks (thus, controllers which require their own internal infinitely precise clocks are not necessary), and (2) a linear (in the number of clocks) number of memory bits. Precisely, we show that for safety objectives, a memory of size $\big(3 \cdot |C| + \lg(|C| + 1)\big)$ bits suffices for winning controller strategies, where C is the set of clocks of the timed automaton game, significantly improving the previous known exponential bound. We also settle the open question of whether winning *region* controller strategies require memory for safety objectives by showing with an example the necessity of memory for region strategies to win for safety objectives.

Categories and Subject Descriptors

F.4 [**Mathematical Logic and Formal Languages**]: Mathematical Logic—*Computability theory*

General Terms

Algorithms, Theory, Verification

Keywords

Timed Automata, Timed Games, Safety Objectives, Control

1. INTRODUCTION

Synthesizing controllers to ensure that a plant stays in a safe set is an important problem in the area of systems

---

*This work has been financially supported in part by the European Community's Seventh Framework Program via project Control for coordination of distributed systems (C4C; Grant Agreement number INFSO-ICT-223844); and by Austrian FWF NFN ARiSE funding.

control. We study the synthesis of *timed* controllers in the present paper. Our formalism is based on timed automata [1], which are models of real-time systems in which states consist of discrete locations and values for real-time clocks. The transitions between locations are dependent on the clock values. The real-time controller synthesis problem is modeled using *timed automaton games*, which are played by two players on timed automata, where player 1 is the "controller" and player 2 the "plant". Obtaining winning strategies for player 1 in such games corresponds to the construction of controllers for real-time systems with desired objectives.

The issue of *time divergence* is crucial in timed games, as a naive control strategy might simply block time, leading to "zeno" runs. The following approaches have been proposed to avoid such invalid zeno solutions: (1) discretize time so that players can only take transitions at integer multiples of some fixed time period, e.g. in [12]; (2) put syntactic restrictions on the timed game structure so that zeno runs are not possible (the syntactic restriction is usually presented as the *strong non-zenoness* assumption where the obtained controller synthesis algorithms are guaranteed to work correctly only on timed automaton games where every cycle is such that in it some clock is reset to 0 and is also greater than an integer value at some point, e.g. in [3, 4, 14]); (3) require player 1 to ensure time divergence (e.g. by only taking transitions if player 2 can never take transitions in the future from the current location, as in [10, 5]); (4) give the controller access to an extra (infinitely precise) clock which measure global time and require that player 1 wins if either its moves are chosen only finitely often, or if the ticks of this extra clock are seen infinitely often while satisfying the desired objective, e.g, in [9, 2].

The above approaches are not optimal in certain cases and below we point out some drawbacks. Discretizing the system blows up the state space; and might not be faithful to the real-time semantics. Putting syntactic restrictions is troublesome as it can lead to disallowing certain system models. For example, consider the timed automaton game $\mathcal{T}$ in Figure 1. The details of the game are not important and are omitted here for the sake of brevity. In the figure, the edges are labelled as a_1^j for actions controlled by player 1; and by a_2^j for actions controlled by player 2. The safety objective is to avoid the location "Bad" (player 1 can satisfy this objective without blocking time). One can easily show that zeno runs are possible in this timed automaton game, mainly, due to the edges a_2^0 and a_2^1. The game $\mathcal{T}$ can be made to be non-zeno syntactically by changing the guards of the

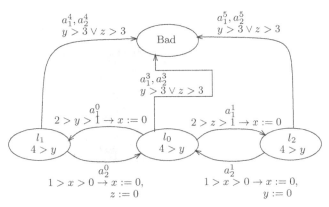

Figure 1: A timed automaton game.

edges a_2^0 and a_2^1 to $1 > x > d$, where d is some conservative constant (say 0.001 time units , where it is assumed that the plant takes at least 0.001 time units to transition out of l_1 and l_2). This change unfortunately blows up the finite state *region abstraction* of the timed automaton game (the region abstraction is used in every current solution to the real-time controller synthesis problem for timed automaton games). If the constant d is 0.001, then the number of states in the region abstraction blows up from roughly $2.5 * 10^5$ for the original game to $2.5 * 10^5 * 10^9$; a blow up by a *factor* of 10^9. Admittedly however, on-the-fly algorithms for controller synthesis may help mitigate the situation in some cases ([6]) by not explicitly constructing the full graph of the region abstraction.

Requiring player 1 to guarantee time divergence by only taking transitions if player 2 cannot take transitions from the current location is too conservative. If we consider the game in Figure 1, this approach would prevent player 1 from taking *any* of the actions, making the system uncontrollable. Finally, adding an extra infinitely precise clock to measure time, and making it observable to the controller amounts to giving unfair and unrealistic power to the controller in many situations.

In the present paper, we avoid the shortcomings of the previous approaches by using two techniques. First, we use *receptive* [2, 15], player-1 strategies, which, while being required to not prevent time from diverging, are not required to ensure time divergence. Receptiveness is incorporated by using the more general, semantic and fully symmetric formalism of [9] for dealing with the issue of time divergence. This setting places no syntactic restriction on the game structure, and gives both players equally powerful options for advancing time, but for a player to win, it must not be *responsible* for causing time to converge. Formally, our timed games proceed in an infinite sequence of rounds. In each round, both players simultaneously propose moves, with each move consisting of an action and a time delay after which the player wants the proposed action to take place. Of the two proposed moves, the move with the shorter time delay "wins" the round and determines the next state of the game. Let a set Φ of runs be the desired objective for player 1. Then player 1 has a *winning* strategy for Φ if it has a strategy to ensure that, no matter what player 2 does, one of the following two conditions hold: (1) time diverges and the resulting run belongs to Φ, or (2) time does not diverge but player-1's moves are chosen only finitely often (and thus it is not to be blamed for the convergence of time). Second, in the current work, the controller only uses the sys-

tem clocks of the model (unlike [9] which makes available to the controller an extra infinitely precise clock to measure time), ensuring that the controller bases its actions only on the variables corresponding to the physical processes of the system (namely, the system clocks and locations). Time divergence is *inferred* from the history of certain predicates of the system clocks, rather than from an extra infinitely precise clock that the controller has to keep in memory.

Contributions. Our current work significantly improves the results of [7]. In [7] we showed that finite-memory receptive strategies suffice for safety objective in timed automaton games; the problem of establishing a memory bound was left open. In this paper, we first show that a basic analysis, using *Zielonka trees*, of the characterization of receptive strategies of [7] leads to the requirement of an *exponential* number of bits for the memory bound (in the number of clocks) for the winning strategies. We then present an improved new characterization of receptive strategies for safety objectives which allows us to obtain a memory bound for winning strategies that is *linear* in the number of bits. Precisely, we show that a memory of size $(3 \cdot |C| + \lg(|C| + 1))$ bits suffices for winning receptive strategies for safety objectives, where C is the set of clocks of the timed automaton game, considerably improving the exponential bound obtained from the results of [7]. Finally, we settle the open question of whether winning *region* strategies for controllers require memory for safety objectives: we show with an example the necessity of memory for region strategies to win. The proofs are omitted due to lack of space, and are available in [8].

2. TIMED GAMES

2.1 Timed Game Structures

In this subsection we present the definitions of timed game structures, runs, objectives, strategies and the notions of winning in timed game structures.

Timed game structures. A *timed game structure* is a tuple $\mathcal{G} = \langle S, A_1, A_2, \Gamma_1, \Gamma_2, \delta \rangle$ with the following components.
- S is a set of states.
- A_1 and A_2 are two disjoint sets of actions for players 1 and 2, respectively. We assume that $\perp_i \notin A_i$, and write $A_i^\perp$ for $A_i \cup \{\perp_i\}$. The set of *moves* for player i is $M_i = \mathbb{R}_{\geq 0} \times A_i^\perp$. Intuitively, a move $\langle \Delta, a_i \rangle$ by player i indicates a waiting period of Δ time units followed by a discrete transition labeled with action a_i. The move $\langle \Delta, \perp_i \rangle$ is used to represent the move of player i where player-i just lets time elapse for Δ time units without taking any of the discrete actions from A_i.
- $\Gamma_i : S \mapsto 2^{M_i} \setminus \emptyset$ are two move assignments. At every state s, the set $\Gamma_i(s)$ contains the moves that are available to player i. We require that $\langle 0, \perp \rangle \in \Gamma_i(s)$ for all states $s \in S$ and $i \in \{1, 2\}$. Intuitively, $\langle 0, \perp_i \rangle$ is a time-blocking stutter move.
- $\delta : S \times (M_1 \cup M_2) \mapsto S$ is the transition function. We require that for all time delays $\Delta, \Delta' \in \mathbb{R}_{\geq 0}$ with $\Delta' \leq \Delta$, and all actions $a_i \in A_i^\perp$, we have (1) $\langle \Delta, a_i \rangle \in \Gamma_i(s)$ iff both $\langle \Delta', \perp_i \rangle \in \Gamma_i(s)$ and $\langle \Delta - \Delta', a_i \rangle \in \Gamma_i(\delta(s, \langle \Delta', \perp_i \rangle))$; and (2) if $\delta(s, \langle \Delta', \perp_i \rangle) = s'$ and $\delta(s', \langle \Delta - \Delta', a_i \rangle) = s''$, then $\delta(s, \langle \Delta, a_i \rangle) = s''$.

The game proceeds as follows. If the current state of the game is s, then both players simultaneously propose moves $\langle \Delta_1, a_1 \rangle \in \Gamma_1(s)$ and $\langle \Delta_2, a_2 \rangle \in \Gamma_2(s)$. The move with the shorter duration "wins" in determining the next state of the game. If both moves have the same duration, then player 2 determines whether the next state will be determined by its move, or by the move of player 1. We use this setting as our goal is to compute the winning set for player 1 against all possible strategies of player 2. Formally, we define the *joint destination function* $\delta_{\mathsf{jd}} : S \times M_1 \times M_2 \mapsto 2^S$ by $\delta_{\mathsf{jd}}(s, \langle \Delta_1, a_1 \rangle, \langle \Delta_2, a_2 \rangle) =$

$$
\begin{cases}
\{\delta(s, \langle \Delta_1, a_1 \rangle)\} & \text{if } \Delta_1 < \Delta_2; \\
\{\delta(s, \langle \Delta_2, a_2 \rangle)\} & \text{if } \Delta_2 < \Delta_1; \\
\{\delta(s, \langle \Delta_2, a_2 \rangle), \delta(s, \langle \Delta_1, a_1 \rangle)\} & \text{if } \Delta_2 = \Delta_1.
\end{cases}
$$

The time elapsed when the moves $m_1 = \langle \Delta_1, a_1 \rangle$ and $m_2 = \langle \Delta_2, a_2 \rangle$ are proposed is given by $\mathsf{delay}(m_1, m_2) = \min(\Delta_1, \Delta_2)$. The boolean predicate $\mathsf{blame}_i(s, m_1, m_2, s')$ indicates whether player i is "responsible" for the state change from s to s' when the moves m_1 and m_2 are proposed. Denoting the opponent of player i by $\sim i = 3 - i$, for $i \in \{1, 2\}$, we define $\mathsf{blame}_i(s, \langle \Delta_1, a_1 \rangle, \langle \Delta_2, a_2 \rangle, s') = \left(\Delta_i \leq \Delta_{\sim i} \ \wedge \ \delta(s, \langle \Delta_i, a_i \rangle) = s' \right)$.

Runs. A *run* of the timed game structure $\mathcal{G}$ is an infinite sequence $r = s_0, \langle m_1^0, m_2^0 \rangle, s_1, \langle m_1^1, m_2^1 \rangle, \ldots$ such that $s_k \in S$ and $m_i^k \in \Gamma_i(s_k)$ and $s_{k+1} \in \delta_{\mathsf{jd}}(s_k, m_1^k, m_2^k)$ for all $k \geq 0$ and $i \in \{1, 2\}$. For $k \geq 0$, let $\mathsf{time}(r, k)$ denote the "time" at position k of the run, namely, $\mathsf{time}(r, k) = \sum_{j=0}^{k-1} \mathsf{delay}(m_1^j, m_2^j)$ (we let $\mathsf{time}(r, 0) = 0$). By $r[k]$ we denote the $(k+1)$-th state s_k of r. The run prefix $r[0..k]$ is the finite prefix of the run r that ends in the state s_k. Let Runs be the set of all runs of $\mathcal{G}$, and let FinRuns be the set of run prefixes.

Objectives. An *objective* for the timed game structure $\mathcal{G}$ is a set $\Phi \subseteq \mathsf{Runs}$ of runs. We will be interested in the classical safety objectives. Given a set of states Y, the *safety* objective consists of the set of runs that stay within Y, formally, $\mathsf{Safe}(Y) = \{ r \mid \text{for all } i \text{ we have } r[i] \in Y \}$. To solve timed games for safety objectives, we shall need to solve for certain ω-regular objectives that will be explained later (see [16] for the definition of ω-regular sets).

Strategies. A *strategy* for a player is a recipe that specifies how to extend a run. Formally, a *strategy* π_i for player $i \in \{1, 2\}$ is a function π_i that assigns to every finite run prefix $r[0..k]$ a move m_i in the set of moves available to player i at the state $r[k]$. For $i \in \{1, 2\}$, let Π_i be the set of strategies for player i.

Outcomes. Given two strategies $\pi_1 \in \Pi_1$ and $\pi_2 \in \Pi_2$, the set of possible *outcomes* of the game starting from a state $s \in S$ is denoted $\mathsf{Outcomes}(s, \pi_1, \pi_2)$. We let $\mathsf{Outcomes}_k(s, \pi_1, \pi_2)$ denote the set of finite runs $r[0..k-1]$ which are possible according to the two strategies given the initial state s.

Receptive strategies. We will be interested in strategies that are meaningful (in the sense that they do not block time). To define them formally we first present the following two sets of runs.

- A run r is *time-divergent* if $\lim_{k \to \infty} \mathsf{time}(r, k) = \infty$. We denote by Timediv the set of all time-divergent runs.
- The set $\mathsf{Blameless}_i \subseteq \mathsf{Runs}$ consists of the set of runs in which player i is responsible only for finitely many transitions. A run $s_0, \langle m_1^0, m_2^0 \rangle, s_1, \langle m_1^1, m_2^1 \rangle, \ldots$ belongs to the set $\mathsf{Blameless}_i$, for $i = \{1, 2\}$, if there exists a $k \geq 0$ such that for all $j \geq k$, we have $\neg \, \mathsf{blame}_i(s_j, m_1^j, m_2^j, s_{j+1})$.

A strategy π_i is *receptive* if for all strategies $\pi_{\sim i}$, all states $s \in S$, and all runs $r \in \mathsf{Outcomes}(s, \pi_1, \pi_2)$, either $r \in$ Timediv or $r \in \mathsf{Blameless}_i$. Thus, no what matter what the opponent does, a receptive strategy of player i cannot be responsible for blocking time. Strategies that are not receptive are not physically meaningful. A timed game structure $\mathcal{G}$ is *well-formed* if both players have receptive strategies. We restrict our attention to well-formed timed game structures. We denote Π_i^R to be the set of receptive strategies for player i. Note that for $\pi_1 \in \Pi_1^R, \pi_2 \in \Pi_2^R$, we have $\mathsf{Outcomes}(s, \pi_1, \pi_2) \subseteq$ Timediv.

Winning sets. Given an objective Φ, let $\mathsf{WinTimeDiv}_1^{\mathcal{G}}(\Phi)$ denote the set of states s in $\mathcal{G}$ such that player 1 has a receptive strategy $\pi_1 \in \Pi_1^R$ such that for all receptive strategies $\pi_2 \in \Pi_2^R$, we have $\mathsf{Outcomes}(s, \pi_1, \pi_2) \subseteq \Phi$. The strategy π_1 is said to be a winning strategy. A player-2 *spoiling strategy* π_2 against a player-1 strategy π_1 is such that we have $\mathsf{Outcomes}(s, \pi_1, \pi_2) \not\subseteq \Phi$. In computing the winning sets, we shall quantify over *all* strategies, but modify the objective to take care of time divergence. Given an objective Φ, let $\mathsf{TimeDivBl}_1(\Phi) = (\mathsf{Timediv} \cap \Phi) \cup (\mathsf{Blameless}_1 \setminus \mathsf{Timediv})$, i.e., $\mathsf{TimeDivBl}_1(\Phi)$ denotes the set of runs such that either time diverges and Φ holds, or else time converges and player 1 is not responsible for time to converge. Let $\mathsf{Win}_1^{\mathcal{G}}(\Phi)$ be the set of states in $\mathcal{G}$ such that for all $s \in \mathsf{Win}_1^{\mathcal{G}}(\Phi)$, player 1 has a (possibly non-receptive) strategy $\pi_1 \in \Pi_1$ such that for all (possibly non-receptive) strategies $\pi_2 \in \Pi_2$, we have $\mathsf{Outcomes}(s, \pi_1, \pi_2) \subseteq \Phi$. The strategy π_1 is said to be winning for the non-receptive game. The following result establishes the connection between Win and WinTimeDiv sets.

Theorem 1 ([13]). *For all well-formed timed game structures $\mathcal{G}$, and for all ω-regular objectives Φ, we have $\mathsf{Win}_1^{\mathcal{G}}(\mathsf{TimeDivBl}_1(\Phi)) = \mathsf{WinTimeDiv}_1^{\mathcal{G}}(\Phi)$.*

We observe here that $\mathsf{TimeDivBl}_1(\Phi)$ is *not* equivalent to $(\neg \, \mathsf{Blameless}_1) \to \mathsf{Timediv} \cap \Phi$. Player 1 loses even if it does not get moves infinitely often, provided time diverges and the run does not belong to Φ.

2.2 Timed Automaton Games

In this subsection we first define a special class of timed game structures, namely, timed automaton games, and the notion of region equivalence. We then present the computation of winning sets for timed automaton games based on the framework of [9], and derive various basic properties of winning strategies.

Timed automaton games. Timed automata [1] suggest a finite syntax for specifying infinite-state timed game structures. A *timed automaton game* is a tuple $\mathcal{T} = \langle L, C, A_1, A_2, E, \gamma \rangle$ with the following components:

- L is a finite set of locations.
- C is a finite set of clocks.
- A_1 and A_2 are two disjoint sets of actions for players 1 and 2, respectively.
- $E \subseteq L \times (A_1 \cup A_2) \times \mathsf{Constr}(C) \times L \times 2^C$ is the edge relation, where the set $\mathsf{Constr}(C)$ of *clock constraints* is generated by the grammar $\theta ::= x \leq d \mid d \leq x \mid \neg\theta \mid \theta_1 \wedge \theta_2$ for clock variables $x \in C$ and nonnegative integer constants d. For an edge $e = \langle l, a_i, \theta, l', \lambda \rangle$, the clock constraint θ acts as a guard on the clock values which specifies when the edge e can be taken, and by taking the edge e, the clocks in the set $\lambda \subseteq C$ are reset to 0. The setting of clocks to 0 is called the *reset map*. We require that for all edges $\langle l, a_i, \theta', l', \lambda' \rangle, \langle l, a_i, \theta'', l'', \lambda'' \rangle \in E$ with $l' \neq l''$, the conjunction $\theta' \wedge \theta''$ is unsatisfiable. This

requirement ensures that a state and a move together uniquely determine a successor state.

- $\gamma : L \mapsto \mathsf{Constr}(C)$ is a function that assigns to every location an invariant for both players. All clocks increase uniformly at the same rate. When at location l, each player i must propose a move out of l before the invariant $\gamma(l)$ expires. Thus, the game can stay at a location only as long as the invariant is satisfied by the clock values.

A *clock valuation* is a function $\kappa : C \mapsto \mathbb{R}_{\geq 0}$ that maps every clock to a nonnegative real. The set of all clock valuations for C is denoted by $K(C)$. Given a clock valuation $\kappa \in K(C)$ and a time delay $\Delta \in \mathbb{R}_{\geq 0}$, we write $\kappa + \Delta$ for the clock valuation in $K(C)$ defined by $(\kappa + \Delta)(x) = \kappa(x) + \Delta$ for all clocks $x \in C$. For a subset $\lambda \subseteq C$ of the clocks, we write $\kappa[\lambda := 0]$ for the clock valuation in $K(C)$ defined by $(\kappa[\lambda := 0])(x) = 0$ if $x \in \lambda$, and $(\kappa[\lambda := 0])(x) = \kappa(x)$ if $x \notin \lambda$. A clock valuation $\kappa \in K(C)$ *satisfies* the clock constraint $\theta \in \mathsf{Constr}(C)$, written $\kappa \models \theta$, if the condition θ holds when all clocks in C take on the values specified by κ. A *state* $s = \langle l, \kappa \rangle$ of the timed automaton game $\mathcal{T}$ is a location $l \in L$ together with a clock valuation $\kappa \in K(C)$ such that the invariant at the location is satisfied, that is, $\kappa \models \gamma(l)$. Let S be the set of all states of $\mathcal{T}$. In a state, each player i proposes a time delay allowed by the invariant map γ, together either with the action $\bot$, or with an action $a_i \in A_i$ such that an edge labeled a_i is enabled after the proposed time delay. We require that for $i \in \{1, 2\}$ and for all states $s = \langle l, \kappa \rangle$, if $\kappa \models \gamma(l)$, either $\kappa + \Delta \models \gamma(l)$ for all $\Delta \in \mathbb{R}_{\geq 0}$, or there exist a time delay $\Delta \in \mathbb{R}_{\geq 0}$ and an edge $\langle l, a_i, \theta, l', \lambda \rangle \in E$ such that (1) $a_i \in A_i$ and (2) $\kappa + \Delta \models \theta$ and for all $0 \leq \Delta' \leq \Delta$, we have $\kappa + \Delta' \models \gamma(l)$, and (3) $(\kappa + \Delta)[\lambda := 0] \models \gamma(l')$. This requirement is necessary (but not sufficient) for well-formedness of the game. Given a timed automaton game $\mathcal{T}$, the definition of a associated timed game structure $[\![\mathcal{T}]\!]$ is standard [9].

Clock region equivalence. Timed automaton games can be solved using a region construction from the theory of timed automata [1]. *Clock region equivalence*, denoted as $\cong$ is an equivalence relation on states of timed automaton games. The equivalence classes of the relaton are called *regions*, and induce a time abstract bisimulation on the timed game structure. There are finitely many clock regions; more precisely, the number of clock regions is bounded by $|L| \cdot \prod_{x \in C}(c_x + 1) \cdot |C|! \cdot 4^{|C|}$. The formal definition of the region relation is standard and omitted here for lack of space, see *e.g.* [8] for the formal definition.

Region strategies and objectives. For a state $s \in S$, we write $\mathsf{Reg}(s) \subseteq S$ for the clock region containing s. For a run r, we let the *region sequence* $\mathsf{Reg}(r) = \mathsf{Reg}(r[0]), \mathsf{Reg}(r[1]), \cdots$. Two runs r, r' are region equivalent if their region sequences are the same. An ω-regular objective Φ is a region objective if for all region-equivalent runs r, r', we have $r \in \Phi$ iff $r' \in \Phi$. A strategy π_1 is a *region strategy*, if for all runs r_1 and r_2 and all $k \geq 0$ such that $\mathsf{Reg}(r_1[0..k]) = \mathsf{Reg}(r_2[0..k])$, we have that if $\pi_1(r_1[0..k]) = \langle \Delta, a_1 \rangle$, then $\pi_1(r_2[0..k]) = \langle \Delta', a_1 \rangle$ with $\mathsf{Reg}(r_1[k] + \Delta) = \mathsf{Reg}(r_2[k] + \Delta')$. The definition for player 2 strategies is analogous. Two region strategies π_1 and π_1' are region-equivalent if for all runs r and all $k \geq 0$ we have that if $\pi_1(r[0..k]) = \langle \Delta, a_1 \rangle$, then $\pi_1'(r[0..k]) = \langle \Delta', a_1 \rangle$ with $\mathsf{Reg}(r[k] + \Delta) = \mathsf{Reg}(r[k] + \Delta')$. An ω-regular objective Φ is a location ω-regular objective if, whenever a run $s_0, \langle m_1^0, m_2^0 \rangle, s_1, \langle m_1^1, m_2^1 \rangle \ldots$ is in Φ. then any run

$\ddot{s}_0, \langle \ddot{m}_1^0, \ddot{m}_0 \rangle, \ddot{s}_1, \langle \ddot{m}_1^1, \ddot{m}_2^1 \rangle \ldots$ such that the locations of s_j and $\ddot{s}_j$ match for each j, is in Φ. Region objectives can similarly be defined. Henceforth, we shall restrict our attention to region and location objectives.

Encoding time-divergence by enlarging the game structure. Given a timed automaton game $\mathcal{T}$, consider the enlarged game structure $\widehat{\mathcal{T}}$ (based mostly on the construction in [9]) with the state space $S^{\widehat{\mathcal{T}}} \subseteq S \times \mathbb{R}_{[0,1)} \times \{\text{TRUE}, \text{FALSE}\}^2$, and an augmented transition relation $\delta^{\widehat{\mathcal{T}}} : S^{\widehat{\mathcal{T}}} \times (M_1 \cup M_2) \mapsto S^{\widehat{\mathcal{T}}}$. In an augmented state $\langle s, \mathfrak{z}, tick, bl_1 \rangle \in S^{\widehat{\mathcal{T}}}$, the component $s \in S$ is a state of the original game structure $[\![\mathcal{T}]\!]$, $\mathfrak{z}$ is value of a fictitious clock z which gets reset to 0 every time it crosses 1 (i.e., if κ' is the clock valuation resulting from letting time Δ elapse from an initial clock valuation κ, then, $\kappa'(z) = (\kappa(z) + \Delta) \mod 1$), *tick* is true iff z crossed 1 at last transition and bl_1 is true if player 1 is to blame for the last transition (ie., blame_1 is true for the last transition). Note that any strategy π_i in $[\![\mathcal{T}]\!]$, can be considered a strategy in $\widehat{\mathcal{T}}$: the values of the clock z, *tick* and bl_1 correspond to the values each player keeps in memory in constructing its strategy. Given any initial value of $\mathfrak{z} = \mathfrak{z}^*, tick = tick^*, bl_1 = bl_1^*$; any run r in $\mathcal{T}$ has a corresponding unique run $\widehat{r}$ in $\widehat{\mathcal{T}}$ with $\widehat{r}[0] = \langle r[0], \mathfrak{z}^*, tick^*, bl_1^* \rangle$ such that r is a projection of $\widehat{r}$ onto $\mathcal{T}$. For an objective Φ, we can now encode time-divergence as the LTL objective: $\mathsf{TimeDivBl}_1(\Phi) = (\Box \Diamond \ tick \to \Phi) \wedge (\neg \Box \Diamond \ tick \to \Diamond \Box \neg \ bl_1)$, where $\Box$ and $\Diamond$ are the standard LTL modalities ("always" and "eventually" respectively), the combinations $\Box \Diamond$ and $\Diamond \Box$ denoting "infinitely often" and "all but for a finite number of steps" respectively.

A μ-calculus formulation for describing the sure winning sets. Given an ω-regular objective $\widehat{\Phi}$ of the enlarged game structure $\widehat{\mathcal{T}}$, a μ-calculus formula φ to describe the winning set $\mathsf{Win}_1^{\widehat{\mathcal{T}}}(\widehat{\Phi})$ is given in [9]. The μ-calculus formula uses the *controllable predecessor* operator for player 1, $\mathsf{CPre}_1 : 2^{\widehat{S}} \mapsto 2^{\widehat{S}}$ (where $\widehat{S} = S^{\widehat{\mathcal{T}}}$), defined formally by $\widehat{s} \in \mathsf{CPre}_1(Z)$ iff $\exists m_1 \in \Gamma_1^{\widehat{\mathcal{T}}}(\widehat{s}) \ \forall m_2 \in \Gamma_2^{\widehat{\mathcal{T}}}(\widehat{s}) . \delta_{\mathrm{jd}}^{\widehat{\mathcal{T}}}(\widehat{s}, m_1, m_2) \subseteq Z$. Informally, $\mathsf{CPre}_1(Z)$ consists of the set of states from which player 1 can ensure that the next state will be in Z, no matter what player 2 does. The operator CPre_1 preserves regions of $\widehat{\mathcal{T}}$ (this follows from the results of Lemma 1). It was also shown in [9] that only unions of regions arise in the μ-calculus iteration for ω-regular location objectives.

Regions suffice for determining winning moves. Let Y, Y_1', Y_2' be regions. We show in Lemma 1 that one of the following two conditions hold: (a) for all states in Y there is a move for player 1 with destination in Y_1', such that against all player 2 moves with destination in Y_2', the next state is guaranteed to be in Y_1'; or (b) for all states in Y for all moves for player 1 with destination in Y_1' there is a move of player 2 to ensure that the next state is in Y_2'; or (c) if $Y_1' = Y_2'$ (except for the bl_1 component), then player 2 can pick the same time delay as player 1 and hence there are two resulting states.

Lemma 1 ([8]). *Let $\mathcal{T}$ be a timed automaton game, and let Y, Y_1', Y_2' be regions in the corresponding enlarged timed game structure $\widehat{\mathcal{T}}$. Suppose player-i has a move $\langle \Delta_i, \bot_i \rangle$ from some $\widehat{s} \in Y$ to $\widehat{s}_i \in Y_i'$, for $i \in \{1, 2\}$. Then, for all states $\widehat{s} \in Y$ and for all player-1 moves $m_1^{\widehat{s}} = \langle \Delta_1, a_1 \rangle$ with $\widehat{s} + \Delta_1 \in Y_1'$, one of the following cases must hold.*
1. *$Y_1' \neq Y_2'$ and for all moves $m_2^{\widehat{s}} = \langle \Delta_2, a_2 \rangle$ of*

player-2 with $\widehat{s} + \Delta_2 \in Y_2'$, we have $\Delta_1 < \Delta_2$ (and hence $\mathsf{blame}_1(\widehat{s}, m_1^{\widehat{s}}, m_2^{\widehat{s}}, \widehat{\delta}(\widehat{s}, m_1^{\widehat{s}})) = $ TRUE and $\mathsf{blame}_2(\widehat{s}, m_1^{\widehat{s}}, m_2^{\widehat{s}}, \widehat{\delta}(\widehat{s}, m_2^{\widehat{s}})) = $ FALSE).

2. $Y_1' \neq Y_2'$ and for all player-2 moves $m_2^{\widehat{s}} = \langle \Delta_2, a_2 \rangle$ with $\widehat{s} + \Delta_2 \in Y_2'$, we have $\Delta_2 < \Delta_1$ (and hence $\mathsf{blame}_2(\widehat{s}, m_1^{\widehat{s}}, m_2^{\widehat{s}}, \widehat{\delta}(\widehat{s}, m_2^{\widehat{s}})) = $ TRUE and $\mathsf{blame}_1(\widehat{s}, m_1^{\widehat{s}}, m_2^{\widehat{s}}, \widehat{\delta}(\widehat{s}, m_1^{\widehat{s}})) = $ FALSE).

3. $Y_1' = Y_2'$ and there exists a player 2 move $m_2^{\widehat{s}} = \langle \Delta_2, a_2 \rangle$ with $\widehat{s} + \Delta_2 \in Y_2'$ such that $\Delta_1 = \Delta_2$ (and hence $\mathsf{blame}_1(\widehat{s}, m_1^{\widehat{s}}, m_2^{\widehat{s}}, \widehat{\delta}(\widehat{s}, m_1^{\widehat{s}})) = $ TRUE and $\mathsf{blame}_2(\widehat{s}, m_1^{\widehat{s}}, m_2^{\widehat{s}}, \widehat{\delta}(\widehat{s}, m_2^{\widehat{s}})) = $ TRUE).

Region strategies suffice. We now present a lemma which states that region strategies suffice for winning for region ω-regular objectives, and that all strategies region-equivalent to a region winning strategy are also winning.

Lemma 2 ([7])**.** *Let $\mathcal{T}$ be a timed automaton game and $\widehat{\mathcal{T}}$ be the corresponding enlarged game structure. Let $\widehat{\Phi}$ be an ω-regular region objective of $\widehat{\mathcal{T}}$. Then, (1) there exists a region winning strategy for $\widehat{\Phi}$ from $\mathsf{Win}_1^{\widehat{\mathcal{T}}}(\widehat{\Phi})$, and (2) if π_1' is a strategy that is region-equivalent to a region winning strategy π_1, then π_1' is a winning strategy for $\widehat{\Phi}$ from $\mathsf{Win}_1^{\widehat{\mathcal{T}}}(\widehat{\Phi})$.*

Infinite memory required to maintain global clock. Note that there is an infinitely precise global clock z in the enlarged game structure $\widehat{\mathcal{T}}$. If $\mathcal{T}$ does not have such a global clock, then strategies in $\widehat{\mathcal{T}}$ correspond to strategies in $\mathcal{T}$ where player 1 (and player 2) maintain the value of the infinitely precise global clock in memory. This requires infinite memory in general. An example can be constructed where an unbounded number of unique global values arise in a run (see [8]), thus requiring infinite memory to distinguish between them.

3. RECEPTIVE STRATEGIES FOR SAFETY OBJECTIVES

In this section we show the existence of finite-memory winning strategies for safety objectives in timed automaton games, and study the memory requirements of these strategies. The encoding of time-divergence in the previous section required an infinitely precise clock which had to be kept in the memory of player 1, thus requiring infinite memory. In this section, we derive an alternative characterization of receptive strategies which does not requires this extra clock. The characterization of receptive strategies is used to derive receptive strategies for safety objectives. We then analyze the memory requirements of these strategies using *Zielonka trees* (see [11]).

3.1 Analyzing Spoiling Strategies of Player 2

In this subsection we analyze the spoiling strategies of player 2. This analysis will be used in characterizing the receptive strategies of player 1.

Adding predicates to the game structure. We add some predicates to timed automaton games; the predicates will be used later to analyze receptive safety strategies. Given a timed automaton game $\mathcal{T}$ and a state s of $\mathcal{T}$, we define two functions $V_{>0} : C \mapsto \{$TRUE, FALSE$\}$ and $V_{\geq 1} : C \mapsto \{$TRUE, FALSE$\}$. We obtain $2 \cdot |C|$ predicates based on the two functions. For a clock x, the values of the predicates $V_{>0}(x)$ and $V_{\geq 1}(x)$ indicate if the value of clock x was greater than 0, or greater than or equal to 1 respectively, at the transition point, just before the reset map. For example,

for a state $s^p = \langle l^p, \kappa^p \rangle$ and $\delta(s^p, \langle \Delta, a_1 \rangle) = s$, the predicate $V_{>0}(x)$ is TRUE at state s iff $\kappa'(x) > 0$ for $\kappa' = \kappa^p + \Delta$. Consider the enlarged game structure $\widetilde{\mathcal{T}}$ with the state space $\widetilde{S} = S \times \{$TRUE, FALSE$\} \times \{$TRUE, FALSE$\}^C \times \{$TRUE, FALSE$\}^C$ and an augmented transition relation $\widetilde{\delta}$. A state of $\widetilde{\mathcal{T}}$ is a tuple $\langle s, bl_1, V_{>0}, V_{\geq 1} \rangle$, where s is a state of $\mathcal{T}$, the component bl_1 is TRUE iff player 1 is to be blamed for the last transition, and $V_{>0}, V_{\geq 1}$ are as defined earlier. The clock region equivalence relation can be lifted to states of $\widetilde{\mathcal{T}}$: $\langle s, bl_1, V_{>0}, V_{\geq 1} \rangle \cong_{\widetilde{\mathcal{T}}} \langle s', bl_1', V_{>0}', V_{\geq 1}' \rangle$ iff $s \cong_{\mathcal{T}} s'$, $bl_1 = bl_1'$, $V_{>0} = V_{>0}'$ and $V_{\geq 1} = V_{\geq 1}'$. We next present a finite state concurrent game $\widetilde{\mathcal{T}}^{\mathsf{F}}$ based on the regions of $\widetilde{\mathcal{T}}$ which will be used to analyze spoiling strategies of player 2.

Finite state concurrent game $\widetilde{\mathcal{T}}^{\mathsf{F}}$ based on the regions of $\widetilde{\mathcal{T}}$. We first show that there exists an finite state concurrent game $\widetilde{\mathcal{T}}^{\mathsf{F}}$ which can be used to obtain winning sets and winning strategies of $\widetilde{\mathcal{T}}$. The two ideas behind $\widetilde{\mathcal{T}}^{\mathsf{F}}$ are that (1) only region sequences are important for games with ω-regular location objectives, and (2) only the destination regions of the players are important (due to Lemma 1). Formally, the game $\widetilde{\mathcal{T}}^{\mathsf{F}}$ is defined as the tuple $\langle S^{\mathsf{F}}, M_1^{\mathsf{F}}, M_2^{\mathsf{F}}, \Gamma_1^{\mathsf{F}}, \Gamma_2^{\mathsf{F}}, \delta^{\mathsf{F}} \rangle$ where

- S^{F} is the set of states of $\widetilde{\mathcal{T}}^{\mathcal{F}}$, and is equal to the set of regions of $\widetilde{\mathcal{T}}$.
- M_i^{F} for $i \in \{1, 2\}$ is the set of moves of player-i.

$$M_1^{\mathsf{F}} = \{ \langle \widetilde{R}, a_1 \rangle \mid \widetilde{R} \text{ is a region of } \widetilde{\mathcal{T}}, \text{ and } a_1 \in A_1^\perp \}$$

$$M_2^{\mathsf{F}} = \left\{ \begin{array}{l} \langle \widetilde{R}, a_2, i \rangle \mid \widetilde{R} \text{ is a region of } \widetilde{\mathcal{T}}, \text{ with} \\ i \in \{1, 2\}, \text{ and } a_2 \in A_2^\perp \end{array} \right\}$$

Intuitively, the moves of player-i denote which region it wants to let time pass to, and then take the discrete action $a_i^\perp$. In addition, for player 2, the "i" denotes which player's move will be chosen should the two players propose moves to the same region.

- Γ_i^{F} for $i \in \{1, 2\}$ is the move assignment function. Given a state $\widetilde{R} \in S^{\mathsf{F}}$, we have $\Gamma_i^{\mathsf{F}}(\widetilde{R})$ to be the set of moves available to player i at state $\widetilde{R}$.

$$\Gamma_1^{\mathsf{F}}(\widetilde{R}) = \left\{ \begin{array}{l} \langle \widetilde{R}', a_1 \rangle \mid \exists \widetilde{s} \in \widetilde{R} \text{ such that player 1 has a move} \\ \langle \Delta, a_1 \rangle \text{ in } \widetilde{\mathcal{T}} \text{ from } \widetilde{s} \text{ with } \mathsf{Reg}(\widetilde{s} + \Delta) = \widetilde{R}' \end{array} \right\}$$

$$\Gamma_2^{\mathsf{F}}(\widetilde{R}) = \left\{ \begin{array}{l} \langle \widetilde{R}', a_2, i \rangle \mid \exists \widetilde{s} \in \widetilde{R} \text{ such that player 2 has a} \\ \text{move } \langle \Delta, a_2 \rangle \text{ in } \widetilde{\mathcal{T}} \text{ from } \widetilde{s} \text{ with} \\ \mathsf{Reg}(\widetilde{s} + \Delta) = \widetilde{R}' \text{ and } i \in \{1, 2\} \end{array} \right\}$$

- The transition function δ^{F} is specified as $\delta^{\mathsf{F}}(\widetilde{R}, \langle \widetilde{R}_1, a_1 \rangle, \langle \widetilde{R}_2, a_2, i \rangle) =$

$$\begin{cases} \widetilde{R}' & \text{if } \widetilde{R}_1 \neq \widetilde{R}_2, \widetilde{R}_2 \text{ is a time successor of } \widetilde{R}_1, \text{ and} \\ & \exists \widetilde{s}_1 \in \widetilde{R}_1 \text{ such that } \delta^{\widetilde{\mathcal{T}}}(\widetilde{s}_1, \langle 0, a_1 \rangle) \in \widetilde{R}' \\ \widetilde{R}' & \text{if } \widetilde{R}_1 \neq \widetilde{R}_2, \widetilde{R}_1 \text{ is a time successor of } \widetilde{R}_2, \text{ and} \\ & \exists \widetilde{s}_2 \in \widetilde{R}_2 \text{ such that } \delta^{\widetilde{\mathcal{T}}}(\widetilde{s}_2, \langle 0, a_2 \rangle) \in \widetilde{R}' \\ \widetilde{R}' & \text{if } \widetilde{R}_1 = \widetilde{R}_2, i = 1 \text{ and} \\ & \exists \widetilde{s}_1 \in \widetilde{R}_1 \text{ such that } \delta^{\widetilde{\mathcal{T}}}(\widetilde{s}_1, \langle 0, a_1 \rangle) \in \widetilde{R}' \\ \widetilde{R}' & \text{if } \widetilde{R}_1 = \widetilde{R}_2, i = 2 \text{ and} \\ & \exists \widetilde{s}_2 \in \widetilde{R}_2 \text{ such that } \delta^{\widetilde{\mathcal{T}}}(\widetilde{s}_2, \langle 0, a_2 \rangle) \in \widetilde{R}' \end{cases}$$

Note that given player-1 and player-2 strategies $\pi_1^{\widetilde{\mathcal{T}}^{\mathsf{F}}}$ and $\pi_2^{\widetilde{\mathcal{T}}^{\mathsf{F}}}$, and any state $\widetilde{R}$, we have only one run in $\mathsf{Outcomes}(\widetilde{R}, \pi_1^{\widetilde{\mathcal{T}}^{\mathsf{F}}}, \pi_2^{\widetilde{\mathcal{T}}^{\mathsf{F}}})$.

Mapping runs and states in $\widetilde{\mathcal{T}}$ to those in $\widetilde{\mathcal{T}}^{\mathsf{F}}$ using RegMap() and RegStates(). Given a run $\widetilde{r} = \widetilde{s}_0, \langle m_1^0, m_2^0 \rangle, \widetilde{s}_1, \langle m_1^1, m_2^1 \rangle, \ldots$ of $\widetilde{\mathcal{T}}$, we let $\mathsf{RegMap}(\widetilde{r})$ be the corresponding run in $\widetilde{\mathcal{T}}^{\mathsf{F}}$ such that the states in $\widetilde{r}$ are mapped to their regions, and the moves of $\widetilde{\mathcal{T}}$ are mapped to corresponding moves in $\widetilde{\mathcal{T}}^{\mathsf{F}}$. Formally, $\mathsf{RegMap}(\widetilde{r})$ is the run $\mathsf{Reg}(\widetilde{s}_0), \langle m_1^{0,\mathsf{F}}, m_2^{0,\mathsf{F}} \rangle, \mathsf{Reg}(\widetilde{s}_1), \langle m_1^{1,\mathsf{F}}, m_2^{1,\mathsf{F}} \rangle, \ldots$ in $\widetilde{\mathcal{T}}^{\mathsf{F}}$ such that for $m_1^j = \langle \Delta_1^j, a_1^j \rangle$ and $m_2^j = \langle \Delta_2^j, a_1^j \rangle$ we have (1) $m_1^{j,\mathsf{F}} = \langle \mathsf{Reg}\left(\widetilde{s}_j + \Delta_1^j\right), a_1^j \rangle$, and (2) $m_2^{j,\mathsf{F}} = \langle \mathsf{Reg}\left(\widetilde{s}_j + \Delta_2^j\right), a_2^j, i \rangle$ with $i = 1$ if $\Delta_1^j < \Delta_2^j$, or $\Delta_1^j = \Delta_2^j$ and $\widetilde{s}_{j+1} = \delta(\widetilde{s}_j, m_1^j)$ (i.e., the move of player 1 gets picked in round j); otherwise $i = 2$. Given a set of regions X of $\widetilde{\mathcal{T}}$ (i.e., X is a set of states of $\widetilde{\mathcal{T}}^{\mathsf{F}}$), let $\mathsf{RegStates}(X) = \{\widetilde{s} \mid \widetilde{s} \in \bigcup X\}$. We have the following lemma which states the equivalence of the games $\widetilde{\mathcal{T}}^{\mathsf{F}}$ and $\widetilde{\mathcal{T}}$.

Lemma 3. *Let $\mathcal{T}$ be a timed automaton game, $\widetilde{\mathcal{T}}$ the enlarged game structure as described above, and $\widetilde{\mathcal{T}}^{\mathsf{F}}$ the corresponding finite state concurrent game structure. Let $\widetilde{\Phi}$ be an ω-regular location objective of $\widetilde{\mathcal{T}}$ (and naturally also of $\widetilde{\mathcal{T}}^{\mathsf{F}}$). We have* $\mathsf{Win}_1^{\widetilde{\mathcal{T}}}(\widetilde{\Phi}) = \mathsf{RegStates}\left(\mathsf{Win}_1^{\widetilde{\mathcal{T}}^{\mathsf{F}}}(\widetilde{\Phi})\right)$.

Obtaining a class of player-2 spoiling strategies in $\widetilde{\mathcal{T}}$ using the game structure $\widetilde{\mathcal{T}}^{\mathsf{F}}$. We use the finite state game $\widetilde{\mathcal{T}}^{\mathsf{F}}$ to analyze the spoiling strategies of player 2 for any given player-1 strategy π_1 in $\widetilde{\mathcal{T}}$. To do this analysis, we (1) map any player-1 strategy π_1 in $\widetilde{\mathcal{T}}$ to a corresponding player-1 strategy π_1^{F} in $\widetilde{\mathcal{T}}^{\mathsf{F}}$; and (2) map any player-2 spoiling strategies in $\widetilde{\mathcal{T}}^{\mathsf{F}}$ against π_1^{F} to a class of player-2 spoiling strategies in $\widetilde{\mathcal{T}}$, all of which will be spoiling against π_1. These mappings are described next.

Mapping player-1 strategies in $\widetilde{\mathcal{T}}$ to player-1 strategies in $\widetilde{\mathcal{T}}^{\mathsf{F}}$. Let $\mathsf{FinRuns}^{\widetilde{\mathcal{T}}^{\mathsf{F}}}$ be the set of finite runs of $\widetilde{\mathcal{T}}^{\mathsf{F}}$. A set of finite runs $\mathcal{O}$ of $\widetilde{\mathcal{T}}$ is said to *cover* $\mathsf{FinRuns}^{\widetilde{\mathcal{T}}^{\mathsf{F}}}$ if for every (finite) run $\widetilde{r}^{\mathsf{F}} \in \mathsf{FinRuns}^{\widetilde{\mathcal{T}}^{\mathsf{F}}}$, there exists a *unique* finite run $\widetilde{r} \in \mathcal{O}$ such that $\mathsf{RegMap}(\widetilde{r}) = \widetilde{r}^{\mathsf{F}}$. There exists at least one such run-cover $\mathcal{O}$ ([8]). Abusing notation, we let $\mathcal{O}(\widetilde{r}^{\mathsf{F}})$ denote the unique run $\widetilde{r} \in \mathcal{O}$ such that $\mathsf{RegMap}(\widetilde{r}) = \widetilde{r}^{\mathsf{F}}$. Given a player-1 strategy π_1 in $\widetilde{\mathcal{T}}$, and a run-cover $\mathcal{O}$ of $\mathsf{FinRuns}^{\widetilde{\mathcal{T}}^{\mathsf{F}}}$, we obtain the mapped player-1 strategy in $\widetilde{\mathcal{T}}^{\mathsf{F}}$, denoted, $\mathbb{F}^{\mathcal{O}}(\pi_1)$, as follows.

$$\left(\mathbb{F}^{\mathcal{O}}(\pi_1)\right)(\widetilde{r}^{\mathsf{F}}) = \begin{cases} \langle \widetilde{R}, a_1 \rangle & \text{such that } \pi_1\left(\mathcal{O}\left(\widetilde{r}^{\mathsf{F}}\right)\right) = \langle \Delta_1, a_1 \rangle, \\ & \text{and } \mathsf{Reg}\left(\mathcal{O}\left(\widetilde{r}^{\mathsf{F}}\right)[k] + \Delta_1\right) = \widetilde{R} \text{ (where} \\ & \mathcal{O}\left(\widetilde{r}^{\mathsf{F}}\right)[k] \text{ is the last state in } \mathcal{O}\left(\widetilde{r}^{\mathsf{F}}\right)) \end{cases}$$

Intuitively, the strategy $\mathbb{F}^{\mathcal{O}}(\pi_1)$, on the finite run $\widetilde{r}^{\mathsf{F}}$, acts like π_1 on the finite run $\mathcal{O}(\widetilde{r}^{\mathsf{F}})$ (i.e., the move is to the same region, with the same discrete action). A run $\widetilde{r}^{\mathsf{F}}$ of the untimed game $\widetilde{\mathcal{T}}^{\mathsf{F}}$ corresponds to several different runs of the timed game $\widetilde{\mathcal{T}}$, where for each such timed run $\widetilde{r}$ we have $\mathsf{RegMap}(\widetilde{r}) = \widetilde{r}^{\mathsf{F}}$. The function $\mathbb{F}^{\mathcal{O}}$ specifies which specific timed run from $\widetilde{\mathcal{T}}$ to base the player-1 move in $\widetilde{\mathcal{T}}^{\mathsf{F}}$ on.

Mapping player-2 strategies in $\widetilde{\mathcal{T}}^{\mathsf{F}}$ to player-2 strategies in $\widetilde{\mathcal{T}}$. We present a procedure for mapping a player-2 strategy $\pi_2^{\widetilde{\mathcal{T}}^{\mathsf{F}}}$ in $\widetilde{\mathcal{T}}^{\mathsf{F}}$ to player-2 strategies in $\widetilde{\mathcal{T}}$. This mapping will depend on a given player-1 pure strategy π_1 in $\widetilde{\mathcal{T}}$ (the strategy π_1 will be given as a parameter). We next formally

define a *set*, denoted $\mathbb{T}\mathsf{Set}_{\pi_1}(\pi_2^{\widetilde{\mathcal{T}}^{\mathsf{F}}})$, of player-2 strategies in $\widetilde{\mathcal{T}}$; given a player-2 strategy $\pi_2^{\widetilde{\mathcal{T}}^{\mathsf{F}}}$ in $\widetilde{\mathcal{T}}^{\mathsf{F}}$, and a player-1 strategy π_1 in $\widetilde{\mathcal{T}}$, against π_1.

Definition 1. The set $\mathbb{T}\mathsf{Set}_{\pi_1}(\pi_2^{\widetilde{\mathcal{T}}^{\mathsf{F}}})$ of player-2 spoiling strategies, is defined as containing all player-2 strategies π_2 in $\widetilde{\mathcal{T}}$ satisfying the following conditions Given any run prefix $\widetilde{r}[0..k]$ in $\widetilde{\mathcal{T}}$, with $\pi_1(\widetilde{r}[0..k]) = \langle \Delta_1, a_1 \rangle$, we have $\pi_2(\widetilde{r}[0..k]) = \langle \Delta_2, a_2 \rangle$ such that

- $\Delta_2 < \Delta_1$ and $\mathsf{Reg}(\widetilde{r}[k] + \Delta_2) = \widetilde{R}_2$ if
  (1) $\pi_2^{\widetilde{\mathcal{T}}^{\mathsf{F}}}(\mathsf{RegMap}(\widetilde{r}[0..k])) = \langle \widetilde{R}_2, a_2, i \rangle$; and
  (2) $\mathsf{Reg}(\widetilde{r}[k] + \Delta_1)$ is a time successor of $\widetilde{R}_2$.
- $\Delta_2 > \Delta_1$ and $\mathsf{Reg}(\widetilde{r}[k] + \Delta_2) = \widetilde{R}_2$ if
  (1) $\pi_2^{\widetilde{\mathcal{T}}^{\mathsf{F}}}(\mathsf{RegMap}(\widetilde{r}[0..k])) = \langle \widetilde{R}_2, a_2, i \rangle$; and
  (2) $\widetilde{R}_2$ is a time successor of $\mathsf{Reg}(\widetilde{r}[k] + \Delta_1)$.
- $\Delta_2 \geq \Delta_1$ and $reg(\widetilde{r}[k] + \Delta_2) = \widetilde{R}_2$ if
  (1) $\pi_2^{\widetilde{\mathcal{T}}^{\mathsf{F}}}(\mathsf{RegMap}(\widetilde{r}[0..k])) = \langle \widetilde{R}_2, a_2, 2 \rangle$; and
  (2) $\mathsf{Reg}(\widetilde{r}[k] + \Delta_1) = \widetilde{R}_2$.

Intuitively, a strategy π_2 in $\mathbb{T}\mathsf{Set}_{\pi_1}(\pi_2^{\widetilde{\mathcal{T}}^{\mathsf{F}}})$ picks a move of time duration bigger than that of π_1 if the strategy $\pi_2^{\widetilde{\mathcal{T}}^{\mathsf{F}}}$ in $\widetilde{\mathcal{T}}^{\mathsf{F}}$ allows a corresponding player-1 move $\langle \mathsf{Reg}(\widetilde{r}[k] + \Delta_1), a_1 \rangle$. Otherwise, the strategies π_2 pick a move of shorter duration.

The mapped Player-2 spoiling strategies set $\mathsf{Spoil}^{\mathcal{O}}(\pi_1, \pi_2^{\widetilde{\mathcal{T}}^{\mathsf{F}}, \mathcal{O}, \pi_1})$ in $\widetilde{\mathcal{T}}$. Given a player-1 strategy π_1 in $\widetilde{\mathcal{T}}$, we now obtain a specific *set* of player-2 spoiling strategies in $\widetilde{\mathcal{T}}$ against π_1 (for some ω-regular location objective $\widetilde{\Phi}$ of $\widetilde{\mathcal{T}}$). The set is denoted as $\mathsf{Spoil}^{\mathcal{O}}(\pi_1, \pi_2^{\widetilde{\mathcal{T}}^{\mathsf{F}}, \mathcal{O}, \pi_1})$, where $\mathcal{O}$ is a runcover of $\mathsf{FinRuns}^{\widetilde{\mathcal{T}}^{\mathsf{F}}}$, and $\pi_2^{\widetilde{\mathcal{T}}^{\mathsf{F}}, \mathcal{O}, \pi_1}$ is a given player-2 spoiling strategy against $\mathbb{F}^{\mathcal{O}}(\pi_1)$ in $\widetilde{\mathcal{T}}^{\mathsf{F}}$ for the same objective $\widetilde{\Phi}$, if one such exists. The set $\mathsf{Spoil}^{\mathcal{O}}(\pi_1, \pi_2^{\widetilde{\mathcal{T}}^{\mathsf{F}}, \mathcal{O}, \pi_1})$ of player-2 spoiling strategies for π_1 is defined as $\mathbb{T}\mathsf{Set}_{\pi_1}(\pi_2^{\widetilde{\mathcal{T}}^{\mathsf{F}}, \mathcal{O}, \pi_1})$.

The next Lemma shows that the player-2 strategies of $\widetilde{\mathcal{T}}$ in $\mathsf{Spoil}^{\mathcal{O}}(\pi_1, \pi_2^{\widetilde{\mathcal{T}}^{\mathsf{F}}, \mathcal{O}, \pi_1})$ all spoil the player-1 strategy π_1 from winning in $\widetilde{\mathcal{T}}$. The intuition behind the Lemma is that given a state $\widetilde{s} \notin \mathsf{Win}_1^{\widetilde{\mathcal{T}}}(\widetilde{\Phi})$, we have that $\mathsf{Reg}(\widetilde{s}) \notin \mathsf{Win}_1^{\widetilde{\mathcal{T}}^{\mathsf{F}}}(\widetilde{\Phi})$; and that player-2 can obtain spoiling strategies for any player-1 strategy π_1 in $\widetilde{\mathcal{T}}$ by prescribing moves to the same regions as the player-2 spoiling strategy in $\widetilde{\mathcal{T}}^{\mathsf{F}}$, which spoils $\mathbb{F}^{\mathcal{O}}(\pi_1)$ (for some suitably chosen $\mathcal{O}$). This result will be used in the next subsection to show that receptive player-1 strategies *must* satisfy certain requirements.

Lemma 4. *Let $\mathcal{T}$ be a timed automaton game, $\widetilde{\mathcal{T}}$ the enlarged game structure, and $\widetilde{\mathcal{T}}^{\mathsf{F}}$ the corresponding finite state concurrent game structure. Given an ω-regular location objective $\widetilde{\Phi}$ of player 1 (in $\widetilde{\mathcal{T}}$ and $\widetilde{\mathcal{T}}^{\mathsf{F}}$), the following assertions hold.*

1. *$\widetilde{s} \in \mathsf{Win}_1^{\widetilde{\mathcal{T}}}(\widetilde{\Phi})$ iff $\mathsf{Reg}(\widetilde{s}) \in \mathsf{Win}_1^{\widetilde{\mathcal{T}}^{\mathsf{F}}}(\widetilde{\Phi})$.*

2. *Let $\widetilde{s} \notin \mathsf{Win}_1^{\widetilde{\mathcal{T}}}(\widetilde{\Phi})$. Given any player-1 strategy π_1 in $\widetilde{\mathcal{T}}$; there exists a runcover $\mathcal{O}$ of $\mathsf{FinRuns}^{\widetilde{\mathcal{T}}^{\mathsf{F}}}$ such that for any player-2 spoiling strategy $\pi_2^{\widetilde{\mathcal{T}}^{\mathsf{F}}}$ against $\mathbb{F}^{\mathcal{O}}(\pi_1)$ in $\widetilde{\mathcal{T}}^{\mathsf{F}}$ from $\mathsf{Reg}(\widetilde{s})$ for the objective $\widetilde{\Phi}$ (such spoiling strategies exist by the previous part of the lemma), we have that every player-2 strategy in $\mathsf{Spoil}^{\mathcal{O}}(\pi_1, \pi_2^{\widetilde{\mathcal{T}}^{\mathsf{F}}})$ is a spoiling strategy against π_1 in the structure $\widetilde{\mathcal{T}}$ for the objective $\widetilde{\Phi}$ from the state $\widetilde{s}$.*

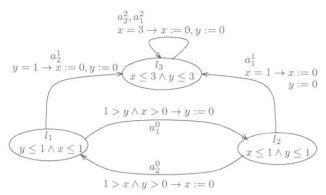

Figure 2: A time automaton game $\mathcal{T}_1$ with player-1 receptive strategies.

3.2 Characterizing Receptive Strategies Without Using Extra Clocks

In this subsection we present characterizations of receptive strategies in timed automaton games, and show that receptiveness can be expressed as an LTL condition on the states of $\widetilde{\mathcal{T}}$, from which it follows that receptive strategies require finite memory in timed automaton games. First, we consider the case where all clocks are bounded in the game (i.e., location invariants of the form $\bigwedge_{x\in C} x \leq d_x$ can be put on all locations).

Lemma 5 ([7]). *Let $\mathcal{T}$ be a timed automaton game in which all clocks are bounded (i.e., for all clocks x we have $x \leq d_x$, for constants d_x in all reachable states). Let $\widetilde{\mathcal{T}}$ be the enlarged game structure obtained from $\mathcal{T}$. Then player 1 has a receptive strategy from a state s of $\mathcal{T}$ iff $\langle s, \cdot\rangle \in \mathsf{Win}_1^{\widetilde{\mathcal{T}}}(\Phi)$, where $\Phi = \bigl(\Box\Diamond(bl_1 = \mathrm{TRUE})\bigr) \rightarrow \Phi_*$, and $\Phi_* =$*

$$
\left(\bigwedge_{x\in C}\Box\Diamond(x=0)\right)\wedge\left\{
\begin{array}{c}
\Box\Diamond\Bigl((bl_1 = \mathrm{TRUE}) \wedge \bigwedge_{x\in C}(V_{>0}(x) = \mathrm{TRUE})\Bigr)\\
\vee\\
\Box\Diamond\Bigl((bl_1 = \mathrm{FALSE}) \wedge \bigvee_{x\in C}(V_{\geq 1}(x) = \mathrm{TRUE})\Bigr)
\end{array}\right\}
$$

The proof of the above Lemma is given in [8], and uses the player-2 spoiling strategies from Lemma 4 of the previous subsection. The formula Φ of the Lemma states that if player-1 moves are chosen infinitely often then (1) all clocks must be 0 infinitely often; and either (2a) infinitely often player-1 moves are chosen such that the resulting states have $V_{>0}(x) = \mathrm{TRUE}$ for *every* clock x, or (2b) infinitely often player-2 moves are chosen such that for *some* clock x, we have $V_{\geq 1}(x) = \mathrm{TRUE}$ in the resulting state. We next present a couple of examples to demonstrate the role of the various clauses in the the formula Φ of Lemma 5.

Example 1. Consider the timed automaton game in Figure 1. The edges a_1^j are player-1 edges and a_2^j player-2 edges. The annotation $y = 1 \rightarrow x := 0, y := 0$ for the transition edge labelled a_2^1 denotes that a_2^1 is the action, with the guard of the transition being $y = 1$, and that the reset map of the transition sets clocks x and y to 0. The edges a_2^2 and a_1^1 have the same guards and reset maps. The objective of player 1 is simply TRUE (there are no bad states), that is, the objective of player 1 is simply to play with any receptive strategy. We show why this is not trivial, and the utility of Lemma 5. It is clear that player 1 has a receptive strategy when at location l_3; it repeatedly takes (or tries to take) the edge a_1^2. Let us hence focus our attention on plays which consist of (l_1, l_2) cycles (i.e., player 2 picks the edge a_2^0 from location

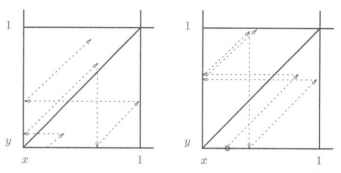

Figure 3: Two trajectories of the cycle (a_1^0, a_2^0) traversing through two zones of $\mathcal{T}_1$.

l_2, and player 1 takes the edge a_1^0 from location l_1). Let the starting state satisfy $(x < 1) \wedge (y < 1)$. In a run which consists of (l_1, l_2) cycles, we have that (1) both clocks are reset infinitely often, and (2) both clocks are greater than 0 infinitely often when the edge a_1^0 is taken (this is because the condition on the edge a_2^0 ensures that clock y is greater than 0 when at location l_1, and the edge condition on a_1^0 further ensure $x > 0$ when edge a_1^0 is taken). Thus, a run of (l_1, l_2) cycles satisfies the formula Φ of Lemma 5. We next illustrate why such a run would be time-divergent, *when player 1 chooses the appropriate moves for the edge a_1^0.*

Observe that after one (l_1, l_2) cycle, the states always satisfy $1 > x > y \geq 0$ when at l_2, and $1 > y > x \geq 0$ when at l_1. Figure 1 illustrates two paths through these two zones after at least one (l_1, l_2) cycle. Note that the transitions from the zone $1 > x > y \geq 0$ are controlled by player 2 (via edge a_2^0), and those from $1 > y > x \geq 0$ are controlled by player 1 (via edge a_1^0). In the second trajectory, player 1 is *not* able to take transitions which make the clock x more than $1/2$; but it is able to ensure that the clock y is more than $1/2$ infinitely often via its a_1^0 transitions. Since the clock y is more than $1/2$ infinitely often and is also reset infinitely often, time diverges (we will present a more formal proof of time divergence of the run shortly). It is easy to construct another timed automaton $\mathcal{T}_*$ in which player 1 can only ensure that clock x is more than $1/2$ infinitely often via player-1 actions. It can then be seen that the automatons $\mathcal{T}_1$ and $\mathcal{T}_*$ can be "combined" by a player-2 action so that player 1 can only ensure that *some* clock is more than $1/2$ infinitely often via its actions; it cannot ensure that any one *particular* clock will satisfy this property. To ensure time divergence, player 1 hence also needs to ensure that *all* clocks are reset infinitely often (as it does not know which clock will be more than $1/2$ infinitely often).

We now formally show time divergence of the runs shown in Figure 1. Let the duration of the j-th player 2 move be Δ_2^j. The value of the clock y is then Δ_2^j when location l_1 is entered for the j-th time, after the j-th a_2^0 move. Player 1 can pick its j-th a_1^0 move to be of duration $1 - \Delta_2^j - \varepsilon$. Thus, in one cycle time passes by $1 - \varepsilon$ time units. With $\varepsilon < 1$, it can be seen that time diverges. □

Example 2. We now illustrate why we require in the formula Φ of Lemma 5 that if $\Box\Diamond\bigl((bl_1 = \mathrm{TRUE}) \wedge \bigwedge_{x\in C}(V_{>0}(x) = \mathrm{TRUE})\bigr)$ does not hold, then $\Box\Diamond\bigl((bl_1 = \mathrm{FALSE}) \wedge \bigvee_{x\in C}(V_{\geq 1}(x) = \mathrm{TRUE})\bigr)$ must hold. Consider the timed automaton game $\mathcal{T}_2$ in Figure 4. The objective of player 1 is to simply play with a receptive strategy (as in Example 1). The edges a_1^j are player-1 edges and a_2^j player-2 edges. The edges a_2^2 and

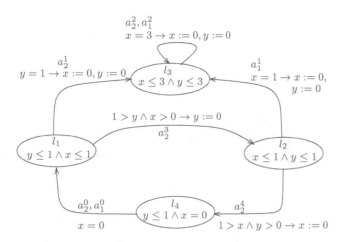

Figure 4: A timed automaton game $\mathcal{T}_2$ without player-1 receptive strategies.

a_1^2 have the same guards and reset maps. It is clear that player 1 has a receptive strategy when at location l_3; it repeatedly takes (or tries to take) the edge a_1^2. Hence, player 2 keeps the game in the (l_1, l_2, l_4). For the j-th a_2^3 and the j-th a_2^4 move, player 2 chooses a time duration of $1/2^j$. Player 1 is forced to take the move a_1^0 (of time duration 0) when at location l_4. In this cycle with such a strategy by player 2, we have that (1) all clocks are reset infinitely often, (2) the moves of player 1 are picked infinitely often, and (3) all clock values are greater than 0 infinitely often (i.e., $\Box\Diamond \bigwedge_{x \in C}(V_{>0}(x) = \text{TRUE})$ holds). But, time converges in such a run (and thus player 1 does not have a receptive strategy). The states in l_1, l_2, l_4 (with $x < 1 \wedge y < 1$) do *not* satisfy Φ of Lemma 5 because even though $\Box\Diamond \bigwedge_{x \in C}(V_{>0}(x) = \text{TRUE})$ holds, $\Box\Diamond (bl_1 = \text{TRUE}) \wedge \bigwedge_{x \in C}(V_{>0}(x) = \text{TRUE})$ does not hold. As this example shows, if player 2 picks moves to satisfy $\bigwedge_{x \in C}(V_{>0}(x) = \text{TRUE})$, then it can choose arbitrarily small moves. That is why require that if we are considering player 2 moves, then $\bigvee_{x \in C}(V_{\geq 1}(x) = \text{TRUE})$ must hold infinitely often. $\qquad\Box$

Old characterization of receptive strategies for general timed automaton games([7]). Lemma 5 was generalized to all timed automaton games in the following lemma presented in [7]. The idea of the generalization is to identify the subset of clocks which "escape" to infinity; and then to take a disjunction over all such possible subsets. Note that once a clock x becomes more than c_x, then its actual value can be considered irrelevant in determining regions. If only the clocks in $X \subseteq C$ have escaped beyond their maximum tracked values, the rest of the clocks still need to be tracked.

Lemma 6 ([7]). *Let $\mathcal{T}$ be a timed automaton game, and $\widetilde{\mathcal{T}}$ be the corresponding enlarged game. Then player 1 has a receptive strategy from a state s iff* $\langle s, \cdot \rangle \in \text{Win}_1^{\widetilde{\mathcal{T}}}(\Phi^*)$, *where* $\Phi^* = (\Box\Diamond(bl_1 = \text{TRUE})) \to \bigvee_{X \subseteq C} \phi_X^1 \wedge \phi_X^2 \wedge \phi_X^2$, *and* $\phi_X^1 = (\bigwedge_{x \in X} \Diamond\Box(x > c_x))$, *and* $\phi_X^2 = (\bigwedge_{x \in C \setminus X} \Box\Diamond(x = 0))$, *and* $\phi_X^3 = $
$$\left(\begin{array}{c} \Box\Diamond\left((bl_1 = \text{TRUE}) \wedge \bigwedge_{x \in C \setminus X}(V_{>0}(x) = \text{TRUE})\right) \\ \vee \\ \Box\Diamond\left((bl_1 = \text{TRUE}) \wedge \bigvee_{x \in C \setminus X}(V_{\geq 1}(x) = \text{TRUE})\right) \end{array} \right)$$

New characterization of receptive strategies for general timed automaton games. We shall see later that player-1 strategies which win for the objective Φ^* of Lemma 6 have a bound of $(|C+1|)^{2^{|C|}}$ for the number of memory states required. We present a new characterization of receptive strategies for which we can prove a memory bound of only $(|C|+1)$. First, we need to add $|C|$ predicates to the game structure $\widetilde{\mathcal{T}}$. For a state s of $\mathcal{T}$, we define another function $V_{>\max}^* : C \mapsto \{\text{TRUE}, \text{FALSE}\}$. The value of the predicate $V_{>\max}^*(x)$ for a clock $x \in C$ is TRUE at a state s iff the value of clock x is more than c_x, and was more than c_x in the previous state. That is, if a state $s^p = \langle l^p, \kappa^p \rangle$ and $\delta(s^p, \langle \Delta, a_1 \rangle) = s = \langle l, \kappa \rangle$, then at the state s, the predicate $V_{>\max}^*(x)$ is TRUE iff $\kappa^p(x) > c_x$ and $\kappa(x) > c_x$ (thus, $V_{>\max}^*(x)$ being true implies that the clock x has stayed greater than c_x throughout the transition). Let $\ddot{\mathcal{T}}$ be the enlarged game structure similar to $\widetilde{\mathcal{T}}$ with the state space being enlarged to also have $V_{>\max}^*$ values (in addition to $V_{>0}$ and $V_{\geq 1}$ values): $\ddot{S} = S \times \{\text{TRUE}, \text{FALSE}\} \times \{\text{TRUE}, \text{FALSE}\}^C \times \{\text{TRUE}, \text{FALSE}\}^C \times \{\text{TRUE}, \text{FALSE}\}^C$. A state of $\widetilde{\mathcal{T}}$ is a tuple $\langle s, bl_1, V_{>0}, V_{\geq 1}, V_{>\max}^* \rangle$, where s is a state of $\mathcal{T}$, the component bl_1 is TRUE iff player 1 is to be blamed for the last transition, and $V_{>0}, V_{\geq 1}, V_{>\max}^*$ are as defined earlier. A finite state concurrent game $\ddot{\mathcal{T}}^\mathsf{F}$ analogous to $\widetilde{\mathcal{T}}^\mathsf{F}$ can be constructed, and results analogous to Lemmas 3 and 4 hold for the structures $\ddot{\mathcal{T}}$ and $\ddot{\mathcal{T}}^\mathsf{F}$.

The next Lemma presents the new characterization of receptive strategies when clocks may be unbounded in $\mathcal{T}$. The proof of the Lemma can be found in [8], and uses the player 2 strategies from Lemma 4 of the previous subsection.

Lemma 7. *Let $\mathcal{T}$ be a timed automaton game, and $\ddot{\mathcal{T}}$ be the corresponding enlarged game. Then player 1 has a receptive strategy from a state s of $\mathcal{T}$ iff $\langle s, \cdot \rangle \in \text{Win}_1^{\ddot{\mathcal{T}}}(\Phi^\dagger)$, where*
$$\Phi^\dagger = \left(\Box\Diamond(bl_1 = \text{TRUE})\right) \to \left(\left(\Psi_1^\dagger \wedge \left(\Psi_2^\dagger \vee \Psi_3^\dagger\right)\right) \vee \Psi_4^\dagger\right)$$
$$\Psi_1^\dagger = \bigwedge_{x \in C} \Box\Diamond\left((x = 0) \vee (V_{>\max}^*(x) = \text{TRUE})\right)$$
$$\Psi_2^\dagger = \Box\Diamond\left(\begin{array}{c}(bl_1 = \text{TRUE}) \wedge \left(\bigwedge_{x \in C}(V_{>0}(x) = \text{TRUE})\right) \wedge \\ \left(\bigvee_{x \in C}(V_{>\max}^*(x) = \text{FALSE})\right)\end{array}\right)$$
$$\Psi_3^\dagger = \Box\Diamond\left((bl_1 = \text{FALSE}) \wedge \bigvee_{x \in C}\left((V_{\geq 1}(x) = \text{TRUE}) \wedge (V_{>\max}^*(x) = \text{FALSE})\right)\right)$$
$$\Psi_4^\dagger = \bigwedge_{x \in C} \Diamond\Box(V_{>\max}^*(x) = \text{TRUE}).$$

3.3 Memory Requirement of Receptive Strategies

In this subsection we deduce memory bounds on player-1 receptive strategies using Zielonka tree analysis (see [11] for details on Zielonka trees). We first deduce a bound that allows player 1 to win in the finite state concurrent game $\ddot{\mathcal{T}}^\mathsf{F}$. A player-1 winning strategy in $\ddot{\mathcal{T}}^\mathsf{F}$ can be mapped to a player-1 winning strategy in $\ddot{\mathcal{T}}$ by letting $\pi_1^{\ddot{\mathcal{T}}}(\ddot{r}[0..k]) = \langle \Delta, a_1 \rangle$ such that $\pi_1^{\ddot{\mathcal{T}}^\mathsf{F}}(\text{Reg}(\ddot{r}[0..k])) = \langle \ddot{R}, a_1 \rangle$ for $\text{Reg}(\ddot{r}[k] + \Delta) = \ddot{R}$. Thus, the memory requirement for a player-1 winning strategy in $\ddot{\mathcal{T}}$ is not more than as for in the finite game $\ddot{\mathcal{T}}^\mathsf{F}$. We note that Zielonka tree analysis holds only for turn based games, but since concurrent games with sure winning conditions reduce to concurrent games in which both players may use only pure strategies, which in turn reduce to turn based games, the Zielonka tree analysis is valid for game $\ddot{\mathcal{T}}^\mathsf{F}$ with sure winning conditions.

Zielonka tree analysis. Let AP be a set of atomic propositions, and let AP_N be AP together with the negations of the propositions, i.e., $AP \cup \{\neg P \mid P \in AP\}$. We say a set $\mathcal{B} \subseteq AP_N$ is *consistent* with respect to AP iff for all propositions $P \in AP$, either $P \in \mathcal{B}$, or $\neg P \in \mathcal{B}$ (or both belong to $\mathcal{B}$). A *Muller* winning condition $\mathcal{F}$ is a consistent subset of 2^{AP_N}. An infinite play satisfies the Muller condition $\mathcal{F}$ iff the set of propositions (or the negation of propositions) occurring infinitely often in the play belongs to $\mathcal{F}$. Given $\mathcal{B} \subseteq AP_N$, let $\mathcal{F} \upharpoonright \mathcal{B}$ denote the set $\{D \in \mathcal{F} \mid D \subseteq \mathcal{B}\}$. The *Zielonka tree* $\mathcal{Z}_{\mathcal{F},\mathcal{B}}$ of a Muller condition $\mathcal{F}$ over AP with $\mathcal{B} = AP_N$ is defined inductively as follows:

1. If $\mathcal{B} \in \mathcal{F}$, then the root of $\mathcal{Z}_{\mathcal{F},\mathcal{B}}$ is labelled with $\mathcal{B}$. Let $\mathcal{B}_1, \ldots, \mathcal{B}_k$ be all the maximal sets in: $\{\mathcal{B}^* \notin \mathcal{F} \mid \mathcal{B}^* \subseteq \mathcal{B}, \text{ and } \mathcal{B}^* \text{ consistent with respect to } AP\}$. The root of $\mathcal{Z}_{\mathcal{F},\mathcal{B}}$ then has as children the Zielonka trees $\mathcal{Z}_{\mathcal{F} \upharpoonright \mathcal{B}_i, \mathcal{B}_i}$ of $\mathcal{F} \upharpoonright \mathcal{B}_i$ for $1 \leq i \leq k$.

2. If $\mathcal{B} \notin \mathcal{F}$, then $\mathcal{Z}_{\mathcal{F},\mathcal{B}} = \mathcal{Z}_{\overline{\mathcal{F}},\mathcal{B}}$, where $\overline{\mathcal{F}} = \{D \in 2^{\mathcal{B}} \mid D \notin \mathcal{F} \text{ and } D \text{ is consistent with respect to } AP\}$.

A node of the Zielonka tree $\mathcal{Z}_{\mathcal{F},\mathcal{B}}$ is a *Good* node if it is labelled with a set from $\mathcal{F}$, otherwise it is a *Bad* node.

Equivalent definition of Zielonka trees. We now present an equivalent definition (which suffices for our purposes) of the Zielonka tree $\mathcal{Z}_{\mathcal{F},AP_N}$ of a Muller condition $\mathcal{F}$ over AP. Every node of the Zielonka tree $\mathcal{Z}_{\mathcal{F},AP_N}$ with is labelled with a consistent subset $\mathcal{B} \subseteq AP_N$. A node of the Zielonka tree $\mathcal{Z}_{\mathcal{F},AP_N}$ is a *Good* node if it is labelled with a set from $\mathcal{F}$, otherwise it is a *Bad* node. The root is labelled with AP_N. The children of a node v are defined inductively as follows:

1. Suppose v is a Good node labelled with $\mathcal{B}_v$. Let $\mathcal{B}_1, \ldots, \mathcal{B}_k$ be all the maximal sets in: $\{\mathcal{B}^* \notin \mathcal{F} \mid \mathcal{B}^* \subseteq \mathcal{B}, \text{ and } \mathcal{B}^* \text{ consistent with respect to } AP\}$. The node v then has k children (that are all Bad) labelled with $\mathcal{B}_1, \ldots, \mathcal{B}_k$.

2. Suppose v is a Bad node labelled with $\mathcal{B}_v$. Let $\mathcal{B}_1, \ldots, \mathcal{B}_k$ be all the maximal sets in: $\{\mathcal{B}^* \in \mathcal{F} \mid \mathcal{B}^* \subseteq \mathcal{B}, \text{ and } \mathcal{B}^* \text{ consistent with respect to } AP\}$. The node v then has k children (that are all Good) labelled with $\mathcal{B}_1, \ldots, \mathcal{B}_k$.

The number $m_{\mathcal{F}}$ of a Muller condition. Let $\mathcal{F}$ be a Muller condition that is a consistent subset of 2^{AP_N}. Consider the Zielonka tree $\mathcal{Z}_{\mathcal{F},AP_N}$ of $\mathcal{F}$. We define a number $m_{\mathcal{F}}^v$ for each node v of $\mathcal{Z}_{\mathcal{F},AP_N}$ inductively.

$$m_{\mathcal{F}}^v = \begin{cases} 1 & \text{if } v \text{ is a leaf,} \\ \sum_{i=1}^{k} m_{\mathcal{F}}^{v_i} & \text{if } v \text{ is a Good node and} \\ & \text{has children } v_1, \ldots, v_k, \\ \max\{m_{\mathcal{F}}^{v_1}, \ldots, m_{\mathcal{F}}^{v_k}\} & \text{if } v \text{ is a Bad node and} \\ & \text{has children } v_1, \ldots, v_k. \end{cases}$$

The number $m_{\mathcal{F}}$ of the Muller condition $\mathcal{F}$ is defined to be $m_{\mathcal{F}}^{v_r}$ where v_r is the root of the Zielonka tree $\mathcal{Z}_{F,AP_N}$.

Lemma 8 ([11]). *Let $\mathcal{G}^f$ be a finite state turn based game. If player 1 has a winning strategy for a Muller objective $\mathcal{F}$ from a state s in $\mathcal{G}^f$, then it has a winning strategy from s that requires at most $m_{\mathcal{F}}$ memory states.*

Now we can use Zielonka tree analysis to deduce memory requirements of receptive strategies. The proof of the following Lemma can be found in [8], and uses Zielonka tree analysis to obtain the memory bounds.

Lemma 9. 1. *Let $\phi_1 = (\Diamond\Box F_1) \vee (\Diamond\Box F_2) \vee \bigwedge_{i \leq n}(\Box\Diamond I_j)$, where F_1, F_2, I_j are boolean predicates on states of a finite state game $\mathcal{G}^f$. Player 1 has a winning strategy from $\mathsf{Win}_1(\phi_1)$ in $\mathcal{G}^f$ that requires at most n memory states for the objective ϕ_1.*

2. *Let $\phi_2 = (\Diamond\Box F) \vee \bigvee_{\alpha \leq m}\left(\Diamond\Box F_\alpha \wedge \left(\bigwedge_{i \leq n}\Box\Diamond I_{\alpha,i}\right) \wedge \Box\Diamond I_\alpha\right)$, where $F, F_\alpha, I_{\alpha,i}, I_\alpha$ are boolean predicates on states of a finite state game $\mathcal{G}^f$. Player 1 has a winning strategy from $\mathsf{Win}_1(\phi_2)$ in $\mathcal{G}^f$ that requires at most $(n+1)^m$ memory states for the objective ϕ_2.*

Corollary 1. *Let $\mathcal{T}$ be a timed automaton game with the clocks C, and let $\ddot{\mathcal{T}}$ be the corresponding enlarged game.*

1. *Let $\Phi^\dagger$ be as in Lemma 7. Player 1 has a winning strategy in $\ddot{\mathcal{T}}$ from $\mathsf{Win}_1(\Phi^\dagger)$ that requires at most $(|C|+1)$ memory states.*

2. *Let Φ^* be as in Lemma 6. Player 1 has a winning strategy in $\ddot{\mathcal{T}}$ from $\mathsf{Win}_1(\Phi^*)$ that requires at most $(|C| + 1)2^{|C|}$ memory states.*

3.4 Memory Requirement of Receptive Strategies for Safety Objectives

Player 1 can ensure it stays in a set Y in a receptive fashion if it uses a receptive strategy that only plays moves to Y states at each step. The next theorem uses this fact to characterize safety strategies.

Theorem 2. *Let $\mathcal{T}$ be a timed automaton game and $\ddot{\mathcal{T}}$ be the corresponding enlarged game. Let Y be a union of regions of $\mathcal{T}$. Then the following assertions hold.*

1. *$\mathsf{WinTimeDiv}_1^{\ddot{\mathcal{T}}}(\Box Y) = \mathsf{Win}_1^{\ddot{\mathcal{T}}}\big((\Box Y) \wedge \Phi^\sharp\big)$, where $\Phi^\sharp = \Phi^*$ (as defined in Lemma 6), or $\Phi^\sharp = \Phi^\dagger$ (as defined in Lemma 7).*

2. *Player 1 has a finite-memory, receptive, region strategy in $\ddot{\mathcal{T}}$ that is winning for the safety objective $\mathsf{Safe}(Y)$ at every state in $\mathsf{WinTimeDiv}_1^{\ddot{\mathcal{T}}}(\Box Y)$, that requires at most $(|C|+1)$ memory states (where $|C|$ is the number of clocks in $\mathcal{T}$).*

3. *Player 1 has a finite-memory, receptive, strategy in $\mathcal{T}$ that is winning for the safety objective $\mathsf{Safe}(Y)$ at every state in $\mathsf{WinTimeDiv}_1^{\mathcal{T}}(\Box Y)$, that requires at most $(|C| + 1) \cdot 2^{3 \cdot |C|}$ memory states, i.e. $\big(\lg(|C| + 1) + 3 \cdot |C|\big)$ bits of memory (where $|C|$ is the number of clocks in $\mathcal{T}$).*

3.5 Memory Requirement of Receptive Region Strategies for Safety Objectives

We now show memoryless *region* strategies for safety objectives do not suffice (where the regions are as classically defined for timed automata).

Example 3 (Memory necessity of winning region strategies for safety). Consider the timed automaton game $\mathcal{T}_3$ in Figure 5. The edges a_1^j are player-1 edges and a_2^j player-2 edges. The safety objective of player-1 is to avoid the location "Bad". It is clear that to avoid the bad location, player-1 must ensure that the game keeps cycling around the locations l_0, l_1, l_2, and that the clock value of y never exceeds 1. Cycling around only in l_0, l_1 cannot be ensured by a receptive player-1 strategy as player 2 can take smaller and smaller time steps to take the a_2^0 transition. Cycling around only in l_0, l_2 also cannot be ensured by a receptive player-1 strategy as the clock value of would always need to stay below 1 without being reset, implying that more than 1 time unit does not pass. Thus, any receptive player-1 strat-

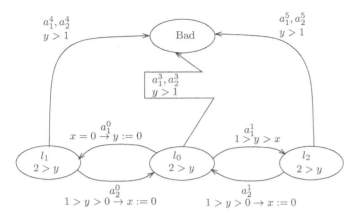

Figure 5: A time automaton game $\mathcal{T}_3$ where player-1 does not have receptive region strategies for the safety objective.

egy which avoids the bad location must cycle infinitely often between l_0, l_1, and also between l_0, l_2.

Suppose a player-1 *memoryless* region strategy π_1^* exists for avoiding the bad location, starting from a state in the region $R = \langle l_0, x = 0 \wedge 0 < y < 1 \rangle$. Suppose π_1^* always proposes the transition a_1^0 from the region R_1. Then, player 2 can take the a_2^0 transitions with smaller and smaller time delays and ensure that the region is R after each a_2^0 transition. This will make time converge, and player 1 will not be blameless, thus π_1^* is not a receptive strategy. Suppose π_1^* always proposes the transition a_1^1 from the region R_1 (or proposes a non-zero time delay move, which has the equivalent effect of disabling the a_1^0 transition). In this case, player 2 can take the a_2^1 transition to again ensure that the region is R after the a_2^1 transition. This will result in the situation where the l_0, l_2 cycle is always taken, time is not divergent, and player 1 is not blameless; thus π_1^* is again not a receptive strategy.

We now demonstrate that a finite-memory (actually memoryless in this case) receptive player-1 strategy $\pi_1^\dagger$ exists from states in the region $R = \langle l_0, x = 0 \wedge 0 < y < 1 \rangle$ for avoiding the bad location. If the current state is in the region R with the clock value of y being less than $1/2$, then player 1 proposes the a_1^1 transition with a delay which will make make clock y have a value greater than $1/2$. If the current state is in the region R with the clock value of y being greater than or equal to $1/2$, then player 1 proposes to take the a_1^0 transition (immediately). This strategy ensures that against any player-2 receptive strategy: (1) the game will cycle infinitely often between l_0, l_1, and also between l_0, l_2, and (2) the clock y will be at least $1/2$ infinitely often, and also be reset infinitely often, giving us time divergence. Thus, $\pi_1^\dagger$ is a receptive memoryless player-1 winning strategy.

Finally, we demonstrate a player-1 finite-memory receptive *region* strategy $\pi_1^\ddagger$ for avoiding the bad location, starting from a state in the region $R = \langle l_0, x = 0 \wedge 0 < y < 1 \rangle$. The strategy acts as follows when at region R. If the previous cycle was to l_1, the strategy $\pi_1^\ddagger$ proposes to take the edge a_1^1 with a delay which will make make clock y have a value greater than $1/2$. If the previous cycle was to l_2, the strategy $\pi_1^\ddagger$ proposes to take the edge a_1^0 (immediately). It can be verified that the strategy $\pi_1^\ddagger$ requires only one memory state, and is a player-1 winning receptive region strategy. □

Theorem 3 (Memory necessity of winning region strategies

for safety). *There is a timed automaton game $\mathcal{T}$, a union of regions Y of $\mathcal{T}$, and a state s such that player 1 does not have a winning memoryless receptive region strategy from s, but has a winning receptive region strategy from s that requires at most $(3 \cdot |C| + \lg(|C| + 1))$ bits of memory states (where $|C|$ is the number of clocks in $\mathcal{T}$), for the safety objective of staying in the set Y.*

4. REFERENCES

[1] R. Alur and D. L. Dill. A theory of timed automata. *Theor. Comput. Sci.*, 126(2):183–235, 1994.

[2] R. Alur and T. A. Henzinger. Modularity for timed and hybrid systems. In *CONCUR 97*, LNCS 1243, pages 74–88. Springer, 1997.

[3] E. Asarin and O. Maler. As soon as possible: Time optimal control for timed automata. In *HSCC 99*, LNCS 1569, pages 19–30. Springer, 1999.

[4] P. Bouyer, E. Brinksma, and K. G. Larsen. Staying alive as cheaply as possible. In *HSCC 04*, LNCS 2993, pages 203–218. Springer, 2004.

[5] P. Bouyer, D. D'Souza, P. Madhusudan, and A. Petit. Timed control with partial observability. In *CAV 03*, LNCS 2725, pages 180–192. Springer, 2003.

[6] F. Cassez, A. David, E. Fleury, K.G. Larsen, and D. Lime. Efficient on-the-fly algorithms for the analysis of timed games. In *CONCUR 05*, pages 66–80. Springer, 2005.

[7] K. Chatterjee, T. A. Henzinger, and V. S. Prabhu. Trading infinite memory for uniform randomness in timed games. In *HSCC 08*, LNCS 4981. Springer, 2008.

[8] K. Chatterjee and V. S. Prabhu. Synthesis of memory-efficient real-time controllers for safety objectives. *CoRR*, abs/1101.5842, 2011.

[9] L. de Alfaro, M. Faella, T A. Henzinger, R. Majumdar, and M. Stoelinga. The element of surprise in timed games. In *CONCUR 03*, LNCS 2761, pages 144–158. Springer, 2003.

[10] D. D'Souza and P. Madhusudan. Timed control synthesis for external specifications. In *STACS 02*, LNCS 2285, pages 571–582. Springer, 2002.

[11] S. Dziembowski, M. Jurdziński, and I. Walukiewicz. How much memory is needed to win infinite games? In *LICS 97*, pages 99–110. IEEE Computer Society, 1997.

[12] T. A. Henzinger and P. W. Kopke. Discrete-time control for rectangular hybrid automata. *Theoretical Computer Science*, 221:369–392, 1999.

[13] T. A. Henzinger and V. S. Prabhu. Timed alternating-time temporal logic. In *FORMATS 06*, LNCS 4202, pages 1–17. Springer, 2006.

[14] A. Pnueli, E. Asarin, O. Maler, and J. Sifakis. Controller synthesis for timed automata. In *Proc. System Structure and Control*. Elsevier, 1998.

[15] R. Segala, R. Gawlick, J.F. Søgaard-Andersen, and N. A. Lynch. Liveness in timed and untimed systems. *Inf. Comput.*, 141(2):119–171, 1998.

[16] W. Thomas. Languages, automata, and logic. In *Handbook of Formal Languages*, volume 3, Beyond Words, chapter 7, pages 389–455. Springer, 1997.

Hybrid Controllers for Tracking of Impulsive Reference State Trajectories: A Hybrid Exosystem Approach

Manuel Robles
University of Arizona
Tucson, AZ 85721
robles86@u.arizona.edu

Ricardo G. Sanfelice
University of Arizona
Tucson, AZ 85721
sricardo@u.arizona.edu

ABSTRACT

We study the problem of designing controllers to track state trajectories for plants with jumps in the state that are given by constrained differential equations capturing the continuous dynamics and constrained difference equations (or inclusions) capturing the discrete dynamics. The reference trajectories consist of signals having intervals of flow and instantaneous jumps, and are generated via a known hybrid exosystem. The class of controllers considered are hybrid and are designed to guarantee that the jump times of the plant coincide with those of the given reference trajectories. By recasting the tracking problem as the stabilization of a set and using asymptotic stability tools for time-invariant hybrid systems, we derive sufficient conditions for the closed-loop system that guarantee tracking of reference trajectories.

Categories and Subject Descriptors

C.1.m [**Miscellaneous**]: Hybrid systems; I.2.8 [**Problem Solving, Control Methods, and Search**]: Control theory—*hybrid control*.

General Terms

Algorithms, Design, Performance, Theory, Verification.

Keywords

hybrid dynamical systems, tracking control, exosystem, asymptotic stability.

1. INTRODUCTION

We consider plants with state jumps given in terms of a constrained flow equation

$$\dot{\xi} = f_p(\xi, u) \qquad (\xi, u) \in C_p \qquad (1)$$

and a constrained jump inclusion

$$\xi^+ \in G_p(\xi, u) \qquad (\xi, u) \in D_p. \qquad (2)$$

For this class of systems, which are hybrid systems due to exhibiting both continuous and discrete behavior, we address the problem of designing a controller that assigns the input u and measures the plant's state ξ to enforce that the set of points where the state ξ and the reference trajectory r coincide is asymptotically stable. In the hybrid system setting being considered, the reference trajectory may exhibit intervals of continuous evolution or flow as well as instantaneous jumps. Without being precise about the concept of asymptotic stability at this point, a well-posed controller solving such a tracking problem will guarantee that, when the initial condition of the plant coincides with the initial value of the reference r to track, r itself is a solution to the plant. As a consequence, from such initial conditions, the controller has to enforce flows and trigger jumps in the plant over the same flow intervals and jumps instants that the reference trajectory does, while from other initial conditions, it needs to guarantee that the trajectories do not go far away from and converge to the set of points where ξ and r coincide.

A purpose of this paper is to bring to the attention of the hybrid systems community the difficulties to tracking control design for hybrid systems. As it will be illustrated in Section 2, even for very basic plants and reference trajectories, it is not a trivial task to enforce that the error between the state and the reference, i.e., the tracking error, is well behaved. While the tracking problem is highly relevant in numerous engineering applications, such as bipedal locomotion and juggling systems, only a handful of contributions to the solution of the general tracking problem of interest have been proposed in the literature. Noteworthy contributions on tracking control are the time-warping approach for mechanical systems undergoing impacts proposed in [1, 2, 3], the methods for systems modeled as measure differential inclusions used in [4, 5], the techniques for mechanical systems with unilateral constraints in [6, 7, 8], and the hybrid approach for juggling systems in [9]. In this paper, we consider reference trajectories exhibiting flows and jumps that can be generated via a known hybrid exosystem. We recast the tracking problem as the asymptotic stabilization of a time-invariant set of the state space involving the plant, controller, and exosystem. In this setting, the set to stabilize may not be compact, but only closed. The proposed approach leads to the characterization of a class of hybrid controllers guaranteeing that the jump times of the plant coincide with those of the given reference trajectories and that the set of points with tracking error equal zero is asymptotically stable.

The remainder of this paper is organized as follows. In Section 2 we present key difficulties in solving tracking control problems with impulsive reference trajectories and outline the proposed approach. Results to establish asymptotic stability of closed sets in hybrid systems are presented in Section 3. A general tracking control problem is presented in Section 4 and then specialized to the full information case in Section 5, where tools for design of hybrid tracking controllers are proposed. Finally, in Section 6, we exercise in examples the utility of the controller characterization of Section 5.

Notation: We summarize the notation used throughout the paper. $\mathbb{R}^n$ denotes n-dimensional Euclidean space; $\mathbb{R}$ real numbers; $\mathbb{R}_{\geq 0}$ nonnegative real numbers; $\mathbb{N}$ natural numbers including 0; $\mathbb{B}$ the closed unit ball in a Euclidean space. Given a set S, $\overline{S}$ denotes its closure. Given a vector $x \in \mathbb{R}^n$, $|x|$ denotes the Euclidean vector norm. Given a set $S \subset \mathbb{R}^n$ and a point $x \in \mathbb{R}^n$, $|x|_S := \inf_{y \in S} |x - y|$. A function $\alpha : \mathbb{R}_{\geq 0} \to \mathbb{R}_{\geq 0}$ is said to belong to class-$\mathcal{K}$ ($\alpha \in \mathcal{K}$) if it is continuous, zero at zero, and strictly increasing and to belong to class-$\mathcal{K}_\infty$ ($\alpha \in \mathcal{K}_\infty$) if it belongs to class-$\mathcal{K}$ and is unbounded.

2. MOTIVATIONAL EXAMPLE AND PROPOSED APPROACH

Consider the fully controlled hybrid scalar plant given by

$$\begin{array}{llll} \dot{\xi} & = & u_1 & (\xi, u) \in C_p \subset \mathbb{R} \times \mathbb{R}^3 \\ \xi^+ & = & u_3 & (\xi, u) \in D_p \subset \mathbb{R} \times \mathbb{R}^3, \end{array} \quad (3)$$

where $\xi \in \mathbb{R}$ is the state, $u = [u_1 \ u_2 \ u_3]^\top \in \mathbb{R}^3$ is the control input, the set C_p defines the condition allowing continuous evolution according to $\dot{\xi} = u_1$, and the set D_p defines the condition triggering jumps $\xi^+ = u_3$ in the state. Solutions to (3) can be defined as functions on hybrid time domains, which are subsets of $\mathbb{R}_{\geq 0} \times \mathbb{N}$ and parameterize the trajectories by flow time t and jump time j (see Section 3 for more details). That is, the evaluation of the solution ξ to (3) at t units of time and j jumps is denoted $\xi(t, j)$ (similarly for u). Consequently, the value of the solution after a jump at, say, (t', j') is given by $\xi(t', j' + 1)$, which from (3), $(\xi(t', j'), u(t', j')) \in D_p$ must hold and we will have $\xi(t', j' + 1) = u_3(t', j')$.

Consider the reference trajectory r given by the sawtooth signal shown in Figure 1 and the problem of designing a control law

$$u = \kappa_c(\xi, r)$$

so that ξ tracks r. For the purposes of this motivational discussion, we consider the attractivity property required for tracking, that is, the property that, for every initial condition $\xi(0, 0)$ of (3),

$$\lim_{t+j \to \infty} |\xi(t, j) - r(t, j)| = 0;$$

see Section 4 for a complete definition involving stability. A typical approach used in tracking control for continuous-time and discrete-time plants consists of generating the reference trajectory via an *exosystem*, defining the tracking error, and then analyzing the resulting system. Following this approach, sawtooth reference trajectories r can be generated

via the exosystem

$$\begin{array}{llll} \dot{w} & = & 1 & w \in C_e := [0, 1] \\ w^+ & = & 0 & w \in D_e := \{1\} \\ r & = & w. \end{array} \quad (4)$$

In particular, the reference trajectory in Figure 1 is generated from system (4) with initial condition $w(0, 0) = 0$, which has domain $\bigcup_{j \in \mathbb{N}}([t_j, t_{j+1}], j)$ and is given by

$$r(t, j) = t - t_j \qquad \forall t \in [t_j, t_{j+1}],$$

where $t_j = j, j \in \mathbb{N}$. Following the "classical" approach to tracking control, the dynamics of the tracking error

$$\chi := \xi - r \quad (5)$$

are given as follows:

- Differentiating (5), the continuous dynamics of the tracking error are governed by

$$\dot{\chi} = u_1 - 1$$

when both the flow condition of (3) and of (4) hold simultaneously, that is, when

$$(\chi + r, u) \in C_p \qquad \text{and} \qquad r \in [0, 1] \quad (6)$$

hold, where we have used the fact that $\xi = \chi + r$ from (5) and $r = w$;

- Computing the change of χ when jumps occur, we obtain

$$\chi^+ = g(\chi, u_3, r) \quad (7)$$

when either one of the jump conditions of (3) and of (4) hold, that is, when

$$(\chi + r, u) \in D_p \qquad \text{or} \qquad r = 1 \quad (8)$$

hold. For such points the jump map g is defined as

$$g(\chi, u_3, r) = \begin{cases} u_3 - r & (\chi + r, u) \in D_p, r \in [0, 1) \\ \chi + r & (\chi + r, u) \notin D_p, r = 1 \\ u_3 & (\chi + r, u) \in D_p, r = 1. \end{cases}$$

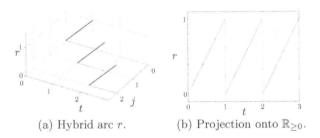

(a) Hybrid arc r.　　　(b) Projection onto $\mathbb{R}_{\geq 0}$.

Figure 1: Reference trajectory for the tracking control problem in Section 2.

First, note that, in general, the constraints (6) and (8), and the jump equation (7) cannot be written in terms of the tracking error solely – hence, the dynamics of the error system are given by (3)-(8). To illustrate this, consider the hybrid plant (3) with

$$C_p = \{(\xi, u) : 0 \leq u_2 \leq 1\}, \quad D_p = \{(\xi, u) : u_2 = 1\} \quad (9)$$

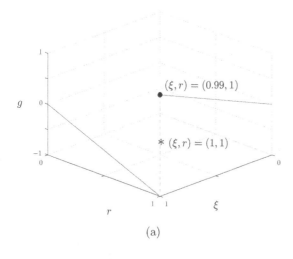

(a)

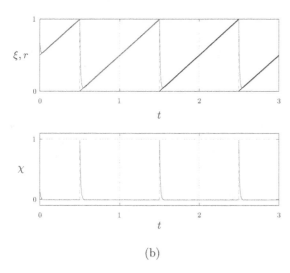

(b)

Figure 2: Jump map g for the error system in tracking control problem in Section 2 and trajectories. The tracking error χ peaks to values close to 1 after jumps.

and the static controller $\mathcal{H}_c$ given by

$$u = \kappa_c(\xi, r) = \begin{bmatrix} 1 \\ r \\ 0 \end{bmatrix}.$$

While in terms of the tracking error χ we have $\dot{\chi} = 0$ and $\chi^+ = 0$, the flow condition is $\xi - \chi \in [0, 1]$ and the jump condition is $\xi - \chi = 1$, which explicitly depend on ξ.

Second, the jump equation may map the tracking error to dramatically different values from nearby points, making it difficult to guarantee that it converges to zero asymptotically. To see this issue, consider the following choice of flow and jump sets:

$$C_p = \{(\xi, u) : 0 \le \xi \le 1\}, \ D_p = \{(\xi, u) : \xi = 1\}. \ (10)$$

Then, picking $u_3 = 0$ would make the jump equation (7) reset the tracking error χ to zero when $(\xi, u) \in D_p$ and $r = 1$, or equivalently, when $\chi = 0$ and $r = 1$. However,

from points χ nearby 0 and $r = 1$, the jump equation (7) updates χ to a value nearby 1, even when u_1 is chosen so that χ converges to zero during flows. Figure 2(a) depicts the map g as a function of ξ and r (solid) for $u_3 = 0$ – at $(\xi, r) = (1, 1)$, g is equal to zero, as $*$ denotes. In fact, for instance, when jumps occur due to $r = 1$ with $\chi = -0.01$, which corresponds to $\xi = 0.99$ as denoted with $\bullet$ in the graph, a much larger tracking error results after the jump ($\chi^+ = 0.99$). Therefore, from points close to $(\xi, r) = (1, 1)$, the value of the error χ after a jump can be nearby 0 or 1, as seen in Figure 2(b). This "peaking phenomenon" imposes a difficulty in guaranteeing that the norm of χ converges to zero as the attractivity property requires.

We propose to design tracking controllers that circumvent such a challenging issue by ensuring that jumps of the plant occur at the same instant as the jumps of the reference trajectories. For the illustrative example above, a controller designed with the said approach will assign u so that the jumps of the plant and exosystem occur jointly. For this purpose, we recast the tracking control problem as the stabilization of a closed, perhaps unbounded, set and exploit sufficient conditions for asymptotic stability of time-invariant hybrid systems already available in the literature. With the proposed approach, the obtained results are a first step in solving the tracking control problem for general hybrid systems, and it is the hope that they will spark the interest of the hybrid systems community.

3. STABILITY OF CLOSED SETS FOR HYBRID SYSTEMS

3.1 Modeling Framework

A hybrid system $\mathcal{H}$ with state x, input u, and output y is modeled as

$$\mathcal{H} \begin{cases} \dot{x} & = & f(x, u) & (x, u) \in C \\ x^+ & \in & G(x, u) & (x, u) \in D \\ y & = & h(x), \end{cases} \quad (11)$$

where $\mathbb{R}^n$ is the space for the state x, $\mathcal{U} \subset \mathbb{R}^m$ is the space for inputs u, the set $C \subset \mathbb{R}^n \times \mathcal{U}$ is the *flow set*, the function $f : C \to \mathbb{R}^n$ is the *flow map*, the set $D \subset \mathbb{R}^n \times \mathcal{U}$ is the *jump set*, the set-valued map $G : D \rightrightarrows \mathbb{R}^n$ is the *jump map*, and $h : \mathbb{R}^n \to \mathbb{R}^p$ is the *output map*. The data of the hybrid system $\mathcal{H}$ is given by (C, f, D, G, h), and at times we use the shorthand notation $\mathcal{H} = (C, f, D, G, h)$. Note that the state x can contain both continuous and discrete state components. That is, the state x can be given by $x := [\xi^\top \ q]^\top$ where $\xi \in \mathbb{R}^{n-1}$ is the continuous state and $q \in \{1, 2, \dots, N\} \subset \mathbb{R}$ is the discrete (or logic) state. Moreover, as illustrated in [10, 11], hybrid automata can be modeled in the framework (11).

We remark that the presentation is focused on single-valued flow maps f due to the control application of interest; however, the general stability results below also hold for the case when f is replaced by a set-valued mapping.

DEFINITION 3.1 (HYBRID TIME DOMAIN) *A set $E \subset \mathbb{R}_{\ge 0} \times \mathbb{N}$ is a compact hybrid time domain if*

$$E = \bigcup_{j=0}^{J-1} ([t_j, t_{j+1}], j)$$

for some finite sequence of times $0 = t_0 \leq t_1 \leq t_2 ... \leq t_J$. *It is a hybrid time domain if for all* $(T, J) \in E$, $E \cap ([0, T] \times \{0, 1, ... J\})$ *is a compact hybrid time domain.*

This definition of time domain has similarities with hybrid time trajectories in [12] and hybrid time sets in [13]. Solutions to hybrid systems $\mathcal{H}$ will be given in terms of hybrid arcs and hybrid inputs. These are parameterized by pairs (t, j), where t is the ordinary-time component and j is the discrete-time component that keeps track of the number of jumps.

DEFINITION 3.2 (HYBRID ARC AND INPUT). *A function* $\phi : \text{dom} \, \phi \to \mathbb{R}^n$ *is a hybrid arc if* $\text{dom} \, \phi$ *is a hybrid time domain and, for each* $j \in \mathbb{N}$, *the function* $t \mapsto \phi(t, j)$ *is absolutely continuous on the interval* $\{t : (t, j) \in \text{dom} \, \phi \}$. *A function* $u : \text{dom} \, u \to \mathcal{U}$ *is a hybrid input if* $\text{dom} \, u$ *is a hybrid time domain and, for each* $j \in \mathbb{N}$, *the function* $t \mapsto u(t, j)$ *is Lebesgue measurable and locally essentially bounded on the interval* $\{t : (t, j) \in \text{dom} \, u \}$.

Purely continuous inputs $t \mapsto \tilde{u}(t)$ can be converted to a hybrid input u on a given hybrid time domain S by defining $u(t, j) = \tilde{u}(t)$ for each $(t, j) \in S$.

With the definitions of hybrid time domain, and hybrid arc and input in Definitions 3.1 and 3.2, respectively, we define a concept of solution for hybrid systems $\mathcal{H}$.

DEFINITION 3.3 (SOLUTION). *Given a hybrid input* $u : \text{dom} \, u \to \mathcal{U}$, *a hybrid arc* $\phi : \text{dom} \, \phi \to \mathbb{R}^n$ *defines a solution pair* (ϕ, u) *to the hybrid system* $\mathcal{H} = (C, f, D, G, h)$ *if the following conditions hold:*

(S0) $(\phi(0, 0), u(0, 0)) \in \overline{C} \cup D$ *and* $\text{dom} \, \phi = \text{dom} \, u$;

(S1) For each $j \in \mathbb{N}$ *such that* $I_j := \{t : (t, j) \in \text{dom}(\phi, u) \}$ *has nonempty interior* $\text{int}(I_j)$,

$$(\phi(t, j), u(t, j)) \in C \text{ for all } t \in \text{int}(I_j),$$

and, for almost all $t \in I_j$,

$$\frac{d}{dt} \phi(t, j) = f(\phi(t, j), u(t, j));$$

(S2) For each $(t, j) \in \text{dom}(\phi, u)$ *such that* $(t, j+1) \in \text{dom}(\phi, u)$,

$$(\phi(t, j), u(t, j)) \in D$$

and

$$\phi(t, j+1) \in G(\phi(t, j), u(t, j)).$$

A solution pair (ϕ, u) to $\mathcal{H}$ is said to be *complete* if $\text{dom}(\phi, u)$ is unbounded, *Zeno* if it is complete but the projection of $\text{dom}(\phi, u)$ onto $\mathbb{R}_{\geq 0}$ is bounded, *discrete* if the domain is $\{0\} \times \mathbb{N}$, and *maximal* if there does not exist another pair (ϕ', u') such that (ϕ, u) is a truncation of (ϕ', u') to some proper subset of $\text{dom}(\phi', u')$.

3.2 Stability and Sufficient Conditions

In preparation for the analysis of closed-loop systems resulting in tracking control, we define stability and Lyapunov

functions for closed hybrid systems (no inputs and outputs) given, with some abuse of notation, by

$$\mathcal{H} \begin{cases} \dot{x} & = & f(x) & x \in C \\ x^+ & \in & G(x) & x \in D. \end{cases} \tag{12}$$

The following definition introduces stability for subsets of the state space, e.g., equilibrium points and attractors. Given $\phi^0 \in \mathbb{R}^n$, $\mathcal{S}_{\mathcal{H}}(\phi^0)$ denotes the set of maximal solutions ϕ to $\mathcal{H}$ with $\phi(0, 0) = \phi^0$.

DEFINITION 3.4 (STABILITY). *A set* $\mathcal{A} \subset \mathbb{R}^n$ *is said to be*

- *uniformly globally stable if there exists* $\alpha \in \mathcal{K}_\infty$ *such that each solution* $\phi \in \mathcal{S}_{\mathcal{H}}(\phi(0, 0))$ *satisfies* $|\phi(t, j)|_{\mathcal{A}} \leq \alpha(|\phi(0, 0)|_{\mathcal{A}})$ *for all* $(t, j) \in \text{dom} \, \phi$;

- *uniformly globally pre-attractive if for each* $\varepsilon > 0$ *and* $r > 0$ *there exists* $N > 0$ *such that, for any solution* $\phi \in \mathcal{S}_{\mathcal{H}}(\phi(0, 0))$ *with* $|\phi(0, 0)|_{\mathcal{A}} \leq r$, $(t, j) \in \text{dom} \, \phi$ *and* $t + j \geq N$ *imply* $|\phi(t, j)|_{\mathcal{A}} \leq \varepsilon$;

- *uniformly globally pre-asymptotically stable if it is both uniformly globally stable and uniformly globally pre-attractive.*

The definition of pre-attractivity above does not impose that every solution is complete, though it implies their boundedness relative to $\mathcal{A}$. Completeness of solutions will be of interest in the tracking problem studied here and will need to be guaranteed separately from asymptotic stability. When $\mathcal{A}$ is uniformly globally pre-asymptotically stable and every maximal solution to $\mathcal{H}$ is complete, the set $\mathcal{A}$ is said to be *uniformly globally asymptotically stable.*

The next result for asymptotic stability of closed sets from [14] will be instrumental in characterizing hybrid controllers for tracking; see also [15] and [16] for related sufficient conditions. It is essentially a Lyapunov stability theorem for hybrid systems for asserting stability that is uniform with respect to initial conditions. Its proof follows the main ideas of the standard classical Lyapunov theorem for continuous-time systems [17] (see also [18]) and is omitted; see [14] for details.

THEOREM 3.5. *(Lyapunov theorem) Let* $\mathcal{H} = (C, f, D, G)$ *be a hybrid system and let* $\mathcal{A} \subset \mathbb{R}^n$ *be closed. If there exist a function* $V : \mathbb{R}^n \to \mathbb{R}$ *that is continuously differentiable on an open set containing* $\overline{C}$, *functions* $\alpha_1, \alpha_2 \in \mathcal{K}_\infty$, *and a continuous positive definite function* ρ *such that*

$$\alpha_1(|x|_{\mathcal{A}}) \leq V(x) \leq \alpha_2(|x|_{\mathcal{A}}) \qquad \forall x \in C \cup D \cup G(D) \tag{13a}$$

$$\langle \nabla V(x), f(x) \rangle \leq -\rho(|x|_{\mathcal{A}}) \qquad \forall x \in C \tag{13b}$$

$$V(g) - V(x) \leq -\rho(|x|_{\mathcal{A}}) \qquad \forall x \in D, \, g \in G(x) \tag{13c}$$

then $\mathcal{A}$ *is globally uniformly pre-asymptotically stable for* $\mathcal{H}$.

The following result introduces relaxed Lyapunov conditions (see [14]). It states that if each solution jumps an arbitrarily large number of times or if it flows for an infinite amount of time, then the conditions in Theorem 3.5 can be relaxed.

COROLLARY 3.6. *(relaxed Lyapunov conditions) Let $\mathcal{H} = (C, f, D, G)$ be a hybrid system and let $\mathcal{A} \subset \mathbb{R}^n$ be closed. If there exist a function $V : \mathbb{R}^n \to \mathbb{R}$ that is continuously differentiable on an open set containing $\overline{C}$, functions $\alpha_1, \alpha_2 \in \mathcal{K}_\infty$, and a continuous positive definite function ρ such that (13a) and either A) or B) below holds:*

A) *Condition (13c) holds,*

$$\langle \nabla V(x), f(x) \rangle \;\leq\; 0 \quad \forall x \in C , \qquad (14)$$

and for each $r > 0$ there exist $\gamma_r \in \mathcal{K}_\infty$ and $N_r \geq 0$ such that for each maximal solution ϕ to $\mathcal{H}$, $|\phi(0,0)|_\mathcal{A} \in (0, r]$, $(t, j) \in \operatorname{dom}\phi$, $t + j \geq N$ imply $j \geq \gamma_r(N) - N_r$;

B) *Condition (13b) holds,*

$$V(g) - V(x) \;\leq\; 0 \quad \forall x \in D, \ g \in G(x) , \qquad (15)$$

and for each $r > 0$ there exist $\gamma_r \in \mathcal{K}_\infty$ and $N_r \geq 0$ such that for each maximal solution ϕ to $\mathcal{H}$, $|\phi(0,0)|_\mathcal{A} \in (0, r]$, $(t, j) \in \operatorname{dom}\phi$, $t + j \geq N$ imply $t \geq \gamma_r(N) - N_r$;

then $\mathcal{A}$ is uniformly globally pre-asymptotically stable.

4. PROBLEM STATEMENT

We consider plants modeled by hybrid systems $\mathcal{H}_p$ with state $\xi \in \mathbb{R}^{n_p}$, input $u \in \mathbb{R}^{m_p}$, and output $y \in \mathbb{R}^{s_p}$ given by

$$\mathcal{H}_p \begin{cases} \dot{\xi} &=& f_p(\xi, u) & (\xi, u) \in C_p \\ \xi^+ &\in& G_p(\xi, u) & (\xi, u) \in D_p \\ y &=& h_p(\xi), \end{cases} \qquad (16)$$

with data $(C_p, f_p, D_p, G_p, h_p)$. We consider hybrid arcs $r : \operatorname{dom} r \to \mathbb{R}^{s_e}$ defining reference trajectories to be tracked. These are generated via hybrid exosystems $\mathcal{H}_e$ of the form

$$\mathcal{H}_e \begin{cases} \dot{w} &=& f_e(w) & w \in C_e \\ w^+ &\in& G_e(w) & w \in D_e \\ r &=& h_e(w) \end{cases} \qquad (17)$$

with state $w \in \mathbb{R}^{n_e}$, output $r \in \mathbb{R}^{s_e}$, and data $(C_e, f_e, D_e, G_e, h_e)$. The following class of tracking hybrid controllers with state $\eta \in \mathbb{R}^{n_c}$ and data $(C_c, f_c, D_c, G_c, \kappa_c)$ is considered:

$$\mathcal{H}_c \begin{cases} \dot{\eta} &=& f_c(\eta, y, r) & (\eta, y, r) \in C_c \\ \eta^+ &\in& G_c(\eta, y, r) & (\eta, y, r) \in D_c \\ u &=& \kappa_c(\eta, y, r). \end{cases}$$

$$(18)$$

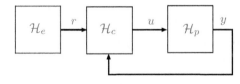

Figure 3: The interconnection of $\mathcal{H}_p$, $\mathcal{H}_c$, and $\mathcal{H}_e$ results in the closed-loop system $\mathcal{H}_{cl}$.

The input to $\mathcal{H}_c$ has been assigned to (y, r) and its output to the input of the plant $\mathcal{H}_p$. Figure 3 depicts the closed-loop system obtained from the interconnection of $\mathcal{H}_p$, $\mathcal{H}_c$, and $\mathcal{H}_e$. It is denoted $\mathcal{H}_{cl}$, has state

$$x := (\xi, w, \eta) \in \mathbb{R}^{n_p} \times \mathbb{R}^{n_e} \times \mathbb{R}^{n_c},$$

and is given by

$$\begin{aligned} \dot{\xi} &=& f_p(\xi, \kappa_c(\eta, y, r)) \\ \dot{w} &=& f_e(w) \\ \dot{\eta} &=& f_c(\eta, y, r) \end{aligned} \left.\begin{array}{c} \\ \\ \end{array}\right\} \begin{array}{c} (\xi, \kappa_c(\eta, y, r)) \in C_p \\ \text{and } w \in C_e \\ \text{and } (\eta, y, r) \in C_c \end{array}$$

$$\begin{aligned} \xi^+ &\in& G_p(\xi, \kappa_c(\eta, y, r)) \\ w^+ &=& w \\ \eta^+ &=& \eta \end{aligned} \left.\begin{array}{c} \\ \\ \end{array}\right\} \ (\xi, \kappa_c(\eta, y, r)) \in D_p$$

$$\begin{aligned} \xi^+ &=& \xi \\ w^+ &\in& G_e(w) \\ \eta^+ &=& \eta \end{aligned} \left.\begin{array}{c} \\ \\ \end{array}\right\} \ w \in D_e$$

$$\begin{aligned} \xi^+ &=& \xi \\ w^+ &=& w \\ \eta^+ &\in& G_c(\eta, y, r) \end{aligned} \left.\begin{array}{c} \\ \\ \end{array}\right\} \ (\eta, y, r) \in D_c$$

$$(19)$$

with $y = h_p(\xi)$ and $r = h_e(w)$. Note that the flow set for $\mathcal{H}_{cl}$ is given by the intersection of the flow sets of $\mathcal{H}_p$, $\mathcal{H}_c$, and $\mathcal{H}_e$, while the jump set is given by the union of the individual jump sets. In this way, flows are enabled when all of the flow conditions are satisfied while jumps are enabled when any of the individual jump conditions hold.

With the above definitions, given a hybrid plant $\mathcal{H}_p$ and a hybrid exosystem $\mathcal{H}_e$, a general tracking control problem consists of designing the controller's data

$$(C_c, f_c, D_c, G_c, \kappa_c)$$

such that the set

$$\{x \ : \ h_p(\xi) = h_e(w) \}$$

is both stable and attractive. In this paper, we consider the case when the function h_p is the identity, that is, the entire state of the plant is available for control. In this case, $n_p = s_p = s_e$ and the reference signals are state trajectories. We insist on rendering the said set uniformly globally asymptotically stable. More precisely, we focus on the following problem:

A State Trajectory Tracking Control Problem ($\star$):
Given a hybrid plant $\mathcal{H}_p$ and a hybrid exosystem $\mathcal{H}_e$ generating the reference trajectories to track design the data $(C_c, f_c, D_c, G_c, \kappa_c)$ of the controller $\mathcal{H}_c$ such that the set

$$\{x \ : \ \xi = h_e(w) \} \qquad (20)$$

is uniformly globally asymptotically stable.

REMARK 4.1. *The attractivity property of the set (20) in Problem ($\star$) implies that each solution to $\mathcal{H}_{cl}$ satisfies*

$$\lim_{t+j \to \infty} |\xi(t, j) - r(t, j)| = 0.$$

Moreover, it implies that solutions to the plant with initial conditions $\xi(0,0) = w(0,0)$, if they exist, satisfy

$$\xi(t, j) = r(t, j) \qquad \text{for all } (t, j) \in \operatorname{dom}\xi.$$

Note that even after intersecting it with the region of operation of the closed-loop system, in general, the set (20) is not necessarily bounded. The asymptotic stability property required in Problem ($\star$) implies completeness of solutions to the closed-loop system, and hence, completeness of the reference trajectories generated by the hybrid exosystem.

5. A CLASS OF HYBRID CONTROLLERS FOR TRACKING KNOWN REFERENCE STATE TRAJECTORIES

5.1 Main Approach

The proposed approach consists of generating the reference trajectories via an exosystem and designing hybrid controllers $\mathcal{H}_c$ that, with the knowledge of the reference trajectories, guarantee that the jumps of the plant and of the reference trajectory happen simultaneously. For instance, for the hybrid plant in (3) with data as in (10) and reference trajectory r generated by (4), a controller $\mathcal{H}_c$ guaranteeing that jumps of the plant and r occur simultaneously enforces, in particular,

$$\xi(t,j) = 1 \iff r(t,j) = 1,$$

$(t,j) \in \operatorname{dom} x$. In general, with a hybrid controller designed so that the jumps of the plant and of the reference trajectory happen simultaneously, the closed-loop system $\mathcal{H}_{cl}$ becomes[1]

$$
\begin{aligned}
\left.
\begin{array}{rcl}
\dot{\xi} &=& f_p(\xi, \kappa_c(\eta, \xi, h_e(w))) \\
\dot{w} &=& f_e(w) \\
\dot{\eta} &=& f_c(\eta, \xi, h_e(w))
\end{array}
\right\}
&
\begin{array}{l}
(\xi, \kappa_c(\eta, \xi, h_e(w))) \in C_p \\
\text{and } w \in C_e \\
\text{and } (\eta, \xi, h_e(w)) \in C_c
\end{array}
\\[1em]
\left.
\begin{array}{rcl}
\xi^+ &\in& G_p(\xi, \kappa_c(\eta, \xi, h_e(w))) \\
w^+ &\in& G_e(w) \\
\eta^+ &=& \eta
\end{array}
\right\}
&
\begin{array}{l}
(\xi, \kappa_c(\eta, \xi, h_e(w))) \in D_p \\
\text{and } w \in D_e
\end{array}
\\[1em]
\left.
\begin{array}{rcl}
\xi^+ &=& \xi \\
w^+ &=& w \\
\eta^+ &\in& G_c(\eta, \xi, h_e(w))
\end{array}
\right\}
&
\quad (\eta, \xi, h_e(w)) \in D_c
\end{aligned}
$$

$$(21)$$

We denote this closed-loop system as $\mathcal{H}_{cl}^\star$. Its data is given by

$$
\begin{aligned}
C &:= \{x : (\xi, \kappa_c(\eta, \xi, h_e(w))) \in C_p, \\
&\qquad w \in C_e, (\eta, \xi, h_e(w)) \in C_c\},
\end{aligned}
$$

$$
f(x) := \begin{bmatrix} f_p(\xi, \kappa_c(\eta, \xi, h_e(w))) \\ f_e(w) \\ f_c(\eta, \xi, h_e(w)) \end{bmatrix},
$$

$$
\begin{aligned}
D &:= D_1 \cup D_2, \\
D_1 &:= \{x : (\xi, \kappa_c(\eta, \xi, h_e(w))) \in D_p, w \in D_e\}, \\
D_2 &:= \{x : (\eta, \xi, h_e(w)) \in D_c\},
\end{aligned}
$$

---

[1]Note that for such a hybrid controller, the jump conditions

$$(\xi, \kappa_c(\eta, \xi, h_e(w))) \in D_p$$

and

$$w \in D_e$$

are equivalent. For completeness, both conditions are included in the closed-loop system (21).

$$
G(x) :=
\begin{cases}
G_1(x) := \begin{bmatrix} G_p(\xi, \kappa_c(\eta, \xi, h_e(w))) \\ G_e(w) \\ \eta \end{bmatrix} & \\
& x \in D_1 \setminus D_2, \\[1.5em]
G_2(x) := \begin{bmatrix} \xi \\ w \\ G_c(\eta, \xi, h_e(w)) \end{bmatrix} & x \in D_2 \setminus D_1, \\[1.5em]
\{G_1(x), G_2(x)\} & x \in D_1 \cap D_2.
\end{cases}
$$

Then, asymptotic stability of the set

$$\mathcal{A} := \{x \in \overline{C} \cup D : \xi = h_e(w)\}$$

can be established using the sufficient conditions in Theorem 3.5 and Corollary 3.6. After more details on the hybrid exosystem, these sufficient conditions applied to our tracking problem are presented in Section 5.3. A discussion on conditions guaranteeing the simultaneous jump property as well as properties involving the solution sets of $\mathcal{H}_e$ and $\mathcal{H}_{cl}^\star$ are also given in Section 5.3.

5.2 Hybrid Exosystem

The data $(C_e, f_e, D_e, G_e, h_e)$ of the exosystem $\mathcal{H}_e$ in (17) can be defined to generate the reference trajectories to be tracked. For instance:

- The system in (4) generates a unique periodic sawtooth reference signal that starts at $w(0,0)$ and oscillates between 0 and 1;

- The hybrid exosystem

$$
\begin{aligned}
\left.
\begin{array}{rcl}
\dot{w}_1 &=& w_2 \\
\dot{w}_2 &=& -\gamma
\end{array}
\right\} & \quad w_1 \geq 0 \\
\left.
\begin{array}{rcl}
w_1^+ &=& w_1 \\
w_2^+ &=& -w_2
\end{array}
\right\} & \quad w_1 = 0, w_2 \leq 0 \\
r &=& w_1,
\end{aligned}
$$

$$(22)$$

where $\gamma > 0$ is the gravity constant, generates a unique periodic reference signal corresponding to the height of a ball bouncing on the ground without energy dissipation at bounces.

Note that the solutions to $\mathcal{H}_e$ may not be unique, in particular, when capturing a family of reference signals. For example, the hybrid exosystem

$$
\begin{aligned}
\left.
\begin{array}{rcl}
\dot{w}_1 &=& 1 \\
\dot{w}_2 &=& 0
\end{array}
\right\} & \quad w_1 \in [0, \Delta] \\
\left.
\begin{array}{rcl}
w_1^+ &=& 0 \\
w_2^+ &=& 1 - w_2
\end{array}
\right\} & \quad w_1 \in [\delta, \Delta] \\
r &=& w_2
\end{aligned}
$$

$$(23)$$

with initial condition $w_1(0,0) = 0$, $w_2(0,0) \in \{0, 1\}$ generates a family of square signals with semi-period taking value in the set $[\delta, \Delta]$, where $0 < \delta < \Delta$.

To guarantee that achieving completeness of the solutions to the closed-loop system, which is required in Problem $(\star)$, is not prevented by the hybrid exosystem itself, we impose the following condition on $\mathcal{H}_e$.

ASSUMPTION 5.1. *Every maximal solution to $\mathcal{H}_e$ is complete.*

It is straightforward to check that the construction of the hybrid exosystems in (4), (22), and (23) are such that Assumption 5.1 holds. In fact, completeness of the solutions to these systems follows from [19, Proposition 2.4] since maximal solutions from $C_e \cup D_e$ are bounded and can either flow for some finite time or jump and stay in it.

5.3 Characterization of Hybrid Controllers for Reference Tracking

The data $(C_c, f_c, D_c, G_c, \kappa_c)$ of $\mathcal{H}_c$ is designed to guarantee the following three properties, which, when satisfied, provide a solution to Problem $(\star)$.

5.3.1 Matching jumps of reference and plant

Motivated by the discussion in Section 2, the proposed approach consists of designing a controller $\mathcal{H}_c$ that, with full knowledge of the exosystem $\mathcal{H}_e$, guarantees that the jumps of the reference and of the plant occur simultaneously. As pointed out in Section 5.1, for this property to hold, it is required to have that the state is in D_e if and only if it is in D_p when the plant is controlled by $\mathcal{H}_c$, that is,

$$(\xi, \kappa_c(\eta, \xi, h_e(w))) \in D_p \iff w \in D_e, \qquad (24)$$

and that from points in

$$\overline{C}_e \cap D_e$$

or

$$\overline{\{x \ : \ (\xi, \kappa_c(\eta, \xi, h_e(w))) \in C_p \}} \cap \\ \{x \ : \ (\xi, \kappa_c(\eta, \xi, h_e(w))) \in D_p \} =: \widetilde{X}_p$$

only jumps are possible. The conditions insisting on having jumps only from the intersection of the respective flow and jump sets guarantee that from points in D_e and D_p, respectively, only jumping is possible.

To illustrate the approach, consider the hybrid scalar plant in (3) with sets given by (9). Let a static state-feedback controller assign $u_1 = 1$ and $u_2 = r(= w)$, which define the first two components of the vector-valued function κ_c. Then, the plant $\mathcal{H}_p$ under the effect of such a controller is given by

$$\begin{array}{lll} \dot{\xi} & = & 1 \qquad\qquad 0 \le r \le 1 \\ \xi^+ & = & u_3 \qquad\qquad r = 1. \end{array}$$

Trivially, the jumps of the plant and of the exosystem (4) occur simultaneously. Moreover, flows from points in

$$\overline{C}_e \cap D_e \ = \ \{w \ : \ w = 1\}$$

or

$$\widetilde{X}_p \ = \ \{x \ : \ w = 1\}$$

are not possible since the respective flow maps point outward the interior of these sets.

5.3.2 Preservation of reference trajectory

The interconnection between $\mathcal{H}_p$, $\mathcal{H}_e$, and $\mathcal{H}_c$ has to be such that a hybrid arc $w : \operatorname{dom} w \to \mathbb{R}^{s_e}$ is the component of a solution of the interconnection if and only if it is a solution to the hybrid exosystem $\mathcal{H}_e$ itself. This property will be guaranteed when the plant dynamics are such that, due to the action of the controller, the given reference trajectory is induced in the plant. It amounts to guarantee the following two conditions:

P1) For each initial condition

$$x(0, 0) = (\xi(0, 0), w(0, 0), \eta(0, 0))$$

and each solution

$$x = (\xi, w, \eta) \in \mathcal{S}_{\mathcal{H}_{cl}^\star}(x(0, 0))$$

we have that

$$\widetilde{w} \in \mathcal{S}_{\mathcal{H}_e}(w(0, 0)),$$

where $\widetilde{w}$ is the hybrid arc w without the jumps due only to the controller, that is, given the w-component of a solution x to $\mathcal{S}_{\mathcal{H}_{cl}^\star}$, $\widetilde{w}$ is constructed from w by removing the points $w(t', j')$ such that $x(t', j') \in D_2 \setminus D_1$;

P2) For each initial condition $w(0, 0)$ and each solution

$$w \in \mathcal{S}_{\mathcal{H}_e}(w(0, 0))$$

we have that, for each $\xi(0, 0)$ and $\eta(0, 0)$,

$$x = (\xi, w', \eta) \in \mathcal{S}_{\mathcal{H}_{cl}^\star}(x(0, 0)),$$

where $x(0, 0) = (\xi(0, 0), w(0, 0), \eta(0, 0))$ and w' is equal to w when the jumps due only to the controller are removed.

Conditions P1) and P2) establish an equivalence between the solutions to $\mathcal{H}_{cl}^\star$ and $\mathcal{H}_e$ as they imply that the w-components of the solutions to the closed-loop system $\mathcal{H}_{cl}^\star$ are solutions to the exosystem $\mathcal{H}_e$, and that every solution to the exosystem is the w-component of a solution to the closed-loop system (after appropriate matching of the jumps potentially added by the controller).

5.3.3 Asymptotic Stability of Closed-loop System

The hybrid controller $\mathcal{H}_c$ will be designed to guarantee that the closed set $\mathcal{A}$ is uniformly globally asymptotically stable. Following Theorem 3.5, this property can be asserted with a function $V : \mathbb{R}^{n_p} \times \mathbb{R}^{n_e} \times \mathbb{R}^{n_c} \to \mathbb{R}$ that is continuously differentiable on an open set containing $\overline{C}$, functions $\alpha_1, \alpha_2 \in \mathcal{K}_\infty$, and a continuous positive definite function ρ such that

$$\alpha_1(|x|_{\mathcal{A}}) \le V(x) \le \alpha_2(|x|_{\mathcal{A}}) \\ \forall x \in C \cup D \cup G(D); \qquad (25)$$

$$\langle \nabla V(x), f(x) \rangle \le -\rho(|x|_{\mathcal{A}}) \\ \forall x \in C; \qquad (26)$$

$$V(g) - V(x) \le -\rho(|x|_{\mathcal{A}}) \\ \forall x \in D_1 \setminus D_2, \ g \in G_1(x); \qquad (27)$$

$$V(g) - V(x) \le -\rho(|x|_{\mathcal{A}}) \\ \forall x \in D_2 \setminus D_1, \ g \in G_2(x); \qquad (28)$$

$$V(g) - V(x) \le -\rho(|x|_{\mathcal{A}}) \\ \forall x \in D_1 \cap D_2, \ g \in \{G_1(x), G_2(x)\}. \qquad (29)$$

While the conditions above could have been expressed in terms of the tracking error χ, it is rarely the case that its dynamics can be written as a function of χ and η only – see Section 2. The data of the hybrid controller has to be chosen so that (25)-(29) hold. In particular, condition (26) depends on f_c, C_c and κ_c; (27) depends on κ_c; and (28) depends on G_c and D_c.

The conditions above imply that solutions to the closed-loop system are such that $|x(t, j)|_{\mathcal{A}} \to 0$, that is, $|\xi(t, j) - h_e(w(t, j))| \to 0$. This includes all possible solutions generated by $\mathcal{H}_e$. Furthermore, when combined with the conditions in Section 5.3.2, it implies that $\xi(t, j) = h_e(w(t, j))$ on the domain of definition of solutions starting from $\xi(0,0) = h_e(w(0,0))$, that is, the controller induces the reference trajectory as a solution to the plant.

The following theorem summarizes the result outlined above.

THEOREM 5.2. *Let Assumption 5.1 hold and let the solutions ϕ_e to $\mathcal{H}_e$ generate the reference trajectories to track. If there exists a hybrid controller $\mathcal{H}_c$ guaranteeing that the jumps of $\mathcal{H}_e$ and $\mathcal{H}_p$ occur simultaneously, and conditions P1) and P2) hold, and if there exist a Lyapunov function candidate $V : \mathbb{R}^{n_p} \times \mathbb{R}^{n_e} \times \mathbb{R}^{n_c} \to \mathbb{R}$ for $\mathcal{H}_{cl}^\star$ with respect to*

$$\mathcal{A} = \left\{ x \in \overline{C} \cup D \ : \ \xi = h_e(w) \right\},$$

functions $\alpha_1, \alpha_2 \in \mathcal{K}_\infty$, and a positive definite and continuous function ρ such that (25)-(29) hold, then $\mathcal{A}$ is uniformly globally pre-asymptotically stable. Moreover, if the closed-loop system is such that every maximal solution is complete then $\mathcal{H}_c$ provides a solution to Problem $(\star)$.

REMARK 5.3. *The stability conditions in Theorem 5.2 can be relaxed according to items A) and B) of Corollary 3.6.*

6. EXAMPLES

Below we exercise the proposed techniques in academic examples. For a first order hybrid plant, we consider the problem of tracking a sawtooth signal and a square wave signal, while, for a double-integrator system with impulsive input, we consider the problem of tracking a triangular signal for the "position" state, which leads to tracking a square wave for the "velocity" state. Due to the reference trajectories having jumps, tools for tracking control design in the literature are not applicable.

• **First-order hybrid plant:** Consider the system

$$\begin{aligned} \dot{\xi} &= a\xi + b + u_1 & (\xi, u) \in C_p \\ \xi^+ &= c + u_3 & (\xi, u) \in D_p \end{aligned} \tag{30}$$

defining the hybrid plant to control, where $a, b, c \in \mathbb{R}, \xi \in \mathbb{R}$, $u = [u_1 \ u_2 \ u_3]^\top$,

$$C_p = \left\{ (\xi, u) \ : \ 0 \le u_2 \le 1 \right\}, \ D_p = \left\{ (\xi, u) \ : \ u_2 = 1 \right\}$$

and the static controller $\mathcal{H}_c$ given by

$$u = \kappa_c(\xi, r) = \begin{bmatrix} \lambda_1 + \lambda_2\xi + \lambda_3 r \\ r \\ \lambda_4 + \lambda_5\xi + \lambda_6 r \end{bmatrix}. \tag{31}$$

To track sawtooth reference trajectories generated by the exosystem (4), let $\lambda_1 = 1 - b$, $\lambda_2 < -a$, $\lambda_3 = -a - \lambda_2$, $\lambda_4 = -c$, $|\lambda_5| \in (0, 1)$, and $\lambda_6 = -\lambda_5$. This static control law forces the jumps of the plant when the reference jumps. It results in a closed-loop system with state $x := (\xi, w)$ and dynamics

$$\left. \begin{aligned} \dot{\xi} &= (a + \lambda_2)\xi + b + \lambda_1 + \lambda_3 w \\ \dot{w} &= 1 \end{aligned} \right\} \ 0 \le w \le 1 \tag{32}$$

$$\left. \begin{aligned} \xi^+ &= c + \lambda_4 + \lambda_5\xi + \lambda_6 w \\ w^+ &= 0 \end{aligned} \right\} \ w = 1. \tag{33}$$

Assumption 5.1 holds. In fact, note that the jump map G_e evaluated at D_e takes the state back to points in C_e from

where flow is possible until D_e is reached, which implies completeness of maximal solutions to $\mathcal{H}_e$. Furthermore, the solutions to the closed-loop system (32),(33) are bounded. Conditions P1) and P2) hold by inspection.

To show that the set

$$\mathcal{A} = \left\{ (\xi, w) \in \overline{C} \cup D \ : \ \xi = w \right\}$$

is uniformly globally asymptotically stable, take

$$V(x) = \frac{1}{2}(\xi - w)^2,$$

which satisfies $V(\mathcal{A}) = 0$, $V((\overline{C} \cup D) \setminus \mathcal{A}) > 0$. Note that $\mathcal{H}_{cl}^\star$ has $C = \mathbb{R} \times [0, 1]$, f given by the right-hand side of (32), $D = \mathbb{R} \times \{1\}$, and g given by the right-hand side of (33). Then, we have

$$\langle \nabla V(x), f(x) \rangle = -2\lambda_3 V(x) \qquad \forall x \in C \tag{34}$$

and

$$V(g(x)) - V(x) = -(1 - \lambda_5^2)V(x) \qquad \forall x \in D. \tag{35}$$

Since $\lambda_3 > 0$ and $|\lambda_5| \in (0, 1)$, and the properties above, Theorem 5.2 implies that $\mathcal{A}$ is uniformly globally asymptotically stable. Figure 4 depicts a simulation of the closed-loop system with the proposed controller. Figure 4(a) shows a plant trajectory converging to the reference asymptotically. The tracking error decreases both during flows and jumps as Figure 4(b) indicates.

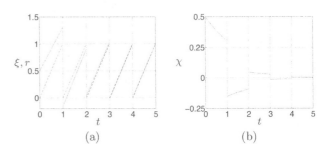

(a)　　　　　　　　(b)

Figure 4: Plant's state ξ (solid) and reference trajectory r (dashed) to the closed-loop system (32)-(33). Initial conditions: $\xi(0,0) = 0.5$, $w(0,0) = 0$. Parameters: $a = -1$, $b = \frac{1}{2}$, $c = 1$, $\lambda_1 = \lambda_2 = \lambda_3 = \lambda_5 = \frac{1}{2}$, $\lambda_4 = -1$, and $\lambda_6 = -\frac{1}{2}$. In (a), the state of the plant is shown in solid lines and the reference trajectory is shown in dotted lines. The tracking error converges to zero asymptotically as (b) shows.

Now, for the same hybrid plant, consider the problem of tracking square reference trajectories generated by the exosystem

$$\begin{aligned} \left. \begin{aligned} \dot{w}_1 &= 0 \\ \dot{w}_2 &= 1 \end{aligned} \right\} & \ 0 \le w_2 \le T \\ \left. \begin{aligned} w_1^+ &= -w_1 \\ w_2^+ &= 0 \end{aligned} \right\} & \ w_2 = T \\ r &= w_1, \end{aligned}$$

where $w_2 \in [0, T]$ is a timer used to change the sign of the discrete state $w_1 \in \{-1, 1\}$. The timer w_2 is reset when reaches the parameter $T > 0$, which denotes the semi-period of the square wave reference signal.

We consider the static controller in (31) with parameters $\lambda_1 = -b$, $\lambda_2 < -a$, $\lambda_3 = -a - \lambda_2$, $\lambda_4 = -c$, $|\lambda_5| \in (0,1)$, and $\lambda_6 = -\lambda_5 - 1$. This static control law forces the jumps of the plant when the reference jumps, and results in a closed-loop system with state $x := (\xi, w)$ and dynamics given by

$$
\left.\begin{array}{rcl}
\dot{\xi} &=& (a + \lambda_2)\xi + b + \lambda_1 + \lambda_3 w_1 \\
\dot{w}_1 &=& 0, \quad \dot{w}_2 = 1
\end{array}\right\} \quad 0 \le w_2 \le T \quad (36)
$$

$$
\left.\begin{array}{rcl}
\xi^+ &=& c + \lambda_4 + \lambda_5 \xi + \lambda_6 w_1 \\
w_1^+ &=& -w_1, \quad w_2^+ = 0
\end{array}\right\} \quad w_2 = T. \quad (37)
$$

As above, it can be shown that Assumption 5.1, conditions P1)-P2) hold, and that every maximal solution to the closed-loop system is bounded and complete.

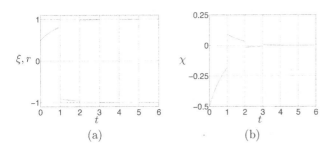

(a)   (b)

Figure 5: Plant's state ξ (solid) and reference trajectory r (dashed) to the closed-loop system (36)-(37). Initial conditions: $\xi(0,0) = 0.5$, $w(0,0) = (1,0)$. Parameters: $T = 1$, $a = 1$, $b = 1$, $c = 1$, $\lambda_1 = -1$, $\lambda_2 = -2$, $\lambda_3 = 1$, $\lambda_4 = -1$, $\lambda_5 = -\frac{1}{2}$ and $\lambda_6 = -\frac{1}{2}$. In (a), the state of the plant is shown in solid lines and the reference trajectory is shown in dotted lines. The tracking error converges to zero asymptotically as (b) shows.

To prove that the set

$$
\mathcal{A} = \left\{ (\xi, w) \in \overline{C} \cup D \; : \; \xi = w_1 \right\}
$$

is uniformly globally asymptotically stable, consider the Lyapunov function

$$
V(x) = \frac{1}{2}(\xi - w_1)^2,
$$

which satisfies the conditions $V(\mathcal{A}) = 0$, $V((\overline{C} \cup D) \setminus \mathcal{A}) > 0$. Then, (34) and (35) hold with this Lyapunov function and the resulting data of the closed-loop system. Then, Theorem 5.2 implies that $\mathcal{A}$ is uniformly globally asymptotically stable. Figure 5 depicts a simulation of the closed-loop system with the proposed controller showing convergence of the plant trajectory to the reference.

- **Second-order impulsive plant:** Consider the plant

$$
\dot{\xi}_1 = \xi_2, \qquad \dot{\xi}_2 = a\xi_1 + b\xi_2 + u_1 \qquad (38)
$$

where $a, b \in \mathbb{R}$ and the input $u \in \mathbb{R}$ has a Lebesgue integrable part u_1 and an impulsive part u_2. This plant can be modeled as a hybrid system with flows given by (38) and with jumps governed by

$$
\xi_1^+ = \xi_1, \qquad \xi_2^+ = \xi_2 + u_2
$$

at every time instant that the input u has an impulse. Suppose that the goal is to have $\xi = [\xi_1 \; \xi_2]^\top$ track the reference

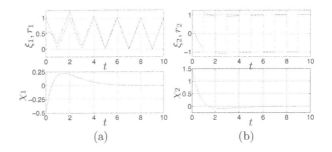

(a)   (b)

Figure 6: Plant's state ξ (solid) and reference trajectory r (dashed) to the closed-loop system (39)-(40). Initial conditions: $\xi(0,0) = (0.5, 0.5)$, $w(0,0) = (1,-1,0)$. Parameters: $T = 1$, $a = 1$, $b = 1$, $\lambda_1 = -1$, $\lambda_2 = -2$. The top graphs show the state of the plant in solid lines and the reference trajectory in dotted lines. The bottom graphs show the tracking error converging to zero asymptotically.

signal $r = [w_1 \; w_2]^\top$ generated by a hybrid exosystem

$$
\left.\begin{array}{rcl}
\dot{w}_1 &=& w_2, \quad \dot{w}_2 = 0, \\
\dot{w}_3 &=& 1
\end{array}\right\} \quad w \in C_e := \{ w \; : \; 0 \le w_3 \le T \}
$$

$$
\left.\begin{array}{rcl}
w_1^+ &=& w_1, \quad w_2^+ = -w_2 \\
w_3^+ &=& 0
\end{array}\right\} \quad w \in D_e := \{ w \; : \; w_3 = T \},
$$

which defines a triangular wave for w_1 and a square wave for w_2 with semi-period $T > 0$. Consider the case when the jumps of the plant are triggered by the jumps of the exosystem and the controller is given by

$$
u = \kappa_c(\xi, r) = \begin{bmatrix} -a\xi_1 - b\xi_2 + \lambda_1(\xi_1 - r_1) + \lambda_2(\xi_2 - r_2) \\ -2r_2 \end{bmatrix},
$$

where $\lambda_1, \lambda_2 \in \mathbb{R}$. To stabilize the set

$$
\mathcal{A} = \left\{ (\xi, w) \in \overline{C} \cup D \; : \; \xi = [w_1 \; w_2]^\top \right\},
$$

pick λ_1 and λ_2 so that $\begin{bmatrix} 0 & 1 \\ \lambda_1 & \lambda_2 \end{bmatrix}$ is Hurwitz. It is straightforward to show that for the resulting closed-loop system, which is given by

$$
\left.\begin{array}{rcl}
\dot{\xi}_1 &=& \xi_2 \\
\dot{\xi}_2 &=& \lambda_1(\xi_1 - w_1) + \lambda_2(\xi_2 - w_2) \\
\dot{w}_1 &=& w_2, \quad \dot{w}_2 = 0, \quad \dot{w}_3 = 1
\end{array}\right\} \quad 0 \le w_3 \le T \quad (39)
$$

$$
\left.\begin{array}{rcl}
\xi_1^+ &=& \xi_1, \quad \xi_1^+ = \xi_2 - 2w_2 \\
w_1^+ &=& w_1, \quad w_2^+ = -w_2, \quad w_3^+ = 0
\end{array}\right\} \quad w_3 = T, \quad (40)
$$

we have that for

$$
V(x) = \frac{1}{2}\left(\xi - \begin{bmatrix} w_1 \\ w_2 \end{bmatrix}\right)^\top P \left(\xi - \begin{bmatrix} w_1 \\ w_2 \end{bmatrix}\right),
$$

$P = P^\top > 0$, there exists $Q > 0$ such that, for all $x \in C$,

$$
\langle \nabla V(x), f(x) \rangle = -\left(\xi - \begin{bmatrix} w_1 \\ w_2 \end{bmatrix}\right)^\top Q \left(\xi - \begin{bmatrix} w_1 \\ w_2 \end{bmatrix}\right) < 0
$$

and $V(g(x)) - V(x) = 0$ for all $x \in D$. Then, using Theorem 5.2 and Corollary 3.6, the set $\mathcal{A}$ is uniformly globally asymptotically stable. Plant trajectories showing that asymptotic tracking is achieved are shown in Figure 6.

7. FINAL REMARKS

We stated a general tracking control problem for impulsive reference signals. For the full information case, sufficient conditions useful for the design of tracking controllers were proposed. These rely on an approach consisting of generating the reference trajectories via an exosystem and designing a control algorithm that guarantees that jumps of the reference system and plant match. The sufficient conditions for asymptotic stability were obtained from Lyapunov stability theorems for time-invariant hybrid systems in [14].

The results in Section 5 are applicable to hybrid plants and exosystems for which a (potentially hybrid) tracking controller inducing simultaneous jumps can be designed. The academic examples in Section 6 suggest that this is possible when the control input enters through the flow and jump set. The challenge in such cases is essentially to stabilize the resulting error system, for which Lyapunov functions that are function of the error seem suitable.

A more explicit (and tighter) set of conditions to those in Section 5.3.1 can be derived. This is possible by writing (24) as a set condition and by computing the sets of points from where, under the effect of the controller, flows of $\mathcal{H}_p$ and $\mathcal{H}_e$ are possible, and then insisting on these to be disjoint from the respective jump sets. These conditions will involve the given data of $\mathcal{H}_p$ and $\mathcal{H}_e$, and of $\mathcal{H}_c$, which is to be determined, in particular, the tangent cones of the flow sets.

The obtained results are an initial step in solving the tracking control problem for general hybrid systems. The authors hope that the difficulties to tracking control design for hybrid systems pointed out in this paper will spark interest in the hybrid systems community and lead to general design methods.

8. ACKNOWLEDGMENTS

The authors would like to thank J. J. B. Biemond, N. van de Wouw, and W. P. M. H. Heemels for insightful discussions on the tracking problem and the anonymous reviewers for helpful comments on the results.

9. REFERENCES

[1] L. Menini and A. Tornambe, "Asymptotic tracking of periodic trajectories for a simple mechanical system subject to nonsmooth impacts," *IEEE Transactions on Automatic Control*, vol. 46, pp. 1122–1126, 2001.

[2] F. Martinelli, L. Menini, and A. Tornambe, "State estimation for a class of linear mechanical systems that become observable thanks to non-smooth impacts," in *Proceedings of the 41st IEEE Conference on Decision and Control*, vol. 4, 2002, pp. 3608–3613.

[3] S. Galeani, L. Menini, and A. Tornambe, "A local observer for linearly observable nonlinear mechanical systems subject to impacts," in *Proceedings of the American Control Conference*, vol. 6, 2003, pp. 4760–4765.

[4] R. I. Leine and N. van de Wouw, "Uniform convergence of monotone measure differential inclusions: with application to the control of mechanical systems with unilateral constraints," *International Journal of Bifurcation and Chaos*, vol. 18, no. 5, pp. 1435–1457, 2008.

[5] N. van de Wouw and R. Leine, "Tracking control for a class of measure differential inclusions," in *Proc. 47th IEEE Conference on Decision and Control*, 2008, pp. 2526–2532.

[6] J. M. Bourgeot and B. Brogliato, "Tracking control of complementarity Lagrangian systems," *Int. J. Bifurcation and Chaos*, vol. 15, no. 6, pp. 1839–1866, 2005.

[7] B. Brogliato, S. Niculescu, and P. Orhant, "On the control of finite-dimensional mechanical systems with unilateral constraints," *IEEE Transactions on Automatic Control*, vol. 42, no. 2, pp. 200–215, 1997.

[8] B. Brogliato, S. Niculescu, and M. Monteiro-Marques, "On tracking control of a class of complementary-slackness hybrid mechanical systems," *Systems and Control Letters*, vol. 39, no. 4, pp. 255–266, 2000.

[9] R. Sanfelice, A. R. Teel, and R. Sepulchre, "A hybrid systems approach to trajectory tracking control for juggling systems," in *Proc. 46th IEEE Conference on Decision and Control*, 2007, pp. 5282–5287.

[10] R. Sanfelice, R. Goebel, and A. Teel, "Generalized solutions to hybrid dynamical systems," *ESAIM: Control, Optimisation and Calculus of Variations*, vol. 14, no. 4, pp. 699–724, 2008.

[11] R. Goebel, R. Sanfelice, and A. Teel, "Hybrid dynamical systems," *IEEE Control Systems Magazine*, pp. 28–93, 2009.

[12] J. Lygeros, K. Johansson, S. Simić, J. Zhang, and S. S. Sastry, "Dynamical properties of hybrid automata," *IEEE Transactions on Automatic Control*, vol. 48, no. 1, pp. 2–17, 2003.

[13] P. Collins, "A trajectory-space approach to hybrid systems," in *Proceedings of the 16th International Symposium on Mathematical Theory of Network and Systems*, 2004.

[14] R. Goebel, R. G. Sanfelice, and A. R. Teel, *Hybrid Dynamical Systems: Modeling, Stability, and Robustness*. Accepted for Publication in Princeton University Press, 2010.

[15] R. Sanfelice, R. Goebel, and A. Teel, "Invariance principles for hybrid systems with connections to detectability and asymptotic stability," *IEEE Transactions on Automatic Control*, vol. 52, no. 12, pp. 2282–2297, 2007.

[16] R. G. Sanfelice and A. R. Teel, "Asymptotic stability in hybrid systems via nested Matrosov functions," *IEEE Transactions on Automatic Control*, vol. 54, no. 7, pp. 1569–1574, 2009.

[17] A. Lyapunov, "The general problem of the stability of motion," Math. Society of Kharkov, English translation: International Journal of Control, vol. 55, 1992, 531-773.

[18] H. Khalil, *Nonlinear Systems*, 3rd ed. Prentice Hall, 2002.

[19] R. Goebel and A. Teel, "Solutions to hybrid inclusions via set and graphical convergence with stability theory applications," *Automatica*, vol. 42, no. 4, pp. 573–587, 2006.

Variational Formulation and Optimal Control of Hybrid Lagrangian Systems

Kathrin Flaßkamp
University of Paderborn
Department of Mathematics
Warburger Str. 100
33098 Paderborn, Germany
kathrinf@math.uni-paderborn.de

Sina Ober-Blöbaum
University of Paderborn
Department of Mathematics
Warburger Str. 100
33098 Paderborn, Germany
sinaob@math.uni-paderborn.de

ABSTRACT

In this contribution, we introduce an optimal control problem for hybrid Lagrangian control systems. The dynamics of these systems and their discrete approximations are derived via a hybrid variational principle which is based on the Lagrange-d'Alembert principle for continuous Lagrangian control systems. Optimal control problems are stated in a continuous and in a numerically treatable, i.e. discrete version. For the implementation, we present a two layer approach which decouples the continuous parts of a hybrid trajectory. Additionally, this enables us to solve the optimal control problem also for multiple objectives. The approach is exemplified for a simple mechanical system.

Categories and Subject Descriptors

G.1.6 [**Numerical Analysis**]: Optimization—*constrained optimization, nonlinear programming*; I.6 [**Simulation and Modeling**]: Miscellaneous

General Terms

Algorithms

Keywords

Discrete mechanics, hybrid Lagrangian system, multiobjective optimization, optimal control, variational mechanics

1. INTRODUCTION

Extensive research has been made in the field of hybrid systems in the last decades, since it is well known that the behavior of modern technical systems can only be modeled appropriately by a combination of *continuous dynamics* and *discrete events* models. Structural changes in the environment (e.g. different road surfaces of a driving car) as well as changing coupling structures of dynamical systems (e.g. a switch from an open to a closed kinematic chain) require to switch between different physical and mathematical models on different domains in state space. However, not only physical influences, but also the neglect of physical influences, e.g. the use of reduced or simplified models on different domains lead to hybrid systems. An adequate way to model hybrid systems has been found to be *hybrid automata* (see e.g. [10, 16]).

A great interest lies in the *optimal control of hybrid systems* since this includes not only the computation of optimal control trajectories for the continuous parts but also an optimization of the discrete variables such as the switching sequence. It is possible to derive necessary optimality conditions for hybrid optimal control problems (see e.g. [2, 15, 17]). The numerical treatment of hybrid optimal control problems results in mixed-integer optimal control problems. In [3] (and the references therein) two solution techniques are presented, which are e.g. combinations of numerical direct collocation methods for continuous dynamic optimization and branch and bound techniques. The mode sequence optimization can be also performed by a relaxation as proposed in [4]. A number of other solution approaches are based on a problem splitting such that e.g. hierarchical optimization algorithms can be applied (see [8] and the references therein).

In this contribution, we focus on hybrid mechanical systems comprised by *Lagrangian systems* for which the dynamics can be described via a *variational principle*. In contrast to [1] we do not restrict ourselves to one single Lagrangian and a single domain but we consider a family of Lagrangian which lead to different dynamics in different parts, i.e. *domains* of the state space. While a number of recent publications combines optimization techniques for the mode sequences and switching times mainly based on indirect approaches (e.g. [2, 4, 6]), we focus on the combination of direct optimal control methods with switching time optimization.

The goal of this contribution is twofold: First of all, we derive a variational formulation of a hybrid mechanical system. To this end, we define a hybrid variational principle for which the influence of the discrete events is taken into account. Besides variations of the continuous variables also variations of the time instants when a discrete event occurs have to be considered. For the variational approach we use similar concepts as in [7] where a nonsmooth variational principle for the modeling of collisions of particles and solid bodies is proposed. A discrete version of the hybrid principle is used to derive a discrete (i.e. discrete-time) variational formulation as an approximation of the hybrid Lagrangian system and its flow. In this procedure, the hybrid optimal control problem is transformed into a constrained optimization problem as it is done for ordinary optimal control problems by the optimal control technique DMOC (*Discrete Mechanics and Optimal Control*, [13]).

Secondly, we present a two layer approach to numerically solve a hybrid multiobjective optimal control problem. Here, we restrict ourselves to the case of a fixed switching sequence. This approach is based on a decoupling of the continuous parts of a trajectory of the hybrid system, such that the coupling constraints, i.e. the switching times and states can be optimized in an upper layer. This resembles the approach presented in [18]. Similar implementations have also been used in [9] for the optimal control of formation flying satellites and in [14] for a bipedal robot, where the solution strategies are based on DMOC. In this contribution, we also use DMOC for solving the optimal control problems of the continuous parts in the lower layer. Furthermore, instead of only one objective as mainly considered in the context of hybrid optimal control theory, we propose a method to optimize several conflicting objectives at the same time. To this aim, we combine our approach with multiobjective optimization techniques used in the upper layer. In this way, we are able to determine the set of optimal compromises, the so called *Pareto set*, of the multiobjective optimal control problem for hybrid systems. For an overview of standard multiobjective optimization methods see e.g. [12].

The outline of this contribution is as follows: In Section 2 we give a brief introduction into Lagrangian formulations and continuous and discrete variational principles since our approach is based on these concepts. In Sections 3 and 4 we define the hybrid Lagrangian control system and derive a hybrid variational principle describing its dynamics for both, the continuous and the discretized system. The corresponding continuous and discrete (multiobjective) optimal control problems are formulated in Section 5. In particular, we present a solution strategy based on a two layer formulation. In Section 6 the performance of the method is demonstrated by means of a simple mechanical example, the hybrid single-mass oscillator. A conclusion and an outlook for future work and directions are given in Section 7.

2. VARIATIONAL MECHANICS

In the following section we give a brief introduction into continuous and discrete variational mechanics. For more details we refer to [11, 13]. Let Q be an n-dimensional configuration manifold with tangent bundle TQ and cotangent bundle T^*Q. Consider a mechanical system with time-dependent configuration vector $q(t) \in Q$ and velocity vector $\dot{q}(t) \in T_{q(t)}Q$, $t \in [0, T]$, whose dynamical behavior is described by the Lagrangian $L : TQ \to \mathbb{R}$. Typically, the Lagrangian L consists of the difference of the kinetic and potential energy. In addition, there is a force $f : Q \times U \to T^*Q$ acting on the system. This force depends on a time dependent control parameter $u(t) \in U \subseteq \mathbb{R}^m$ that influences the system's motion. The equations of motion can be described via a variational principle. Define the *action map* $\mathfrak{S} : C^2([0, T], Q) \to \mathbb{R}$ as

$$\mathfrak{S}(q) = \int_0^T L(q(t), \dot{q}(t)) \, dt.$$

Then the *Lagrange-d'Alembert principle* seeks curves $q \in C^2([0, T], Q)$ with fixed initial $q(0)$ and fixed final value $q(T)$

satisfying

$$\delta \int_0^T L(q, \dot{q}) \, dt + \int_0^T f(q, \dot{q}, u) \cdot \delta q \, dt = 0 \qquad (1)$$

for all variations $\delta q \in T_q C^2([0, T], Q)$ and the second integral in (1) is the *virtual work* acting on the mechanical system via the force f. This yields

$$0 = \int_0^T \left(\frac{\partial L}{\partial q} \cdot \delta q + \frac{\partial L}{\partial \dot{q}} \cdot \delta \dot{q} \right) dt + \int_0^T f \cdot \delta q \, dt$$

$$= \int_0^T \left[\frac{\partial L}{\partial q} - \frac{d}{dt} \left(\frac{\partial L}{\partial \dot{q}} \right) + f \right] \cdot \delta q \, dt + \frac{\partial L}{\partial \dot{q}} \cdot \delta q \Big|_0^T$$

using integration by parts[1]. The expression $\frac{\partial L}{\partial \dot{q}}$ is the conjugate momentum denoted by p and given via the *Legendre transform* $\mathbb{F}L : TQ \to T^*Q$, $(q, \dot{q}) \mapsto (q, \frac{\partial L}{\partial \dot{q}}) = (q, p)$. Since the boundaries are fixed, it holds $\delta q(0) = \delta q(T) = 0$, and (1) is equivalent to the *forced Euler-Lagrange equations*

$$\frac{\partial L}{\partial q}(q, \dot{q}) - \frac{d}{dt} \left(\frac{\partial L}{\partial \dot{q}}(q, \dot{q}) \right) + f(q, \dot{q}, u) = 0.$$

For simplicity we introduce the notation

$$D_{EL} L = \frac{\partial L}{\partial q} - \frac{d}{dt} \left(\frac{\partial L}{\partial \dot{q}} \right).$$

The energy of the system is given as $E = p \cdot \dot{q} - L(q, \dot{q})$.

The goal of discrete variational mechanics is to derive discrete approximations of the solutions of the forced Euler-Lagrange equations that inherit the same qualitative behavior as the continuous solution. To this aim, a discrete variational principle is formulated and the above derivations can be performed in an analogous way. Consider two positions q_0 and q_1 as two points on a curve at time h apart so that $q_0 \approx q(0)$ and $q_1 = q(h)$. A discrete Lagrangian $L_d(q_0, q_1, h)$ approximates the action integral along the curve segment between q_0 and q_1 as $L_d(q_0, q_1, h) \approx \int_0^h L(q(t), \dot{q}(t)) \, dt$.[2] Considering a discrete curve of points $\{q_k\}_{k=0}^N$ the *discrete action* is given by the sum of the discrete Lagrangian on each adjacent pair $\mathfrak{S}_d = \sum_{k=0}^{N-1} L_d(q_k, q_{k+1})$. Similarly, the virtual work can be approximated via $\sum_{k=0}^{N-1} f_d^-(q_k, q_{k+1}, u_k) \cdot \delta q_k + f_d^+(q_k, q_{k+1}, u_k) \cdot \delta q_{k+1}$ with the left and right discrete forces $f_d^\pm(q_k, q_{k+1}, u_k) := f_k^\pm$, where u_k denotes the control parameter guiding the system from q_k to q_{k+1}. The *discrete Lagrange-d'Alembert principle* seeks discrete curves of points $\{q_k\}_{k=0}^N$ satisfying

$$\delta \mathfrak{S}_d + \sum_{k=0}^{N-1} f_k^- \cdot \delta q_k + f_k^+ \cdot \delta q_{k+1} = 0 \qquad (2)$$

for all variations δq_k vanishing at the endpoints. This gives after rearranging the summation

$$0 = \sum_{k=1}^{N-1} \left(D_1 L_d(q_k, q_{k+1}) + D_2 L_d(q_{k-1}, q_k) + f_k^- + f_{k-1}^+ \right) \cdot \delta q_k$$

$$+ \left(D_1 L_d(q_0, q_1) + f_0^- \right) \cdot \delta q_0 + \left(D_2 L_d(q_{N-1}, q_N) + f_{N-1}^+ \right) \cdot \delta q_N,$$

where D_i denotes the derivative w.r.t. the i-th argument. Analogous to the continuous case, the boundary terms can

---

[1] For a function $g : [0, T] \to \mathbb{R}^n$ we use the notation $g|_0^T = g(T) - g(0)$.

[2] We will neglect the h dependence of L_d except where it is important.

be rewritten as $p_0^- \cdot \delta q_0 + p_N^- \cdot \delta q_N$ with the discrete momenta p_k^- and p_{k+1}^+ given by the *forced discrete Legendre transforms* $\mathbb{F}^{\pm} L_d : Q \times Q \to T^*Q$ (cf. [11]) as

$$(q_k, q_{k+1}) \mapsto (q_k, p_k^-) = -D_1 L_d(q_k, q_{k+1}) - f_k^-$$

$$(q_k, q_{k+1}) \mapsto (q_{k+1}, p_{k+1}^+) = D_2 L_d(q_k, q_{k+1}) + f_k^+.$$

With $\delta q_0 = \delta q_N = 0$, (2) is equivalent to the *discrete forced Euler-Lagrange equations*

$$D_1 L_d(q_k, q_{k+1}) + D_2 L_d(q_{k-1}, q_k) + f_k^- + f_{k-1}^+ = 0 \quad (3)$$

for each $k = 1, \dots, N-1$. The discrete energy of the discrete system is given as $E_d(q_k, q_{k+1}) = -D_3 L_d(q_k, q_{k+1}, h)$.

The use of discrete variational mechanics plays an important role for the numerical treatment of Lagrangian systems. The equations (3) provide a time stepping scheme for the simulation of the mechanical system which is called a *variational integrator* (cf. [11]). Since these integrators, derived in a variational way, are structure-preserving, important properties of the continuous system are preserved (or change consistently with the applied forces), such as symplecticity or momentum maps induced by symmetries. In addition, they have an excellent long-time energy behavior. The approximation order depends on the quadrature rule used to approximate the relevant integrals (e.g. second order using a midpoint rule approximation).

3. THE HYBRID LAGRANGIAN CONTROL SYSTEM

In this section, we introduce the notation of a hybrid Lagrangian control system. The equations of motion and the corresponding flow of the hybrid system are described by means of a hybrid variational principle.

Definition 1. A *hybrid Lagrangian control system* is a 9-tuple $\mathcal{H} = (Q, \Gamma, \mathcal{E}, \mathcal{D}, \mathcal{L}, \mathcal{G}, \mathcal{R}, U, \mathcal{F})$, where

- Q is an n-dimensional configuration manifold with associated state space (tangent bundle) given by TQ and cotangent bundle T^*Q,

- Γ is a (countable) set of discrete states,

- $\mathcal{E} \subset \Gamma \times \Gamma$ is a collection of discrete transitions,

- $\mathcal{D} = \{D_i\}_{i \in \Gamma}$ is a collection of domains where $D_i = \{i\} \times TV_i \subset \{i\} \times TQ$ with $V_i \subset Q$ for all $i \in \Gamma$,

- $\mathcal{L} = \{L_i : D_i \to \mathbb{R}\}_{i \in \Gamma}$ is a family of hyperregular Lagrangian,

- $\mathcal{G} = \{G_e\}_{e \in \mathcal{E}}$ is a collection of guards with $G_e \subset D_i$ for $e = (i, j) \in \mathcal{E}$,

- $\mathcal{R} = \{R_e\}_{e \in \mathcal{E}}$ is a collection of reset maps $R_{(i,j)} : G_{(i,j)} \subset D_i \to D_j$,

- $U \subseteq \mathbb{R}^m$ is the set of admissible controls, and

- $\mathcal{F} = \{f_i : D_i \times U \to T^*V_i\}_{i \in \Gamma}$ is a collection of Lagrangian forces.

Definition 1 indicates that for a given $\mathcal{H}$ the motion of a mechanical system in the domain D_i is determined by the Lagrangian L_i and the Lagrangian force f_i. If a guard $G_{(i,j)}$ is reached, a switch (via the reset map $R_{(i,j)}$) between the

according domains and corresponding Lagrangian descriptions occurs assuming determinism. Whenever it is clear from the context, we will drop the discrete part of the domains, e.g. $L_i : D_i \subset TQ \to \mathbb{R}$. Note that Definition 1 does not include discrete controls, which can be added for a more general setting.

For the hybrid control system $\mathcal{H}$ we introduce an appropriate concept of time that includes a description of the time instants of the discrete transitions between different domains. To this aim, we define the hybrid interval as follows.

Definition 2. Let $\Lambda = \{0, 1, \dots, N\} \subset \mathbb{N}$ be an indexing set. A *hybrid interval* $\mathcal{I} = \{I_\lambda\}_{\lambda \in \Lambda}$ is a sequence of intervals such that $I_\lambda = [\tau_\lambda, \tau_{\lambda+1}]$ for all $\lambda = 0, \dots, N-1$, and $I_N = [\tau_N, \tau_{N+1}]$ or $I_N = [\tau_N, \tau_{N+1})$. Here, $\tau_\lambda \in \mathbb{R}$, $\lambda = 0, \dots, N+1$ and $\tau_\lambda \leq \tau_{\lambda+1}$.

In order to keep track of the actual discrete state and thus the according domain of the mechanical system, a switching sequence is introduced as follows.

Definition 3. For an indexing set Λ, a *switching sequence* is determined by a map $\gamma : \Lambda \to \Gamma$ governing the visited discrete states in a hybrid interval.

Based on γ the spaces of admissible state and control functions can be defined. For a fixed hybrid interval $\mathcal{I} = \{I_\lambda\}_{\lambda \in \Lambda}$ and a corresponding switching sequence γ, let

$$C_\gamma = \{c = \{c_\lambda\}_{\lambda \in \Lambda} \mid c_\lambda \in C^1(I_\lambda, D_{\gamma(\lambda)}), \ c_\lambda(\tau_{\lambda+1}) \in$$
$$G_{\gamma(\lambda), \gamma(\lambda+1)}, \ R_{\gamma(\lambda-1), \gamma(\lambda)}(c_{\lambda-1}(\tau_\lambda)) = c_\lambda(\tau_\lambda)\}$$

with $c_0(\tau_0)$ and $c_N(\tau_{N+1})$ arbitrary for $\lambda = 0$ and $\lambda = N$, respectively, denote the collection of all piecewise continuously differentiable functions on $\mathcal{I}$ consistent with the sequence of discrete states. Then, the *space of all admissible state functions* $\mathcal{C} = \bigcup_{\gamma \in \Omega} C_\gamma$ is a union of function spaces C_γ with Ω being the set of all switching sequences compatible with the graph structure of the underlying transition graph that is given by $(\Lambda, \mathcal{E})$, i.e. $\Omega = \{\gamma \mid (\gamma(\lambda), \gamma(\lambda+1)) \in \mathcal{E} \ \forall \lambda \in \{0, 1, \dots, N\}\}$. We use the notation $c_\lambda = (q_\lambda, \dot{q}_\lambda)$, where $q_\lambda = \pi_Q \circ c_\lambda$ is a C^2-function on I_λ whose first time derivative $\dot{q}_\lambda$ is a C^1-function on I_λ, where $\pi_Q : TQ \to Q$ is the canonical projection from the tangent bundle TQ to the configuration space Q. Thus, we can define the *space of all admissible curves* $\mathcal{Q} = \bigcup_{\gamma \in \Omega} \{q = \{q_\lambda\}_{\lambda \in \Lambda} \mid q_\lambda \in C^2(I_\lambda, V_{\gamma(\lambda)}), \ q_\lambda = \pi_Q \circ c_\lambda, c \in \mathcal{C}\}$ as a projection of $\mathcal{C}$ with $T_q\mathcal{Q}$ being its tangent space. Furthermore, let $\mathcal{U}_\gamma = \{u = \{u_\lambda\}_{\lambda \in \Lambda} \mid u_\lambda \in C(I_\lambda, U)\}$ be the space of piecewise continuous functions (continuous on $(\tau_\lambda, \tau_{\lambda+1})$), where $u(t)$ is the *hybrid control function* consistent with the switching sequence γ. We denote the *space of all admissible hybrid control functions* by $\mathcal{U} = \bigcup_{\gamma \in \Omega} \mathcal{U}_\gamma$ being the space of control functions according to the set of all compatible switching sequences γ.

3.1 Hybrid Variational Formulation

The goal of this section is to derive a hybrid version of the Lagrange-d'Alembert principle as introduced in Section 2. Hamilton's principle states that for any variation of a curve q, the action is stationary. The extension for forced systems, the Lagrange-d'Alembert principle, states that for any variation of a curve q the change in the action can be expressed by means of the virtual work due to the forces acting on the system. To provide a hybrid version of the Lagrange-d'Alembert principle, further extensions have to be taken

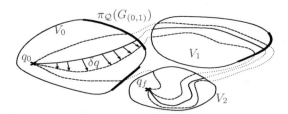

Figure 1: Variations of a curve $q \in \mathcal{Q}$ may lead to different switching sequences γ.

into account: 1. We consider all curves that connect the initial and final configuration q_0 and q_f, where the switching sequence γ may change for different curves (see Figure 1). 2. A variation δq in q may also lead to a variation $\delta \tau_\lambda$ in the switching times τ_λ, $\lambda = 1, \ldots, N$ that has to be taken into account. 3. Due to the reset maps and different Lagrangian descriptions, variations $\delta q(\tau_\lambda)$ and $\delta \tau_\lambda$, $\lambda = 1, \ldots, N$, in the switching points lead to an artificial momentum and energy change in the overall hybrid system that influences the change of the action for varied curves.

To realize 1., we consider the switching sequence γ as a function $\bar{\gamma}$ of a curve $c \in C^1(I_\lambda, TQ)$ in the following way: $\bar{\gamma} : C^1(I_\lambda, TQ) \rightarrow \Gamma$ with $\bar{\gamma}(c_\lambda) = \gamma(\lambda)$, and for $c \in \mathcal{C}$ $\bar{\gamma}(c) := \{\bar{\gamma}(c_\lambda)\}_{\lambda \in \Lambda} = \{\gamma(\lambda)\}_{\lambda \in \Lambda}$, i.e. a given curve c determines a switching sequence γ and thus a variation in the curve q may influence γ. To take the switching time variance (2.) into account, we consider separate variations $\delta q(t) \in T_{q(t)}\mathcal{Q}$ and $\delta \tau_\lambda \in \mathbb{R}$, $\lambda = 1, \ldots, N$, where for $t = \tau_\lambda$ the variation in configuration and time is denoted by $\bar{\delta}q(\tau_\lambda) := \delta q(\tau_\lambda) + \dot{q}(\tau_\lambda)\delta\tau_\lambda$. To describe the instantaneous change in momentum and energy (3.) for varied curves and varied switching times, we define the *impulsive momentum effect* Φ and the *impulsive energy effect* $\mathfrak{E}$ as

$$\Phi := \sum_{\lambda \in \Lambda \setminus \{N\}} \Delta_{(\lambda, \lambda+1)}, \quad \text{and} \quad \mathfrak{E} := \sum_{\lambda \in \Lambda \setminus \{N\}} \mathfrak{d}_{(\lambda, \lambda+1)}, \quad (4)$$

$$\Delta_{(\lambda, \lambda+1)} := p_{\lambda+1}(\tau_{\lambda+1}) \cdot \bar{\delta}q_{\lambda+1}(\tau_{\lambda+1}) - p_\lambda(\tau_{\lambda+1}) \cdot \bar{\delta}q_\lambda(\tau_{\lambda+1})$$
$$\mathfrak{d}_{(\lambda, \lambda+1)} := (E_{\gamma(\lambda+1)}(\tau_{\lambda+1}) - E_{\gamma(\lambda)}(\tau_{\lambda+1})) \cdot \delta\tau_{\lambda+1}$$

with p_λ the conjugate momentum given by the Legendre transform $\mathbb{F}L_{\gamma(\lambda)} : D_{\gamma(\lambda)} \subset TQ \rightarrow T^*Q$ and with the energy $E_{\gamma(\lambda)} = p_\lambda \cdot \dot{q}_\lambda - L_{\gamma(\lambda)}$.

In the next step, for the hybrid interval $\mathcal{I}$, we define the *hybrid action map* $\mathfrak{S} : \mathcal{Q} \rightarrow \mathbb{R}$ to be the sum of the action maps on each interval I_λ

$$\mathfrak{S}(q) = \sum_{\lambda \in \Lambda} \int_{I_\lambda} L_{\gamma(\lambda)}(q_\lambda(t), \dot{q}_\lambda(t)) \, dt, \quad (5)$$

where Λ is the index set of $\mathcal{I}$ and where $\gamma(\lambda)$ is assumed to be $\bar{\gamma}(c_\lambda)$, thus depent on the actual trajectory c_λ. Furthermore, the *hybrid virtual work* $W : \mathcal{Q} \times \mathcal{U} \rightarrow \mathbb{R}$ is given by the sum of the virtual work on each interval I_λ

$$W(q, u) = \sum_{\lambda \in \Lambda} \int_{I_\lambda} f_{\gamma(\lambda)}(q_\lambda(t), \dot{q}_\lambda(t), u(t)) \cdot \delta q_\lambda(t) \, dt \quad (6)$$

with variations $\{\delta q_\lambda\}_{\lambda \in \Lambda} \in T_q\mathcal{Q}$.

By means of the hybrid action map, the hybrid virtual work and the impulsive effects, the dynamics of the hybrid system $\mathcal{H}$ can be determined by a hybrid version of the Lagrange-d'Alembert principle.

Proposition 1. Let $\mathcal{H}$ be a hybrid Lagrangian control system and $\mathcal{I} = \{I_\lambda\}_{\lambda \in \Lambda}$ a hybrid interval with index set Λ. A curve $q \in \mathcal{Q}$ joining $q(\tau_0) = q_0$ and $q(\tau_{N+1}) = q_f$ satisfies the forced Euler-Lagrange equations on each interval I_λ if and only if the *hybrid Lagrange-d'Alembert principle* is fulfilled that is defined as follows: Seek for curves $q \in \mathcal{Q}$ such that for all variations $\delta q \in T_q\mathcal{Q}$ and $\delta \tau_\lambda \in \mathbb{R}$, $\lambda = 1, \ldots, N$ it holds

$$\delta \mathfrak{S}(q) + W(q, u) + \Phi = \mathfrak{E}. \quad (7)$$

PROOF. In the following we omit the argument t for $q_\lambda, \dot{q}_\lambda$ and u_λ in the integral formulation. Taking separate variations δq in configurations and $\{\delta\tau_\lambda\}_{\lambda \in \Lambda \setminus \{0\}}$ in switching times with $\delta q_0(\tau_0) = \delta q_N(\tau_{N+1}) = 0$, using integration by parts, and rearranging sums of the resulting boundary terms gives

$$\mathfrak{E} = \sum_{\lambda \in \Lambda} \int_{I_\lambda} \left(D_{EL} L_{\gamma(\lambda)} + f_{\gamma(\lambda)} \right) \delta q_\lambda \, dt$$
$$+ \sum_{\lambda \in \Lambda \setminus \{N\}} \left(\frac{\partial L_{\gamma(\lambda)}}{\partial \dot{q}_\lambda} \cdot \delta q_\lambda - \frac{\partial L_{\gamma(\lambda+1)}}{\partial \dot{q}_{\lambda+1}} \cdot \delta q_{\lambda+1} \right) \bigg|_{\tau_{\lambda+1}}$$
$$+ \sum_{\lambda \in \Lambda \setminus \{N\}} \left(L_{\gamma(\lambda)}(\tau_{\lambda+1}) - L_{\gamma(\lambda+1)}(\tau_{\lambda+1}) \right) \cdot \delta\tau_{\lambda+1} + \Phi.$$

By substituting the relation between configuration and time variation $\bar{\delta}q_\lambda(\tau_{\lambda+1}) = \delta q_\lambda(\tau_{\lambda+1}) + \dot{q}_\lambda \delta\tau_{\lambda+1}$ this results in

$$\mathfrak{E} = \sum_{\lambda \in \Lambda} \int_{I_\lambda} \left(D_{EL} L_{\gamma(\lambda)} + f_{\gamma(\lambda)} \right) \delta q_\lambda \, dt$$
$$+ \sum_{\lambda \in \Lambda \setminus \{N\}} \left(\frac{\partial L_{\gamma(\lambda)}}{\partial \dot{q}_\lambda} \cdot \bar{\delta}q_\lambda - \frac{\partial L_{\gamma(\lambda+1)}}{\partial \dot{q}_{\lambda+1}} \cdot \bar{\delta}q_{\lambda+1} \right) \bigg|_{\tau_{\lambda+1}}$$
$$+ \sum_{\lambda \in \Lambda \setminus \{N\}} \left(E_{\gamma(\lambda+1)}(\tau_{\lambda+1}) - E_{\gamma(\lambda)}(\tau_{\lambda+1}) \right) \cdot \delta\tau_{\lambda+1} + \Phi$$

where the last two sums (the boundary terms arising from the integration by parts of the Lagrangian) together with the impulsive effects Φ and $\mathfrak{E}$ add up to zero. Thus the principle is fulfilled if and only if $D_{EL} L_{\gamma(\lambda)} + f_{\gamma(\lambda)} = 0$ on I_λ for all $\lambda \in \Lambda$, since the integrands are continuous and δq_λ is arbitrary on $(\tau_\lambda, \tau_{\lambda+1})$. $\square$

Note that we can not simply assume vanishing variations $\delta q_\lambda(\tau_{\lambda+1}) = \delta q_{\lambda+1}(\tau_{\lambda+1}) = 0$ at the switching points, since $q_\lambda(\tau_{\lambda+1})$ and $q_{\lambda+1}(\tau_{\lambda+1})$ are not assumed to be fixed. In [7] the variations $\bar{\delta}q(\tau_\lambda)$ are restricted to be in $T\partial C$, where ∂C is some contact surface. This restriction of the variation leads to extra conditions for the momenta and the energy at the switching time such that the state after the switch is uniquely determined. Rather than deriving the correct reset maps as in [7], for the hybrid Lagrangian system the reset maps are already given and uniquely determine the transformation of the states. In turn, the state transformation together with the Lagrangian uniquely determine the change in momentum and energy at the switching time. Including this change directly into the principle in terms of the impulsive effects and considering arbitrary (instead of restricted) variations $\bar{\delta}q(\tau_\lambda)$ and $\delta\tau_\lambda$, the hybrid principle gives the Euler-Lagrange equations without any additional conditions on the transformation at the switching times.

The role of the impulsive effects becomes clearer if we consider the following three cases. Firstly, assume that the reset maps act trivially, i.e. for all discrete transitions $(\gamma(\lambda), \gamma(\lambda+$

1)) we have $G_{(\gamma(\lambda),\gamma(\lambda+1))} \subset D_{\gamma(\lambda+1)}$ and $R_{(\gamma(\lambda),\gamma(\lambda+1))} = Id$ with Id being the identity on $G_{(\gamma(\lambda),\gamma(\lambda+1))}$. Then the variations of the boundaries, $\bar{\bar{\delta}}q_{\lambda+1}(\tau_{\lambda+1})$ and $\bar{\delta}q_\lambda(\tau_{\lambda+1})$, are equal. Hence, we can simply write $\Phi = \sum_{\lambda \in \Lambda \setminus \{N\}} (p_\lambda(\tau_{\lambda+1}) - p_{\lambda+1}(\tau_{\lambda+1})) \cdot \bar{\delta}q_\lambda(\tau_{\lambda+1})$ and the change in momentum is only influenced by the difference of the Lagrangian. Secondly, assume additionally that the Lagrangian differ only by a term depending on the configurations q, i.e. $L_{\lambda+1}(q,\dot{q}) - L_\lambda(q,\dot{q}) = H(q)$ for some function $H : Q \to \mathbb{R}$. Then, not only the boundary variations coincide but also the momenta $p_\lambda(\tau_{\lambda+1})$ and $p_{\lambda+1}(\tau_{\lambda+1})$ are the same. In this case, $\Phi = 0$ holds and hence, the boundary terms arising from the integration by parts cancel out automatically. In other words, this is the case when a discrete event does not change the momentum in the switching point of the system, i.e. the momentum is preserved in the transition point. Thirdly, if the Lagrangian $L_{\gamma(\lambda)}$ and $L_{\gamma(\lambda+1)}$ even match in the transition point, then it also holds that $E_{\gamma(\lambda)}(\tau_{\lambda+1}) = E_{\gamma(\lambda+1)}(\tau_{\lambda+1})$. Hence, the impulsive energy effect $\mathfrak{E}$ is zero and a variation $\delta\tau_{\lambda+1}$ does not lead to an energy change, i.e. the energy is preserved in the transition point.

3.2 Hybrid Flow

Now we can define a hybrid version of a Lagrangian flow, that is a flow induced by a Lagrangian vector field.

Definition 4. A *hybrid flow* is a 4-tuple $(\mathcal{I},\gamma,u,c)$ where

- $\mathcal{I}$ is a hybrid time interval,

- $\gamma : \Lambda \to \Gamma$ is a switching sequence,

- $u = \{u_\lambda \in C(I_\lambda, U)\}_{\lambda \in \Lambda}$ is a hybrid control function,

- $c = \{c_\lambda = (q_\lambda, \dot{q}_\lambda) : I_\lambda \to D_{\gamma(\lambda)}\}_{\lambda \in \Lambda}$ is a collection of C^1-maps each of which is a solution of the forced Euler-Lagrange equations on I_λ

$$\frac{d}{dt}\frac{\partial L_{\gamma(\lambda)}}{\partial \dot{q}_\lambda} - \frac{\partial L_{\gamma(\lambda)}}{\partial q_\lambda} = f_{\gamma(\lambda)}(q_\lambda, \dot{q}_\lambda, u_\lambda) \quad (8)$$

such that for all $\lambda \in \Lambda \setminus \{N\}$ it holds that (i) $(\gamma(\lambda), \gamma(\lambda + 1)) \in \mathcal{E}$, i.e. $(\gamma(\lambda), \gamma(\lambda + 1))$ is an admissible discrete transition, (ii) $(q_\lambda, \dot{q}_\lambda)(\tau_{\lambda+1}) \in G_{(\gamma(\lambda),\gamma(\lambda+1))} \subset D_{\gamma(\lambda)}$, and (iii) $R_{(\gamma(\lambda),\gamma(\lambda+1))}((q_\lambda,\dot{q}_\lambda)(\tau_{\lambda+1})) = (q_{\lambda+1},\dot{q}_{\lambda+1})(\tau_{\lambda+1})$.

Thus, for a fixed switching sequence and a fixed hybrid control function $u(t)$, a solution on each interval I_λ of the forced Euler-Lagrange equations can be uniquely determined by the initial conditions $(q_\lambda, \dot{q}_\lambda)(\tau_\lambda)$ that are in turn uniquely determined by the reset map $R_{(\gamma(\lambda-1),\gamma(\lambda))}$.

4. THE DISCRETE HYBRID LAGRANGIAN CONTROL SYSTEM

For the numerical simulation and optimization of the hybrid control system we derive a discrete formulation by means of discrete variational mechanics. To this aim, the definitions and the variational derivation introduced in Section 3 have to be reformulated in a discrete way. We replace the state space TQ by $Q \times Q$ and define

Definition 5. A *discrete hybrid mechanical control system* is a 9-tuple $\mathcal{H}_d = (Q, \Gamma, \mathcal{E}, \mathcal{D}_d, \mathcal{L}_d, \mathcal{G}_d, \mathcal{R}_d, U, \mathcal{F}_d)$, where

- Q is an n-dimensional manifold,

- Γ is a (countable) set of discrete states,

- $\mathcal{E} \subset \Gamma \times \Gamma$ is a collection of discrete transitions,

- $\mathcal{D}_d = \{D_{d,i}\}_{i \in \Gamma}$ is a collection of domains where $D_{d,i} = \{i\} \times V_i \times V_i \subset \{i\} \times Q \times Q$ with $V_i \subset Q$,

- $\mathcal{L}_d = \{L_{d,i} : D_{d,i} \to \mathbb{R}\}_{i \in \Gamma}$ is a family of discrete Lagrangian,

- $\mathcal{G}_d = \{G_{d,e}\}_{e \in \mathcal{E}}$ is a collection of guards with $G_{d,e} \subset D_{d,i}$ for $e = (i,j) \in \mathcal{E}$,

- $\mathcal{R}_d = \{R_{d,e}\}_{e \in \mathcal{E}}$ is a collection of discrete reset maps $R_{d,(i,j)} : G_{d,(i,j)} \subset D_{d,i} \to D_{d,j}$,

- $U \subseteq \mathbb{R}^m$ is the set of admissible controls, and

- $\mathcal{F}_d = \{f_{d,i}^\pm : D_{d,i} \times U \to T^*V_i\}_{i \in \Gamma}$ is a collection of left and right discrete Lagrangian forces.

To discretize the hybrid interval $\mathcal{I}$ we consider the grids $\{t_\lambda^k\}_{k=0}^{M(\lambda)}$ with $t_\lambda^{k+1} - t_\lambda^k = h \in \mathbb{R}^+ \, \forall \lambda \in \Lambda$, where $M(\lambda)$ indicates that the number of discrete times per subinterval in $I_\lambda \subset \mathcal{I}$ may vary for different λ (see Figure 2). We also fix $\alpha_\lambda \in [0,1]$, and we let $\tilde{\tau}_\lambda = t_{\lambda-1}^{M(\lambda)} + \alpha_\lambda h = t_\lambda^0 - (1-\alpha_\lambda)h$, $\lambda = 1, \ldots, N$, denote the switching times (corresponding to τ_λ from the continuous model). Define the grid $\Delta t = \{\Delta t_\lambda\}_{\lambda \in \Lambda}$ with $\Delta t_\lambda = \{\tilde{\tau}_\lambda, t_\lambda^0, \ldots, t_\lambda^{M(\lambda)}, \tilde{\tau}_{\lambda+1}\}$ with $\tilde{\tau}_0 = t_0^0 = \tau_0$ for $\lambda = 0$ and $\tilde{\tau}_{N+1} = t_N^{M(N)} = \tau_{N+1}$ for $\lambda = N$, where each Δt_λ discretizes I_λ. We assume only that the steps at which the switches occur are known but not the switching times $\tilde{\tau}_\lambda$, $\lambda = 1, \ldots, N$.

On each interval I_λ we replace a path $q_\lambda : I_\lambda \to Q$ by a discrete path $q_{d,\lambda} : \Delta t_\lambda \to Q$ with $q_\lambda^k = q_{d,\lambda}(t_\lambda^k)$, $k = 0, \ldots, M(\lambda)$, and $\tilde{q}_\lambda^0 = q_{d,\lambda}(\tilde{\tau}_\lambda)$, $\tilde{q}_\lambda^{M(\lambda)} = q_{d,\lambda}(\tilde{\tau}_{\lambda+1})$. We view $q_{d,\lambda} = \{\tilde{q}_\lambda^0, q_\lambda^0, \ldots, q_\lambda^{M(\lambda)}, \tilde{q}_\lambda^{M(\lambda)}\}$ as an approximation to $q_\lambda(t)$, $t \in I_\lambda$. Similarly, we replace the control path $u_\lambda : I_\lambda \to U$ by the discrete control path $u_{d,\lambda} : \Delta t_\lambda \to U$. We define $u_\lambda^k = u_{d,\lambda}(t_\lambda^k)$ on $[t_\lambda^k, t_\lambda^{k+1}]$ as the constant value of the control parameter guiding the system from q_λ^k to q_λ^{k+1} for $k = 0, \ldots, M(\lambda) - 1$, and $\tilde{u}_\lambda^{M(\lambda)} = u_{d,\lambda}(t_\lambda^{M(\lambda)}) = u_{d,\lambda}(\tilde{\tau}_{\lambda+1})$ the control parameter guiding the system from $q_\lambda^{M(\lambda)}$ to $\tilde{q}_\lambda^{M(\lambda)}$ and $\tilde{u}_\lambda^0 = u_{d,\lambda}(\tilde{\tau}_\lambda)$ the control parameter guiding the system from $\tilde{q}_\lambda^0$ to q_λ^0 for all $\lambda \in \Lambda$.

The space of discrete paths $q_d : \Delta t \to Q$ on the hybrid interval $\mathcal{I}$ consistent with the sequence of discrete states is then defined by

$$\mathcal{Q}_{d,\gamma} = \{q_d = \{q_{d,\lambda}\}_{\lambda \in \Lambda} \mid q_{d,\lambda} : \Delta t_\lambda \to V_{\gamma(\lambda)},$$
$$(q_\lambda^{M(\lambda)}, \tilde{q}_\lambda^{M(\lambda)}) \in G_{d,(\gamma(\lambda),\gamma(\lambda+1))},$$
$$R_{d,(\gamma(\lambda-1),\gamma(\lambda))}(q_{\lambda-1}^{M(\lambda-1)}, \tilde{q}_{\lambda-1}^{M(\lambda-1)}) = (\tilde{q}_\lambda^0, q_\lambda^0)\}$$

with $q_N^{M(N)} = \tilde{q}_N^{M(N)}$ ($\lambda = N$) and $\tilde{q}_0^0 = q_0^0$ ($\lambda = 0$) arbitrary. The discrete time interval as well as the discrete path q_d are sketched in Figure 2. Similarly, the space of *discrete hybrid control paths* $u_d : \Delta t \to U$ on the hybrid interval $\mathcal{I}$ is defined by $\mathcal{U}_{d,\gamma} = \{u_d = \{u_{d,\lambda}\}_{\lambda \in \Lambda} \mid u_{d,\lambda} : \Delta t_\lambda \to U\}$. Analogous to the continuous case,

$$\mathcal{Q}_d = \bigcup_{\gamma \in \Omega} \mathcal{Q}_{d,\gamma} \quad \text{and} \quad \mathcal{U}_d = \bigcup_{\gamma \in \Omega} \mathcal{U}_{d,\gamma}.$$

consist of all *admissible discrete paths* q_d and *all admissible discrete hybrid control paths* u_d each consistent with any switching sequence γ compatible with the graph structure.

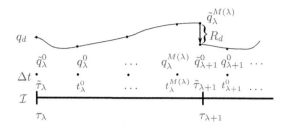

Figure 2: The grid Δt discretizes the hybrid interval $\mathcal{I}$. A discrete path $q_{d,\lambda}$ is defined on each interval $[\tau_\lambda, \tau_{\lambda+1}]$. The transition in the switching point τ_λ is given by the reset map R_d.

4.1 Discrete Hybrid Variational Formulation

In the next step, all continuous quantities, i.e. the action, the virtual work, the reset maps, the guards, and the impulsive effects have to be approximated by their discrete counterparts. For the reset maps and guards a discrete version can be defined based on the continuous and discrete Legendre transforms as $G_{d,(i,j)} := (\mathbb{F}^+L_{d,i})^{-1} \circ \mathbb{F}L_i(G_{(i,j)})$ and $R_{d,(i,j)} := (\mathbb{F}^-L_{d,j})^{-1} \circ \mathbb{F}L_j \circ R_{(i,j)} \circ (\mathbb{F}L_i)^{-1} \circ \mathbb{F}^+L_{d,i}$.

The *discrete hybrid action map* $\mathfrak{S}_d : \mathcal{Q}_d \times [0,1]^N \to \mathbb{R}$ is defined as

$$
\mathfrak{S}_d(q_d, \{\alpha_\lambda\}_{\lambda=1}^N) = \sum_{\lambda \in \Lambda} \sum_{k=0}^{M(\lambda)-1} L_{d,\gamma(\lambda)}(q_\lambda^k, q_\lambda^{k+1}, h)
$$
$$
+ \sum_{\lambda \in \Lambda \setminus \{N\}} \Big[L_{d,\gamma(\lambda)}(q_\lambda^{M(\lambda)}, \tilde{q}_\lambda^{M(\lambda)}, \alpha_{\lambda+1}h)
$$
$$
+ L_{d,\gamma(\lambda+1)}(\tilde{q}_{\lambda+1}^0, q_{\lambda+1}^0, (1-\alpha_{\lambda+1})h) \Big].
$$

The *discrete hybrid virtual work* $W_d : \mathcal{Q}_d \times \mathcal{U}_d \to \mathbb{R}$ is defined in an analogous way as

$$
W_d(q_d, u_d) = \sum_{\lambda \in \Lambda} \sum_{k=0}^{M(\lambda)-1} \Big(f_{\gamma(\lambda)}^{k,-} \cdot \delta q_\lambda^k + f_{\gamma(\lambda)}^{k,+} \cdot \delta q_\lambda^{k+1} \Big)
$$
$$
+ \sum_{\lambda \in \Lambda \setminus \{N\}} \Big[\Big(\tilde{f}_{\gamma(\lambda)}^{M(\lambda),-} \cdot \delta q_\lambda^{M(\lambda)} + \tilde{f}_{\gamma(\lambda)}^{M(\lambda),+} \cdot \delta \tilde{q}_\lambda^{M(\lambda)} \Big)
$$
$$
+ \Big(\tilde{f}_{\gamma(\lambda+1)}^{0,-} \cdot \delta \tilde{q}_{\lambda+1}^0 + \tilde{f}_{\gamma(\lambda+1)}^{0,+} \cdot \delta q_{\lambda+1}^0 \Big) \Big].
$$

where $\delta q_\lambda^k, \delta \tilde{q}_\lambda^0, \delta \tilde{q}_\lambda^{M(\lambda)}$ are variations of the discrete configurations $q_\lambda^k, \tilde{q}_\lambda^0, \tilde{q}_\lambda^{M(\lambda)}$, respectively, $f_{\gamma(\lambda)}^{k,\pm} := f_{d,\gamma(\lambda)}^{k,\pm}(q_\lambda^k, q_\lambda^{k+1}, u_\lambda^k)$ for $k = 0, \ldots, M(\lambda) - 1 \forall \lambda \in \Lambda$, and $\tilde{f}_{\gamma(\lambda+1)}^{0,\pm} := f_{d,\gamma(\lambda+1)}^{0,\pm}(\tilde{q}_{\lambda+1}^0, q_{\lambda+1}^0, \tilde{u}_{\lambda+1}^0)$ and $\tilde{f}_{\gamma(\lambda)}^{M(\lambda),\pm} := f_{d,\gamma(\lambda)}^{M(\lambda),\pm}(q_\lambda^{M(\lambda)}, \tilde{q}_\lambda^{M(\lambda)}, \tilde{u}_\lambda^{M(\lambda)})$ for all $\lambda \in \Lambda \setminus \{N\}$. Here, each term of the sums approximates the integrals of the Lagrangian and the virtual work on two neighboring points of the grid Δt as described in Section 2.

The discrete versions Φ_d and $\mathfrak{E}_d$ of the impulsive effects (4) are given as

$$
\Phi_d := \sum_{\lambda \in \Lambda \setminus \{N\}} \Delta_{d,(\lambda,\lambda+1)} \quad \text{and} \quad \mathfrak{E}_d := \sum_{\lambda \in \Lambda \setminus \{N\}} \eth_{d,(\lambda,\lambda+1)}
$$

with

$$
\Delta_{d,(\lambda,\lambda+1)} := \tilde{p}_{\lambda+1}^{0,-} \cdot \delta \tilde{q}_{\lambda+1}^0 - \tilde{p}_\lambda^{M(\lambda),+} \cdot \delta \tilde{q}_\lambda^{M(\lambda)},
$$
$$
\eth_{d,(\lambda,\lambda+1)} := \big(E_{d,\gamma(\lambda+1)}(\tilde{q}_{\lambda+1}^0, q_{\lambda+1}^0)
$$
$$
- E_{d,\gamma(\lambda)}(q_\lambda^{M(\lambda)}, \tilde{q}_\lambda^{M(\lambda)}) \big) \cdot \delta \tilde{\tau}_{\lambda+1},
$$

with $\delta \tilde{\tau}_{\lambda+1} = h \delta \alpha_{\lambda+1}$, i.e. a variation of $\alpha_{\lambda+1} \in [0,1]$ leads to a variation of the switching time $\tilde{\tau}_{\lambda+1}$. The discrete momenta $\tilde{p}_\lambda^{k,\pm}$ are given by the discrete Legendre transforms $\mathbb{F}^\pm L_{d,\gamma(\lambda)} : D_{d,\gamma(\lambda)} \to T^*V_{\gamma(\lambda)}$ and $E_{d,\gamma(\lambda)}$ is the discrete energy. Having the discrete versions of the action, the virtual work, and the impulsive effects, we derive the discrete equations of motions via a discrete version of the hybrid Lagrange-d'Alembert principle. Let $\mathcal{H}_d$ be a discrete hybrid Lagrangian control system, $\mathcal{I} = \{I_\lambda\}_{\lambda \in \Lambda}$ a hybrid interval with index set Λ, and Δt a time grid discretizing the hybrid interval $\mathcal{I}$. We formulate the *discrete hybrid Lagrange-d'Alembert principle* as follows: Find discrete paths q_d such that for all variations $\delta q_d \in T_{q_d}\mathcal{Q}_d$ and $\delta \alpha_\lambda \in \mathbb{R}$ for $\lambda = 1, \ldots, N$ with fixed boundaries (i.e. $\delta q_0^0 = \delta q_N^{M(N)} = 0$), one has

$$
\delta \mathfrak{S}_d(q_d) + W_d(q_d, u_d) + \Phi_d = \mathfrak{E}_d. \tag{9}
$$

By taking variations and rearranging sums, the terms arising from the discrete variations corresponding to the switching points $\delta \tilde{q}_{\lambda+1}^0, \delta \tilde{q}_\lambda^{M(\lambda)}, \delta \alpha_{\lambda+1} \forall \lambda \in \Lambda \setminus \{N\}$ are

$$
\sum_{\lambda \in \Lambda \setminus \{N\}} \Big(D_1 L_{d,\gamma(\lambda+1)}(\tilde{q}_{\lambda+1}^0, q_{\lambda+1}^0, (1-\alpha_{\lambda+1})h) + \tilde{f}_{\gamma(\lambda+1)}^{0,-} \Big) \cdot \delta \tilde{q}_{\lambda+1}^0
$$
$$
+ \Big(D_2 L_{d,\gamma(\lambda)}(q_\lambda^{M(\lambda)}, \tilde{q}_\lambda^{M(\lambda)}, \alpha_{\lambda+1}h) + \tilde{f}_{\gamma(\lambda)}^{M(\lambda)-1,+} \Big) \cdot \delta \tilde{q}_\lambda^{M(\lambda)}
$$
$$
+ \Big(D_3 L_{d,\gamma(\lambda)}(q_\lambda^{M(\lambda)}, \tilde{q}_\lambda^{M(\lambda)}, \alpha_{\lambda+1}h)
$$
$$
- D_3 L_{d,\gamma(\lambda+1)}(\tilde{q}_{\lambda+1}^0, q_{\lambda+1}^0, (1-\alpha_{\lambda+1})h) \big) \cdot h \delta \alpha_{\lambda+1}.
$$

These cancel with the discrete impulsive effects. Setting the remaining parts to zero results in the standard discrete forced Euler-Lagrange equations that have to be fulfilled on each discrete grid $\Delta t_\lambda \forall \lambda \in \Lambda$ reading

$$
D_1 L_{d,\gamma(\lambda)}(q_\lambda^k, q_\lambda^{k+1}) + D_2 L_{d,\gamma(\lambda)}(q_\lambda^{k-1}, q_\lambda^k)
$$
$$
+ f_{\gamma(\lambda)}^{k,-} + f_{\gamma(\lambda)}^{k-1,+} = 0 \tag{10}
$$

for $k = 1, \ldots, M(\lambda) - 1$ and all $\lambda \in \Lambda$ and

$$
D_1 L_{d,\gamma(\lambda)}(q_\lambda^{M(\lambda)}, \tilde{q}_\lambda^{M(\lambda)}, \alpha_{\lambda+1}h) + \tilde{f}_{\gamma(\lambda)}^{M(\lambda),-}
$$
$$
+ D_2 L_{d,\gamma(\lambda)}(q_\lambda^{M(\lambda)-1}, q_\lambda^{M(\lambda)}, h) + f_{\gamma(\lambda)}^{M(\lambda)-1,+} = 0 \tag{11}
$$
$$
D_1 L_{d,\gamma(\lambda+1)}(q_{\lambda+1}^0, q_{\lambda+1}^1, h) + f_{\gamma(\lambda+1)}^{0,-}
$$
$$
+ D_2 L_{d,\gamma(\lambda+1)}(\tilde{q}_{\lambda+1}^0, q_{\lambda+1}^0, (1-\alpha_{\lambda+1})h) + \tilde{f}_{\gamma(\lambda+1)}^{0,+} = 0 \tag{12}
$$

for all $\lambda \in \Lambda \setminus \{N\}$.

4.2 Discrete Hybrid Flow

The discrete hybrid flow is thus defined as follows.

Definition 6. A *discrete hybrid flow* is a 5-tuple $(\mathcal{I}, \Delta t, \gamma, u_d, q_d)$ where

- $\mathcal{I}$ is a hybrid time interval,
- Δt is a time grid discretizing the hybrid interval $\mathcal{I}$,

- $\gamma : \Lambda \to \Gamma$ is a switching sequence,

- $u_d = \{u_{d,\lambda} : \Delta t_\lambda \to U\}_{\lambda \in \Lambda}$ a discrete hybrid control path, and

- $q_d = \{q_{d,\lambda} : \Delta t_\lambda \to Q\}_{\lambda \in \Lambda}$ is a discrete path where each $q_{d,\lambda}$ is a solution of the discrete forced Euler-Lagrange equations (10), (11), and (12)

such that for all $\lambda \in \Lambda \setminus \{N\}$ it holds that (i) $(\gamma(\lambda), \gamma(\lambda + 1)) \in \mathcal{E}$, i.e. $(\gamma(\lambda), \gamma(\lambda + 1))$ is an admissible discrete transition, (ii) $(q_\lambda^{M(\lambda)}, \tilde{q}_\lambda^{M(\lambda)}) \in G_{d,(\gamma(\lambda),\gamma(\lambda+1))} \subset D_{d,\gamma(\lambda)}$, and (iii) $R_{d,(\gamma(\lambda),\gamma(\lambda+1))}(q_\lambda^{M(\lambda)}, \tilde{q}_\lambda^{M(\lambda)}) = (\tilde{q}_{\lambda+1}^0, q_{\lambda+1}^0)$.

As for the continuous flow, for a fixed switching sequence and a fixed discrete control path u_d the discrete solution q_d is uniquely determined by the discrete forced Euler-Lagrange equations and the initial conditions given by the discrete reset map after each switching point.

5. OPTIMAL CONTROL OF A HYBRID LAGRANGIAN CONTROL SYSTEM

Optimal control aims for optimizing a cost functional while taking into account the dynamics of the system to be controlled. In Section 3 we have described the dynamics of a hybrid Lagrangian control system based on a variational principle. In the next step, we introduce a cost functional for these hybrid systems. Let us denote $\mathcal{C}_\lambda = C^1(I_\lambda, D_{\gamma(\lambda)})$ and $\mathcal{U}_\lambda = C(I_\lambda, U)$.

The cost functional $J : \mathcal{C} \times \mathcal{U} \to \mathbb{R}$ of a hybrid flow $(\mathcal{I}, \gamma, u, c)$ is given by the sum of the cost functionals $J_{\gamma(\lambda)} : \mathcal{C}_\lambda \times \mathcal{U}_\lambda \to \mathbb{R}$ of each continuous part of the switching sequence:

$$J(c, u) = \sum_{\lambda \in \Lambda} J_{\gamma(\lambda)}(c_\lambda, u_\lambda) = \sum_{\lambda \in \Lambda} \int_{I_\lambda} B_{\gamma(\lambda)}(c_\lambda(t), u_\lambda(t)) \, dt \tag{13}$$

with $B_{\gamma(\lambda)} : D_{\gamma(\lambda)} \times U \to \mathbb{R}$ being continuously differentiable. In the following formulation of the optimal control problem we assume that the number of interval parts, $N+1$, and the switching sequence γ are fixed, while the intermediate switching times and in particular the final time τ_{N+1} are considered as optimization variables.

Problem 1. An *optimal control problem* for a hybrid Lagrangian control system $\mathcal{H}$ with switching sequence γ is given by

$$\min_{\mathcal{I},(c,u)\in\mathcal{C}_\gamma \times \mathcal{U}_\gamma} J(c, u) = \sum_{\lambda \in \Lambda} J_{\gamma(\lambda)}(c_\lambda, u_\lambda) \tag{14}$$

subject to

$$\tau_0 = 0, c_0(\tau_0) = c_{\text{start}}, c_N(\tau_{N+1}) = c_{\text{final}} \tag{15}$$

$$\frac{d}{dt}\frac{\partial L_{\gamma(\lambda)}}{\partial \dot{q}_\lambda} - \frac{\partial L_{\gamma(\lambda)}}{\partial q_\lambda} = f_{\gamma(\lambda)}(c_\lambda, u_\lambda) \tag{16}$$

$$\text{for } t \in I_\lambda \text{ and } \forall \lambda \in \Lambda$$

$$\mathcal{I} = \{I_\lambda\}_{\lambda \in \Lambda} \text{ with } \tau_\lambda \leq \tau_{\lambda+1}. \tag{17}$$

Note that in this formulation the constraints with regard to the guards and the reset maps are included due to the definition of the function space $\mathcal{C}_\gamma \ni c$. [3]

---

[3] We assume existence of solutions with higher regularity as in general optimal control theory.

5.1 Multiobjective Optimal Control of a Hybrid Lagrangian System

In typical applications, there arise more than one cost functional. This leads to an extension of Problem 1.

Problem 2. A *multiobjective optimal control problem* for a hybrid Lagrangian control system $\mathcal{H}$ with switching sequence γ is given by

$$\min_{\mathcal{I},(c,u)\in\mathcal{C}_\gamma \times \mathcal{U}_\gamma} \mathbf{J}(c, u) \tag{18}$$

subject to the constraints (15)-(17), where $\mathbf{J}(c, u) : \mathcal{C} \times \mathcal{U} \to \mathbb{R}^k$ is a vector of cost functionals $\mathbf{J}(c, u) = (J^1(c, u), \ldots, J^k(c, u))$ with each $J^l(c, u)$, $l = 1, \ldots, k$ being of form (13).

The minimization of the vector valued functional $\mathbf{J}(c, u)$ is defined by the partial order $<_p$ on $\mathbb{R}^k$. Let $v, w \in \mathbb{R}^k$, then the vector v is *less than* w ($v <_p w$), if $v_i < w_i$ for all $i \in \{1, \ldots, k\}$. The relation $\leq_p$ is defined analogously. By this relation, we can introduce the concept of dominance and Pareto optimality (see also [12]).

Definition 7. (a) A solution $(\mathcal{I}, c, u)$ of Problem 2 is called *admissible* if it satisfies the constraints (15)-(17).

(b) A solution $(\mathcal{I}, c, u)$ is *dominated* by a solution $(\mathcal{I}^*, c^*, u^*)$ w.r.t. (18) if $\mathbf{J}(c^*, u^*) \leq_p \mathbf{J}(c, u)$ and $\mathbf{J}(c, u) \neq \mathbf{J}(c^*, u^*)$, otherwise $(\mathcal{I}, c, u)$ is non-dominated by $(\mathcal{I}^*, c^*, u^*)$.

(c) A solution $(\mathcal{I}^*, c^*, u^*)$ is called *(Pareto) optimal* if there exists no $(\mathcal{I}, c, u)$ which dominates $(\mathcal{I}^*, c^*, u^*)$.

(d) The set of all Pareto optimal solutions $(\mathcal{I}^*, c^*, u^*)$ is called the *Pareto set* and its image under $\mathbf{J}$ the *Pareto front*.

5.2 The Discrete Optimal Control Problem

For the numerical solution of the (multiobjective) optimal control problem we need a discretized version of Problem 1 and 2. Here, we follow the philosophy of DMOC [13], where the discrete forced Euler-Lagrange equations serve as equality constraints for a discretized cost functional. We therefore introduce the discrete cost function

$$J_d(q_d, u_d) = \sum_{\lambda \in \Lambda} J_{d,\gamma(\lambda)}(q_{d,\lambda}, u_{d,\lambda}) \tag{19}$$

$$= \sum_{\lambda \in \Lambda} \sum_{k=0}^{M(\lambda)-1} B_{d,\gamma(\lambda)}(q_\lambda^k, q_\lambda^{k+1}, u_\lambda^k)$$

$$\sum_{\lambda \in \Lambda \setminus \{N\}} \Big[B_{d,\gamma(\lambda)}(q_\lambda^{M(\lambda)}, \tilde{q}_\lambda^{M(\lambda)}, \tilde{u}_\lambda^{M(\lambda)})$$

$$+ B_{d,\gamma(\lambda+1)}(\tilde{q}_{\lambda+1}^0, q_{\lambda+1}^0, \tilde{u}_{\lambda+1}^0) \Big],$$

where we approximate the cost functionals on the time slice $[kh, (k+1)h]$ by

$$B_{d,\gamma(\lambda)}(q_\lambda^k, q_\lambda^{k+1}, u_\lambda^k) \approx \int_{kh}^{(k+1)h} B_{\gamma(\lambda)}(q_\lambda(t), \dot{q}_\lambda(t), u_\lambda(t)) \, dt$$

for $k = 0, \ldots, M(\lambda) - 1 \forall \lambda \in \Lambda$ and correspondingly for the intervals at the switching points. As for the continuous case, if several discrete cost functions come into play, they can be collected in a vector $\mathbf{J}_d(q_d, u_d)$. Using the discrete description of a hybrid Lagrangian control system as derived in Section 4 leads to the following discrete version of Problem 2.

Problem 3. The *discrete optimal control problem* for a discrete hybrid system $\mathcal{H}_d$ with switching sequence γ is given by

$$\min_{\mathcal{I},(q_d,u_d)\in\mathcal{Q}_{d,\gamma}\times\mathcal{U}_{d,\gamma}} \mathbf{J}_d(q_d,u_d)$$

subject to the initial and final conditions

$$\mathbb{F}^- L_{d,\gamma(0)}(q_0^0,q_0^1) = \mathbb{F}L_{\gamma(0)}(c_{\text{start}}),$$

$$\mathbb{F}^+ L_{d,\gamma(N)}(q_N^{M(N)-1},q_N^{M(N)}) = \mathbb{F}L_{\gamma(N)}(c_{\text{final}}),$$

and the discrete Euler-Lagrange equations (10)–(12).

5.3 Two Layer Formulation

The discretization of the hybrid optimal control problem is the first crucial step to derive a formulation that is numerically treatable. As a second step, we split up the large optimization problem into two layers (as also proposed in [18]), see Figure 3 for a sketch.

This can be done by moving the switching constraints, i.e. the switching times and reset states (the final states of each interval I_λ, $\lambda \in \Lambda \setminus \{N\}$) into the upper layer such that in the lower layer there remain uncoupled "ordinary" optimal control problems.

To this end we add the reset state values as auxiliary optimization variables, $\mathcal{A} = \{a_\lambda\}_{\lambda\in\Lambda}$ and $\mathcal{B} = \{b_\lambda\}_{\lambda\in\Lambda}$. By construction they coincide with the boundary values of each continuous part of the hybrid path, i.e. $\mathcal{A}$ and $\mathcal{B}$ are the sets of initial and final states of $I_\lambda \, \forall \lambda \in \Lambda$, respectively. In the lower layer we search on each interval I_λ, $\lambda \in \Lambda$ for an optimal solution $(c_\lambda, u_\lambda) \in \mathcal{C}_\lambda \times U_\lambda$ with fixed boundaries given by a_λ and b_λ. This results in $N+1$ decoupled subproblems, each defined on I_λ, $\lambda \in \Lambda$. Let $\mathcal{S} = \{\tau_1,\dots,\tau_N\}$ denote the set of switching times. In the upper layer the switching times $\mathcal{S}$ and reset state values $\mathcal{A},\mathcal{B}$ are optimized with respect to the cost functional (14) or the vector of cost functionals (18) that are evaluated based on the optimal solutions of the problems in the lower layer. Using similar concepts as in [9] one can show that the two layer formulation is equivalent to the original problem.

Problem 4. The optimal control problems of the lower layer are of the form:

$$\min_{(c_\lambda,u_\lambda)\in\mathcal{C}_\lambda\times U_\lambda} J_{\gamma(\lambda)}(c_\lambda,u_\lambda) = \int_{I_\lambda} B_{\gamma(\lambda)}(c_\lambda(t),u_\lambda(t))\,dt$$

subject to $c_\lambda(\tau_\lambda) = a_\lambda$, $c_\lambda(\tau_{\lambda+1}) = b_\lambda$ and

$$\frac{d}{dt}\frac{\partial L_{\gamma(\lambda)}}{\partial \dot{q}_\lambda} - \frac{\partial L_{\gamma(\lambda)}}{\partial q_\lambda} = f_{\gamma(\lambda)}(c_\lambda,u_\lambda)\,\forall t \in I_\lambda$$

for all $\lambda \in \Lambda$ with $a_0 = c_{\text{start}}$ and $b_N = c_{\text{final}}$.

While Problem 4 consists of ordinary optimal control problems with fixed boundary conditions, in Problem 5 we optimize w.r.t. the reset states and switching times.

Problem 5. The optimization problem of the upper layer is of the form:

$$\min_{\mathcal{S},\mathcal{A},\mathcal{B}} \mathbf{J}(c^*,u^*)$$

subject to $\tau_\lambda \leq \tau_{\lambda+1} \, \forall \lambda \in \Lambda \setminus \{N\}$ and

$$b_\lambda \in G_{(\gamma(\lambda),\gamma(\lambda+1))}, \quad a_{\lambda+1} = R_{(\gamma(\lambda),\gamma(\lambda+1))}(b_\lambda),$$

where $(c^*,u^*) = \{(c_\lambda^*,u_\lambda^*)\}_{\lambda\in\Lambda}$ are the optimal solutions of Problem 4.

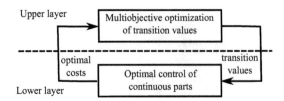

Figure 3: The computation of a multiobjective optimal control problem for hybrid systems is split up into two layers.

For the numerical treatment we use a discrete formulation of the lower layer problems as in Problem 3. Then $(q_{d,\lambda},u_{d,\lambda})$ approximates (c_λ,u_λ) with order dependent on the approximation order of the discrete Lagrangian and discrete forces (cf. [13]), e.g. we get an approximation error of order 2 using the midpoint rule for the approximation of the integrals. The splitting into two layers enables us to choose numerical tools for both layers almost independently, as long as an exchange of information is possible.

We solve the discretized optimal control problems in the lower layer by using a sparse SQP optimization algorithm that is implemented in the routine `nag_opt_nlp_sparse` of the NAG library[4]. The numerical tool for the optimization of the upper layer is chosen depending on the type of problem. If only one objective has to be minimized, we use a state of the art nonlinear optimization algorithm for which no derivative information is required. In case of multiobjective optimal control problems, we are interested in the entire Pareto set, i.e. all optimal compromises. Here, we use the set-oriented methods implemented in the software package GAIO (*Global Analysis of Invariant Objects*, [5]). These methods allow an approximation of the entire Pareto set using a box covering of the parameter space and are either based on gradient information or on function evaluations only.

6. NUMERICAL EXAMPLE

We apply the presented approach to a simple hybrid mechanical system which we call a hybrid single-mass oscillator (see Figure 4). The system consists of two masses which are firmly connected (that justifies the term single-mass oscillator) and a suspension system. The suspension consists of one linear and one nonlinear spring which are mounted parallel but differ in their lengths in the unloaded case. The parameters are chosen such that in the equilibrium state of the first mass, only one spring is tensioned whereas both masses load the second spring as well (cf. Figure 4).

In detail, the hybrid dynamical system $\mathcal{H}_{SMO} = (Q,\Gamma,\mathcal{E},\mathcal{D},\mathcal{L},\mathcal{G},\mathcal{R},U,\mathcal{F})$ is modeled with $Q = \mathbb{R}$, $TQ = \mathbb{R}^2$, $\Gamma = \{0,1\}$ and $\mathcal{E} = \{(0,1),(1,0)\}$. The set of domains is $\mathcal{D} = \{D_0,D_1\}$ with $D_0 = \{(q,\dot{q}) \in \mathbb{R}^2 | q \leq q^s\}$, $D_1 = \{(q,\dot{q}) \in \mathbb{R}^2 | q \geq q^s\}$ and the family of Lagrangian is $\mathcal{L} = \{L_0,L_1\}$ with $L_0(q,\dot{q}) = \frac{1}{2}m\dot{q}^2 - \frac{1}{2}c_1 q^2 - mgq$ and $L_1(q,\dot{q}) = \frac{1}{2}m\dot{q}^2 - \frac{1}{2}c_1 q^2 - \frac{1}{4}c_2(q-q^s)^4 - mgq$ with $m = m_1 + m_2$. The guards and resets are defined to be $\mathcal{G} = \{G_{(0,1)},G_{(1,0)}\}$ with $G_{(0,1)} = \{(q,\dot{q}) \in D_0 | q = q^s, \dot{q} > 0\}$, $G_{(1,0)} = \{(q,\dot{q}) \in D_1 | q = q^s, \dot{q} < 0\}$ and $\mathcal{R} = \{R_{(0,1)},R_{(1,0)}\}$ with $R_{(0,1)} = R_{(1,0)} = Id$. The set of admissible controls and the Lagrangian forces are $U = \{u \in \mathbb{R} | -m_2 g \leq u \leq 0\}$ and

[4]http://www.nag.com

248

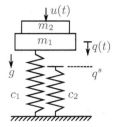

Parameter	Value
1st mass (m_1)	1,500
2nd mass (m_2)	500
1st spring (c_1)	10,000
2nd sping (c_2)	20,000
gravity (g)	10
initial point (q_0)	1.5
final point (q_f)	1.96
switching point (q^s)	1.7

Figure 4: Sketch of the single-mass oscillator and parameters used for the numerical computations.

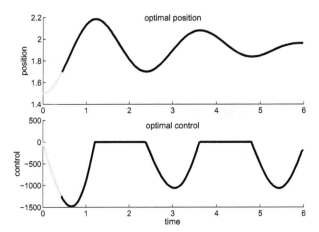

Figure 5: Hybrid optimal solution of the two layer implementation for one switching and fixed final time $t_f = 5.98$. The optimal switching time is $\tau_1 = 0.44$, the optimal switching velocity $\dot{q}^s = 0.79$ and the optimal costs are $J = 2.75 \cdot 10^6$.

$\mathcal{F} = \{f_0, f_1\}$ with $f_0(q, \dot{q}, u) = f_1(q, \dot{q}, u) = u$. We assume to start the optimal control of the system right after the second mass is put on top of the stationary first mass. Since no damping effects are taken into account, the uncontrolled system would start to oscillate forever. However, by applying a control force as depicted in Figure 4, it is possible to steer the system into the equilibrium state of both masses. At first, we are interested in the optimal solution for a hybrid trajectory consisting of two parts and with fixed final time t_f, hence we take $\Lambda = \{0, 1\}$, $\mathcal{I} = \{[0, \tau_1], [\tau_1, t_f]\}$ and $\gamma(0) = 0, \gamma(1) = 1$. The cost functional for the optimal control problem 1 of $\mathcal{H}_{SMO}$ is the control effort given as

$$J(c, u) = J_0(c_0, u_0) + J_1(c_1, u_1) = \int_{I_0} u_0^2(t)\, dt + \int_{I_1} u_1^2(t)\, dt. \quad (20)$$

Using the two layer formulation as presented in Section 5.3, we search for optimal state and control trajectories with initial state $c_{\text{start}} = c_0(0) = (q_0, 0)$ and final state $c_{\text{final}} = c_1(t_f) = (q_f, 0)$, as well as the optimal switching time τ_1 and switching velocity $\dot{q}^s$. The optimal trajectories (position and control) for the fixed final time $t_f = 5.98$ are shown in Figure 5.

In the next step, we take a second objective function into account, that is the duration T of the maneuver. Intuitively, this is contradictory to minimizing the control effort (20) and hence, we are faced with a multiobjective optimal control problem. For the approximation of the Pareto front,

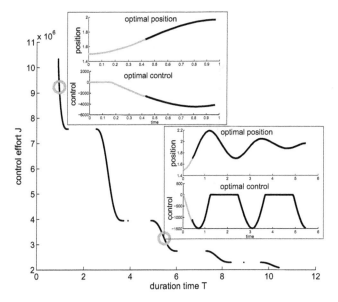

Figure 6: An approximation of the Pareto front with exemplary Pareto optimal solutions.

we use algorithms provided in the software package GAIO (cf. Section 5.3). Figure 6 shows the resulting approximation of the Pareto front. The shape of the Pareto front shows the typical trade-off between control effort and duration, i.e. if one objective improves, the other one gets worse. The holes in the Pareto front emerge, because the corresponding solutions are dominated by Pareto optimal solutions which have the same control effort but reach the final state faster.

Finally, we investigate the optimal solutions with regard to the number of switches. In this example, only an odd number of switches is reasonable, so we consider the control problem for $N = 3$ and $N = 5$ switches. The resulting Pareto fronts are depicted in Figure 7. While for $N = 1$ only three parameters $(\tau_1, T, \dot{q}^s)$ have to be optimized in the upper layer, the problems with $N = 3$ and $N = 5$ lead to seven and eleven parameters, respectively. It can be seen in Figure 7 that – as expected – for longer duration times $(T > 8)$, solutions with several switches become better, i.e. cheaper w.r.t. the control effort than solutions with only one switch.

7. CONCLUSION AND OUTLOOK

An optimal control problem for hybrid Lagrangian control systems has been stated and solution strategies have been proposed. We derive the equations of motion by a hybrid variational principle. The corresponding discrete hybrid variational principle is based on a discrete approximation of the hybrid Lagrangian system leading to discrete forced Euler-Lagrange equations and to the definition of a discrete hybrid flow. Theses equations form equality constraints for the hybrid optimal control problem stated in Section 5. Additional constraints are given by the underlying graph structure of the hybrid system and the hybrid time interval. The two layer approach proposed for an implementation could be successfully applied to the quite simple example of a single-mass oscillator. In the future, further applications have to reveal advantages and disadvantages of this approach. Another useful step is the combination of our

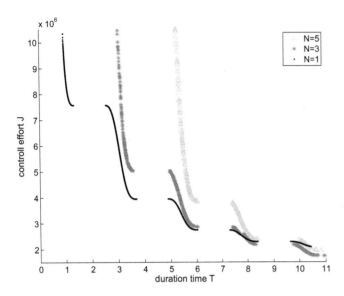

Figure 7: Approximation of the Pareto fronts for hybrid solutions with $N = 1$, $N = 3$ **and** $N = 5$ **switches.**

approach with discrete optimization techniques (as e.g. in [3]) such that also the switching sequence can be optimized. In order to use nonlinear optimization algorithms that work with gradient information, the techniques presented in [4, 6, 18] or similar approches could be used to compute derivatives w.r.t. the switching times. In particular, we want to focus on multi-body systems where discrete events correspond to e.g. impacts and where different Lagrangian correspond to different topologies (e.g. different coupling structures of the multi-body system). For these kinds of systems, reset maps could be derived based on a hybrid variational formulation similar to [7], where the relation between velocity before and after a single impact emerges from a nonsmooth variational principle.

8. ACKNOWLEDGMENTS

This contribution was developed and published in the course of the Collaborative Research Centre 614 "Self-Optimizing Concepts and Structures in Mechanical Engineering" funded by the German Research Foundation (DFG) under grant number SFB 614. The authors gratefully acknowledge the helpful suggestions of S. Hage-Packhäuser and the assistance of A. Seifried in the numerical computations.

9. REFERENCES

[1] A. Ames and S. Sastry. Hybrid Routhian Reduction of Lagrangian Hybrid Systems. *American Control Conference*, 2006.

[2] S. Attia, V. Azhmyakov, and J. Raisch. On an Optimization Problem for a Class of Impulsive Hybrid Systems. *Discrete Event Dynamic Systems*, 20:215–231, 2010.

[3] M. Buss, M. Glocker, M. Hardt, O. von Stryk, R. Bulirsch, and G. Schmidt. Nonlinear Hybrid Dynamical Systems: Modeling, Optimal Control, and Applications. In S. Engell, G. Frehse, and E. Schnieder, editors, *Modelling, Analysis, and Design of Hybrid Systems*, volume 279 of *Lecture Notes in Control and Information Sciences*, pages 311–335. Springer Berlin, 2002.

[4] T. Caldwell and T. Murphey. Switching mode generation and optimal estimation with application to skid-steering. *Automatica*, In Press, 2010.

[5] M. Dellnitz, O. Schütze, and T. Hestermeyer. Covering Pareto Sets by Multilevel Subdivision Techniques. *J. of Optimization Theory and Application*, 124 (1):113–136, 2005.

[6] M. Egerstedt, Y. Wardi, and H. Axelsson. Transition-time optimization for switched-mode dynamical systems. *IEEE Transactions on Automatic Control*, 51(1):110 – 115, 2006.

[7] R. C. Fetecau, J. E. Marsden, M. Ortiz, and M. West. Nonsmooth Lagrangian Mechanics and Variational Collision Integrators. *SIAM J. Applied Dynamical Systems*, 2(3):381–416, 2003.

[8] H. Gonzales, R. Vasudevan, M. Kamgarpour, S. Sastry, R. Bajcsy, and C. J. Tomlin. A Descent Algorithm for the Optimal Control of Constrained Nonlinear Switched Dynamical Systems. *13th ACM International Conference on Hybrid Systems: Computation and Control*, 2010.

[9] O. Junge, J. E. Marsden, and S. Ober-Blöbaum. Optimal Reconfiguration of Formation Flying Spacecraft - a decentralized approach. In *IEEE Conference on Decision and Control*, pages 5210–5215, San Diego, CA, USA, 2006.

[10] J. Lygeros, K. H. Johansson, S. N. Simić, J. Zhang, and S. Sastry. Dynamical Properties of Hybrid Automata. *IEEE Transactions on automatic control*, 48(1), 2003.

[11] J. E. Marsden and M. West. Discrete mechanics and variational integrators. *Acta Numerica*, 10:357–514, 2001.

[12] K. Miettinen. *Nonlinear Multiobjective Optimization*. Kluwer Academic Publishers, 1999.

[13] S. Ober-Blöbaum, O. Junge, and J. E. Marsden. Discrete mechanics and optimal control: an analysis. *Control, Optimisation and Calculus of Variations*, 2010. DOI: 10.1051/cocv/2010012.

[14] D. Pekarek, A. D. Ames, and J. E. Marsden. Discrete Mechanics and Optimal Control Applied to the Compass Gait Biped. In *46th IEEE Conference on Decision and Control*, New Orleans, LA, USA, 2007.

[15] M. Shaikh and P. Caines. On the hybrid optimal control problem: Theory and algorithms. *IEEE Transactions on Automatic Control*, 52(9):1587 –1603, 2007.

[16] S. N. Simić, K. H. Johansson, S. Sastry, and J. Lygeros. Towards a Geometric Theory of Hybrid Systems. *Dynamics of Continuous, Discrete and Impulsive Systems, Series B: Applications & Algorithms*, 12(5-6):649–687, 2005.

[17] H. J. Sussmann. A maximum principle for hybrid optimal control problems. In *38th IEEE Conference on Decision and Control*, Phoenix, Arizona, 1999.

[18] X. Xu and P. J. Antsaklis. Optimal control of switched systems via non-linear optimization based on direct differentiations of value functions. *Int. J. of Control*, 75(16):1406–1426, 2002.

A Stochastic Reach-Avoid Problem with Random Obstacles[*] [†]

Sean Summers
Automatic Control Laboratory
ETH Zürich
Physikstrasse 3
8092 Zürich, Switzerland
summers@control.ee.ethz.ch

Maryam Kamgarpour
Department of Mechanical
Engineering
UC Berkeley
Berkeley CA 94720, USA
maryamka@eecs.berkeley.edu

Claire Tomlin
Department of Electrical Engineering
and Computer Sciences
UC Berkeley
Berkeley CA 94720, USA
tomlin@eecs.berkeley.edu

John Lygeros
Automatic Control Laboratory
ETH Zürich
Physikstrasse 3
8092 Zürich, Switzerland
lygeros@control.ee.ethz.ch

ABSTRACT

We present a dynamic programming based solution to a stochastic reachability problem for a controlled discrete-time stochastic hybrid system. A sum-multiplicative cost function is introduced along with a corresponding dynamic recursion which quantifies the probability of hitting a target set at some point during a finite time horizon, while avoiding an obstacle set during each time step preceding the target hitting time. In contrast with earlier works which consider the reach and avoid sets as both deterministic and time invariant, we consider the avoid set to be both time-varying and probabilistic. Optimal reach-avoid control policies are derived as the solution to an optimal control problem via dynamic programming. A computational example motivated by aircraft motion planning is provided.

Categories and Subject Descriptors

I.6.4 [**Simulation and modeling**]: Model Validation and Analysis

[*]This work was partially supported by the European Commission under the project iFly, FP6-TREN-037180, and the MoVeS project, FP7-ICT-2009-257005.

[†]The work of M. Kamgarpour was supported by AFOSR under grant FA9550-06-1-0312 and by National Science and Engineering Research Council of Canada (NSERC).

General Terms

Theory, Verification

Keywords

reachability, dynamic programming, Markov processes, random sets

1. INTRODUCTION

Reachability analysis of deterministic dynamical systems constitutes a practically important and intensely researched area in control theory. Over the years, methods and numerical tools for reachability of continuous time deterministic systems have been well researched (see [5, 17, 19, 28] and the references therein). In particular, the reachability problems considered are often solved via dynamic programming [17, 20]. Additionally, reachability problems for deterministic hybrid systems and uncertain hybrid systems have been addressed using computational methods based on dynamic programming [26] and nonsmooth analysis [13].

Stochastic Hybrid System (SHS) models have become a common mechanism for the analysis and design of complex systems given their ability to capture the variable temporal and spatial behavior often found in realistic systems. In the continuous time setting, early contributions to SHS theory include the works of [11, 14, 16], with [9] establishing a theoretical foundation for the measurability of events for reachability problems. Given that technical issues such as measurability are easier to resolve in the discrete-time setting, consideration of discrete-time stochastic hybrid systems (DTSHS) [4] has also attracted considerable attention. Based on a theoretical foundation for the solution of stochastic optimal control problems of general discrete-time systems of [7], probabilistic reachability of DTSHS has been addressed in [2, 22, 25].

In this paper we extend the recent results of [2, 25] in the area of probabilistic reachability of DTSHS. We consider a probabilistic reach-avoid problem where the objective is to maximize or minimize the probability that a system starting at a specific initial condition will hit a target set while avoiding an unsafe set over a finite time horizon. However, in contrast with [2,25], we consider that the unsafe set (or obstacle set) in the reach-avoid problem may be time-dependent and random. In particular, that it can be accurately modeled by a time-indexed sequence of random closed sets [18,21].

Following the methods of [2, 25], we formulate the reach-avoid problem with random obstacles as a finite horizon stochastic optimal control problem with a sum-multiplicative cost-to-go function. Specifically, we consider two distinct possibilities for the random set-valued obstacle process. In the first, we consider the random set process to be an independent stochastic process, and thus decoupled from the evolution of the DTSHS. In the second case, we consider the obstacle process as a set-valued Markov process that can be expressed through an appropriate parameterization. In both cases, dynamic programming is used to compute the Markov control policy that maximizes or minimizes the cost of the optimal control problem. A numerical example motivated by aircraft motion planning under uncertain weather predictions is provided.

The rest of the work is arranged as follows. In Section 2.1 we briefly recall the DTSHS model of [2]. In Section 2.2, we recall the basic theory of random closed sets and define a set-valued stochastic process as a model for the dynamic obstacle set. In Section 3, we introduce the notion of probabilistic reach-avoid over a finite time horizon with dynamic and stochastic obstacle sets, and develop a mechanism to determine optimal Markov control policies based on dynamic programming. In Section 4 we provide a numerical example.

2. MATHEMATICAL BACKGROUND

Here we recall the DTSHS model and associated semantics introduced in [2] and aspects of the theory of random closed sets [18, 21].

2.1 DTSHS

Throughout, given a Borel set K, $\mathcal{B}(K)$ denotes the Borel σ−algebra of K.

DEFINITION 1. *A discrete-time stochastic hybrid system,* $\mathcal{H} = (\mathcal{Q}, n, \mathcal{A}, T_v, T_q, R)$, *comprises*

- *A discrete state space* $\mathcal{Q} := \{q_1, q_2, ..., q_m\}$, *for some* $m \in \mathbb{N}$;

- *A map* $n : \mathcal{Q} \to \mathbb{N}$ *which assigns to each discrete state value* $q \in \mathcal{Q}$ *the dimension of the continuous state space* $\mathbb{R}^{n(q)}$. *The hybrid state space is then given by* $X := \bigcup_{q \in \mathcal{Q}} \{q\} \times \mathbb{R}^{n(q)}$;

- *A compact Borel space* $\mathcal{A}$ *representing the control space;*

- *A Borel-measurable stochastic kernel on* $\mathbb{R}^{n(\cdot)}$ *given* $X \times \mathcal{A}$, $T_v : \mathcal{B}(\mathbb{R}^{n(\cdot)}) \times X \times \mathcal{A} \to [0,1]$, *which assigns to each* $x = (q, v) \in X$ *and* $a \in \mathcal{A}$ *a probability measure* $T_v(\cdot|x, a)$ *on the Borel space* $(\mathbb{R}^{n(q)}, \mathcal{B}(\mathbb{R}^{n(q)}))$;

- *A discrete stochastic kernel on* $\mathcal{Q}$ *given* $X \times \mathcal{A}$, $T_q : \mathcal{Q} \times X \times \mathcal{A} \to [0,1]$, *which assigns to each* $x \in X$ *and* $a \in \mathcal{A}$ *a probability distribution* $T_q(\cdot|x, a)$ *over* $\mathcal{Q}$;

- *A Borel-measurable stochastic kernel on* $\mathbb{R}^{n(\cdot)}$ *given* $X \times \mathcal{A} \times \mathcal{Q}$, $R : \mathcal{B}(\mathbb{R}^{n(\cdot)}) \times X \times \mathcal{A} \times \mathcal{Q} \to [0,1]$, *which assigns to each* $x \in X$, $a \in \mathcal{A}$, *and* $q' \in \mathcal{Q}$ *a probability measure* $R(\cdot|x, a, q')$ *on the Borel space* $(\mathbb{R}^{n(q')}, \mathcal{B}(\mathbb{R}^{n(q')}))$.

Consider the DTSHS evolving over the finite time horizon $k = 0, 1, ..., N$ with $N \in \mathbb{N}$. We specify the initial state as $x_0 \in X$ at time $k = 0$, and define the notion of a Markov policy.

DEFINITION 2. *A Markov Policy for a DTSHS,* $\mathcal{H}$, *is a sequence* $\mu = (\mu_0, \mu_1, ..., \mu_{N-1})$ *of universally measurable maps* $\mu_k : X \to \mathcal{A}$, $k = 0, 1, ..., N - 1$. *The set of all admissible Markov policies is denoted by* $\mathcal{M}_m$.

Let $\tau_v : \mathcal{B}(\mathbb{R}^{n(\cdot)}) \times X \times \mathcal{A} \times \mathcal{Q} \to [0,1]$ be a stochastic kernel on $\mathbb{R}^{n(\cdot)}$ given $X \times \mathcal{A} \times \mathcal{Q}$, which assigns to each $x \in X$, $a \in \mathcal{A}$, and $q' \in \mathcal{Q}$, a probability measure on the Borel space $(\mathbb{R}^{n(q')}, \mathcal{B}(\mathbb{R}^{n(q')}))$ given by

$$\tau_v(dv'|(q,v), a, q') = \begin{cases} T_v(dv'|(q,v), a), & \text{if } q' = q \\ R(dv'|(q,v), a, q'), & \text{if } q' \neq q. \end{cases}$$

Based on τ_v we introduce the kernel $Q : \mathcal{B}(X) \times X \times \mathcal{A} \to [0,1]$:

$$Q(dx'|x, a) = \tau_v(dv'|x, a, q')T_q(q'|x, a).$$

DEFINITION 3. *Consider the DTSHS,* $\mathcal{H}$, *and time horizon* $N \in \mathbb{N}$. *A stochastic process* $\{x_k, k = 0, ..., N\}$ *with values in* X *is an execution of* $\mathcal{H}$ *associated with a Markov policy* $\mu \in \mathcal{M}_m$ *and an initial condition* $x_0 \in X$ *if and only if its sample paths are obtained according to the DTSHS Algorithm.*

Algorithm 1 DTSHS Algorithm

Require: Sample Path $\{x_k, k = 0, ..., N\}$
Ensure: Initial hybrid state $x_0 \in X$ at time $k = 0$, and Markov control policy $\mu = (\mu_0, \mu_1, ..., \mu_{N-1}) \in \mathcal{M}_m$
1: **while** $k < N$ **do**
2:     Set $a_k = \mu_k(x_k)$
3:     Extract from X a value x_{k+1} according to $Q(\cdot|x_k, a_k)$
4:     Increment k
5: **end while**

Equivalently, the DTSHS $\mathcal{H}$ can be described as a Markov control process with state space X, control space $\mathcal{A}$, and controlled transition probability function Q. Further, given a specific control policy $\mu \in \mathcal{M}_m$ and initial state $x_0 \in X$, the execution $\{x_k, k = 0, ..., N\}$ is a time inhomogeneous stochastic process defined on the canonical sample space $\Omega = X^{N+1}$, endowed with its product σ−algebra $\mathcal{B}(\Omega)$. The probability measure $P_{x_0}^\mu$ is uniquely defined by the transition kernel Q, the Markov policy $\mu \in \mathcal{M}_m$, and the initial condition $x_0 \in X$ (see [7]). From now on, we will use interchangeably the notation $Q(\cdot|x, \mu_k(x))$ and $Q_x^{\mu_k}(\cdot|x)$ to represent the one-step transition kernel.

2.2 Random Sets

For the hybrid state space X one can select a metric d such that (X, d) becomes a complete separable metric space (see e.g. [11]). Let $\mathcal{K}$ denote the set of all closed subsets of the hybrid state space X and let d_H denote the Hausdorff

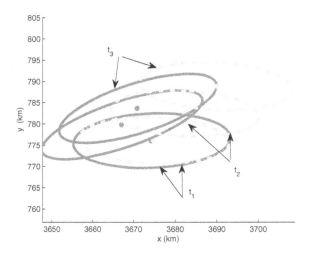

Figure 1: A trajectory of forecasted and realized weather obstacles over a 15 minute horizon according to [12] for a section of airspace centered at latitude 30° and longitude 86.5^o, near the gulf coast of Florida, on 01/07/2009. The forecast storms are shown in dashed lines. The horizon is shown with the labels $t_1 = 5$ minutes, $t_2 = 10$ minutes, and $t_3 = 15$ minutes.

metric. It follows that $(\mathcal{K}, d_H)$ is also a complete separable metric space where the corresponding open subsets generate a σ-algebra on $\mathcal{K}$ [21], i.e. the Borel σ-algebra $\mathcal{B}(\mathcal{K})$ corresponding to the Hausdorff metric of $\mathcal{K}$.

DEFINITION 4. *A random closed set is a measurable function Ξ from a probability space $(\Omega, \mathcal{F}, P)$ into the measure space $(\mathcal{K}, \mathcal{B}(\mathcal{K}))$.*

The distribution of a random closed set Ξ, is given by the probabilities

$$P\{\Xi \cap F \neq \emptyset\}$$

for $F \in \mathcal{K}$. For $F = \{x\} \in X$, the probability $P\{x \in \Xi\} = P\{\omega \in \Omega : x \in \Xi(\omega)\}$ is obtained which satisfies the expression

$$P\{x \in \Xi\} = 1 - P\{x \notin \Xi\}.$$

We refer to the function $p_\Xi(x) = P\{x \in \Xi\}$ as the covering function. For some set $K \subseteq X$, let $\mathbf{1}_K(\cdot) : X \to \{0,1\}$ denote the indicator function. The covering function can also be interpreted as the mean of the indicator function $\mathbf{1}_\Xi$, i.e.

$$p_\Xi(x) = E[\mathbf{1}_\Xi(x)].$$

Note that the covering function is a universally measurable function [18] and takes values between 0 and 1.

We now define a stochastic set-valued process to be used as a model for obstacle movement. For $k = 0, 1, 2, \ldots, N$, let G_k be a Borel-measurable stochastic kernel on $\mathcal{K}$ given $\mathcal{K}$, $G_k : \mathcal{B}(\mathcal{K}) \times \mathcal{K} \to [0,1]$, which assigns to each $K \in \mathcal{K}$ a probability measure $G_k(\cdot|K)$ on the Borel space $(\mathcal{K}, \mathcal{B}(\mathcal{K}))$. That is, let G_k represent a collection of probability measures on $(\mathcal{K}, \mathcal{B}(\mathcal{K}))$ parametrized by the elements of $\mathcal{K}$ and indexed by time k. A discrete-time time-inhomogeneous set-valued Markov process $\boldsymbol{\Xi} = (\Xi_k)_{k \in \mathbb{N}_0}$ taking values in the

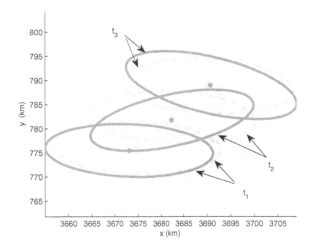

Figure 2: A trajectory of the random obstacles based on forecast data is shown. The forecasted ellipses are represented by dashed lines and the realization of the random ellipses is represented as solid lines.

Borel space $\mathcal{K}$ is described by the stochastic kernel G_k. This general model for the set-valued Markov process includes as special cases time-homogeneous set-valued Markov processes and independent distributions of random sets taking values according to time-indexed random closed set stochastic kernels.

In most cases the characterization of a stochastic set valued process and the computation of associated functions (e.g. the covering function) is difficult due to the size of $\mathcal{K}$. Yet, methods have been suggested in the literature that alleviate the complexity of these processes and the related functions [10,24]. For example, random closed set processes are often characterized by families of closed subsets of $\mathcal{K}$ which are parametrized by real parameters (referred to as morphological transformations in [18]).

DEFINITION 5. *A parameterization of a discrete-time set-valued Markov process $\boldsymbol{\Xi}$ is a discrete-time Markov process $\xi = (\xi_k)_{k \in \mathbb{N}_0}$ with parameter space $\mathcal{O}$ and transition probability function $T_k : \mathcal{B}(\mathcal{O}) \times \mathcal{O} \to [0,1]$ together with a function $\gamma : \mathcal{O} \to \mathcal{K}$ such that*

$$\boldsymbol{\Xi} = (\Xi_k)_{k \in \mathbb{N}_0} = (\gamma(\xi_k))_{k \in \mathbb{N}_0}.$$

From now on we restrict our attention to parameterized discrete-time set-valued Markov processes. Analysis of the associated functions is often completed via Monte Carlo methods. Consider the following example.

2.3 Example - Vertically Integrated Liquid

In aircraft path planning, the ability to identify and characterize regions of hazardous weather is vitally important. One factor that can be used to determine the safety of a region of the airspace for an aircraft to fly through is the Vertically Integrated Liquid (VIL) water content measurement [3], which represents the level of precipitation in a column of the airspace. This measurement has proven useful in the detection of severe storms and short-term rainfall forecasting [8], and hence can be used as an indicator for establishing a no-fly zone for aircraft: A region of the airspace

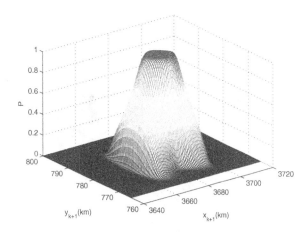

Figure 3: An example covering function for the set-valued Markov process of Section 2.2.

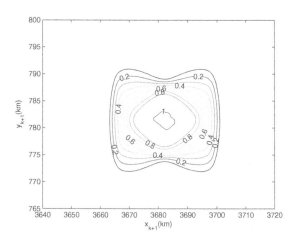

Figure 4: Contour plot of the covering function shown in Figure 3.

with a VIL measurement level above a certain threshold is recommended as a no-fly zone for aircraft.

Regions with VIL levels above the safety threshold come in numerous shapes and sizes. To simplify the expression of these regions, one method of representation is to enclose the no-fly region by minimum volume ellipsoids. Simplifying the problem further, we consider a constant flight level for the aircraft and consequently focus on ellipses in two dimensions. Consider the forecast data at time instance k, from which we obtain elliptical obstacles $\{\mathcal{E}_k^l\}$, where $l = 1, 2, \ldots, L_k$, and L_k denotes the number of obstacles at discrete time k. Each ellipse $\mathcal{E}_k^l$ is parameterized by its center $m_k^l \in \mathbb{R}^2$ and its positive definite eccentricity matrix M_k^l:

$$x \in \mathcal{E}_k^l(m_k^l, M_k^l) \iff (x - m_k^l)^T M_k^l (x - m_k^l) \leq 1 \quad (6)$$

For simplicity, in the sequel we assume that there exists $l = 1$ elliptical no-fly region and denote the ellipse and its representative parameters at time k by $\mathcal{E}_k$, m_k, and M_k.

Uncertainties associated with forecast data naturally exist, and become more prominent with the horizon length of the forecast. For example, Figure 1 shows forecast storm regions versus actual storm regions over a 15 minute horizon at samples of 5 minutes. Figure 1 clearly illustrates the effect that uncertainties can have on aircraft path planning. We account for these uncertainties by modeling the hazardous regions (i.e. obstacles) as a sequence of random closed sets whose distribution is determined from forecast data and associated statistics. In this formulation, the subsets of the airspace that are unsafe to fly through become probabilistic.

As stated in Section 2.2, mathematically characterizing the uncertainty in the forecast, and hence modeling the hazardous regions as random closed sets, is difficult in general. However, given the ellipse parameterization introduced in (6), we can model the dynamic obstacles as random closed sets by considering the parameters of the ellipse representation (e.g. the ellipse center) as random variables distributed according to the forecast data. Assuming that the forecast data is available in 5 minute increments and is given in terms of the expected ellipse center m_k and eccentricity matrix M_k, we consider two mathematical formulations for

the model of the random sets. In the first, we consider that the evolution of the ellipse centers is Markovian, and hence the obstacle process is a set-valued Markov process (i.e. the distribution of $\mathcal{E}_{k+1}$ is a function of $\mathcal{E}_k$). In the second, we consider the distribution of the obstacle at time $k + 1$ to be independent of $\mathcal{E}_k$. In the current section we consider the Markov model for the set-valued process, although both special cases will be addressed in Sections 3 and 4.

Consider a storm region characterized by its minimum-volume ellipse and let m_k and m_{k+1} be the forecast center of the ellipse at time k and time $k+1$ respectively. Then the term $\mu_k \in \mathbb{R}^2$ defined as $\mu_k = m_{k+1} - m_k$, for $k \in \mathbb{N}$ denotes the incremental motion of the storm. Given that there are uncertainties in the forecast, we assume that the true center is a random variable whose motion is described as

$$\xi_{k+1} = \xi_k + \mu_k + \eta_k, \quad (7)$$

where $\eta_k \sim \mathcal{N}(0, \Sigma)$. In addition, we approximate the variation in the ellipse eccentricity according to the expression

$$C_k = R(\theta_k)^T M_k R(\theta_k), \quad (8)$$

where M_k is the eccentricity obtained from the forecast at time k, $R(\cdot)$ is a rotation matrix, and the angle of rotation θ_k at time k is a random variable with uniform distribution over an interval $[-\alpha, \alpha]$. Note that in this formulation only the ellipse centers are Markov, although one could also consider the angle of rotation to be Markov as well. The noise parameters Σ and α are best determined from the quality of the forecast and the rate of movement of the storms. A trajectory of the forecast ellipses and a realization of the random ellipses over a horizon of 15 minutes is shown in Figure 2. Based on the analysis of the storm movements [12], the noise parameters were set to $\Sigma = I_{2 \times 2}$ and $\alpha = \frac{\pi}{6}$.

Given that the objective is to evaluate the safety of an aircraft path with respect to hazardous weather, the covering function for the random closed sets is of immediate interest. That is, for an aircraft position $x_k \in \mathbb{R}^2$ at some time k, the probability of being in the hazardous region $\Xi_k = \mathcal{E}_k(\xi_k, C_k)$ is given by

$$P\{x_k \in \Xi_k\} = P\{(x_k - \xi_k)^T C_k (x_k - \xi_k) \leq 1\}.$$

Unfortunately, the calculation of the above probability is

not analytically possible. For the case in which the eccentricity of the ellipse is assumed to be deterministic, the above probability obeys a Chi-squared distribution and can be approximated using statistical computational tools. For the more general case, one can use Monte Carlo simulations to approximate the probabilities of hitting an obstacle. For example, consider the Markov set-valued obstacle process introduced above with $\xi_k = [3675 \ 775]^T$, $\mu_k = [7.1 \ 6.4]^T$, and $M_{k+1} = [0.0028 \ 0; \ 0 \ 0.0278]$. The covering function $p_{\Xi_{k+1}}(x_{k+1})$ at time $k+1$ is approximated over the region $[3640 \ 3720] \times [765 \ 800]$ using 10^5 Monte Carlo samples and a 101×101 grid discretization. The result is displayed in Figure 3 and Figure 4.

3. FINITE HORIZON REACH-AVOID

Let $K, K'_k \in \mathcal{B}(X)$, with $K \subseteq K'_k$ for all $k = 0, 1, \ldots, N$. We define the stopping time associated with hitting K as $\tau := \inf\{j \geq 0 | x_j \in K\}$, and the stopping time associated with hitting $X \setminus K'_k$ as $\tau' := \inf\{j \geq 0 | x_j \in X \setminus K'_j\}$; if a set is empty we set its infimum equal to $+\infty$. Our goal is to evaluate the probability that the execution of the Markov control process associated with the Markov policy $\mu \in \mathcal{M}_m$ and the initial condition x_0 will hit K before hitting $X \setminus K'_k$ during the time horizon N. We assume that the initial avoid set $\Xi_0 = X \setminus K'_0$ is known and $\Xi_k = X \setminus K'_k$ for $k = 1, \ldots, N$ is an execution of the stochastic set-valued process G. The probability that the system initialized at $x_0 \in X$, with control policy $\mu \in \mathcal{M}_m$ and initial avoid set $\Xi_0 \in \mathcal{K}$, reaches K while avoiding $X \setminus K'_k$ for all $k = 0, 1, \ldots, N$ is given by

$$
\begin{aligned}
r^\mu_{(x_0, \Xi_0)}(K) &:= P^\mu_{(x_0, \Xi_0)}\{\exists j \in [0, N] : x_j \in K \ \wedge \\
&\qquad\qquad \forall i \in [0, j-1] \ x_i \in K'_i \setminus K\}, \\
&= P^\mu_{(x_0, \Xi_0)}\{\{\tau < \tau'\} \wedge \{\tau \leq N\}\},
\end{aligned}
$$

where $\wedge$ denotes the logical AND, and we operate under the assumption that the requirement on i is automatically satisfied when $x_0 \in K$; subsequently we will use a similar convention for products, i.e. $\prod_{i=k}^{j}(\cdot) = 1$ if $k > j$.

As in [25], consider

$$
\sum_{j=0}^{N}\left(\prod_{i=0}^{j-1} \mathbf{1}_{K'_i \setminus K}(x_i)\right)\mathbf{1}_K(x_j) =
$$

$$
\begin{cases} 1, & \text{if } \exists j \in [0, N] : x_j \in K \wedge \\ & \quad \forall i \in [0, j-1] \ x_i \in K'_i \setminus K \\ 0, & \text{otherwise.} \end{cases}
$$

Hence $r^\mu_{(x_0, \Xi_0)}(K)$ can be expressed as the expectation

$$
r^\mu_{(x_0, \Xi_0)}(K) = E^\mu_{(x_0, \Xi_0)}\left[\sum_{j=0}^{N}\left(\prod_{i=0}^{j-1} \mathbf{1}_{K'_i \setminus K}(x_i)\right)\mathbf{1}_K(x_j)\right].
$$

Analytical (and computational) evaluation of $r^\mu_{(x_0, \Xi_0)}(K)$ can be separated into two distinct classes of problems. In the first class of problems we consider the dynamics of the DTSHS and the set-valued obstacle process to be decoupled. We assume that the set-valued obstacle process is described (or can be fairly approximated) by a time-indexed independent distribution of random sets. Further, we assume the control actions (optimal or otherwise) do not depend on the state of the set-valued obstacle process at each step in time.

It follows (in Section 3.1) that a property of this model class is the ability to separate the computational burden of the obstacle process from the computational burden of the DT-SHS.

In the second class of problems, we consider the dynamics of the DTSHS and the set-valued obstacle process to be coupled. We assume that the set-valued obstacle process is modeled as a set-valued Markov process and that the control policy can depend on both the state of the DTSHS and the state of the obstacle process. The second class of systems subsumes the first class of systems, hence a more general set of models and problems is considered. However, the approach for the second class of systems (introduced in Section 3.2) requires that the state space of the Markov process be augmented, consequently restricting the size of problems that can be approximated numerically due to the Curse of Dimensionality [6]. As a result, it is sometimes necessary to approximate a coupled system with a decoupled system (in both the model dynamics and the space of control policies). While computationally prudent, this approximation will lead to inferior success rates since the available control policy cannot make use of the state of the obstacle at each time step.

In Sections 3.1 and 3.2 we assume that the obstacle process can be characterized according to Definition 5. In Section 3.1 this results in the ability to computationally evaluate the covering functions via Monte Carlo analysis. In Section 3.2 this results in the ability to augment the state space of the DTSHS with the parameters of the obstacle process.

3.1 Decoupled Markov Process

Assume that the set-valued obstacle process is described (or can be fairly approximated) by a time-indexed independent distribution of random sets. It follows that the product measure of the obstacle process is equal to (or well approximated by) the product measure of time-indexed independent stochastic kernels, i.e. for $N \in \mathbb{N}$

$$
\prod_{j=0}^{N-1} G_j(d\Xi_j | \Xi_{j-1}) \approx \prod_{j=0}^{N-1} G_j(d\Xi_j).
$$

Note that since the initial state of the obstacle Ξ_0 is assumed known, we define $G_0(d\Xi_0 | \Xi_{0-1}) = G_0(d\Xi_0) = \delta_{\Xi_0}(d\Xi_0)$.

For a DTSHS with independent set-valued obstacle processes, it can be shown that

$$
\begin{aligned}
r^\mu_{(x_0, \Xi_0)}(K) &= E^\mu_{(x_0, \Xi_0)}\left[\sum_{j=0}^{N}\left(\prod_{i=0}^{j-1} \mathbf{1}_{K'_i \setminus K}(x_i)\right)\mathbf{1}_K(x_j)\right], \\
&= E^\mu_{x_0}\left[\sum_{j=0}^{N}\left(\prod_{i=0}^{j-1} p_{K'_i \setminus K}(x_i)\right)\mathbf{1}_K(x_j)\right],
\end{aligned}
$$

where the covering function notation is used liberally for the simplifying expression $p_{K'_i \setminus K}(x_i) = \mathbf{1}_{X \setminus K}(x_i) - p_{\Xi_i}(x_i)$ (since $K'_i \setminus K$ is not necessarily a random closed set). Clearly, the covering functions are defined

$$
p_{\Xi_i}(x_i) = E\left[\mathbf{1}_{\Xi_i}(x_i)\right] = \int_{\mathcal{K}} \mathbf{1}_{\Xi_i}(x_i)G_i(d\Xi_i).
$$

A proof (by Fubini's Theorem [23]) of the preceding claim

follows:

$$r^{\mu}_{(x_0,\Xi_0)}(K) = E^{\mu}_{(x_0,\Xi_0)}\left[\sum_{j=0}^{N}\left(\prod_{i=0}^{j-1}\mathbf{1}_{K'_i\setminus K}(x_i)\right)\mathbf{1}_K(x_j)\right]$$

$$= \int_{X^N\times\mathcal{K}^N}\left[\sum_{j=0}^{N}\left(\prod_{i=0}^{j-1}(\mathbf{1}_{X\setminus K}(x_i)-\mathbf{1}_{\Xi_i}(x_i))\right)\right.$$
$$\left.\mathbf{1}_K(x_j)\right]\prod_{j=0}^{N-1}Q^{\mu_j}(dx_{j+1}|x_j)G_j(d\Xi_j)$$

$$= \int_{X^N}\int_{\mathcal{K}^N}\left[\sum_{j=0}^{N}\left(\prod_{i=0}^{j-1}(\mathbf{1}_{X\setminus K}(x_i)-\mathbf{1}_{\Xi_i}(x_i))\right)\right.$$
$$\left.\mathbf{1}_K(x_j)\right]\prod_{j=0}^{N-1}G_j(d\Xi_j)\prod_{j=0}^{N-1}Q^{\mu_j}(dx_{j+1}|x_j)$$

$$= \int_{X^N}\left[\sum_{j=0}^{N}\left(\int_{\mathcal{K}^j}\prod_{i=0}^{j-1}(\mathbf{1}_{X\setminus K}(x_i)-\mathbf{1}_{\Xi_i}(x_i))\right.\right.$$
$$\left.\left.\prod_{i=0}^{j-1}G_i(d\Xi_i)\right)\mathbf{1}_K(x_j)\right]\prod_{j=0}^{N-1}Q^{\mu_j}(dx_{j+1}|x_j)$$

$$= \int_{X^N}\left[\sum_{j=0}^{N}\left(\prod_{i=0}^{j-1}(\mathbf{1}_{X\setminus K}(x_i)-\right.\right.$$
$$\left.\left.\int_{\mathcal{K}}\mathbf{1}_{\Xi_i}(x_i)G_i(d\Xi_i))\right)\mathbf{1}_K(x_j)\right]\prod_{j=0}^{N-1}Q^{\mu_j}(dx_{j+1}|x_j)$$

$$= \int_{X^N}\left[\sum_{j=0}^{N}\left(\prod_{i=0}^{j-1}p_{K'_i\setminus K}(x_i)\right)\mathbf{1}_K(x_j)\right]$$
$$\prod_{j=0}^{N-1}Q^{\mu_j}(dx_{j+1}|x_j)$$

$$= E^{\mu}_{x_0}\left[\sum_{j=0}^{N}\left(\prod_{i=0}^{j-1}p_{K'_i\setminus K}(x_i)\right)\mathbf{1}_K(x_j)\right].$$

For a fixed Markov policy $\mu \in \mathcal{M}_m$, let us define the functions $V^{\mu}_k : X \to [0,1]$, $k = 0,\ldots,N$ as

$$V^{\mu}_N(x) = \mathbf{1}_K(x) , \tag{9}$$
$$V^{\mu}_k(x) = \mathbf{1}_K(x)+$$
$$p_{K'_k\setminus K}(x)\int_{X^{N-k}}\sum_{j=k+1}^{N}\left(\prod_{i=k+1}^{j-1}p_{K'_i\setminus K}(x_i)\right)$$
$$\mathbf{1}_K(x_j)\prod_{j=k+1}^{N-1}Q^{\mu_j}(dx_{j+1}|x_j)Q^{\mu_k}(dx_{k+1}|x). \tag{10}$$

Note that

$$V^{\mu}_0(x_0) = E^{\mu}_{x_0}\left[\sum_{j=0}^{N}\left(\prod_{i=0}^{j-1}p_{K'_i\setminus K}(x_i)\right)\mathbf{1}_K(x_j)\right]$$
$$= r^{\mu}_{(x_0,\Xi_0)}(K).$$

Let $\mathcal{F}$ denote the set of functions from X to $\mathbb{R}$ and define

the operator $H : X \times \mathcal{A} \times \mathcal{F} \to \mathbb{R}$ as

$$H(x,a,Z) := \int_X Z(y)Q(dy|x,a). \tag{11}$$

The following lemma shows that $r^{\mu}_{(x_0,\Xi_0)}(K)$ can be computed via a backwards recursion.

LEMMA 12. *Fix a Markov policy* $\mu = (\mu_0,\mu_1,\ldots\mu_{N-1}) \in \mathcal{M}_m$. *The functions* $V^{\mu}_k : X \to [0,1]$, $k = 0,1,\ldots,N-1$ *can be computed by the backward recursion:*

$$V^{\mu}_k(x) = \mathbf{1}_K(x) + p_{K'_k\setminus K}(x)H(x,\mu_k(x),V^{\mu}_{k+1}), \tag{13}$$

initialized with $V^{\mu}_N(x) = \mathbf{1}_K(x)$, $x \in X$.

PROOF 14. *By induction. First, due to the definition of (9) and (10), we have that*

$$V^{\mu}_{N-1}(x) = \mathbf{1}_K(x) +$$
$$p_{K'_{N-1}\setminus K}(x)\int_X V^{\mu}_N(x_N)Q^{\mu_{N-1}}(dx_N|x),$$

so that (13) is proven for $k = N-1$. *For* $k < N-1$ *we can separate the terms associated with* x_{k+1} *as follows*

$$V^{\mu}_k(x) = \mathbf{1}_K(x) +$$
$$p_{K'_k\setminus K}(x)\int_X\left(\mathbf{1}_K(x_{k+1}) + p_{K'_{k+1}\setminus K}(x_{k+1})\right.$$
$$\int_{X^{N-k-1}}\sum_{j=k+2}^{N}\left(\prod_{i=k+2}^{j-1}p_{K'_i\setminus K}(x_i)\right)\mathbf{1}_K(x_j)$$
$$\left.\prod_{j=k+2}^{N-1}Q^{\mu_j}(dx_{j+1}|x_j)Q^{\mu_{k+1}}(dx_{k+2}|x_{k+1})\right)$$
$$Q^{\mu_k}(dx_{k+1}|x)$$
$$= \mathbf{1}_K(x) +$$
$$p_{K'_k\setminus K}(x)\int_X V^{\mu}_{k+1}(x_{k+1})Q^{\mu_k}(dx_{k+1}|x)$$

which concludes the proof. $\square$

DEFINITION 15. *Let* $\mathcal{H}$ *be a Markov control process,* $\Xi = (\Xi_k)_{k\in\mathbb{N}_0}$ *a random closed set stochastic process,* $K \in \mathcal{B}(X)$, $K'_k \in \mathcal{B}(X)$, *with* $K \subseteq K'_k$ *and* $K'_k = X \setminus \Xi_k$ *for all* $k = 0,1,2,\ldots,N$. *A Markov policy* μ^* *is a maximal reach-avoid policy if and only if* $r^{\mu^*}_{(x_0,\Xi_0)}(K) = \sup_{\mu\in\mathcal{M}_m} r^{\mu}_{(x_0,\Xi_0)}(K)$, *for all* $x_0 \in X$.

THEOREM 16. *Define* $V^*_k : X \to [0,1]$, $k = 0,1,\ldots,N$, *by the backward recursion:*

$$V^*_k(x) = \sup_{a\in\mathcal{A}}\{\mathbf{1}_K(x) + p_{K'_k\setminus K}(x)H(x,a,V^*_{k+1})\} \tag{17}$$

initialized with $V^*_N(x) = \mathbf{1}_K(x)$, $x \in X$. *Then,* $V^*_0(x_0) = \sup_{\mu\in\mathcal{M}_m} r^{\mu}_{(x_0,\Xi_0)}(K)$, $x_0 \in X$ *and* $\Xi_0 \in \mathcal{K}$. *If* $\mu^*_k : X \to \mathcal{A}$, $k \in [0,N-1]$, *is such that for all* $x \in X$

$$\mu^*_k(x) = \arg\sup_{a\in\mathcal{A}}\{\mathbf{1}_K(x) + p_{K'_k\setminus K}(x)H(x,a,V^*_{k+1})\} \tag{18}$$

then $\mu^* = (\mu^*_0,\mu^*_1,\ldots,\mu^*_{N-1})$ *is a maximal reach-avoid policy. A sufficient condition for the existence of* μ^* *is that* $U_k(x,\lambda) = \{a \in \mathcal{A}|H(x,a,V^*_{k+1}) \geq \lambda\}$ *is compact for all* $x \in X$, $\lambda \in \mathbb{R}$, $k \in [0,N-1]$.

PROOF 19. *For all $k = 0, 1, ..., N$, the covering function p_{Ξ_k}, and therefore $p_{K'_k \setminus K}$, is a universally measurable function [18]. Hence, we apply the proof of Theorem 6 in [25] with $p_{K'_k \setminus K}$ replacing $\mathbf{1}_{K' \setminus K}$ everywhere.* $\square$

Note that Theorem 16 gives a sufficient condition for the existence of an optimal nonrandomized Markov policy. While the consideration of randomized Markov policies is indeed interesting in the event that an optimal nonrandomized Markov policy does not exist, in most cases the "best" policy can be taken to be nonrandomized [7]. In light of this fact and in the interest of space, we do not consider randomized Markov policies in the present work and urge the interested reader to consider Chapter 8 in [7] for additional details.

3.2 Coupled Markov Process

Assume that the sequence of obstacles is modeled as a set-valued Markov process. It follows that the product measure of the obstacle process is equal to the product measure of the stochastic kernel G, i.e. for $N \in \mathbb{N}$ the product measure is

$$\prod_{j=0}^{N-1} G(d\Xi_j | \Xi_{j-1}).$$

Note that since the initial state of the obstacle Ξ_0 is assumed known, we define $G(d\Xi_0 | \Xi_{0-1}) = \delta_{\Xi_0}(d\Xi_0)$.

By Definition 5, we have that an equivalent characterization of the set-valued Markov process $\mathbf{\Xi}$ with transition kernel G is given by the discrete-time Markov process $\xi = (\xi_k)_{k \in \mathbb{N}_0}$ with parameter space $\mathcal{O}$ and transition probability function T along with the function γ. Let $\bar{x} \in \bar{X}$ be the augmented state of the DTSHS, where $\bar{x} = \begin{bmatrix} x^T, & \xi^T \end{bmatrix}^T$ and $\bar{X} = X \times \mathcal{O}$ is the augmented state space of the DTSHS. Further, let us define the stochastic kernel $\bar{Q} : \mathcal{B}(\bar{X}) \times \bar{X} \times \mathcal{A} \to [0,1]$:

$$\bar{Q}(d\bar{x}' | \bar{x}, a) = Q(dx' | x, a) T(d\xi' | \xi).$$

We call the resulting process an augmented DTSHS (ADT-SHS) $\bar{\mathcal{H}}$.

DEFINITION 20. *A Markov Policy for an augmented DT-SHS, $\bar{\mathcal{H}}$, is a sequence $\mu = (\mu_0, \mu_1, ..., \mu_{N-1})$ of universally measurable maps $\mu_k : \bar{X} \to \mathcal{A}$, $k = 0, 1, ..., N - 1$. The set of all admissible Markov policies is denoted by $\bar{\mathcal{M}}_m$.*

Hence, the stochastic reach-avoid problem with time-varying probabilistic obstacles is transformed into a stochastic reach-avoid problem with deterministic obstacles, and thus can be solved using the nominal reach-avoid methods in [25].

4. AIRCRAFT PATH PLANNING

Here we consider a discrete-time stochastic hybrid model of aircraft motion inspired by the work [27]. We model the aircraft motion as a simple point mass unicycle with three modes of operation; straight flight, right turn, and left turn. The discrete-time continuous dynamics of the aircraft are given by

$$x_{k+1}^1 = x_k^1 + t_e a_k^1 \cos(x_k^3) + w_k^1 \qquad (21)$$
$$x_{k+1}^2 = x_k^2 + t_e a_k^1 \sin(x_k^3) + w_k^2$$
$$x_{k+1}^3 = x_k^3 + t_e a_k^2 + w_k^3$$

where $x = \begin{bmatrix} x^1, & x^2, & x^3 \end{bmatrix}^T \in \mathbb{R}^3$ are the states of the system, $a = \begin{bmatrix} a^1, & a^2 \end{bmatrix}^T \in \mathcal{A}$ are the control variables for the system, $w = \begin{bmatrix} w^1, & w^2, & w^3 \end{bmatrix}^T \sim \mathcal{N}(0, \Sigma_w)$ is the process noise of the system, and t_e is the sampling time according to a Euler discretization of the continuous time model in [27]. In the model, $\begin{bmatrix} x^1, & x^2 \end{bmatrix}^T \in \mathbb{R}^2$ denotes the position of the aircraft in two dimensions and $x^3 \in [-\pi, \pi]$ denotes the heading angle of the aircraft. The linear velocity of the aircraft takes values between the minimum and maximum aircraft velocity, i.e. $a^1 \in [v_{\min}, v_{\max}]$, with $v_{\min} \in \mathbb{R}$ and $v_{\max} \in \mathbb{R}$. The angular velocity of the aircraft takes one of three possible values, corresponding to the three modes of operation of the DTSHS, i.e. $a^2 \in \{0, -u, u\}$ where $u \in \mathbb{R}$ is the angular velocity of the aircraft when in turning mode. In the following we consider a sampling time of $t_e = 1$ minute, aircraft speed $a^1 = 7.1$ km per minute, angular velocity $u = 0.3$ radians per minute, and disturbance variance $\Sigma_w \in \mathbb{R}^{3 \times 3}$ defined by $\Sigma_w(1,1) = 0.25$, $\Sigma_w(2,2) = 0.25$, $\Sigma_w(3,3) = 0.05$, and $\Sigma_w(i,j) = 0$ if $i \neq j$. As in [27], the protected zone of the aircraft is an 8 km cylindrical block in the state space x that should not intersect the weather obstacle.

In the following examples, we consider the maximization and verification of aircraft trajectory safety and success given a probabilistic hazard forecast. In the first example, we model the probabilistic obstacles as a sequence of independent random closed sets and evaluate the probability of the aircraft attaining a target region while avoiding the hazardous regions. In the second example, we augment the state space of the aircraft DTSHS (21) with a parameterization of the hazardous regions and evaluate the probability of safety of the aircraft.

In both examples, set-valued obstacle processes are constructed according to the forecast product [12], which provides VIL numbers in a 1 km by 1 km gridded form for the entire United States over a time horizon of two hours. We consider a section of airspace centered at latitude $30°$ and longitude $86.5°$, near the Florida gulf coast, on 01/07/2009, a day in which storms were observed in the region under consideration. Evaluating the data from [12] for the location and time above, we have extracted a thirty minute forecast comprising centers m_k and eccentricities M_k at one minute increments, i.e. $k \in \{0, ..., 30\}$. Figure 5 represents the deterministic forecast over a thirty minute period. Executions of the random sets are a function of the deterministic forecast in both the decoupled and coupled examples. Figure 6 shows an aircraft path and obstacle location over a 10 minute period for a flight on the same day. While the aircraft path avoids the deterministic forecast, it intersects the hazardous region which is shown by the true obstacle location obtained from the weather data.

4.1 Decoupled Process

Consider the region $\bar{K} = [3600, 3800] \times [750, 850] \times [-\pi, \pi]$ with target set $K = [3742, 3768] \times [752, 778] \times [-\pi, \pi]$ and safe set

$$K'_k = \bar{K} \setminus \Xi_k.$$

Given forecast data extracted from [12] in the form of expected centers m_k and expected eccentricities M_k, we consider a time-indexed probabilistic model of the ellipse pa-

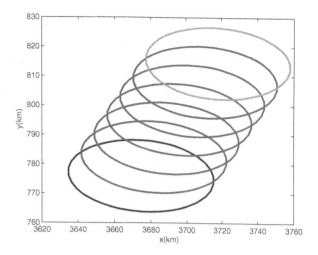

Figure 5: Deterministic weather forecast over a thirty minute period, given at 5 minute increments.

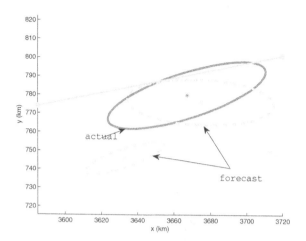

Figure 6: Path of aircraft around the forecasted and actual weather obstacles for a 5-minute portion of the flight. Notice that the path avoids the forecasted ellipse but intersects with the actual storm.

rameters

$$\xi_k \sim \mathcal{N}(m_k, \Sigma_k), \qquad (22)$$

$$C_k = R(\theta_k)^T M_k R(\theta_k), \qquad (23)$$

where Σ_k is a covariance matrix, $R(\cdot)$ is a rotation matrix, and the angle of rotation θ_k at time k is a random variable with uniform distribution over an interval $[-\alpha, \alpha]$. Σ_k and α are defined as in Section 2.3 and the initial parameter values are $\xi_0 = [3675, 776]^T$ and $\theta_0 = 0$ (see Figure 5). Accounting for the hazardous weather regions of Section 2.3 and the protected zone of the aircraft, the random closed set Ξ_k is consequently defined

$$\Xi_k = \mathcal{E}(\xi_k, C_k) \oplus \mathcal{C}(0, 8) \times [-\pi, \pi]$$

where $\mathcal{C}(c, r)$ is a circle (representing the protected zone of the aircraft) defined by its center c and radius r and $\oplus$ denotes the Minkowski sum.

Considering the DTSHS for aircraft motion (21) and the model of probabilistic obstacles, characterized through (22) and (23), we would like to evaluate (and subsequently maximize) the probability that an aircraft attains K while avoiding the hazardous regions over a horizon of thirty minutes (i.e. $k \in \{0, \ldots, 30\}$). All numerical computations were performed on a $201 \times 101 \times 40$ grid according to the methods in [1]. The optimal value function, which represents the maximum probability of attaining the target region safely at some point during the time horizon, is shown in Figure 7 for an initial heading angle of $x_0^3 = -0.0785$ radians. For example, the DTSHS initialized at $x_0 = [3620, \ 830, \ -0.0785]^T$ has a maximum probability of success of 93.3 percent according to the optimal value function. An example execution of the process from $x_0 = [3620, \ 830, \ -0.0785]^T$ is shown in Figure 8.

4.2 Coupled Process

Here we consider the DTSHS for aircraft motion (21) and the set-valued Markov process for obstacle (hazardous weather) movement given in Section 2.3 by the equations (7) and (8). According to Section 3.2 we augment the state of the DTSHS with the state of the obstacle such that the

state of the coupled Markov process is

$$[x^T, \xi^T, \theta]^T \in \mathbb{R}^6.$$

Solving a dynamic program in 6 dimensions is intractable due to the Curse of Dimensionality. We therefore make the following modifications. We assume that $\theta_k = 0$ and $M_k = M$ for all k, thereby removing θ as a state. Additionally, we form a new state corresponding to the relative coordinate of the aircraft and obstacle location

$$[x^1, x^2]^T - \xi \in \mathbb{R}^2$$

and remove the states x^1, x^2, ξ^1, and ξ^2. The resulting state of the coupled process is

$$\bar{x} = \begin{bmatrix} x^1 - \xi^1 \\ x^2 - \xi^2 \\ x^3 \end{bmatrix} = \begin{bmatrix} \bar{x}^1 \\ \bar{x}^2 \\ \bar{x}^3 \end{bmatrix}.$$

Combining equations (21) and (7), the process dynamics of the augmented system are given by the difference equations

$$\bar{x}_{k+1}^1 = \bar{x}_k^1 + t_e a_k^1 \cos(\bar{x}_k^3) + w_k^1 - \mu_k^1 - \eta_k^1 \qquad (24)$$
$$\bar{x}_{k+1}^2 = \bar{x}_k^2 + t_e a_k^1 \sin(\bar{x}_k^3) + w_k^2 - \mu_k^2 - \eta_k^2$$
$$\bar{x}_{k+1}^3 = \bar{x}_k^3 + t_e a_k^2 + w_k^3.$$

By defining an augmented system for the coupled process, which combines the dynamics of the DTSHS and the dynamics of the random obstacle process, we can now define a reach-avoid problem in the spirit of [25]. Consider $\bar{K}_1 = \mathbb{R} \times \mathbb{R} \times [-\pi, \pi]$ and $\bar{K}_2 = [-69, 89] \times [-24, 40] \times [-\pi, \pi]$. We define the target region $K = \bar{K}_1 \setminus \bar{K}_2$ and the safe set $K' = \bar{K}_1 \setminus \Xi$, where the obstacle set Ξ is static, deterministic, and defined

$$\Xi = \mathcal{E}(0, M) \oplus \mathcal{C}(0, 8) \times [-\pi, \pi]$$

where $\mathcal{C}(c, r)$ is a circle (representing the protected zone of the aircraft) defined by its center c and radius r and $\oplus$ denotes the Minkowski sum.

Considering the ADTSHS (24), we would like to evaluate (and subsequently maximize) the probability that an aircraft

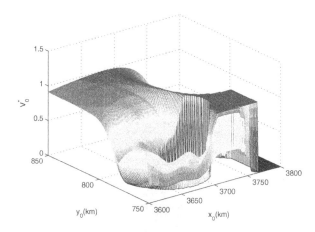

Figure 7: Optimal value function for the aircraft path planning problem with decoupled Markov processes (with initial heading angle $\bar{x}_0^3 = -0.0785$).

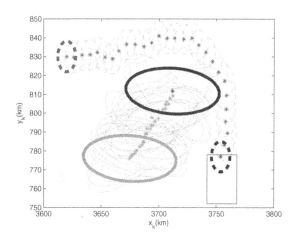

Figure 8: Example process execution for the aircraft path planning problem with decoupled Markov processes.

attains K (i.e. the aircraft escapes a region of the airspace that is considered too close to the hazardous weather region) while avoiding the hazardous region Ξ over a horizon of thirty minutes (i.e. $k \in \{0, \ldots, 30\}$). All numerical computations were performed on a $161 \times 67 \times 20$ grid according to the methods in [1]. The optimal value function, which represents the maximum probability of attaining the target region safely at some point during the time horizon, is shown in Figure 9 for an initial aircraft heading angle of $\bar{x}_0^3 = -0.1571$ radians (note that the value function is shown as $1 - V_0^*$). For example, the DTSHS initialized at $\bar{x}_0 = [-60, \quad -10, \quad -0.1571]^T$ has a maximum probability of success of 83.08 percent according to the optimal value function. The $V_0^* = 0.95$ level set of the optimal value function is shown in Figure 10. Note that all initial conditions which start outside the level set have a success probability greater than 95 percent.

5. CONCLUSION

Extending the methods of [2, 25] and integrating the theory of random closed sets [18, 21], we formulated a reach-avoid problem with random obstacles as a finite horizon stochastic optimal control problem. We considered two possibilities for the random set-valued obstacle process. In the first, we considered the random set process to be an independent stochastic process, and thus decoupled from the evolution of the DTSHS. In the second case, we considered the obstacle process as a set-valued Markov process, equivalently expressed through parameterization. In both cases, it was shown that dynamic programming can be used to compute the Markov control policy that maximizes or minimizes the cost of the optimal control problem. A numerical example motivated by aircraft motion planning under uncertain weather predictions was used to illustrate the effectiveness of the methods introduced.

6. REFERENCES

[1] A. Abate, S. Amin, M. Prandini, J. Lygeros, and S. Sastry. Computational approaches to reachability analysis of stochastic hybrid systems. In A. Bemporad, A. Bicchi, and G. C. Buttazzo, editors, *Hybrid Systems: Computation and Control*, volume 4416 of *Lecture Notes in Computer Science*, pages 4–17. Springer, 2007.

[2] A. Abate, M. Prandini, J. Lygeros, and S. Sastry. Probabilistic reachability and safety for controlled discrete time stochastic hybrid systems. *Automatica*, 44(11):2724–2734, November 2008.

[3] S. A. Amburn and P. L. Wolf. VIL density as a hail indicator. *Weather and Forecasting*, 12(3):473–478, 1997.

[4] S. Amin, A. Abate, M. Prandini, J. Lygeros, and S. Sastry. Reachability analysis for controlled discrete time stochastic hybrid systems. In Hespanha and Tiwari [15], pages 49–63.

[5] J.-P. Aubin. *Viability theory*. Birkhauser Boston Inc., Cambridge, MA, USA, 1991.

[6] R. Bellman. *Dynamic Programming*. Princeton University Press, Princeton, New Jersey, USA, 1957.

[7] D. P. Bertsekas and S. E. Shreve. *Stochastic Optimal Control: The Discrete-Time Case*. Athena Scientific, February 2007.

[8] B. Boudevillain and H. Andrieu. Assessment of vertically integrated liquid (VIL) water content radar measurement. *Journal of Atmospheric and Oceanic Technology*, 20(6):807–819, 2003.

[9] M. L. Bujorianu and J. Lygeros. Reachability questions in piecewise deterministic markov processes. In O. Maler and A. Pnueli, editors, *Hybrid Systems: Computation and Control*, volume 2623 of *Lecture Notes in Computer Science*, pages 126–140. Springer, 2003.

[10] N. Cressie and G. M. Laslett. Random set theory and problems of modeling. *SIAM Review*, 29(4):pp. 557–574, December 1987.

[11] M. Davis. *Markov Models and Optimization*. Chapman & Hall, London, 1993.

[12] J. Evans, K. Carusone, M. Wolfson, B. Crowe, D. Meyer, and D. Klingle-Wilson. The Corridor Integrated Weather System (CIWS). In *10th*

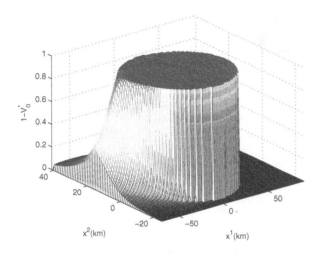

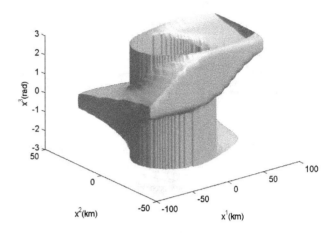

Figure 9: Optimal value function for the aircraft path planning problem with coupled Markov processes (portrayed as $1 - V_0^*$ with initial heading angle $x_0^3 = -0.1571$).

Figure 10: Level set corresponding to $V_0^* = 0.95$ for the aircraft path planning problem with coupled Markov processes.

Conference on Aviation, Range, and Aerospace Meteorology, 2001.

[13] Y. Gao, J. Lygeros, and M. Quincampoix. On the Reachability Problem for Uncertain Hybrid Systems. *IEEE Transactions on Automatic Control*, 52(9):1572–1586, September 2007.

[14] M. K. Ghosh, A. Arapostathis, and S. I. Marcus. Ergodic control of switching diffusions. *SIAM J. Control Optim.*, 35(6):1952–1988, November 1997.

[15] J. P. Hespanha and A. Tiwari, editors. *Hybrid Systems: Computation and Control*, volume 3927 of *Lecture Notes in Computer Science*. Springer, 2006.

[16] J. Hu, J. Lygeros, and S. Sastry. Towards a theory of stochastic hybrid systems. In N. A. Lynch and B. H. Krogh, editors, *Hybrid Systems: Computation and Control*, volume 1790 of *Lecture Notes in Computer Science*, pages 160–173. Springer, 2000.

[17] J. Lygeros. On reachability and minimum cost optimal control. *Automatica*, 40(6):917–927, 2004.

[18] G. Matheron. *Random Sets and Integral Geometry*. Wiley, New York, 1975.

[19] I. Mitchell. The flexible, extensible and efficient toolbox of level set methods. *J. Sci. Comput.*, 35(2-3):300–329, 2008.

[20] I. Mitchell and C. Tomlin. Level set methods for computation in hybrid systems. In *Hybrid Systems: Computation and Control*, pages 310–323, London, UK, 2000. Springer-Verlag.

[21] I. Molchanov. *Theory of Random Sets*. Springer, New York, 2005.

[22] F. Ramponi, D. Chatterjee, S. Summers, and J. Lygeros. On the connections between PCTL and dynamic programming. In *HSCC '10: Proceedings of the 13th ACM international conference on Hybrid systems: computation and control*, pages 253–262, New York, NY, USA, 2010. ACM.

[23] W. Rudin. *Real and complex analysis, 3rd ed.* McGraw-Hill, Inc., New York, NY, USA, 1987.

[24] D. Stoyan. Random sets: Models and statistics. *International Statistical Review*, 66(1):pp. 1–27, April 1998.

[25] S. Summers and J. Lygeros. Verification of discrete time stochastic hybrid systems: A stochastic reach-avoid decision problem. *Automatica*, 46(12):1951 – 1961, 2010.

[26] C. Tomlin, J. Lygeros, and S. Sastry. A Game Theoretic Approach to Controller Design for Hybrid Systems. *Proceedings of IEEE*, 88:949–969, July 2000.

[27] C. Tomlin, G. J. Pappas, and S. Sastry. Conflict resolution for air traffic management: a study in multiagent hybrid systems. *Automatic Control, IEEE Transactions on*, 43(4):509–521, 1998.

[28] C. J. Tomlin, I. Mitchell, A. M. Bayen, and M. Oishi. Computational techniques for the verification of hybrid systems. *Proceedings of the IEEE*, 91(7):986–1001, July 2003.

Impulsive Data Association with an Unknown Number of Targets

Matthew Travers
Northwestern University
Department of Mechanical
Engineering
Evanston, Illinois 60208
botics11@gmail.com

Todd Murphey
Northwestern University
Department of Mechanical
Engineering
Evanston, Illinois 60208
t-
murphey@northwestern.edu

Lucy Pao
University of Colorado
Department of Electrical,
Computer, and Energy
Engineering
Boulder, Colorado
pao@colorado.edu

ABSTRACT

First- and second-order solution methods for the multi-target data association problem with an unknown number of targets are presented. It is shown that by considering a single continuous measurement signal with impulsive switching between measuring the position of different objects, the data association problem can be recast as a continuous optimization over the impulse times and magnitudes. First- and second-order adjoint formulations are derived which reduce the calculation of the either the first- or second-order derivative of the cost function to a single integration (over any number of impulse times and magnitudes). These adjoint formulations as well as a method for estimating the total number of impulses which occur are the main contributions of this work.

Categories and Subject Descriptors

J [**Computer Applications**]: Miscellaneous; J.2 [**Physical Sciences and Engineering**]: Mathematics & Statistical Engineering

General Terms

Measurements & Verification

Keywords

Data Association, Optimization, & Filtering Theory

1. INTRODUCTION

Multi-target data association is a common interest across several different fields [2, 4, 10, 11, 16, 19, 26, 27, 29]. The systems of interest contain multiple targets from which measurements may originate. There is uncertainty in the positions of the targets due to an imperfect system model as well as noise contained in the measurements. The goal of

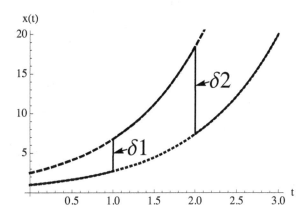

Figure 1: Trajectories of two dynamically identical objects with different initial conditions, represented by the dotted and dashed lines. The solid black line represents the portions of the two trajectories measured over the time horizon $t = [0, 3]$. The change between measuring different trajectories is an impulse in the measurement signal. In this example, the magnitudes of the impulses are δ_1 and δ_2.

the data association problem is to assign the measurements to the correct target.

This paper presents a new method for nonlinear data association referred to as *impulsive data association* (IDA). Figure 1 shows a simple example. In Figure 1, the dotted line represents the trajectory of a pre-specified object of interest (object 1) and the dashed line some other object nearby (object 2). There is a single sensor which produces a single measurement signal that is assumed to be continuous. The solid black line in Figure 1 represents the portions of the two trajectories that the sensor measures over $t = [0, 3]$. The transitions between measuring objects 1 and 2 are impulses, which in this example have magnitudes δ_1 and δ_2. Assuming the object of interest is originally being measured and that the times at which impulses occur in the "measurement trajectory" are known, it is possible to determine the the origin of individual measurements.

The main contribution of this work is the derivation of the first- and second-order adjoint formulations for the derivative of a cost with respect to impulse times, magnitudes, and the second-order coupling between the impulse times and magnitudes. An interesting and computationally useful result, presented in Section 3, is that the same adjoint operators which appear in the first- and second-order derivatives of the cost with respect to the impulse times also appear unmodified in the derivatives with respect to the impulse magnitudes. This result is computationally useful due to the fact that quadratic convergence with respect to the impulse times and magnitudes is achieved by calculating two integrals, regardless of the number of total impulses.

There are several techniques in the literature that address the multi-target data association problem with an unknown number of objects. The most popular approaches are referred to as *maximum a posterior (MAP)* techniques [3]. MAP techniques analytically define the posterior distribution and test various elements from the set of possible solutions, i.e., partitions of the discrete measurements, to determine which one maximizes the posterior . Pre-existing MAP techniques have been shown to perform well in many realistic situations. The main issue facing current MAP techniques is the way in which the solution space is searched. The main theoretical contribution of this work is a method that takes advantage of the maximum principle to search the space of all possible partitions of measurements for an optimum with convergence in quadratic time.

Multiple-hypothesis tracking (MHT) [6, 8, 28] is an example of a MAP technique. The power of MHT is that, given enough time, it will always find the correct answer. The issue with MHT is that it exhaustively tests every element of the solution space at each time step to determine the MAP solution. As the number of measurements increases the MHT algorithm experiences exponential growth in terms of running time. Heuristic methods for dealing with this exponential growth in complexity have been proposed [10, 11, 20, 28], but at the expense of optimality.

More recently, *Markov Chain Monte Carlo data association (MCMCDA)* techniques have been developed [5, 18, 22, 24, 25]. MCMCDA differs from MHT due to the fact that MCMCDA uses MCMC techniques to search the space of possible solutions for areas with nontrivial probability with respect to the posterior distribution. Although this is a more computationally efficient method than MHT, it still experiences run time problems as the number of measurements gets large.

The rest of this paper is organized as follows. The problem is formally defined in Section 2. After defining the problem, the first and second-order derivatives of the cost function (defined in Section 2) with respect to the impulse times and magnitudes are derived (as well as the cross terms for the second derivatives). Finally, a simulated implementation of the IDA algorithm is provided and results analyzed. In Section 5 conclusions and future directions of this work are provided.

2. PROBLEM DEFINITION

In general, impulsive data association is applied to non-linear systems with dynamics of the form

$$\dot{x}^j = f(x^j(t), t), \ x^j(0) = x_0^j,$$
$$z(t) = h(x^j(t), t) + \nu(t), \quad \text{where } j \in \{1, \dots, N\} \quad (1)$$

where N is the total number of objects in the system, each with identical dynamics $f(\cdot)$, and $z(t)$ is the *measurement signal* which contains the noise term $\nu(\cdot)$. In the work presented in this paper we do not assume that N is known.

An assumption that we are making is that we have a way of addressing the uncertainty in the system model, such as the *Kalman-Bucy Filter* [1]. In Section 3 we do not explicitly talk about the trajectory $x(t)$ being the result of a filtered signal, but in practice this is the case. The derivatives of the trajectory that are defined in Section 3 depend only on evaluating the value of the trajectory at certain points. The trajectory $x(t)$ can thus be a filtered signal without loss of generality.

We define a cost function of the form

$$J(\cdot) = \int_{t_0}^{t_f} \ell(x(s), s) ds \quad (2)$$

where, for example $\ell(x(s), s) = (x_d(s) - x(s))^T (x_d(s) - x(s))$, $x_d(\cdot)$ is the reference trajectory (which is the measurement signal in this work), and $x(\cdot)$ is the model of the trajectory. Assume that $\ell(\cdot)$ is C^2, $f(\cdot)$ is C^2, and $x_d(\cdot)$ is C^1.

3. THE DERIVATIVES OF $J(\cdot)$

In this section, we analytically derive the first and second derivatives of the cost function (2) with respect to an arbitrary number of impulse times and magnitudes. The second-order cross terms between the impulse times and magnitudes are also derived. An interesting result is that the same adjoint operators that appear in the first- and second-order derivatives of the cost with respect to the impulse times appear in a similar way when taking derivatives with respect to an arbitrary impulse amplitude.

3.1 First derivatives of $J(\cdot)$

In finding the first derivative of $J(\cdot)$ it is helpful to first derive the derivatives of the trajectory $x(\cdot)$. Note that in Lemmas 1 and 2, the first derivative of the trajectory with respect to the impulse times and magnitudes have the exact same linear form, differing only in initial conditions. Lemma 1 gives the derivative of $x(\cdot)$ with respect to an arbitrary impulse magnitude.

Lemma 1 The first derivative of the trajectory with respect to the impulse magnitudes δ_i is

$$D_{\delta_i} x(t) \circ \partial \delta_i = \begin{cases} 0, & t < \tau_i \\ \Phi(t, \tau_i) \circ \Delta^i, & t \geq \tau_i \end{cases} \quad (3)$$

$$\Delta^i = D_{\delta_i} x(\tau_i) \circ \partial \delta_i$$

where $\Phi(t, \tau_i)$ is the state transition [9] matrix for the system

$$\dot{q} = A(t)q$$

and $A(t) = D_1 f(x(t), t)$ (where $D_1 f$ means the derivative of f with respect to the first argument).

Proof: Using the fundamental theorem of calculus, the trajectory $x(t)$ can be written as

$$x(0) = x_0, \qquad x(t) = x(\tau_i) + \int_{\tau_i}^{t} f(x(s), s) ds, \quad (4)$$

where τ_i can be any time such that $\tau_i < t$. Taking the derivative of (4) with respect to δ_i

$$D_{\delta_i} x(\cdot) \circ \partial \delta_i =$$

$$D_{\delta_i} x(\tau_i) \circ \partial \delta_i + \int_{\tau_i}^{t} D_1 f(x(s), s) \circ D_{\delta_i} x(s) \circ \partial \delta_i ds \quad (5)$$

The fundamental theorem of calculus can be used to rewrite (5) in differential form

$$\frac{\partial}{\partial t} D_{\delta_i} x(t) \circ \partial \delta_i = D_1 f(x(t), t) \circ D_{\delta_i} x(t) \circ \partial \delta_i.$$

with initial condition,

$$D_{\delta_i} x(\tau_i) \circ \partial \delta_i$$

This is a linear differential equation and can thus be represented as a state transition matrix operating on an initial condition as follows:

$$D_{\delta_i} x(t) \circ \partial \delta_i = \Phi(t, \tau_i) \circ D_{\delta_i} x(\tau_i) \circ \partial \delta_i.$$

This result proves the Lemma for $t \geq \tau_i$. For $t < \tau_i$, observe that (4), does not depend on δ_i until time τ_i. ■

The next lemma states the first derivative of the trajectory with respect to an arbitrary impulse time τ_i. Again due to the application of the chain rule, the derivative of the trajectory appears in the derivative of the cost with respect to the impulse times.

Lemma 2 The first derivative of the trajectory with respect to the impulse times τ_i is

$$D_{\tau_i} x(t) \circ \partial \tau_i = \begin{cases} 0, & t < \tau_i \\ \Phi(t, \tau_i) \circ X^i, & t \geq \tau_i \end{cases} \quad (6)$$

$$X^i = (f(x(\tau_i^-), \tau_i^-) - f(x(\tau_i^+), \tau_i^+)) \partial \tau_i$$

where $\Phi(t, \tau_i)$ is the same state transition matrix from Lemma 1, τ_i^- refers to the time right before the i-th impulse, and τ_i^+ the time right after.

Proof: Taking the derivative of (4) with respect to τ_i

$$D_{\tau_i} x(t) \circ \partial \tau_i = D_{\tau_i} x(\tau_i^-) \circ \partial \tau_i - f(x(\tau_i^+), \tau_i^+)$$
$$+ \int_{\tau_i^+}^{t} D_1 f(x(s), s) D_{\tau_i} x(s) \circ \partial \tau_i ds$$
$$= f(x(\tau_i^-), \tau_i^-) - f(x(\tau_i^+), \tau_i^+)$$
$$+ \int_{\tau_i^+}^{t} D_1 f(x(s), s) D_{\tau_i} x(s) \circ \partial \tau_i ds. \quad (7)$$

The first term in (7) is the result of taking the derivative of the initial condition in (4) with respect to its argument. The second term in (7) is the result of applying the Leibniz rule. The fundamental theorem of calculus can be used to rewrite (7) in differential form

$$D_{\tau_i} x(\tau_i) \circ \partial \tau_i = f(x(\tau_i^-), \tau_i^-) - f(x(\tau_i^+), \tau_i^+)$$
$$\frac{\partial}{\partial t} D_{\tau_i} x(t) \circ \partial \tau_i = D_1 f(x(t), t) \circ D_{\tau_i} x(t) \circ \partial \tau_i.$$

This is a linear differential equation and can thus be represented as a state transition matrix operating on an initial condition

$$D_{\tau_i} x(t) = \Phi(t, \tau_i) \circ D_{\tau_i} x(\tau_i) \circ \partial \tau_i.$$

This result is one part of the Lemma. To obtain the second part of the Lemma, go back to Equation (4) and take the derivative of $x(t)$ with respect to an arbitrary τ_k such that $\tau_k < \tau_i$. The initial condition $x(\tau_i^-)$ is constant and thus does not depend on τ_k. For the integral term, $x(t)$ does not depend on τ_k anywhere in the interval $[\tau_i^+, t]$. Thus $D_{\tau_k} x(t) \circ \partial \tau_k = 0$ when $t < \tau_i$. ■

Having derived the first derivatives of the trajectory it is now straightforward to derive the first derivatives of the cost $J(\cdot)$.

Theorem 1 The derivative of the cost function $J(\cdot)$ with respect to the impulse magnitudes δ_i is:

$$D_{\delta_i} J(\cdot) \circ \partial \delta_i = \psi(t_f, \tau_i) \circ \Delta^i. \quad (8)$$

The linear operator $\psi(t_f, \tau_i) : \mathbb{R}^n \to \mathbb{R}$ is found by integrating

$$\psi(t, t) \circ U = 0$$
$$\frac{\partial}{\partial \tau} \psi(t, \tau) \circ U =$$
$$-D_1 \ell(x(\tau), \tau) \circ U - \psi(t, \tau) \circ D_1 f(x(\tau), \tau) \circ U$$

backward along τ from t_f to τ_i.

Proof: Take derivative of (2) with respect to δ_i

$$D_{\delta_i} J(\cdot) \circ \partial \delta_i = \int_{\tau_i}^{t_f} D_1 \ell(x(s), s) D_{\delta_i} x(s) \circ \partial \delta_i ds, \quad (10)$$

and substitute in (3) to obtain

$$D_{\delta_i} J(\cdot) \circ \partial \delta_i = \int_{\tau_i}^{t_f} D_1 \ell(x(s), s) \circ \Phi(s, \tau_i) ds \circ \Delta^i \quad (11)$$

where Δ^i has been taken out of the integral because it does not depend on s. Defining

$$\psi(t, \tau) \circ U = \left(\int_{\tau}^{t} D_1 \ell(x(s), s) \circ \Phi(s, \tau) ds \right) \circ U, \quad (12)$$

(11) can be rewritten as

$$D_{\delta_i} J(\cdot) \circ \partial \delta_i = \psi(t_f, \tau_i) \circ \Delta^i$$

which is the first part of Theorem 1. To obtain the second part of Theorem 1, take the derivative of (12) with respect to τ [12].

$$\frac{\partial}{\partial \tau} \psi(t, \tau) \circ U =$$
$$- D_1 \ell(x(\tau), \tau) \circ U$$
$$- \int_{\tau}^{t} D_1 \ell(x(s), s) \circ \Phi(s, \tau) \circ A(\tau) U ds \quad (13a)$$
$$= -D_1 \ell(x(\tau), \tau) \circ U$$
$$- \left(\int_{\tau}^{t} D_1 \ell(x(s), s) \circ \Phi(s, \tau) ds \right) \circ A(\tau) \circ U \quad (13b)$$
$$= -D_1 \ell(x(\tau), \tau) \circ U - \psi(t, \tau) \circ D_1 f(x(\tau), \tau) \circ U \quad (13c)$$

Equation (13c) along with evaluating ψ in (12) at $\tau = t$ yields the final two parts of Theorem 1. ■

The next theorem states the first derivative of the cost with respect to an arbitrary impulse time. The derivation of

Theorem 2 is similar to the derivation of Theorem 1, except that it relies on the application of Leibniz's rule.

Theorem 2 The derivative of the cost function $J(\cdot)$ with respect to each of the impulse times τ_i is

$$D_{\tau_i} J(\cdot) \circ \partial \tau_i = \Psi(t_f, \tau_i) \tag{14}$$

where $\Psi(t_f, \tau_i) : \mathbb{R}^n \to \mathbb{R}$,

$$\Psi(t_f, \tau_i) = \psi(t_f, \tau_i) \circ X^i + \ell(x(\tau_i^-), \tau_i^-) - \ell(x(\tau_i^+), \tau_i^+).$$

Proof: Take the derivative of the cost function as it is written in Equation (2) with respect to τ_i. The derivative is the sum of three parts, the derivative of the integrand itself along with two terms that come from applying Leibniz's rule. Recall that in Equation (6), $D_{t_i} x(t) = 0$ up until $t = \tau_i$. Therefore the derivative of the integrand only needs to be integrated from τ_i up to t_f. Thus,

$$D_{\tau_i} J(\cdot) \circ \partial \tau_i = \int_{\tau_i}^{t_f} D_1 \ell(x(s), s) \circ D_{\tau_i} x(s) \circ \partial \tau_i ds$$
$$+ \ell(x(\tau_i^-), \tau_i^-) - \ell(x(\tau_i^+), \tau_i^+). \tag{15}$$

Substituting in (6) and observing that X^i is independent of s, we can write

$$D_{\tau_i} J(\cdot) \circ \partial \tau_i = \psi(t, \tau_i) \circ X^i + \ell(x(\tau_i^-), \tau_i^-) - \ell(x(\tau_i^+), \tau_i^+)$$

where $\psi(\cdot)$ is found by integrating (13c) backwards in time. ∎

Note that the adjoint operator $\psi(\cdot)$ appears in both (8) and (14), and thus the calculation of the first derivatives of $J(\cdot)$ with respect to each impulse time as well as impulse magnitude requires only a single integration. This result is independent of the total number of impulses.

3.2 Second Derivatives of $J(\cdot)$

In deriving the second derivatives of the cost $J(\cdot)$, we will proceed in a similar manner to the derivations of the first derivatives, i.e., we will first derive the second derivatives of the trajectory.

The following two lemmas are provided in order to define the initial conditions which appear in the second derivatives of the trajectory and thus in the second derivatives of the cost due to the chain rule.

Lemma 3 For $i \geq j$ and $t \geq \tau_i$, the derivative of $D_{\delta_i} x(t) \circ \partial \delta_i$ with respect to δ_j satisfies the differential equation (with initial condition $\Delta^{i,j}$)

$$\frac{d}{dt} D_{\delta_j} D_{\delta_i} x(t) \circ (\partial \delta_j, \partial \delta_i) =$$
$$D_1 f(x(t), t) \circ D_{\delta_j} D_{\delta_i} x(t) \circ (\partial \delta_j, \partial \delta_i)$$
$$+ D_1^2 f(x(t), t) \circ (D_{\delta_j} x(t) \circ \partial \delta_j, D_{\delta_i} x(t) \circ \partial \delta_i) \tag{16a}$$

$$\Delta^{i,j} = D_{\delta_j} D_{\delta_i} x(\tau_i) \circ (\partial \delta_j, \partial \delta_i) \tag{16b}$$

Proof: Differentiate (5) and apply the fundamental theorem of calculus. This is straightforward so it is not included here. ∎

The next lemma provides an initial condition that appears in the second derivatives of the cross terms between the impulse times and magnitudes.

Lemma 4 For $i \geq j$ and $t \geq \tau_i$, the derivative of $D_{\tau_i} x(t) \circ \partial \tau_i$ with respect to δ_j satisfies the differential equation (with initial condition $\Delta X^{i,j}$)

$$\frac{d}{dt} D_{\delta_j} D_{\tau_i} x(t) \circ (\partial \delta_j, \partial \tau_i) =$$
$$D_1 f(x(t), t) \circ D_{\delta_j} D_{\tau_i} x(t) \circ (\partial \delta_j, \partial \tau_i)$$
$$+ D_1^2 f(x(t), t) \circ (D_{\delta_j} x(t) \circ \partial \delta_j, D_{\tau_i} x(t) \circ \partial \tau_i) \tag{17a}$$

$$\Delta X^{i,j} = D_{\delta_j} D_{\tau_i} x(\tau_i) \circ (\partial \delta_j, \partial \tau_i) =$$
$$D_1 f(x(\tau_i^-), \tau_i^-) \circ D_{\delta_j} x(\tau_i^-) \circ \partial \delta_j$$
$$- D_1 f(x(\tau_i^+), \tau_i^+) \circ D_{\delta_j} x(\tau_i^+) \circ \partial \delta_j \tag{17b}$$

Proof: As before, differentiate $D_{\tau_i} x(t) \circ \partial \tau_i$ and apply the fundamental theorem of calculus. ∎

The following lemma provides the second derivative of the trajectory with respect to two impulse times.

Lemma 5 For $i \geq j$ and $t \geq \tau_i$, the second derivative of $x(t)$ satisfies the differential equation (with initial condition $X^{i,j}$)

$$\frac{d}{dt} D_{\tau_j} D_{\tau_i} x(t) \circ (\partial \tau_j, \partial \tau_i) =$$
$$D_1 f(x(t), t) \circ D_{\tau_j} D_{\tau_i} x(t) \circ (\partial \tau_j, \partial \tau_i)$$
$$+ D_1^2 f(x(t), t) \circ (D_{\tau_j} x(t) \circ \partial \tau_j, D_{\tau_i} x(t) \circ \partial \tau_i) \tag{18a}$$

$$X^{i,j} = D_{\tau_j} D_{\tau_i} x(\tau_i) \circ (\partial \tau_j, \partial \tau_i) =$$
$$\begin{cases} D_1 f(x(\tau_i^+), \tau_i^+) \circ f(x(\tau_i^+), \tau_i^+) \partial \tau_j \partial \tau_i \\ + D_1 f(x(\tau_i^-), \tau_i^-) \circ f(x(\tau_i^-), \tau_i^-) \partial \tau_j \partial \tau_i \\ - 2 D_1 f(x(\tau_i^+), \tau_i^+) \circ f(x(\tau_i^-), \tau_i^-) \partial \tau_j \partial \tau_i \\ + D_2 f(x(\tau_i^-), \tau_i^-) \circ \partial \tau_j \partial \tau_i \\ \quad - D_2 f(x(\tau_i^+), \tau_i^+) \circ \partial \tau_j \partial \tau_i, \qquad i = j \\ (D_1 f(x(\tau_i^-), \tau_i^-) \\ \quad - D_1 f(x(\tau_i^+), \tau_i^+)) \circ \Phi(\tau_i, \tau_j) \circ X^j \partial \tau_i, \quad i > j. \end{cases} \tag{18b}$$

Proof: Differentiate (7) and apply the fundamental theorem of calculus. ∎

Looking at Lemmas 3, 4, and 5, observe that the ODE's for the second derivatives of the trajectory are not linear as in the first derivatives, they are *affine*. From the form of the solution to a linear affine system, the next few lemmas complete our derivations of the second-order derivatives of the trajectories with respect to the impulse times and magnitudes.

Lemma 6 The second derivative $D_{\delta_j} D_{\delta_i} x(t) \circ (\partial \delta_j, \partial \delta_i)$ is

$$D_{\delta_j} D_{\delta_i} x(t) \circ (\partial \delta_j, \partial \delta_i) =$$
$$\Phi(t, \tau_i) \circ \Delta^{i,j} + \phi(t, \tau_i)(\Phi(\tau_i, \tau_j) \circ \Delta^j, \Delta^i) \tag{19}$$

where $\Phi(t,\tau)$ is the state transition matrix from Lemma 1 and the bilinear operator $\phi(t,\tau) : \mathbb{R}^n \times \mathbb{R}^n \to \mathbb{R}^n$ is defined as

$$\phi(t,\tau) \circ (U,V) =$$
$$\int_\tau^t \Phi(t,s) \circ D_1^2 f(x(s),s) \circ (\Phi(s,\tau) \circ U, \Phi(s,\tau) \circ V) ds \tag{20}$$

with $\Delta^{i,j}$ as the initial condition from (16b).

Proof: Notice that (16a) is in the affine form $\dot{x} = A(t)x + B(t)$. Recalling that the solution to an affine system is $x(t) = \Phi(t,t_0) \circ x_0 + \int_\tau^t \Phi(t,s) \circ B(s) ds$ [9], as well as using (3) and (16b), we find that

$$D_{\delta_j} D_{\delta_i} x(t) \circ (\partial\delta_j, \partial\delta_i) =$$
$$= \Phi(t,\tau_i) \circ \Delta^{i,j} + \int_{\tau_i}^t \Phi(t,s) \circ D_1^2 f(x(s),s)$$
$$\circ (D_{\delta_j} x(s) \circ \partial\delta_j, D_{\delta_i} x(s) \circ \partial\delta_i) ds$$
$$= \Phi(t,\tau_i) \circ \Delta^{i,j} + \int_{\tau_i}^t \Phi(t,s) \circ D_1^2 f(x(s),s)$$
$$\circ (\Phi(s,\tau_j) \circ \Delta^j, \Phi(s,\tau_i) \circ \Delta^i) ds$$
$$= \Phi(t,\tau_i) \circ \Delta^{i,j} + \int_{\tau_i}^t \Phi(t,s) \circ D_1^2 f(x(s),s)$$
$$\circ (\Phi(s,\tau_i) \circ \Phi(\tau_i,\tau_j) \circ \Delta^j, \Phi(s,\tau_i) \circ \Delta^i) ds$$
$$= \Phi(t,\tau_i) \circ \Delta^{i,j} + \phi(t,\tau_i) \circ (\Phi(\tau_i,\tau_j) \circ \Delta^j, \Delta^i)$$

where Δ^i and Δ^j have been pulled out of the integral because they do not depend on s. ∎

By direct inspection of (20) we can write

$$\phi(t,t) \circ (U,V) = 0$$
$$\frac{\partial}{\partial\tau} \phi(t,\tau) \circ (U,V) = -\Phi(t,\tau) \circ D_1^2 f(x(\tau),\tau) \circ (U,V)$$
$$- \phi(t,\tau) \circ (D_1 f(x(\tau),\tau) \circ U, V)$$
$$- \phi(t,\tau) \circ (U, D_1 f(x(\tau),\tau) \circ V). \tag{21}$$

In a similar way to the calculation of $\psi(\cdot)$, $\phi(\cdot)$ can be found by integrating (21) backwards in time.

The next lemma derives the second derivative of the trajectory for the cross terms between impulse times and magnitudes.

Lemma 7 The second derivative $D_{\delta_j} D_{\tau_i} x(t) \circ (\partial\delta_j, \partial\tau_i)$ is

$$D_{\delta_j} D_{\tau_i} x(t) \circ (\partial\delta_j, \partial\tau_i) =$$
$$\Phi(t,\tau_i) \circ \Delta X^{i,j} + \phi(t,\tau_i)(\Phi(\tau_i,\tau_j) \circ \Delta^j, X^i) \tag{22}$$

Proof: Using (17a) and (17b) and plugging in (3) and (6)

$$D_{\delta_j} D_{\tau_i} x(t) \circ (\partial\delta_j, \partial\tau_i)$$
$$= \Phi(t,\tau_i) \circ \Delta X^{i,j} + \int_{\tau_i}^t \Phi(t,s) \circ D_1^2 f(x(s),s)$$
$$\circ (D_{\delta_j} x(s) \circ \partial\delta_j, D_{\tau_i} x(s) \circ \partial\tau_i) ds$$
$$= \Phi(t,\tau_i) \circ \Delta X^{i,j} + \int_{\tau_i}^t \Phi(t,s) \circ D_1^2 f(x(s),s)$$
$$\circ (\Phi(s,\tau_j) \circ \Delta^j, \Phi(s,\tau_i) \circ X^i) ds$$
$$= \Phi(t,\tau_i) \circ \Delta X^{i,j} + \int_{\tau_i}^t \Phi(t,s) \circ D_1^2 f(x(s),s)$$
$$\circ (\Phi(s,\tau_i) \circ \Phi(\tau_i,\tau_j) \circ \Delta^j, \Phi(s,\tau_i) \circ X^i) ds$$
$$= \Phi(t,\tau_i) \circ \Delta X^{i,j} + \phi(t,\tau_i) \circ (\Phi(\tau_i,\tau_j) \circ \Delta^j, X^i)$$

where Δ^i and Δ^j have been taken out of the integral because they do not depend on s. ∎

The following lemma provides the second-order derivative of the trajectory with respect to the impulse times τ_i and τ_j.

Lemma 8 The second derivative $D_{\tau_j} D_{\tau_i} x(t) \circ (\partial\tau_j, \partial\tau_i)$ is

$$D_{\tau_j} D_{\tau_i} x(t) \circ (\partial\tau_j, \partial\tau_i) = \Phi(t,\tau_i) \circ X^{i,j}$$
$$+ \phi(t,\tau_i)(\Phi(\tau_i,\tau_j) \circ X^j, X^i) \tag{23}$$

where $\Phi(t,\tau)$ is the state transition matrix from Lemma 1 and $X^{i,j}$ is the intial condition from (18b).

Proof: Using (18a) and (18b)

$$D_{\tau_j} D_{\tau_i} x(t) \circ (\partial\tau_j, \partial\tau_i)$$
$$= \Phi(t,\tau_i) \circ X^{i,j} + \int_{\tau_i}^t \Phi(t,s) \circ D_1^2 f(x(s),s)$$
$$\circ (D_{\tau_j} x(s) \circ \partial\tau_j, D_{\tau_i} x(s) \circ \partial\tau_i) ds$$
$$= \Phi(t,\tau_i) \circ X^{i,j} + \int_{\tau_i}^t \Phi(t,s) \circ D_1^2 f(x(s),s)$$
$$\circ (\Phi(s,\tau_j) \circ X^j, \Phi(s,\tau_i) \circ X^i) ds$$
$$= \Phi(t,\tau_i) \circ X^{i,j} + \int_{\tau_i}^t \Phi(t,s) \circ D_1^2 f(x(s),s)$$
$$\circ (\Phi(s,\tau_i) \circ \Phi(\tau_i,\tau_j) \circ X^j, \Phi(s,\tau_i) \circ X^i) ds$$
$$= \Phi(t,\tau_i) \circ X^{i,j} + \phi(t,\tau_i) \circ (\Phi(\tau_i,\tau_j) \circ X^j, X^i)$$

where X^i and X^j have been pulled out of the integral because they do not depend on s. ∎

Having derived the second derivatives of the trajectory with respect to the impulse times, magnitudes, and cross terms between the impulse times and magnitudes, it is now possible to derive the second derivatives of the cost.

Theorem 3 The second derivative of the cost function $J(\cdot)$ with respect to the impulse magnitude δ_j where $\tau_i \geq \tau_j$ is

$$D_{\delta_j} D_{\delta_i} J(\cdot) \circ (\partial\delta_j, \partial\delta_i) =$$
$$\psi(t_f, \tau_i) \circ \Delta^{i,j} + \Omega(t_f, \tau_i) \circ (\Phi(\tau_i,\tau_j) \circ \Delta^j, \Delta^i)$$

where $\Omega(t,\tau) \circ (U,V) : \mathbb{R}^n \times \mathbb{R}^n \to \mathbb{R}$ is the bilinear operator found by integrating

$$\Omega(t,t) \circ (U,V) = 0_{n \times n} \tag{24a}$$

$$\frac{\partial}{\partial \tau} \Omega(t,\tau) \circ (U,V) = -D_1^2 \ell(x(\tau),\tau) \circ (U,V)$$
$$- \psi(t,\tau) \circ D_1^2 f(x(\tau),\tau) \circ (U,V)$$
$$- \Omega(t,\tau) \circ (D_1 f(x(\tau),\tau) \circ U, V)$$
$$- \Omega(t,\tau) \circ (U, D_1 f(x(\tau),\tau) \circ V) \tag{24b}$$

backwards over τ from t_f to τ_i.

Proof: Take the derivative of (10) with respect to δ_j and plug in (19)

$$D_{\delta_j} D_{\delta_i} J(\cdot) \circ (\partial \delta_j, \partial \delta_i) =$$
$$\int_{\tau_i}^{t_f} D_1 \ell(x(s),s) \circ D_{\delta_j} D_{\delta_i} x(s) \circ (\partial \delta_j, \partial \delta_i)$$
$$+ D_1^2 \ell(x(s),s) \circ (D_{\delta_j} x(s) \circ \partial \delta_j, D_{\delta_i} x(s) \circ \partial \delta_i) ds$$
$$= \int_{\tau_i}^{t_f} D_1 \ell(x(s),s) \circ \Phi(s,\tau_i) \circ \Delta^{i,j}$$
$$+ D_1 \ell(x(s),s) \circ \phi(s,\tau_i) \circ (\Phi(\tau_i,\tau_j) \circ \Delta^j, \Delta^i)$$
$$+ D_1^2 \ell(x(s),s) \circ (\Phi(s,\tau_i) \circ \Phi(\tau_i,\tau_j) \circ \Delta^j, \Phi(s,\tau_i) \circ \Delta^j)$$
$$= \psi(t_f,\tau_i) \circ \Delta^{i,j} + \Omega(t_f,\tau_i) \circ (\Phi(\tau_i,\tau_j) \circ \Delta^j, \Delta^i)$$

where $\Omega(t,\tau)$ is defined as

$$\Omega(t,\tau) \circ (U,V) = \int_{\tau}^{t} D_1 \ell(x(s),s) \circ \Phi(s,\tau) \circ (U,V)$$
$$+ D_1^2 \ell(x(s),s) \circ (\Phi(s,\tau) \circ U, \Phi(s,\tau) \circ V) ds. \tag{25}$$

This provides the first part of the proof. To obtain the second parts, take the derivative of (25) with respect to τ (this calculation is straight forward and is thus omitted here). ∎

The following theorem provides the second-order cross derivatives of the cost with respect to the impulse magnitude δ_j and impulse time τ_i.

Theorem 4 The second order cross derivative of the cost function $J(\cdot)$ with respect to the impulse time τ_i and magnitude δ_j where $\tau_i \geq \tau_j$ is

$$D_{\delta_j} D_{\tau_i} J(\cdot) \circ (\partial \delta_j, \partial \tau_i) =$$
$$D_1 \ell(x(\tau_i^-),\tau_i^-) \circ D_{\delta_j} x(\tau_i^-) \circ \partial \delta_j$$
$$- D_1 \ell(x(\tau_i^+),\tau_i^+) \circ D_{\delta_j} x(\tau_i^+) \circ \partial \delta_j$$
$$+ \psi(t_f,\tau_i) \circ \Delta X^{i,j} + \Omega(t_f,\tau_i) \circ (\Phi(\tau_i,\tau_j) \circ \Delta^j, X^i)$$

Proof: Take the derivative $D_{\tau_i} J(\cdot) \circ \partial \tau_i$ with respect to δ_j and plug in (22)

$$D_{\delta_j} D_{\tau_i} J(\cdot) \circ (\partial \delta_j, \partial \tau_i) =$$
$$D_1 \ell(x(\tau_i^-),\tau_i^-)) \circ D_{\delta_j} x(\tau_i^-) \circ \partial \delta_j$$
$$- D_1 \ell(x(\tau_i^+),\tau_i^+)) \circ D_{\delta_j} x(\tau_i^+) \circ \partial \delta_j$$
$$+ \int_{\tau_i}^{t_f} D_1 \ell(x(s),s) \circ D_{\delta_j} D_{\tau_i} x(s) \circ (\partial \delta_j, \partial \tau_i)$$
$$+ D_1^2 \ell(x(s),s) \circ (D_{\delta_j} x(s) \circ \partial \delta_j, D_{\delta_i} x(s) \circ \partial \delta_i) ds$$

$$= D_1 \ell(x(\tau_i^-),\tau_i^-)) \circ D_{\delta_j} x(\tau_i^-) \circ \partial \delta_j$$
$$- D_1 \ell(x(\tau_i^+),\tau_i^+)) \circ D_{\delta_j} x(\tau_i^+) \circ \partial \delta_j$$
$$+ \int_{\tau_i}^{t_f} D_1 \ell(x(s),s) \circ \Phi(s,\tau_i) \circ \Delta X^{i,j}$$
$$+ D_1 \ell(x(s),s) \circ \phi(s,\tau_i) \circ (\Phi(\tau_i,\tau_j) \circ \Delta^j, X^i)$$
$$+ D_1^2 \ell(x(s),s) \circ (\Phi(s,\tau_i) \circ \Phi(\tau_i,\tau_j) \circ \Delta^j, \Phi(s,\tau_i) \circ X^i)$$
$$= D_1 \ell(x(\tau_i^-),\tau_i^-)) \circ D_{\delta_j} x(\tau_i^-) \circ \partial \delta_j$$
$$- D_1 \ell(x(\tau_i^+),\tau_i^+)) \circ D_{\delta_j} x(\tau_i^+) \circ \partial \delta_j$$
$$+ \psi(t_f,\tau_i) \circ \Delta X^{i,j} + \Omega(t_f,\tau_i) \circ (\Phi(\tau_i,\tau_j) \circ \Delta^j, X^i) ∎$$

The following theorem provides the second-order derivatives of the cost with respect to the impulse times τ_i and τ_j.

Theorem 5 The second derivative of the cost function $J(\cdot)$ with respect to the impulse times τ_j and τ_i where $\tau_i \geq \tau_j$ is

$$D_{\tau_j} D_{\tau_i} J(\cdot) \circ (\partial \tau_j, \partial \tau_i) =$$
$$D_1 \ell(x(\tau_i^-),\tau_i^-) \circ (D_{\tau_j} x_d(\tau_i^-) \circ \frac{\partial \tau_i}{\partial \tau_j} - D_{\tau_j} x(\tau_i^-) \circ \partial \tau_j)$$
$$- D_1 \ell(x(\tau_i^+),\tau_i^+) \circ (D_{\tau_j} x_d(\tau_i^+) \circ \frac{\partial \tau_i}{\partial \tau_j} - D_{\tau_j} x(\tau_i^+) \circ \partial \tau_j)$$
$$- D_1 \ell(x(\tau_i),\tau_i) \circ X^i \partial \tau_i \delta_i^j + \psi(t_f,\tau_i) \circ X^{i,j}$$
$$+ \Omega(t_f,\tau_i) \circ (\Phi(\tau_i,\tau_j) \circ X^j, X^i)$$

Proof: Take the derivative of (15) with respect to τ_j and plugging in (23)

$$D_{\tau_j} D_{\tau_i} J(\cdot) \circ (\partial \tau_j, \partial \tau_i) =$$
$$\frac{\partial}{\partial \tau_j} \left(\int_{\tau_i^+}^{t_f} D_1 \ell(x(s),s) \circ D_{\tau_i} x(s) \circ \partial \tau_i ds \right.$$
$$\left. + \ell(x(\tau_i^-),\tau_i^-) - \ell(x(\tau_i^+),\tau_i^+) \right)$$
$$= D_1 \ell(x(\tau_i^-),\tau_i^-) \circ (D_{\tau_j} x_d(\tau_i^-) \circ \frac{\partial \tau_i}{\partial \tau_j}$$
$$- D_{\tau_j} x(\tau_i^-) \circ \partial \tau_j) - D_1 \ell(x(\tau_i^+),\tau_i^+)$$
$$\circ (D_{\tau_j} x_d(\tau_i^+) \circ \frac{\partial \tau_i}{\partial \tau_j} - D_{\tau_j} x(\tau_i^+) \circ \partial \tau_j)$$
$$- D_1 \ell(x(\tau_i^+),\tau_i^+) \circ D_{\tau_i} x(\tau_i^+) \circ \partial \tau_i \frac{\partial \tau_i}{\partial \tau_j}$$
$$+ \int_{\tau_i^+}^{t_f} (D_1 \ell(x(s),s) \circ D_{\tau_j} D_{\tau_i} x(s) \circ (\partial \tau_j, \partial \tau_i)$$
$$+ D_1^2(x(s),s) \circ (D_{\tau_j} x(s) \circ \partial \tau_j, D_{\tau_i} x(s) \circ \partial \tau_i)) ds$$
$$= D_1 \ell(x(\tau_i^-),\tau_i^-) \circ (D_{\tau_j} x_d(\tau_i^-) \circ \frac{\partial \tau_i}{\partial \tau_j}$$
$$- D_{\tau_j} x(\tau_i^-) \circ \partial \tau_j) - D_1 \ell(x(\tau_i^+),\tau_i^+)$$
$$\circ (D_{\tau_j} x_d(\tau_i^+) \circ \frac{\partial \tau_i}{\partial \tau_j} - D_{\tau_j} x(\tau_i^+) \circ \partial \tau_j)$$

$$- D_1\ell(x(\tau_i^+), \tau_i^+) \circ D_{\tau_i} x(\tau_i^+) \circ \partial\tau_i \frac{\partial\tau_i}{\partial\tau_j}$$

$$+ \int_{\tau_i^+}^{t_f} (D_1\ell(x(s), s) \circ \Phi(s, \tau_i) \circ X^{i,j}$$

$$+ D_1\ell(x(s), s) \circ \phi(s, \tau_i) \circ (\Phi(\tau_i, \tau_j) \circ X^j, X^i))ds$$

$$+ \int_{\tau_i^+}^{t_f} D_1^2\ell(x(s), s) \circ (\Phi(s, \tau_i) \circ \Phi(\tau_i, \tau_j)$$

$$\circ X^j, \Phi(s, \tau_i) \circ X^i)ds$$

$$= D_1\ell(x(\tau_i^-), \tau_i^-) \circ (D_{\tau_j} x_d(\tau_i^-) \circ \partial\tau_j \delta_j^i$$

$$- D_{\tau_j} x(\tau_i^-) \circ \partial\tau_j) - D_1\ell(x(\tau_i^+), \tau_i^+)$$

$$\circ (D_{\tau_j} x_d(\tau_i^+) \circ \partial\tau_j \delta_j^i - D_{\tau_j} x(\tau_i^+) \circ \partial\tau_j)$$

$$- D_1\ell(x(\tau_i^+), \tau_i^+) \circ X^i \partial\tau_j \delta_i^j + \psi(t_f, \tau_i) \circ X^{i,j}$$

$$+ \Omega(t_f, \tau_i) \circ (\Phi(\tau_i, \tau_j) \circ X^j, X^i)$$

where δ_j^i is the Kronecker delta and

$$\frac{\partial\tau_i}{\partial\tau_j} = \begin{cases} \partial\tau_j & i = j \\ 0 & i \neq j. \end{cases} \tag{26}$$

The first two terms in (26) are the derivatives of the two Leibniz terms in (15), $\ell(x(\tau_i^-), \tau_i^-)$ and $\ell(x(\tau_i^+), \tau_i^+)$. It is important to note that the derivative of the reference $x_d(\cdot)$ appears in the second derivative of $J(\cdot)$ with respect to the impulse times τ_i and τ_j for impulsive systems.

To explain the appearance of the derivative of the reference in the second derivatives, return to the Leibniz terms in (15) (recalling that $\ell(x(t), t) = (x_d(t) - x(t))^T (x_d(t) - x(t))$). The Leibniz terms in the first derivatives of $J(\cdot)$ (with respect to the impulse time τ_i) are evaluated at τ_i^- and τ_i^+, respectively, and thus both the reference and the trajectory depend explicitly on τ_i. Thus, when taking the derivatives of these Leibniz terms with respect to τ_i a second time, the derivatives of both the reference and model must be present. Taking the derivatives of the reference is a nontrivial operation. It turns out that evaluating the derivatives of the model that result from the Leibniz terms is also nontrivial. When $i \neq j$ the evaluation of the $D_{\tau_j} x(\tau_i) \circ \partial\tau_j$ is straight forward and can be calculated using (6). When $i = j$ care must be taken due to the fact that the derivative is now being taken with respect to the argument of $x(\cdot)$. In this case $(i = j)$, $D_{\tau_j} x(\tau_i) \circ \partial\tau_j = f(x(\tau_i), \tau_i)$. ∎

4. IMPLEMENTATION

4.1 Trajectory Optimization

It was mentioned earlier that in order to properly implement the impulse optimization, we first need a way in which to estimate the total number of impulses that occur. Nonlinear trajectory optimization is the method proposed as a pre-step used to estimate the number of impulses that occur.

The exact form of the nonlinear trajectory optimization problem being solved is

$$\dot{x} = U(t)f(x, t) = F(x, U) \tag{27}$$

where $U(t)$ is a diagonal matrix function unless otherwise noted and $f(x, t)$ are the same dynamics from Equation (1). Note that for the system in (27) the trajectory optimization will always be nonlinear, even when the dynamics $f(\cdot)$ are linear.

In order to perform the trajectory optimization, we define a cost function $S(\cdot)$ separate from the cost $J(\cdot)$ in Equation (2) such that

$$S(\eta) = \int_0^T \ell(s, x(s), U(s))ds + m(x(T)) \tag{28}$$

where $\eta \in \mathcal{T}$, and $\mathcal{T}$ is the trajectory manifold associated with the dynamics defined in (1). The problem can thus be stated as the constrained optimization

$$\min_{\eta \in \mathcal{T}} S(\eta). \tag{29}$$

Through the use of the projection operator ([7, 13, 14, 15], the constrained optimization in (29) can be rewritten as the unconstrained optimization

$$\min_{\xi \in \mathcal{L}} S(P(\xi))$$

where ξ is an element of the infinite dimensional function space $\mathcal{L}$. This optimization problem is solved using a first-order descent method with a line search. Most of the technical details of the trajectory optimization will be left to several references ([7, 13, 14, 15]) and are thus not provided here.

In standard implementations of trajectory optimization the goal is to find an optimal trajectory with respect to a cost of the form (28). For the purposes of estimating the total number of impulses as well as approximating the times at which the impulses occur, the goal of the trajectory optimization is slightly modified. Using trajectory optimization as a pre-step to impulse optimization, the goal is to find deviations in the control $U(\cdot)$. It is explained further below, but we wish the nominal value of $U(\cdot)$ to be equal to the identity. That is, we want—when possible—to have $\dot{x} = f(x, t)$. The goal of the trajectory optimization is to find deviations away from $1(t)$.

To explain why we would like to constrain $U = I$, consider the 1-D system containing two separate bodies each with dynamics $\dot{x} = x$, but with different initial conditions. This situation is shown in Figure 1, where a single sensor measures the position of one object over the intervals $t = (0, 1)$ and $t = (2, 3)$, and a second object over the interval $t = (1, 2)$. Note that this example is deterministic and intentionally oversimplified for the purposes of clarity.

Figure 2 shows results of applying trajectory optimization to this 1-D system. In Figure 2(a) the solid line represents the desired signal, the dotted line the initial guess in the trajectory optimization, and the dashed line the current guess in the optimization after several iterations.

Figure 2(b) shows the control signal associated with the same step in the trajectory optimization that produced the dashed line in Figure 2(a). Through inspection of Figure 2 we can see that when the reference signal (solid line) is very close to the current trajectory in the optimization procedure (dashed line), the control signal value is close to one. When the difference between the current trajectory and the reference trajectory changes suddenly, the difference is reflected in the control signal as spikes that generate the delta function in the state (as can be seen in Figure 2(b)). The spikes in Figure 2(b) are the deviations from $1(t)$ in the control that are desired. In this example, we would use thresholding [21, 23] to determine that there are two impulse times that we need to optimize over and our initial guess for the impulse optimization would be $(\tau_1, \tau_2) \approx (1, 2)$.

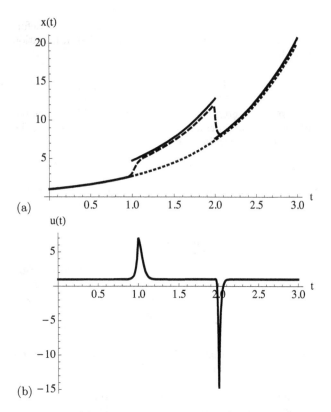

(a)

(b)

Figure 2: (a) Measurement signal (solid line), initial guess (dotted line) for trajectory optimization, and result (dashed line) of applying trajectory optimization to the system with dynamics $\dot{x} = ux$. (b) Control signal obtained as a result of applying trajectory optimization in (a).

4.2 Descent Methods

Having found the derivatives in Section 3 and a method for estimating the total number of impulses in Section 4.1, it is possible to implement both first- and second-order optimizations on (2), such as steepest descent with a line search and Newton's method [17]. Note that in the example in Section 4.3, a combination of a quasi-Newton's method and standard Newton's method are used to produce the convergence results shown in Figure 4. The quasi-Newton's method checks the eigenvalues of the Hessian and replaces any negative eigenvalues with 1, thus performing steepest descent in that subspace.

4.3 Example

The example selected to demonstrate the IDA algorithm has dynamics

$$\dot{x} = v(t)\cos(\theta(t))$$
$$\dot{y} = v(t)\sin(\theta(t))$$
$$\dot{\theta} = \omega(t).$$

where $v(t)$ and $\omega(t)$ are some inputs. Two objects are present in the system. Figure 3 shows an example of the two objects' trajectories for $v(t) = 1$ and $\omega(t) = 1$. In Figure 3(a), the dotted line represents the trajectory of a pre-specified object of interest, the dashed line represents the trajectory of the second nearby object, and the solid line represents the

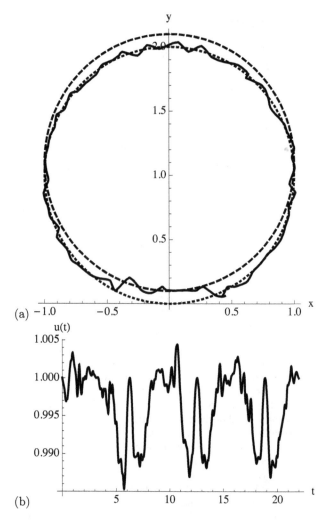

(a)

(b)

Figure 3: (a) Measurements for the two-plane example (solid line). The dotted line represents the trajectory of the object of interest and the dashed line the trajectory of some plane nearby. (b) Result of applying the trajectory optimization of Section 4.1 to the two-plane system when six total impulses occur. The six spikes correspond to the six impulses.

continuous measurement signal, found by interpolating the set of discrete noisy measurements.

It was mentioned earlier that in order to determine the times at which impulses occur in the measurement trajectory, the total number of impulses must first be estimated. Figure 3(b) shows the results of applying the trajectory optimization algorithm of Section 4.1 to the two airplane system over three periods of the circular trajectories shown in Figure 3(a). Note that there are six distinct peaks in the control signal shown in Figure 3(b) that correspond to the six impulses that occur in the measurement trajectory. Note also that we are showing a single component of the matrix $U(\cdot)$ in Figure 3(b) due to the fact that for this example

$$U(t) = \begin{pmatrix} u_1 & 0 & 0 \\ 0 & u_2 & 0 \\ 0 & 0 & u_3 \end{pmatrix}, \tag{30}$$

where $u_1 = u_3 = 1$ because the impulse always occurs in the $y-$direction.

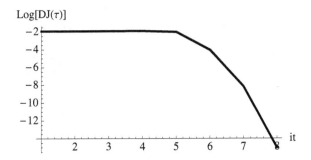

Figure 4: Second-order convergence results of applying impulse optimization over impulse times and magnitudes to the two plane, six impulse system.

Figure 4 presents the second-order convergence that results from optimizing (2) over six impulse times as well as magnitudes. Figure 4 shows that we do in fact achieve quadratic convergence and that we reach a solution, with magnitude of the gradient less than 10^{-14} on a log based 10 scale, within about 8 iterations.

5. CONCLUSIONS AND FUTURE WORK

This paper presented a new method for solving the multi-target data association problem for systems with an unknown number of targets, called impulsive data association. The algorithm is "impulsive" in that the sensor switching between measuring the trajectories of different targets is modeled as an impulse in a continuous measurement signal.

The main contribution of the work presented in this paper is the derivation of first- and second-order adjoint equations for the first- and second-order derivatives of the cost function (2) with respect to impulse times and magnitudes. An interesting result that arises in deriving the derivatives of the cost is that the same adjoint operator appears in the derivatives with respect the both the impulse times and magnitudes. This result means that it is possible to compute the entire second derivative of the cost with respect to an unknown number of impulse times and magnitudes by integrating two separate equations. This helps minimize the computational complexity.

In Section 2, we mentioned the fact that we were ignoring the presence of process noise in the work presented in this paper. In an actual implementation of IDA the uncertainty in the initial distribution as well as the uncertainty associated with the process noise needs to be explicitly addressed. One particular choice of methods to address these two sources of additional uncertainty in the IDA framework is the Kalman-Bucy filter. The resulting method would simultaneously run the filter and IDA (with a moving window in time), where $x(\cdot)$ becomes the estimate calculated by the filter. The implementation of an example that includes this combination of methods is a current direction of future work.

Another direction of future work is addressing the possibility of impulsively switching between the trajectories of two objects that are not the object of interest. In this paper we have made the assumption that the only impulses that occur are those between the object of interest and another

object nearby. A method similar to gating [3] is currently being developed to handle this possibility.

6. ACKNOWLEDGMENTS

The authors would like to thank each of the reviewers as well as Benjamin Tovar for their help with this paper.

7. REFERENCES

[1] B. Anderson and J. B. Moore. *Optimal Control: Linear Quadratic Methods.* Dover Publications, Inc, 1990.

[2] T. Bailey and H. Durrant-Whyte. Simultaneous Localization and Mapping (SLAM): Part II. *Robotics Automation Magazine, IEEE*, 13(3):108 –117, Sep. 2006.

[3] Y. Bar-Shalom and T. E. Fortmann. *Tracking and Data Association.* Academic Press, Inc., 1988.

[4] Y. Bar-Shalom, F. Daum, and J. Huang. The Probabilistic Data Association Filter. *IEEE Control Systems Magazine*, 29(6):82 – 100, 2009.

[5] N. Bergman and A. Doucet. Markov Chain Monte Carlo Data Association Target Tracking Problems. In *IEEE Conference on Decision and Control*, pages 735–742, 2004.

[6] S. S. Blackman. Multiple Hypothesis Tracking for Multiple Target Tracking. *IEEE Aerospace and Electronic Systems Magazine*, pages 5–18, 2004.

[7] T. Caldwell and T. Murphey. Switching Mode Generation and Optimal Estimation with Application to Skid-Steering. *Automatica*, 47(1):50–64, 2011.

[8] K. Chang, S. Mori, and C. Chong. Evaluating a Multiple-Hypothesis Multi-Target Tracking Algorithm. *IEEE Transactions on Aerospace and Electronic Systems*, 30(2):578–590, 1994.

[9] C. T. Chen. *Linear Systems Theory and Design, 3rd ed.* Oxford University Press, 1999.

[10] I. Cox. A Review of Statistical Data Association Techniques for Motion Correspondence. *International Journal of Computer Vision*, 10(1):53–66, 1993.

[11] I. Cox and S. Hingorani. An Efficient Implementation of Reid's Multiple-Hypothesis Tracking Algorithm and its Evaluation for the Purpose of Visual Tracking. *IEEE Transactions on Pattern Analysis and Machine Intelligence*, 18(2):138–150, 1996.

[12] M. Egerstedt, Y. Wardi, and F. Delmotte. Optimal Control of Switching Times in Switched Dynamical Systems. In *IEEE Conference on Decision and Control*, pages 2138–2143, 2003.

[13] J. Hauser. A Projection Operator Approach to the Optimization of Trajectory Functionals. *IFAC World Congress*, 2002.

[14] J. Hauser and D. G. Meyer. The Trajectory Manifold of a Nonlinear Control System. In *IEEE Conference on Decision and Control*, pages 1034–1039, 1998.

[15] J. Hauser and A. Saccon. A Barrier Function Method for the Optimization of Trajectory Functionals with Constraints. In *IEEE Conference on Decision and Control*, pages 864–869, 2006.

[16] M. Kalandros, L. Trailović, L. Y. Pao and Y. Bar Shalom. Tutorial on Multisensor Management and

Fusion Algorithms for Target Tracking. In *American Control Conference (ACC)*, 2004.

[17] C. T. Kelley. *Iterative Methods for Optimization.* Society for Industrial and Applied Mathematics, 1999.

[18] Z. Khan, T. Balch, and F. Dellaert. *Lecture Notes in Computer Science*, chapter : An MCMC-Based Particle Filter for Tracking Multiple Interacting Targets. Springer, Berlin / Heidelberg, Germany, 2004.

[19] Z. Khan, T. Balch, and F. Dellaert. MCMC Data Association and Sparse Factorization Updating for Real Time Multi-target Tracking with Merged and Multiple Measurements. *IEEE Trans. Pattern Anal. Mach. Intell.*, 28(12):1960–72, 2006.

[20] T. Kurien. *Multi-Target Multi-Sensor Tracking: Advanced Applications*, chapter : Issues in the Design of Practical Multi-Target Tracking Algorithms. Y. Bar-Shalom, Ed. Artech House, Norwood, MA, 1990.

[21] M. Mariscotti. A Method for Automatic Identification of Peaks in the Presence of Background and its Application to Spectrum Analysis. *Nucl. Instrum. Methods*, 50:309–320, 1967.

[22] B. Milch, B. Marthi, S. Russell, D. Sontag, D. L. Ong, and A. Kolobov. *BLOG: Probabilistic Models with Unknown Objects*. MIT Press, 1999.

[23] N. Mtetwa and L. S. Smith. Smoothing and Thresholding in Neuronal Spike Detection. *Neurocomput.*, 69(10-12):1366–1370, 2006.

[24] S. Oh, S. Russell, and S. Sastry. Markov Chain Monte Carlo Data Association for General Multiple-Target Tracking Problems. In *IEEE Conference on Decision and Control*, 2004.

[25] S. Oh, S. Russell, and S. Sastry. Markov Chain Monte Carlo Data Association for Multi-Target Tracking. *IEEE Transactions on Automatic Control*, 54(3):481–497, 2009.

[26] R. Powers and L. Pao. Power and Robustness of a Track-Loss Detector Based on Kolmogorov-Smirnov Tests. In *American Control Conference (ACC)*, pages 3757 – 3764, 2006.

[27] K. M. Reichard, E. C. Crow, and D. C. Swanson. Automated Situational Awareness Sensing for Homeland Defense. *SPIE System Diagnosis and Prognosis: Security and Condition Monitoring Issues Conference*, pages 64 – 71, 2003.

[28] D. Reid. An Algorithm for Tracking Multiple Targets. *IEEE Transactions on Automatic Control*, 24.6:843–854, 1979.

[29] B. Schumitsch, S. Thrun, G. Bradski, and K. Olukotun. The Information Form Data Association Filter. In *Conference on Neural Information Processing Systems*, 2005.

Resource Constrained LQR Control
Under Fast Sampling [*]

Jerome Le Ny
Department of Electrical and
Systems Engineering
University of Pennsylvania
200 South 33rd Street
Philadelphia, PA 19104, USA
jeromel@seas.upenn.edu

Eric Feron
School of Aerospace
Engineering
Georgia Institute of
Technology
270 Ferst Drive
Atlanta, GA 30332, USA
feron@gatech.edu

George J. Pappas
Department of Electrical and
Systems Engineering
University of Pennsylvania
200 South 33rd Street
Philadelphia, PA 19104, USA
pappasg@seas.upenn.edu

ABSTRACT

We investigate a state feedback Linear Quadratic Regulation problem with a constraint on the number of actuation signals that can be updated simultaneously. Such a constraint arises for example in networked and embedded control systems, due to limited communication and computation capabilities. Following recent results on the dual problem of scheduling Kalman filters, we first develop a bound on the achievable performance that can be computed efficiently by semidefinite programming. This bound can be approached arbitrarily closely by an analog periodic controller that can switch between control inputs arbitrarily fast. We then discuss implementation issues on digital platforms, i.e., the discretization of the analog controller in the presence of a relatively fast but finite sampling rate.

Categories and Subject Descriptors

C.3 [**Special-Purpose and Application-Based Systems**]: real-time and embedded systems; G.1.10 [**Mathematics of Computing**]: Numerical Analysis—*Applications*; F.2.2 [**Analysis of algorithms and problem complexity**]: Nonnumerical Algorithms and Problems—*Sequencing and Scheduling*

General Terms

Algorithms, Theory

Keywords

Networked and Embedded Control Systems

---

[*]This work was supported by NSF award No. 0931239

1. INTRODUCTION

Modern control systems are increasingly implemented on networked and embedded platforms [10]. This trend raises many interesting questions at the interface of control, computing, and communications, see e.g. [2, 21]. In particular, for many control systems it is important to understand the impact of limited computational and communication resources. Taking this aspect into account as much as possible during the control system design phase can help reduce the overall system cost and provides more flexibility at the implementation and system integration phases [3, 15].

In this paper we consider a control problem for a plant with multiple actuation channels, only one of which can be updated at a time, see Fig. 1. Such a constraint arises for example if the controller is implemented on a platform with limited computational power or executing other computation-intensive tasks. In this case, it might be desirable or necessary to divide the control function into subfunctions with shorter execution times that update only a subset of the control signals when executed [15]. A natural question is then to decide which input signals should be updated more frequently. A similar situation occurs if the control signals must be sent to the actuators via a communication network. In this case, the restriction on the number of simultaneous control signal updates is due to the limited communication capacity constraint between the controller and the plant. Here we limit our discussion to the Linear Quadratic Regulation (LQR) problem [1].

Similar problems at the interface of control and scheduling have been considered at least since Meier et al. studied the dual problem of scheduling measurement systems [18]. More recently much research has been done on linear quadratic control problems and other optimal control problems for switched systems, see e.g. [11, 16, 17, 22, 26], and on joint control and scheduling problems, especially in the context of Networked Control Systems (NCS), see e.g [3,20,21]. In this paper, we consider a continuous-time infinite-horizon linear quadratic control problem under state feedback. Lee [16] considers a closely related but more general output feedback infinite-horizon Linear Quadratic Gaussian (LQG) problem in discrete time, and Zhang et al. [26] study the finite-horizon problem, also under state feedback. The importance of the continuous-time infinite-horizon version of the problem, which is the dual of a Kalman filter scheduling problem considered in [14], comes from the fact that it admits

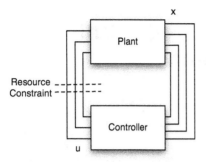

Figure 1: Plant-Controller configuration with an additional resource constraint, due to the presence of a communication network or because the processor can only update one control input at a time.

a simple solution that requires only solving a Linear Matrix Inequality (LMI) whose size is of the order of the plant dimension. In contrast, the available solutions of discrete-time formulations typically involve searching over the set of possible input sequences [18], whose size grows exponentially with the length of the sequence. Moreover, there is no a priori known bound on the size of the search space for infinite-horizon formulations [16]. The continuous-time approach also has the advantage of considering the inter-sample behavior of closed-loop sampled-data systems. The potential drawback of this method is that the controller obtained is a continuous-time (periodic, fast switching) controller, which must be discretized for most applications. Hence an important goal of this paper is to discuss in some details implementation issues of this analog controller on digital platforms for the control of continuous-time plants.

The paper is organized as follows. Section 2 describes our formulation of the resource-constrained LQR control problem. A solution for this problem consists in designing a feedback control signal together with a scheduling policy specifying which control channel to use at each time. These two subproblems can in fact be treated independently, and section 3 describes the optimal analog controller given a scheduling policy, which is simply obtained from the solution of a standard LQR control problem. In section 4 we show how to compute a bound on the performance achievable by any controller and scheduling policy, by solving a semidefinite program. If we assume that we can implement an analog control law, this bound can be approximated arbitrarily closely by certain periodic switching policies described in section 5, with the performance gap simply controlled by the rate at which the policies are allowed to switch between inputs. The rest of the paper is devoted to the discussion of digital implementations of these analog controllers. Two possible implementations, one assuming possibly time-varying sampling rates and the other assuming a time-triggered platform respectively, are described in section 6, and their performance is evaluated by simulations in section 7.

2. PROBLEM FORMULATION

Consider a linear continuous-time plant with dynamics

$$dx = Ax\,dt + B_{\sigma(t)}u(t)\,dt + dw, \qquad (1)$$

where $x \in \mathbb{R}^n$ is the n-dimensional state vector, $u \in \mathbb{R}^m$ is the input vector, and $A \in \mathbb{R}^{n \times n}$. Moreover, $\sigma(t) \in \{1, \dots, N\} =: [N]$ is an additional control parameter used to select the matrix $B_{\sigma(t)} \in \{B_1, \dots, B_N\}$, where $B_i \in \mathbb{R}^{n \times m}$ for all $i \in [N]$. The noise process w is a vector Wiener process with zero mean and incremental covariance $W\,dt$. We assume that we have access to the full state x to design a feedback controller. Also, assume for simplicity that x_0 is deterministic, so that $\Sigma(0) = E[x(0)x(0)]^T = 0$. Finally, assume that the pair $(A, [B_1, \dots, B_N])$ is controllable.

Let us first fix a time horizon T, and consider the problem of designing a switching signal $\sigma(t)$ together with a control input $u(t)$ that jointly minimize the quadratic cost

$$J_T(\sigma, u) = \frac{1}{T} E\left\{ \int_0^T x^T Q x + u^T R_\sigma u \, dt + x^T(T) Q_f x(T) \right\},$$
$$\qquad (2)$$

where $Q \succeq 0$, $Q_f \succ 0$, and $R_i \succ 0$, for all $i \in [N]$. We are mostly concerned with the infinite-horizon version of this problem, where the goal is to design a policy σ and a control law u minimizing

$$J(\sigma, u) = \limsup_{T \to \infty} J_T(\sigma, u). \qquad (3)$$

To study this problem, we assume throughout the paper that $(A, Q^{1/2})$ is observable.

To the switching signal $\sigma(t)$ we associate an N-dimensional signal of binary variables $\beta(t) \in \{0, 1\}^N$ such that

$$\beta_i(t) = 1 \Leftrightarrow \sigma(t) = i.$$

In other words, β is a unit vector with a one at the index σ. The scalars $\beta_i(t)$ corresponding to the signal $\sigma(t)$ are therefore subject to the constraints $\beta_i(t) \in \{0, 1\}$ for all $i \in [N]$ and

$$\sum_{i=1}^N \beta_i(t) \leq 1. \qquad (4)$$

Clearly there is a one-to-one correspondence between σ and β and we employ both notations interchangeably. Note in particular that we can rewrite the dynamics equation (1) as

$$dx = Ax\,dt + \left(\sum_{i=1}^N \beta_i(t) B_i \right) u(t)\,dt + dw.$$

Example 1. Consider the situation of Fig. 1, where there are N control channels $u = [u_1, \dots, u_N]$ available, each possibly multidimensional, with $u_i \in \mathbb{R}^{m_i}$ and $\sum_{i=1}^N m_i = m$. However, only one channel be be used at each time. The control inputs that cannot be updated are set to zero, as in e.g. [25]. In other words, the matrices B_i are of the form

$$B_i = \begin{bmatrix} \mathbf{0}_{n \times m_1} & \cdots & \mathbf{0}_{n \times m_{i-1}} & \hat{B}_i & \mathbf{0}_{n \times m_{i+1}} & \cdots & \mathbf{0}_{n \times m_N} \end{bmatrix}$$

where $\hat{B}_i \in \mathbb{R}^{n \times m_i}$.

3. CONTROLLER DESIGN

Given a switching policy σ, the system (1) evolves as a linear time-varying system. Fixing $T < \infty$, the controller design problem is then a standard linear quadratic regulation problem and so the optimal controller is a static feedback controller

$$u_\sigma(t) = L_\sigma(t; T)x, \qquad (5)$$

where $L_\sigma(t;T) = -R_{\sigma(t)}^{-1} B_{\sigma(t)}^T P_\sigma(t;T)$ and $P_\sigma(t;T)$ satisfies the differential Riccati equation

$$\dot{P}_\sigma(t;T) = - A^T P_\sigma(t;T) - P_\sigma(t;T)A - Q$$
$$+ P_\sigma(t;T)B_\sigma(t)R_\sigma^{-1}B_\sigma^T(t)P_\sigma(t;T), \quad (6)$$
$$P_\sigma(T;T) = Q_f.$$

Note that (6) can be rewritten

$$\dot{P}_\sigma(t;T) = - A^T P_\sigma(t;T) - P_\sigma(t;T)A - Q \quad (7)$$
$$+ P_\sigma(t;T) \left(\sum_{i=1}^{N} \beta_i(t) B_i R_i^{-1} B_i^T \right) P_\sigma(t;T).$$

Moreover, denoting $\Sigma(t) = E[x(t)x^T(t)]$ the correlation matrix of the state vector, and using $u = L_\sigma(t;T)x$, the cost expression (2) becomes

$$J_T(\sigma) = \frac{1}{T} \Big\{ \int_0^T \mathrm{Tr}[(Q + L_\sigma^T(t)R_\sigma L_\sigma(t;T))\Sigma(t)] \, dt$$
$$+ \mathrm{Tr}[Q_f \Sigma(T)] \Big\}.$$

It is useful to further transform these expressions, as follows. First, we have

$$A^T P_\sigma + P_\sigma A - P_\sigma B_\sigma R_\sigma^{-1} B_\sigma^T P_\sigma + Q$$
$$= (A + B_\sigma L_\sigma)^T P_\sigma + P_\sigma(A + BL_\sigma) + L_\sigma^T RL_\sigma + Q,$$

which gives, in the Riccati equation (6)

$$\dot{P}_\sigma(t;T) = -(A + B_\sigma L_\sigma)^T P_\sigma - P_\sigma(A + BL_\sigma) - L_\sigma^T RL_\sigma - Q.$$

We also have a Lyapunov equation describing the dynamics of the covariance matrix $\Sigma(t)$ under the control $u = L_\sigma x$, namely [4]

$$\dot{\Sigma}(t) = (A + B_\sigma L_\sigma)\Sigma(t) + \Sigma(t)(A + B_\sigma L_\sigma)^T + W.$$

Therefore

$$\mathrm{Tr}[(Q + L_\sigma^T(t)R_\sigma L_\sigma(t))\Sigma(t)]$$
$$= -\mathrm{Tr}[(\dot{P}_\sigma + (A + B_\sigma L_\sigma)^T P_\sigma + P_\sigma(A + BL_\sigma))\Sigma(t)]$$
$$= -\mathrm{Tr}[\dot{P}_\sigma \Sigma(t) + P_\sigma(\dot{\Sigma} - W)]$$
$$= \mathrm{Tr}[P_\sigma W] - \mathrm{Tr}\left[\frac{d}{dt}(P_\sigma \Sigma)\right].$$

Finally, we obtain the following expression for the cost function, using the optimal control law u given σ

$$J_T(\sigma) = \frac{1}{T} \Big\{ \int_0^T \mathrm{Tr}[P_\sigma(t;T)W] \, dt - \mathrm{Tr}[P_\sigma(T;T)\Sigma(T)]$$
$$\mathrm{Tr}[P_\sigma(0;T)\Sigma(0)] + \mathrm{Tr}[Q_f \Sigma(T)] \Big\},$$

that is,

$$J_T(\sigma) = \frac{1}{T} \int_0^T \mathrm{Tr}[P_\sigma(t;T)W] \, dt \quad (8)$$

since $P\sigma(T;T) = Q_f$ and $\Sigma(0) = 0$. The remaining goal is then to obtain a scheduling policy σ minimizing (9), or the infinite-horizon cost

$$J(\sigma) = \limsup_{T\to\infty} \frac{1}{T} \int_0^T \mathrm{Tr}[P_\sigma(t;T)W] \, dt. \quad (9)$$

4. PERFORMANCE BOUND

Our approach is to first derive a bound on the performance achievable by any switching policy σ. Let

$$X_\sigma(t;T) = P_\sigma^{-1}(t;T).$$

Then in terms of this new variable, equation (7) becomes

$$\dot{X}_\sigma(t;T) = X_\sigma(t;T)A^T + AX_\sigma(t;T) + X_\sigma(t;T)QX_\sigma(t;T)$$
$$\qquad\qquad (10)$$
$$- \left(\sum_{i=1}^{N} \beta_i(t) B_i R_i^{-1} B_i^T \right), \quad (11)$$
$$X_\sigma(T;T) = Q_f^{-1}.$$

Now define the time averages

$$\tilde{X}_\sigma(T) = \frac{1}{T} \int_0^T X_\sigma(t;T)dt, \quad \tilde{\beta}_i(T) = \frac{1}{T} \int_0^T \beta_i(t)dt.$$

Integrating equation (11) over the interval $[0,T]$, we get

$$\frac{X_\sigma(T;T) - X_\sigma(0;T)}{T} = \tilde{X}_\sigma(T)A^T + A\tilde{X}_\sigma(T)$$
$$- \left(\sum_{i=1}^{N} \tilde{\beta}_i(T) B_i R_i^{-1} B_i^T \right) + \frac{1}{T} \int_0^T X_\sigma(t;T)QX_\sigma(t;T)dt.$$

Now by Jensen's inequality applied to the matrix convex function $X \to XQX$ [7, p. 110], we have

$$\frac{1}{T} \int_0^T X_\sigma(t;T)QX_\sigma(t;T)dt$$
$$\succeq \left(\frac{1}{T} \int_0^T X_\sigma(t;T)dt \right) Q \left(\frac{1}{T} \int_0^T X_\sigma(t;T)dt \right).$$

Together with the fact that $X_\sigma(0;T) \succeq 0$, we get the following convex quadratic inequality

$$\frac{Q_f^{-1}}{T} \succeq \tilde{X}_\sigma(T)A^T + A\tilde{X}_\sigma(T) \quad (12)$$
$$- \left(\sum_{i=1}^{N} \tilde{\beta}_i(T) B_i R_i^{-1} B_i^T \right) + \tilde{X}_\sigma(T)Q\tilde{X}_\sigma(T).$$

Also, note that in the cost function

$$\frac{1}{T} \int_0^T \mathrm{Tr}[P_\sigma(t;T)W]dt = \mathrm{Tr}\left[\left(\frac{1}{T} \int_0^T P_\sigma(t;T)dt \right) W \right],$$

and, again by Jensen's inequality and the matrix convexity of $X \to X^{-1}$ on the positive definite matrices [7, p. 76], we have

$$\left(\frac{1}{T} \int_0^T P_\sigma(t;T)dt \right) = \left(\frac{1}{T} \int_0^T X_\sigma(t;T)^{-1}dt \right) \succeq \tilde{X}_\sigma(T)^{-1}.$$

Finally, we see therefore that for any T, the optimal cost is lower bounded by the quantity

$$\mathrm{Tr}[\tilde{X}_\sigma(T)^{-1}W] \quad (13)$$

where $\tilde{X}_\sigma(T)$ satisfies the constraint (12) and $\tilde{\beta}_i(T)$ is subject to

$$\sum_{i=1}^{N} \tilde{\beta}_i(T) \leq 1, \quad 0 \leq \tilde{\beta}_i(T), i = 1, \ldots, N.$$

A lower bound on the achievable performance for T finite can then be obtained by letting $\tilde{X}_\sigma(T) = X$ in (13), where X is the solution of the following program with variables $X, \{b_i\}_{1 \leq i \leq N}$

$$\min_{X \succ 0, \{b_i\}_{1 \leq i \leq N}} \quad \mathrm{Tr}[X^{-1}W] \qquad (14)$$

$$\text{s.t.} \quad \frac{Q_f^{-1}}{T} \succeq XA^T + AX \qquad (15)$$

$$- \left(\sum_{i=1}^N b_i B_i R_i^{-1} B_i^T \right) + XQX$$

$$\sum_{i=1}^N b_i \leq 1, \quad 0 \leq b_i, i = 1, \ldots, N.$$

For $(A, Q^{1/2})$ observable, this optimization problem also has a solution for $T \to \infty$ (see [14]), where (15) is replaced by

$$0 \succeq XA^T + AX - \left(\sum_{i=1}^m b_i B_i R_i^{-1} B_i^T \right) + XQX. \qquad (16)$$

The resulting optimal value is a bound on the performance achievable by any switching policy for the infinite-horizon problem.

Finally, we can compute this performance bound efficiently by solving a semidefinite program. We consider only the infinite-horizon problem from now on, i.e., with the constraint (16). Introduce the slack variable $Y \succeq X^{-1}$. By taking Schur complements, we then see that the bound can be obtained by solving

$$Z^* = \min_{X, Y, \{b_i\}_{1 \leq i \leq N}} \quad \mathrm{Tr}[YW] \qquad (17)$$

$$\text{s.t.} \quad \begin{bmatrix} Y & I \\ I & X \end{bmatrix} \succ 0$$

$$\begin{bmatrix} XA^T + AX - \left(\sum_{i=1}^N b_i B_i R_i^{-1} B_i^T \right) & XQ^{1/2} \\ Q^{1/2}X & -I \end{bmatrix} \preceq 0$$

$$\sum_{i=1}^N b_i \leq 1, \quad 0 \leq b_i, i = 1, \ldots, N.$$

5. OPTIMAL CONTINUOUS-TIME POLICIES

In the rest of the paper, we discuss certain switching policies approaching the performance bound (17). Note first that one can evaluate empirically the cost of any policy (say, by simulation) and compare this cost to the performance bound to obtain an indication of the policy performance with respect to an optimal policy. Depending on certain implementation issues, one can approach this bound more or less closely. In this section, we start by assuming essentially no limitation on the signals σ, u. In particular, σ is allowed to switch arbitrarily fast, and u can be implemented as the continuous-time function (5). In this context, there are periodic continuous-time policies that approach the lower bound on achievable performance arbitrarily closely, and hence are essentially optimal. These policies can serve as a benchmark to evaluate the performance of policies satisfying more realistic implementation constraints, such as the ones discussed in section 6.

To design optimal continuous-time policies for the infinite-horizon problem (3), we start by solving (17) and thus obtain

a set of optimal parameters $\{b_i\}_{1 \leq i \leq N}$. Let $b_0 = 0$. We then choose a period length ϵ to execute the following policy, for an ϵ-periodic function $t \to P(t)$ to be described next. We divide each period in subintervals as follows

(i) Use control

$$u = L_i(t)x = -R_i^{-1}B_i^T P(t)x \qquad (18)$$

for the interval

$$[\epsilon \sum_{k=0}^{i-1} b_k, \epsilon \sum_{k=0}^i b_k], 1 \leq i \leq N.$$

(ii) If $\sum_{k=0}^N b_k < 1$, then set $u(t) = 0$ for the interval $[\epsilon \sum_{k=0}^N b_k, \epsilon]$.

Hence the policy spends a proportion b_i of each period in mode i, and does not use any actuation signal for a proportion $1 - \sum_{i=1}^N b_i$ of each period (although we typically have $\sum_{i=1}^N b_i = 1$).

We define a periodic switching signal σ such that $\sigma(t) = i$ over the interval $[\epsilon \sum_{k=0}^{i-1} b_k, \epsilon \sum_{k=0}^i b_k]$, and $\sigma(t) = 0$ over the interval $[\epsilon \sum_{k=0}^N b_k, \epsilon]$ in each period, where $\sigma = 0$ signifies that no input signal is sent to the plant. Define also the formal notation $B_0 R_0^{-1} B_0^T := \mathbf{0}_{n \times n}$. The matrix $P(t)$ used in (18) is the unique positive definite stabilizing ϵ-periodic solution of the following periodic Riccati differential equation (PRE)

$$\dot{P}(t) = -A^T P(t) - P(t)A - Q + P(t)B_{\sigma(t)}R_{\sigma(t)}^{-1}B_{\sigma(t)}^T P(t), \qquad (19)$$

see [6,14]. Denote by $Z(\epsilon)$ the value of the infinite-horizon cost $J(\sigma)$ for the ϵ-periodic continuous-time policy described above, and recall that we denoted by Z^* the value of the performance bound (17). The following theorem says that in the limit of arbitrarily fast switching ($\epsilon \to 0$), these policies perform essentially optimally.

THEOREM 1. We have $Z(\epsilon) - Z^* = O(\epsilon)$ as $\epsilon \to 0$. In particular $Z(\epsilon) \to Z^*$ as $\epsilon \to 0$.

The proof of this theorem follows from the results in [14] on scheduling continuous-time Kalman filters by duality. We now turn to evaluating the performance impact of more realistic implementation constraints.

6. DIGITAL IMPLEMENTATION

There are essentially three ways of designing digital controllers for sampled-data systems [5, 8]. We can first discretize the plant dynamics and design the controller in discrete-time, as in much of the recent literature on the optimal control of switched systems [11, 16, 26]. This potentially ignores some behaviors of the closed-loop continuous-time system, such as hidden oscillations [5]. The second approach, adopted here, is to design a continuous-time controller which is then discretized in actual implementations. The last approach is the most challenging and relatively unexplored in the switched systems literature, and consists in modeling the digital implementation explicitly at the continuous-time level [8] prior to control design.

The policies described in the previous section are continuous-time control laws, i.e., they require a continuous update of the control signal u. The function $t \to P(t)$ in (i) can be computed a priori, but this requires a numerical integration method to obtain the stabilizing periodic solution of the Riccati equation (19). Moreover, in practice most digital control implementations use piecewise constant input signals. Another issue with the result of theorem 1 is that in actual implementations the period ϵ is a finite positive constant governed by the rate at which the physical system can switch between modes or sample the state. Hence the purpose of this section is to discuss more practical digital implementation schemes.

6.1 Restrictions on the Switching Times: Time-Triggered Policies

Fixing ϵ a priori imposes that the system following (i), (ii) is able to switch mode at times $b_1\epsilon, (b_1+b_2)\epsilon, (b_1+b_2+b_3)\epsilon,\ldots$ In fact, the result of theorem 1 remains valid for any continuous-time policy σ that spends a proportion b_i of its period in mode i, not necessarily in a single interval as in (i), which gives additional flexibility to the schedule. Let us call a policy or schedule σ where ϵ can be fixed a priori and the system can guarantee a time $b_i\epsilon$ per period in mode i a policy of type (U) (unrestricted). Typical technological limitations impose a bound of the form

$$\min_{1\le i\le N}\{b_i\epsilon\} \ge \Delta,$$

where Δ is here the minimum dwell-time, i.e., the minimum time that the system must remain in any mode, e.g. due to limited sampling rates or to the implementation overhead of the mode switching mechanism.

In some cases, it can be easier in an embedded control system implementation to further restrict the times at which the mode can change to an a priori fixed set of regularly spaced times $k\Delta, k\ge 0$. This is the case for example if a time-triggered protocol is used for scheduling tasks on an embedded processor [12], which can simplify system integration and verification, see e.g. [20, 24]. Let us call a policy σ satisfying this constraint of fixed computing slots of size Δ a policy of type (TT) (time-triggered). Policies of type (U) and type (TT) give somewhat different discretized models, as discussed in the next subsection. We now turn to the problem of approximating a schedule of type (U) by a more restricted policy of type (TT).

For policies of type (TT), we use the following periodic schedules. First, we approximate the parameters $\{b_i\}_{1\le i\le N}$ optimal for (17) by

$$b_i \approx \frac{l_i}{l}, \quad l_i, l \in \mathbb{N}, \tag{20}$$

where l is an admissible length for the period of the overall schedule. For example, if l is of the form $2^p, p \in \mathbb{N}$, then (20) represents a partial binary expansion of the number b_i. In general, increasing l provides a better approximation of the optimal parameters $b_{i1\le i\le N}$ but increases the memory requirements of the implementation. The rounding procedure (20) might have to be adjusted slightly in general, to make sure that the constraint $\sum_{i=1}^N l_i \le l$ is always enforced, corresponding to $\sum_{i=1}^N b_i \le 1$. We then design a cycle of l slots of length Δ, such that in each cycle, mode i is used in l_i time slots. For resource constrained applications as described in example 1, it is best to spread the slots dedicated to each mode as much as possible within each cycle, to avoid channels operating in an open-loop manner for too long. In contrast, in order to minimize the number of mode switches, one can schedule the slots for each given mode consecutively. In the simulations presented in section 7 we choose the l_i positions of the slots for each mode randomly among the l slots of a cycle. Note that the length $l\Delta$ of the period of a schedule (TT) can be much longer than the length ϵ of a period of a schedule (U), which must only satisfy $\epsilon \ge \frac{\Delta}{\min_{1\le i\le N} b_i}$. In general for schedules of type (TT), there is a trade-off between reducing the length of the schedule and obtaining a good approximation of the parameters $\{b_i\}_{1\le i\le N}$.

6.2 Controller Discretization

Let us now fix the parameter Δ, representing dwell-time for schedules (U) and time-slot length for schedules (TT). Let us also assume that the continuous-time control law (18) must be implemented on a digital controller, which can therefore only update the control signal u at discrete times (we ignore quantization effects in this paper). We assume that the controller also samples the state at these times and we neglect the computation interval between the sampling time and the control update time. We assume however that for both types of policies Δ is a lower bound on the inter-sampling times. Note that these conventions simply fix a possible choice of implementation constraints for the rest of the discussion, and other scenarios could be considered, e.g. where the inter-sampling times can be shorter than Δ. Finally, a zero-order hold is assumed to be present between the digital controller and the plant, so that the plant sees a piecewise-constant control signal.

Denoting the sampling and computation times $t_k, k \ge 0$, we have $t_k = k\Delta$ for policies of type (TT). For policies of type (U), Let us assume for simplicity of notation that $b_1 = \min_{1\le i\le N} b_i$. We then make the choice $\epsilon = \frac{\Delta}{\min_{1\le i\le N} b_i} = \Delta/b_1$, so that only one sample per period is used in the mode with shortest length. For $i \ge 2$, we divide the length $b_i\epsilon = b_i\Delta/b_1$ of mode i into $\lambda_i = \lfloor b_i/b_1 \rfloor$ blocks of equal length δ_i, and sample and compute the control input for mode i at times $(b_1+\ldots+b_i+k\delta_i)\epsilon, 0 \le k \le \lambda_i$, within each period. Note that the time interval between two successive samples in mode i is

$$\delta_i = \frac{b_i\epsilon}{\left\lfloor \frac{b_i}{b_1} \right\rfloor} \ge b_1\epsilon = \Delta,$$

hence the constraint on inter-sampling times is satisfied. The sampling times over the whole period for schedules of types (U) are not exactly periodic, which allows us ignore the approximation issue in (20) but could be more complicated to implement. On the other hand, this device allows us to obtain schedules of type (U) of shorter length.

Next, we discuss the time-discretization of the continuous-time function $L_\sigma(t)$ in (18). For simplicity of notation, we discuss this process only for the policies of type (TT). The discretization process for policies of type (U) is similar, except for the slightly different sampling periods in different modes. Hence let us denote by $x_k = x(k\Delta), k \ge 0$, the sample sequence measured by the controller following a schedule (TT). We construct l gain matrices $L_{k,\Delta}, k = 0,\ldots,l-1$, which are stored in the controller memory to compute the discrete-time periodic control sequence $u_k, k \ge 0$. With the zero-order hold assumption, the signal $u(t)$ at the input of the plant is piecewise constant equal to u_k on the interval $[k\Delta, (k+1)\Delta)$. We use the notation $\sigma(k) := \sigma(t_k) = \sigma(k\Delta)$

to represent the mode used at time k by the schedule σ. Because of the l-periodicity of the schedule, we have $\sigma(k) = \sigma(k \bmod l)$.

Rather that performing say a simple first-order Euler integration scheme for the continuous-time Riccati equation (6), we use the discrete-time Riccati equation describing the optimal solution for the discrete-time LQR problem. Since we emphasize the connection between continuous-time and discrete-time models here, it is best to use incremental models of discrete-time systems [9, 19]. Incremental models also offer improved numerical properties for a small sampling period Δ [23], which is the our focus in this paper.

Hence define

$$A_\Delta = \frac{e^{A\Delta} - I}{\Delta}, \quad B_{i,\Delta} = \frac{1}{\Delta} \int_0^\Delta e^{A\tau} d\tau B_i, 1 \le i \le N.$$

Note that $A_\Delta \to A$ and $B_{i,\Delta} \to B_i$ as $\Delta \to 0$. Moreover, the continuous-time cost (2) for the infinite-horizon problem is approximated by the Riemann sum

$$J(\sigma, u) \approx \lim_{T \to \infty} \frac{1}{T} E \left[\sum_{k=0}^{\lceil T/\Delta \rceil} x_k^T (\Delta Q) x_k + u_k^T (\Delta R_{\sigma(k)}) u_k \right]. \tag{21}$$

Then consider the Difference Periodic Riccati Equation (DPRE) in incremental form (see e.g. [23]), associated to the discrete-time LQR problem (21)

$$\frac{P_k - P_{k+1}}{\Delta} = A_\Delta^T P_{k+1} + P_{k+1} A_\Delta - P_{k+1} B_{\sigma(k),\Delta} \times$$
$$(\Delta B_{\sigma(k),\Delta}^T P_{k+1} B_{\sigma(k),\Delta} + R_{\sigma(k)})^{-1} B_{\sigma(k),\Delta}^T P_{k+1}$$
$$+ \gamma(\sigma(k), \Delta), \tag{22}$$

where

$$\gamma(\sigma(k), \Delta) = \Delta \Big\{ A_\Delta^T P_{k+1} A_\Delta - A_\Delta^T P_{k+1} B_{\sigma(k),\Delta}$$
$$\times (\Delta B_{\sigma(k),\Delta}^T P_{k+1} B_{\sigma(k),\Delta} + R_{\sigma(k)})^{-1} B_{\sigma(k),\Delta}^T P_{k+1}$$
$$- P_{k+1} B_{\sigma(k),\Delta} (\Delta B_{\sigma(k),\Delta}^T P_{k+1} B_{\sigma(k),\Delta} + R_{\sigma(k)})^{-1}$$
$$\times B_{\sigma(k),\Delta}^T P_{k+1} A_\Delta$$
$$- \Delta A_\Delta^T P_{k+1} B_{\sigma(k),\Delta} (\Delta B_{\sigma(k),\Delta}^T P_{k+1} B_{\sigma(k),\Delta} + R_{\sigma(k)})^{-1}$$
$$\times B_{\sigma(k),\Delta}^T P_{k+1} A_\Delta \Big\}.$$

Note in particular that $\gamma(\sigma(k), \Delta) = O(\Delta)$, so that (22) can be seen as an approximation of the continuous-time dynamics (19) as $\Delta \to 0$.

Under our controllability and observability assumptions, the DPRE has a unique stabilizing l-periodic solution [6, 14], which is then used to approximate the continuous-time solution $t \to P(t)$ in (18) at the sampling times. We precompute the matrices $P_0, \dots P_{l-1}$ defining this periodic solution. Then we define the gain matrices as

$$L_{k,\Delta} = -(\Delta B_{\sigma(k),\Delta} P_{k+1} B_{\sigma(k),\Delta} + R_{\sigma(k)})^{-1}$$
$$\times B_{\sigma(k),\Delta}^T P_{k+1} (I + \Delta A_\Delta), \quad k = 0, \dots, l-1,$$

with $P_l = P_0$. Note again that $L_{i,\Delta} \to L_i$ as $\Delta \to 0$, where L_i was defined in (18). Then at period $k \ge 0$, the controller computes

$$u_k = \Delta L_{(k \bmod l),\Delta} x_k,$$

recalling that we neglect the time it takes to compute the matrix-vector product $L_{(k \bmod l)} x_k$. The dynamics of the

closed-loop system at the sampling times are then

$$x_{k+1} - x_k = (A_\Delta + L_{\sigma(k \bmod l),\Delta}) x_k \Delta + w_k, \tag{23}$$

approximating the continuous-time closed loop dynamics, with $\{w_k\}_{k \ge 0}$ a zero-mean Gaussian white noise with covariance

$$E[w_i w_j^T] = \int_0^\Delta e^{A\tau} W e^{A^T \tau} d\tau.$$

Note that the discretization process for policies of types (U) produces a discrete-time system with (periodic) time-varying dynamics instead of the time invariant system (23), due to the unequal sampling period in different modes.

7. SIMULATION RESULTS

In this section, we present simulation results illustrating the performance of the discretized versions of the optimal continuous-time policies. First, we consider the impact of increasing the dwell time or time-slot length Δ, which is desirable to reduce implementation costs. In particular, we evaluate the region of values Δ where the performance bound (17) is close to the performance achieved by the discrete periodic controllers. Consider the following matrices

$$A_1 = \begin{bmatrix} -1 & 1 & 0 & 0 & 0 \\ 0 & -1 & 1 & 0 & 0 \\ 0 & 0 & -1 & 1 & 0 \\ 0 & 0 & 0 & -1 & 1 \end{bmatrix}, \quad B = \begin{bmatrix} 0 & 0 & 0 \\ 0 & 0 & 0 \\ 1 & 0 & 0 \\ 0 & 1 & 0 \\ 0 & 0 & 1 \end{bmatrix}$$

$$A_2 = \begin{bmatrix} 1 & 1 & 0 & 0 & 0 \\ 0 & 1 & 1 & 0 & 0 \\ 0 & 0 & 1 & 1 & 0 \\ 0 & 0 & 0 & 1 & 1 \end{bmatrix}, \quad W = 0.1 \, \mathbf{I}_5.$$

Here and throughout this section, Q and $R_i, 1 \le i \le N$, are taken to be identity matrices. We study the behavior of the policies for the controllable stable system (A_1, B) and unstable system (A_2, B), assuming a scenario as described in example 1, where each control channel selects one of the three columns of B.

Solving the semidefinite program (17) provides the optimal parameters $b_1 \approx 0.54, b_2 \approx 0.44, b_3 \approx 0.02$ for $(A1, B)$ and $b_1 \approx 0.74, b_2 \approx 0.12, b_3 \approx 0.14$ for the unstable system (A_2, B). This results in a schedule of type (U) for the stable system where 27 equidistant samples are taken in the first mode, 22 in the second mode, one in the third mode, before repeating the cycle. For the unstable system, we take 6 samples in the first mode, one sample in mode 2, one sample in mode 3 and repeat the cycle (recall that for schedules (U) the sampling periods in different modes can be different). For both systems, we approximate the optimal values b_1, b_2, b_3 to 2 digits of precision in the (TT) schedules, i.e., the length of the schedule is set to $l = 100$. Hence we obtain $l_1 = 54, l_2 = 44, l_3 = 2$ for the stable system and $l_1 = 74, l_2 = 12, l_3 = 14$ for the unstable system. The positions of the slots for each mode are chosen randomly. The performance curves are shown on Fig. 2 for (A_1, B) and Fig. 3 for (A_2, B). The achieved average performance for each value of Δ is evaluated via Monte-Carlo simulations, with the cost averaged over 10^5 samples. We also show the performance of the unconstrained controller obtained by first discretizing the dynamics and then designing the optimal discrete-time LQR controller.

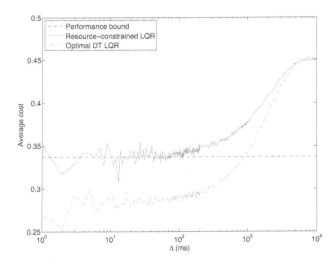

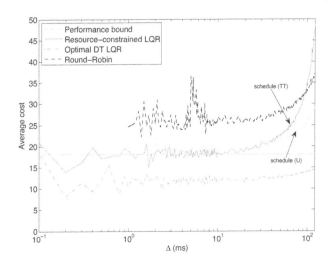

Figure 2: Performance degradation as Δ increases and comparison with the continuous-time performance bound for a stable open-loop system. Note the logarithmic scale on the x-axis (Δ varies from 1ms to 10s). There is no significant performance difference observed between schedules of type (U) and (TT), and only the cost of the later one is shown.

Figure 3: Performance degradation as Δ increases and comparison with the continuous-time performance bound for an unstable open-loop system. The Round-Robin policy simply cycles between the 3 modes by spending one slot in each mode in each cycle.

We can see for these systems that for reasonably small values of Δ (approximately $\Delta \le 100$ ms for (A_1, B) and $\Delta \le 10$ ms for (A_2, B)), the performance of the digital controllers of section 6.2 matches the performance bound (17) (up to the noise in the simulation results), hence these controllers and schedules perform essentially optimally. We observe in general that the bound is tight for an interesting range of values of Δ, and typically quite informative for much of the interval of sampling times that is of practical interest given the system dynamics. There is no significant different between the performance of the schedules of type (U) and (TT) for small values of Δ. Increasing Δ further, we then observe a relatively rapid performance degradation. Nonetheless, the schedules of type (U) show a much better performance than the long schedules of types (TT) in this range for unstable systems (see Fig. 3). Naturally, the value of Δ for which the performance degradation starts to become noticeable depends, among other things, on the degree of stability of the open-loop system (the location of the eigenvalue of A with maximum real part).

Note that the scheduling sequences obtained from the continuous time analysis are not necessarily optimal in general for the problem with large sampling periods Δ, where other effects becomes significant and allow us to distinguish between the performance of schedules (U) and (TT) for example. This can be seen by looking at the unstable system (A_2, B), and comparing the schedule (TT) to the simple Round-Robin policy with $l = 3$, $l_1 = l_2 = l_3 = 1$, see Fig. 3. For $\Delta > 100$ ms, the Round-Robin policy starts to show a better performance, and diverges at a much smaller rate than for the (TT) schedule (but not the (U) schedule) as Δ continues to increase.

Let us investigate in more details the relationship between the range of values Δ where the performance of the digital implementation matches the continuous-time bound and the degree of stability of the open-loop system, measured here

by the maximum real part of the eigenvalues of A. For this purpose, we fix the matrix B to be the 3×2 matrix

$$B = \begin{bmatrix} 0 & 0 \\ 1 & 0 \\ 0 & 1 \end{bmatrix},$$

and generate 3×3 matrices A randomly with entries generated according to a standard normal distribution. For each such matrix, we evaluate empirically the threshold Δ at which the cost becomes greater than 1.2 times the performance bound. Let us denote Δ_{th} this threshold. Define, for a matrix A

$$s(A) := \max\{\mathrm{Re}(\lambda) | \lambda \text{ eigenvalue of } A\}.$$

Fig. 4 shows the variation of Δ_{th} with $s(A)$ for a number of such randomly generated systems, for schedules of type (TT). The thresholds were approximately determined using Monte-Carlo simulations and a Robbins-Monro procedure [13]. As in the previous example, we see that the region where the performance of the digital controller approximately matches the performance bound (17), here within a tolerance of 20%, extends to fairly large values of slot length, with Δ in the tens of milliseconds allowed even for fairly unstable open-loop dynamics. Better results could in fact be expected with schedules of type (U).

Finally, Fig. 5 shows an example of regulation result, using the discretized periodic controller and a schedule of type (TT). The system dynamics is described by a random 10×10 matrix A and a random 10×7 matrix B, and as in example 1 each column of B is associated to a different control channel (i.e. $m_i = 1, i = 1, \ldots, 7$). The regulation responses are compared to a design based on implementing the optimal discrete-time linear-quadratic regulator for the system (A, B), i.e., assuming all input channels can be used simultaneously. As expected, in the presence of the control constraint the response is found to be more sluggish and the

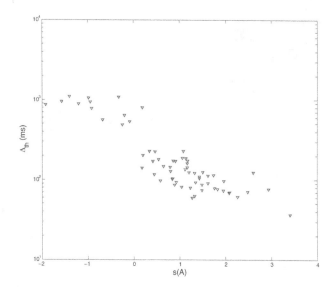

Figure 4: Variation of the threshold value Δ_{th} associated with 20% performance decrease with respect to the performance bound (17) for random linear systems with a three-dimensional state space. On the x-axis we represent the degree of stability of the open loop dynamics, measured by the maximum real part of the eigenvalues of A.

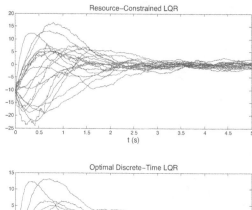

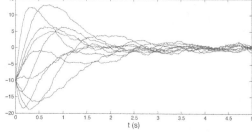

Figure 5: Sample trajectories (plotted at the sample points $k\Delta, k \geq 0$) for a random system with 10 states and 7 inputs, for the digital implementation with schedule (TT) and $\Delta = 10$ ms. One of the 7 inputs can be used in each slot. We also show the response obtained with the standard discrete-time LQR controller (i.e., LQR design after discretizing the dynamics).

noise perturbing the dynamics cannot be filtered as well as with the standard LQR controller.

8. CONCLUSION

We have discussed a linear quadratic control problem under scheduling constraints. It is found that working directly with the continuous-time formulation of the problem allows us to obtain a simple performance bound, which is optimal assuming an analog controller and in the limit of infinitely fast switching rates. Relaxing these assumptions, the bound remains relevant for the characterization of the performance of digital versions of the analog controller under realistic sampling rates. It is interesting to note that for control problems concerning sampled-data systems subject to resource utilization constraints, it seems to be often easier to work directly in continuous-time rather than discretize the dynamics first. This is related to the well-known possibility to "convexify" the set of control inputs in continuous-time optimal control by fast switching. More generally however, continuous-time modeling is also often a richer modeling framework. For example, from the discussion of the schedules of type (U) in section 6, it emerges that time-varying sampling rates can be useful, a fact that would not be noticed by using discrete-time techniques assuming from the start a fixed sampling rate.

The problem considered here is the dual of the Kalman filtering problem with scheduling constraints studied in [14]. Future work will consider the LQG problem under output feedback and constraints on the measurement and control signals.

9. REFERENCES

[1] B. D. O. Anderson and J. Moore. *Optimal Control: Linear Quadratic Methods*. Dover, 2007. Republication of the 1990 edition.

[2] P. J. Antsaklis and J. Bailleul. Special issue on networked control systems technology. *Proceedings of the IEEE*, 95(1):9–28, 2007.

[3] K.-E. Arzen, A. Cervin, and D. Henriksson. Implementation-aware embedded control systems. In D. Hristu-Varsakelis and W. Levine, editors, *Handbook of Networked and Embedded Control Systems*, 2005.

[4] K. J. Åström. *Introduction to Stochastic Control Theory*. Dover, 2006. Republication of the 1970 edition.

[5] K. J. Åström and B. Wittenmark. *Computer-Controlled Systems: Theory and Design*. Prentice Hall, 3rd edition, 1997.

[6] S. Bittanti, P. Colaneri, and G. D. Nicolao. The periodic Riccati equation. In S. Bittanti, A. J. Laub, and J. C. Willems, editors, *The Riccati equation*. Springer-Verlag, 1991.

[7] S. P. Boyd and L. Vandenberghe. *Convex Optimization*. Cambridge University Press, 2006.

[8] T. Chen and B. Francis. *Optimal Sampled-Data Control Systems*. Springer, 1995.

[9] G. C. Goodwin, J. I. Yuz, J. C. Agüero, and M. Cea. Sampling and sampled-data models. In *Proceedings of the American Control Conference*, pages 1–20, Baltimore, MD, 2010.

[10] D. Hristu-Varsakelis and W. S. Levine, editors.

Handbook of Networked and Embedded Control Systems. Springer-Verlag, 2005.

[11] D. Hristu-Varsakelis and L. Zhang. LQG control of networked control systems with access constraints and delays. *International Journal of Control*, 81(8):1266–1280, August 2008.

[12] H. Kopetz. *Real-Time Systems: Design Principles for Distributed Embedded Applications*. Kluwer Academic Publishers, 1997.

[13] H. J. Kushner and G. G. Yin. *Stochastic Approximation and Recursive Algorithms and Applications*. Springer, 2nd edition, 2003.

[14] J. Le Ny, E. Feron, and M. Dahleh. Scheduling continuous-time Kalman filters. *IEEE Transactions on Automatic Control*, 2011. To appear.

[15] J. Le Ny and G. J. Pappas. Robustness analysis for the certification of digital controller implementations. In *Proceedings of the First International Conference on Cyber-Physical Systems (ICCPS)*, Stockholm, Sweden, April 2010.

[16] J.-W. Lee. Infinite-horizon joint LQG synthesis of switching and feedback in discrete time. *IEEE Transactions on Automatic Control*, 54(8):1945–1951, 2009.

[17] B. Lincoln and B. Bernhardsson. LQR optimization of linear system switching. *IEEE Transactions on Automatic Control*, 47(10):1701–1705, October 2002.

[18] L. Meier, J. Peschon, and R. Dressler. Optimal control of measurement systems. *IEEE Transactions on Automatic Control*, 12(5):528–536, 1967.

[19] R. Middleton and G. C. Goodwin. *Digital Control and Estimation. A Unified Approach*. Prentice Hall, Englewood Cliffs, New Jersey, 1990.

[20] T. Nghiem, G. J. Pappas, R. Alur, and A. Girard. Time-triggered implementations of dynamic controllers. *ACM Transactions in Embedded Computing Systems*, 2010. In press.

[21] J. Nilsson. *Real-Time Control Systems with Delays*. PhD thesis, Dept. Automatic Control, Lund Institute of Technology, Lund, Sweden, January 1998.

[22] S. Sager. *Numerical Methods for Mixed-Integer Optimal Control Problems*. PhD thesis, Universität Heidelberg, 2006.

[23] M. Salgado, R. Middleton, and G. Goodwin. Connection between continuous and discrete Riccati equations with applications to Kalman filtering. *Control Theory and Applications, IEE Proceedings D*, 135(1):28–34, 1988.

[24] G. Weiss and R. Alur. Automata based interfaces for control and scheduling. In *Proceedings of the 10th International Conference on Hybrid Systems: Computation and Control*, 2007.

[25] L. Zhang and D. Hristu-Varsakelis. Communication and control co-design for networked control systems. *Automatica*, 42(953-958), 2006.

[26] W. Zhang, J. Hu, and J. Lian. Quadratic optimal control of switched linear stochastic systems. *Systems and Control Letters*, 2010. submitted.

Consensus in Networked Multi-Agent Systems with Adversaries

Heath LeBlanc
heath.j.leblanc@vanderbilt.edu

Xenofon Koutsoukos
xenofon.koutsoukos@vanderbilt.edu

Institute for Software Integrated Systems
Department of Electrical Engineering and Computer Science
Vanderbilt University
Nashville, TN, USA

ABSTRACT

In the past decade, numerous consensus protocols for networked multi-agent systems have been proposed. Although some forms of robustness of these algorithms have been studied, reaching consensus securely in networked multi-agent systems, in spite of intrusions caused by malicious agents, or adversaries, has been largely underexplored. In this work, we consider a general model for adversaries in Euclidean space and introduce a consensus problem for networked multi-agent systems similar to the Byzantine consensus problem in distributed computing. We present the Adversarially Robust Consensus Protocol (ARC-P), which combines ideas from consensus algorithms that are resilient to Byzantine faults and from linear consensus protocols used for control and coordination of dynamic agents. We show that ARC-P solves the consensus problem in complete networks whenever there are more cooperative agents than adversaries. Finally, we illustrate the resilience of ARC-P to adversaries through simulations and compare ARC-P with a linear consensus protocol for networked multi-agent systems.

Categories and Subject Descriptors

C.2.4 [**Computer-Communication Networks**]: Distributed Systems; H.1.1 [**Models and Principles**]: Systems and Information Theory—*General Systems Theory*

General Terms

Algorithms, Security, Theory

Keywords

Consensus, Dynamic agent, Networked multi-agent system, Robustness, Adversary

1. INTRODUCTION

Reaching consensus is a fundamental problem in group coordination. The formal study of consensus has a rich history in management science [5] and distributed computing [18]. More recently, there has been a surge of research in the coordination of multi-agent networks. Within mobile robotics, there are several approaches researching the minimum attributes required to achieve distributed tasks such as gathering [1, 4, 29, 32]. In control, consensus algorithms have been used for the coordination of dynamic agents for group formation [10, 33], conflict resolution [24], and a host of other problems [19, 21]. In sensor networks, consensus has been considered for filtering [30], sensor fusion [36], and distributed hypothesis testing [22].

Various forms of uncertainty have been considered in consensus protocols for multi-agent networks. Reaching average consensus in a wireless network with interference is studied in [34]. Additive channel noise is addressed in [14]. Packet loss in ring networks is studied in [13]. Nonuniform time delays for a class of linear systems is considered in [17]. Contraction analysis is used in [35] to study nonlinear systems and wave variables are used in the communication for robustness to nonuniform constant delays. A virtual layer is used for self-stabilization of a network of robots to a desired curve in [11] whenever there are intermittent disturbances in the network. Robustness in terms of sensitivity to model uncertainty has been addressed in [15].

On the other hand, robustness of consensus protocols in networked multi-agent systems to malicious attacks and failures similar to the Byzantine failures of [16] has only been studied in the last few years. In [25–27], detecting and isolating malicious agents in discrete-time linear consensus networks is considered. Similarly, [31] addresses calculating functions of the initial states of cooperative agents in discrete-time linear consensus networks in the presence of malicious agents. Similar to these works, this paper considers a problem that addresses security of consensus networks; however, the problem introduced here is formulated in continuous-time. Moreover, the algorithm described here is less computationally complex when implemented in discrete-time.

Specifically, in this paper we consider robustness of networked multi-agent systems to adversaries. The multi-agent system is comprised of two classes: cooperative and adversarial agents. The cooperative agents have first-order integrator dynamics and the only assumption about the behavior of the adversarial agents is that their state trajectories are bounded and continuous. Additionally, we assume there is an upper bound on the number of adversarial agents present. The agents exchange scalar state information in an all-to-all manner either through communication or sensing. The goal is for the states of the cooperative agents to asymptotically align to a constant value within the range of their initial states, without knowledge of which agents are adversaries.

The main contribution of this work is the design and analysis of a consensus protocol that is robust to the presence of adversaries.

First, we introduce a system model in the context of ordinary differential equations (ODEs). Then, we define the adversarial agreement problem and present the consensus protocol, which we refer to as the Adversarially Robust Consensus Protocol (ARC-P). ARC-P is inspired by the *ConvergeApproxAgreement* algorithm [6, 18] and the linear consensus protocol (LCP) [23]. We prove that ARC-P yields a unique solution which solves the adversarial agreement problem with an exponential rate of convergence. Then, we show an upper bound on performance that allows for approximate solutions to the problem in finite time. Finally, we provide several simulation results comparing ARC-P to LCP in the presence of adversaries, and illustrate the performance tradeoff incurred by ARC-P for robustness to adversaries.

ARC-P combines ideas from the *ConvergeApproxAgreement* algorithm, which is resilient to Byzantine faults, and LCP. Specifically, it employs the sort and reduce function – which eliminate outlying values to ensure the output is within the range of cooperative agents' states – similarly to [6]. The concept of using a sum of relative state values (i.e., the difference between a neighboring agent's state and the given agent's state) as control input to a first-order integrator agent – in order to drive the agents' states together – is taken from LCP. By combining these ideas from distributed computing and control, we obtain a new consensus protocol that is resilient to adversaries, and can be analyzed using system theoretic techniques. Specifically, we analyze the Lipshcitz continuity of the protocol to ensure the uniqueness of solutions. We use error dynamics to show exponential pairwise convergence of the cooperative agent states. We show that the states of the cooperative agents always converge to a point within the range of their initial conditions using an invariant set argument. Finally, we bound the worst-case convergence time with respect to any arbitrarily small error tolerance, so the protocol can terminate in finite time with an approximate solution.

The paper is organized as follows: Section 2 describes the system model and the adversarial agreement problem. In Section 3 the protocol ARC-P is described. In Section 4, we prove that ARC-P solves the adversarial agreement problem. Section 5 presents simulations illustrating ARC-P in the presence of adversaries, and compares the performance of our solution with LCP. Related work is discussed in Section 6. Finally, Section 7 provides concluding remarks and some ideas for future work.

2. SYSTEM MODEL AND PROBLEM

2.1 System Model

The topology of the networked multi-agent system is described by a labelled strongly connected digraph, $\mathcal{D} = (\mathcal{V}, \mathcal{E})$, where $\mathcal{V} = \{1, \ldots, n\}$ describes the n dynamic agents. Without loss of generality, $\mathcal{V}$ is partitioned into a set of p cooperative agents, $\mathcal{V}_c = \{1, \ldots, p\}$, and a set of q adversarial agents, $\mathcal{V}_a = \{p+1, \ldots, n\}$, with $q = n - p$. The number of adversarial agents in the network is bounded by a constant $F \in \mathbb{Z}^+$, so that $q \leq F$. The edge set $\mathcal{E} \subseteq \mathcal{V} \times \mathcal{V}$ models the information flow between the agents, which is realized either through communication or sensing. For each ordered pair $(i, j) \in \mathcal{E}$, state information flows from agent i to agent j. For loops, $(i, i) \in \mathcal{E}$ represents local state feedback. In this paper, the network is assumed to be complete, i.e., $\mathcal{E} = \{(i, j) | i, j \in \mathcal{V}\}$.

The networked multi-agent system is a composition of two interacting subsytems, i.e., the set of cooperative and adversarial agents. The agents interact in a synchronous manner by sharing state information, as shown in Figure 1. In the figure, $x_c = [x_1, \ldots, x_p]^\mathsf{T} \in \mathbb{R}^p$ represents the states of the cooperative agents. Similarly, $x_a =$

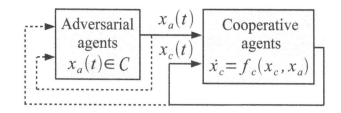

Figure 1: System model.

$[x_{p+1}, \ldots, x_n]^\mathsf{T} \in \mathbb{R}^q$ represents the states of the adversaries. The state feedback to the adversarial agents is shown as dashed lines to indicate that this information may or may not be used to influence the behavior of the adversaries. On the other hand, the cooperative agents must use the state information from all the agents in a similar manner since the cooperative agents are unaware of which agents are adversaries. However, from a global perspective, the states of the adversaries can be viewed as uncertain inputs to the cooperative agents. This is the approach used to analyze the convergence properties of the subsystem of cooperative agents.

2.1.1 Cooperative Agents

Each cooperative agent $i \in \mathcal{V}_c$ has dynamics given by $\dot{x}_i = u_i$, where $u_i = f_i(x_c, x_a)$ is a control input, which is designed in such a way so that the cooperative agents reach consensus in spite of the influence of at most F adversaries. The state of the adversarial agents, x_a, is treated as an uncertain input; however, because there is no prior knowledge concerning which agents are adversarial, the control input must treat the state information from neighboring agents in the same manner. With these clarifications, the dynamics of the system of cooperative agents are given by

$$\dot{x}_c = f_c(x_c, x_a), \quad x_c(0) = x_{c_0}, \, x_a(t) \in \mathcal{C}, \quad (1)$$

where $f_c(x_c, x_a) = [f_1(x_c, x_a) \ \ldots \ f_p(x_c, x_a)]^\mathsf{T}$, $x_{c_0} \in \mathbb{R}^p$ is the vector of initial values of the cooperative agents, and $\mathcal{C} \subset \mathbb{R}^q$ is some fixed compact set.

2.1.2 Adversarial Agents

The adversarial agents are assumed to be designed for the purpose of disrupting the objective of the cooperative agents. The main limitation on the behavior of the adversaries is that the state trajectory of each agent is restricted to bounded continuous functions of time. Specifically, we assume that $x_a(t)$ has a continuous trajectory that remains in some arbitrarily large, but fixed compact set $\mathcal{C} \subset \mathbb{R}^q$ for all $t \geq 0$. Although these assumptions eliminate most unstable systems, the fixed compact set $\mathcal{C}$ can be chosen large enough to include any finite region. One interesting case is when the adversaries are designed to drive the states of the cooperative agents to some unsafe region.

2.2 Adversarial Agreement Problem

Consider a networked multi-agent system consisting of n agents, where a subset of the agents are adversaries. Assume there exists an upper bound F on the number of such agents. Then the *adversarial agreement problem* is defined by two conditions: agreement and validity.

The *agreement condition* states that the pairwise absolute difference between the states of the cooperative agents approaches zero asymptotically, regardless of the adversaries' trajectories. That is, for all $x_c(0) \in \mathbb{R}^p$,

$$\lim_{t \to \infty} |x_i(t) - x_j(t)| = 0, \quad \forall i, j \in \mathcal{V}_c, \, x_a(t) \in \mathcal{C}. \quad (2)$$

Equivalently, the cooperative agents achieve the agreement condition if the state of the cooperative agents, x_c, converges to the agreement space, $\mathcal{A} = \{y \in \mathbb{R}^p | y_i = y_j, \forall i, j \in \mathcal{V}_c\}$.

The *validity condition* states that the limit of the state trajectory of each cooperative agent exists and is contained in the interval formed by the initial states of cooperative agents, regardless of the adversaries' trajectories. That is, if we define the interval

$$\mathcal{I}_0 = [\min_{i \in \mathcal{V}_c} x_i(0), \max_{j \in \mathcal{V}_c} x_j(0)],$$

then the validity condition is formulated as

$$\lim_{t \to \infty} x_i(t) \in \mathcal{I}_0, \quad \forall i \in \mathcal{V}_c, \, x_a(t) \in \mathcal{C}. \tag{3}$$

As in the case of the agreement condition, the validity condition can be stated in terms of x_c. Let $\mathcal{H}_0 = \mathcal{I}_0^p \subset \mathbb{R}^p$ denote the hypercube formed by the Cartesian product of p copies of $\mathcal{I}_0$. Then the validity condition stated in (3) is equivalent to $lim_{t \to \infty} x_c(t) \in \mathcal{H}_0$ for all $x_a(t) \in \mathcal{C}$. Note that if the system satisfies both the agreement and validity conditions, then all of the states of the cooperative agents will converge to a single limit point within $\mathcal{I}_0$.

Example: Consider the linear consensus protocol [23], which we will denote by LCP throughout this paper:

$$\dot{x}_i(t) = \sum_{j \in \mathcal{N}_i} (x_j(t) - x_i(t)), \, x_i(0) = x_{0_i}, \tag{4}$$

where $\mathcal{N}_i = \{j \in \mathcal{V} | (j, i) \in \mathcal{E}, j \neq i\}$. In complete networks, adversaries of the type outlined above cannot prevent the cooperative agents from asymptotically aligning their states to a consensus state. This is because for complete networks, (4) can be written as

$$\dot{x}_i(t) = -nx_i(t) + \sum_{j=1}^{n} x_j(t).$$

Therefore, the pairwise error $e_{ij} = x_i - x_j$ for $i, j \in \mathcal{V}_c$ has dynamics given by

$$\dot{e}_{ij}(t) = -ne_{ij}(t),$$

which converges exponentially to zero with rate n.

However, the validity condition is not satisfied by LCP. Even a single adversary in a complete network can become the leader and drive the state of each cooperative agent to an arbitrary point in the interval $\mathcal{C}$. Although LCP is designed for cooperative agents, the sensitivity to adversarial influence on the cooperative agents is undesirable in cases where security is an issue.

3. CONSENSUS PROTOCOL

This section introduces the Adversarially Robust Consensus Protocol (ARC-P), which is robust to adversaries in complete networks whenever there are more cooperative agents than adversaries, or $n > 2F$. The main idea of the protocol is for each agent to sort the state values and then filter (remove) the F largest and F smallest values so that the remaining values lie within the range of cooperative states. The state of the given agent i is then subtracted from each of the remaining values to form $m = n - 2F$ relative state values. A relative state value is negative if the state of agent i is greater than the filtered state value and nonnegative otherwise. The rate of change of the state of agent i is then the sum of these m relative state values. The result is that the state of agent i increases (decreases) whenever it is smaller (larger) than the average of the m filtered values, and remains constant if it is equal. Intuitively, this process should force the cooperative agents to converge to the average of the filtered values. In the extreme case, whenever $n = 2F + 1$, only the median of the state values remains,

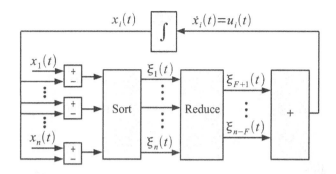

Figure 2: Synchronous data flow model of ARC-P for agent i.

and therefore, the cooperative agents' states are driven toward the median of the state values. Before stating the protocol, we require some definitions.

DEFINITION 1. *Let $m = n - 2F$, and denote the elements of a vector $x \in \mathbb{R}^n$ by x_i, $i = 1, 2, \ldots, n$. Then:*

1. *The concatenation function, $\chi_{p,q} \colon \mathbb{R}^p \times \mathbb{R}^q \to \mathbb{R}^{p+q}$, is defined by[1]*

$$\chi_{p,q}(y, z) = \begin{bmatrix} y \\ z \end{bmatrix}; \tag{5}$$

2. *The (ascending) sorting function on n elements, $\rho_n \colon \mathbb{R}^n \to \mathbb{R}^n$, is defined by $\xi = \rho_n(x)$ such that ξ is a permutation of x satisfying*

$$\xi_1 \leq \xi_2 \leq \cdots \leq \xi_n; \tag{6}$$

3. *The reduce function with respect to $F \in \mathbb{Z}^+$ is defined by $r_F \colon \mathbb{R}^n \to \mathbb{R}^{n-2F}$, $n > 2F$, satisfying*

$$r_F(\xi) = [\xi_{F+1} \, \xi_{F+2} \, \cdots \, \xi_{n-F}]^T; \tag{7}$$

4. *The sum function, $s \colon \mathbb{R}^m \to \mathbb{R}$, is defined by*

$$s(x) = \sum_{i=1}^{m} x_i; \tag{8}$$

5. *The composition of the concatenation, sorting, reduce, and sum functions is defined by $\phi \colon \mathbb{R}^p \times \mathbb{R}^q \to \mathbb{R}$, satisfying for all $(y, z) \in \mathbb{R}^p \times \mathbb{R}^q$,*

$$\phi(y, z) = (s \circ r_F \circ \rho \circ \chi_{p,q})(y, z). \tag{9}$$

The concatenation, sorting, and sum functions are defined in a natural way. The reduce function is intended to be composed with the sorting function – as in the definition of ϕ – so that the resulting operation removes the F smallest and F largest elements.

With these definitions, ARC-P calculates $u_i = f_i(x_c, x_a)$ for each cooperative agent $i \in \mathcal{V}_c$ by

$$f_i(x_c(t), x_a(t)) = -mx_i(t) + \phi(x_c(t), x_a(t)), \tag{10}$$

where $m = n - 2F$. Thus, to ensure $m \geq 1$, we require $n > 2F$.

Figure 2 shows the data flow model of ARC-P for cooperative agent i. In the figure, the state, $x_i(t)$, of the agent, whose dynamics are $\dot{x}_i(t) = u_i(t)$, is subtracted from each of the other states,

---

[1]The concatenation function $\chi_{p,q}$ is essentially the identity function on $\mathbb{R}^{p+q}$. It is used for notational reasons so that x_a can be treated as an uncertain input to the system of cooperative agents.

including itself. The resulting relative state values are sorted and then reduced by eliminating the largest and smallest F elements. Finally, the remaining elements are summed to produce the control input $u_i(t)$ to the integrator agent. It is straightforward to show that this order of operations is equivalent to (10). This implementation is most beneficial in cases where the relative state is sensed directly, so there is no need for a global coordinate system.

4. ANALYSIS

In this section, we analyze the continuity and convergence properties of ARC-P. First, we consider the Lipschitz continuity of the protocol, in order to ensure existence and uniqueness of solutions for all time $t > 0$, over all initial conditions $x_c(0) \in \mathbb{R}^p$, and for all adversarial trajectories, $x_a(t) \in \mathcal{C}$. Next, we show the agreement condition is satisfied and characterize the convergence rate to the agreement space. Then, we prove that the validity condition holds, thereby showing that ARC-P solves the adversarial agreement problem. Finally, we consider a uniform upper bound on convergence in order to obtain ϵ-approximate solutions to the adversarial agreement problem in finite time.

4.1 Lipschitz Continuity

We begin by recalling the definition of Lipschitz continuity. For the purposes of this paper, we restrict ourselves to Euclidean spaces.

DEFINITION 2. *Let $|| \cdot ||$ denote any norm defined on a Euclidean space, and let $g \colon \mathbb{R}^n \to \mathbb{R}^m$. Then g is Lipschitz continuous with Lipschitz constant L if the following condition holds for all $x, y \in \mathbb{R}^n$:*

$$||g(x) - g(y)|| \le L||x - y||.$$

In order to show the Lipschitz continuity of ARC-P, we must show that the sorting function is Lipschitz continuous. First, we consider an interesting property of the sorting function; namely, given any two vectors, then the angle between the vectors will never increase by sorting the vectors. This result is then used to show Lipschitz continuity of the sorting function.

LEMMA 1. *Given $x, x_0 \in \mathbb{R}^n$ with $\xi = \rho_n(x)$ and $\xi_0 = \rho_n(x_0)$, then*

$$\xi^\mathsf{T}\xi_0 = \sum_{i=1}^n \xi_i\xi_{0_i} \ge \sum_{i=1}^n x_i x_{0_i} = x^\mathsf{T} x_0. \tag{11}$$

PROOF. We prove the result by induction on n. The base step ($n = 1$) is obvious (since $\xi = x$, $\xi_0 = x_0$). Now, suppose (11) is true for $1 \le n \le m$, and let $n = m + 1$, with $x, x_0, \xi, \xi_0 \in \mathbb{R}^{m+1}$. Let j (and k) denote the index of the element with minimum value in x (x_0). If there are multiple elements with minimum values in either vector, arbitrarily fix the index to correspond to one of the minimum values. There are two cases: $j \ne k$ and $j = k$.

Case 1, $j \ne k$: Swap the elements x_j and x_k in x so that the minimum values of x and x_0 occur in the same index (k in this case). Remove the kth element from each vector and denote the resulting vectors by $x', x_0' \in \mathbb{R}^m$, and their corresponding sorted versions by ξ' and ξ_0' respectively. Then, by the inductive hypothesis

$$\sum_{i=1}^m \xi_i'\xi_{0_i}' \ge \sum_{i=1}^m x_i' x_{0_i}'. \tag{12}$$

But the terms in (12) are related to the terms in $x^\mathsf{T} x_0$ and $\xi^\mathsf{T}\xi_0$ as follows. For the right-hand side, the only elements altered in x are x_j and x_k, which have been swapped (with x_j removed), and only

x_{0_k} has been removed from x_0, with no other changes to x_0. Thus, we have

$$\sum_{i=1}^m x_i' x_{0_i}' = \sum_{\substack{i=1 \\ i \ne j,k}}^{m+1} x_i x_{0_i} + x_k x_{0_j}. \tag{13}$$

Similarly, for the left-hand side of (12), only one minimum value of each vector has been removed; therefore, the inner product of the resulting sorted vectors $(\xi'^\mathsf{T}\xi_0')$ contain the same terms as $\xi^\mathsf{T}\xi_0$, except for the term $x_j x_{0_k} = \xi_1\xi_{0_1}$. Hence,

$$\sum_{i=1}^m \xi_i'\xi_{0_i}' = \sum_{i=2}^{m+1} \xi_i\xi_{0_i}. \tag{14}$$

Substituting (13) and (14) into (12) and adding $x_j x_{0_k} = \xi_1\xi_{0_1}$ to both sides of the inequality yields

$$\sum_{i=1}^{m+1} \xi_i\xi_{0_i} \ge \sum_{\substack{i=1 \\ i \ne j,k}}^{m+1} x_i x_{0_i} + x_k x_{0_j} + x_j x_{0_k}. \tag{15}$$

Now, since $x_k \ge x_j$ and $x_{0_j} \ge x_{0_k}$, we have

$$(x_k - x_j)(x_{0_j} - x_{0_k}) \ge 0$$
$$\implies x_k x_{0_j} + x_j x_{0_k} \ge x_k x_{0_k} + x_j x_{0_j}.$$

Finally, combining this with (15) produces the desired result

$$\sum_{i=1}^{m+1} \xi_i\xi_{0_i} \ge \sum_{i=1}^{m+1} x_i x_{0_i}, \tag{17}$$

which completes the inductive step.

Case 2, $j = k$: Fix x' and x_0' by removing the kth element (the minimum value) of x and x_0, respectively. Then, (12) is true by the inductive hypothesis. Analogous to Case 1, (14) also holds. In this case, the right-hand side of (12) is given by

$$\sum_{i=1}^m x_i' x_{0_i}' = \sum_{\substack{i=1 \\ i \ne k}}^{m+1} x_i x_{0_i}. \tag{18}$$

Substituting (14) and (18) into (12) and adding $x_k x_{0_k} = \xi_1\xi_{0_1}$ to both sides of the inequality yields (17), which completes the inductive step and the proof. $\square$

LEMMA 2. *The sorting function, $\xi = \rho_n(x) \in \mathbb{R}^n$, defined by (6), is a Lipschitz continuous function of $x \in \mathbb{R}^n$.*

PROOF. Fix $x, x_0 \in \mathbb{R}^n$ and let $\rho_n(x) = \xi$, $\rho_n(x_0) = \xi_0$. Then, using the norm preservation property of permutations and Lemma 1, we have

$$||\xi - \xi_0||_2 = \left(\xi^\mathsf{T}\xi + \xi_0^\mathsf{T}\xi_0 - 2\xi^\mathsf{T}\xi_0\right)^{\frac{1}{2}}$$
$$\le \left(x^\mathsf{T} x + x_0^\mathsf{T} x_0 - 2x^\mathsf{T} x_0\right)^{\frac{1}{2}} = ||x - x_0||_2. \quad \square$$

THEOREM 1. *The function $f_c(x_c, x_a)$ defining the dynamics of the subsystem of cooperative agents (c.f. (1)), with f_i defined in (10), is Lipschitz continuous in x_c and x_a.*

PROOF. The concatenation, reduce, and sum functions are linear maps and are therefore Lipschitz continuous. The sorting function is Lipschitz continuous by Lemma 2. The result then follows since scalar multiplication is Lipschitz continuous and the composition and difference of Lipschitz functions result in a Lipschitz continuous function. $\square$

Since $x_a(t) \in \mathcal{C}$ is a bounded continuous trajectory and f_c is Lipschitz continuous, the system (1) admits a solution $x_c(t)$ which is uniquely defined on $\mathbb{R}^+$ for all $x_c(0) \in \mathbb{R}^p$ and $x_a(t) \in \mathcal{C}$ [2].

COROLLARY 1. *The networked multi-agent system with $n > 2F$ and each cooperative agent's control protocol given by (10), has a unique solution for all $t \geq 0$, $x_c(0) \in \mathbb{R}^p$, and $x_a(t) \in \mathcal{C}$.*

4.2 Agreement

In this section, we prove the agreement condition for ARC-P and characterize the convergence rate to the agreement space.

THEOREM 2. *The networked multi-agent system with $n > 2F$ and each cooperative agent's control protocol given by (10), satisfies the agreement condition (2). Moreover, the convergence to the agreement space is exponential with rate $m = n - 2F$.*

PROOF. For each pair $i, j \in \mathcal{V}_c$, $i \neq j$, we define the pairwise error $e_{ij}(t) = x_i(t) - x_j(t)$. Since $n > 2F$, $\phi(x_c(t), x_a(t))$ is defined. Because the network is complete, $\phi(x_c(t), x_a(t))$ is identical for each agent. Therefore, the ϕ-terms cancel in the error dynamics of $\dot{e}_{ij}(t) = -m e_{ij}(t)$. Now, define $e(t)$ as the column vector containing all $\binom{p}{2}$ pairwise errors of the form $e_{ij}(t)$. Clearly, $e = 0$ is equivalent to $x_c \in \mathcal{A}$. The error dynamics are then

$$\dot{e} = -me \implies e(t) = e(0)\mathrm{e}^{-mt},$$

which proves the agreement condition is satisfied and that the convergence is exponential with rate $m = n - 2F$. $\quad\square$

4.3 Validity

While the agreement condition follows directly from the symmetry provided by the complete network, the validity condition requires an invariant set argument, facilitated by some results from the theory of uncertain systems. As described in Section 2.1, we consider a decomposition of the multi-agent system into cooperative and adversarial agents which interact through sharing state information. The state information from the adversarial agents is viewed as an uncertain input to the subsystem of cooperative agents and may take values in the compact set $\mathcal{C}$. We begin with the definition of robustly positively invariant sets.

DEFINITION 3. *The set $\mathcal{S} \subset \mathbb{R}^p$ is robustly positively invariant for the system given by (1) if for all $x_c(0) \in \mathcal{S}$ and all $x_a(t) \in \mathcal{C}$ the solution is such that $x_c(t) \in \mathcal{S}$ for $t > 0$.*

In order to show that the validity condition holds, we first show that the hypercube $\mathcal{H}_0$, containing all of the initial values of the cooperative agents, is a robustly positively invariant set using an extension of Nagumo's Theorem for uncertain systems. Then we prove that the limit of the cooperative agents' states exists and therefore lies in this hypercube. For notational brevity, we denote

$$x_{\min} = \min_{i \in \mathcal{V}_c} x_i(0) \text{ and } x_{\max} = \max_{i \in \mathcal{V}_c} x_i(0).$$

LEMMA 3. *If each cooperative agent's control protocol is given by (10), then the hypercube $\mathcal{H}_0 = \mathcal{I}_0^p$ defined by*

$$\mathcal{H}_0 = \{y \in \mathbb{R}^p | x_{min} \leq y_i \leq x_{max}\},$$

is robustly positively invariant for the system (1).

PROOF. First we require a definition. For any compact and convex set $\mathcal{S} \subset \mathbb{R}^p$, the tangent cone to $\mathcal{S}$ in y is the set

$$\mathcal{T}_{\mathcal{S}}(y) = \{z \in \mathbb{R}^p | \lim_{h \to 0} \frac{\mathrm{dist}(y + hz, \mathcal{S})}{h} = 0\},$$

where $\mathrm{dist}(y, \mathcal{S}) = \inf_{x \in \mathcal{S}} \|y - x\|_2$. Since $\mathcal{H}_0$ is closed and convex, an extension to Nagumo's Theorem presented in [3, p.106] states that $\mathcal{H}_0$ is robustly positively invariant if and only if

$$f_c(y, x_a) \in \mathcal{T}_{\mathcal{H}_0}(y), \ \forall y \in \mathcal{H}_0 \text{ and } x_a \in \mathcal{C}.$$

For interior points y in $\mathcal{H}_0$, we have $\mathcal{T}_{\mathcal{H}_0}(y) = \mathbb{R}^p$, so we only need to check the boundary of $\mathcal{H}_0$. The boundary, $\partial\mathcal{H}_0$, is given by

$$\partial\mathcal{H}_0 = \{y \in \mathcal{H}_0 | \exists i \in \{1, 2, \ldots, p\} \text{ s.t. } y_i \in \{x_{\min}, x_{\max}\}\}.$$

In other words, points on the boundary have at least one component that is either a minimum or maximum of the initial values.

Fix $y \in \partial\mathcal{H}_0$ and let $\mathcal{I}_{\min}, \mathcal{I}_{\max} \subseteq \{1, 2, \ldots, p\}$ denote the sets of indices such that

$$j \in \mathcal{I}_{\min} \Leftrightarrow y_j = x_{\min} \text{ and } k \in \mathcal{I}_{\max} \Leftrightarrow y_k = x_{\max}.$$

Since $y \in \partial\mathcal{H}_0$, at least one of these index sets is nonempty. Let e_j denote the j-th canonical basis vector. From the geometry of the hypercube, it is sufficient for the following to hold:

$$e_j^{\mathsf{T}} f_c(y, x_a) \geq 0 \quad \forall j \in \mathcal{I}_{\min}, x_a(t) \in \mathcal{C},$$
$$e_k^{\mathsf{T}} f_c(y, x_a) \leq 0 \quad \forall k \in \mathcal{I}_{\max}, x_a(t) \in \mathcal{C}.$$

It can be shown that each component $j \in \{1, 2, \ldots, m\}$ of $\zeta \triangleq (r_F \circ \rho_n \circ \chi_{p,q})(y, x_a)$ satisfies

$$x_{\min} \leq \zeta_j \leq x_{\max}, \ \forall x_a(t) \in \mathcal{C}.$$

Indeed, otherwise $y \notin \mathcal{H}_0$. Therefore, we have that for all $x_a(t) \in \mathcal{C}$, $j \in \mathcal{I}_{\min}$, and $k \in \mathcal{I}_{\max}$,

$$e_j^{\mathsf{T}} f_c(y, x_a) = -m y_j + \sum_{i=1}^m \zeta_i = -m x_{\min} + \sum_{i=1}^m \zeta_i \geq 0,$$

$$e_k^{\mathsf{T}} f_c(y, x_a) = -m y_k + \sum_{i=1}^m \zeta_i = -m x_{\max} + \sum_{i=1}^m \zeta_i \leq 0. \quad\square$$

THEOREM 3. *The networked multi-agent system with $n > 2F$ and each cooperative agent's control protocol given by (10), satisfies the validity condition (3).*

PROOF. Consider $\psi(t) = \max_{i \in \mathcal{V}_c}(x_i(t))$. Define for each $j \in \mathcal{V}_c$,

$$\mathcal{I}_j = \{t \geq 0 \mid x_j(t) = \psi(t)\}.$$

Then, let $\mathcal{S} = \{j \in \mathcal{V}_c | \mathcal{I}_j \neq \emptyset\}$. We claim that elements of $\mathcal{S}$ satisfy the following property: for $i, j \in \mathcal{S}$, $x_i(t) \equiv x_j(t)$, for all $t \geq 0$. To show this, fix $i, j \in \mathcal{S}$, and consider the errors $e_{ij}(t)$ from the proof of Theorem 2. If $x_i(0) > x_j(0)$, then

$$e_{ij}(t) = e_{ij}(0)\mathrm{e}^{-mt} > 0, \quad \forall t \geq 0.$$

This contradicts the fact that $j \in \mathcal{S}$. By symmetry, we cannot have $x_j(0) > x_i(0)$. Therefore, $x_i(0) = x_j(0)$, and $e_{ij}(t) \equiv 0$, which implies $x_i(t) \equiv x_j(t)$, for all $t \geq 0$.

Therefore, $\psi(t)$ uniquely describes the positive trajectory of the subset of cooperative agents with initial value $x_{\max}$. Now, since $\phi(x_c(t), x_a(t)) \leq m\psi(t)$, $\psi(t)$ is nonincreasing. Furthermore, since $\psi(t)$ must remain in $\mathcal{I}_0$ by Lemma 3, it is bounded below by $x_{\min}$; thus $\lim_{t \to \infty} \psi(t) \in \mathcal{I}_0$ exists. Finally, by Theorem 2, all of the states of the cooperative agents converge to $\psi(t)$ and therefore, we have $\lim_{t \to \infty} \psi(t) = \lim_{t \to \infty} x_i(t)$, for all $i \in \mathcal{V}_c$. Since this is independent of $x_a(t)$, the validity condition is satisfied. $\quad\square$

By Theorem 2, ARC-P satisfies the agreement condition. By Theorem 3, ARC-P satisfies the validity condition. Therefore, ARC-P solves the adversarial agreement problem.

THEOREM 4. *The networked multi-agent system with* $n > 2F$ *and each cooperative agent's control protocol given by (10), solves the adversarial agreement problem.*

4.4 Finite Termination

In this section, we derive an upper bound on the performance of ARC-P in order to terminate in finite time while ensuring an ϵ-approximate solution to the adversarial agreement problem. By Theorem 2, the rate of convergence is exponential with rate m. But, the upper bound on convergence can be made precise.

COROLLARY 2. *Consider a networked multi-agent system with* $n > 2F$ *and each cooperative agent's control protocol given by (10). Define* $\beta = \max_{i \in \mathcal{V}} x_i(0) - \min_{i \in \mathcal{V}} x_i(0)$. *Then,*

$$\max_{i,j \in \mathcal{V}_c} e_{ij}(t) \leq \beta e^{-mt}, \forall t \geq 0.$$

PROOF. For all $i, j \in \mathcal{V}_c$,

$$e_{ij}(t) = e_{ij}(0)e^{-mt} \leq \beta e^{-mt}, \forall t \geq 0. \qquad \square$$

Using Corollary 2, an ϵ-approximate solution to the adversarial agreement problem can be obtained in finite time. Specifically, to ensure that the maximum pairwise error between the states of cooperative agents is less than $\epsilon > 0$, Corollary 2 implies we may terminate at any time greater than $\frac{1}{m}|\log(\frac{\epsilon}{\beta})|$ (provided $\beta \neq 0$, in which case $x_c(0) \in \mathcal{A}$).

5. SIMULATIONS

To illustrate the robustness of ARC-P, we consider three examples in which a subset of the agents have been overtaken and redesigned with malicious intent, and a fourth example that illustrates the performance tradeoff incurred for the robustness to adversaries. The first scenario considers the case where two out of the eight agents are adversaries and their goal is to drive the consensus state of the cooperative agents to an unsafe set $\mathcal{U}$. In the second scenario, three of the agents have been redesigned as oscillators in order to force the cooperative agents to oscillate at the desired frequency. Finally, in the third scenario a single adversary in a large network tries to force the other agents to follow a sinusoidal trajectory in the unsafe set.

To motivate the need for a consensus protocol that is robust to adversaries, we compare ARC-P with LCP under the same conditions. It is shown that LCP achieves the agreement condition in spite of the behavior of the adversaries, but not the validity condition. For LCP, the states of the cooperative agents effectively converge to the average of the aversaries' trajectories. Thus, in all three scenarios, the adversaries are able to achieve their goal.

Example 1: Consider a multi-agent network with eight agents, each with unique identifier in $\{1, 2, \ldots, 8\}$, and with initial states equal to their identifier (e.g., for agent 1, $x_1(0) = 1$). Suppose that agents 7 and 8 have been compromised (i.e., $\mathcal{V}_a = \{7, 8\}$). The adversaries are redesigned with

$$\dot{x}_i = -10x_i + 10u_i, \ \forall i \in \mathcal{V}_a,$$

where the reference inputs u_i for the adversarial agents are $u_7 = 25$ and $u_8 = 26$. Therefore, the adversarial agents will converge exponentially to 25 and 26, respectively, with rate 10.

The goal of the adversaries is to drive the states of the cooperative agents into the unsafe set $\mathcal{U} = \{y \in \mathbb{R} | y \geq 20\}$. The results for LCP and ARC-P are shown in Figure 3. The adversaries are able to achieve their goal only with LCP. The cooperative agents equipped with ARC-P achieve both the agreement and validity conditions. Because both of the adversaries always have larger state values, the

consensus process for the cooperative agents is unaffected and the final consensus state is the average of the $m = 4$ initial states of the agents filtered by $\phi(x_c, x_a)$; in this case, 4.5. Also, the rate of convergence is $m = 4$.

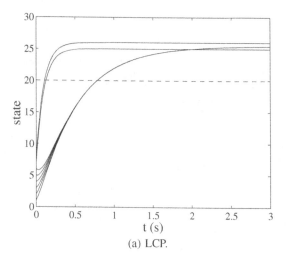

(a) LCP.

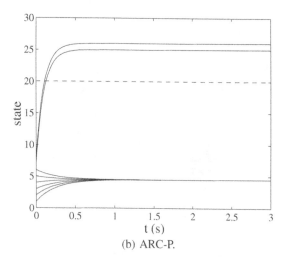

(b) ARC-P.

Figure 3: Adversaries try to drive agents to $\mathcal{U}$.

Example 2: Consider the same multi-agent network as Example 1, with the same initial conditions, but with agents 6, 7, and 8 as adversaries (i.e., $\mathcal{V}_a = \{6, 7, 8\}$). This time the adversaries' dynamics are designed as

$$\ddot{x}_i = -100\pi^2 x_i, \ \forall i \in \mathcal{V}_a.$$

Thus, they are oscillators with natural frequency 5 Hz and amplitude given by their initial state. The goal of the adversaries in this case is to force the cooperative agents' states to oscillate at 5 Hz.

The results for LCP and ARC-P are shown in Figure 4. As can be seen in Figure 4(a), the cooperative agents executing LCP synchronize and begin oscillating at 5 Hz, with a phase lag of 90° with respect to the adversaries. However, for the case of ARC-P, the cooperative agents achieve the agreement and validity conditions. As the adversarial agents move their states inside the range of the filter ϕ, the limit point for the cooperative agents is shifted, which can be seen in Figure 4(b) as a change in the shape of the exponential decay each time the adversaries move through this range.

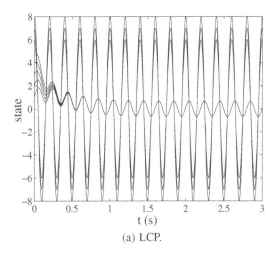

(a) LCP.

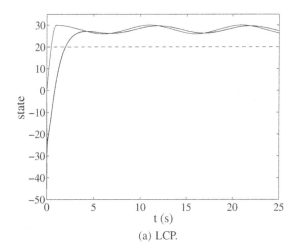

(a) LCP.

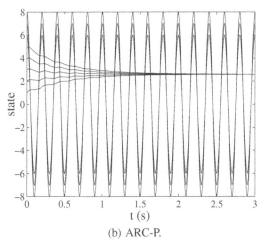

(b) ARC-P.

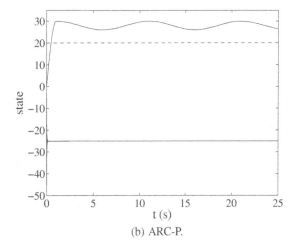

(b) ARC-P.

Figure 4: Adversaries try to force agents to oscillate at 5 Hz.

Figure 5: A single adversary tries to drive 50 cooperative agents to oscillate at 0.1 Hz in the unsafe set $\mathcal{U}$.

This shifts the limit point from 3 to 2.6, without affecting the rate of consensus.

Example 3: Consider a multi-agent network with 51 agents, where only agent 51 is an adversary. The initial states of the cooperative agents are $x_1(0) = -1$, $x_2(0) = -2$, ..., $x_{50}(0) = -50$. The adversary is designed with time-varying dynamics given by the following expressions:

$$\ddot{x}_{51} = -0.25\pi^2 x_{51} \qquad \text{if } 0 \leq t < 1;$$
$$\dot{x}_{51} = -0.4\pi \sin(0.2\pi(t-1)) \quad \text{if } t \geq 1;$$

and initial conditions $x_{51}(0) = 0$, $\dot{x}_{51}(0) = 15\pi$. The resulting trajectory is

$$x_{51}(t) = \begin{cases} 30\sin(0.5\pi t) & t < 1; \\ 2\cos(0.2\pi(t-1)) + 28 & t \geq 1. \end{cases}$$

The objective of the adversary is to bring the states of the cooperative agents into the unsafe set $\mathcal{U}$ (as in Example 1), and force them to oscillate at a frequency of 0.1 Hz. The results for LCP and ARC-P are shown in Figure 5. In this case, the convergence rates for LCP and ARC-P are 51 and 49, respectively, so the pairwise difference between the states of cooperative agents becomes negligible by 0.1 s (approximately five time constants) into the sim-

ulation. The result is that the trajectory of the cooperative agents appears to be a single curve in the figure. Clearly, the adversary is able to achieve its objective only with LCP. The consensus limit point for ARC-P is -25, i.e., the average of $x_1(0), \ldots, x_{49}(0)$.

Example 4: We consider the performance tradeoff required for robustness to adversaries. For this purpose, consider the same multi-agent network of Example 1, but with no adversaries. As shown in Theorem 2, the rate of exponential convergence is $m = n - 2F$. The change in the rate of convergence with F is illustrated in Figure 6 for the eight agent network. Note that ARC-P reduces to LCP in the case $F = 0$. It is also important to note that although the limit point observed in Figure 6 is the same in all four cases, this would not be the case with asymmetries in the initial conditions.

By scaling ARC-P in (10) by the factor $\frac{n}{m}$, we may eliminate the tradeoff in performance for robustness to adversaries, and make ARC-P perform as well as LCP. However, LCP may also be scaled to improve its rate of convergence. Moreover, scaling ARC-P will incur a reduction in robustness to time delays as it does with LCP. Indeed, scaling LCP scales the largest eigenvalue of the Laplacian, which reduces the robustness to time delays [20]. Investigation of the robustness of ARC-P to time delays is left for future work.

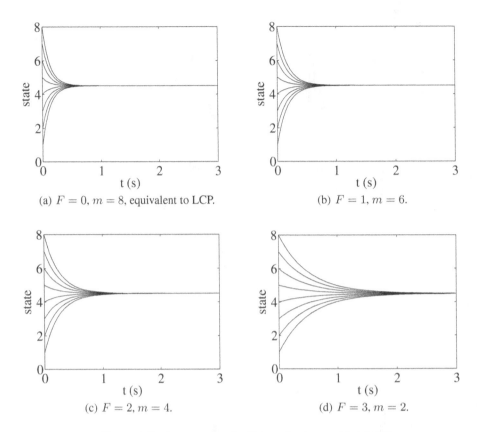

(a) $F = 0$, $m = 8$, equivalent to LCP.

(b) $F = 1$, $m = 6$.

(c) $F = 2$, $m = 4$.

(d) $F = 3$, $m = 2$.

Figure 6: Performance tradeoff for robustness with ARC-P.

6. RELATED WORK AND DISCUSSION

Some of the earliest work on reaching consensus in the presence of adversaries can be found in [28] and [16], where the Byzantine agreement problem was first introduced. The notion of adversary described in this paper is similar to Byzantine failures, and has the same intent, i.e., to consider robustness to some sense of worst-case behavior. However, the restriction on continuity of the trajectory and the requirement to share true state information make our model more limited than the Byzantine fault model. Byzantine faulty processors are allowed to change their state to any valid state, and are allowed to send different messages to different processors. But within the environment of Euclidean space and with evolution in continuous-time, the restriction of continuity is sensible and even required in many physical situations. Furthermore, if the state information is obtained through sensing or broadcast communication, it is reasonable to assume the same state information is received by all agents (under the assumption of perfect sensors).

The Byzantine agreement problem has been generalized to allow approximate solutions for both synchronous and asynchronous systems whenever the values are arbitrary real numbers [6]. The filter ϕ of this work is modeled after the class of approximation functions considered in [6]. In contrast to [6], we construct ϕ on the topology of Euclidean space as opposed to using multisets. Moreover, the select function is not needed because the adversaries cannot hide their true state. This is why we are able to loosen the constraint on the ratio of adversaries to $n \geq 2F + 1$ (from $n \geq 3F + 1$). An interesting consequence of the results presented here is that the class of approximation functions considered in [6] are Lipschitz continuous (the proof of the Lipschitz continuity of the select function follows because the select function is linear).

More recently, there has been work in mobile robotics, which views coordination problems from a distributed computing perspective. The goal of the research is to characterize the weakest set of assumptions required for achieving a certain coordinated task in finite time [29]. One of the tasks considered is for a group of robots to gather at a single point in space (i.e., rendezvous) in finite time [32]. In order to consider the weakest assumptions on the capabilities of the robots, it is common to assume that the robots are indistinguishable, have different local coordinate frames, and are oblivious (which means they do not remember past observations or computations performed in previous steps). In some cases, it is assumed that all robots are able to obtain the exact position of all other robots, which is similar to the case considered in this paper.

The robot gathering problem has also been studied in the presence of faults [1], including Byzantine faults [4]. From a distributed computing perspective, the issue of stability of the robot dynamics is ignored, and the evolution of the robots occur in computational cycles (e.g., the *Wait-Look-Compute-Move* cycle) [29]. The common computational models used are the ATOM model [32], where the full cycle is executed instantaneously and atomically, and the CORDA model [29], where each stage requires a (nonzero) finite amount of time to execute any given stage, and any non-null move action will result in a (nonzero) finite distance moved.

The gathering problem with Byzantine failures shares some similarities with the adversarial agreement problem. As mentioned above, in some cases all-to-all sharing of state information is assumed [4]. Also, the cooperative agents are oblivious, which is also the case in our work because the control input is static. Additionally, in ARC-P the cooperative agents are indistinguishable, although this is not explicitly required in our model. There are also some key differences between the gathering and adversarial agreement problems. In our model, it is implicitly assumed there is a global inertial coordinate frame, unless the relative states are the quantities sensed. Also, in the gathering problem, consensus to a point must be achieved in finite time, as opposed to the asymptotic requirement of the adversarial agreement problem. However, because the dynamics of the robots are not considered, stability is assumed, whereas in the adversarial agreement problem, the dynamics of the individual agent and of the subsystem of cooperative agents is fundamental. A consequence of modeling the dynamics as ODEs is that the system model is synchronous. This differs from the gathering problem with Byzantine faults, where the system is often assumed to be asynchronous [4].

Another closely related research problem is the issue of security of consensus in multi-robot systems [9]. A distributed intrusion detection system (IDS) has been developed using a hybrid model of robotic agents that monitors neighboring agents to detect non-cooperative agents using only local information [9]. The distributed IDS has been extended to deal with the case where some of the monitors provide false information [8], and improved by using past information [7]. The distributed IDS approach differs from ARC-P because the agents executing ARC-P are oblivious, so they do not require past information for robustness to adversaries. Moreover, the approximation functions can be computed in linear time, so they are very efficient. However, the adversarial agreement problem is not concerned with issues such as collision avoidance, which makes the agents executing ARC-P susceptible in this case. On the other hand, avoiding collisions with misbehaving agents has been considered for the distributed IDS [8].

Also, the issue of security of linear consensus networks has been studied. In [25], the issue of a single malicious agent is considered. The same authors later extended the work to characterize the connectivity of the network required to tolerate misbehaving agents and non-colluding agents in [26]. A computationally expensive but exact algorithm was presented in [26] to detect and isolate up to k misbehaving agents in networks with connectivity at least $2k + 1$. Additionally the structure of the entire network was necessary for the exact algorithm [26]. In [27], two approaches were considered to reduce the computational complexity and require only partial network information. The first assumes the network is comprised of weakly interconnected subcomponents and restricts the behavior of the misbehaving agents. The second imposes a hierarchical structure to detect and isolate the misbehaving agents. The problem considered in these papers requires detecting and isolating the misbehaving agents, and therefore results in more complex algorithms than ARC-P.

Another approach is considered in [31], where the feasibility of reaching consensus on any function of the initial states is considered in the presence of malicious agents. In this case, the linearity of the protocol is exploited to calculate the initial values exactly in at most n steps, where n is the number of nodes. Similar to our model for adversary, the malicious agents send the same information to all their neighbors. However, the malicious agents are modeled in discrete-time, so there is no continuity restriction.

Finally, [12] presents a framework for determining the robustness of distributed algorithms for discrete-time, synchronous algorithms on undirected graphs. The robustness of an algorithm is defined with respect to a fault model and is measured using a cost function. The cost function is a formal model of the cooperative task, and is defined on a domain consisting of all state and input trajectories, initial states, and environmental variables of the networked multi-agent system. The minimizer of the cost function should occur on a subset of the domain where the networked multi-

agent system achieves the task. By extending the framework of [12] to continuous-time and by defining the cost function

$$
C = \lim_{t \to \infty} \left(\sum_{i,j=1}^{n} (x_i(t) - x_j(t))^2 + \sum_{i=1}^{n} \mathrm{dist}(x_i(t), \mathcal{H}_0)^2 \right),
$$

for the adversarial agreement problem, it is straightforward to show using the results of this paper that ARC-P is worst-case robust to adversaries up to $F = \lfloor \frac{n-1}{2} \rfloor$ agents in complete networks. This shows that our interpretation of robustness is conformable to that of [12].

7. CONCLUSIONS AND FUTURE WORK

In this paper, we provide a general model for adversaries in Euclidean space and propose the adversarial agreement problem, a consensus problem in the context of adversaries. Then, we introduce the Adversarially Resilient Consensus Protocol (ARC-P) that combines ideas from distributed computing and control consensus protocols. We analyze the convergence properties of ARC-P, and show that it solves the adversarial agreement problem. Additionally, we show how to obtain approximate solutions to the problem in finite time. Then we provide several simulations to illustrate the resilient behavior of ARC-P to adversaries and the performance tradeoff required for the robustness to adversaries. Finally, we describe the related work and show how ARC-P is worst-case robust [12] to adversaries in complete networks whenever the ratio of adversaries to total agents is less than one-half.

Currently, we assume an "ideal" network, where all agents are able to obtain instantaneous, real-valued state information from every other agent. This assumption isolates the network uncertainties from the issue of adversaries. In future work, we would like to relax these idealized assumptions by simultaneously considering adversaries with other network uncertainties, such as time delays, packet loss, channel noise, and quantization. Also, for the protocol to be more useful, we intend to generalize the protocol for arbitrary network topologies by using a local bound on the number of adversaries amongst each agent's neighbors.

8. ACKNOWLEDGMENTS

The authors would like to thank the reviewers for their comments, which have improved the paper. This work is supported in part by the National Science Foundation (CNS-1035655, CCF-0820088), the U.S. Army Research Office (ARO W911NF-10-1-0005), and Lockheed Martin. The views and conclusions contained herein are those of the authors and should not be interpreted as necessarily representing the official policies or endorsements, either expressed or implied, of the U.S. Government.

9. REFERENCES

[1] N. Agmon and D. Peleg. Fault-tolerant gathering algorithms for autonomous mobile robots. *SIAM J. Comput.*, 36(1):56–82, 2006.

[2] F. Blanchini. Set invariance in control. *Automatica*, 35(11):1747 – 1767, 1999.

[3] F. Blanchini and S. Miani. *Set-Theoretic Methods in Control*. Birkhauser, Boston, Massachusetts, 2008.

[4] Z. Bouzid, M. Gradinariu Potop-Butucaru, and S. Tixeuil. Byzantine convergence in robot networks: The price of asynchrony. In *Int. Conf. on Principles of Distributed Systems*, pages 54–70, December 2009.

[5] M. H. DeGroot. Reaching a consensus. *Journal of the American Statistical Association*, 69(345):118–121, 1974.

[6] D. Dolev, N. A. Lynch, S. S. Pinter, E. W. Stark, and W. E. Weihl. Reaching approximate agreement in the presence of faults. *Journal of the ACM*, 33(3):499 – 516, 1986.

[7] A. Fagiolini, F. Babboni, and A. Bicchi. Dynamic distributed intrusion detection for secure multi-robot systems. In *Int. Conf. of Robotics and Aut.*, pages 2723 – 2728, May 2009.

[8] A. Fagiolini, A. Bicchi, G. Dini, and I. Savino. Tolerating malicious monitors in detecting misbehaving robots. In *IEEE Int. Workshop on Safety, Security, and Rescue Robotics*, pages 108 – 114, October 2008.

[9] A. Fagiolini, M. Pellinacci, M. Valenti, G., G. Dini, and A. Bicchi. Consensus-based distributed intrusion detection for multi-robot systems. In *Int. Conf. on Robotics and Aut.*, pages 120 – 127, May 2008.

[10] J. A. Fax and R. M. Murray. Information flow and cooperative control of vehicle formations. *IEEE Trans. on Aut. Control*, 49(9):1465 – 1476, 2004.

[11] S. Gilbert, N. Lynch, S. Mitra, and T. Nolte. Self-stabilizing robot formations over unreliable networks. *ACM Trans. Auton. Adapt. Syst.*, 4(3):1–29, 2009.

[12] V. Gupta, C. Langbort, and R. Murray. On the robustness of distributed algorithms. In *IEEE Conf. on Decision and Control*, December 2006.

[13] P. Hovareshti, J. Baras, and V. Gupta. Average consensus over small world networks: A probabilistic framework. In *IEEE Conf. on Decision and Control*, December 2008.

[14] M. Huang and J. Manton. Stochastic Lyapunov analysis for consensus algorithms with noisy measurements. In *American Control Conf.*, July 2007.

[15] Q. Hui, W. Haddad, and S. Bhat. On robust control algorithms for nonlinear network consensus protocols. *Int. J. Robust Nonlinear Control*, 20(3):269 – 284, 2010.

[16] L. Lamport, R. Shostak, and M. Pease. The Byzantine generals problem. *ACM Trans. Program. Lang. Syst.*, 4(2):382–401, 1982.

[17] D. Lee and M. Spong. Agreement with non-uniform information delays. In *American Control Conf.*, pages 756 – 761, June 2006.

[18] N. A. Lynch. *Distributed Algorithms*. Morgan Kaufmann Publishers Inc., San Francisco, California, 1997.

[19] M. Mesbahi and M. Egerstedt. *Graph Theoretic Methods in Multiagent Networks*. Princeton University Press, Princeton, New Jersey, 2010.

[20] R. Olfati-Saber. Ultrafast consensus in small-world networks. In *American Control Conf.*, pages 2371 – 2378, June 2005.

[21] R. Olfati-Saber, J. A. Fax, and R. M. Murray. Consensus and cooperation in networked multi-agent systems. *Proceedings of the IEEE*, 95(1):215–233, 2007.

[22] R. Olfati-Saber, E. Franco, E. Frazzoli, and J. Shamma. Belief consensus and distributed hypothesis testing in sensor networks. In *Networked Embedded Sensing and Control*, volume 331 of *Lecture Notes in Control and Information Sciences*, pages 169–182. Springer Berlin / Heidelberg, 2006.

[23] R. Olfati-Saber and R. M. Murray. Consensus problems in networks of agents with switching topology and time-delays. *IEEE Trans. on Aut. Control*, 49(9):1520 – 1533, September 2004.

[24] L. Pallottino, V. G. Scordio, E. Frazzoli, and A. Bicchi. Decentralized cooperative policy for conflict resolution in multi-vehicle systems. *IEEE Trans. on Robotics*, 23(6):1170–1183, 2007.

[25] F. Pasqualetti, A. Bicchi, and F. Bullo. Distributed intrusion detection for secure consensus computations. In *IEEE Conf. on Decision and Control*, pages 5594 –5599, December 2007.

[26] F. Pasqualetti, A. Bicchi, and F. Bullo. On the security of linear consensus networks. In *IEEE Conf. on Decision and Control*, December 2009.

[27] F. Pasqualetti, R. Carli, A. Bicchi, and F. Bullo. Identifying cyber attacks via local model information. In *IEEE Conf. on Decision and Control*, December 2010.

[28] M. Pease, R. Shostak, and L. Lamport. Reaching agreement in the presence of faults. *J. ACM*, 27(2):228–234, 1980.

[29] G. Prencipe. CORDA: Distributed coordination of a set of autonomous mobile robots. In *Proc. 4th European Research Seminar on Advances in Distributed Systems*, pages 185–190, May 2001.

[30] D. Spanos, R. Olfati-Saber, and R. Murray. Approximate distributed Kalman filtering in sensor networks with quantifiable performance. In *4th Int. Symp. on Information Processing in Sensor Networks*, pages 133 – 139, April 2005.

[31] S. Sundaram and C. Hadjicostis. Distributed function calculation via linear iterations in the presence of malicious agents; part II: Overcoming malicious behavior. In *American Control Conf.*, pages 1356 –1361, June 2008.

[32] I. Suzuki and M. Yamashita. Distributed anonymous mobile robots: Formation of geometric patterns. *SIAM Journal on Computing*, 28:1347–1363, 1999.

[33] H. Tanner, G. Pappas, and V. Kumar. Leader-to-formation stability. *IEEE Trans. on Robotics and Automation*, 20(3):443 – 455, 2004.

[34] S. Vanka, V. Gupta, and M. Haenggi. Power-delay analysis of consensus algorithms on wireless networks with interference. *Int. J. Syst., Control Commun.*, 2(1/2/3):256–274, 2010.

[35] W. Wang and J.-J. Slotine. Contraction analysis of time-delayed communications and group cooperation. *IEEE Trans. on Aut. Control*, 51(4):712–717, April 2006.

[36] L. Xiao, S. Boyd, and S. Lall. A scheme for robust distributed sensor fusion based on average consensus. In *4th Int. Symp. on Information Processing in Sensor Networks*, pages 63–70, April 2005.

Reputation-Based Networked Control with Data-Corrupting Channels

Shreyas Sundaram
Department of Electrical and
Computer Engineering
University of Waterloo
Waterloo, ON, Canada
ssundara@uwaterloo.ca

Jian Chang
Computer and Information
Science Department
University of Pennsylvania
Philadelphia, PA
jianchan@cis.upenn.edu

Krishna K. Venkatasubramanian
Computer and Information
Science Department
University of Pennsylvania
Philadelphia, PA
vkris@cis.upenn.edu

Chinwendu Enyioha
Department of Electrical and
Systems Engineering
University of Pennsylvania
Philadelphia, PA
cenyioha@ee.upenn.edu

Insup Lee
Computer and Information
Science Department
University of Pennsylvania
Philadelphia, PA
lee@cis.upenn.edu

George J. Pappas
Department of Electrical and
Systems Engineering
University of Pennsylvania
Philadelphia, PA
pappasg@ee.upenn.edu

ABSTRACT

We examine the problem of reliable networked control when the communication channel between the controller and the actuator periodically drops packets and is faulty *i.e.,* corrupts/alters data. We first examine the use of a standard triple modular redundancy scheme (where the control input is sent via three independent channels) with majority voting to achieve mean square stability. While such a scheme is able to tolerate a single faulty channel when there are no packet drops, we show that the presence of lossy channels prevents a simple majority-voting approach from stabilizing the system. Moreover, the number of redundant channels that are required in order to maintain stability under majority voting increases with the probability of packet drops. We then propose the use of a reputation management scheme to overcome this problem, where each channel is assigned a reputation score that predicts its potential accuracy based on its past behavior. The reputation system builds on the majority voting scheme and improves the overall probability of applying correct (stabilizing) inputs to the system. Finally, we provide analytical conditions on the probabilities of packet drops and corrupted control inputs under which mean square stability can be maintained, generalizing existing results on stabilization under packet drops.

Categories and Subject Descriptors

B.4.5 [**Reliability, Testing, and Fault-Tolerance**]: Redundant Design; G.1.0 [**Numerical Analysis**]: General—*Stability (and instability)*; K.6.m [**Miscellaneous**]: Security

General Terms

Reliability, Security, Design, Algorithms, Theory

Keywords

Networked Control, Reputation, Majority Voting, Stability Analysis

1. INTRODUCTION

Networked control systems are spatially distributed systems where the communication between the sensors, actuators and controllers takes place through a network [12]. The presence of a network within the control loop can adversely affect the performance of the system due to the inherent unreliability of the underlying channel. For instance, the network can drop packets with a certain probability or introduce delays in transmission. Recent years have seen much work on the problem of control in the presence of such imperfections [9, 21, 8, 13, 10, 19, 2].

While the issue of stability over relatively benign packet-dropping channels has been well studied, the topic of maintaining stability under faults or attacks (*i.e.,* data alterations or corruptions) in the *network* has not yet received much attention. The potential for such disruptions is becoming greater as control networks become increasingly integrated with standard corporate and data networks [23], and an unmitigated fault or attack could have disastrous consequences in safety-critical applications. The paper [1] takes a step towards addressing this problem, studying the stability of networked control systems under a malicious denial-of-service attack where an attacker stops packets from reaching the controller or actuator for an extended period of time. Other recent investigations of this topic can be found in [20].

In this work, we study the problem of reliable networked control with channels that are both packet dropping and data-corrupting. We first examine the use of a simple majority voting scheme with multiple redundant channels between the controller and the actuator. Each of these channels may drop packets or modify the data that they carry. We show that a straightforward implementation of triple modular redundancy *may not* be sufficient to ensure stability, due to the presence of packet drops in the network. Specifically, we show that the number of redundant channels required in order to ensure stability increases with the probability of packet drops by each channel. It is worth noting that [19] also studied the use of multiple packet-dropping channels to obtain stability in a networked control system. However, the channels in that paper were not assumed to corrupt the data, and redundancy was added only to increase the probability of receiving a packet at the actuator. In contrast, we introduce redundant channels in our setup only to deal with data-corruptions in certain channels, and show that the combination of data corruptions and packet drops imposes a lower bound on the number of redundant channels that are required to stabilize the system (even if a single channel is sufficient to stabilize the system under non-faulty conditions).

To address the problem of stabilization under packet drops and data corruptions, we introduce the use of *reputation management* into the networked control setting. At its core, this involves placing a computational element (called the *reputation manager*) in the feedback loop, which examines the values that it receives from the redundant channels and uses the *history* of the channels' (data corruption) behaviors to assign to them a quantitative reputation 'score'. This score is then used to switch between the available channels (or to apply no input at all). Reputation management schemes have been well studied in the computer science community for the past decade for applications such as file sharing [24], peer-to-peer networks [17], spam detection [25], and inter-domain routing [6]. The characterization of "good" versus "bad" behavior with regard to maintaining reputation depends on the particular application and domain. While reputation management systems have traditionally been used to characterize and control virtual processes in the computing domain, in this work, we apply the concept to a *cyber-physical* system revolving around the control of a plant. Consequently, we tie the performance of the reputation manager to a *physical process*, namely the stability of the plant (encapsulated by the square of the largest eigenvalue of the system matrix, as we will show).

The principal contributions of the paper are: (1) an analytical characterization of the probabilities of packet drops and data corruptions under which the networked control system can maintain mean-square-stability, and (2) a demonstration of the capability of reputation management to improve upon triple modular redundancy schemes to provide mean square stability (under certain conditions). In addition to the specific contributions listed above, our analysis and evaluations are intended to lay a foundation for further integration of reliability and security mechanisms developed by the computer science community into feedback control settings.

The paper is organized as follows, Section 2 presents the necessary background on networked control systems and the fault model for this work. This is followed by Section 3, where we describe the system model for our reliable control scheme, and Section 4 delves into the majority voting approach and its pitfalls. Section 5 presents the reputation management scheme. In Section 6, we present a detailed stability analysis for systems with a faulty channel. Finally, Section 7 concludes the paper.

Notation: For a given matrix $\mathbf{A}$, the notation $\mathbf{A}'$ indicates the transpose of the matrix. For any vector $\mathbf{a}$, the notation $\|\mathbf{a}\|$ denotes the Euclidean norm of the vector. More generally, for a given positive definite matrix $\mathbf{P}$ and any vector $\mathbf{a}$,

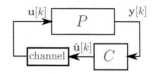

Figure 1: Networked Control

the notation $\|\mathbf{a}\|_{\mathbf{P}}^2$ denotes $\mathbf{a}'\mathbf{P}\mathbf{a}$. The eigenvalues of maximum and minimum modulus of a matrix $\mathbf{A}$ are denoted by $\lambda_{max}(\mathbf{A})$ and $\lambda_{min}(\mathbf{A})$, respectively. The notation $E[\cdot]$ denotes the expectation of a random variable. The notation $\mathbb{N}$ denotes the set of nonnegative integers.

2. BACKGROUND AND FAULT MODEL

Consider the networked control system shown in Fig. 1. The plant P is given by the dynamical system

$$\mathbf{x}[k+1] = \mathbf{A}\mathbf{x}[k] + \mathbf{B}\mathbf{u}[k], \quad \mathbf{y}[k] = \mathbf{C}\mathbf{x}[k] , \qquad (1)$$

where $\mathbf{x} \in \mathbb{R}^n$ is the system state vector, $\mathbf{u} \in \mathbb{R}^m$ is the system input vector, $\mathbf{y} \in \mathbb{R}^p$ is the system output vector, and k denotes the time-step of the system. The matrices $\mathbf{A}, \mathbf{B}$ and $\mathbf{C}$ are real-valued matrices of appropriate dimensions.

To obtain the desired behavior from the plant, the output $\mathbf{y}[k]$ is sent to a controller C, as shown in Fig. 1. Based on $\mathbf{y}[k]$ (or perhaps the entire history of the outputs), the controller computes an input $\hat{\mathbf{u}}[k]$ to apply to the plant. This value is sent through a channel (or more generally, a network), which results in a value $\mathbf{u}[k]$ to be applied at the actuators. There are multiple ways to model the channel; for example, it may drop packets according to some probability, introduce time-varying delays, or have dynamics that cause the input and output to be related in complex ways. For the Bernoulli packet-drop model commonly studied in the literature (*e.g.,* [12, 10, 22]), at each time-step k, we have

$$\mathbf{u}[k] = \begin{cases} \hat{\mathbf{u}}[k] & \text{with probability } 1-p, \\ \emptyset & \text{with probability } p, \end{cases}$$

where $\emptyset$ denotes that no signal is received. In the latter case, we assume that the actuators simply apply the input $\mathbf{u}[k] = 0$. This case has been studied extensively, leading to conditions on the plant and success probability p under which the system will be stable. Due to the probabilistic nature of the applied inputs, the following notion of stability is considered in the literature.

Definition 1. The system is said to be *mean square stable* if $E\left[\|x[k]\|^2\right] < \infty, \ \forall k \in \mathbb{N}$.

Various conditions for ensuring mean square stability have been obtained for general plants [13, 21, 8]. When the input matrix $\mathbf{B}$ is square and full rank (*i.e.,* the system is fully actuated), the following result has been established.

THEOREM 1 ([9, 13, 10]). *Suppose that $\mathbf{B}$ is square and full rank, and let p be the packet drop probability. Then, there exists a linear controller C such that the closed loop system is stable if and only if $p|\lambda_{max}(A)|^2 < 1$.*

EXAMPLE 1. *Consider the first-order plant*

$$x[k+1] = 4x[k] + u[k], \quad y[k] = x[k] ,$$

where $x, u, y \in \mathbb{R}$. Suppose we wish to control this plant via the standard networked control architecture shown in Fig. 1, where the channel has drop probability $p = 0.05$. Since $p|\lambda_{max}(A)|^2 = 0.05(4)^2 = 0.8 < 1$, one can find a controller (e.g., $u[k] = -4y[k]$) such that the system is mean square stable.

In this paper, we expand the discussion on networked control systems to the case where the channel is capable of *modifying* the data that it carries, in addition to dropping it with a certain probability. This can happen, for example, if an attacker takes control of one of the intermediate nodes in the network, and executes *man-in-the-middle* attacks [3]. It can also happen due to accidental faults in the network (*e.g.*, when encoding or decoding during transmissions). Regardless of the cause of the error, we will refer to channels that modify the data that they carry as faulty.

Definition 2. Let $\hat{\mathbf{u}}[k]$ be the input to the channel at time-step k, and let $\mathbf{u}[k]$ be the output of the channel. The channel is said to be *faulty* at time-step k if the channel outputs a value that is not equal to the input (*i.e.*, $\mathbf{u}[k] \neq \hat{\mathbf{u}}[k]$ and $\mathbf{u}[k] \neq \emptyset$).

REMARK 1. *Note that the above definition allows the faulty channel to change the control input arbitrarily. Also, note that if the channel drops the packet (i.e., $\mathbf{u}[k] = \emptyset$), it is irrelevant whether or not the data was modified en-route; thus, we will use the term faulty to refer only to channels that actually deliver a modified packet at a given time-step k.*

REMARK 2. *In this paper, we will make the assumption that the controller is located at the sensor (without an intermediate channel), and that the controller has full access to the state (i.e., $\mathbf{y}[k] = \mathbf{x}[k]$). The latter assumption can be relaxed if there is an acknowledgment mechanism in the network, whereby the actuator can inform the controller of the input value that is applied [13]. We will leave the full treatment of more general scenarios for future work; as we will see, even this scenario offers various challenges for control.*

REMARK 3. *One can readily incorporate random noise into system (1) without affecting the property of mean square stability that we discuss in this paper. In order to avoid introducing additional variables and to keep the exposition clear, we will leave noise out of our discussion.*

REMARK 4. *One can also consider the inclusion of cryptographic mechanisms into the feedback loop to protect the data that is being transmitted over the channels. However, such mechanisms are primarily intended to protect the confidentiality of the data, and do not typically address the issue of what happens when the entire data packet is corrupted (i.e., even if the actuator is able to determine that the integrity of the transmitted data is compromised, it has no way to know what input to apply to the system). In other words, the real-time nature of feedback control requires a mechanism by which data is reliably delivered. As we will see in the next section, we will achieve this by transmitting the control inputs along multiple independent channels, and enforce the assumption that the packets on at most one of the channels can be corrupted by the attacker. In return, we do not have to assume that the attacker is computationally bounded (which is a typical assumption in cryptographic systems). This assumption of an attacker restricted by the topology of the network, as opposed to bounded in the computations that he/she can perform, is also considered in the literature on Byzantine fault tolerance [18] and information theoretic security [15].*

3. NETWORKED CONTROLLER MODEL AND ASSUMPTIONS

As a first step to addressing the problem of a faulty channel in Fig. 1, suppose we implement a standard triple modular redundancy scheme by simply sending the controller inputs through three different channels; for example, this could represent disjoint paths through the communication network between the controller and the actuator. Such an implementation is shown in Fig. 2.

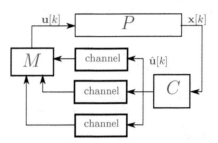

Figure 2: Networked Control with Redundant Channels.

The *manager* block (denoted by M in Fig. 2) compares the different channel values and makes a decision as to what input $\mathbf{u}[k]$ (if any) to apply (i.e., it switches between the available channels, or the input $\mathbf{0}$). We assume that each channel drops its packet with probability p, and that during the course of operation, no more than one[1] of the channels can be faulty (i.e., if a channel is faulty at some time-step, a different channel cannot be faulty at another time-step). We also assume that each channel is independent of the other channels. If the probability of packet drops is $p = 0$, the manager M receives the outputs of all three channels at each time-step, and can simply choose the value that is specified by the majority of the channels as the input $\mathbf{u}[k]$. The more interesting case occurs when we have a nonzero probability of packet drop, and the manager must adopt some rule to choose which input to apply among the ones it does receive. We study this case next.

4. MAJORITY VOTING WITH PACKET DROPS

When the manager does not receive all three signals at each time-step, one can adopt the following natural extension of the standard majority-voting mechanism: if the manager receives at least two signals that match, then that signal is applied. Otherwise, the input $\mathbf{u}[k] = \mathbf{0}$ is applied. We denote the number of received signals at time-step k by R_k. Note that R_k is a binomial random variable with parameters $(3, p)$. Thus, $R_k = t$ with probability $\binom{3}{t}p^{3-t}(1-p)^t$, for $t \in \{0, 1, 2, 3\}$. The probability of receiving two matching signals will depend on whether one of the channels is *faulty*.

4.1 No Faulty Channels

When there are no faulty channels, the correct input is applied whenever at least two signals are received, and $\mathbf{u}[k] = 0$ is applied otherwise. This latter case occurs with probability $\bar{p}_1 = p^3 + 3p^2(1-p)$. One can verify that $\bar{p}_1 \leq p$ for $p \in [0, 0.5]$, and $\bar{p}_1 \geq p$ for $p \in [0.5, 1]$. Thus, when there are

---

[1]In general, if f faulty/malicious channels are to be tolerated, one requires $2f + 1$ channels in total.

no faulty channels, this mechanism is equivalent to a single non-faulty channel with drop probability $\bar{p}_1$; there will be an overall improvement when p is less than 0.5, and a degradation when p is greater than 0.5.

4.2 One Faulty Channel

When one of the channels is faulty (*i.e.*, its output is not equal to the signal generated by the controller), the probability that the manager receives at least two matching signals is equal to the probability that it receives signals from both of the non-faulty channels. This probability is given by $(1-p)^2$, and thus the drop probability is given by

$$\bar{p}_2 = 1 - (1-p)^2 = -p^2 + 2p. \qquad (2)$$

One can verify that $\bar{p}_2 \geq p$ for $p \in [0, 1]$. The presence of a faulty channel therefore results in an overall degradation in performance (*i.e.*, the majority voting mechanism functions as a single non-faulty channel with drop probability $\bar{p}_2 \geq p$). This may result in the overall system no longer being stable, as the following example illustrates.

EXAMPLE 2. *Suppose we would like to protect the plant described in Example 1 against channel faults (or attacks) by implementing the modular redundancy scheme shown in Fig. 2. If one of the channels is faulty, the probability that the manager M does not receive two equal values (equal to $-4y[k]$) at any time-step k is given by equation (2), namely $\bar{p}_2 = -(0.05)^2 + 2(0.05) = 0.0975$. However, $\bar{p}_2|\lambda_{max}(A)|^2 = 0.0975(4)^2 = 1.56 > 1$, and this violates the bound in Theorem 1. Thus the plant cannot be stabilized via this majority voting scheme when one of the channels is faulty.*

The above example shows that when one considers the possibility of both faulty and packet-dropping channels, a simple triple-modular redundancy scheme with majority voting may not be sufficient to maintain stability, *even* when the probability of packet drop is low enough that a single non-faulty channel with the same drop probability would suffice.

One way to restore the performance of the networked control scheme would be to increase the number of redundant channels. Specifically, suppose that we consider $N+1$ channels, where one channel is allowed to be faulty. Now, the probability that the manager does not get two (non-empty) matching signals is:

$$\bar{p} = p^N + Np^{N-1}(1-p) . \qquad (3)$$

For any desired drop probability p^*, we would like to find the value of N for which $\bar{p} = p^*$ (*i.e.*, the number of channels that are required in order to obtain the desired performance). This is given by the following theorem.

THEOREM 2. *Consider a majority voting scheme with $N+1$ channels, each of which can drop packets with a probability p. Furthermore, suppose up to one of the channels can be faulty, whereby it changes the value of the data that it carries. The number of redundant channels N required to maintain a drop probability of p^* is given by*

$$N = \frac{1}{\ln p} W\left(\frac{p^* \ln p}{1-p} p^{\frac{1}{1-p}}\right) - \frac{p}{1-p},$$

where $W(\cdot)$ is the Lambert W-function.[2]

[2]This is the inverse of the function xe^x, and cannot be written in terms of more elementary functions [7].

PROOF. In order to have $\bar{p} = p^*$, we use (3) to obtain

$$p^N + Np^{N-1}(1-p) = p^*$$

$$\Leftrightarrow (p + N(1-p))p^{N-1} = p^*$$

$$\Leftrightarrow (p + N(1-p))e^{(N-1)\ln p} = p^*$$

$$\Leftrightarrow (1-p)\left((N-1) + \frac{1}{1-p}\right)e^{(N-1)\ln p} = p^*$$

$$\Leftrightarrow \left((N-1) + \frac{1}{1-p}\right)(\ln p)e^{\left((N-1)+\frac{1}{1-p}\right)\ln p} = \frac{p^* \ln p}{1-p}e^{\frac{\ln p}{1-p}}.$$

Letting $f(x) = xe^x$, the above equation becomes

$$f\left((N-1)\ln p + \frac{\ln p}{1-p}\right) = \frac{p^* \ln p}{1-p}p^{\frac{1}{1-p}} .$$

The solution to the above equation is given by the expression in the theorem. $\square$

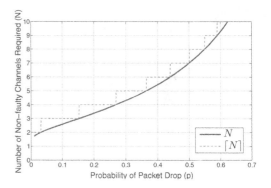

Figure 3: Number of non-faulty channels (N) required in a majority voting scheme in order to tolerate a single faulty channel and maintain a drop probability of $p^* = \frac{1}{4^2}$.

A plot of N versus p is shown in Fig. 3 for the case where $p^* = \frac{1}{4^2}$ (i.e., this is the maximum drop probability able to stabilize the scalar plant in Example 1); since the number of channels must be integer-valued, the value of $\lceil N \rceil$ is also shown in the plot. Note that for $p = \frac{1}{4^2}$, one requires $N = 3$ additional channels (for a total of four channels) in order to tolerate one faulty channel and still maintain the same probability of packet drop as a single non-faulty channel. As the packet drop probability increases, the number of channels required also increases rapidly.[3] Note that, even though we assume that at most one channel is faulty over all times, the analysis presented in this section will hold for the more general case where different channels can be faulty at different time-steps (as long as at most one channel is faulty in any given time-step). This is because the above majority voting scheme is memoryless, and decisions in the present are made without regard to past behavior of the channels.

In the next section, we will present a mechanism to address the shortcomings of modular redundancy in control loops with packet drops. For certain types of faults that afflict the channels, this mechanism will provide mean-square

[3]This result can potentially be used by an intelligent attacker to execute a subtle "jamming" or denial-of-service attack on the network, whereby he or she increases the rate of packet drops to the point where four disjoint channels are no longer sufficient to maintain stability.

stability without requiring the use of more than two redundant channels *i.e.*, it will address the problems described above with regard to the loss of stability in triple modular redundancy.

5. MAJORITY VOTING WITH A REPUTATION MANAGER

In practice, the behavior of a channel at any time-step k will be correlated with past and future behavior. For example, if a channel is compromised by an attacker, it is likely that the channel will misbehave for several time-steps (or perhaps permanently). Similarly, if the channel has an internal state, and if the state gets disturbed from its normal operating condition for some reason, it will take several time-steps for this error to die away. To formally capture such behavior, we will first need a metric to evaluate "good" versus "bad" behavior by a channel. Let $C = \{c_1, c_2, c_3\}$ be the set of channels between the controller and reputation manager and $I_{c_i,k}$ be the input received from a channel $c_i \in C$ at time-step k (this value can be $\emptyset$ if the input is dropped by the channel during the time-step).

Definition 3. An input $I_{c_i,k}$ is *verifiably correct* if the reputation manager receives at least one additional signal from any channel $c_j \in C$, $i \neq j$ and $I_{c_i,k} = I_{c_j,k} \neq \emptyset$. An input $I_{c_u,k}$ is *verifiably incorrect*, if the reputation manager receives two other signals c_i and c_j, such that $c_i, c_j \in C$, $i \neq j \neq u$, $I_{c_i,k} = I_{c_j,k} \neq \emptyset$ and $I_{c_u,k} \neq I_{c_i,k}$.

Let $G_{c_i,k}$ and $B_{c_i,k}$ represent the sets of verifiably correct and verifiably incorrect inputs, respectively, received from a channel $c_i \in C$ up to time-step k.

Definition 4. The reputation of a channel $c_i \in C$ at a time-step k is a map $Rep : G_{c_i,k} \times B_{c_i,k} \to [0,1] \cup \{\phi\}$, where ϕ denotes an *unknown* reputation.

To simplify the notation, we use $Rep(c_i)$ to denote the reputation of a channel c_i at a given time-step. The exact map that is used will depend on the types of faulty behavior that are assumed. The reputation values for each channel will be administered and updated by a *reputation manager* (RM), which we describe next.

5.1 Reputation Manager

The reputation manager is incorporated into the block labeled M in Fig. 2, and operates in four stages: (1) observe the inputs from the channels at a given time-step, (2) verify the correctness (or incorrectness) of the received inputs, (3) update the reputation of the channels whose inputs can be verified, and (4) use the reputation to choose between the available channels (or apply no input at all). As we shall see, the use of reputation management outperforms the majority voting mechanism by increasing the probability of applying a correct control input and reducing the probability of applying no input.

When there is only one faulty channel and the RM receives at least two matching signals, it is *certain* that the input is correct. However, there may be situations when the RM is *uncertain* about the quality of the signal; this can happen, for example, if two non-matching signals are received, and the reputations of the two channels are identical. If the RM randomly chooses one of the two inputs to apply, there is a chance that an *incorrect* value will be applied. It would be prudent in this case to ensure that the input applied by the RM is *bounded* in some sense, so that a faulty channel

---

Algorithm 1 Reputation Management Scheme

Require: Controller injects $-\mathbf{K}\mathbf{x}[k]$ and $\mathbf{u}^b[k]$ into each of the three channels. Both of these signals travel together in one packet. Let R_k denote the number of packets received by the RM at time-step k.

begin
if $R_k = 3$ **then**
 RM applies the value $-\mathbf{K}\mathbf{x}[k]$ specified by the majority of the packets, and updates the reputation of all three channels accordingly
else if $R_k = 2$ and they both match **then**
 RM applies $-\mathbf{K}\mathbf{x}[k]$ and updates the reputation of the two channels.
else if $R_k = 2$ and they do not match **then**
  if $Rep(c_i) \geq \Theta$ and $0 \leq Rep(c_j) < \Theta$ **then**
   RM applies the value $-\mathbf{K}\mathbf{x}[k]$ specified by c_i, and increases the reputation of c_i while decreasing that of c_j.
  else if $Rep(c_i) \geq \Theta$ and $Rep(c_j) \geq \Theta$ **then**
   RM randomly chooses one of the two channels and applies the input $\mathbf{u}^b[k]$ specified by it.
  else if $Rep(c_i) \geq \Theta$ and $Rep(c_j) = \phi$ **then**
   RM applies $\mathbf{u}^b[k]$ specified by c_i.
  else if $(Rep(c_i) = \phi$ and $Rep(c_j) = \phi)$ or $(Rep(c_i) < \Theta$ and $Rep(c_j) < \Theta)$ **then**
   RM applies $\mathbf{u}[k] = 0$
  end if
else if $R_k = 1$ **then**
  if $Rep(c_i) \geq \Theta$ **then**
   RM applies $\mathbf{u}^b[k]$ specified by c_i.
  else if $Rep(c_i) = \phi$ or $Rep(c_i) < \Theta$ **then**
   RM applies $\mathbf{u}[k] = 0$
  end if
else if $R_k = 0$ **then**
 RM applies $\mathbf{u}[k] = 0$
end if
end

---

is not able to immediately increase the system state to a large value. Rather than having the RM scale the received inputs itself, we will instead have the controller inject *two* values into each channel: one value is a state feedback input $-\mathbf{K}\mathbf{x}[k]$, and the other is a *bounded* input $\mathbf{u}^b[k]$ satisfying $\|\mathbf{u}^b[k]\|_{\mathbf{P}} < b$, for some matrix $\mathbf{K}$, some positive definite matrix $\mathbf{P}$ and some positive real number b. We will later discuss conditions under which $\mathbf{K}, \mathbf{P}, b$ and $\mathbf{u}^b[k]$ can be chosen to ensure mean square stability, even when bounded incorrect inputs are applied with a certain probability. Note that a faulty channel can corrupt both the state feedback and bounded input injected by the controllers, but if the corruption causes the norm of the bounded input to increase above b, this can be immediately detected and identified by the reputation manager.

The above ideas are formally presented as Algorithm 1. Here, i and j are indices of channels from which input is received. The reputation value for each channel is initially set to ϕ (*i.e.*, unknown), and gets updated as verifiably correct or incorrect inputs are received from the channels. The parameter $\Theta \in [0,1]$ is a *threshold* value, and input from a channel with reputation below Θ is not applied. A variety of mappings/functions can be used to obtain the actual value of the reputation $Rep(c_i)$ based on the observed history $H_{c_i,k} = G_{c_i,k} \cup B_{c_i,k} \cup (\cup_{g=1}^{k} I_{c_i,g} - \{G_{c_i,k} \cup B_{c_i,k}\})$ at any time-step k. In this section we describe two such mappings that prove to be simple yet more effective than majority voting. Before delving into the details, we present the fault model that we are considering (the analysis of other fault models will be left as an extension for future work).

5.1.1 Fault Model

We assume that, over all time-steps, at most one channel is faulty, and without loss of generality we take this to be c_3

(of course, this is unknown to the manager).[4] Furthermore, we assume that c_3 becomes faulty at an arbitrary time-step. In this work, we take the faultiness of the channel to be probabilistic and expressed as d_{c_i}, which is the probability that channel c_i is *fault-free* at any given time-step. Thus, we consider the case where $d_{c_1} = 1$, $d_{c_2} = 1$ and $0 \leq d_{c_3} \leq 1$ at any given time-step. Finally, we assume that the reputation manager knows that at most one channel is faulty, but it does not know *a priori* that c_3 is the faulty channel nor does it know the probability of being c_3 being faulty (*i.e.*, $1 - d_{c_3}$).

5.2 Stratified Reputation

We will start by considering a simple reputation function that switches between a finite set of reputation values. Initially all the channels have an unknown reputation ϕ.

Definition 5. (Stratified Reputation) Let $G_{c_i,k}$ and $B_{c_i,k}$ represent the number of verifiably correct and verifiably incorrect inputs, respectively, received from channel c_i up to time-step k. Then, the reputation of the channel c_i at time-step k is defined as:

$$Rep(c_i) = \begin{cases} 1 & G_{c_i,k} > 0 \ \& \ B_{c_i,k} = 0, \\ 0 & B_{c_i,k} > 0, \\ \phi & B_{c_i,k} = 0 \text{ and } G_{c_i,k} = 0. \end{cases}$$

Using the stratified reputation function, the moment we verify that a channel is faulty, it is "black-listed" and no unbounded input provided by that channel is ever applied again. Since reputations can only be either 0 or 1 (when not unknown), we can use any $0 < \Theta \leq 1$ as the threshold in Algorithm 1. While simple, this reputation function does not provide us with much information about the extent of faultiness of the channel (which could be useful for identifying the cause of the fault). We will next consider a slightly more complex reputation function that still provides an improvement over majority voting, and also provides an estimate of the fault probability of the channel.

5.3 Bayesian Reputation

To estimate the fault probability of the channel, we use a *Bayesian Reputation* function [14]. To understand this, note that when channel c_i becomes faulty, each verifiable input provided by c_i is correct or incorrect with a Bernoulli distribution, given by the (unknown) probability d_{c_i}. Consequently, given a set of verifiably correct and incorrect inputs from a channel, the quantity d_{c_i} satisfies a *beta distribution*, which is the probability distribution of seeing a particular combination of correct and incorrect inputs from a channel [14]. The expected value of the beta distribution forms the reputation and is given by the ratio of the number of correct values received to the number of total values received. This is also the required estimate of the fault-free probability of c_3 (*i.e.*, d_{c_3}).

Definition 6. (Bayesian Reputation) Let $G_{c_i,k}$, $B_{c_i,k}$ and $S_{c_i,k}$ represent the number of verifiably correct, verifiably incorrect and total number of verifiable inputs, respectively, received from the channel c_i up to time-step k. Then, the reputation of the channel c_i at time-step k is defined as:

$$Rep(c_i) = \frac{G_{c_i,k}}{S_{c_i,k}}, where \ \ S_{c_i,k} = G_{c_i,k} + B_{c_i,k}. \quad (4)$$

---

[4]For instance, in a multi-hop network, the three channels could represent three node-disjoint paths in the network between the controller and the manager, and misbehavior on the part of nodes on only one of these paths would leave the other paths (channels) reliable.

Note that, it might be preferable in some situations to give weight to more recent behaviors than older ones, *e.g.*, by introducing time decay functions for the reputation calculation. We will leave a treatment of such reputation functions for future work.

5.3.1 Discussion

Even though both reputation functions can detect the faulty channel, the Bayesian reputation function allows one to estimate the probability of the channel's faultiness through its reputation value. This helps system administrators perform better fault diagnosis. For example, if the reputation of channel c_i is lower than a certain value, it might imply that the root cause of the faulty signal could be serious network outage or a malicious attack, rather than random network operation errors. Consequently, unlike the stratified function, with the Bayesian reputation function, the reputation value is constantly updated even when the RM is aware which channel might be faulty.

Note that, these two are not the only reputation functions that can be used here. Their choice is motivated by the desire to demonstrate simple functions that can provide improvements over majority voting (as we will show below). The choice of the reputation function is largely dictated by the fault model assumed in the system. Identifying different criteria for choosing appropriate reputation functions is the focus of ongoing research.

5.4 Evaluation Metrics

From Algorithm 1 it can be seen that each of the inputs (i.e., a full state-feedback input, a bounded correct input, a bounded faulty input, and $\mathbf{0}$) will be applied with a certain probability. We will characterize the probabilities of each of these events by $\mathcal{C}, \mathcal{C}_b, \mathcal{E}$ and $\mathcal{N} = 1 - \mathcal{C} - \mathcal{C}_b - \mathcal{E}$, respectively. In the rest of the paper we refer to these collectively as ECN metrics. Furthermore, we use subscripts NR and R with the ECN metrics, to indicate that the scenario being considered *does not* or *does* use reputation, respectively.

5.4.1 Performance Without Reputation Manager

In the absence of a reputation manager (*i.e.*, the standard triple-modular-redundancy scheme with majority voting from the previous section), it is easy to see that $\mathcal{E}_{NR} = 0$, $\mathcal{C}_{b_{NR}} = 0$, $\mathcal{C}_{NR} = (1-p)^2 + 2p(1-p)^2 d_{c_3}$ and $\mathcal{N}_{NR} = 1 - [(1-p)^2 + 2p(1-p)^2 d_{c_3}]$.

5.4.2 Performance With Reputation Manager

Considering our setup with reputation, Table 1 shows the ECN metrics for different combinations of possible channel reputation values. We omit the value of $\mathcal{C}_R$, since $\mathcal{C}_R = 1 - \mathcal{C}_{b_R} - \mathcal{E}_R - \mathcal{N}_R$. The following example demonstrates how the ECN probabilities are calculated.

EXAMPLE 3. *Assume initially all channels behave as expected without modifying the controller inputs. This will eventually cause the reputations of c_1, c_2 and c_3 to become 1 (i.e., perfect). Now suppose c_3 starts behaving in a faulty manner with a probability $1 - d_{c_3}$. As the reputation of c_3 is perfect, there is a non-zero probability of applying an incorrect input to the plant.*

A bounded incorrect controller input is applied to the plant under two conditions: (1) a value from c_3 alone was received, the probability of which is given by: $p^2(1-p)$, or (2) values were received from c_1 and c_3 or c_2 and c_3 and in either case, c_3 was chosen at random, given by the probability $\frac{1}{2}p(1-$

Reputation of Channel			Metrics		
c_1 (Good)	c_2 (Good)	c_3 (Faulty)	$\mathcal{E}_R$	$\mathcal{N}_R$	$\mathcal{C}_{b_R}$
1	1	$\geq \Theta$	$(1-p)p(1-d_{c_3})$	p^3	$p(1-p)(1+p+d_{c_3}(2p-1))$
ϕ	1	1	$\frac{1}{2}p(1-p)(1-d_{c_3})(3-p)$	$p^3+(1-p)p^2$	$\frac{1}{2}p(1-p)(1+p+2d_{c_3})$
1	ϕ	1	$\frac{1}{2}p(1-p)(1-d_{c_3})(3-p)$	$p^3+(1-p)p^2$	$\frac{1}{2}p(1-p)(1+p+2d_{c_3})$
1	1	ϕ or $< \Theta$	0	$p^3+(1-p)p^2$	$2p(1-p)$
ϕ	ϕ	ϕ	0	$p^3+3(1-p)p^2+2(1-p)^2p(1-d_{c_3})$	0

$p)^2 + \frac{1}{2}p(1-p)^2 = p(1-p)^2$. *Furthermore, the probability that c_3 is faulty, given that its value was received, is given by $(1-d_{c_3})$. Given these values, we have $\mathcal{E}_R = (p(1-p)^2 + p^2(1-p))(1-d_{c_3}) = p(1-p)(1-d_{c_3})$.*

Similarly, given the perfect reputation values, no input will be applied to the plant only when no value is received, i.e., all three channels drop their packets. The probability of this happening is given by $\mathcal{N}_R = p^3$.

Finally, a bounded correct controller input is applied to the plant under three conditions: (1) a value from c_1 or c_2 is received, the probability of which is given by: $2p^2(1-p)$; (2) a value was received from c_1 and c_3 or c_2 and c_3, the values do not match, and c_1 and c_2 are chosen at random, respectively. The probability of this scenario is given as: $p(1-p)^2(1-d_{c_3})$; and (3) the value from c_3 alone is received and it is not faulty, which is given by $p^2(1-p)d_{c_3}$. Given these values, we have $\mathcal{C}_{b_R} = p(1-p)(1+p+d_{c_3}(2p-1))$.

Given that c_1 and c_2 are never faulty, and their reputation is updated only when majority of the inputs are identical (2 out of 3), the reputation c_1 and c_2 can therefore only take the values 1 or ϕ. In the case of channel c_3 however, the reputation can be either ϕ or $0 \leq Rep(c_3) \leq 1$ depending upon the faultiness of the channel, and the packet drop characteristics of the three channels.

5.5 Simulation Results

We simulated the triple modular redundancy scheme in a networked control system with one faulty channel. The principal goal of the simulation was to evaluate whether the reputation manager improves the probability of applying correct inputs to the plant. Each simulation run was set to execute for 1000 time-steps with different p and d_{c_3} values. Furthermore, in order to compensate for the variation of individual simulation cycles, the entire simulation process was executed 10,000 times for each combination of p and d_{c_3} values and the averages of the values are reported. Our simulations use the Bayesian reputation function with the threshold value of $\Theta = 1$. However, with this choice, one can verify that Algorithm 1 performs identically under both the Stratified and Bayesian reputation functions.

Fig. 4 shows the variation in the probability of applying a correct state-feedback input ($\mathcal{C}$). The probability of applying a correct input to the plant when reputation is used is always higher than when it is not used (*i.e.*, in the case where majority voting alone is used). As the packet drop rate increases, however, this difference is reduced considerably. Note that all of the curves corresponding to the cases where reputation is used are almost identical, which demonstrates an important property of using reputation: performance does not significant degrade even when the value of d_{c_3} drops. This is a consequence of our fault model, where at most one of the three channels is faulty. Consequently, once the RM has identified the good channels, it automatically does not consider the input from the third one (given the chosen threshold value of $\Theta = 1$).

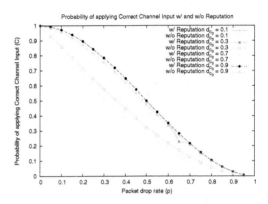

Figure 4: Effect on $\mathcal{C}$ in a system with dynamic channel failure due to reputation

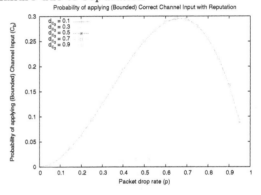

Figure 5: Effect on $\mathcal{C}_b$ in a system with dynamic channel failure due to reputation

Fig. 5 shows the probability of applying a bounded correct controller input ($\mathcal{C}_b$). At lower packet drop rates, the probability of applying a full (correct) state-feedback input to the plant is much higher, resulting in low $\mathcal{C}_b$ values. As the drop rate increases, not all inputs arrive at the reputation manager. Consequently, the reputation manager has to increasingly apply bounded inputs, resulting in the increase in the value of $\mathcal{C}_b$. As the packet drop rate increases further, not enough inputs arrive which causes $\mathcal{C}_b$ to drop again. Again, all curves are nearly identical.

Fig. 6 shows the probability of applying no controller input ($\mathcal{N}$). This value is always lower when reputation is used, but predictably increases with the packet drop rate. Once again, all curves corresponding to the cases where reputation is used are almost identical.

Fig. 7 shows the probability of applying a bounded incorrect input ($\mathcal{E}$). The values are very close to zero when the packet drop rate is low, irrespective of the value of d_{c_3}. This is because 2 or more identical inputs are frequently received, causing the reputation of c_1 and c_2 to reach 1 very quickly. For higher packet drop rates, $\mathcal{E}$ is greater for channels with higher d_{c_3}. The reasons are two-fold: (1) c_3 amasses a bet-

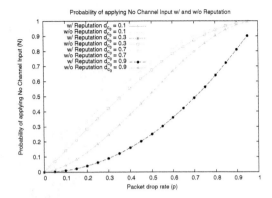

Figure 6: Effect on $\mathcal{N}$ in a system with dynamic channel failure due to reputation

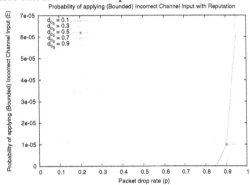

Figure 7: Effect on $\mathcal{E}$ in a system with dynamic channel failure due to reputation

ter reputation than in the case where d_{c_3} is lower, and (2) with the high packet drop rate, a relatively large number of time-steps will have only c_3's input showing up, potentially increasing the number of incorrect inputs applied to the plant. Note that after channel c_3 becomes faulty and the system has run for some time, the reputations reach their steady state values and the probability of applying an incorrect (bounded) input $\mathcal{E}$ is always 0. In Fig. 7, the small non-zero value of $\mathcal{E}$ is due to the few incorrect (bounded) inputs that are applied during the transient period (while the reputations are converging). The convergence rate of the reputations can potentially be analyzed by associating an appropriate Markov chain with the RM and examining its mixing time; this is the subject of ongoing research.

6. STABILITY UNDER INTERMITTENT FAULTY INPUTS

In the previous section, we showed that a reputation manager can improve the probability of applying correct inputs to the plant under a majority voting scheme with packet drops. However, in situations where the RM is uncertain about the quality of the inputs that it receives, it can potentially apply incorrect (bounded) inputs to the plant with a certain probability. While this probability is zero in steady state for the fault model (only one channel can be faulty for all time) and reputation functions that we consider in this paper, the same is not true for more general fault models (e.g., where multiple channels can be faulty). In this section, we will generalize existing results on stability under packet dropping channels to the case where incorrect (but bounded) inputs are periodically applied to the system. We

will then verify that for the fault model considered in this paper, reputation management is able to satisfy the resulting conditions for mean square stability better than majority voting.

To this end, suppose that system (1) operates as follows; at each time-step k, the input is:

$$\mathbf{u}[k] = \begin{cases} -\mathbf{K}\mathbf{x}[k] & \text{with probability } \mathcal{C} \\ \mathbf{u}^b[k] & \text{with probability } \mathcal{C}_b \\ \mathbf{d}[k] & \text{with probability } \mathcal{E} \\ \mathbf{0} & \text{with probability } 1 - \mathcal{C} - \mathcal{C}_b - \mathcal{E}. \end{cases} \quad (5)$$

In the above, $\mathcal{C}$, $\mathcal{C}_b$, and $\mathcal{E}$ are the probabilities of applying a stabilizing state-feedback input, a bounded stabilizing input or a bounded incorrect input, respectively. The bounds on the inputs $\mathbf{u}^b[k]$ and $\mathbf{d}[k]$ are of the form $\|\mathbf{u}^b[k]\|_{\mathbf{P}} \leq b$ and $\|\mathbf{d}[k]\|_{\mathbf{P}} \leq b$ for some positive definite matrix $\mathbf{P}$ (which we will specify below) and for some positive value b. The input $\mathbf{u}^b[k]$ is taken to be *designed* (*i.e.,* chosen to improve the stability of the system), and the input $\mathbf{d}[k]$ is a *faulty* input (perhaps chosen maliciously), satisfying the above bound. The matrix $\mathbf{K}$ will be chosen to obtain mean square stability. Under these conditions, the following theorem provides a sufficient condition for the probabilities of applying state-feedback control inputs and applying bounded faulty inputs that will maintain mean square stability of the system.

THEOREM 3. *Consider a system* $(\mathbf{A}, \mathbf{B})$, *and let* $\mathcal{C}$ *be the probability of applying the state feedback input* $-\mathbf{K}\mathbf{x}[k]$, *and* $\mathcal{E}$ *be the probability of applying an incorrect, but bounded, input to the plant. If there exists a positive definite matrix* $\mathbf{P}$ *and a matrix* $\mathbf{K}$ *such that*

$$(\mathbf{A} - \mathbf{BK})'\mathbf{P}(\mathbf{A} - \mathbf{BK})\mathcal{C} + \mathbf{A}'\mathbf{PA}(1 - \mathcal{C} + \mathcal{E}) < \mathbf{P}, \quad (6)$$

the system is mean square stable.

PROOF. For system (1) with inputs (5), we will examine the quantity

$$\sigma_{k+1} \triangleq E[\|\mathbf{x}[k+1]\|_{\mathbf{P}}^2]$$
$$= E\left[E[\|\mathbf{x}[k+1]\|_{\mathbf{P}}^2 \mid \mathbf{x}[k], \mathbf{d}[k], \mathbf{u}^b[k]]\right]$$
$$= E[\|(\mathbf{A} - \mathbf{BK})\mathbf{x}[k]\|_{\mathbf{P}}^2]\mathcal{C} + E[\|\mathbf{A}\mathbf{x}[k] + \mathbf{B}\mathbf{u}^b[k]\|_{\mathbf{P}}^2]\mathcal{C}_b$$
$$+ E[\|\mathbf{A}\mathbf{x}[k] + \mathbf{B}\mathbf{d}[k]\|_{\mathbf{P}}^2]\mathcal{E}$$
$$+ E[\|\mathbf{A}\mathbf{x}[k]\|_{\mathbf{P}}^2](1 - \mathcal{C} - \mathcal{C}_b - \mathcal{E}).$$

Now, note that $\|\mathbf{A}\mathbf{x} + \mathbf{B}\mathbf{u}\|_{\mathbf{P}}^2 = \|\mathbf{A}\mathbf{x}\|_{\mathbf{P}}^2 + \|\mathbf{B}\mathbf{u}\|_{\mathbf{P}}^2 + 2\mathbf{x}'\mathbf{A}'\mathbf{PB}\mathbf{u}$ for any vectors $\mathbf{x}$ and $\mathbf{u}$. Substituting this into the above expression, we get

$$\sigma_{k+1} = E[\|(\mathbf{A} - \mathbf{BK})\mathbf{x}[k]\|_{\mathbf{P}}^2]\mathcal{C} + E[\|\mathbf{A}\mathbf{x}[k]\|_{\mathbf{P}}^2](1 - \mathcal{C})$$
$$+ E[2\mathbf{x}'[k]\mathbf{A}'\mathbf{PB}\mathbf{u}^b[k] + \|\mathbf{B}\mathbf{u}^b[k]\|_{\mathbf{P}}^2]\mathcal{C}_b \quad (7)$$
$$+ E[2\mathbf{x}'[k]\mathbf{A}'\mathbf{PB}\mathbf{d}[k] + \|\mathbf{B}\mathbf{d}[k]\|_{\mathbf{P}}^2]\mathcal{E}.$$

Suppose now that the input $\mathbf{u}^b[k]$ is chosen to minimize[5]

$$2\mathbf{x}'[k]\mathbf{A}'\mathbf{PB}\mathbf{u}^b[k] + \|\mathbf{B}\mathbf{u}^b[k]\|_{\mathbf{P}}^2,$$

subject to the constraint $\|\mathbf{u}^b[k]\|_{\mathbf{P}} \leq b$; note that the minimum is guaranteed to be nonpositive (since choosing $\mathbf{u}^b[k] = 0$ produces a value of 0). Furthermore, note that since $\|\mathbf{a} - \mathbf{b}\|_{\mathbf{P}}^2 = \|\mathbf{a}\|_{\mathbf{P}}^2 - 2\mathbf{a}'\mathbf{Pb} + \|\mathbf{b}\|_{\mathbf{P}}^2 \geq 0$ for any vectors $\mathbf{a}$

---

[5]This is a Quadratically Constrained Quadratic Program (QCQP), which can be solved numerically [4], but difficult to solve analytically.

and $\mathbf{b}$, we have $E[\mathbf{a}'\mathbf{Pb}] \leq \frac{1}{2}(E[\|\mathbf{a}\|_{\mathbf{P}}^2] + E[\|\mathbf{b}\|_{\mathbf{P}}^2])$. Thus, we can write

$$E[2\mathbf{x}'[k]\mathbf{A}'\mathbf{PBd}[k]] \leq E[\|\mathbf{Ax}[k]\|_{\mathbf{P}}^2] + E[\|\mathbf{Bd}[k]\|_{\mathbf{P}}^2] \ .$$

Substituting these facts into (7), we get

$$\begin{aligned}
\sigma_{k+1} &\leq E[\|(\mathbf{A} - \mathbf{BK})\mathbf{x}[k]\|_{\mathbf{P}}^2]\mathcal{C} + E[\|\mathbf{Ax}[k]\|_{\mathbf{P}}^2](1 - \mathcal{C} + \mathcal{E}) \\
&\quad + 2E[\|\mathbf{Bd}[k]\|_{\mathbf{P}}^2]\mathcal{E} \\
&\leq E[\|(\mathbf{A} - \mathbf{BK})\mathbf{x}[k]\|_{\mathbf{P}}^2]\mathcal{C} + E[\|\mathbf{Ax}[k]\|_{\mathbf{P}}^2](1 - \mathcal{C} + \mathcal{E}) \\
&\quad + 2\mathcal{E}\lambda_{max}(\mathbf{B}'\mathbf{PB})E[\|\mathbf{d}[k]\|^2] \\
&\leq E[\|(\mathbf{A} - \mathbf{BK})\mathbf{x}[k]\|_{\mathbf{P}}^2]\mathcal{C} + E[\|\mathbf{Ax}[k]\|_{\mathbf{P}}^2](1 - \mathcal{C} + \mathcal{E}) \\
&\quad + 2\mathcal{E}\frac{\lambda_{max}(\mathbf{B}'\mathbf{PB})}{\lambda_{min}(\mathbf{P})}b^2. \qquad (8)
\end{aligned}$$

Now, examine the quantity

$$\|(\mathbf{A} - \mathbf{BK})\mathbf{x}[k]\|_{\mathbf{P}}^2\mathcal{C} + \|\mathbf{Ax}[k]\|_{\mathbf{P}}^2(1 - \mathcal{C} + \mathcal{E}) =$$
$$\mathbf{x}'[k]\left((\mathbf{A} - \mathbf{BK})'\mathbf{P}(\mathbf{A} - \mathbf{BK})\mathcal{C} + \mathbf{A}'\mathbf{PA}(1 - \mathcal{C} + \mathcal{E})\right)\mathbf{x}[k].$$

If we can choose the positive definite matrix $\mathbf{P}$ and matrix $\mathbf{K}$ to satisfy

$$(\mathbf{A} - \mathbf{BK})'\mathbf{P}(\mathbf{A} - \mathbf{BK})\mathcal{C} + \mathbf{A}'\mathbf{PA}(1 - \mathcal{C} + \mathcal{E}) < \mathbf{P},$$

then we would obtain $E[\|(\mathbf{A} - \mathbf{BK})\mathbf{x}[k]\|_{\mathbf{P}}^2\mathcal{C} + \|\mathbf{Ax}[k]\|_{\mathbf{P}}^2(1 - \mathcal{C} + \mathcal{E})] < E[\|\mathbf{x}[k]\|_{\mathbf{P}}^2]$, and (8) would become

$$\sigma_{k+1} < \sigma_k + 2\mathcal{E}\frac{\lambda_{max}(\mathbf{B}'\mathbf{PB})}{\lambda_{min}(\mathbf{P})}b^2 < \alpha\sigma_k + 2\mathcal{E}\frac{\lambda_{max}(\mathbf{B}'\mathbf{PB})}{\lambda_{min}(\mathbf{P})}b^2,$$

for some $0 \leq \alpha < 1$. This signifies a stable system (provided that $b < \infty$), which proves mean square stability. $\square$

Equation (6) in the above theorem can be readily transformed to a *linear matrix inequality* (LMI), which can then be solved to determine whether there exists a feasible pair $(\mathbf{P}, \mathbf{K})$ [4]. When the matrix $\mathbf{B}$ is square and full rank, we can easily choose $\mathbf{K} = \mathbf{B}^{-1}\mathbf{A}$, in which case (6) becomes

$$\mathbf{A}'\mathbf{PA}(1 - \mathcal{C} + \mathcal{E}) - \mathbf{P} < 0. \qquad (9)$$

This is a Lyapunov function, and admits a positive definite matrix $\mathbf{P}$ if and only if the matrix $\sqrt{1 - \mathcal{C} + \mathcal{E}}\mathbf{A}$ has all eigenvalues inside the unit circle [11]. This immediately leads to the following result.

COROLLARY 1. *Consider a system* $(\mathbf{A}, \mathbf{B})$ *with matrix* $\mathbf{B}$ *being square and full rank. Let* $\mathcal{C}$ *be the probability that the input* $-\mathbf{B}^{-1}\mathbf{Ax}[k]$ *is applied to the system, and let* $\mathcal{E}$ *be the probability that an incorrect (but bounded) input is applied to the system. Then, the system is mean square stable if*

$$(1 - (\mathcal{C} - \mathcal{E}))|\lambda_{max}(A)|^2 < 1 \ . \qquad (10)$$

It is instructive to compare this condition to that in Theorem 1: the term $\mathcal{C} - \mathcal{E}$ indicates that incorrect inputs essentially 'cancel out' a certain number of the correct inputs. As expected, this is worse than a channel that simply drops inputs, which only limits the number of correct inputs that are applied. Note that the value of the bound b does not impact the mean square stability of the system as long as $b < \infty$; the proof of the theorem reveals, however, that the bound on the second moment of the state increases with b.

Finally, to verify that the reputation manager described in Section 5 improves the metric $1 - \mathcal{C} + \mathcal{E}$ for stability from equation (10), Fig. 8 shows the variation in the value of

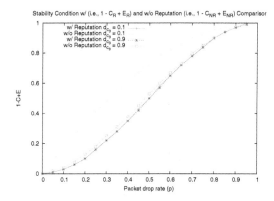

Figure 8: Stability condition $(1 - \mathcal{C} + \mathcal{E})$ with and without reputation for different p and d values

$1 - \mathcal{C} + \mathcal{E}$ with respect to p, for a few representative d_{c_3} values. The main point to note in the graph is that the value of $1 - \mathcal{C} + \mathcal{E}$ with no reputation is invariably higher compared to reputation-based networked control. The difference in performance decreases with the increase in packet-drop rates. Note that reputation management is not guaranteed to stabilize the system for all drop probabilities; this work is intended to demonstrate that reputation management can provide mean square stability under larger drop probabilities than majority voting. In summary, these results collectively demonstrate the utility of using even a very simple reputation scheme to augment networked controller with modular redundancy to compensate for faults.

7. CONCLUSIONS & FUTURE WORK

In this paper we studied a reliable networked control scheme to ensure mean square stability when the channel between the controller and actuator is faulty, in addition to dropping packets. To achieve this, we first studied the use of triple modular redundancy, but showed that due to the potential for packet drops, a straightforward application of majority voting in such a scheme may not be sufficient to ensure stability. We characterized the amount of additional redundancy that would be required in order to rectify this situation. We then provided a *reputation management* scheme to reduce the amount of redundancy required. The scheme builds on majority voting and improves the probability of applying correct inputs to the system, but potentially injects (bounded) incorrect inputs as well. We then generalized existing results on networked control to show that mean square stability will be maintained as long as the bounded incorrect inputs are applied to the plant infrequently enough.

Having introduced a reputation management scheme for networked control systems, there are a variety of avenues for future research. First, we intend to study the effects of other control policies on the ability to stabilize the system in the presence of data-corrupting channels (e.g., allowing the manager to apply the previously applied input when unverifiable values are received, instead of applying an input of zero). Second, we intend to perform a detailed analysis of the convergence of the channel reputations to their steady states (i.e., how quickly the reputations go from being 'unknown' to providing an accurate representation of the channels' reliability). One approach would be to model each possible combination of channel reputations as the states of a Markov Chain, and then to analyze the mixing time of the chain. We also intend to study more general networked control architectures, where the plant's sensors are no longer located at a single point, but instead geographically dispersed. This would necessitate the use of different state-estimators or controllers at each of those locations, and would require a scheme to fuse these different values appropriately (perhaps by making connections with recent work on trust-based distributed Kalman filtering [16]). In cases where the values received by the reputation manager are noisy, or are transmitted asynchronously through the network. A new metric would have to be defined in order to determine which values "agree", and which value is sufficiently different from the others that it can be tagged as incorrect. The work on bounded-delay and threshold majority voting from [5] would be of interest in this regard. Finally, the chosen reputation functions were well-suited for the fault model assumed for this work. However, more elaborate fault models may require more elaborate reputation functions to ensure stability. It would be interesting to see what specific mathematical properties reputation functions should possess to ensure the plant is stabilized. Ultimately, our goal is to build a complete foundational framework for reputation-based networked control.

8. ACKNOWLEDGMENT

This research was supported in part by a grant from NSERC, ONR MURI N00014-07-1-0907, NSF CNS-0834524 and NSF CNS-0931239. Chinwendu Enyioha is supported by a Ford Fellowship administered by the National Research Council of the National Academies.

9. REFERENCES

[1] S. Amin, A. Cardenas, and S. Sastry. Safe and secure networked control systems under denial-of-service attacks. In R. Majumdar and P. Tabuada, editors, *Hybrid Systems: Computation and Control*, volume 5469 of *Lecture Notes in Computer Science*, pages 31–45. 2009.

[2] A. Bemporad, W. Heemels, and M. J. (eds). *Networked Control Systems*, volume 406. Lecture Notes in Control and Information Sciences, Springer-Verlag, 2010.

[3] M. Bishop. *Introduction to Computer Security*. Pearson Education Inc., 2005.

[4] S. Boyd and L. Vandenberghe. *Convex Optimization*. Cambridge University Press, 2004.

[5] P. Caspi and R. Salem. Threshold and bounded-delay voting in critical control systems. In M. Joseph, editor, *Formal Techniques in Real-Time and Fault-Tolerant Systems*, volume 1926 of *Lecture Notes in Computer Science*, pages 70–81. 2000.

[6] J. Chang, K. Venkatasubramanian, A. G. West, S. Kannan, I. Lee, B. Loo, and O. Sokolsky. AS-CRED: Reputation service for trustworthy inter-domain routing. In *University of Pennsylvania Technical Report, MS-CIS-10-17*, April 2010.

[7] R. M. Corless, G. H. Gonnet, D. E. G. Hare, D. J. Jeffrey, and D. E. Knuth. On the Lambert W function. *Advances in Computational Mathematics*, 5(1):329–359, Dec. 1996.

[8] V. Gupta, A. F. Dana, J. Hespanha, R. M. Murray, and B. Hassibi. Data transmission over networks for estimation and control. *IEEE Transactions on Automatic Control*, 54(8):1807–1819, Aug. 2009.

[9] V. Gupta and N. C. Martins. On stability in the presence of analog erasure channels. In *Proc. of the 47th IEEE Conference on Decision and Control*, pages 429–434, 2008.

[10] C. N. Hadjicostis and R. Touri. Feedback control utilizing packet dropping network links. In *Proc. of the 41st IEEE Conference on Decision and Control*, pages 1205–1210, 2002.

[11] J. P. Hespanha. *Linear Systems Theory*. Princeton University Press, 2009.

[12] J. P. Hespanha, P. Naghshtabrizi, and Y. Xu. A survey of recent results in networked control systems. *Proc. of the IEEE*, 95(1):138–162, Jan. 2007.

[13] O. C. Imer, S. Yuksel, and T. Basar. Optimal control of LTI systems over unreliable communication links. *Automatica*, 42(9):1429–1439, Sep. 2006.

[14] R. Ismail and A. Josang. The beta reputation system. In *the 15th BLED Electronic Commerce Conference*, page 41, 2002.

[15] K. Jain. Security based on network topology against the wiretapping attack. *IEEE Wireless Communications*, 11(1):68–71, Feb. 2004.

[16] T. Jiang, I. Matei, and J. S. Baras. A trust based distributed Kalman filtering approach for mode estimation in power systems. In S. S. Sastry and M. McQueen, editors, *Proc. of the First Workshop on Secure Control Systems*. 2010.

[17] S. D. Kamvar, M. T. Schlosser, and H. Garcia-Molina. The Eigentrust algorithm for reputation management in P2P networks. In *Proc. of the 12th Int. Conf. on the World Wide Web*, pages 640–651, 2003.

[18] N. A. Lynch. *Distributed Algorithms*. Morgan Kaufmann Publishers, Inc., 1996.

[19] A. Mesquita, J. Hespanha, and G. Nair. Redundant data transmission in control/estimation over wireless networks. In *Proc. of the 2009 American Control Conference*, pages 3378–3383, 2009.

[20] S. S. Sastry and M. McQueen. Proceedings of the first workshop on secure control systems, 2010.

[21] L. Schenato, B. Sinopoli, M. Franceschetti, K. Poolla, and S. S. Sastry. Foundations of control and estimation over lossy networks. *Proc. of the IEEE*, 95(1):163–187, Jan. 2007.

[22] P. Seiler and R. Sengupta. Analysis of communication losses in vehicle control problems. In *Proc. of the 2001 American Control Conference*, pages 1491–1496, 2001.

[23] K. Stouffer, J. Falco, and K. Scarfone. Guide to industrial control systems (ICS) security. Technical Report 800-82, National Institute of Standards and Technology, Sep. 2008.

[24] K. Walsh and E. G. Sirer. Experience with an object reputation system for peer-to-peer filesharing. In *NSDI'06: Proc. of the 3rd conference on Networked Systems Design & Implementation*, pages 1–1. USENIX Association, 2006.

[25] A. G. West, A. J. Aviv, J. Chang, and I. Lee. Spam mitigation using spatio-temporal reputations from blacklist history. In *Proceedings of ACSAC 2010*, pages 161–170, 2010.

Almost Sure Stability of Networked Control Systems under Exponentially Bounded Bursts of Dropouts

Michael Lemmon
Department of Electrical Engineering
University of Notre Dame
Notre Dame, IN 46556, USA
lemmon@nd.edu

Xiaobo Sharon Hu
Dept. of Computer Science and Engineering
University of Notre Dame
Notre Dame, IN 46556, USA
shu@nd.edu

ABSTRACT

A wireless networked control systems (NCS) is a control system whose feedback path is realized over a wireless communication network. The stability of such systems can be problematic given the random way in which wireless channels drop feedback messages. This paper establishes sufficient conditions for the almost sure stability of NCS under random dropouts. These conditions relate the burstiness in the dropout process to the nominal response of the controlled system. In particular, this means that the burstiness of the dropout process provides a convenient quality-of-service (QoS) constraint on the wireless channel that can be used to adaptively reconfigure the control system in a manner that guarantees the almost sure stability of the NCS. We also show how a probabilistic extension of the network calculus can be used to reconfigure multi-hop communication networks so this paper's sufficient stability condition is not violated.

Categories and Subject Descriptors

J.7 [**Computers in Other Systems**]: Command and control

General Terms

Theory

Keywords

Wireless, Networked Control Systems, almost sure stability, dropout process, burstiness, network calculus

1. INTRODUCTION

In recent years there has been considerable interest in using wireless communication networks to support the monitoring and management of geographically distributed systems [6]. This interest has been driven by a wireless network's low deployment cost and ease of reconfiguration. Significant concerns arise, however, when wireless networks are suggested for use in time-critical control applications. These concerns stem from the probabilistic nature of message delivery in wireless networks. At any point in time, there is a finite probability that the wireless network will fail to deliver a given packet and this means it is impossible for such networks to meet the *hard* real-time quality-of-service (QoS) constraints usually expected by control applications. So while wireless communication technologies may be inexpensive and easy to deploy, it is still unclear if this technology can be used for control applications expecting hard real-time guarantees on message delivery.

Control applications have traditionally demanded hard real-time QoS guarantees from network infrastructure. This expectation could be satisfied by wireline networks, but it is an unreasonable expectation for wireless networks. Do these control applications really need to meet hard real-time deadlines? Are there any control applications that can tolerate firm or event soft real-time guarantees? Firm/soft real-time guarantees may be sufficient if one is willing accept stochastic guarantees on control system stability. Past papers (for example, see [8]) have clearly demonstrated that networked control systems will be mean square stable provided the average rate of dropped feedback data is sufficiently bounded. Hard real-time constraints, therefore, may not be a prerequisite for all control applications.

The problem with mean square stability, however, is that it only requires the variance of the state process to be asymptotically bounded. This means that sample state trajectories of mean square stable processes have a finite probability of being arbitrarily far from the system's equilibrium point. So while the system may be well-behaved on the average, there is always the possibility of a large transient occurring and for many applications this sort of behavior is unacceptable. Demanding applications, therefore, should satisfy a much stronger stability concept than mean square stability. One such stability concept is *almost sure stability*.

Almost sure stability requires the probability of the system's largest excursion from equilibrium for all times $k > T$ go to zero as T goes to infinity. In other words, the further out one goes along the state trajectory, the probability of being arbitrarily far from the equilibrium becomes vanishingly small. This stronger form of stochastic stability is difficult to guarantee in dynamical systems with external disturbances. In fact, it has been shown that if the system disturbance is uniformly bounded, then the resulting system will be almost surely unstable [10]. On the basis of these results, therefore, it may appear highly unlikely that wireless communication should ever be used for highly critical control applications.

The reason for this instability is the wireless channel's propensity for generating a long string or *burst* of dropped data packets [10]. This observation suggests that if one were able to limit the probability of long bursts of dropouts, it may still be possible to ensure the almost sure stability of the process. This fact was recently exploited in [9] to show that if no dropout bursts greater than a given length occur, then quantized control systems will be almost sure practically stable in the presence of uniformly bounded disturbances. While this finding is encouraging, it is of little value in building wireless NCS since there is always a probability that a burst of dropouts may occur.

In wireless NCS it is impossible to require the probability of long dropout bursts to be zero. It may, however be possible to compensate for a long burst by reconfiguring the controller or network. This is the viewpoint adopted in this paper. In particular, this paper shows that if the dropout process has exponentially bounded burstiness, then one can guarantee the almost sure stability of the control system provided its response to an admissible disturbance satisfies certain bounds. In particular, let's characterize a channel's burstiness by a *burst exponent*, γ. This parameter is chosen to be inversely proportional to the probability of a long burst of dropouts. Let's also characterize a system's disturbance rejection in terms of an exponent, s, which parameterizes the closed-loop system's disturbance rejection ability. The main finding in this paper shows that provided the product $s\gamma$ is large enough, then we can guarantee the almost sure stability of the closed-loop process. This sufficient condition, therefore, suggests that it may be possible to adaptively reconfigure the controller and network to assure almost-sure stability. Controller reconfiguration selects a controller whose disturbance rejection renders the parameter s sufficiently large. Network reconfiguration adjusts the communication channel to increase the burst exponent γ.

The remainder of this paper is organized as follows. Section 2 introduces the system model under study. The main result of this paper will be found in section 3. This result is the aforementioned sufficient condition for almost sure stability in networked control systems with single hop wireless networks. Experimental results validating this condition will be found in section 4. This experimental section also presents preliminary simulation results concerning the adaptive reconfiguration of the controller in response to changes in the wireless channel's state. Section 5 extends the result in section 3 multi-hop wireless networks. This section presents a network optimization problem that seeks to minimize overall network energy consumption subject to a constraint on the burstiness in individual network links. Final remarks will be found in section 6.

2. SYSTEM MODEL

The system under study is shown in figure 1. In this figure, one is interested in stabilizing a discrete-time dynamical system called the *plant*. The plant accepts two real-valued inputs; a positive external disturbance, $\{w_k\}_{k=0}^{\infty}$, and a controlled input, $\{u_k\}_{k=0}^{\infty}$. In response to these inputs, the plant generates a real valued *state*, $\{x_k\}_{k=0}^{\infty}$, that satisfies the following equation,

$$x_{k+1} = \alpha x_k + u_k + w_k \qquad (1)$$

for $k = 0, 1, \ldots, \infty$ where $\alpha > 1$ and the initial condition $0 \leq x_0 \in \mathbb{R}$ is given. The only thing known about the

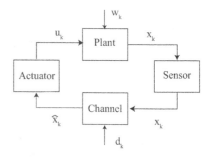

Figure 1: System Model

disturbance is that it is bounded in a sense to be specified in the next section. The control input at time k is assumed to satisfy the following equation,

$$u_k = (\beta - \alpha)\hat{x}_k \qquad (2)$$

for $k = 0, 1, \ldots, \infty$ where $0 < \beta < 1$ and $\hat{x}_k$ is the output of another system called the *channel*.

The channel is a memoryless system that accepts two inputs. Physically this channel is a single-hop wireless communication network. The first input is the plant's state, x_k, measured by a perfect *sensor*. The second input is a binary valued stochastic process, $\{d_k\}_{k=0}^{\infty}$ called the *dropout process*. The relationship between the channel's output, $\hat{x}_k$, and its two inputs is

$$\hat{x}_k = \begin{cases} x_k & \text{if } d_k = 0 \\ 0 & \text{if } d_k = 1 \end{cases} \qquad (3)$$

for all $k \geq 0$. The dropout process, d_k, therefore takes the value of 0 if the channel successfully transport the sensor's measurement, x_k. If $d_k = 1$, then the channel is said to have *dropped* or *erased* the sensor's measurement and it outputs the value 0.

If one combines equations (1)-(3), then the closed-loop system's dynamics are characterized by the following switched difference equation,

$$x_{k+1} = \begin{cases} \alpha x_k + w_k & \text{if } d_k = 1 \\ \beta x_k + w_k & \text{if } d_k = 0 \end{cases} \qquad (4)$$

for all $k = 0, 1, \ldots, \infty$. Since $1 < \alpha$ and $0 < \beta < 1$, the undriven system's state is increasing if sensor data is dropped and it is decreasing asymptotically to zero if there are no dropouts. In particular, if we let $x(k; x_0)$ denote the system state at time k assuming initial input x_0, then at time $k \geq 0$ the system state can be shown to be

$$x(k; x_0) = \alpha^{d_{0,k}} \beta^{k - d_{0,k}} + \sum_{\ell=1}^{k} \alpha^{d_{\ell,k}} \beta^{k-\ell-d_{\ell,k}} w_{\ell-1} \quad (5)$$

where

$$d_{\ell,k} = \sum_{j=\ell}^{k-1} d_j \qquad (6)$$

is the total number of packets that were dropped over the time interval $[\ell, k-1]$ and $d_{k,k} = 0$. We let $\rho_{\ell,k} = \frac{d_{\ell,k}}{k-\ell}$ denote the *local average dropout rate* over time interval $[\ell, k-1]$.

The dropout process $\{d_k\}_{k=0}^{\infty}$ is a stochastic process. The process will be *Bernoulli* with dropout rate $\lambda \geq 0$ if the probability of a dropout at any time $k > 0$ is λ. We will also consider dropout processes that have *exponentially bounded burstiness* or EBB [15]. In particular, given two constants $\rho, \gamma > 0$, we say the process $\{d_k\}_{k=0}^{\infty}$ is (ρ, γ)-EBB if and only if for all $\sigma > 0$ and all $0 \leq \ell < k$,

$$\Pr\{d_{\ell,k} > \rho(k-\ell) + \sigma\} \leq e^{-\gamma\sigma}. \tag{7}$$

In the above equation, ρ may be viewed as a long term dropout rate and σ may be viewed as the length of a dropout burst (i.e. a dropout burst consists of a several consecutive dropouts). A dropout process is therefore (ρ, γ)-EBB if the probability of the total number of dropouts over an interval Δ being greater than $\rho\Delta + \sigma$ can be exponentially bounded as a function of the *dropout burst length*, σ. Throughout this paper, γ will be referred to as the process' *burst exponent*.

This paper studies the *almost sure* stability of the process in equation (4) when the dropouts form a stochastic process. Let the event $A_k^{\epsilon}(x_0)$ be defined as

$$A_k^{\epsilon} = \{|x(k; x_0)| > \epsilon\} \tag{8}$$

then the system is *almost sure asymptotically stable* if and only if for all $\epsilon > 0$

$$\Pr\left\{\limsup_k A_k^{\epsilon}(x_0)\right\} = 0. \tag{9}$$

It is *almost sure practically stable* if there exists $\epsilon > 0$ such that equation (9) holds. Finally the process is said to be *almost sure unstable* if there exists $\epsilon > 0$ such that

$$\Pr\left\{\limsup_k A_k^{\epsilon}(x_0)\right\} = 1. \tag{10}$$

3. SINGLE-HOP NETWORKS

This section examines the almost sure stability of the control system shown in figure 1 assuming that the dropout process has exponentially bounded burstiness. The feedback channel in this system can be considered a single-hop network. The following theorem shows that any Bernoulli process has exponentially bounded burstiness. A version of this theorem was proven in [15]. The following theorem differs from the earlier version in that it explicitly characterizes the burst exponent γ.

THEOREM 3.1. *Let $\{d_k\}_{k=0}^{\infty}$ be a Bernoulli process with parameter λ. For any $\rho > \lambda$ there exists a constant $\gamma > 0$ such that*

$$\Pr\{d_{\ell,k} \geq \rho(k-\ell) + \sigma\} \leq e^{-\gamma\sigma} \tag{11}$$

for all $k > \ell \geq 0$ where

$$\gamma = \sup\left\{z \in \mathbb{R}^+ : \lambda e^z + 1 - \lambda \leq e^{\rho z}\right\}. \tag{12}$$

PROOF. The Markov inequality implies that

$$\Pr\{d_{\ell,k} \geq \rho(k-\ell) + \sigma\} = \Pr\left\{e^{zd_{\ell,k}} \geq e^{z(\rho(k-\ell)+\sigma)}\right\}$$
$$\leq \mathbb{E}\left[e^{zd_{\ell,k}}\right] e^{-z(\rho(k-\ell)+\sigma)}$$

for any $z > 0$. $\mathbb{E}[e^{zd_{\ell,k}}]$ is the moment generating function for the random variable $d_{\ell,k}$. Since d_k is an independent and identically distributed process we can see that

$$\mathbb{E}[e^{zd_{\ell,k}}] = (\lambda e^z + 1 - \lambda)^{k-\ell}.$$

The probability in equation (11) may therefore be bounded as

$$\Pr\{d_{\ell,k} \geq \rho(k-\ell) + \sigma\} \leq f(z)^{k-\ell} g(-z)^{k-\ell} e^{-z\sigma}, \tag{13}$$

where

$$f(z) = (\lambda e^z + 1 - \lambda)$$
$$g(z) = e^{\rho z}.$$

Note that

$$f(0) = g(0) = 1$$
$$f'(0) = \lambda < \rho = g'(0).$$

This means that $1 \leq f(z) \leq g(z)$ for all $z \in [0, \gamma]$ for the γ given in the theorem's statement. We can use this fact in equation (13) to obtain

$$\Pr\{d_{\ell,k} \geq \rho(k-\ell) + \sigma\} \leq g(z)^{k-\ell} g(-z)^{k-\ell} e^{-z\sigma}$$
$$\leq e^{-z\sigma}$$

for any $z \in [0, \gamma]$ which completes the proof. $\square$

A number of papers (see [8] for instance) characterize the mean square stability of a linear process under Bernoulli dropouts. The following theorem establishes similar results for homogeneous (i.e. no input) versions of the linear systems in equation (4) assuming the dropout process has exponentially bounded burstiness.

THEOREM 3.2. *Assume the dropout process is (ρ, γ)-EBB where*

$$0 < \rho < \rho^* = -\frac{\log\beta}{\log\alpha - \log\beta} < 1. \tag{14}$$

If the input $w_k = 0$ for all $k \geq 0$, then the system in equation (4) is almost sure asymptotically stable.

PROOF. Under the assumption that $w_k = 0$ for all k, equation (5) reduces to

$$x_k = \alpha^{d_{0,k}} \beta^{k-d_{0,k}} x_0. \tag{15}$$

Now assume that a particular instance of the dropout process satisfies the inequality

$$d_{\ell,k} \leq \rho(k-\ell) + \sigma \tag{16}$$

for $0 < \ell < k$ and for a given $\sigma > 0$. With this particular dropout process the above equation (15) can be bounded as

$$x_k \leq \mu^k \left(\frac{\alpha}{\beta}\right)^{\sigma} x_0$$

for all $k \geq 0$ where $\mu = \alpha^{\rho}\beta^{1-\rho}$. The right hand side of the above inequality can be bounded by a polynomial function of k. In particular, for any $\gamma > 0$ there exists a positive $C > 0$ such that

$$\mu^k < Ck^{-\frac{2}{\gamma}\log(\alpha/\beta)}$$

for all $k \geq 1$. This allows us to bound x_k with a polynomial function of k,

$$x_k \leq Ck^{-\frac{2}{\gamma}\log(\alpha/\beta)} \left(\frac{\alpha}{\beta}\right)^{\sigma} x_0. \tag{17}$$

So if the dropout sequence d_k satisfies equation (16) then the system state is bounded above as in equation (17).

So let's consider the event $A_k^{\epsilon}(x_0)$. From equation (17) we know that if $x_k > \epsilon$ and if $d_{\ell,k} \leq \rho(k-\ell) + \sigma$ for some choice of σ, then

$$\epsilon < Ck^{-\frac{2}{\gamma}\log(\alpha/\beta)}(\alpha/\beta)^{\sigma} x_0.$$

Taking the log of both sides and solving for σ provides a bound on σ of the form,

$$\sigma > \frac{\log(\epsilon/Ck^{-\frac{2}{\gamma}\log(\alpha/\beta)}x_0)}{\log(\alpha/\beta)} = \sigma^*(\epsilon). \qquad (18)$$

The right hand side of equation (18) is a lower bound on the burst length σ giving rise to the system state, $x_k > \epsilon$. In other words, if $x_k > \epsilon$, then the dropout process must have had burst length $\sigma > \sigma^*(\epsilon)$. Since we assumed the process is (ρ, γ)-EBB, the probability of this event occurring must be less than $e^{-\gamma\sigma^*}$. We may, therefore, bound the probability of $A_k^\epsilon(x_0)$ as

$$
\begin{aligned}
\Pr\{A_k^\epsilon(x_0)\} &\leq \Pr\{d_{\ell,k} > \rho(k-\ell) + \sigma^*(\epsilon)\} \\
&\leq \exp\left(-\gamma\left(\frac{\log(\epsilon/Ck^{-\frac{2}{\gamma}\log(\alpha/\beta)}x_0)}{\log(\alpha/\beta)}\right)\right) \\
&= C_1 k^{-2}
\end{aligned}
$$

where $C_1 = \left(\frac{Cx_0}{\epsilon}\right)^{\frac{\gamma}{\log(\alpha/\beta)}}$.

If we now sum these probabilities over all $k \geq 1$, we obtain

$$\sum_{k=1}^\infty \Pr\{A_k^\epsilon(x_0)\} \leq C_1 \sum_{k=1}^\infty k^{-2} = \frac{C_1\pi^2}{6}.$$

This sum is clearly bounded for any finite ϵ so by the first Borel-Cantelli lemma we can conclude that

$$\Pr\left\{\limsup_k A_k^\epsilon(x_0)\right\} = 0$$

for any $\epsilon > 0$ which means the system is almost sure asymptotically stable. $\square$

It is well known that homogeneous systems under Bernoulli dropouts are almost sure stable [8] [11]. The result in theorem 3.2 extends that prior result to the larger class of dropout processes with exponentially bounded burstiness. We now extend these results to inhomogeneous systems. In general, it is well known that the system in equation (4) is almost surely unstable when driven by a *uniformly bounded* disturbance [10]. This is because at any time k there is a finite probability of having a dropout burst whose length is great enough to force x_k to exceed ϵ. One way to get around this issue is to simply require that no burst occurs with length beyond the critical threshold of σ^*. This approach was adopted in [9] for quantized feedback control systems with random dropouts. The following theorem adopts a less heavy-handed approach. In particular, we assume the dropout process is (ρ, γ)-EBB and identify a class of inputs for which the system is almost sure asymptotically stable.

The following theorem assumes that the dropout process is (ρ, γ)-EBB with a given ρ that is less than the bound ρ^* defined in equation (14) of theorem 3.2. The other parameter, γ, is the *burst exponent* of the dropout process. Since we already know that this system is almost sure unstable under uniformly bounded dropouts, we relax the uniform bound and require that the response of an "averaged" closed-loop system is bounded above Ck^{-s} where $s > 0$. We refer to s as the system's *response exponent*. The larger s is, the faster the system rejects the input disturbance w. The following theorem asserts that if the product $s\gamma$ is greater than $\log\alpha - \log\beta$, then the closed-loop system will be almost surely asymptotically stable.

THEOREM 3.3. *Assume the dropout process is (ρ, γ)-EBB where*

$$\rho < \rho^* = -\frac{\log\beta}{\log\alpha - \log\beta} < 1. \qquad (19)$$

Assume that the system input, w is such that there exist positive real constants C and s such that

$$\mu^k x_0 + \sum_{j=0}^{k-1}\mu^j w_{k-j-1} \leq Ck^{-s} \qquad (20)$$

where $\mu = \alpha^\rho\beta^{1-\rho}$. If s and γ satisfy,

$$s\gamma > \log\alpha - \log\beta \qquad (21)$$

then the driven system in equation (4) is almost sure asymptotically stable.

PROOF. From equation (5) we know

$$x_k = \alpha^{d_{0,k}}\beta^{k-d_{0,k}}x_0 + \sum_{\ell=1}^k \alpha^{d_{\ell,k}}\beta^{k-\ell-d_{\ell,k}}w_{\ell-1}.$$

Consider a dropout sequence that satisfies the bound,

$$d_{\ell,k} \leq \rho(k-\ell) + \sigma \qquad (22)$$

for all $0 < \ell < k$ where $\sigma > 0$ is given. This implies that

$$
\begin{aligned}
x_k &\leq \alpha^{\rho k+\sigma}\beta^{(1-\rho)k-\sigma}x_0 \qquad (23)\\
&\quad + \sum_{\ell=1}^k \alpha^{\rho(k-\ell)+\sigma}\beta^{(1-\rho)(k-\ell)-\sigma}w_{\ell-1} \\
&= \left(\mu^k x_0 + \sum_{j=0}^{k-1}\mu^j w_{k-j-1}\right)\left(\frac{\alpha}{\beta}\right)^\sigma \qquad (24)
\end{aligned}
$$

where $\mu = \alpha^\rho\beta^{1-\rho}$. Under the assumption in equation (19), it can be readily shown that $0 < \mu < 1$. Moreover, the other assumption in equation (20) and the above relation in equation (24) imply that

$$x_k \leq Ck^{-s}\left(\frac{\alpha}{\beta}\right)^\sigma. \qquad (25)$$

So if the dropout sequence, d_k, satisfies equation (22), then the system state x_k must be bounded as in equation (25).

Now consider the event $A_k^\epsilon(x_0)$ so that $x_k > \epsilon$ at time instant k. If the dropout sequence also satisfies equation (22) then we can use equation (25) to infer that

$$\epsilon \leq Ck^{-s}\left(\frac{\alpha}{\beta}\right)^\sigma.$$

Taking the log of both sides and solving for σ yields the bound

$$\sigma > \frac{\log(\epsilon/Ck^{-s})}{\log(\alpha/\beta)} = \sigma^*(\epsilon) \qquad (26)$$

The right hand side of equation (26) is a lower bound on the burst length, σ, that gives rise to $x_k > \epsilon$. Since the dropout process was assumed to be (ρ, γ)-EBB, the probability of this event occurring is $e^{-\gamma\sigma^*}$. We may therefore bound the probability of event $A_k^\epsilon(x_0)$ as

$$
\begin{aligned}
\Pr\{A_k^\epsilon(x_0)\} &\leq \Pr\{d_{\ell,k} > \rho(k-\ell) + \sigma^*(\epsilon)\} \\
&\leq \exp\left(-\gamma\left(\frac{\log(\epsilon/Ck^{-s})}{\log(\alpha/\beta)}\right)\right) \\
&= C_1 k^{-\frac{s\gamma}{\log(\alpha/\beta)}}
\end{aligned}
$$

where $C_1 = \left(\frac{C}{\epsilon}\right)^{\frac{\gamma}{\log(\alpha/\beta)}}$.

So as we did in the proof for theorem 3.1, we sum these probabilities over $k = 1$ to ∞ to obtain

$$\sum_{k=1}^{\infty} \Pr\{A_k^\epsilon(x_0)\} \leq C_1 \sum_{k=1}^{\infty} k^{-\frac{\gamma}{\log(\alpha/\beta)}}.$$

This sum is convergent if $\frac{s\gamma}{\log(\alpha/\beta)}$ is greater than one and that sum is given by the Riemann zeta function, $\zeta(s)$. This is precisely the condition assumed in equation (21). We can therefore conclude that the sum of these probabilities is bounded for any choice of ϵ we can make. By the first Borel-Cantelli lemma this implies that $\Pr\{\limsup_k A_k^\epsilon\} = 0$ which means the system is almost sure asymptotically stable. $\square$

Unlike the earlier result in [10] where the disturbance was uniformly bounded, theorem 3.3 restricts the input to asymptotically approach zero in the manner prescribed by equation (20). We may take, however, the polynomial exponent, s, on the right hand side of equation (20) to be arbitrarily close to zero so that in the limiting case one approaches the uniformly bounded case. To guarantee almost sure asymptotic stability as $s \to 0$, one would then need to have the burst exponent γ go to infinity as well, which means that the probability of any burst essentially goes to zero. If, however, we are only interested in assuring almost sure practical stability (for a specified ϵ), then these limiting conditions require that the probability of a burst of length greater than $\sigma^*(\epsilon)$ goes to zero. This finding is consistent with recent results in [9].

The results in this section confined their attention to linear scalar systems in which the state x_k is always going to be positive. This may, at first, appear to be a significant limitation. We view the system in equation (4), however, as a Lyapunov comparison system. In other words x_k corresponds to the Lyapunov function of the system at time k. This is precisely what was done for so-called *noise-to-state* stable (NSS) systems [4]. In this regard, we believe it may be possible to extend these results to characterize stability in probability for discrete-time nonlinear systems with exponential dropouts.

4. EXPERIMENTAL RESULTS

This section presents experimental results from preliminary Monte Carlo simulations examining how tight the sufficient condition in theorem 3.3 might be. We consider the system in equation (4) with Bernoulli dropouts having parameter λ. The closed-loop dynamic is characterized by the parameter $\beta = .2$. The open-loop dynamic constant α takes values of 5, 2, and 1.25. These are the three systems examined in these simulation studies.

We first study these systems by fixing the input to the system and then varying the dropout parameter λ. We select a parameter $\rho < \rho^*$ where

$$\rho^* = -\frac{\log\beta}{\log\alpha - \log\beta},$$

was defined in equation (14) of theorem 3.2. We selected $\rho = 0.8\rho^*$. We then drove the system in equation (4) with an input function $w_k = \frac{1}{\sqrt{k}}$ for $k \geq 1$ and determined a response ,exponent $\bar{s}$, which over bounds the system's actual response x_k with the function $Ck^{-\bar{s}}$. The value of $\bar{s}$ was then used in equation (21) to determine an upper bound, γ^*, on

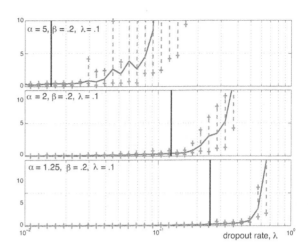

Figure 2: Simulation results where system input was held constant and dropout process parameter, λ, was varied between 0.01 **to** 0.5

the burst exponent γ such that the overall system would be almost surely asymptotically stable. From this γ^*, we then computed the dropout parameter λ^* using the relation

$$\lambda^* e^{\gamma^*} + 1 - \lambda^* = e^{\rho\gamma^*}$$

in equation (12). The dropout parameter λ^* computed above represents a threshold level above which we can expect the driven system to be almost surely unstable.

To test this hypothesis, we simulated the driven system with the given input where the dropout rate λ was varied between 0.01 and 0.5. For each λ, the system was simulated 10 times over the time interval from 0 to 10000. The maximum excursion of the state after time instant 5000 was recorded for each simulation run. The mean, maximum, and minimum value of each collection was recorded and then plotted. Figure 2 shows this plot for the three systems in which $\alpha = 5$, 2, or 1.25. The x-axis in the plot represents the dropout parameter λ on a log scale. The y-axis represents the mean, maximum, and minimum values that x_k achieved over the 10 simulation runs at the specified dropout parameter. The dark blue line shows the mean value. The max and min values are marked by the dashed error bars. For each simulation, the computed threshold, λ^*, is marked by the black vertical line.

The parameters of these three systems are shown in table 1. The threshold dropout rates for the three systems are $\lambda^* = 0.015$, 0.125, and 0.25, respectively. These thresholds are marked by the dark vertical line in each plot in figure 2. The plots show that to the right of the threshold, the variation in the state, x_k, increases dramatically. To the left of this threshold, the state remains close to zero for all times after 5000. This is precisely the behavior one would expect if the systems were almost sure asymptotically stable for $\lambda < \lambda^*$. These results, therefore, seem to confirm the findings in theorem 3.3.

We also studied this system from the standpoint of varying the response exponent s in equation (20). For these experiments, we used the same three systems and we fixed the dropout rate $\lambda = 0.1$. Under these assumptions we can then compute ρ^* as before and selected $\rho = 0.8\rho^*$. For this value of ρ we computed the bound, γ^*, on the burst expo-

	α	β	ρ^*	γ^*	λ^*
system 1	5	.2	.5	6.19	0.015
system 2	2	.2	.69	4.42	0.125
system 3	1.25	.2	.88	3.52	0.25

Table 1: Parameter Values for Simulation Experiment Varying Dropout Rate λ

	α	β	λ	γ^*	s^*
system 1	5	.2	.1	3.4	-0.95
system 2	2	.2	.1	5.1	-0.45
system 3	1.25	.2	.1	7.7	-0.24

Table 2: Parameters for simulation where system responsiveness was varied

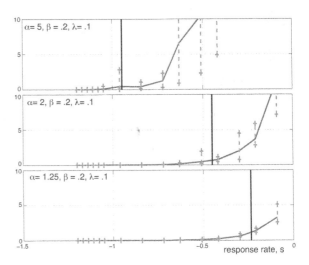

Figure 3: Simulation results where the dropout rate was held to $\lambda = 0.1$ and the response exponent, s, for the system was varied between -1.5 and $-.1$.

nent using equation (12). This value of γ^* was then used in equation (21) to determine the threshold s^* for the system's response exponent. For systems whose actual response exponent, $\overline{s} > s^*$, we expect the system to no longer be almost surely asymptotically stable. The computed values for γ^* and s^* are shown in table 2

The results from this simulation experiment are plotted in figure 3 for all three systems. The x-axis is the response exponent, s, that was actually seen in the simulation. As before in figure 2, the y-axis is the mean, max, and min values achieved by the system state after time instant 5000 over 10 independent simulation runs. The dark vertical line in each plot marks the threshold, s^*, computed for table 2. For simulation runs to the right of these vertical lines, we expect a large increase in the variation of the state (as shown by the larger error bars and larger mean values). To the left of this line, the system states are relatively small. We take this behavior as indicative of the dividing line between almost sure stability and instability as suggested in theorem 3.3. Once again, therefore, these simulation results seem to predict over what range of burst exponents, γ, and response exponents, s, we can expect these systems to be almost sure asymptotically stable.

The simulation results in figures 2 and 3 support theorem 3.3's assertion regarding a tradeoff between the system's response exponent, s, and the dropout process' burst exponent, γ. This suggests a strategy for reconfiguring the control law in wireless networked control systems. It is well known that radio communication channels with Ralyeigh fading can be modeled as two state Markov chains whose two states represent the channel state [14]. The channel state can be detected quickly by monitoring the signal to noise ratio (SNR) at the receiver. From the SNR, one may compute the bit error rate and thereby predict the dropout rate expected at the receiver. What this means is that the channel state and the channel's dropout parameter can be observed and may be used by the control application to reconfigure its controller and thereby guarantee the almost sure stability of the process, even if the channel state randomly changes according to a Markov chain.

A simple simulation was devised to test this idea. We assume a dropout process whose dropout parameter λ switches between 0.01 and 0.1 according to a two-state Markov chain. The Markov chain state, q is either $q_1 = $ GOOD with $\lambda = 0.01$ and $q_2 = $ BAD with $\lambda = 0.1$. The transition probability $\Pr(q_i : q_j)$ is 0.99 if $i = j$ and is 0.01 if $i \neq j$. The sys-

tem under study has a state x_k that satisfies the following difference equation

$$x_{k+1} = \begin{cases} 5x_k + w_k & \text{if } d_k = 1 \text{ (dropout)} \\ 0.75x_k + w_k & \text{if } d_k = 0 \text{ and } q = \text{GOOD} \\ 0.5x_k + w_k & \text{if } d_k = 0 \text{ and } q = \text{BAD} \end{cases}$$

where $w_k = \frac{1}{k^{0.8}}$. In this case, the control system's gain is a function of the channel state. The top plot in figure 4 shows a state trajectory for this system. This system is almost surely asymptotically stable since the pulses due to packet dropouts grow less frequent as time increases. As a point of comparison, the second plot in figure 4 shows the state trajectory of a system in which the control gain is independent of channel state. In particular, we let $x_k = 0.75x_k + w_k$ when $d_k = 0$, regardless of the channel condition. For this case, one sees very large pulses (some on the order of 10^4) arbitrarily far out in time. So without this channel aware switching the system loses almost sure asymptotic stability.

5. MULTI-HOP NETWORKS

The results in section 3 pertain to a networked control system in which the feedback communication channel has a single hop. The last section suggested that if the burst exponent of this single hop is too small, then we can modify the controller to still assure closed-loop stability. We may, however, also consider this from the standpoint of controlling the network. In other words, for a given response exponent, how might we reconfigure the network to enforce the sufficient stability condition in theorem 3.3? This section examines that question with regard to a multi-hop communication network (rather than single-hop). In particular, we use a probabilistic extension of the network calculus to identify an optimization problem whose solution yields the burst exponents of individual network links whose end-to-end quality of service enforces the almost sure stability condition in theorem 3.3.

Consider a networked control system whose feedback information is transported over the network shown in figure 5. This network consists of N wireless nodes connected in

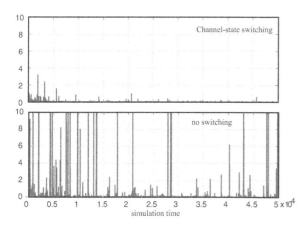

Figure 4: Simulation history of a wireless networked control system whose dropouts following a two-state Markov chain. The top plot shows the response of a channel-aware switching controller and the bottom plot shows the response of a controller that does not switch on channel state.

$$A^{\text{net}} = A^1 \rightarrow \boxed{\text{Node 1}} \xrightarrow{D^1 = A^2} \boxed{\text{Node 2}} \xrightarrow{D^2 = A^3} \cdots \xrightarrow{A^N} \boxed{\text{Node N}} \xrightarrow{D^N = D^{\text{net}}}$$

Figure 5: Traffic flow through N forwarding nodes

series. The first node receives sensor data and transports this over a multi-hop network to a destination node that is connected to the control system's actuator.

Rather than assuming that each link's dropout process is Bernoulli, we adopt the viewpoint used in probabilistic extensions of the network calculus [15] where each link provides its arriving packets with a *statistical service curve*. Using recent results in [1], we use these link service curves to identify a *network* statistical service curve and then show that this network service curve results in an end-to-end dropout process that has exponentially bounded burstiness. The main finding is that the network's burst exponent can be related back to burst exponents for each link in the network. This relationship takes the form of a constraint on the link burst exponents which guides the reconfiguration of the overall network. The following subsections first review some basic results in [1] regarding statistical service curves. We then go on to use these results to characterize the network's burst exponent.

Probabilistic Network Calculus: The network calculus [2, 3, 7] uses a min-plus algebra to relate the end-to-end quality of service (latency) in deterministic networks to the QoS of each link. Our problem needs a probabilistic extension of the network calculus that relates the burstiness of each link to the network's end-to-end burstiness [15]. If one uses the techniques in [15] to bound end-to-end network latency, one finds that this bound grows as $O(N^3)$ where N is the number of network nodes. A recent alternative approach [1] allows one to obtain an upper bound on a network's latency that grows as $O(N \log N)$. The following discussion uses this later method to bound the network's burst exponent. The method makes use of *statistical service curves*

To define a statistical service curve, it will be useful to introduce some notational conventions. In particular, we let

$(x)_+ = \max(0, x)$ where x is any real number. We let $x \wedge y = \min\{x, y\}$ for any real x and y. Given a function $S(\cdot) : \mathbb{Z} \to \mathbb{R}$, we let S_δ be the function where $S_\delta(k) = S(k) + \delta k$. Given two functions $A(\cdot) : \mathbb{Z} \to \mathbb{R}$ and $S(\cdot) : \mathbb{Z} \to \mathbb{R}$, we define the min-plus convolution of A and S as the function $A * S$ that takes values

$$(A * S)(k) = \inf_{0 \le \ell \le k} \{A(k - \ell) + S(\ell)\} \qquad (27)$$

for all $k \ge 0$. With these notational conventions, we can now introduce the concept of a statistical service curve.

Consider a network node whose input is a stochastic process $A = \{A(k)\}_{k=0}^{\infty}$ called the *arrival process*. $A(k)$ denotes the total number of packets received by the node over the time interval $[0, k]$. We let the output of the node be a stochastic process $D = \{D(k)\}_{k=0}^{\infty}$ called the *departure process*. $D(k)$ represents the total number of packets that have departed the node over the interval $[0, k]$. Given a function $S(\cdot) : \mathbb{Z} \to \mathbb{R}$, we say that S is a *statistical service curve* for the node provided for any real $\sigma > 0$,

$$\Pr\{D(k) < (A * (S - \sigma)_+)(k)\} < \epsilon(\sigma) \qquad (28)$$

where $\epsilon(\cdot) : \mathbb{R} \to \mathbb{R}$ is a non-increasing function called the *error function*.

Theorem 1 in [1] characterizes the network service curve for the network shown in figure 5. Assuming that node i for $i = 1, 2, \ldots, N$ provides a statistical service curve $S^i(\cdot) : \mathbb{Z} \to \mathbb{R}$ with error function $\epsilon^i(\cdot)$, then for any $\delta > 0$, the function

$$S^{\text{net}} = S^1 * S^2_{-\delta} * S^3_{-2\delta} * \cdots * S^N_{(N-1)\delta} \qquad (29)$$

is a statistical service curve for the network with an error function

$$\epsilon^{\text{net}}(\sigma) = \inf_{\sigma_1 + \cdots + \sigma_N = \sigma} \left[\epsilon^N(\sigma_N) + \sum_{j=1}^{N-1} \int_{\sigma_j}^{\infty} \epsilon^j(u) du \right]. \quad (30)$$

The proof for the above result will be found in [1].

Network Burst Exponent: We now use the result in equations (29) and (30) to bound the network's end-to-end burst exponent. We start by showing that a single node with a statistical service curve will have a dropout process that has exponentially bounded burstiness. The proof of this assertion requires the following technical lemma.

LEMMA 5.1. *if $S(k) = ((1 - \rho)k)_+$ for all k and $A(k)$ is such that $A(k + \ell) - A(k) < \ell$, then for all k*

$$(A * (S - \sigma)_+)(k) \ge (A(k) - (\rho k + \sigma))_+. \qquad (31)$$

PROOF. We use the bound $(A(k+\ell) - A(k) < \ell)$ to bound the min-plus convolution,

$$(A * (S - \sigma)_+)(k)$$
$$= \inf_{0 \le \ell \le k} \{A(k - \ell) + ((1 - \rho)\ell - \sigma)_+\}$$
$$= \inf_{0 \le \ell \le k} \{A(k) + (A(k - \ell) - A(k)) + ((1 - \rho)\ell - \sigma)_+\}$$
$$\ge \inf_{0 \le \ell \le k} \{A(k) - \ell + ((1 - \rho)\ell - \sigma)_+\}$$
$$\ge \inf_{0 \le \ell \le k} \{(A(k) - (\rho\ell + \sigma))_+\}$$
$$= (A(k) - (\rho k + \sigma))_+.$$

The lemma follows since this holds for all $k \ge 0$. $\square$

We use lemma 5.1 to establish the following theorem. This theorem asserts under relatively mild assumptions that if the wireless node provides a statistical service curve with error function $e^{-\gamma\sigma}$, then the dropout process has exponentially bounded burstiness with the same error function.

THEOREM 5.2. *Consider a node with arrival process $A(k)$ and departure process $D(k)$. Suppose there exist constants $0 < \rho < 1$ and $\gamma > 0$ such that for all $\sigma > 0$, the node provides a statistical service curve $S(k) = ((1-\rho)k)_+$ to the arrival process with error function $\epsilon(\sigma) = e^{-\gamma\sigma}$. Then the dropout process $d_{0,k} = A(k) - D(k)$ is (ρ, γ)-EBB.*

PROOF. Under the assumptions we know that

$$\Pr\left\{D(k) < (A * (S - \sigma)_+)(k)\right\} < e^{-\gamma\sigma}.$$

Equation (31) implies that the following events satisfy

$$\{D(k) < (A(k) - (\rho k + \sigma))_+\} \subset \{D(k) < (A * (S - \sigma)_+)(k)\}.$$

So the probability of the left hand event must be less than the probability of the right hand event which is, in turn, less than $e^{-\gamma\sigma}$, thereby completing the proof. $\square$

Theorem 5.2 can now be used to establish the main result of this section. Return to the network shown in figure 5 and assume that this is the feedback channel used by the networked control system in equation (4). Further assume that the ith node in this network ($i = 1, 2, \ldots, N$) provides a statistical service curve $S^i(k) = ((1 - \rho^i)k)_+$ with error function $\epsilon^i(\sigma) = e^{-\gamma^i\sigma}$ for all $\sigma > 0$, some $\gamma^i > 0$ and some $\rho^i > -\frac{\log\beta}{\log\alpha - \log\beta}$. From theorem 1 in [1], we know that the network service curve, S^{net}, in equation (29) has the error function

$$\epsilon^{\text{net}}(\sigma) = \inf_{\sigma_1 + \cdots + \sigma_N = \sigma} \left[e^{-\gamma^N\sigma_N} + \sum_{j=1}^{N-1} \frac{1}{\delta\gamma^j} e^{-\gamma^j\sigma_j} \right]$$

for any $\delta > 0$. We can select a specific partition of the delays $\sigma_i = \frac{\sigma}{N}$ so that

$$\epsilon^{\text{net}}(\sigma) \leq \left[e^{-\gamma^N\sigma/N} + \sum_{j=1}^{N-1} \frac{1}{\delta\gamma^j} e^{-\gamma^j\sigma/N} \right].$$

Since the network gives the arrival process into the first node a statistical service curve S^{net}, we know from theorem 5.2 that the end-to-end dropout process has exponentially bounded burstiness. In addition to this, theorem 3.3 suggests that for a given response exponent, s, the end-to-end process burst exponent, γ^{net}, should be greater than

$$\gamma^* = \frac{\log\alpha - \log\beta}{s}.$$

So we need to select the individual link exponents, γ^i for $i = 1, 2, \ldots, N$ such that the error function $\epsilon^{\text{net}}(\sigma)$ is less than the burst error function $e^{-\gamma^*\sigma}$. If this is done we expect the networked control system to remain almost sure stable. In other words, we need to select the link exponents γ^i so that

$$\epsilon^{\text{net}}(\sigma) \leq \left[e^{-\gamma^N\sigma/N} + \sum_{j=1}^{N-1} \frac{1}{\delta\gamma^j} e^{-\gamma^j\sigma/N} \right] \leq e^{-\gamma^*\sigma} \quad (32)$$

In light of theorem 3.3, if the end-to-end ρ of the network's service curve still satisfies the required bound in equation

(14), then equation (32) represents an inequality constraint on the burst exponents of each link, that must be satisfied to assure the almost sure stability of the networked control system.

We suggest that the inequality in equation (32) can be used to adaptively reconfigure the wireless nodes in the network. Consider a wireless radio node in which a single message packet consists of M information bits. For this packet to be successfully received, all M information bits must be received. Assume that the node transmits $L > M$ bits at R bits/second with power w. If the bit error rate is known then we can compute the probability that M information bits will be received within a specified deadline $D > LR$. If this probability is too small we can take steps to decrease the bit error rate (increase broadcast power) or we can increase the bit transmission rate R. In either case one may, within realistic limits, formulate an optimization problem whose solution would generate a set of link burst exponents that satisfy the inequality constraint in equation (32). The decision variable in this problem would be either the link's bit rate or transmission power.

For example, consider a scenario in which the radio nodes adjust their transmitted bit rate, R_i, for $i = 1, 2, \ldots, N$. Let $E(R_i)$ denote the energy each node expends in transmitting a packet at this bit rate. We may then pose the following problem that seeks to minimize the summed energy of all network nodes subject to the end-to-end burstiness constraint required to achieve almost sure stability. This optimization problem could take the form,

$$
\begin{array}{ll}
\text{minimize} & \sum_{i=1}^{N} E(R_i) \\
\text{with respect to:} & R_1, R_2, \cdots, R_N \\
\text{subject to:} & R_i < \overline{R}_i, \quad (i = 1, 2, \ldots, N) \\
& \left[e^{-\gamma^N\sigma/N} + \sum_{j=1}^{N-1} \frac{1}{\delta\gamma^j} e^{-\gamma^j\sigma/N} \right] \leq e^{-\gamma^*\sigma}
\end{array}
$$

where γ^i (the link's burst exponent) is a function of the node's transmission rate R_i, $\overline{R}_i$ is an upper bound on the ith node's maximum allowable transmission rate, and δ is a tuning parameter.

This section has suggested how probabilistic extensions of the network calculus might be used in conjunction with theorem 3.3 to adaptively reconfigure wireless networks to ensure almost sure stability in networked control systems. The basic approach involves using the network calculus to form an optimization problem whose solution minimizes overall network energy consumption while ensuring the link transmission rates satisfy a bound on the end-to-end burstiness of the network. The resulting optimization problem appears to be separable so that any one of a number of distributed optimization algorithms might be used to solve this problem [12, 5, 13].

6. CONCLUDING REMARKS

This paper studied the almost sure stability of discrete-time linear systems under dropout processes that have exponentially bounded burstiness. The main finding is theorem 3.3 which provides a sufficient characterization for almost sure stability in systems whose driven response decays to zero at an arbitrarily slow rate. The sufficient condition establishes a tradeoff between the system's rate of decay and the dropout process' burst exponent. Preliminary simulation experiments suggest that the bound in theorem 3.3 is reasonably tight.

Theorem 3.3 suggests a method by which one can adaptively reconfigure a networked control system to maintain almost sure stability over wireless networks with random dropouts. If, for instance, one has no control over the wireless channel, then it may be possible to adjust the controller to increase the system's rate of decay for a bounded class of inputs. This may be used to compensate for temporary increases in dropout burstiness. Another approach for reconfiguring the system focuses on the links in a multi-hop network. Using results from the probabilistic network calculus we identified a constraint on the network's burst exponents that could be used to guide the adaptation of link bit rates to try and meet the almost sure stability conditions presented in theorem 3.3.

The results in this paper, therefore, suggest a promising direction for adaptively reconfiguring a wireless sensor-actuator network to guarantee the almost sure stability of the controlled process. The results in theorem 3.3 provide guidance on how to adjust either the control application or the network's communication infrastructure to achieve these goals.

7. ACKNOWLEDGEMENT

The authors acknowledge the partial financial support of the National Science Foundation NSF-CNS-0931195.

8. REFERENCES

[1] F. Ciucu, A. Burchard, and J. Liebeherr. Scaling properties of statistical end-to-end bounds in the network calculus. *Information Theory, IEEE Transactions on*, 52(6):2300–2312, 2006.

[2] R. Cruz. A calculus for network delay. I. Network elements in isolation. *Information Theory, IEEE Transactions on*, 37(1):114–131, 1991.

[3] R. Cruz. A calculus for network delay, part II: Network analysis. *IEEE Transactions on Information theory*, 37(1):132–141, 1991.

[4] H. Deng, M. Krstic, and R. Williams. Stabilization of stochastic nonlinear systems driven by noise of unknown covariance. *IEEE Transactions on Automatic Control*, 46(8):1237–1253, 2001.

[5] B. Johansson, M. Rabi, and M. Johansson. A randomized incremental subgradient method for distributed optimization in networked systems. *SIAM Journal on Optimization*, 20(3):1157–1170, 2009.

[6] K. Koumpis, L. Hanna, M. Andersson, and M. Johansson. Wireless industrial control and monitoring beyond cable replacement. In *Proceedings of the 2nd PROFIBUS International Conference*. 2005

[7] J.-Y. Le Boudec and P. Thiran. *Network Calculus: A Theory of Deterministic Queuing Systems for the Internet*, volume 2050 of *Lecture notes in computer science*. Springer, 2001.

[8] Q. Ling and M. Lemmon. Soft real-time scheduling of networked control systems with dropouts governed by a Markov chain. In *American Control Conference, 2003. Proceedings of the 2003*, volume 6, pages 4845–4850. IEEE, 2003.

[9] Q. Ling and M. Lemmon. A necessary and sufficient feedback dropout condition to stabilize quantized linear control system with bounded noise. to appear in IEEE Transactions on Automatic Control, 2010.

[10] A. Matveev and A. Savkin. Comments on" control over noisy channels" and relevant negative results. *IEEE Transactions on Automatic Control*, 50(12):2105–2110, 2005.

[11] L. Montestruque and P. Antsaklis. Stability of model-based networked control systems with time-varying transmission times. *IEEE Transactions on Automatic Control*, 49(9), 2004.

[12] A. Nedic and A. Ozdaglar. Distributed subgradient methods for multi-agent optimization. *IEEE Transactions on Automatic Control*, 54(1):48–61, 2009.

[13] P. Wan and M. Lemmon. Distributed network utility maximization using event-triggered augmented lagrangian methods. In *Proceedings of the American Control Conference*, 2009.

[14] H. Wang and N. Moayeri. Finite-state Markov channel–a useful model for radio communication channels. *IEEE Transactions on Vehicular Technology*, 44(1):163–171, 1995.

[15] O. Yaron and M. Sidi. Performance and stability of communication networks via robust exponential bounds. *IEEE/ACM Transactions on Networking (TON)*, 1(3):385, 1993.

SHAVE — Stochastic Hybrid Analysis of Markov Population Models

[Tool Presentation]

Maksim Lapin
Saarland University,
Saarbrücken, Germany
lapin@cs.uni-saarland.de

Linar Mikeev
Saarland University,
Saarbrücken, Germany
mikeev@cs.uni-saarland.de

Verena Wolf
Saarland University,
Saarbrücken, Germany
wolf@cs.uni-saarland.de

ABSTRACT

We present a tool called SHAVE that approximates the transient distribution of a continuous-time Markov population process by combining moment-based and state-based representations of probability distributions. As an intermediate step, SHAVE constructs a stochastic hybrid model from the original process which is then solved numerically.

Categories and Subject Descriptors: G.3 [Mathematics of Computing]: PROBABILITY AND STATISTICS [Queueing theory, Markov Processes]

General Terms: Experimentation

Keywords: Markov processes, Tool, Transient analysis

1. INTRODUCTION

Markov processes are an omnipresent modeling approach in the applied sciences. Often, they describe *population processes*, that is, they operate on a multidimensional discrete state space, where each dimension of a state represents the number of individuals of a certain type. Depending on the application area, "individuals" refers to customers in a queuing network, molecules in a chemically reacting volume, servers in a computer network, etc.

Here, we are particularly interested in dynamical models of biochemical reaction networks, such as signaling pathways, gene regulatory networks, and metabolic networks. They are an important emerging application area of continuous-time Markov processes and operate on an abstraction level where a state of the system is given by an n-dimensional vector of chemical populations, that is, the system involves n different types of molecules and the i-th coordinate represents the number of molecules of type i. Molecules collide randomly and may undergo chemical reactions, which change the state of the system. Classical modeling approaches in biochemistry are based on a system of ordinary differential equations that assume a continuous deterministic change of chemical concentrations. Over the last decade, however, various experimental results have shown that the discreteness and randomness of the chemical reactions need to be taken into account. Thus, discrete-state continuous-time Markov processes have gained in importance for describing the dynamics in the cell [8]. From the transient probability distribution of such a population process different measures of interest can be derived such as the distribution of switching delays [8] or the distribution of the time of DNA replication initiation at different origins [9]. Moreover, many parameter estimation methods require the computation of the posterior distribution because means and variances do not provide enough information to calibrate parameters [4].

The SHAVE tool approximates the transient distribution of a population process which requires the solution of the Kolmogorov differential equations (KDE) that provide for each population vector $x \in \mathbb{Z}_+^n$ the time-derivative of the probability that the process is in state x at time t. In most cases no tight upper and lower bounds on the population sizes are known a-priori and, thus, the number of possible states is large (or even infinite) which renders the direct numerical integration of the KDE infeasible. Existing tools designed for large state spaces use symbolic representations based on MTBDDs [7] or Kronecker representations [1]. These techniques try to reduce the amount of memory needed at a certain step of the computation but not the number of states that are considered. Therefore they perform slowly for population processes of realistic size. Moreover, they exploit that the same transition rates occur several times in the underlying Markov process. This, however, is rarely the case in population models since the transition rates are usually density-dependent, that is, proportional to the relative abundance of certain species.

In contrast, the SHAVE software uses a dynamical truncation of the state space as well as moment-based representations of distributions which leads to a stochastic hybrid model. The latter is then solved numerically and its solution yields an approximation of the transient distribution of the population process [5]. SHAVE is written in C++ and available at http://alma.cs.uni-saarland.de/shave.

2. MODEL SPECIFICATION

The input of the SHAVE tool is a population process that is specified using guarded commands, that is, commands of the form "guard |- rate -> update" are used to specify classes of transitions. The guard is a boolean expression over the state variables that represent the different populations. The transition is only possible in states where the guard is true. The rate (function) is a polynomial expression over the state variables that specifies the transition rate. The update

HSCC'11, April 12–14, 2011, Chicago, Illinois, USA.
Copyright 2011 ACM 978-1-4503-0629-4/11/04 ...$10.00.

describes the change of the state variables if the transition occurs. SHAVE is restricted to rate functions that are rational functions of at most order two. Moreover, we only allow constant updates, i.e., if x is the current state then the successor state of the transition is $x + v \in \mathbb{Z}_+^n$ for some constant vector $v \in \mathbb{Z}^n$.

EXAMPLE 2.1. *Consider two genes that share a common promotor region at which only a single protein can bind. Let P_i be the protein produced by gene i at rate g_i and assume that if a protein of type P_i binds to the promotor (which happens at rate b_i for a selected protein) then only gene i can be transcribed but not gene j ($j \in \{1, 2\}$, $j \neq i$). Let $x = (x_1, x_2, x_3)$ denote a state of the underlying Markov process where x_i denotes the population size of P_i for $i \in \{1, 2\}$. The entry $x_3 \in \{1, 2, 3\}$ denotes the state of the promotor where $x_3 = i$ represents the case where P_i is bound to the promotor and if $x_3 = 3$ then no molecule is bound to the promotor. The corresponding guarded commands are*

```
(x_3=3 ∨ x_3=i) |- g_i      -> x_i=x_i+1;
(x_i>0)         |- d_i*x_i -> x_i=x_i-1;
(x_3=3 ∧ x_i>0) |- b_i*x_i -> x_i=x_i-1,x_3=i;
(x_3=i)         |- u_i      -> x_i=x_i+1,x_3=3;
```

where $i \in \{1, 2\}$ and g_i, d_i, b_i, u_i are positive constants. The second command describes degradation of P_i proteins and the last command describes unbinding from the promoter.

3. NUMERICAL METHODS

The SHAVE tool combines two strategies that work well for the transient analysis of population processes [2, 5].

- If a population is large, the corresponding marginal probability distribution can accurately be approximated by using a moment-based representation, that is, instead of integrating the probability of each state over time, we only integrate the first two moments of the distribution as well as the covariances [3, 10]. This approximation is known to be exact if the population sizes tend to infinity in the scaled process [6].

- For small populations we integrate the corresponding KDE directly using an explicit fourth-order Runge-Kutta method. We apply a dynamical truncation of the state space during the computation, i.e., whenever the probability of a state becomes less than a certain threshold, we remove the state from the current state space. When a new state is found we add it to our current state space. If the threshold is chosen small (e.g. 10^{-15}), this method has shown to yield accurate approximations [2].

Since in most applications we have both large and small populations, we construct a stochastic hybrid model where the modes represent the states of the small populations and each mode has a system of differential equations for the dynamics of the large populations. Instead of integrating the global moments of the large populations, we integrate the moments conditioned on the mode.

EXAMPLE 3.1. *Assume that in Ex. 2.1 the populations of P_1 and P_2 are large. Then population x_3 will be represented by three modes where we assign two differential equations to each mode that describe the dynamics of the conditional expectations of the P_i populations. For instance, in mode $x_3 = 1$ we have $\dot{x}_1 = g_1 - d_1 x_1 + u_1$ and $\dot{x}_2 = d_2 \cdot x_2$. The probability p_1 of having $x_3 = 1$ is then given by the "reduced KDE" $\dot{p}_1 = b_1 x_1 p_3 - u_1 p_1$ where p_3 is the probability of $x_3 = 3$.*

Our numerical method integrates the probabilities of the modes and the differential equations of each mode for a small time interval $[t, t+h]$. Then we approximate the conditional expectations at time $t + h$ by taking into account the probability that the mode changes during $[t, t+h]$. If an order-two representation is chosen for the large populations then the (co-)variances are calculated in a similar way. A detailed description of the approximation has been presented in [5].

4. EXPERIMENTAL RESULTS

We compute an approximation of the transient distribution of the process in Ex. 2.1 and compare our results with a purely state-based solution as well as a purely moment-based solution. For the parameters $g_1 = 5$, $g_2 = 0.5$, $d_1 = 0.0005$, $d_2 = 0.005$, $b_1 = b_2 = 0.1$, $u_1 = u_2 = 0.005$, a purely stochastic solution for a time horizon of $t = 500$ takes nearly 5 hours whereas the hybrid method needs only 25 seconds. The relative errors of the first three moments (averaged over all populations) are 0.06, 0.08, and 0.09 for the hybrid method while a purely moment-based solution of order one has a relative error of 0.45. For further experimental results we refer to [5].

5. CONCLUSIONS

SHAVE is a tool for the approximate solution of Markov processes with an underlying population structure. It efficiently computes an accurate approximation of the transient probability distribution based on a stochastic hybrid model that combines moment-based and state-based representations of distributions.

References

[1] P. Buchholz and P. Kemper. Numerical analysis techniques in the APNN toolbox. In *Workshop on Formal Methods in Performance Evaluation and Applications*, pages 1–6, 1999.

[2] F. Didier, T. A. Henzinger, M. Mateescu, and V. Wolf. Fast adaptive uniformization of the chemical master equation. *IET Systems Biology Journal*, 2010. To appear.

[3] S. Engblom. Computing the moments of high dimensional solutions of the master equation. *Appl. Math. Comput.*, 180:498–515, 2006.

[4] D. Henderson, R. Boys, C. Proctor, and D. Wilkinson. Linking systems biology models to data: a stochastic kinetic model of p53 oscillations. In *Handbook of Appl. Bayesian Analysis*. Oxford University Press, 2009.

[5] T. A. Henzinger, M. Mateescu, L. Mikeev, and V. Wolf. Hybrid numerical solution of the chemical master equation. In *Proc. of CMSB'10*. ACM Digital Library, 2010.

[6] T. G. Kurtz. Strong approximation theorems for density dependent Markov chains. *Stochastic Processes Appl.*, 6(3):223–240, 1977/78.

[7] M. Kwiatkowska, G. Norman, and D. Parker. Prism: Probabilistic model checking for performance and reliability analysis. *ACM SIGMETRICS Performance Evaluation Review*, 36(4):40–45, 2009.

[8] H. H. McAdams and A. Arkin. Stochastic mechanisms in gene expression. *PNAS, USA*, 94:814–819, 1997.

[9] P. Patel, B. Arcangioli, S. Baker, A. Bensimon, and N. Rhind. DNA replication origins fire stochastically in fission yeast. *Mol. Biol. Cell*, 17:308–316, 2006.

[10] A. Singh and J. P. Hespanha. Approximate moment dynamics for chemically reacting systems. *IEEE Trans. on Automat. Contr.*, 2010. To appear.

TuLiP: A Software Toolbox for Receding Horizon Temporal Logic Planning

Tichakorn Wongpiromsarn*, Ufuk Topcu**, Necmiye Ozay**, Huan Xu**, and Richard M. Murray**

* Singapore-MIT Alliance for Research and Technology, Singapore
** California Institute of Technology, Pasadena, CA
{nok, utopcu, necmiye, mumu, murray}@cds.caltech.edu

ABSTRACT

This paper describes TuLiP, a Python-based software toolbox for the synthesis of embedded control software that is provably correct with respect to an expressive subset of linear temporal logic (LTL) specifications. TuLiP combines routines for (1) finite state abstraction of control systems, (2) digital design synthesis from LTL specifications, and (3) receding horizon planning. The underlying digital design synthesis routine treats the environment as adversary; hence, the resulting controller is guaranteed to be correct for any admissible environment profile. TuLiP applies the receding horizon framework, allowing the synthesis problem to be broken into a set of smaller problems, and consequently alleviating the computational complexity of the synthesis procedure, while preserving the correctness guarantee.

Categories and Subject Descriptors

D.2.4 [**Software Engineering**]: Software/Program Verification—*Formal methods*; D.2.10 [**Software Engineering**]: Design—*Methodologies*

General Terms

Design, Verification

Keywords

Linear temporal logic, receding horizon control

1. INTRODUCTION

To achieve higher levels of autonomy, modern embedded control systems need to reason about complex, uncertain environments and make decisions that enable complex missions to be accomplished safely and efficiently. To this end, linear temporal logic (LTL) is widely used as a specification language to precisely define system correctness properties. The embedded control software needs to be able to integrate discrete and continuous decision-making and provide correctness guarantee with respect to a given specification. Furthermore, since the environment may be dynamic and unknown a priori, it is important that the controller ensures proper response to all the admissible environment profiles.

A common approach to embedded control software synthesis is to construct a finite transition system that serves as an abstract model of the physical system and synthesize a strategy, represented by a finite state automaton, satisfying the given properties based on this model. Software packages based on this procedure include LTLCon [4], conPAS2 [11], Pessoa [5] and LTLMoP [1]. LTLCon and conPAS2 handle affine systems and piecewise affine systems, respectively, and arbitrary LTL specifications. Pessoa admits nonlinear and switched dynamics but only a very limited class of LTL specifications. However, these three tools do not handle the adversarial nature of the environment. Hence, the controller is only provably correct with respect to an a priori known and fixed environment. In contrast, LTLMoP accounts for adversaries but only considers fully actuated systems operating in the Euclidean plane. To keep the synthesis problem tractable, LTLMoP only admits the GR[1] fragment of LTL. A sampling-based method has been proposed for μ-calculus specifications in [3] but does not provide a correctness guarantee for all the admissible environments.

This paper introduces TuLiP, a Python-based toolbox for embedded control software synthesis. Similar to LTLMoP, TuLiP models the environment as an adversary and only considers GR[1] formulas. This often leads to the state explosion problem since all the admissible environment profiles need to be taken into consideration in the synthesis process. TuLiP alleviates this complexity by integrating a receding horizon framework [9]. Additionally, it admits general affine dynamics with bounded disturbances.

2. TuLiP FEATURES

We now summarize two key features of TuLiP (available at http://www.cds.caltech.edu/tulip).

2.1 Embedded Control Software Synthesis

TuLiP deals with systems that comprise the plant, i.e., the physical component regulated by the controller, and its potentially dynamic and a priori unknown environment. Note that the environment does not only include the factors that are external to the plant but it also includes the factors over which the system does not have control, e.g., hardware failure. The plant may contain both continuous (e.g. physical) and discrete (e.g. computational) components. TuLiP models the embedded control software synthesis problem as a game between the plant and the environment. Given the model of the plant and system specification φ in LTL, TuLiP provides a function that automatically synthesizes a controller that ensures system correctness with respect to φ for any admissible environment, if such a controller exists (i.e., φ is realizable). If φ is unrealizable, TuLiP also provides counter examples, i.e., initial states starting from which the environment can falsify φ regardless of controller's actions.

The synthesis feature relies on (1) generating a proposition preserving partition of the continuous state space, (2) continuous state space discretization based on the evolution of the continuous state [8], and (3) digital design synthesis. JTLV [6] is used as the underlying synthesis routine.

Currently, TuLiP handles the case where the continuous state of the plant evolves according to discrete-time linear time-invariant dynamics: for $t \in \{0, 1, 2, \ldots\}$, $s[t+1] = As[t] + Bu[t] + Ed[t]$, $u[t] \in \mathcal{U}$, $d[t] \in \mathcal{D}$, $s[0] \in \mathcal{S}$, where $\mathcal{S} \in \mathbb{R}^n$ is the continuous state space, $\mathcal{U} \in \mathbb{R}^m$ and $\mathcal{D} \in \mathbb{R}^p$ are the sets of admissible control inputs and exogenous disturbances, $s[t], u[t], d[t]$ are the continuous state, the control signal and the exogenous disturbance, respectively, at time t. $\mathcal{U}, \mathcal{D}, \mathcal{S}$ are assumed to be bounded polytopes.

The specification φ is assumed to be of the form

$$\varphi = \left(\psi_{init} \wedge \Box\psi_e \wedge \bigwedge_{i \in I_f} \Box\Diamond\psi_{f,i}\right) \implies \left(\Box\psi_s \wedge \bigwedge_{i \in I_g} \Box\Diamond\psi_{g,i}\right),$$

known as GR[1]. Here $\psi_{init}, \psi_e, \psi_{f,i}, i \in I_f, \psi_s$ and $\psi_{g,i}, i \in I_g$ are propositional formulas. ψ_{init}, ψ_e and $\psi_{f,i}, i \in I_f$ essentially describe the assumptions on the initial state of the system and the environment. ψ_s and $\psi_{g,i}, i \in I_g$ describe the desired behavior of the system. See [9] for more details.

2.2 Receding Horizon Framework

For systems with a certain structure, the computational complexity of the planner synthesis can be alleviated by solving the planning problems in a receding horizon fashion, i.e., compute the plan or strategy over a "shorter" horizon, starting from the current state, implement the initial portion of the plan, and recompute the plan. This approach essentially reduces the problem into a set of smaller problems. Certain sufficient conditions ensure that this "receding horizon" strategy preserves the desired system-level properties.

Given a specification in the form of φ above, TuLiP first constructs a finite state abstraction of the physical system. Then, for each $i \in I_g$, we organize the system states into a partially ordered set $\mathcal{P}^i = (\{\mathcal{W}_j^i\}, \preceq_{\psi_{g,i}})$ where $\mathcal{W}_0^i$ are the set of states satisfying $\psi_{g,i}$. For each j, we define a short-horizon specification Ψ_j^i associated with $\mathcal{W}_j^i$ as

$$\begin{aligned} \Psi_j^i &= \left((\nu \in \mathcal{W}_j^i) \wedge \Phi \wedge \Box\psi_e \wedge \bigwedge_{k \in I_f} \Box\Diamond\psi_{f,k}\right) \\ &\implies \left(\Box\psi_s \wedge \Box\Diamond(\nu \in \mathcal{F}^i(\mathcal{W}_j^i)) \wedge \Box\Phi\right). \end{aligned}$$

Here Φ is a propositional formula that describes receding horizon invariants and $\mathcal{F}^i : \{\mathcal{W}_j^i\} \to \{\mathcal{W}_j^i\}$ defines the intermediate goal for starting in $\mathcal{W}_j^i$. Let $\mathcal{V}$ be the entire state space of the system. As described in [9], a sufficient condition for the receding horizon strategy to lead to correct execution with respect to φ is that for all $i \in I_g$, (1) $\mathcal{W}_0^i \cup \mathcal{W}_1^i \cup \ldots \cup \mathcal{W}_p^i = \mathcal{V}$, (2) $\mathcal{W}_0^i \prec_{\psi_{g,i}} \mathcal{W}_j^i, \forall j \neq 0$, (3) $\mathcal{F}^i(\mathcal{W}_j^i) \prec_{\psi_{g,i}} \mathcal{W}_j^i, \forall j \neq 0$, (4) $\psi_{init} \implies \Phi$ is a tautology, and (5) Ψ_j^i is realizable $\forall j$.

Given the plant model, φ, $\{\mathcal{W}_j^i\}$, $\mathcal{F}^i$ and Φ, TuLiP automatically constructs the short horizon specification Ψ_j^i for each i, j. It includes functions for verifying that there exists a partial order $\preceq_{\psi_{g,i}}$ and that the sufficient condition above is satisfied; and automatically computing the receding horizon invariant Φ if one exists or report an error otherwise.

3. APPLICATIONS AND DISCUSSIONS

We have demonstrated the successful applications of TuLiP in multiple applications, including autonomous driving [9], vehicle management systems in avionics [10] and multi-target

tracking. Other simpler examples are included in the current release of the toolbox. For the autonomous driving problem, the receding horizon framework needs to be applied for the car to be able to drive a reasonable distance. Due to the state explosion problem, TuLiP cannot automatically find the receding horizon invariant Φ for this specific application. Nevertheless, it provides useful guidelines for the user to easily manually construct Φ. Once Φ is constructed, TuLiP successfully checks that the sufficient condition for applying the receding horizon strategy is satisfied.

TuLiP constructs Φ roughly by starting $\Phi = True$ and iterating between (1) checking the realizability of each Ψ_j^i, and (2) updating Φ to be the conjunction of current Φ and the negation of the counter examples of unrealizable Ψ_j^i (if any). This process stops when $\psi_{init} \implies \Phi$ is no longer a tautology or all the Ψ_j^i are realizable. Since the counter examples are given as the enumeration of all the infeasible initial states, the size of Φ quickly increases. An extension of the current version of TuLiP is to implement a procedure for reducing the counter examples into a small formula. We also plan to integrate various existing software packages into TuLiP including a user-friendly simulation environment such as Player/Stage [2] and a state space discretization procedure that admits a more general class of systems (e.g. one based on approximate simulations and bisimulations as discussed in [7] and implemented in [5]).

4. REFERENCES

[1] C. Finucane, G. Jing, and H. Kress-Gazit. LTLMoP. http://code.google.com/p/ltlmop/.

[2] B. Gerkey, R. Vaughan, and A. Howard. The Player/Stage project: Tools for multi-robot and distributed sensor systems. In *Conf. on Advanced Robotics*, 2003.

[3] S. Karaman and E. Frazzoli. Sampling-based motion planning with deterministic μ-calculus specifications. In *IEEE CDC*, 2009.

[4] M. Kloetzer and C. Belta. LTLCon. http://iasi.bu.edu/~software/LTL-control.htm.

[5] M. Mazo, A. Davitian, and P. Tabuada. Pessoa: A tool for embedded controller synthesis. In T. Touili, B. Cook, and P. Jackson, editors, *CAV*, volume 6174 of *LNCS*, pages 566–569. Springer, 2010.

[6] N. Piterman, A. Pnueli, and Y. Sa'ar. Synthesis of reactive(1) designs. In *Verification, Model Checking and Abstract Interpretation*, volume 3855 of *LNCS*, pages 364 – 380. Springer, 2006. http://jtlv.sourceforge.net/.

[7] P. Tabuada. *Verification and Control of Hybrid Systems: A Symbolic Approach*. Springer, 2009.

[8] T. Wongpiromsarn, U. Topcu, and R. M. Murray. Automatic synthesis of robust embedded control software. In *AAAI SS on Embedded Reasoning: Intelligence in Emb'd Systems*, pages 104–111, 2010.

[9] T. Wongpiromsarn, U. Topcu, and R. M. Murray. Receding horizon control for temporal logic specifications. In *HSCC*, pages 101–110, 2010.

[10] T. Wongpiromsarn, U. Topcu, and R. M. Murray. Formal synthesis of embedded control software: Application to vehicle management systems. In *AIAA Infotech@Aerospace*, 2011. submitted.

[11] B. Yordanov and C. Belta. conPAS2. http://hyness.bu.edu/conPAS2.html.

Pessoa 2.0: A Controller Synthesis Tool for Cyber-Physical Systems

Pritam Roy
UC Los Angeles
pritam@ee.ucla.edu

Paulo Tabuada
UC Los Angeles
tabuada@ee.ucla.edu

Rupak Majumdar
MPI-SWS and UCLA
rupak@cs.ucla.edu

ABSTRACT

We introduce PESSOA 2.0, a tool that automatically synthesizes controllers for cyber-physical systems based on correct-by-design methodology. PESSOA 2.0 accepts a cyber-physical system represented by a set of smooth differential equations and automata and a specification in a fragment of Linear Temporal Logic that is expressive enough to describe interesting properties but simple enough to avoid Safra's construction. It outputs, if possible, a controller for the system that enforces the specification up to an abstraction parameter. We report on examples illustrating the expressiveness of the fragment and the controllers synthesized by the tool.

Categories and Subject Descriptors I.2.2 Automatic Programming. I.2.8 Problem Solving, Control Methods, and Search.

General Terms Design, verification

Keywords Cyber-physical systems, Controller synthesis, Symbolic Algorithms, Linear temporal logic

1. INTRODUCTION

We present PESSOA 2.0, an extension of the controller synthesis tool for cyber-physical systems PESSOA [3]. PESSOA 2.0 takes as input a cyber-physical system modeled by smooth differential equations and automata, and a specification consisting of two parts: a *safety* part in safe linear temporal logic (LTL), and a *liveness* part in an easily determinizable fragment of LTL. It outputs a correct-by-design controller for the specification. PESSOA 2.0 is available for download at www.cyphylab.ee.ucla.edu/pessoa.

Internally, PESSOA 2.0 computes a discrete symbolic abstraction of the continuous system that is ϵ-bisimilar to the original system, and solves automata-theoretic games on the discrete abstraction. It avoids Safra's construction, a difficult step in automata-theoretic synthesis, by using a subset construction for determinizing the safety part [1], and using an easily-determinizable subset of LTL for the liveness part. It symbolically computes maximal strategies for the safety requirements, and in a second step, computes a strategy to enforce the liveness requirement while also enforcing the safety specification. Internally, the abstraction is represented using MTBDDs and games are solved symbolically. We report preliminary results on the use of PESSOA 2.0 in robotics examples,

showing the applicability of the tool. The formal details of the algorithms underlying PESSOA 2.0 are available in the companion technical report [6].

2. TOOL DETAILS

Figure 1 shows the overall architecture of PESSOA 2.0. PESSOA 2.0 takes as input a controlled differential equation modeling the physical components, an automaton modeling the cyber components, and a specification consisting of two parts: a safety part in safe-LTL and an easily determinizable liveness part. It outputs, if possible, a software controller that ensures that cyber and physical components composed with the controller satisfy the specification up to precision $\varepsilon \in \mathbb{R}^+$, a parameter also specified by the user. The controller is refined to a Simulink block for closed-loop simulation.

Abstraction: In the first step PESSOA 2.0 computes a finite abstraction of the continuous part of the system, parameterized by a user-provided precision ε. Technically, the abstraction is either an approximate alternating bisimulation or an approximate alternating simulation with precision ε, see [3]. Internally, the abstraction is modeled as a state transition system with inputs and outputs, and represented using an MTBDD. The input actions are provided by the synthesized controller. The finite abstractions can be composed with automata modeling the cyber components.

Specification: Specifications are given as a conjunction of a safe-LTL formula and an easily-determinizable liveness formula in LTL. The set of *LTL* formulae [4] is generated by the grammar:

$$\varphi := p \mid \neg p \mid \varphi \vee \varphi \mid \varphi \wedge \varphi \mid \bigcirc \varphi \mid \varphi \, \mathsf{W} \, \varphi \mid \varphi \, \mathsf{U} \, \varphi$$

where p ranges over atomic propositions. Safe-LTL formulae are the subset of LTL formulae without the occurrence of any subformula of the form $\varphi_1 \mathsf{U} \varphi_2$. We use $\Diamond \varphi$ (resp. $\Box \varphi$) to denote **true** U φ (resp. φ W **false**). For the liveness part, PESSOA 2.0 currently accepts a *templated formula* of the form $\bigwedge_{i=1}^{n} \psi_i$, where ψ_i, for $i \in \{1, \ldots, n\}$, is a formula of the type:

$$\oplus (M_i \implies (M_i \wedge \neg T_i) \, \mathsf{U} \, ((M_i \wedge T_i) \, \mathsf{W} \, \neg M_i)). \quad (1)$$

where $\oplus$ is an optional $\Box$ operator, and M_i and T_i are propositional formulae. Each formula M_i defines a *mode* and each formula T_i defines the *target* for mode i. We assume that the modes are mutually exclusive and exhaustive. The formula $\bigwedge_{i=1}^{n} \psi_i$ is satisfied when in mode i, *i.e.* when M_i holds, the system either changes to a different mode, *i.e.* M_i no longer holds, or T_i is eventually satisfied and continues to hold. These templates can be directly converted to a deterministic generator that has linear space complexity with respect to the number of modes and they express many important properties in practice. A simple example is reach and stay at T_1 that is obtained by setting $n = 1$, $M_1 = $ **true**, T_1 to the target to be reached, and omitting the $\Box$ operator.

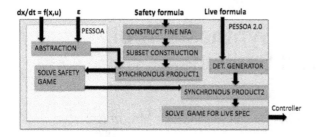

Figure 1: Overall architecture

Fine Automaton: The set of *fine* prefixes for a safe-LTL formula φ is the set of finite prefixes that are sufficient to prove that the computation is unsafe. Kupferman and Vardi [1] show that an automaton that accepts fine prefixes for φ can be constructed from φ. For a formula φ of length n, the size of the automaton can be at most 2^n. The algorithm may produce a non-deterministic finite-state automaton (NFA) which is determinized using the subset construction. Subset construction, in theory, can have an exponential blowup, however the determinization of fine automaton works well in practice (even reduces the number of states).

Controllers as Winning Strategies: We construct controllers as *winning strategies* in games [2], working in two steps.

1) Controller Synthesis for Safety: In the first step, we construct maximal memoryless winning strategies for the safety part of the specification. Internally, the tool computes the synchronous product of the abstracted system with a deterministic automaton that is fine for the safe LTL specification. Then, it solves a safety game on the product structure where the controller enforces that the game always stays in the non-final states of the automaton. The maximal memoryless winning strategy of the controller gives the possible inputs the controller can provide the system to ensure that the safety specification holds for the closed loop system. Construction of winning sets and maximal strategies is performed symbolically, using BDD manipulations.

2) Controller Synthesis for Liveness: In the second step, PESSOA 2.0 constructs a deterministic Büchi automaton for the liveness part of the specification. Then, it solves a Büchi game on the product of the system and the Büchi automaton, while restricting available inputs for the controller to those allowed by the maximal winning strategy computed in Step 1 for the safety objective. A winning strategy for the controller computed at this step ensures that the controller ensures both the safety and the liveness specifications on the abstract system. Again, the construction of winning sets and strategies is performed symbolically using BDD manipulations.

Refinement: So far we have focused on the synthesis of discrete strategies enforcing LTL formulae over the finite abstraction of a physical component possibly composed with automata describing cyber components. The results in [5] guarantee that any such discrete strategy can be refined to a controller acting on the original physical component and enforcing the desired specification up to a precision of ε, a design parameter chosen by the user. PESSOA 2.0 synthesizes the refined controller in the form of a Simulink block for closed-loop simulation.

3. CASE STUDY: ROBOT CONTROLLER

We consider a nonholonomic robot described by the following differential equations: $\dot{x} = v\cos\theta$, $\dot{y} = v\sin\theta$, $\dot{\theta} = \omega$ where (x, y) denotes the robot's position and θ its orientation. The inputs v and ω correspond to the linear and angular velocity of the robot, respectively. We consider several specifications for this robot.

Until: The objective of reaching the target set T, while avoiding the obstacle set O is naturally expressed by the LTL formula $(\neg O) \cup T$. This formula can be decomposed as $((\neg O) \operatorname{W} T) \wedge \Diamond T$. We first solve the safety problem specified by $((\neg O) \operatorname{W} T)$ and then we solve the reachability problem specified by $\Diamond T$.

Mode Switch: The model switching problem can be described using the template formula (1). This pattern frequently occurs when specifying requirements for autonomous vehicles and other cyber-physical domains.

Fault Tolerance: We assume that the communication between the several sensors onboard of the robot with the microprocessor running the control code is governed by a protocol that reports if communication is successful or not. The main microprocessor may fail to receive sensor measurements more than once and needs to react to such failures. One possible way of encoding this objective as a safety property is to require that if sensor measurements are not received k times during last n consecutive control cycles, the robot should stop and remain at its current location. We can encode this objective in the safe LTL formula $\varphi = \Box(fail_{n,k} \rightarrow \bigcirc^n stop)$ where $\bigcirc^n \phi$ is a shorthand of n $\bigcirc$-operators applied to ϕ, $fail_{n,k}$ denotes k is the number of sensor failures in n consecutive readings. . For example, $fail_{2,1} = (f \wedge \neg \bigcirc f) \vee (\neg f \wedge \bigcirc f)$ and the predicate f holds whenever sensor fails. The predicate $stop$ is true when the input v is equal to zero.

Results: The tool computes a finite abstraction of the robot model with a precision of $\varepsilon = 0.1$. The inputs v, ω are restricted to take values in the sets $\{0, 0.2, 0.4\}$, $\{-0.2, 0, 0.2\}$ respectively. In the following table, the columns $|\neg\varphi|$, $|N_\varphi|$ and $|D_\varphi|$ denote the length of the negation of formula, the size of nondeterministic and deterministic generators of the formula. The columns Gen and $Synth$ indicate the time (in seconds) to build DFA with respect to formula and to synthesize the controller respectively.

Property	Params	$|\neg\varphi|$	$|N_\varphi|$	Gen	$|D_\varphi|$	$Synth$
Until	-	4	6	0.001	3	73
Mode Switch	mode = 1	-	-	0.01	3	32
	mode = 2	-	-	0.02	6	58
	mode = 3	-	-	0.01	8	62
	mode = 4	-	-	0.01	10	63
Fault Tolerance	n=4, k=1	13	861	10	15	49
	n=5, k=1	16	2717	112	21	92
	n=5, k=3	16	3685	368	35	120
	n=6, k=1	19	7933	2095	28	129

Acknowledgement. We thank Manuel Mazo Jr. for answering our queries on PESSOA. This research was sponsored in part by the NSF award 0953994 and the DARPA award HR0011-09-1-0037.

4. REFERENCES

[1] O. Kupferman and M. Y. Vardi. Model checking of safety properties. *Form. Methods Syst. Des.*, 19(3):291–314, 2001.

[2] O. Maler, A. Pnueli, and J. Sifakis. On the synthesis of discrete controllers for timed systems. In *STACS'95*, LNCS 900, pages 229–242. Springer-Verlag, 1995.

[3] M. Mazo, A. Davitian, and P. Tabuada. Pessoa: A tool for embedded controller synthesis. In *CAV*, volume 6174 of *LNCS*, pages 566–569. Springer, 2010.

[4] A. Pnueli. The temporal logic of programs. In *FOCS 77*, pages 46–57, 1977.

[5] G. Pola, A. Girard, and P. Tabuada. Approximately bisimilar symbolic models for nonlinear control systems. *Automatica*, 44(10):2508–2516, 2008.

[6] P. Roy, P. Tabuada, and R. Majumdar. Safety-guarantee controller synthesis for cyber-physical systems. *CoRR*, abs/1010.5665, 2010.

A Step Towards Verification and Synthesis from Simulink/Stateflow Models*

Karthik Manamcheri
manamch1@illinois.edu

Sayan Mitra
mitras@illinois.edu

Stanley Bak
sbak2@illinois.edu

Marco Caccamo
mcaccamo@illinois.edu

ABSTRACT

This paper describes a toolkit for synthesizing hybrid supervisory control systems starting from the popular Simulink/Stateflow modeling environment. The toolkit provides a systematic strategy for translating Simulink/Stateflow models to hybrid automata and a discrete abstraction-based algorithm for synthesizing supervisory controllers.

Categories and Subject Descriptors

D.2.4 [**Software Engineering**]: Software/Program Verification—*Model Checking*

General Terms

Algorithms, Verification

Keywords

simulink, stateflow, hybrid automata

1. INTRODUCTION

A key barrier towards applying hybrid system design and analysis techniques to engineering problems is the steep learning curve of the existing hybrid tools. To address this issue, we are building a toolkit that connects Mathwork's popular Simulink/Stateflow (SLSF) environment with new and existing hybrid system tools. While our overarching goals are similar to those of several other projects (see, for example, [9, 8, 6]), the technical approaches are distinct. For example, the synthesis algorithms in [6] rely on approximate bisimulations while ours is based on traditional simulation. Checkmate [8] provides custom modeling blocks within SLSF, whereas our approach allows the use of commonly used Simulink blocks. Deferring a systematic comparison of these approaches

*This project is partially funded by John Deere Technology Innovation Center, Champaign, IL and National Science Foundation (NSF) under the grant CNS-1016791.

for a later paper, here we introduce two core components of our toolkit: (a) HyLink: a tool that transforms a restricted class of SLSF models to hybrid automata for verification, and (b) SimplexGen: a tool that generates supervisory controller logic for a class of SLSF models. These tools can be downloaded from [1].

Figure 1(a) shows the workflow of our tool suite. The SLSF environment provides the front-end for developing and simulating the models which are read by HyLink to produce an internal representation called Hybrid Intermediate Representation (HIR). Different code generation modules analyze the HIR and generate specifications for analysis and synthesis tools, such as HyTech [5], UPPAAL [4], and SimplexGen [2].

2. HYLINK: SLSF TO HYBRID AUTOMATA

A Simulink model describes a dynamical system as a collection of interconnected functional blocks. Stateflow is used to model hierarchical state machines with discrete transitions and continuous flows, and they can be interconnected with other Simulink blocks.

An Example. We consider the model of an autonomous waypoint tracking system (WTS). A *planner* provides the vehicle *controller* a sequence of way points on the two-dimensional plane. The controller periodically sets the acceleration and steering of the vehicle based on the current position, heading, and the waypoint. The vehicle moves according to certain differential equations with the above inputs. The system is said to be *safe* if the vehicle remains within a specified bound of the line joining the current and the previous waypoints (see Figure in 1(c)). Figure 1(b) shows the Simulink model of the vehicle, which is a part of the SLSF model for the complete system. The output of the sub-system in Figure 1(b), reflects the functions which calculate the coordinates x and y of the vehicle.

We have identified a restricted class of Stateflow models which can be translated to hybrid automata. These models requires, for example, that (a) the transition guards to be closed sets, (b) the reset map for an incoming transition at a discrete state (location) and the invariant for that location to overlap. Stateflow models do not have explicit invariants associated with locations. However, as all transitions are urgent, an implicit invariant is derived as follows: the closure of the complement of each outgoing transition guard is computed and the intersection of these sets gives the invariant. Establishing the formal relationship between the set of executions produced by the original Stateflow model and that of our translated hybrid automaton is ongoing work.

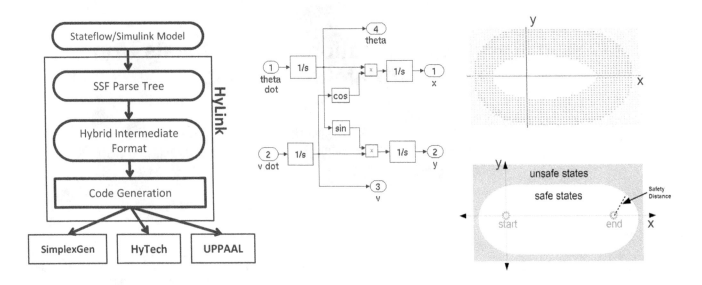

Figure 1: (a) Toolkit work flow, (b) Simulink model of the WTS, (c) Top: Unsafe regions (green points) and switching set (red points) generated by SimplexGen for WTS, Bottom: Single-waypoint system with unsafe and safe regions

For Simulink models that do not have any Stateflow components, HyLink translation to hybrid automaton proceeds as follows: For a model with n switch blocks, a hybrid automaton with 2^n locations is produced—one for each possible configuration of the switches. For each location, the differential and algebraic equations guiding the evolution of the (continuous) variables are derived by composing the functions of the individual blocks that are relevant in the corresponding switch configuration. The invariant condition for the locations are derived by composing the functions of the blocks which drive the switches. The generated hybrid automata have switched modes with invariants but transition conditions and resets cannot be extracted from the Simulink model.

SLSF models do not permit nondeterminism which is a serious restriction in modeling concurrent systems. We address this by developing an idiom: introduction of redundant input variables. For the purposes of simulation these variables are connected to *nondeterminism resolver blocks (NRB)* which set the inputs either according to some deterministic function or using random numbers. During tranalsation, these inputs are ignored to produce a nondeterminisitic hybrid automata.

2.1 Acknowledgments

We thank Daniel Grier for developing the translator from HIR to HyTech, which revealed to us several subtleties in Stateflow semantics.

3. GENERATING SWITCHING CONTROLLER

The Simplex Architecture [7] provides a way to safely sandbox untrusted controllers [3]. The *complex controller* (CC), is combined with a safety wrapper which monitors the state of the plant and can switch to a simpler *safety controller* (SC) before the system state becomes unrecoverable. The switching boundary where the Simplex system should switch from complex to safety controller is given by [2, 3]:

$$G_c = \delta\text{-}BackReach_{\mathcal{HC}}(BackReach_{\mathcal{HS}}(\mathcal{U})),$$

where $\mathcal{U}$ is unsafe set, $BackReach_{\mathcal{HS}}$ is the backwards reach-

able set with the SC and δ-$BackReach_{\mathcal{HC}}$ is the bounded backwards reachable set with the CC within δ time, and δ is the period of the logic.

The SimplexGen tool overapproximates these *BackReach* sets by constructing a discrete abstraction of the hybrid automaton and using bounds on the derivatives for each variable also extracted from the automaton model. The maximum and minimum derivatives are then be used in the discrete abstraction to overapproximate the reach set of the original automata. SimplexGen takes as input the hybrid automata for the CC and SC, an equation describing the unsafe set, and outputs the safety wrapper, namely the switching logic of the supervisory controller. In Figure 1(c)(top), the computed switching set (red) and the unsafe set (green) are shown for a fixed velocity and heading for the WTS example.

4. REFERENCES

[1] Hylink tool. http://hsver.crhc.illinois.edu.
[2] S. Bak, A. Greer, and S. Mitra. Hybrid cyberphysical system verification with simplex using discrete abstractions. In *RTAS '10*, 2010.
[3] S. Bak, K. Manamcheri, S. Mitra, and M. Caccamo. Sandboxing controllers for cyber-physical systems. In *ICCPS'11*, 2011.
[4] J. Bengtsson, K. G. Larsen, F. Larsson, P. Pettersson, and W. Yi. UPPAAL in 1995. In *TACAS*, pages 431–434, 1996.
[5] T. A. Henzinger, P.-H. Ho, and H. Wong-Toi. Hytech: A model checker for hybrid systems. In *CAV '97*.
[6] M. M. Jr., A. Davitian, and P. Tabuada. Pessoa: A tool for embedded control software synthesis. In *CAV*, 2010.
[7] L. Sha. Using simplicity to control complexity. *IEEE Software*, 18:20–28, 2001.
[8] B. I. Silva, K. Richeson, B. H. Krogh, and A. Chutinan. Modeling and verification of hybrid dynamical system using checkmate. In *ADPM*, 2000.
[9] A. Tiwari. Formal semantics and analysis methods for Simulink Stateflow models. Technical report, SRI International, 2002.

Author Index

www.ingramcontent.com/pod-product-compliance
Lightning Source LLC
Chambersburg PA
CBHW080549060326
40689CB00021B/4793